Causality in the Sciences

... ma ... gael i chi hyd y dyddiad
... d oni bai fod darllen... d

Causality in the Sciences

Edited by

Phyllis McKay Illari
Research Fellow, University of Kent
Federica Russo
Research Associate, University of Kent
Jon Williamson
Professor of Reasoning, Inference and Scientific Method, University of Kent

OXFORD
UNIVERSITY PRESS

OXFORD
UNIVERSITY PRESS

Great Clarendon Street, Oxford OX2 6DP

Oxford University Press is a department of the University of Oxford.
It furthers the University's objective of excellence in research, scholarship,
and education by publishing worldwide in

Oxford New York

Auckland Cape Town Dar es Salaam Hong Kong Karachi
Kuala Lumpur Madrid Melbourne Mexico City Nairobi
New Delhi Shanghai Taipei Toronto

With offices in

Argentina Austria Brazil Chile Czech Republic France Greece
Guatemala Hungary Italy Japan Poland Portugal Singapore
South Korea Switzerland Thailand Turkey Ukraine Vietnam

Oxford is a registered trademark of Oxford University Press
in the UK and in certain other countries

Published in the United States
by Oxford University Press Inc., New York

First published 2011

British Library Cataloguing in Publication Data

Data available

Library of Congress Cataloging in Publication Data

Library of Congress Control Number: 2011922687

Typeset by SPI Publisher Services, Pondicherry, India
Printed in Great Britain
on acid-free paper by
CPI Group (UK) Ltd., Croydon, CR0 4YY

ISBN 978–0–19–957413–1

3 5 7 9 10 8 6 4 2

Contents

List of Contributors

Ahn, Woo-kyoung Department of Psychology, Yale University, woo-kyoung.ahn@yale.edu

Aizawa, Ken Department of Philosophy, Centenary College of Louisiana, ken.aizawa@gmail.com

Aliferis, Constantin Center for Health Informatics and Bioinformatics, New York University, constantin.aliferis@nyumc.org

Baetu, Tudor M. ARHU-Philosophy, University of Maryland, tbaetu@hotmail.com

Bogen, Jim HPS Department, University of Pittsburgh, rtjbog@comcast.net

Boniolo, Giovanni FOM, Firc Institute of Molecular Oncology & Dipartimento di Medicina, Chirurgia e Odontoiatria, Università di Milano, giovanni.boniolo@ifom-ieo-campus.it

Broadbent, Alex Department of Philosophy, University of Johannesburg, a.b.broadbent@gmail.com

Cartwright, Nancy Department of Philosophy, Logic and Scientific Method, London School of Economics, and Department of Philosophy, University of California, San Diego, N.L.Cartwright@lse.ac.uk

Cooper, Gregory Department of Biomedical Informatics, University of Pittsburgh, gfc@pitt.edu

D'Abramo, Flavio Department of Philosophical and Epistemological Studies, University of Rome 'La Sapienza', and Department of Epidemiology and Public Health, Imperial College London, flavio.dabramo@gmail.com

Dawid, A. Philip Centre for Mathematical Sciences, University of Cambridge, A.P.Dawid@statslab.cam.ac.uk

Dowe, Phil Philosophy, University of Queensland, p.dowe@uq.edu.au

Elisseeff, André Nhumi Technologies, Zurich, andre@nhumi.com

Faraldo, Rossella Department of Physics, University of Padova, and INFN Padova Section, faraldo@pd.infn.it

Fennell, Damien Centre for Philosophy of Natural and Social Science, London School of Economics and Political Science, D.J.Fennell@lse.ac.uk

Fitelson, Branden Department of Philosophy, Rutgers University, branden@fitelson.org

Geneletti, Sara Department of Statistics, London School of Economics and Political Science, S.Geneletti@lse.ac.uk

Gillett, Carl Department of Philosophy, Northern Illinois University, carl.gillett@gmail.com

Gillies, Donald Department of Science and Technology Studies, University College London, donald.gillies@ucl.ac.uk

Glennan, Stuart Department of Philosophy and Religion, Butler University, sglennan@butler.edu

Groff, Ruth Department of Political Science, Saint Louis University, rgroff@slu.edu

Guyon, Isabelle Clopinet, Berkeley, California, guyon@clopinet.com

Gyenis, Balázs Department of History and Philosophy of Science, University of Pittsburgh, gyepi@hps.elte.hu

Hassani, Hossein School of Mathematics, Cardiff University, HassaniH@cf.ac.uk

Hitchcock, Christopher Division of Humanities and Social Sciences, California Institute of Technology, cricky@its.caltech.edu

Holmes, Dawn E. Department of Statistics and Applied Probability, University of California, holmes@pstat.ucsb.edu

Hope, Lucas Bayesian Intelligence, 2/21 The Parade, Clarinda, VIC 3169, Australia, lucas.r.hope@gmail.com

Hoover, Kevin D. Department of Economics and Department of Philosophy, Duke University, kd.hoover@duke.edu

Illari, Phyllis McKay Philosophy, School of European Culture and Languages, University of Kent, p.mckay@kent.ac.uk

Khan, Aneire MRC/HPA Centre for Environment and Health, School of Public Health, Imperial College London, aneire.khan@imperial.ac.uk

Kincaid, Harold Departments of Philosophy and Epidemiology, University of Alabama at Birmingham, hkincaid@me.com

Kleinberg, Samantha Computer Science Department, New York University, samantha@cs.nyu.edu

Korb, Kevin B. School of Info Tech, Monash University, kbkorb@gmail.com

Kuhlmann, Meinard Institute of Philosophy, University of Bremen, meik@uni-bremen.de

Lagnado, David Division of Psychology and Language Sciences, University College London, d.lagnado@ucl.ac.uk

Lappi, Otto Cognitive Science, Institute of Behavioural Sciences, University of Helsinki, olappi@mappi.helsinki.fi

Lemeire, Jan ETRO Department, Vrije Universiteit Brussel, jlemeire@etro.vub.ac.be

Leuridan, Bert Centre for Logic and Philosophy of Science, Ghent University, Bert.Leuridan@UGent.be

Little, Daniel Philosophy, University of Michigan-Dearborn, dlittle30@gmail.com

Luhmann, Christian Department of Psychology, Stony Brook University, christian.luhmann@stonybrook.edu

Machamer, Peter History and Philosophy of Science, University of Pittsburgh, pkmach@pitt.edu

Markus, Keith A. John Jay College of Criminal Justice of The City University of New York, kmarkus@aol.com

Millstein, Roberta L. Department of Philosophy, University of California, Davis, RLMillstein@ucdavis.edu

Mishra, Bud Computer Science Department, New York University, mishra@nyu.edu

Mouchart, Michel Institut de statistique, biostatistique et sciences actuarielles (ISBA), Université catholique de Louvain, Belgium, Michel.Mouchart@uclouvain.be

Nyberg, Erik P. History and Philosophy of Science Program, The University of Melbourne, e.nyberg@pgrad.unimelb.edu.au

Patterson, Kerry Professor of Econometrics, School of Economics, University of Reading, k.d.patterson@reading.ac.uk

Pearl, Judea Cognitive Systems Lab, Computer Science Department, University of California, judea@cs.ucla.edu

Pellet, Jean-Philippe IBM Zurich Research Labs, Zurich, jppellet@gmail.com

Pendergraft, Garrett Philosophy Department, Pepperdine University, garrett.pendergraft@pepperdine.edu

Psillos, Stathis Department of Philosophy and History of Science, University of Athens, psillos@phs.uoa.gr

Rédei, Miklós Department of Philosophy, Logic and Scientific Method, London School of Economics, M.Redei@lse.ac.uk

Reiss, Julian Faculty of Philosophy, Erasmus University Rotterdam, reiss@fwb.eur.nl

Rottman, Benjamin Psychology, Yale University, benjamin.rottman@yale.edu

Rusanen, Anna-Mari Philosophy of Science Group, Department of Philosophy, History, Culture and Arts Studies, University of Helsinki, anna-mari.rusanen@helsinki.fi

Russo, Federica Philosophy, School of European Culture and Languages, University of Kent, f.russo@kent.ac.uk

Saggion, Antonio Department of Physics, University of Padova, saggion@pd.infn.it; antonio.saggion@unipd.it

Sarkar, Sahotra Department of Philosophy and Section of Integrative Biology, University of Texas at Austin, sarkar@mail.utexas.edu

Silva, Ricardo Department of Statistical Science, University College London, ricardo@stats.ucl.ac.uk

Sloman, Steven A. Cognitive, Linguistic, and Psychological Sciences, Brown University, steven_sloman@brown.edu

Soofi, Abdol S. Department of Economics, University of Wisconsin-Platteville, Soofi@uwplatt.edu

Spirtes, Peter Department of Philosophy, Carnegie Mellon University, ps7z@andrew.cmu.edu

Statnikov, Alexander Center for Health Informatics and Bioinformatics, New York University, Alexander.Statnikov@med.nyu.edu

Steenhaut, Kris ETRO Department, Vrije Universiteit Brussel, kris.steenhaut@vub.ac.be

Thompson, R. Paul Institute for the History and Philosophy of Science and Technology, and Department of Ecology and Evolutionary Biology, University of Toronto, p.thompson@utoronto.ca

Touhafi, Abdellah ETRO Department, Vrije Universiteit Brussel, abdellah.touhafi@vub.ac.be

Vineis, Paolo MRC/HPA Centre for Environment and Health, School of Public Health, Imperial College London, p.vineis@imperial.ac.uk

Walsh, Clare R. School of Psychology, University of Plymouth, clare.walsh@plymouth.ac.uk

Weber, Erik Centre for Logic and Philosophy of Science, Ghent University, Erik.Weber@UGent.be

Williamson, Jon Philosophy, School of European Culture and Languages, University of Kent, j.williamson@kent.ac.uk

Zhigljavsky, Anatoly School of Mathematics, Cardiff University, ZhigljavskyAA@cf.ac.uk

PART I
Introduction

·1·

Why look at causality in the sciences? A manifesto

Phyllis McKay Illari, Federica Russo, and Jon Williamson

Abstract

This introduction to the volume begins with a manifesto that puts forward two theses: first, that the sciences are the best place to turn in order to understand causality; second, that scientifically-informed philosophical investigation can bring something to the sciences too. Next, the chapter goes through the various parts of the volume, drawing out relevant background and themes of the chapters in those parts. Finally, the chapter discusses the progeny of the papers and identifies some next steps for research into causality in the sciences.

1.1 A manifesto

One might think that the sciences are the last place one should look to gain insights about causality. This is because, due to influential arguments of Karl Pearson, Ernst Mach and Bertrand Russell at the turn of the twentieth century, research scientists have for a long time taken great pains to eradicate causal talk from their research papers and to talk instead of associations, correlations, risk factors and other ephemeral properties of data. Thus the traditional home of the study of causality has been within the field of metaphysics in philosophy – a field that has in its turn been treated sceptically by many scientists.

Our first thesis is that, on the contrary, the sciences are the best place to turn in order to understand causality. We maintain this thesis for a variety of reasons.

First, as explained in Section 1.2.5, causal talk became more respectable in the sciences at the turn of the twenty-first century, thanks to attempts to mathematize the notion of cause. It is now becoming clearer that causal reasoning is of central concern to scientists in many fields, as well as to philosophers, and it is fruitfully pursued as a project of mutual concern.

Second, although causal talk was unfashionable in the twentieth century, causality never really went away: scientists' claims were always intended to inform policy, experiment and technology, and such applications require causation, rather than mere association which tells us nothing about what happens when we intervene to change the world.

Third, the concept of cause is changing, and the sciences are at the forefront of these changes. In Aristotle's time causality was understood as explanation in general: the search for causes was a search for 'first principles', which were meant to be explanatory. However, now causal explanation is usually thought of as just one kind of explanation. In the modern era, causality became tied up with the notion of determinism, the prevailing scientific view of the world in Newtonian times. But determinism fell out of favour in science due to the advent of quantum mechanics. Moreover, a non-deterministic notion of cause became increasingly relevant to science (in medicine, for example, claims like 'smoking causes cancer', where the cause is not sufficient to ensure the effect, became quite acceptable), and causality lost its deterministic connotations. If attempts within science to mathematize the notion of cause should prove successful (though this is controversial), the current concept of cause may be replaced by some formal explication, as happened so systematically with the concept of probability. It is science that is driving change in the concept of cause.

Fourth, the field of metaphysics generally benefits immeasurably from interactions with the sciences. Our understanding of time and space, for example, is derived from the use of these notions in physics, just as our understanding of what an organism is (and could be) is derived from the biological study of organisms. It is part of the job of any scientific field to decide what the constituents of its field are, whether that is four-dimensional space-time, bacteria, or market transactions. This is the same question that is faced at a higher degree of abstraction by the metaphysician concerned with what the constituents of the world are. It is bizarre to try to answer those questions without looking at how the same questions are dealt with in the sciences.

Our second thesis is that scientifically-informed philosophical investigation can bring something valuable to the sciences, too. As can be seen in this volume, many scientific fields are wrestling with the same methodological problems concerning causality. Different sciences use different languages and different paradigmatic examples, which can obscure the fact that they are facing the very same problems. But philosophers of science are in a natural position to identify common ground in the methods they encounter across the sciences. These philosophers are becoming increasingly well-informed about the sciences and so able to exploit that position in order to identify best practice. Of course philosophers are also well placed to identify any conceptual problems that they encounter in the methods developed in the sciences and to clarify the very concept of cause that these methods appeal to.

We think, then, that the most promising way forward in understanding causality and making methodological progress is as a mutual project between philosophically-minded scientists and scientifically-informed philosophers. We hope that this volume is testimony to the fruitfulness of this way of looking at causality in the sciences.

1.2 The core issues

1.2.1 Health sciences

While biomedical issues have long been a concern of ethicists and phenomenologists, only very recently have the health sciences become prominent in the debates of philosophers of science and philosophers of causality. It is now clear that the health sciences are an inspiring source for methodological, epistemological and metaphysical issues concerning causation. The chapters in this part of the volume testify to the increasing awareness of both philosophers and practising scientists that biomedical research shares with other domains a number of concerns, from the conditions for inferring causation from correlational data to the definition, use, and role of mechanisms. What triggered philosophers to pay more attention to this domain has been the rise of the so-called *evidence-based medicine* (EBM) movement. Although the first works by the epidemiologist Archie Cochrane going in this direction date from the early 1970s, the term was coined and started to be customarily used only in the early 1990s. The main result has been the production of the so-called 'evidence-hierarchy', i.e. a list ranking methods for causal inference from the strongest (notably, meta-analyses of randomized controlled trials) to the weakest (notably, expert opinion). *Evidence*, it seems, is the pillar of science and the tenets of EBM are well-entrenched. But these strongholds have been under attack for the last 10 years at least. The battle to set the debate straight is happening in this volume too.

For instance, in *Causality, theories, and medicine* Paul Thompson argues against RCTs as the gold standard of causal inference in medicine. Ultimately, Thompson's critical target is statistical methods *alone* as reliable tools for causal inference. His argument largely hinges upon the crucial differences between trials in biomedical contexts and in agricultural settings, where Fisher first developed the methods of randomization. He thus emphasises the role of theory and of background knowledge in establishing causal claims. Thompson's emphasis on the role of 'non-statistical' elements in causal inference is also shared by Alex Broadbent in *Inferring causation in epidemiology: mechanisms, black boxes, and contrasts* and by Harold Kincaid in *Causal modelling, mechanism, and probability in epidemiology*. They turn attention to the contentious issue of whether causal claims in epidemiology are supported by mechanisms and, if so, how. Broadbent in particular opposes the 'mechanistic stance' and the 'black box stance' in epidemiology. He thoroughly discusses pros and cons of taking mechanisms as necessary or sufficient to establish causal claims. He also investigates assumptions and consequences of taking mechanistic considerations in causal assessment to be descriptive or normative. Kincaid, on the other hand, focuses on the use of mechanisms, hoping to make observational studies in epidemiology 'more formal' and consequently stronger.

In *The IARC and mechanistic evidence*, Bert Leuridan and Erik Weber focus on yet another aspect of using mechanisms. Their philosophical considerations about causality and mechanisms are more specifically applied to the procedures for evaluating carcinogenicity of agents by the International Agency for Research on Cancer (IARC). They argue for an evidential role of mechanisms. Mechanisms help in excluding confounding, that is when one or more variables interfere and confound the 'real' causal relations. This may lead IARC panels to conclude that an agent is carcinogenic when it is not, and vice versa. A more theoretical contribution is that of Donald Gillies in *The Russo-Williamson thesis and the question of whether smoking causes heart disease*. Gillies specifically addresses the thesis, put forward in Russo and Williamson 2007, that evidence of both difference-making and mechanisms are needed to establish causal claims. Using examples from the studies on smoking and heart disease, Gillies refines the thesis, requiring that mechanisms be 'plausible' rather than 'confirmed' or 'well established'.

The leitmotif of the chapters of this part seems to be that (*pace* EBM partisans) there is more to causation in health contexts than simply statistics. This, as we shall see next, is a thread followed also in the investigations on causality in psychology. Likewise, chapters in the psychology part share concerns about the role and import of difference-making and mechanistic information for disease causation or causal assessment. Another relevant aspect highlighted by this sample of works in the health sciences is that debates on conceptual issues such as mechanisms are not pursued in abstract terms but are meant to positively contribute to the discussions about the 'use' of causality, for instance in IARC procedures.

1.2.2 Psychology

Psychology has a history of paying serious attention to the philosophical literature and of valuing rigorous philosophical clarification of the basic concepts and distinctions of psychology. Philosophy has not always returned the compliment. As a result, many philosophers will be unaware of the explosion of work in psychology on all aspects of causal reasoning. This fascinating work should be of interest not only to philosophy but also to *any* area of science that is wrestling with causality.

Psychologists test empirically how people *do* reason, not directly how people *should* reason. Nevertheless, empirical results are of direct interest to other fields. Psychologists test 'folk intuitions' on causal reasoning, which is a useful check on whether philosophers' intuitions are systematically different from those of the unphilosophical folk. For the rest of science, primarily concerned with the *normative* aspect of causal reasoning, psychology can find out which weaknesses and fallacies we are susceptible to in our causal reasoning, and in which circumstances we do better – or worse.

The first three chapters in this part give a taster of this growing psychological literature. In *Causal thinking*, David Lagnado brings to bear a body of empirical work to criticize the usual practice in psychology of separating the study of causal learning (learning *about* a causal structure) and causal reasoning (reasoning on the basis of a *known* causal structure). Lagnado argues that studying both aspects together – on the basis of a psychological process account of causal reasoning – will be superior. In *When and how do people reason about unobserved causes?*, Benjamin Rottman, Woo-kyoung Ahn and Christian Luhmann similarly use a body of empirical work to argue that people's reasoning about unobserved causes is more sophisticated than has been recognized. Reasoning about unobserved causes is a big problem for inferring causation from correlation – a concern of almost any science. In particular they examine patterns that people infer in data that deviates from simple correlations and conclude that there is a dynamic interplay between observed and unobserved causes that any attempt to explain causal learning must consider. Clare Walsh and Steven Sloman, in *Counterfactual and generative accounts of causal attribution*, argue that there is evidence that people think about both counterfactuals and mechanisms in forming causal judgements. They go on to examine reasoning about *prevention* or causing an *absence*, noting that there is considerably less consensus on prevention than on positive causation.

The remaining chapters are more philosophical, and illustrate how integrating psychological and philosophical work can benefit both disciplines. In *The autonomy of psychology in the age of neuroscience*, Ken Aizawa and Carl Gillett examine the issue psychologists or neuroscientists face when they discover more than one neurological realizer for what was initially treated as a single psychological phenomenon: do they keep a single psychological phenomenon, with multiple realizers, or do they decide that after all there was more than one psychological phenomenon? Aizawa and Gillett argue, with reference to the discovery of the neural realizers of colour vision, that the higher-level theory plays an essential role in this choice. It is worth noting that Baetu's chapter in Section 1.2.4 examines the same theme with regard to classical genetics and molecular biology. Otto Lappi and Anna-Mari Rusanen, in *Turing machines and causal mechanisms in cognitive science*, argue that explanation using abstract representations in Turing machines illustrates limitations of the account of mechanistic explanation put forward in recent years in the philosophical literature on mechanisms. Finally, in *Real causes and ideal manipulations: Pearl's theory of causal inference from the point of view of psychological research methods*, Keith Markus sets out a detailed examination of Pearl's account of causal reasoning (see Section 1.2.5), when applied to psychology. Markus discusses ways in which Pearl's formalism should be interpreted and argues that it has certain limitations in the context of psychology.

This part thus develops themes arising in psychology itself and from examination of psychology – but which are also of vital interest to other sciences. If it is true that reasoning to a causal structure and reasoning from a causal structure influence each other, as Lagnado argues, then that is of concern to the many scientists whose work is related to either or both forms of reasoning. The issue of scientific taxonomy, or how a field should chop up its domain, is of wide concern, as is examination of the limitations of highly successful inference methods such as those based on Pearl. Finally, mechanisms and their position in causal judgements, and explanations, are clearly of increasing interest. The use of mechanisms in causal reasoning is now a substantial debate in psychology, that has come from philosophy, and – on the basis of much of this volume – is rapidly becoming a debate of interest right across the sciences.

1.2.3 Social sciences

The social sciences are another area that took time to attract the interest of philosophers of causality and of philosophically-minded scientists. It has perhaps been methodological advancements, especially in quantitative research, that have enabled the social sciences to shake off an inferiority complex with respect to the hard sciences. Arguably, the social sciences still cannot establish the same kind of laws as physics, but the use of more rigorous methods has allowed a much deeper understanding of many social phenomena, and more accurate predictions and well-informed interventions through social policy. Moreover, recent debates in philosophy give room for rethinking even traditional debates in social science.

This is the case, for instance, in the contributions *Causal mechanisms in the social realm* by Daniel Little and *Getting past Hume in the philosophy of social science* by Ruth Groff. On the one hand, Little endorses causal realism and asks what ontology is to be developed for the social realm. He argues for mechanisms, but within a microfoundations approach: in social contexts 'causal mechanisms are constituted by the purposive actions of agents within constraints'. Little also makes clear that such an ontology in the social context is overtly anti-Humean, because causation is not in regularities but in mechanisms. Humeanism is also the critical target of Groff. Notably, she discusses how the tacit Humean metaphysics can be by-passed in social science and touches on issues related to methodological individualism and causal powers. Although Groff does not offer any definite positive arguments, she nicely builds bridges between the traditional philosophy of science literature, stances in analytic philosophy, and the methodology of social science. The kind of anti-Humeanism argued for in these two chapters concerns metaphysics, namely whether or not all there *is* about causation is the regular sequence of effects following causes in time. The line of argument of Little and Groff may also be extended to epistemological considerations, namely whether or not in order to *know* about causal relations all we have to do is to track regular sequences of effects following causes in time. An attempt to challenge 'epistemological Humeanism' has been carried out by Russo and Williamson (2009a, b) for the social sciences and for epidemiology. Russo has argued that causal epistemology hinges upon the notion of variation. Simply put, model building and model testing is about meaningful joint *variations* between variables of interest; conditions of invariance of parameters or regularity of occurrence are instead constraints to ensure that variations are causal rather than spurious or accidental. Arguments given in Section 1.2.2 seem to suggest that psychologists also track variations rather than regularities.

In the next group of chapters, two main issues come up: mechanism and structure. In *Causal explanation: recursive decompositions and mechanisms*, Michel Mouchart and Federica Russo tackle the problem of causal explanation in social science research, especially quantitative-oriented research. They present the structural modelling approach as a means to causally explain a social phenomenon and advance the view that the core formal tool – i.e. the recursive decomposition – needs to be interpreted in mechanistic terms. In *Counterfactuals and causal structure* by Kevin Hoover, structural modelling has a slightly different facet. 'Structural' does not refer to the structure or *mechanism* that the recursive decomposition represents, but to the structural *equations*. Hoover's structural account hinges on Simon's notion of causal ordering, and the key aspect is the invariant parametrization of the system. The two chapters have in common, though, that structural modelling is an alternative to a manipulationist or interventionist account *à la* Woodward (Woodward, 2003). Simply put, manipulationist accounts hold that x is a cause of y if, and only if, were we to manipulate or intervene on x, some change in y would accordingly follow (with the usual caveats of holding fixed any other factor liable to interfere in the relation between x and y). Interestingly enough, manipulability theories now enter philosophical discussions in a different way. It seems that the importance of the notion of manipulation is not so much in providing an explication of the concept of causation, but rather in explicating other notions, e.g. that of 'constitutive relevance' used by Craver (2007). In *The error term and its interpretation in structural models in econometrics*, Damien Fennell also considers structural models based on Herbert Simon's notion of causal ordering, and in particular examines issues related to the error term in the equations. The goal of the chapter is mainly expository, in making those who use these kinds of models in econometrics aware of conceptual issues that can hinder successful and meaningful results. In the last chapter *A comprehensive causality test based on the singular spectrum analysis*,

Hossein Hassani, Anatoly Zhigljavsky, Kerry Patterson and Abdol S. Soofi discuss a new statistical method for testing causal relations not in the tradition of structural modelling, but rather in the tradition of Granger's approach. In this approach causality does not lie in the *structures* or *mechanisms* identified in the joint probability distributions, but, to put things very simply, it lies in the power of a (set of) cause-variables to convey information in order to predict the effect-variable.

The second group of chapters, and in particular Hoover's, is closely related to issues also addressed by Judea Pearl and Nancy Cartwright in Section 1.2.5: counterfactuals and structural models, structural models and external validity, going beyond statistics in drawing causal inferences. It is also worth mentioning that most chapters again deal, either directly or indirectly, with mechanisms. A possible explanation is that one may require more than probabilities to give a satisfying conceptual analysis of causation. Perhaps probabilities are not enough even from a methodological point of view: large parts of these methodological chapters invoke, albeit in different ways, mechanisms. The fact that so much emphasis is given to mechanisms may be due to a shift of focus from probabilities to mechanisms. This does not necessarily mean, of course, that probabilities do not play any role in the explication of causation.

1.2.4 Natural sciences

The natural sciences, and particularly physics, are the traditional source for philosophers of science, and for many years natural scientists have taken an interest in the philosophical literature on their field. This part begins with chapters representing the established but growing interest of philosophers in the biological sciences. These engage with topics from mechanism discovery in molecular biology to mathematical modelling in evolutionary biology. The increasing diversity of engagement between philosophy and the natural sciences is also represented by work on the far newer climate science. The part closes with two chapters demonstrating the cutting edge of work on causality emanating from physics.

In *Mechanism schemas and the relationship between biological theories*, Tudor Baetu looks at the relationship between classical genetics and molecular biology, and argues that there are cases where the accommodation of data from molecular biology results in better *classical* explanations. For example, Marfan, Loeys-Dietz and Ehlers-Danlos syndromes can be confused as a single genetic disease, but they are different – and this has implications for their treatment. Baetu argues that this means that classical genetics and molecular biology are not merely parallel explanatory projects, but related. He offers an account of this relation in terms of mechanism schemas. Note that this chapter is thematically linked to the chapter by Aizawa and Gillett in Section 1.2.2. The common concern is with when a difference in lower-level realiser (for Baetu, biochemical molecules; for Aizawa and Gilett, neural systems) matters to the higher-level theory (for Baetu, classical genetics; for Aizawa and Gillett, colour vision), when it does not, and why.

Roberta Millstein turns to our concept of *chance* in *Chances and causes in evolutionary biology: how many chances become one chance.* Millstein argues that at least seven colloquial uses of chance in evolutionary biology can all be translated into the Unified Chance Concept (UCC) by specifying the types of causes that are taken into account (i.e. considered), the types of causes that are ignored or prohibited, and the possible types of outcomes. The UCC is useful, Millstein argues, because it makes it easier to translate between the colloquial chance concepts, and also from them to more formal probabilistic language. In *Drift and the causes of evolution,* Sahotra Sarkar takes a very different approach. Drift is an explanation for evolutionary outcomes which are *not* due to natural selection, mutation, migration or the other recognized causes of evolution. There is always deviation from expected outcomes due to these causes, and this is drift. Sarkar works in the framework of mathematical modelling of evolutionary processes. He distinguishes between the constitutive and the facultative assumptions of a model. The constitutive assumptions define the model, and cannot be changed without changing the system, while the facultative assumptions can vary. So the facultative assumptions give you the causes which act against the background conditions that are given in the constitutive assumptions. Sarkar takes whether the initial size of a population is finite or infinite as a constitutive assumption, and builds a simple mathematical model to show that this models drift, satisfying the usual conditions for drift. But drift is in no facultative assumption of this model. All that is required for drift is that the population be of a finite size; this finite size is part of the conditions under which the evolutionary causes – selection and mutation – operate. Sarkar concludes that drift is not a cause of evolution.

In the chapter, *In defense of a causal requirement on explanation,* Garrett Pendergraft examines whether equilibrium explanations, which explain an observed equilibrium state of a dynamical system by providing a range of possible initial states and possible causal trajectories of the event being explained, violate Pendergraft's Causal Factors Requirement: an explanation of an event must provide information about the causal factors that influenced whether or not that event occurred. Pendergraft argues that equilibrium explanations satisfy this, since they do provide information about causal factors. In so far as drift is an explanation of evolutionary outcomes in terms of chance, the question of whether or not it is a cause of evolution is a link between Millstein and Sarkar's work, and Pendergraft's.

Paolo Vineis, Aneire Khan and Flavio D'Abramo, in *Epistemological issues raised by research on climate change,* examine some of the epistemological challenges faced by climate change research. This is an area with special challenges for coming to causal conclusions, since randomized experiments cannot be done, but only experiments on microenvironments artificially constructed in the laboratory, where the results don't always extrapolate to the real world, and some highly speculative attempts to control real weather, such as to make rain by seeding clouds. This chapter considers particularly the example of rising levels of certain diseases that can clearly be traced to rising salt levels in Bangladesh, and whether we can say that climate change caused these diseases.

One interesting account of causality that emanates from the traditional engagement of philosophy with physics is the process theory of causality (Reichenbach 1956; Salmon 1998; Dowe 2000). Reichenbach's seminal idea, taken up and developed by Salmon, held a process to be causal if it is capable of transmitting a mark. Salmon and Dowe later adopted a version of the theory according to which a causal process is one that transmits or possesses a conserved physical quantity such as charge or angular momentum. In *Explicating the notion of 'causation': the role of extensive quantities*, Giovanni Boniolo, Rosella Faraldo and Antonio Saggion present a development of the process theory, in which conserved quantities are replaced by *extensive quantities*. An extensive quantity is defined as a quantity whose value is given by the volume integral of some function defined over space-time points. Extensive quantities include conserved quantities like angular momentum and charge, but also quantities such as volume and entropy.

For Reichenbach, causal relationships were also characterized *probabilistically*. His probabilistic theory was based around the *common cause principle*, which says roughly that if two events are probabilistically dependent but neither causes the other, then there is some set of common causes of the two events that screens off the dependence (i.e. the two events are probabilistically independent conditional on the common causes). Miklós Rédei and Balázs Gyenis, in *Causal completeness of probability theories – results and open problems*, investigate the question of when the common cause principle is satisfiable. It turns out that in some probability spaces it is possible to satisfy the principle but in others it is not. Their chapter considers both classical and non-classical probability spaces and presents the state-of-the-art concerning what is known about this problem.

On the surface these chapters are very different, arising from different concerns from different scientific fields. But there are some common themes at work here, and in the rest of the volume. The concern of Vineis, Khan and D'Abramo over difficulties with randomized experiments also arises in Section 1.2.1, on the health sciences, and in Section 1.2.5, on computer science, probability and statistics. The issue of mechanisms arises here, as elsewhere. For Baetu, understanding mechanisms and mechanism discovery is vital to understanding the relation between theories, while for Pendergraft the challenge is better to understand different approaches to explanation. The overall project of better understanding explanation is also reflected in the chapter by Lappi and Rusanen examining mechanistic explanation in Section 1.2.2.

The work on mechanisms is, on the face of it, very different from the work on causal processes originating in physics, but there are commonalities in the role of mechanisms and processes in causal explanation and inference, as developed in Section 1.2.6.

1.2.5 Computer science, probability and statistics

As discussed earlier, in the face of criticisms from Mach, Pearson and Russell, in the twentieth century research scientists largely avoided explicit discussion of the causal claims that were implicit in their papers. But certain developments at the turn of the millennium have helped to rehabilitate explicit talk of causality in the sciences, and now 'causality' is no longer a dirty word. It is in the context of these developments that the chapters of this part of the book should be placed.

The 1980s saw the beginning of a revolution in the use of causal methods in the sciences, stemming from interest amongst computer scientists and statisticians in probabilistic and graphical methods for reasoning with causal relationships. Of course revolutions don't just pop out of thin air, and there were several – rather disjoint – lines of thought that led to these important advances. Notably, philosophers of science attempted to characterize causal relationships in terms of patterns of probabilistic dependencies and independencies, and represent them graphically using 'causal nets' (Reichenbach 1956); computer scientists used graphs that chart probabilistic dependencies and independencies to construct computationally tractable representations of probability distributions (see, e.g. Chow and Liu 1968); statisticians were also using graphical models to represent dependence and independence relationships in the analysis of contingency tables (Darroch *et al.* 1980). In the 1980s these advances led to *Bayesian net* methods for causal reasoning (Pearl 1988). Here causal relationships are represented by a directed acyclic graph and causality is tied to probability via the *causal Markov condition*, which says that each variable in the network is probabilistically independent of its non-effects, conditional on its direct causes (see, e.g. Williamson 2005). In the 1990s these methods were reconciled with the use of structural equation models to handle causal relationships – a formalism, stemming from work in the 1920s, that is essentially very similar to the Bayesian net approach (Pearl 2000). As can be seen from the chapters in this part of the book, the Bayesian net approach, and more generally the approach to causality stemming from recent developments in computer science, probability and statistics, remains a thriving area of interesting research questions and lively debate.

In *Causality workbench*, Isabelle Guyon, Constantin Aliferis, Gregory Cooper, André Elisseeff, Jean-Philippe Pellet, Peter Spirtes and Alexander Statnikov focus their attention on methods for the automated learning of causal models directly from data. Hitherto, the field of machine learning in computer science has primarily concerned itself with the task of generating models that are predictively accurate. Broadly speaking, predictive accuracy merely requires that the model adequately capture the underlying probability distribution. Recently, however, there has been some demand for models that are explicitly causal, in order to predict the effects of interventions. Thus a supermarket may wish to use shopping data not only to predict which aisles will need stocking most regularly but also to determine where to move a particular product in order to increase sales of that product. *Causality workbench* presents and discusses an exciting new testbed for computer systems that attempt to learn causal relations directly from data.

The standard approach to learning causal relationships from data is to find a Bayesian net with the least number of arrows from all those that fit the data, and to interpret the arrows in the graph of that net as characterizing the causal relationships. In *When are graphical causal models not good models?*, Jan Lemeire, Kris Steenhaut and Abdellah Touhafi argue that this approach may be unsatisfactory. By appealing to ideas concerning *Kolmogorov complexity*, used widely in computer science in the context of data compression, they argue that the correct causal model may not be a minimal Bayesian net.

Under the standard machine learning approach, the probabilities of a Bayesian net that has been learnt from data are usually simply the frequencies induced by the data. But Bayesian nets were originally conceived of as *belief networks*: the probabilities in the net were supposed to represent degrees of belief that would be appropriate for an agent to adopt given the evidence of the data (Pearl 1988). In *Why making Bayesian networks objectively Bayesian makes sense*, Dawn Holmes argues for a return to the Bayesian, degree of belief interpretation. But rather than advocating the usual subjective Bayesian approach, according to which degrees of belief are subject to rather loose constraints and are largely a question of personal choice, Holmes advocates objective Bayesianism, which holds that degrees of belief are typically subject to tight constraints that leave little or no room for personal choice (Jaynes 1957; Williamson 2010). The key question is: given certain causal and probabilistic evidence, which Bayesian net best represents appropriate degrees of belief? This question has been tackled by Williamson (2005) and Schramm and Fronhöfer (2005), as well as in a distinct line of work culminating in Holmes' chapter.

Bayesian nets are normally construed as representing causal relationships in a qualitative way, via the arrows in the graph of the net. But one might suspect that causality is a matter of degree, in which case the question arises as to how one could measure the extent to which one variable causes another. This question is taken up by the next two chapters. *Probabilistic measures of causal strength*, by Branden Fitelson and Christopher Hitchcock, presents a detailed comparative analysis of a plethora of measures of causal strength that have been put forward in the literature on causality. Kevin Korb, Erik Nyberg and Lucas Hope, in their chapter, *A new causal power theory*, argue that a good measure of degree of causal power can be constructed by appealing to concepts from information theory in computer science – in particular to the concept of mutual information, a concept that is very natural in this context and which underpins, for example, the approach of Chow and Liu (1968) alluded to above.

A quantitative view of causal relationships also forms the backdrop of *Multiple testing of causal hypotheses* by Samantha Kleinberg and Bud Mishra. Their chapter seeks to use methods from computer science and statistics to determine those causal hypotheses that are significant in the statistical sense. Rather than framing their approach in the Bayesian net formalism, which can struggle to cope with the kind of time-series data under consideration in this chapter, Kleinberg and Mishra develop a framework using other methods from computer science, in particular temporal logic and model checking. They apply their approach to microarray data, to neural spike trains, and also to data concerning political speeches and job approval ratings.

Machine learning methods for constructing Bayesian nets can be categorized according to whether or not they attempt to discover latent variables, i.e. variables which are not themselves measured in the data but which are causes of two or more variables that are measured. Latent variables are important to many sciences, not least to psychology which typically uses factor analysis to discover unmeasured common causes (cf. the chapter of Rottman *et al.* discussed in Section 1.2.2). In *Measuring latent causal structure*, Riccardo Silva presents an approach to learning causal relationships that explicitly represents latent variables as nodes in the graph of the Bayesian net. This approach is applied to an example concerning democracy and industrialization and to an example concerning depression.

Judea Pearl, in *The structural theory of causation*, continues his programme of providing a mathematical formalism for causality that unifies approaches to causal reasoning that are extant in the sciences. After explaining the core features of his new theory – which extends the Bayesian net approach of Pearl (1988) and the structural equation approach of Pearl (2000) – Pearl discusses how his new theory can underwrite counterfactual conditionals (conditionals whose antecedents are false), a topic already encountered in Hoover's chapter earlier in the volume. Pearl argues that his account supersedes attempts by philosophers of science to provide a probabilistic analysis of causality, and should be preferred to the *potential-outcomes* (also called *potential response*) approach that emerges from work by Neyman and Rubin (Neyman, 1923; Rubin, 1974).

The potential response approach is also discussed by Sara Geneletti and Philip Dawid in *Defining and identifying the effect of treatment on the treated*. They argue that their decision-theoretic version of the Bayesian net approach can be viewed as a generalization of the potential response approach. Moreover, they argue that their approach can be used to formulate and measure the *effect of treatment on the treated*, an important measure of causal strength that applies to cases where those who are treated are to some extent self-selected.

Statistical and machine learning methods examine data involving a sample of individuals and make general causal claims on the basis of this data. (As Geneletti and Dawid emphasize, one needs to be very careful not to overgeneralize at this stage.) Then policy makers need to apply the general causal claims to a group of individuals who require remedial action in order to identify the most effective interventions. This two-stage process is the focus of Nancy Cartwright's chapter, *Predicting 'it will work for us': (Way) beyond statistics*. Cartwright argues that statistical methods alone will not guarantee the success of either stage of the process. The second stage, Cartwright maintains, needs to be informed by case-specific causal models, concerning the group of individuals who will be treated, and this requires local knowledge about that group that goes well beyond the original dataset. On Cartwright's account, general causal claims are claims about tendencies or capacities and the first stage needs to be backed up by theoretical knowledge of the domain – knowledge of the mechanisms that are responsible for the regularities in the data. (This latter view accords with Thompson's chapter, discussed in Section 1.2.1.)

1.2.6 Causality and mechanisms

This final part of the volume examines mechanisms and their relationship to causality. Mechanisms are important to causal *explanation*, as one way of explaining a phenomenon is to point out the mechanism responsible for it. As we see in the parts on Health Sciences, Section 1.2.1, and Social Sciences, Section 1.2.3, mechanisms are of increasing importance in causal *inference*. (See also Russo and Williamson, 2007; Russo and Williamson, 2011; Illari, 2011). As we see in the part on Psychology, Section 1.2.2, mechanisms are also important in causal *reasoning* (reasoning from a known causal structure). It seems that mechanisms are of interest to every aspect of thinking about causality. The widespread feeling that investigating mechanisms is a fruitful avenue to explore is illustrated in the sheer number of chapters in this volume that touch on the methodology, epistemology and metaphysics of mechanisms in some way or another.

This is a very clear case where philosophical theorizing about the methodology of science is of interest right across scientific disciplines. And while these chapters are more theoretical than those in other sections, since they are not examining an issue arising in a single scientific field, they are all aiming to contribute to scientific work, and are scientifically informed. These chapters illustrate the sheer breadth of interesting work concerning causality and mechanisms, stretching from the very idea of mechanism, their metaphysics, and the applicability of particular conceptions of mechanism across scientific domains.

In *The idea of mechanism* Stathis Psillos disentangles two historical ideas of mechanism. The first is the mechanical conception of mechanism, that mechanisms are configurations of matter in motion subject to mechanical laws. Psillos examines Poincaré's critique that such mechanical mechanisms are too easy to envisage to be informative, because if there is any possible configuration of matter in motion that can underpin a set of phenomena, then there is an infinity of such configurations. The second idea of mechanism is the quasi-mechanical conception of mechanisms, where a mechanism is any arrangement of parts into wholes in such a way that the behaviour of the whole depends on the properties of the parts and their mutual interactions, where this is what constitutes their unity. Psillos discusses Hegel's critique that the unity that such mechanisms possess is external to them, because of the need to identify a *privileged* decomposition out of those available, and so the idea that all explanation is mechanical in this sense is devoid of content. Psillos argues on the basis of these two critiques that mechanisms are not the building blocks of nature, so undermining the *metaphysics* of mechanisms, but that nevertheless the search for mechanism is *epistemologically* and *methodologically* useful. This is a valuable critical historical introduction to the idea of mechanism, against which many of the other chapters can be seen as developing a distinct *new* notion of mechanism.

The first two chapters engage with the metaphysics of mechanisms. In *Singular and general causal relations: a mechanist perspective,* Stuart Glennan examines the relation between singular and general causal relations – the difference between Fred's taking penicillin curing *him,* and penicillin in general curing certain forms of infection. Glennan argues that the simplest reason for preferring singularism from a *mechanista's* perspective (the perspective of someone promoting mechanisms for at least one of the three purposes outlined above) is because mechanisms are particulars – particular things. Glennan then argues for singular determination, which is the view that any causal interaction is a singular case of causal determination, where any causal generalisation is true merely in virtue of a pattern of such singular instances. For Glennan, this is the best metaphysical view of the fundamental *components* of mechanisms since it offers a unified singularist view of these, with the singularist view of mechanisms *themselves* as particulars. Phyllis McKay Illari and Jon Williamson, in *Mechanisms are real and local,* examine the implications of two widely shared premises concerning mechanistic explanation: that mechanistic explanation offers a welcome alternative to traditional laws-based explanation, and that there are two senses of mechanistic explanation: epistemic and physical explanation. They argue that in mechanistic explanation, mechanisms are treated as both real and local, and argue that reality and locality require an *active* metaphysics for the components of mechanisms, illustrated using Cartwright's capacities approach.

The next two chapters both address the idea of a causal *link*, or causal continuity. In *Mechanistic information and causal continuity*, Jim Bogen and Peter Machamer set out to give a novel account of causal continuity in terms of mechanistic information. They use examples of Crick's early conception of gene expression and a sensory-motor reflex in the leech to argue that mechanistic information can be understood in terms of *goals* served by mechanisms, and the *reach* or strength and independence of influence of initial stages of the mechanism on the final stages. Information is ineliminable because the continuity of some mechanisms is a function of their teleological structure, i.e. the goal of the mechanism, and so without attention to the teleological structure, the vital continuity is lost. This chapter has potential implications for the epistemology and methodology of mechanisms, along with their metaphysics. In *The causal-process-model theory of mechanisms*, Phil Dowe addresses the issue of the applicability of causal process theories – such as his own view, mentioned in Section 1.2.4, that causal processes involve the maintenance of conserved quantities – to areas of science other than physics. Dowe considers the need for an account of what it is that scientists look for when they look for something that underlies correlations as an important motivation for his account. If processes involve a spatiotemporally continuous link between cause and effect, then processes cannot involve absences, which would be a gap in a causal process. But absences are sometimes cited as cause or effect, such as in: 'my failure to water the plants caused their death'. An *absence* of watering is said to cause a positive outcome. Dowe offers an account of causal relevance in mechanisms, which can incorporate his theory that causation involves causal processes understood in terms of conserved quantities, but which also allows *absences* in causal explanation.

Meinard Kuhlmann, in *Mechanisms in dynamically complex systems*, examines whether the concept of mechanism can be extended to cover systems that are not just *compositionally* complex, but exhibit complex *dynamics* – what he calls 'dynamically complex systems'. These dynamics arise from the interaction of the system's parts, but are largely irrespective of many properties of these parts. Kuhlmann uses detailed examples of dynamical systems in analysis of heart beat, and financial markets, to argue that dynamically complex systems are not sufficiently covered by the available conceptions of mechanisms. He explores how the notion of a mechanism has to be modified to accommodate this case.

Julian Reiss, in *Third time's a charm: causation, science, and Wittgensteinian pluralism*, examines pluralism about causality: the claim that there is no single correct account of what *cause* means, but instead multiple concepts of cause. Reiss examines three different accounts that all reject any attempt to define 'cause' in terms of necessary and sufficient conditions. Instead they regard different instances of causal relationships such as 'pulling', 'pushing', 'breaking' or 'binding' as sharing family resemblances at best: pushing and pulling clearly share something in common, as do breaking and binding, but there is no single property shared by all instances of such causal terms. This is a pluralist tradition inspired by Wittgenstein and shared by Anscombe, Cartwright, and Machamer, Darden and Craver, and is a form of pluralism about causality that interests many working on mechanisms. Reiss argues for the third form of pluralism, which he says is a form of inferentialism: the method of verifying a causal claim – of evidentially supporting it – determines with what other claims it is inferentially related.

In different ways these chapters are attempting to give an account of mechanisms suitable to their place in causal explanation, inference and reasoning. It is their place in explanation that drives Glennan's emphasis on singularism, and Illari and Williamson's related examination of locality, while Machamer and Bogen, and Dowe's very different attempts to give an account of the causal link – if successful – are important to the usefulness of mechanisms to causal inference. One ambition is also for a single account of mechanism that is applicable across scientific disciplines. Ultimately, the hope is for a general account of mechanisms – the first glimmerings of which can be seen here – which fruitfully addresses all three *methodological* uses of mechanisms in all scientific *disciplines*. This is ambitious, and it remains an open question whether it will be possible.

1.3 Whence and whither?

Progeny of the chapters.

Some of the chapters in this volume were invited contributions, but most were submissions to an open call for papers. Within the broad remit of *Causality in the Sciences*, all authors chose their own topics and titles, and all papers were refereed. Many submissions were received from participants in two events of the *Causality in the Sciences Conference Series* (`http://www.kent.ac.uk/reasoning/cits`): *Causality Study Fortnight* held at the University of Kent in September 2008, and *Mechanisms and Causality in the Sciences* held at the University of Kent in September 2009.

Next steps.

The individual chapters in this volume indicate a plethora of open questions for research on causality. Here we highlight just a few topics for future research that stand out as particularly pressing.

From the volume it is clear that there is a mature field of research centred on the question of the relationship between causality and probability (see Section 1.2.5). But the volume also indicates that there is also a newer, rapidly developing area of research, exploring the relationship between causality and mechanisms (see in particular Section 1.2.6). However, we received very few papers on *all* three: on causality *and* probability *and* mechanisms, and the question of how probabilistic accounts of causality can mesh with mechanistic accounts of causality desperately needs answering. This suggests that a first hot topic for future research will be on causality, probability and mechanisms, bridging the causality-probability agenda on the one hand, and the causality-mechanisms agenda on the other hand.

Although successful formalisms exist for handling aspects of causal reasoning using probabilities, few are explicitly designed for handling mechanisms (see however the discussions of the possible mechanistic interpretations of models in social research in Section 1.2.3). Indeed, a detailed formal understanding of causal reasoning using mechanisms is sorely lacking. So a second hot topic is likely to concern formalisms for handling mechanisms, particularly in causal inference and reasoning. Such formalisms may emerge from the existing formalisms for reasoning using probabilities (e.g. Bayesian nets or multilevel models), or they may need to be entirely new – tailor-made methods for handling causal mechanisms.

A number of chapters invoke mechanisms as evidence for causal tasks, e.g. for the assessment of carcinogenicity. Interestingly, in biomedical and social contexts alike scientists are suggesting that the 'mechanistic picture' is more complicated than it may look at first sight. They are thus moving towards 'ecological views', namely approaches that aim to include both biological and socio-economic factors in the *same* mechanism. This suggests a third hot topic will be to develop *pan-scientific* causal methods. In particular, we are in need of accounts where (i) the *concept* of mechanism permits the inclusion of factors of different natures, (ii) factors of different natures can provide multiple points of *epistemic access* to the same mechanism, and (iii) *formal models* can handle factors of different natures.

Having presented three questions that are likely to feature in future research, we should make some cautionary remarks about how these questions might be solved. We suggested in our manifesto (Section 1.1) that theorizing about causality is best pursued as a collaborative project involving both philosophers of science and scientists from different disciplines and fields. But such a broad project poses two related challenges.

Causality is at the crux of metaphysical, epistemological and methodological issues in the sciences. And different participants in the debate have different primary concerns. The first challenge in theorising about causality is to avoid blurring these three kinds of issue, remaining explicit about which kind is being addressed, and how. For example, the question above of how to integrate ontologically different factors in the same mechanism has metaphysical, epistemological and methodological facets. Yet giving a *methodological* answer to someone concerned about the *metaphysics* of this question, or vice versa, will not help them.

Nevertheless, the metaphysics, the epistemology and the methodology of causality are not wholly distinct. We should expect answers to any one of the three kinds of issue to have implications for the other two kinds. The second challenge is to produce an understanding of causality that successfully addresses all three kinds of issue in a unified way, *without* blurring the distinctions between metaphysics, epistemology and methodology. To make progress on this requires making explicit how metaphysics, epistemology and methodology impact on each other. This is challenging. Note that Cartwright (2007) is pioneering in this regard, urging that questions of metaphysics, methods and *use* cannot be successfully addressed in isolation. Cartwright makes it clear that she thinks an *understanding* of causality that does not help us address how causal claims inform *policy* will never be adequate.

In an era of concern about the 'impact' of research, philosophers have to make the effort to explain why and how philosophical discussions of causality have a bearing on policy and other questions of intervention and control. But scientists also need to make an effort to step back and think of the coherence of the foundations of their work: a 'methodological salad' – an eclectic mix of methods – will inspire no confidence at all unless unifying foundations can be found for the ingredient methods.

In sum, while a sound understanding of causality can best be gained though a mutual project involving the sciences and philosophy, care must be taken not to make progress on metaphysics at the expense of epistemology and methodology, or vice versa.

Acknowledgements

This volume is a product of two research projects hosted by the Centre for Reasoning at the University of Kent: *Mechanisms and causality*, funded by the Leverhulme Trust, and *Causality across the levels: biomedical mechanisms and public health policies*, funded by the British Academy. We are also grateful to the funders who supported the *Causality in the Sciences* conferences that gave rise to some of the papers of this volume: the British Academy, the Mind Association, the British Society for the Philosophy of Science, the Aristotelian Society, the Kent Institute for Advanced Study in the Humanities, the School of European Culture and Literature and the Centre for Reasoning at the University of Kent. We also wish to thank the other members of the steering committee of the *Causality in the Sciences* conference series: Julian Reiss and Erik Weber, and all the participants at those conferences for the lively debates. Finally, we are deeply indebted to all the authors and referees for their hard work on this volume, and to Elizabeth Hannon, Dewi Jackson, Louise Kane, Keith Mansfield, Victoria Mortimer and Mike Nugent at Oxford University Press.

References

Cartwright, N. (2007). *Hunting Causes and Using Them: Approaches in Philosophy and Economics*. Cambridge University Press, Cambridge.

Chow, C. K. and Liu, C. N. (1968). Approximating discrete probability distributions with dependence trees. *IEEE Transactions on Information Theory*, **IT-14**, 462–467.

Craver, Carl (2007). *Explaining the Brain*. Clarendon Press, Oxford.

Darroch, J. N., Lauritzen, S. L., and Speed, T. P. (1980). Markov-fields and log-linear models for contingency tables. *Annals of Statistics*, **8**, 522—539.

Dowe, Phil (2000). *Physical Causation*. Cambridge University Press, Cambridge.

Illari, Phyllis McKay (2011). Mechanistic evidence: Disambiguating the Russo-Williamson thesis.

Jaynes, E. T. (1957). Information theory and statistical mechanics. *The Physical Review*, **106(4)**, forthcoming.

Neyman, Jerzy (1923). On the application of probability theory to agricultural experiments. Essay on principles. Section 9. *Statistical Science*, **5**(4), 465–472. 1990. Translated by Dorota M. Dabrowska and Terence P. Speed.

Pearl, Judea (1988). *Probabilistic Reasoning in Intelligent Systems: Networks of Plausible Inference*. Morgan Kaufmann, San Mateo CA.

Pearl, Judea (2000). *Causality: Models, Reasoning, and Inference*. Cambridge University Press, Cambridge.

Reichenbach, Hans (1956). *The Direction of Time* (1971 edn). University of California Press, Berkeley and Los Angeles.

Rubin, Donald B. (1974). Estimating causal effects of treatments in randomized and nonrandomized studies. *Journal of Educational Psychology*, **66**(5), 688–701.

Russo, Federica (2009*a*). *Causality and Causal Modelling in the Social Sciences. Measuring Variations*. Methodos Series. Springer, New York.

Russo, Federica (2009*b*). Variational causal claims in epidemiology. *Perspectives in Biology and Medicine*, **52**(4), 540–554.

Russo, Federica and Williamson, Jon (2007). Interpreting causality in the health sciences. *International Studies in Philosophy of Science*, **21**(2), 157–170.

Russo, Federica and Williamson, Jon (2011). Generic versus single-case causality: the case of autopsy. *European Journal for Philosophy of Science*, **forthcoming**.

Salmon, Wesley C. (1998). *Causality and Explanation*. Oxford University Press, Oxford.

Schramm, Manfred and Fronhöfer, Bertram (2005). Completing incomplete Bayesian networks. In *Proceedings of the Workshop on Conditionals, Information and Inference* (ed. G. Kern-Isberner, W. Rödder, and F. Kulmann), Lecture Notes in Artificial Intelligence 3301, pp. 200–218. Springer, Berlin.

Williamson, Jon (2005). *Bayesian Nets and Causality: philosophical and Computational Foundations*. Oxford University Press, Oxford.

Williamson, Jon (2010). *In Defence of Objective Bayesianism*. Oxford University Press, Oxford.

Woodward, J. (2003). *Making Things Happen: a Theory of Causal Explanation*. Oxford University Press, Oxford.

PART II
Health sciences

·2·

Causality, theories and medicine

R. Paul Thompson

Abstract

Randomized controlled trials (RCTs) are pervasive in clinical medical research, which stands in stark contrast to other sciences such as physics, chemistry and biology. Most clinical researchers that use RCTs regard them as uncovering causal connections. R. A. Fisher best articulated the rationale for this position in 1935. According to Fisher, if randomization, blocking and replication demonstrated a connection between an intervention and an outcome, that connection is causal. In this paper, I argue that RCTs in clinical medicine do not reveal causal connections. Causal claims in clinical medicine, as in the rest of science, are justified by reference to a robust theory, not RCTs. Part of the argument rests on crucial differences between Fisher's use of RCTs in agriculture and the current use of RCTs in clinical medicine. Two key differences are: the different role of randomization and the legitimacy of assuming homogeneity of the intervention and control entities. A more significant part rests on the integrative power of robust theories; causal attributions are justified by demonstrating that they are, or can be, embedded in a large well-confirm framework. RCTs, by contrast, at best provide isolated input-output connections. A secondary thesis of the paper is that robust theories also allow causal claims to be well-confirmed.

2.1 Introduction

R.A. Fisher was a brilliant mathematician whose contributions to the mathematical foundations of statistics were deep, elegant and robust. His applications of statistics to agricultural research and to Mendelian-based population dynamics were, and remain, transformative. In medicine, the nearly ubiquitous acceptance of the mantra that Randomised Controlled Trials (RCTs) are the gold standard of evidence is traceable to Fisher – although most who espouse the mantra seem not to know its Fisherian origins. Embedded in Fisher's conceptions of experimental design and statistical inference is a conception of causality, and of the role and power of randomization. Both these conceptions I shall argue are untenable, especially in the context of medicine.

RCTs may have considerable utility in a number of research contexts since they do provide some support for a scientific theory (a dynamical (and mechanistic) system which is asserted to model (mathematically) the ontology[1] and dynamics of phenomena). They, however, are far from a gold standard. Contrary to Fisher, they do not provide a causal account of the phenomena under study; nor is randomization essential or, in some contexts, desirable. In what follows, I shall explore what kind of empirical support RCTs provide and why in medicine that support is problematic. I shall also argue that causal attributions are only possible when a robust scientific theory underwrites the attributions; RCTs fall far short of providing a basis for causal attributions. Notwithstanding the ubiquitous use of RCTs in clinical medicine, impressive and robust theoretical underpinnings exist for a substantial array of medical knowledge and causal attributions; it is on these, and not RCTs, that explanation and prediction rest.

2.2 Causality and randomized controlled trials

Outcomes from RCTs constitute the basis for many knowledge claims in medicine and many clinical decisions – especially pharmacological interventions and lifestyle interventions (changes in diet, exercise, and so on). During the last half-century, RCTs have risen in number and authority; government regulators, many epidemiologists and the media have come to regard RCTs as the gold standard of evidence in clinical medicine. This rise in number and authority began with R. A. Fisher who claimed that RCTs provide the basis for discovering and justifying causal connections.[2] According to Fisher, randomization, replication and control (for him, the method of pairwise blocking) guarantee that the intervention is the **cause** of the difference between the outcome in the experimental group (those receiving an intervention) and that in the control group.

Fisher developed his views on experimental design while working at Rothamsted Experimental Station, which he joined as a statistician in 1919. The Station, by that time, had been engaged in agricultural research for more than 70 years. By 1935 when his *Design of Experiments* was published his adherence to the principles of randomization, replication and blocking were

[1] The ontology of a system is the collection of entities postulated to exist (e.g. DNA, proteins, amino acids, leukocytes, and the like) and their properties and their physical relationships to each other (e.g. proteins are strings of amino acids). Dynamical relationships are expressed by a set of equations with these entities as variables.

[2] Randomized trials existed prior to Fisher. James Lind, for example, conducted controlled trials in the eighteenth century – not quite randomized but employing the same reasoning. The most well known is his controlled trial involving the causes of scurvy – a small sample size but manifesting the principles of current RCTs (see: Lind, 1753). Nonetheless, Fisher looms large in the history of RCTs and he contributed significantly to dogma that RCTs provide a strong basis for asserting causal connections and that they should be the basis for experimental design in agriculture and other fields.

firmly entrenched in his method of experimental design. These principles proved powerful in the context of plant agriculture where replication involved sectioning a field into plots, blocking was pairing adjacent plots that were assumed to be identical in all relevant respects (nutrient content and soil composition for example) and randomization was the designation of one of each pair as the experimental plot by a random process such as flipping a coin. Since there were numerous paired plots, the experimental intervention was replicated many times within a single experiment. If a consistent similar difference was found between each experimental plot and its paired control, constant conjunction of intervention and outcome could be assumed. Since paired plots were assumed to be identical in all relevant respects except for the intervention, the outcome could be declared as *caused* by the intervention. Randomization bolstered the assumption of plot-pair identity by removing any systematic bias in favour of one plot of a pair – such as, always choosing the left plot as the experimental plot or favouring lower elevation.

This kind of agricultural research presents an ideal state of affairs for applying Fisher's method of experimental design. The values of the relevant variables in a modest-sized field, as is most frequently used in research, are reasonably homogeneous (soil Ph, organic-to-inorganic material ratio, clay content, and so on). Moreover, dividing the field into small plots and pairing adjacent plots reduces what little heterogeneity might be found in non-adjacent parts of the entire field. Genetic diversity and trait diversity can be managed in agricultural plants to ensure minimal heterogeneity. The adjacent plots remain adjacent throughout the trial and external factors (such as rainfall, hours of sunlight) will be virtually identical for both plots. In short, homogeneity of relevant factors exists naturally or can be easily produced.

In clinical medicine, things are very different. First, a population of individuals is an extremely heterogeneous collection: a wide variety of genotypes, different environmental histories, different physiological dynamics, different interpersonal contacts, and the like. Second, unlike two adjacent plots of land, pairs of individuals or paired groups of individuals do not remain together during a trial; hence, each individual is exposed to different environmental factors. Consider, for example, the simple difference of diet and timing of meals, which may or may not be relevant and often we do not really know their relevance. In the world of RCTs, the Fisherian principles of experimental design are supposed to tame this heterogeneity. The paired groups of individuals are, ideally, random samples from a population; randomization, in principle, results in groups with identical heterogeneity; for each individual in one group, there will be an individual in the other group with the same characteristics. Hence, it is assumed that although any two individuals chosen randomly can be expected to differ in relevant respects, when a large number of individuals are assigned randomly to two different groups, the groups, taken as collectives, will be homogeneous.

Most texts on statistical methods define 'random sample' as: 'A sample of *n* individuals from the population [the whole set or collection of items about which we want information] chosen in such a way that all possible sets of *n* individuals are equally likely to occur' (Wetherill, 1967). As an assumption underpinning aspects of experimental design and statistical analysis, this definition is essential and potent. In the messy world of RCTs, its potency evaporates. First, even the most careful actual random sampling cannot be known to satisfy the requirement of the definition. Second, for financial, ethical and/or trial management reasons, most 'random' sampling in clinical experiments are not from 'the whole set or collection of items about which we want information' (the population) but an already much reduced collection – those, for example, willing and available to participate, those located near the research centre, those who have a relevant disorder, those who do not have signs or symptoms that suggest the experiment might put them at risk, and the list goes on.

Third, even if two groups satisfied the statistical definition of random samples, internal heterogeneity undercuts the assumption that the groups are identical in relevant respects. Most traits of individuals are quantitative traits (traits that differ in the amount or degree) such as blood pressure and lung capacity. These traits are usually the product of many genes and, importantly, are somewhat environmentally sensitive (for example, the effect of exercise on lung capacity and muscle strength).[3]

Suppose only 10 traits are relevant in a particular RCT, the traits vary little (say, six possible values for each) and the experimental and control groups each contain 10,000 individuals and the sampling was random (i.e. the sampling was such that from the population all possible sets of 10,000 individuals were equally likely to occur). Even this does not warrant the conclusion that the groups are identical in relevant respects. The possible combinations of 10 traits each with six possible values entails, assuming independence, equals 6^{10} (60,466,176). Hence, no two individuals in a population of 60 million are the same.

If the number of relevant factors is 11, the possible combinations equal 362,797,056. The US Census Bureau estimated the population of the United States as of August 8, 2009 to be 307,118,070. Consequently, two groups of 10,000 randomly sampled from the entire population of the United States will not be identical in relevant respects. There are almost always more

[3] Traits determined by multiple genes and subject to environmental influences are known as multifactorial traits. Separating the genetic determinants from the environmental ones is challenging. Also, even with stable environmental factors, understanding the genetic transmission of polygenic traits (those caused by more than one gene) is complex. These traits do not manifest simple Mendelian transmission. Although the transmission is Mendelian, it is a complex process which is the subject of the field of quantitative genetics. Within that field, numerous statistical tools have been developed to deal with this complexity (see: Hartl and Clark, 1989).

than 11 relevant variables (factors) and many more than six possible values for those variables and the population being sampled is always dramatically less than 300,000,000 (often fewer than 50,000). In addition, although the values between individuals can be assumed independent, those in a particular individual are rarely independent of each other – such that altering one will alter one or more of the others – and the specific dynamics are frequently idiosyncratic, which further increases the heterogeneity.

To complicate the situation further, the dynamics involved in multivariable interacting systems, such as those involved in the human endocrine, immune and other such systems are usually chaotic; the trajectory of the system is highly sensitive on initial conditions. Since those initial conditions frequently will be different for different individuals, the trajectory of the systems will vary widely. In short, the homogeneity found in agricultural RCTs is entirely illusory and elusive in clinical medicine; in clinical medicine heterogeneity is ubiquitous.[4] This is why some clinical researchers have endorsed the utility of *n* of 1 trials (i.e. where a single patient is the entire trial).[5]

Fisher's motivation for insisting on randomization was rooted in a requirement for the application of mathematical statistics and the definition of causality he adopted. Randomization is required in order to apply the statistical tools of analysis that underpinned his methods and his causal claims. Given the near impossibility of random sampling in clinical trials, Fisher's methods are not applicable;[6] agricultural trials do not rest on random sampling, they involve the toss of a coin (or some other random binary process) to determine which plot of a pair is the experimental plot. There are, of course, other experimental design methodologies that can be applied in clinical medicine but to successfully avoid the aforementioned challenges their validity must be independent of random sampling. Many other critiques of randomised controlled trials have been offered (see: Cartwright, 2007a; Urbach, 1993; Worrall, 2002 and 2007).[7] For the most part they focus on challenging whether randomization is necessary. I agree with those critiques but here have argued that, whether necessary or not, it is unachievable.

The definition of causality that appears to underlie Fisher's methods is elementary. A cause is that which makes a difference. If two states of affairs at t_1 are identical except for one element E and the states differ at t_2, E can

[4] Of course, there are some cases where the required homogeneity exists. In such cases, the relevant variables are few – usually one or two – and the values are constrained. For example, placing a 10% solution of iodine on the skin will result in discolouration of the contacted area.

[5] See, for example, Guyatt, GH, Keller, JL *et al.* (1990) and Avins, AL, Bent, S, *et al.* (2005).

[6] Fisher, with support from the Rockefeller Foundation, did some work in medicine on blood. He contributed significantly to the understanding of the Rh factor in blood groups.

[7] Many defences have been offered as well (see, for example, Papineau, 1994 and Suppes, 1982). Those defences strike me as inadequate against the criticisms of Cartwright, Urbach and Worrall and do not undermine the criticism I have articulated.

be declared the cause of the difference. In Fisher's agricultural trials, the homogeneity of the states being compared (plots adjacent to each other) virtually ensured identity. By having numerous paired plots (paired states) and randomly selecting which plot from each pair would receive the intervention (E), the validity of the claim that the two plots being compared in each case are identical is certain or near certain. Since, the only difference between compared plots is E, it can with certainty be declared 'the cause' of any difference that arises. The upshot of all this is that Fisher's experimental methods do not provide any basis for discovering or justifying causal claim made on the basis of RCTs in clinical medicine.

2.3 Scientific theories

The conclusion of Section 2.2 is that RCTs in clinical medicine provide no, or at best the weakest possible, support for causal claims.[8] I'll return later to the question, 'For what, if anything, do they provide evidence?' Before doing that, in this section I turn to a positive thesis with respect to clinical medicine – namely, what can and does underwrite causal claims.

Physics and chemistry have a rich toolkit of methods which have teased from nature a large and deep body of knowledge. Noteworthy is the fact that RCTs are not among their methods. There is no shortage of recourse to probability and statistics; indeed they employ the entire domain of mathematical knowledge and techniques (e.g. the infinitesimal calculus, topology, and linear and nonlinear algebra). Physics and chemistry employ probability and statistics in contexts where the phenomena are considered truly random or where, even though a system is held to be deterministic, the current understanding of the system admits of uncertainty. In the latter case, the ultimate quest

[8] It might be thought that, in clinical medicine, RCTs are just testing hypotheses without any clear causal structure. That, in fact, is a role that RCTs could play under the auspices of a causal theory, as I indicate below. That, however, is not how medical epidemiologists view them. With few exceptions, books on clinical epidemiology are quite explicit about the causal goals. Consider, for example, Haynes *et al.*'s book, *Clinical Epidemiology*, in which they claim, 'Our key point is this: RCTs provide the best evidence for causation, so don't give up on doing an RCT to settle a causal issue just because it may be difficult or contentious to do so.' (p. 360). Elwood, in his *Critical Appraisal of Epidemiological Studies and Clinical Trials*, devotes an entire chapter to causality. In the section 'A direct test of causation', he claims, 'If a causal relationship exists, the frequency of the defined outcome will be higher in the group exposed to the causal factor. A study design which uses this approach is the randomized trial; that is, the assignment of the treatment for each subject is made by a random or chance procedure.' (p. 7). Further, Rothman and Greenland in *Modern Epidemiology* also have an entire chapter on causality – one of the most nuanced accounts I have found in epidemiological writings. Uncovering causal relationships is clearly on the minds of epidemiologists who engage in RCTs.

is to diminish the uncertainty; the need to use probability and statistics is unsatisfying, though necessary.

Engineering – an applied endeavour similar to clinical medicine – draws heavily on physics and chemistry. Indeed, much of the confidence we have in the claims, predictions and explanations in engineering rest on the confidence we have in the theories, models and knowledge in physics and chemistry. One fundamental logical feature of engineering reasoning – and reasoning in physics and chemistry – is the use and justification of counterfactual claims (typically expressed as conditional – if, then – claims in which the antecedent – the if part – has not occurred or is not known to be true). For example, the claim, 'If my computer keyboard were in motion relative to me, then it would be shorter in the dimension of travel than it was when stationary relative to me,' is a counterfactual claim since the keyboard is in fact currently stationary relative to me. No physicist, however, would doubt the truth of the claim because its truth rests on Einstein's special theory of relativity; hence, to doubt the truth of the claim is to doubt the validity of that theory.

The reason theories support counterfactuals is that they unify and integrate a large body of knowledge into a connected web.[9] The logical structure of this web is such that explanation and, importantly, prediction rest on a wealth of interrelated knowledge claims. Predictions made on the basis of a theory are possible because the integrated wealth of knowledge claims comprising the theory can be used to justify confidence in the predictions. Prediction is an instance of a counterfactual claim.

Although I hold a view of the structure of theories that understands them as a certain sort of mathematical model – a view I will set out below – the logical empiricist conception of theories as interconnected statements (formalized in first-order predicate logic) is a heuristically useful entry point for uncovering the underlying logic of explanation and prediction. Some statements in a theory are extremely general and cannot be deduced from other statements in the collection; these are the axioms of the theory. In a fully developed theory, all the other statements in the collection can be deduced from the axioms. It is that deductive connectivity that integrates the wealth of knowledge claims; it also justifies confidence in predictions and explanations because they are deductive consequences of the theory.[10] Some claims deduced from the axioms are still very general. The further down the deductive hierarchy one moves, the less general the claims; at the lowest level of generality are claims about specific causes of specific effects (see Figure 2.1 for a stylized schematic diagram).

[9] Quine and Ullian also employ the metaphor of a web in a way analogous to mine (see Quine & Ullian 1970). Kuhn's holistic view of theories also treats them as a web (see Kuhn, 1962).

[10] Deduction is the ideal in deterministic systems and theories describing them but frequently the connections are probabilistic such that the truth of a claim is highly probable based on a collection of other claims in the theory, but not a deductive certainty.

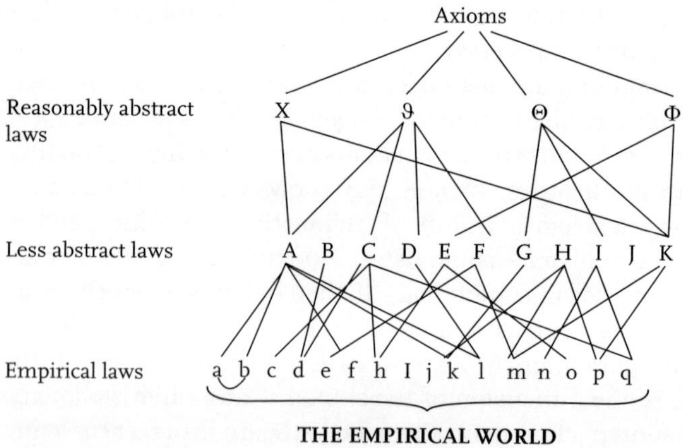

Fig. 2.1 Simple schematic of the deductive web of a theory.

Views differ on how the axioms are generated (discovered). Simplistic empiricism assumes that the first step is the generation of the empirical laws from empirical observation – perhaps by induction, perhaps hypothetico-deductively. Among these empirical laws patterns occur that suggest that a number of empirical laws can be usefully subsumed under a more general claim. Among these more general claims patterns occur that suggest subsets of these more general claims can be subsumed under even more general laws. The process continues until the axioms emerge as the most general claims. Simplistic rationalism assumes the axioms are generated by rational thinking and are subsequently justified by deducing the consequences of accepting the axioms, consequences which are then empirically tested. The history of science suggests that a mixture of these methods is usually involved.

Returning to the claim, 'If my computer keyboard were in motion relative to me, then it would be shorter in the dimension of travel than it was when stationary relative to me', this claim, as noted, is counterfactual. It, however, can be accepted with confidence because it is deducible from the axioms of the special theory of relativity. The degree of confidence, of course, is proportional to the confidence one has in that theory.

The logical empiricist view of theories just sketched assumes that the language of science is first-order predicate logic with identity (symbolic or mathematical logic). The view of theories that I, along with many others,[11] have promoted allows any appropriate mathematical domain to be the language of a theory (e.g. set theory, probability, topology, string theory, and so on). Following Galileo (1623), and three centuries before him Bradwardine (1330),[12] this

[11] See: Suppe (1967, 1972, 1989), Suppes (1957, 1961, 1962, 1967, 1968), Lloyd (1984, 1986, 1988), Thompson (1983, 1986, 1987, 1989, 2007) and van Fraassen (1967, 1969, 1970, 1972).

[12] See: Weisheipl (1967).

view sees mathematics broadly as the language of science. Consequently, it is not a sentential (linguistic) view of the language of science. Theories are not deductively connected statements formalized in symbolic logic but mathematical formulations of dynamical systems. This is still a deductive framework; the deductive structure and techniques of the domain of mathematics used are fully available. Unlike the logical empiricist's view, however, this view understands theories to specify an ontology and the dynamics of a physical system,[13] which is achieved by identifying variables and their range of magnitudes, and specifying, mathematically, the relations among the variables and how the variables change over time (e.g. using transition functions such as $x_{t+1} = f(x_t)$: $f(x_t)$ might be rx_t where r is the rate of population growth (births minus deaths) and x is population size). The thesis of this paper does not rest on which view of theories one accepts. Counterfactual claims are deductive consequences of the theory on both views and confidence in them rests on that deducibility.

2.4 Theories, RCTs and causality

A central role of scientific enquiry and scientific knowledge is to answer '**why**' **questions**, and the perceived importance of uncovering causal connections is motivated by the view that knowledge of cause and effect relationships allows 'why' questions to be answered.[14] Consider the question, 'why, at the onset of the luteal phase in the menstrual cycle, does the level of plasma gonadotropins decrease'. The answer found in *Harrison's Principles of Internal Medicine* (a leading medical resource) is, 'A secondary rise in estrogen **causes** a further gonadotropine suppression' (Braunwald, 2001 p. 2157, emphasis added). Of course, the entire answer to the why question is much more complicated and the additional elements will require the citing additional causes.

In this section, I explore what I argue is a more fundamental element in answering why questions, namely the role of theories. As indicated in

[13] See note 2.

[14] Silvain Bromberger, over 40 years ago, renewed philosophical attention on the importance and role of why questions (Bromberger, 1966). Significant criticisms of Bromberger's specific account have been proffered. I find van Fraassen's early criticism compelling (van Fraassen 1980, pp. 126–130) but do not believe that it diminishes the centrality of why questions; it simply identifies the difficulties with a particular account of the connection between explanation and why questions and the canonical form prescribed by Bromberger. Almost all philosophers of science, and van Fraassen is among this majority, accept that a why question is a request for an explanation and such requests are central to the scientific enterprise. That is all the arguments of this paper require. Of course, van Fraassen and others have put forward a compelling case for the importance of context in determining which theory and/or parts of a theory are relevant to the sought after explanation. I take this as undeniable; a fact that complicates explanation but does not undermine the central arguments of this paper.

previous section, what underwrites confidence in counterfactual claims (and, hence, predictions) is that they can be deduced from relevant parts of a theory. Confidence in the theory comes from the interconnected, and inseparable, nature of the regularities it codifies and the countless predictions deduced using it which have continually been found to be in accordance with the behaviour of the empirical world as we experience it. No isolated claim of a regularity has that robustness and predictions made on the basis of an isolated claim of regularity lack the support and credibility of predictions deduced from regularities embedded in an interconnected web of a vast number and array of regularities. What is true of prediction is equally true of explanation. Although there is no tight symmetry between explanation and prediction, they are two faces of the same logical coin. A robust explanation of an event requires deducing it from a theory, just as a robust prediction requires such a deduction.

To be clear, I am not denying that isolated regularities provide accounts of events; they do. A social worker who asks, 'Why does 8 year old Susan have bruises on her head and shoulder?' may be attempting to determine whether child abuse has occurred. The explanation that Susan fell off a swing the previous day addresses the concern; third-party witnesses make it a compelling explanation. Of course, the social worker has to know that falling off a swing normally results in bruises. For the social worker, this need be no more than an observed regularity – a belief based on the constant conjunction, temporal contiguity and order of the two events. Explanation is pragmatic; the **purpose** of posing a 'why question' determines the relevance of the answer. My claim, as will become clear, is that the purpose of **scientific** research and theorising is to provide an account of observed phenomena, not just a description of them. From the point of view of scientific explanation, the observation that trauma to the skin normally results in a discolouration of the skin in the area of the trauma is simply describing an observation, not explaining it. The goal of scientific[15] research and theorising is to uncover the mechanism that explains the observed phenomenon; in this case, a part of physiological theory.

This goes to the heart of a problem with RCTs. To the extent that they uncover any regularities, they uncover isolated ones unless they can be shown to be among those embedded in, and hence deducible from, a theory. But if such a deduction can be made, it is not the RCT that establishes or justifies acceptance of the regularity; it is the theory. This is not to say that RCTs have no methodological or logical role in scientific enquiry. Fisher's use of them in agriculture is often cast as establishing and justifying causal assertions.

[15] I, of course, am using 'scientific' in a narrower sense that some others might. I think my use accords with standard usage in philosophy of science. Nonetheless, little hangs on this. If forced to broaden the scope of the phrase 'scientific explanation', I would use 'robust scientific explanation' in its place.

But that is the wrong way to view things. RCTs divorced from a theory provide, at best, knowledge of isolated regularities (for Fisher a cause and effect relationship) and, as already argued, isolated regularities lack the robustness required in providing compelling predictions or explanations. What RCTs, used as Fisher used them, can do is provide a method for testing predictions made using a theory – in effect, they are a way of empirically testing a theory. In this way, they confirm or call into question some feature of the theory, such as one or more of its axioms, the validity of a particular deduction from it or some interpretation of its ontology or dynamics, and the like. This is an important role but one that can **only** be played in the context of a theory. Moreover, this role is not an explanatory one.[16] Furthermore, in the absence of a theory, determining whether and in what way an RCT should be conducted is doomed to failure.[17]

An example from immunology illustrates the distinction I am drawing between an isolated result from an RCT (or several RCTs that produce similar results) and a robust explanation of the result that embeds it in a theory. In the eighteenth century, protection from smallpox was discovered to occur in some individuals after inoculation with the dried material from a smallpox pustule. Regrettably, about three in 100 people developed a severe case of smallpox and died. Edward Jenner, in 1796, discovered that inoculation with material from cowpox pustules (the bovine form of smallpox) also conferred protection without causing severe cases. He called this inoculation process vaccination (from *vacca* –Latin for cow: also the origin of vaccinia for the virus that causes cowpox); Pasteur (honouring Jenner) extended this term to cover all inoculations which provide protection from infectious agents. Jenner had no knowledge of the infectious agent; he only knew that protection from smallpox followed vaccination with cowpox, with only a small number of individuals developing serious disease. The vaccine was modified during the following century and a half (e.g. attenuated versions of the smallpox virus were developed).

Jenner's experiment to demonstrate the efficacy of inoculation with cowpox falls significantly short of the 'proof' required today and would completely fail an ethics test. Jenner inoculated an eight-year old boy, James Phipps, with cowpox material. He waited six weeks and inoculated him with fluid extracted from an active smallpox pustule. Phipps did not contract smallpox. On the basis of this 'experiment', he published his success in 1798. As one

[16] Nancy Cartwright, with somewhat different arguments and purposes, has made a similar point in her compelling and insightful recent book, *Hunting Causes and Using Them* (Cartwright, 2007), as also have numerous philosophers over the last 50 or so years (see also Cartwright, 2008).

[17] This is a point elegantly made by Hempel (1966, pp. 10–18). His example and argument are entirely independent of his logical empiricist philosophy; it applies equally to other views of theories and their role in science.

would expect, smallpox vaccines approved in the latter half of the twentieth century were subjected to RCTs. Clearly, an RCT would have provided a higher level of confidence in the efficacy of Jenner's vaccine. It would not, however, have added anything to Jenner's description of the connection of the events. RCTs on more recent smallpox vaccines provide evidence that vaccination is followed by protection against smallpox – knowledge that is, without question, valuable in clinical medicine[18] – but an explanation of why there is a connection is not provided by an RCT.

What does provide an explanation is immunological theory; it provides an account of how the vaccine results in immunity to the *variola major* virus (the virus causing smallpox); and theories in virology and physiology provide an account of how the virus causes the clinical manifestations of smallpox. A comprehensive account of the explanation is complex and more technical than appropriate for this paper but the skeleton can be easily provided. The virus in Jenner's vaccine has the same antigenic determinants (epitopes) as the variola virus but is not a viable pathogen in humans. Lymphocytes (a kind of white blood cell) are produced in the bone marrow. A common lymphoid progenitor gives rise to lymphocytes. There are two major kinds: B lymphocytes (which mature in the bone marrow – hence B) and T lymphocytes (which mature in the thymous – hence T). These are more commonly known as B cells and T cells. Lymphocytes recognise specific sites that are present on antigens – foreign material. A large variety of site-specific T lymphocytes are produced, each recognizing a different epitope. The presence of the vaccine virus is detected by a specific lymphocyte whose receptor matches the virus' epitope; that detection results in the production of a large number of lymphocytes that are site-specific for the virus. Through a complicated biochemical process, the production of armed effector T cells with that site-specific receptor is initiated. These effector T cells inactivate the virus by binding to the epitope. As part of the process, memory T cells and B cells are produced; these provide the observed long-term protection. The same process can be initiated by inoculation with attenuated variola (smallpox) virus.

What makes this a robust explanation of the observed phenomena is the rich body of generalizations on which it draws and the rich ontology involved (hematopoietic stem cells, neutrophils, B cells, T cells, antibodies, basophils, and so on). The deductive network of generalizations at a variety of levels of generality integrates this single-case connection of events (vaccine and protection) in a large framework. A framework that also explains why lymphocytes do not bind to the bodies own tissues, and why the major histocompatability complex (MHC) of genes is important to the production of armed effector

[18] Smallpox vaccination has resulted in one of the great successes of clinical medicine. The last reported case of smallpox was in Somalia in December 1977. On 9 December 1997, the World Health Organization declared that smallpox had been eradicated.

T cells, and why B cells bearing surface CD5 express a distinctive repertoire, and why HIV produces an autoimmune response, and so on and so on. The connection of vaccine and protection is imbedded in this complex dynamical system; a system by means of which we can provide a rich multilayered explanation of the observed connection of events.

Importantly, immunological theory explains the heterogeneity of individuals and explains the heterogeneity of responses to interventions. The explanations will appeal, for example, to genetic differences, such as differences in the MHC group of genes, to compromises to the system, such as low leukocyte counts, to deficiencies in critical precursor elements, such as cytokines, and so on. This explanatory power of heterogeneity is a feature of all theories in medicine.

By contrast, RCTs, independent of this theoretical framework, focus on an isolated connection of events. Even a meta-analysis, which examines and analyses numerous RCTs, focuses on an isolated connection of events. A meta-analysis may provide even stronger evidence that there is a connection but it stills fails to explain why there is a connection. And, the heterogeneity of individuals and their responses to interventions bedevils RCTs divorced from a theory.

Returning now to Fisher to further support my thesis, as noted, for the most part, Fisher's method of experimental design was focused on agricultural research. He did, however, with support from the Rockefeller Foundation, do some work in medicine on blood groups during the period 1935–1943. This work contributed significantly to the early understanding of the Rh factor. It, however, drew heavily on population genetics and evolution, both are robust theories.[19] What Fisher demonstrated in 1943 (see Fisher 1943 and 1944), using the theory of population genetics, was the role that three linked loci with specific allelic combinations could play in the explanation of the puzzling experimental results with the Rh factor. He also predicted, again on the basis of population genetical theory, the existence of antibodies not known to exist at that point. Within the next five years his prediction and explanation received independent empirical support – further confirming the theories. Consequently, Fisher's work in clinical medicine, far from demonstrating the value of RCTs in that domain, elegantly demonstrated the value of a robust theory in providing explanations and making predictions.

Fisher's work on population genetics and evolution and his use of them in medical explanation and prediction makes clear that probability and statistics play an important role in science, outside of RCTs. The domain of mathematics employed as the language of his genetical theory of natural selection is probability and statistics. Using probability and statistics in this way – i.e. as

[19] Fisher contributed significantly to their development (Fisher, 1930).

the language of theory – is common in physics, chemistry and biology; its use in RCTs is not.[20]

What is being questioned here is the appropriateness of the use of probability and statistics in RCTs in medicine. In agriculture, many of the presuppositions on which a legitimate use of probability and statistics are based are met; this is the case in medicine. In medicine, randomization is almost always gerrymandered (sampling is not from the entire relevant population, some individuals assigned to a sample are removed after the fact, samples are adjusted to eliminate relevant differences observed after sampling or known to be likely from past experience – difference in age profile or imbalanced gender, for example – and so on). In addition, the assumption of homogeneity that is reasonably robust in Fisher's agricultural work is absent in medicine,[21] which in part accounts for 'side effects' which are often more prevalent that the target effect, the heterogeneity of outcomes[22] and the constant publication of contradictory findings about the same intervention. The heterogeneity in the population and in outcomes undermines any chance of justifying causal claims. In Fisher's agricultural trials, justifying a causal claim on the basis of the trial is plausible. The problem in this case, as I have argued, is that the causal claim is isolated and, hence, cannot provide a basis for explanation; for that a theory is required. Unlike Fisher's agricultural trials, medical trials do not come remotely close to even justifying the assertion of a causal claim.

In agriculture, randomization is restricted to choosing one of a matched pair of plots – not as in medicine to sampling from a population; paired plots can with a high level of confidence be assumed homogeneous for all relevant factors – contrary to the situation in medicine; and the multiple match-plots which are part of each trial provide replication within the trial – in medicine, replication requires new trials which will be few in number and almost certainly dissimilar to the original trial in important ways.

[20] To avoid any confusion, let me be clear that, as the forgoing use by Fisher makes abundantly clear, the importance or power of probability and statistics in science is not in question. Fisher's use of that domain in population genetics, evolution and medicine parallels its use in physics (e.g. statistical mechanics (a deterministic sphere) and quantum mechanics (an indeterministic sphere)), chemistry (e.g.) and biology (e.g. population genetics). Patrick Suppes used it in a compelling way to develop a probabilistic theory of causality (Suppes, 1970). It worth noting in passing that Suppes is quite clear about the role and importance of theory in his account, 'The analysis of causes and their identification must always be relative to a conceptual framework [what I take a currently accepted theory to be], and there is no successful argument apparently that can show that a particular conceptual framework represents some ultimate and correct view about the structure of the world' (Suppes, 1970, pp. 90–91). There are also a host of other ways in which probability is used in science – from determining goodness of fit between predictions deduced from a theory and the experimental data obtained to describing the distribution of chance events.

[21] See: Upshur (2005).

[22] See: Kravitz *et al.* (2004).

2.5 Causality, theories and an eliminative thesis

Causality has had a rough couple of millennia. Aristotle identified four causes (efficient, formal, final and material). Today 'cause' is only associated with his efficient cause; the other three are held in various states of derision. Even efficient cause has been, and still is, under constant attack. It was, for example, pummelled by David Hume and outright rejected by Bertrand Russell. Responses and counter-responses abound. Patrick Suppes opens his excellent *A Probabilistic Theory of Causality* by quoting the relevant passages from Russell's, 'On the Notion of Cause'; he then argues that Russell's position is a relic of a superseded period of physical science – a period in which 'the fundamental physical phenomena in question were felt to be much better understood at a fundamental level than they are today'. Since, I shall provide a neo-Russellian account repeating the quotation here provides an appropriate starting point.

All philosophers, of every school, imagine that causation is one of the fundamental axioms or postulates of science, yet, oddly enough, in advanced sciences such as gravitational astronomy, the word 'Cause' never occurs.... The law of causality, I believe, like much that passes muster among philosophers, is a relic of a bygone age, surviving, like the monarchy, only because it is erroneously supposed to do no harm.... The principle 'same cause, same effect,' which philosophers imagine to be vital to science, is therefore utterly otiose. As soon as the antecedents have been given sufficiently fully to enable the consequent to be calculated with some exactitude, the antecedents have become so complicated that it is very unlikely they will ever recur. Hence, if this were the principle involved, science would remain utterly sterile.... No doubt the reason why the old 'law of causality' has so long continued to pervade the books of philosophers is simply that the idea of a function is unfamiliar to most of them, and therefore they seek an unduly simple statement. There is no question of repetitions of the 'same' cause producing the 'same' effect; it is not in any sameness of causes and effects that the constancy of scientific laws consists, but in 'sameness of relations'. And even 'sameness of relations' is too simple a phrase; 'sameness of differential equations' is the only correct phrase.

Suppes is correct to point out that the natural sciences at the time he was writing (1970) were more conceptually and theoretically complex than in the period Russell was writing (1910–15); the full impact of Einsteinian relativity (his special theory of relativity was published in 1905 and his general theory of relativity was presentation to the Prussian Academy of Science in 1915) had not occurred and quantum theory only began to coalesce in the 1920s. Since Suppes' observation, the natural sciences have been influenced by chaotic dynamical systems, fractal mathematics and computer simulation to mention but a few important factors. These changes make Suppes' observation more apt. Notwithstanding, however, this correct observation of Suppes', I contend that Russell uncovered the kernel of a profound reinterpretation of causality.

The essence of the argument can be sketched by reflecting on the history of teleology in physics and biology. In the period from Galileo to Newton, the role of teleological accounts of phenomena shrank in physics and astronomy. After Newton, its role was miniscule to non-existent – bursts of attempted resuscitations were unsuccessful. There are cases where the language used appears to invoke a teleological account. For example, it might be said that a particular missile is 'seeking' a fighter aircraft. The missile changes trajectory as the target moves. That phenomenon is a function of an internal positive/negative feedback mechanism; the language of 'seeking' is simply a shorthand expression that can readily be replaced by the mechanistic one. No physicist or engineer – indeed no moderately educated person – would really believe that the missile was 'seeking' the aircraft and was directing its behaviour to achieving that goal. It is simply behaving in accordance with its internal structure and program. If anything can be identified as a goal, it is a humanly constructed goal of incapacitating the aircraft within a larger military goal. Whether this 'goal' is irreducible to a mechanism is complex and irrelevant to the goal-directed language of the physical entity and its behaviour; the physical entity and its behaviour have no intrinsic goals.

In biology, teleological accounts of phenomena were alive and well until the second half of the nineteenth century; the publication of Darwin's *Origin of Species* began the slow decline in the use of teleological accounts. It is not that the use of teleological language was purged from biology. Indeed, today it is easy to find such language in biological books and articles – especially those dealing with the behaviour of organisms. What changed with Darwin was how this language was understood.

When a biologist remarks, 'Hymenoptera perform this dance **in order to** communicate the direction and distance of a food source to other workers,' there is no attribution of intentions. The use of 'in order to' is shorthand for a mechanistic understanding. The behaviour has a genetic basis and is, therefore, biologically programmed and heritable, and the genes responsible for the behaviour have become ubiquitous in that species because they enhance reproductive success. There is no goal of communication, in any teleological sense – though information is in fact conveyed. To the extent anything can be considered a goal, it is the reproductive success of the individual organisms.

Although Suppes' (1970) observation that causal language is pervasive in modern sciences remains true today, I contend that it is, nonetheless, like the use of teleological language; it is a shorthand expression, which owes any meaning and validity to the existence of models and theories. It can, and often should, be eliminated in favour of a mechanistic theoretical account. The claim 'A caused B' is shorthand for 'The claim, whenever the system is in state A, the next state of the system will be B (either always or with $\Pr(x)$) can be deduced from a currently accepted and well-confirmed dynamical theory.'

On this interpretation of 'cause', even in agricultural trials, RCTs do not justify causal claims; only a theory can do that. In addition, this interpretation goes to the heart of a diagnosis of the problem with RCTs in medicine. The problem identified above is that in the absence of a theory, RCTs do not provide explanations or allow predictions. That is a hefty shortcoming, and were not so much at stake it would make the claim that RCTs are a gold standard risible. They do not provide explanations and predictions because the results, unless connected in ways I have described to a theory, stand isolated; an isolation, this reinterpretation of 'cause' suggests, renders causal claims made on the basis of RCTs vacuous.

In physics, chemistry and biology, Russell overstated the case with his claim, 'The law of causality, I believe, like much that passes muster among philosophers, is a relic of a bygone age, surviving, like the monarchy, only because it is erroneously supposed to do no harm.' In those contexts, using the shorthand language of causality almost never results in researchers failing to grasp the importance and role of a theory. Those researchers quite naturally – almost unconsciously – appeal to theories in making predications and providing explanations. Russell's claim, however, has considerable validity in clinical medicine that focuses on RCTs. An impoverished interpretation of causality that divorces it from theories results in significant methodological, epistemological and logical harms. In turn, these result in the harm of suspect findings and claims, and poorly or improperly understood interventions. At the heart of the harm is the undermining of the validity of explanations, predictions and clinical treatments; and that is far from a trivial harm.

The conclusion to draw from all this is not that RCTs and the results of RCTs have no role in medicine; they do. The appropriate conclusion is that their role is dependent on being integrated into (indeed subservient to) a theory. Fortunately, in spite of the emphasis on RCTs, robust theories, that can be used to ground RCTs, abound in medicine; from immunology through to physiology and endocrinology to neurosciences robust theories are found. It is these that have provided the solid, lasting basis for medical explanation, prediction, diagnosis and treatment. Consider, for example, the compendium on immunobiology by Charles Janeway Jr.[23] or the text on medical genetics by Margaret Thompson *et al.*[24] Their entire treatment of their subject is experimental and theoretical, and, most notably, RCTs play no role in the evidence, explanations and predictions provided.[25]

[23] Janeway (1997).

[24] Thompson *et al.* (1996).

[25] The same is true in many other research fields in medicine as a perusal of medical texts in medical genetics, human physiology, neurosciences will reveal.

To appropriate Dobzansky's famous claim that, 'Nothing in biology makes sense except in the light of evolution',[26] 'nothing makes sense in medicine except in the light of a theory'.

Acknowledgements

I am grateful to two reviewers for providing suggestions for strengthening the expositions and arguments of this paper.

References

Avins, A.L., Bent, S., *et al.* (2005). Use of an embedded N-of-1 trial to improve adherence and increase information from a clinical study. *Contemporary Clinical Trials* 26 (3): 397–401.

Bradwardine, T. (*circa* 1330). *Tractatus de Continuo.* [A Latin text based on the three extant manuscripts can be found in J.E. Murdoch (1957) Geometry and the continuum in the fourteenth century: a philosophical analysis of the Thomas Bradwardine's Tractatus de continuo (PhD dissertation: University of Wisconsin)].

Braunwald, Eugene *et al.* (eds.) (2001). *Harrison's Principles of Internal Medicine,* 15th edition, New York: McGraw-Hill.

Bromberger, S. (1966). Why-questions, in R.G. Colodny (ed.) *Mind and the Cosmos.* Pittsburgh: University of Pittsburgh Press, pp. 86–108.

Cartwright, N. (2007a). Are RCTs the gold standard? *BioSocieties,* 2: 11–20.

Cartwright, N. (2007b). *Hunting Causes and Using Them.* New York: Cambridge University Press.

Cartwright, N. (with Andrew Goldfinch and Jeremy Howick) (2008). Evidence-based policy: Where is our theory of evidence? in A. Beckermann, H. Tetens and S. Walter (eds), *Philosophy: Foundations and Applications.* Paderborn: Mentis-Verlag.

Dobzhansky, Theodosius (1964). Biology, molecular and organismic. *American Zoologist* 4: 443–452

Dobzhansky, Theodosius (1973). 'Nothing in Biology Maths sense except in the light of evolution *American Biology Teacher,* 35, No. 3(March): 125–12.

Elwood, M. (1998). *Critical Appraisal of Epidemiological Studies and Clinical Trials* 2nd edition. Oxford: Oxford University Press.

Fisher, R.N. (1930). *The Genetical Theory of Natural Selection.* London: Oxford University Press.

Fisher, R.A. (1935). *The Design of Experiments.* Edinburgh: Oliver & Boyd.

Fisher R.A. and Fraser Roberts, J.A. (1943). A sex difference in blood-group frequencies. *Nature,* 151: 640–641.

[26] Dobzhansky (1964) p. 449, see also, Dobzhansky (1973). Dobzhansky meant by 'evolution' both the fact that it occurred and, most importantly for him, the modern synthetic theory of evolution.

Fisher, R.A., Race, R.R. and Taylor, G.L. (1944). Mutation and the Rhesus reaction. *Nature*, 153: 106.
Galileo, G. (1623). *Il saggiatori* (reprinted in English translation in C.D. O'Malley and S. Drake (eds) *Controversy on the Comets of 1618*, Philadelphia: 1960.
Guyatt, G.H., Keller, J.L., Jaeschke, R., Rosenbloom, D., Adachi, J.D., Newhouse, M.T. (1990). The *n*-of-1 randomized controlled trial: clinical usefulness. Our three-year experience. *Annals of Internal Medicine* 112 (4): 293–299.
Hartl, D.L. and Clark, A.G. (1989). *Principles of Population Genetics*. Sunderland, Massachusetts: Sinauer Associates, Inc.
Haynes, R.B., Sackett, D.L., *et al.* (2006). *Clinical Epidemiology: How to Do Clinical Practice Research* (3rd edition). Philadelphia, PA: Lippincott Williams and Wilkins.
Hempel, C.G. (1967). *The Philosophy of Natural Science*. Englewood Cliffs: Prentice Hall.
Janeway, Jr., C.A., Travers, P. *et al.* (1997). *Immunobiology: The Immune System in Health and Disease* (3rd edition). New York: Garland Publishing, Inc.
Kuhn, Thomas S. (1962). *The Structure of Scientific Revolutions*. Chicago: University of Chicago Press.
Kravitz, Richard L. *et al.* (2004). Evidence-based medicine, heterogeneity of treatment effects, and the trouble with averages. *The Milbank Quarterly* 82: 661–687.
Lind, J. (1753). *A Treatise on Scurvy*. Edinburgh: Sand, Murray and Cochan.
Quine, W.V. and Ullian, J.S. (1970). *The Web of Belief*. New York: Random House.
Rothman, K.J. and Greenland, S. (eds.) (1998). *Modern Epidemiology* (2nd edition). Philadelphia, PA: Lippincott Williams and Wilkins.
Suppe, F. (1967). *On the Meaning and Use of Models in Mathematics and the Exact Sciences*. Ann Arbor: University Microfilms International (Ph D dissertation).
Suppe, F. (1972). What's wrong with the received view on the structure of scientific theories? *Philosophy of Science* 39: 1–19.
Suppe, F. (1989). *The Semantic Conception of Theories and Scientific Realism*. Urbana: University of Illinois Press.
Suppes, P. (1957). *Introduction to Logic*. Princeton: Van Nostrand.
Suppes, P. (1961). A Comparison of the meaning and Uses of Models in Mathematics and the Empirical Sciences, in Hans Freudenthal (ed.) The concept and the Role of th Model in Mathematics and Natural and social sciences Dordrecht: D, Reidel, (1961) pp. 163–177.
Suppes, P. (1962). Models of data, in E. Nagel, P. Suppes and A. Tarski (eds.) *Logic, Methodology and Philosophy of Science: Proceedings of the 1960 International Congress*. Stanford: Stanford University Press, pp. 252–261.
Suppes, P. (1967). What is a scientific theory? in S. Morgenbesser (ed.) *Philosophy of Science Today*. New York: Basic Books, pp. 55–67.
Suppes, P. (1968). The desirability of formalization in science, *Journal of Philosophy* 65: 651–664.
Thompson, M.W., McInnis, R.R., *et al.* (1966). *Genetics in Medicine* (5th edition). Philadelphia, Pennsylvania: W. B. Saunders Company.
Thompson, P. (1983). The structure of evolutionary theory: A semantic approach. *Studies in History and Philosophy of Science* 14: 215–229.

Thompson, P. (1986). The interaction of theories and the semantic conception of evolutionary theory. *Philosophica* 37: 73–86.

Thompson, P. (1987). A defence of the semantic conception of evolutionary theory. *Biology and Philosophy* 2: 26–32.

Thompson, P. (1989). *The Structure of Biological Theories*. Albany: State University of New York Press.

Thompson, P. (2007). Formalisations of evolutionary biology, in M. Matthen and C. Stephens (eds.) *Philosophy of Biology*. Amsterdam: Elsevier, pp. 485–523.

Upshur, Ross E.G. (2005). Looking for rules in a world of exceptions: Reflections on evidence-based practice. *Perspectives in Biology and Medicine* 48: 477–489.

Urbach, P. (1993). The value of randomization and control in clinical trials. *Statistical Medicine*, 12: 1421–1431.

van Fraassen, B.C. (1967). Meaning relations among predicates. *Nous* 1: 161–179.

van Fraassen, B.C. (1969). Meaning relations and modalities. *Nous* 3: 155–167.

van Fraassen, B.C. (1970). On the extension of beth's semantics of physical theories. *Philosophy of Science*, 37: 325.

van Fraassen, B.C. (1972). A formal approach to philosophy of science, in R.E. Colodny (ed.) *Paradigms and Paradoxes*, Pittsburgh: The University of Pittsburgh Press.

van Fraassen, B.C. (1980). *The Scientific Image*. Oxford: Clarendon Press.

Weisheipl, J.A. (1967). Galileo and his precursors, in E. McMullin (ed.) *Galileo: Man of Science*. New York: Basic Books.

Worrall, J. (2002). What evidence in evidence-based medicine? *Philosophy of Science* 69: S316–S330.

Worrall, J. (2007). Why there's no cause to randomize, *British Journal for the Philosophy of Science* 58 (3): 451–488.

·3·

Inferring causation in epidemiology: Mechanisms, black boxes, and contrasts

Alex Broadbent

Abstract

This chapter explores the idea that causal inference is warranted if and only if the mechanism underlying the inferred causal association is identified. This *mechanistic stance* is discernible in the epidemiological literature, and in the strategies adopted by epidemiologists seeking to establish causal hypotheses. But the exact opposite methodology is also discernible, the *black box stance*, which asserts that epidemiologists can and should make causal inferences on the basis of their evidence, without worrying about the mechanisms that might underlie their hypotheses. I argue that the mechanistic stance is indeed a bad methodology for causal inference. However, I detach and defend a mechanistic interpretation of causal generalisations in epidemiology as existence claims about underlying mechanisms.

3.1 Causal hypotheses in epidemiology

What does it take to establish a causal hypothesis in epidemiology? What standards need to be met? Or, if establishment comes in degrees, degrees of what?

The most obvious aspect of this problem concerns inferring causation in a particular study. A study reveals a statistical association between smoking and lung cancer, or a certain gene and obesity. Statistical analysis reveals a low *p*-value – a measure of the chance that the association is due to chance. Study design controls for confounding variables (what philosophers would call common causes of the putative cause and effect). Can it be inferred that, for *this* group, a causal relationship exists between smoking and lung cancer, or having that gene and obesity?

Oddly enough, this is not a question that epidemiologists like to answer. A single study would not normally be considered a sufficient basis for a causal inference. Replication is a guiding epidemiological principle. From a methodological point of view this is extremely interesting. Epidemiologists' credence in a causal hypothesis about Study Group A increases when the effect is replicated in Study Group B. Explaining (or, I suppose, refuting) this attitude is a central task for any methodological analysis.

A second difficulty concerns the inference from a study, or a collection of studies, to a wider population. Epidemiologists are centrally concerned with extrapolating from the people they study to people they have not studied. Replication is important here too, because one way to argue that differences between the population studied and the target population are causally irrelevant is to replicate the study among people who are drawn from the target population. However, replication cannot solve the problem of generalisation. Often the study group will *already* be drawn from the target population: for example, when generalizing from the Whitehall studies to the population of Britain.[1] Differences between those studied and those not studied will always remain; the difficulty is working out when these differences make a difference. On other occasions, studies on a subset of the target population may be impractical: for an obvious example, consider future populations. Quite generally, a central purpose of epidemiology is to get more for less: to learn something not only about those who *have* been studied, but about those who *have not*.

Epidemiologists make efforts to be precise about the scope of their claims, by explicitly stating whether they are intended to apply to the group studied, or to a wider population, and if the latter what conditions the wider population are to meet. For example, instead of saying 'genetic influences cause paediatric obesity', they might say:

> Genetic influences on BMI and abdominal adiposity are high in children born since the onset of the pediatric obesity epidemic.
>
> (Wardle *et al.* 2008, p. 398)

However, even this is an incomplete specification. The children studied were British, but the obesity epidemic affects Europe and America too. Are these results evidence for high genetic influences on BMI and adiposity in children in Britain only, or in Europe and America too? The incompleteness of the specification is not necessarily a failing of the authors of the study. It reflects the genuine difficulty of deciding how to generalize.

Note that replicating the study in European or American children is a way to *avoid* the question, not to answer it. Replication cannot tell us whether a generalisation from this study to American children would be *warranted*, only whether it would be *correct*. In circumstances where we can replicate, that may

[1] The Whitehall studies concerned various aspects of social status and health among British civil servants (Marmot 2006). The aim was to identify social determinants of health, especially the role of purely social differences in explaining differences in health, by studying a group of people whose absolute wealth was such that their basic biological needs were met.

be the best strategy; but for reasons I have already given, we cannot universally substitute replication for generalisation.

Thirdly, there is a difficulty interpreting general claims, even when their scope is fixed. A great deal of philosophical attention has been directed towards singular causal claims, such as 'Jones' smoking caused his lung cancer'. But epidemiologists are almost exclusively concerned with general claims, such as 'smoking causes lung cancer'. Does the latter express a relationship between smoking and lung cancer, or is it a generalisation over individual causal relationships – along the lines of '*in X% of cases*, smoking causes lung cancer'?

This difficulty is a relatively familiar one to philosophers, but it is perhaps not the most pressing one for epidemiologists. In practice, epidemiological hypotheses are explicitly exception-ridden. Accordingly they are framed not as universal generalisations, but as measures of the *influence* of one factor on an outcome, or measures of the *strength* of an association, or of the *proportion* of an effect that is due to a particular factor or group of factors. These sorts of claims raise what is fundamentally the same problem, but in a slightly different way. For example, saying 'Genetic influences... are high' makes it clear that the generalisation is not *exceptionless*. But it still does not make clear exactly how the degree of influence is to be interpreted. Is the claim that, in each individual, the genetic influence is high? – This would amount to a universal generalisation attributing a certain genetic influence to each individual. Such an interpretation is hard to make sense of on either the effect side or the cause side. On the effect side, obesity might be absent in some of the individuals studied. Genetic factors cannot then be said to influence it. Switching from a qualitative property (obesity) to a quantitative one (such as bodyweight) will not always be straightforward: the absence of effects such as lung cancer, diabetes, and suicide are hard to interpret as zero degrees on any quantitative scale. Similarly on the cause side, it makes little sense to attribute some degree of influence to a factor that is absent. This is not clear in the example I have picked, since 'genetic influences' are always present in people, but it is obvious when we consider single-gene conditions. When somebody lacks the gene but has the trait in question, it makes no sense to attribute the trait's presence to the influence of the absent gene in *any* degree.

Another interpretation would see measures of influence, proportion, strength of association, and similar, as measures of the *proportion of cases in which a factor is causal*. (This interpretation is akin to the generalisation-over-singular-causation view of universal causal generalisations.) This view is easy enough to make sense of, but there is a case that it reflects metaphysical commitments rather than epidemiological evidence. Take a measure such as heritability, which is the proportion of a given trait in a given population that is due to genetic factors. The idea that causes are either present or absent, strictly speaking, and the view that we cannot quantify the contribution of a *particular* cause to a *particular* effect, are widespread among philosophers from John Stuart Mill to David Lewis (Mill 1843; Lewis 1973). On such a view, heritability expresses the proportion of the *population* in which the trait is caused by genetic factors – but in each *individual*, the trait either is or is not caused by genetic factors. But this interpretation is not stable, because on this metaphysical picture, causation is not exclusive. Saying that a trait is caused by genetic factors in an individual is compatible with saying that it is caused by non-genetic factors: events have many causes. In the study I have been using as an example, the aim is to measure the *contribution* of genetic influence to obesity in a population. To interpret this as a claim about the proportion of individuals in whom genetic factors cause obesity would be bizarre, since genetic factors are part of the causal history in 100% of cases of obesity. Indeed, every trait is both 100% genetic and 100% environmental, on this interpretation (Rothman and Greenland 2005, S146). Better, we could see it as a claim about the proportion of individuals in whom genetic factors *make the difference* between being obese and not. This interpretation might be made to work; but that would be no trivial philosophical achievement. The interpretation of heritability is a topic of considerable dispute (e.g. Schonemann 1997; Sesardic 2005).[2]

Two lines of response to this bundle of difficulties may be discerned in the contemporary methodological-epidemiological literature. These lines of response are in tension. One is the *mechanistic stance*: the view that causal inference in epidemiology aims at discovering mechanisms: that discovering mechanisms is necessary and sufficient for establishing a causal hypothesis. The other is the *black box stance*: the view that epidemiology is primarily concerned with statistical analysis of associations, and only incidentally concerned with uncovering mechanisms. In Sections 3.2 and 3.3 I will describe and evaluate each stance, and in Section 3.4 I will propose a resolution.

My terms 'stance', 'line of thought', and similar are intended to avoid commitment on the question of whether any actual epidemiologist wholeheartedly asserts any of the views discussed. I rather doubt that any does. Nonetheless, these are not straw men: these stances are evident in the methodological writings of actual epidemiologists, and there is value in seeking to draw them out into the light for explicit evaluation, even though – indeed, partly because –

[2] Heritability is not to be confused with heredity. An individual can inherit a trait from her parents, for example her eye colour. The disposition to develop blue eyes given a certain broadly congenial environment is hereditary; favourite colour probably is not. The heritability of a trait is defined with reference to a population, and makes no sense applied to an individual. It is meant to be a measure of the *relative* contribution of genes and environment, and is not fixed for a given trait. Hence Wardle's interest in showing that obesity is *still* highly heritable despite an increase in the availability of calories.

nobody would endorse these views when stated explicitly and taken to their logical conclusions.

3.2 Mechanisms

There has been a surge of interest in mechanisms in recent philosophy of science. One well-known definition is this:

> A mechanism for a behavior is a complex system that produces that behavior by the interaction of a number of parts, where the interactions between parts can be characterized by direct, invariant, change-relating generalisations.
>
> (Glennan 2002, S344)

The word 'mechanism' is less common in epidemiology than it is in some other biological sciences. Nevertheless, it would be interesting if a similar idea could be identified in actual epidemiological practice.

I think it that it can. Epidemiologists, like neuroscientists, use the term 'mechanism', and they do so in revealing ways, as I shall argue. But in addition, I suggest that it is plausible to take the common epidemiological phrase 'causal pathway' as referring to a mechanism. Perhaps a causal pathway is not exactly the same thing as a mechanism: for one thing, it may be longer, and include several 'mechanisms' in the sense intended by Glennan. Nevertheless, causal pathways probably do meet the proffered criterion for mechanisms, since they typically will be postulated to explain a 'behaviour', and will plausibly constitute a 'complex system' whose parts interact according to 'direct, invariant, change-relating generalisations'. Moreover, identifying a mechanism in neuroscience, and identifying a mechanism in epidemiology, satisfy the same goal: they both explain *how something works*. For these reasons, it is plausible to see the search for causal pathways in epidemiology as a search for what Glennan and others call 'mechanisms', even if the causal pathways identified in epidemiology are not terribly similar to the mechanisms in neuroscience.

The notion of mechanism neatly captures one methodological story that may be discerned in the epidemiological literature. That story has two parts. Initially, associations are identified between variables and health outcomes. By a sort of process of elimination, it is established that it is very unlikely that the association is due to chance, or to confounding variables (common causes, to philosophers) or other biases. A variable for which such an association has been established is called a *risk factor*. This first part of the process provides a good reason to think that the risk factor is causal, but a second stage is required for a conclusive case. The *mechanism* for the operation of the risk factor must be identified. Perhaps not immediately; but if in the fullness of time no mechanism is identified, the credibility of the hypothesis will suffer.

Moreover establishing a mechanism is sufficient for proving that a causal hypothesis is correct: showing *how* A actually causes B is conclusive proof that A *does* cause B.[3] The *mechanistic stance* is the methodological position that identifying the underlying mechanism is both necessary and sufficient for warranted inference to a causal hypothesis.

As a description of contemporary epidemiological methodology, the mechanistic stance is appealing. It has plenty of illustrations. Genetic epidemiology is an extremely good fit. Consider this extract from the introduction of another high profile clinical study in the genetics of obesity:

> The genetic contribution to body weight has been established through family studies, investigation of parent–offspring relationships, and the study of twins and adopted children ... As is the case for height ... environmentally driven changes in body weight in the population occur against a background of susceptibility to weight gain that is determined by genetic factors. Thus, genetic approaches can be applied to understand both the molecular and physiological mechanisms involved in human obesity.
>
> (Farooqi and O'Rahilly 2006, p. 710)

The rationale appears to follow exactly the lines of the mechanistic stance I have described. There is considerable evidence for a causal link between genetics and weight; the next step is to understand the mechanisms involved.

A similar sentiment is evident at what might be seen as the other end of the epidemiological spectrum, among those who work on social determinants of health. They, too, see it as crucial to identify plausible pathways for the determinants they identify. Here is an extract from another introductory rationale, this time from a chapter by two leading proponents of social determinants of health research:

> Psychosocial factors and their influences on health are active areas of research ... There is now enough evidence to suggest that this is an important field for those concerned with improving public health ... Plausible mechanisms linking psychosocial factors to health are described in the first half of this chapter. We then look to the evidence from both human and animal literature to illustrate the ways in which social organisation can influence our biology and, therefore, the health of individuals and populations.
>
> (Brunner and Marmot 2006, p. 8)

This passage mixes the mechanistic stance as a purely methodological stance with the natural companion view that identifying mechanisms is a good way to improve public health. Setting that aside, the mechanistic stance is clearly discernible. There is already evidence that psychosocial factors influence health; the purpose of the chapter is to work out how, by postulating 'plausible mechanisms' and presenting 'evidence from both human and animal litera-

[3] Showing that a given mechanism is how A *actually* causes B must not, of course, be confused with exhibiting a mechanism by which A *could* cause B. The entailment goes through only when what has been shown is how A *actually* causes B.

ture' in support. Whether or not this is the authors' intention, they certainly give the impression that providing a mechanism a key element in the case for the hypothesis that social status is a determinant of health.

In each of these cases, the identification of mechanisms is seen as important: important enough to devote an entire paper to. Why? It *could* be simple scientific curiosity: the desire to find out how things work, just for the sake of it. But I think there is more to it than that. Identifying mechanisms is presented as important not only to complete the scientific picture, but also to seal the case for existing causal hypotheses. This is especially clear in the social determinants of health literature, where the identification of pathways is seen by proponents and detractors alike as crucial for the case that the socioeconomic factors identified as health determinants really do have the effects claimed for them. Michael Marmot's Whitehall studies provide evidence for a causal link between social status and health, among British civil servants – a population whose basic biological needs (food, water, shelter) are amply met. Marmot's efforts to confirm this hypothesis have not focused merely on replicating the results in different populations (though that is of course one area of activity). A considerable amount of effort has also been devoted to identifying mechanisms by which social status might affect health.

This suggests a methodological thesis: that identifying an underlying mechanism is both necessary and sufficient for establishing a causal hypothesis.[4] I am not suggesting that Marmot or anyone else endorses this thesis: indeed it may be that no actual epidemiologist would sign up to it, in that blunt form. But identifying mechanisms clearly *is* relevant to establishing causal hypotheses in epidemiology, and setting up a somewhat extreme but clear stance may be a useful technique for exploring that relevance.[5]

It is not hard to see why identifying a mechanism might be considered necessary and sufficient for establishing a causal hypothesis, at least in epidemiology. An argument for its being necessary might appeal to the fact that epidemiology is clearly not a fundamental science. Causal associations identified in epidemiology presumably exist in virtue of the way that things are organized at a more fundemental level. The existence of causal associations at the population level is naturally seen as arising from certain regularities in the way that the members of that population are organised, and regularities in their

[4] Where there is more than one distinct mechanism underlying a given association, presumably all the underlying mechanisms would have to be identified to satisfy the spirit of this requirement.

[5] The mechanistic stance is *not* supposed to be the view that identifying mechanisms is the goal of epidemiological research: only of causal inference. Causal inference may have other goals, such as public health intervention, or indeed informing further causal inferences in an iterative process.

environment. Identifying a mechanism is just identifying the properties and activities of the population's members and environment which together give rise to the population-level association. If no mechanism can be identified, then the status of the population-level causal association remains mysterious. Failure to identify such properties and activities is not always evidence that they do not exist, of course, and perhaps this is why Austin Bradford Hill famously urges that 'biological plausibility' be treated with caution (Hill 1965). Nevertheless, a hypothesis for which no mechanism is *remotely* plausible, or for which no mechanism is discovered after a long period of time, remains at best tentative.

It is similarly obvious why identifying a mechanism underlying a causal association might be considered sufficient for establishing the corresponding causal hypothesis. If the events and activities giving rise to an association are identified, then it presumably follows that the variables which the hypothesis asserts are causally linked at least *can* be causally connected. There is an interesting twist here, however. Showing that a mechanism exists by which, say, stress can cause poor health, does not bear directly on the claim that stress *does* cause poor health in any particular population. The mechanisms identified by Marmot are ways in which the results of the Whitehall studies *might* have come about. This explains his two-part strategy outlined in the excerpt quoted previously, of first identifying mechanisms and then arguing for their actual operation in humans and animals. So identifying a mechanism is not on its own sufficient for establishing a causal hypothesis; a further inference is required to the claim that the identified mechanism is indeed the explanation of the causal association asserted by the hypothesis in question. What is sufficient, then, to establish a causal hypothesis, is the identification of the mechanism *actually* responsible for the association, not merely a mechanism which physically *could* be responsible for it. I take it that the fundamental motivation for this stance is that it follows from the claim that A causes B in a particular way, that A causes B simpliciter. Showing *how* A causes B is only possible if, or in other words entails that, A does in fact cause B.

The mechanistic stance responds as follows to the difficulties facing causal inference in epidemiology which I identified in Section 3.1. In answer to the question when we are justified in inferring causation for a studied group, the mechanistic answer is presumably, 'Not until a mechanism has been identified'. A single epidemiological study will not usually identify a mechanism, perhaps explaining why single studies typically provoke replication and further research, rather than a causal inference. But if epidemiology seeks to uncover mechanisms, then there may be another, more subtle reason that replication is important. Replicating in Study B an association observed in Study A provides evidence that the same mechanism underlies *both* associations; it seems prima facie less likely that two different mechanisms gave rise to the two associations, than that one did.[6] This explains why replication by Study B can confirm a hypothesis about Study A: because Study B provides further reason to think that some mechanism underlies both studies, and thus provides further reason to think that some mechanism underlies the association first observed in Study A. And showing that some mechanism underlies an association is sufficient for showing that the association is indeed causal, on the mechanistic stance. This methodological stance therefore explains and vindicates the reluctance of epidemiologists to make causal inferences on the basis of individual studies, and the importance they attach to replication.

The mechanistic stance also provides useful clarification of the other difficulties we identified, the problem of generalizing from studies to population, and the question of how to interpret general causal hypotheses in epidemiology. Generalizing to a wider population is typically safer when the mechanism underlying a causal association has been identified, because knowledge of the mechanism yields detailed knowledge of what differences are relevant to the association. For example, our knowledge of the mechanism underlying the analgesic effect of paracetamol enables us to identify the circumstances relevant to this effect, and thus to say whether the effect will hold in a very wide range of circumstances. Moreover, it also gives us a lot of other useful knowledge, for example, about *other* associations – other effects of taking paracetamol. These uncontroversial facts motivate a methodological idea: that the generalisation of an association observed in a particular study to a wider population is only really warranted when the mechanism underlying the association has been identified. generalisation before a mechanism has been identified may of course be required in some circumstances, but until the mechanism has been identified, generalizing remains a sort of guessing, according to the mechanistic stance: because we can't be sure exactly *when* our causal generalisations will hold, until we know *why* they hold.

This suggests an interpretation of general causal claims, such as 'smoking causes lung cancer' or 'genetic influences on BMI and abdominal adiposity are high in children born since the onset of the pediatric obesity epidemic'. Such claims are to be interpreted, first, as asserting the existence of a mechanism – or perhaps several – linking the causal and effect variables, and second, as claiming that the operation of this mechanism(s) is (are) in fact what underlies the causal association between the two variables. The mechanism may or may not be known. When it is, this interpretation is very natural. 'Paracetamol causes pain relief' is naturally precisified, not by specifying more exactly the probability that an analgesic effect will be observed in various circumstances, but rather by saying more about the way paracetamol works. Detailed description of the mechanism *yields* information about whether an analgesic effect

[6] This is an application of what is sometimes called the Common Cause Principle.

will be observed in a wide range of circumstances, in an efficient way. When a mechanism is not known, the causal generalisation is a sort of stand-in: a claim that some unknown mechanism does link the variables.

The exception-ridden nature of epidemiological generalisations is unproblematic, on this view. Suppose a causal generalisation is just a claim that a mechanism underlies the association between two variables C and E. That is to say, in individual cases where an individual c leads to an individual e, a mechanism connects c and e. The exception-ridden nature of the generalisation reflects the fact that sometimes, C-events occur, but the mechanism by which they give rise to E-events is absent, and so E-events fail to occur. Likewise, the fact that C-events can by a certain mechanism give rise to E-events does not preclude E-events from coming about some other way. Causal generalisations are claims about the actual linking of individual C-events and E-events by actual instances of a mechanism; such claims are entirely compatible with C-events occurring without the E-producing mechanism, or E-events occurring as the result of some other mechanism. The content of the causal generalisation is that a certain mechanism *is responsible for the causal association between the identified variables*; it is not a claim about the association itself, but about how the association arises. Further support for this interpretation of exception-ridden causal generalisations derives from one of the points argued in the last paragraph, that when a mechanism is known, it is natural to precisify a causal generalisation by specifying the underlying mechanism, rather than offering more statistical detail about the circumstances in which the association holds. Whether or not it is a model for other sciences, it seems to be a good fit for epidemiology.

The one trouble previously identified which the mechanistic stance does not seem to handle is the worry about measures of proportion of influence, such as heritability. It can handle causal generalisations amounting to associations of less than 100%, such as 'smoking causes lung cancer', in the manner already indicated. But a claim such as 'pediatric obesity is over 70% heritable' does not submit to a mechanistic interpretation, since presumably mechanisms linking genes to weight exist in every human being. I am, however, inclined to think that this reflects badly on the concept of heritability, and similar attempts to apportion causal responsibility; such concepts are hard to make sense of in any analysis. It may be that they have no clear sense, or that they need further clarification (cf. Lewontin 1974; Schonemann 1997). Accordingly I propose to set them aside, and focus on interpreting those epidemiological hypotheses that clearly do make sense.

The mechanistic stance has a lot to recommend it, then: it provides a neat interpretation of several tricky features of actual epidemiological practice, and thus vindicates that practice. The idea that epidemiologists ought to first identify risk factors, and then look for the mechanisms underlying them, sounds like both sensible methodological advice and a fair description of actual epidemiological methodology. In the next section, however, I will identify and amplify some doubts that some epidemiologists have expressed about the mechanistic stance.

3.3 Black boxes

Notwithstanding the foregoing, there are some reasons to doubt that the mechanistic stance provides a good methodology for causal inference in epidemiology. Both the necessity and the sufficiency of discovering a mechanism for inferring causation are open to criticisms of principle, and these criticisms can be illustrated by actual episodes in epidemiology.

Let us start with the idea that discovering underlying mechanisms is necessary for the inference of general causal hypotheses. The motivation identified in the previous section for this methodological principle is that the existence of a mechanism is necessary for the existence of a general causal relationship. On the mechanistic interpretation, a causal relationship exists at the population level between two variables only if the particular instances of those variables are related by a mechanism. I suggested that the general causal claim could be interpreted as a claim about the existence of an underlying mechanism. If so, it is natural enough to require that we *discover* this mechanism when we assert a general causal claim. So an inference to a general causal claim requires, as warrant, the discovery of an underlying mechanism – sooner or later.

This 'sooner or later' indicates a difficulty, though. Our failure to discover an underlying mechanism does not mean the mechanism doesn't exist. Moreover, when we are entirely ignorant of a mechanism, it seems that we are often quite incredulous about the existence of one. Combined with the view that a mechanism must underlie causal associations, this native incredulity can lead, and on occasion has led, to an unreasonable prejudice against hypotheses when we do not see how the hypothesised causal link might work. The view that mechanisms need to be identified for a causal inference to be secure can colour our assessment of the evidence for new hypotheses or against accepted hypotheses. I will suggest that this is because it is not really a methodological principle: it tells us what to aim at (discovery of mechanisms) but not how to achieve that goal. As a consequence, it tilts the balance in favour of existing knowledge, and inhibits what it recommends – the discovery of mechanisms about which we do not yet know.

Several well-known episodes from the history of epidemiology illustrate these claims. Perhaps most famously, the miasma theory of the nineteenth century offered a mechanism for the transmission of disease, based on the movements of 'miasms' (roughly, bad air). The fact that the theory purported to give a mechanism for disease transmission was its principle virtue.

Nineteenth century epidemiological heroes such as John Snow and Ignaaz Semmelweis were criticized for failing to identify plausible mechanisms for their causal hypotheses, leading to unnecessary loss of life in both cases. Snow argued, on the basis of incredibly careful door-to-door inquiries, that a causal connection existed between water supply and cholera (Cameron and Jones 1983). Semmelweis that differential childbed fever rates between two wards in a Vienna hospital were caused by the dirty hands of medical students, who worked in one ward but not the other (Carter 1994). Both hypotheses were resisted by the authorities and many doctors (although Snow was somewhat more persuasively successful than Semmelweis), and the principal reason given was that no plausible mechanism for the transmission of disease along these vectors had been identified. It was not until some decades later that a mechanism which might plausibly underlie their respective hypotheses was identified, in the shape of microbial theory (Carter 2003, esp. Ch. 3, 4).

J.P. Vandebroucke argues that Snow had a 'contagionist' hypothesis about the mechanism of disease transmission, and thus that this is not an example of 'black box' epidemiology (Vandenbroucke 1988). Nevertheless, it is doubtful that Snow had what we now view as a *correct* theory of cholera transmission (and contagionism is certainly false as a *general* theory, since it is possible to contract diseases other than from another diseased person, e.g. tetanus; puerperal fever). It is important to distinguish psychological claims about what leads scientists to their theories, from methodological claims about what justifies or otherwise warrants those theories. Snow's results are impressive *to modern epidemiologists,* even though we now believe he was wrong about the mechanism of disease transmission. His proposed mechanism cannot, therefore, be part of the warrant which we *now* accept for his conclusions (whatever he thought); yet we still regard those conclusions as warranted to a high degree by the evidence he procured. Semmelweis also hypothesised a disease mechanism, namely the resorption of animal-organic matter leading to the decaying of the blood (Gillies 2005). But similarly, acceptance of Semmelweis's resorption/decaying blood mechanism surely cannot be part of *our* reason for thinking that any of Semmelweis's causal hypotheses were well-founded, since we believe the resorption/decaying blood mechanism is not what underlay the associations he identified, for instance between disinfecting hands and reductions in differences in mortality between the two wards. Yet even though we reject his account of the underlying mechanism, this does not prevent us from accepting Semmelweis's causal hypothesis that disinfecting hands caused the reduction. Vandenbroucke's argument stumbles on exactly the point I am trying to make, confusing the discovery of mechanism as a *goal* of causal inference, with the discovery of mechanism as *method.* I am suggesting that it is a good goal, but a lousy method. It is not their theories of the mechanism of disease transmission which have elevated Snow and Semmelweis almost to hero status in the eyes of many modern epidemiologists; and it is not the failure of their contemporaries to appreciate these mechanism-theories that is so often lamented. Rather, it is the way they procured evidence to support their causal hypotheses and to refute counter-hypotheses, which is so widely admired; and the way that evidence was ignored that is lamented.

Thus these episodes cast doubt on the usefulness of a methodological principle stating that discovery of mechanism is necessary for warranted causal inference. At least three doubts are distinguishable. First, the obvious logical point that has already been mentioned: that the inference from no known mechanism to no mechanism is a tricky one. Second, demanding that a mechanism be identified before a causal inference is accepted simply seems to be an unreasonable position, because it seems possible to have excellent evidence for a causal link, without understanding how the link works. Even if the mechanistic *interpretation* is plausible, and general causal claims are to be interpreted as existence claims about underlying mechanisms, it does not follow that a general causal claim is only warranted when the underlying mechanism is identified. It is possible, in an epidemiological context, to know that there is a causal link – and therefore that a mechanism must exist, on the mechanistic stance – yet not know what that mechanism is. The opponents of Snow and Semmelweis are generally considered to have been *unreasonable* to doubt the extremely convincing evidence for a causal link; this is difficult to explain if warrant for causal inference requires the identification of a mechanism.

Of course, in these cases, a mechanism *was* eventually discovered. But (and this is the third doubt) the discovery came *after* the causal hypothesis was well-established. The mechanistic methodology gets things the wrong way round, in these cases. The discovery of a mechanism *can* of course help to confirm a causal hypothesis, but a causal hypothesis can also be solidly confirmed well before the underlying mechanism is known. Therefore discovery of an underlying mechanism is not a necessary condition on warranted causal inference.

It must be remarked that drawing morals from historical episodes is a delicate business, because it is possible for different commentators to see different lessons in the same episode. For example, while most commentators (of a moral-drawing sort) would agree that something went wrong in the Semmelweis episode, Federica Russo and Jon Williamson see the episode differently. They insist that, to establish a causal claim, it *is* necessary to identify an underlying mechanism. But they do not adopt the mechanistic stance as I have outlined it, because they do not think that identifying a mechanism is sufficient for causal inference: they hold that it is also necessary to provide what they variously refer to as 'statistical' and 'probabilistic' evidence. In support, they cite episodes where causal hypotheses were supported by evidence of one kind but not of the other, and were rejected on that basis. Thus they cite the Semmelweis episode to illustrate the claim that 'the relation between contamination and puerperal fever ... was not accepted until backed up by mechanistic evidence, i.e. until the germ theory had been developed' (2007, p. 11), as an instance of their more general thesis that identifying mechanisms is necessary (though not sufficient) for causal inference. They go on to propose a theory of causation which is intended to fit this methodological picture.

This line of argument makes me uneasy, because I am unsure whether it is meant as a descriptive account of causal inference (then and now), or as a normative account of the standards which *ought* to be used when deciding whether to infer causation. Suppose we grant (for the argument) Russo–Williamson's descriptive claim, that Semmelweis's contemporaries rejected his theories because of a lack of 'mechanistic evidence', which I take to mean a lack of any then-acceptable theory about what the underlying mechanism for the proposed causal association might be. In this sense, indeed, Semmelweis did fail to establish his various causal hypotheses: he failed to provide evidence which was *in fact* compelling, as demonstrated by the fact that his evidence did not compel many of his contemporaries. But in another sense he clearly *succeeded* in establishing (at least some of) his causal hypotheses: he provided evidence which, in the eyes of most modern epidemiologists, his colleagues *ought* to have taken more seriously. In particular, the evidence for the efficacy of his proposed intervention – disinfecting (not merely washing) hands – is extremely strong. And replication would have made it stronger, without necessarily advancing knowledge of underlying mechanisms.

Unfortunately a purely descriptive reading of Russo and Williamson's claim renders it largely irrelevant from a methodological point of view, and does not justify or explain why they themselves treat it as a motivation to seek a theory of causation, apt for the health sciences. If the claim is merely descriptive, then we may conclude that Semmelweis's contemporaries were simply *wrong* to insist that a mechanism be identified before they accepted any causal connection. (And so we do not need a new theory of causation that would be compatible with this insistence.) Suppose, then, that Russo and Williamson intend their claim to be normative. Then they are making a normative claim that mechanistic (as well as statistical) evidence is necessary for *good, rational, warranted* causal inference. On this reading, the lesson Russo and Williamson draw from the Semmelweis episode is that Semmelweis's theories *ought* not to have been accepted until knowledge of underlying mechanisms was obtained. (This explains why they offer a theory of causation intended to justify this stance.)

But if this is indeed what Russo and Williamson are claiming, then I fear they need a far stronger argument than any they supply. For then they are committed to the startling view that, had germ theory not come along and the underlying mechanism remained a mystery, we today would be rational to dismiss Semmelweis's work, no matter how much evidence we had gathered in the meantime about the efficacy of disinfecting hands. That is not a view which many modern epidemiologists would share. Modern epidemiologists set very high store in some methods, such as the randomized control trial, which involve no requirement to identify underlying mechanisms. Empirical evidence suggests that this view would cost lives if it were adopted by modern epidemiologists, and that indeed it did cost lives if it was in fact the reason that Semmelweis's claims were not accepted more promptly. In short, the descriptive Russo–Williamson thesis does not (without further argument) support any claims about causation, nor bear on questions about how causation *ought* to be inferred; while the normative thesis is false by the methodological standards prevalent in epidemiology, and also, arguably, in light of the empirical evidence from episodes such as the Semmelweis case itself.

In fact, on the mechanistic interpretation, it is not surprising that a causal hypothesis can be well confirmed before the underlying mechanism is discovered. On the mechanistic interpretation I have suggested, a causal generalisation asserts that there *is* some mechanism that is responsible for the association in question. We could have good reason to believe that there is some mechanism, yet not know what it is.[7] Moreover, this order of events suggests a plausible story about how we discover mechanisms when we previously had no idea about them. We make warranted inferences to causal generalisations; these generalisations imply the existence of underlying mechanisms; and we then conduct further research to find the mechanisms. We *know* where to look. It seems, then, that the mechanistic metaphysics does not after all motivate the corresponding methodological principle that the discovery of underlying mechanism is necessary for warranted causal inference.

Stepping for a moment beyond the confines of the methodology of epidemiology, there seems to be little intuitive support for the idea that causal inferences require knowledge of mechanisms as warrant. There are many everyday cases where we take ourselves to have knowledge of causal relations, without having the slightest idea about the mechanisms underlying them. I know that the clear Turkish liquor, rakı, goes cloudy when water is added, but I don't know what the underlying mechanism is. Presumably someone does, and it might be suggested that I can have a warranted causal belief just as long as I have recourse to an expert who can explain the mechanism to me. But the Turks have known for centuries that rakı goes cloudy when water is added. It is absurd to insist that, for centuries before the underlying mechanism was known, the Turks did not know that mixing water and rakı in roughly equal quantities caused the cloudiness they routinelys witnessed.

Moreover, to insist as a general matter that underlying mechanisms must be identified before a causal inference is warranted raises a dilemma. As we go

[7] This is an instance of the more general fact that we can know *that* something exists without knowing what it is. For example, I know that there *is* something holding the floor up, but I don't know what it is; you may know that there *is* something in the cave, but not what, etc.

on uncovering underlying mechanisms, either we will come up against causal relations for which no underlying mechanism can be discovered, or we will not. If we do, then none of our causal knowledge will be secured, because we will have reached a point where we are unable to discover underlying mechanisms and therefore unable to obtain the warrant we sought for our higher-level causal inferences.[8] If we do not, then we will never finish uncovering underlying mechanisms, and thus again we will never obtain the warrant we seek for our causal inferences. As a general methodological principle, then, the requirement that underlying mechanisms be identified, before a causal inference is warranted, is a guarantee that we can never have causal knowledge. If it has any applicability then it must be confined to particular domains, such as causal inference in epidemiology; but historical episodes previously alluded to suggest that it does not work even so confined.

These famous historical episodes have more recent echoes. In an influential (but not uncontroversial) report for the US government, Richard Doll and Richard Peto argued that many environmental causes of cancer could be identified from careful analysis of epidemiological evidence (Doll and Peto 1981). This approach suggests that epidemiologists should treat diseases as 'black boxes' (Peto 1984), and that the identification of a causal mechanism is not necessary for a warranted causal inference. Accordingly, epidemiologists need not concern themselves with the discovery of mechanisms, but can directly attack causal questions without worrying about the mechanisms underlying the hypotheses they generate. As in the cases of Snow and Semmelweis, the pragmatic benefits of this approach are evident. If Doll and Peto are correct, then labouring to uncover mechanisms may well prove to be a waste of time and money, from a public health point of view. Especially where environmental causes are concerned (smoking being the best-known example), refusing to make a causal inference until a mechanism is known can be seriously detrimental to public health. Doll and Peto's recommendation typifies what I will call the *black box stance*.

A great deal of contemporary research is, I think, undecided between the mechanistic and the black box stances. On the one hand, epidemiological research largely proceeds by identifying associations and applying various statistical tests and methodological principles to form a view about whether these associations are causal. On the other hand, the explosion of identified risk factors has not produced a corresponding increase in the scope of our

[8] Might it be objected that warrant of a higher-level causal hypothesis can be conferred by knowledge of an underlying mechanism, even if that underlying mechanism contains causal links for which yet lower-level underlying mechanisms are not known? No: because (on this view) we are not warranted in believing that the causal links in the mechanism underlying our higher-level hypothesis, until we have identified the mechanisms underlying them in turn. Without that warrant, we do not know (on this view) that they are causal links: for all we know (on this view), our putative mechanism may be a coincidental dance of its parts.

understanding of the conditions studied, nor has it been accompanied by a corresponding explosion in public health or medical interventions. This is not just a case of technology lagging: it is also due to the fact that the causal hypotheses in question do not seem terribly reliable. For example, studies seemed to show that hormone replacement therapy reduced risk of heart disease, and public health policies were implemented on this basis, before subsequent studies found the opposite effect (Rutter 2007). A slightly more subtle problem is that directly translating a causal hypothesis into a public health intervention may have unintended consequences (an instance of the generalisation problem). For example, beta blockers administered after surgery appeared to reduce the risk of death by heart attack, but subsequent studies showed that they increased the risk of death overall by increasing risk of death by stroke and other conditions (Devereaux *et al.* 2008). This, of course, is grist to the mill of the mechanistic stance, because one of the chief benefits of discovering the mechanism underlying an association is that it often comes with information about *other* associations, and so makes unintended consequences of this sort less likely.

Requiring that a mechanism be identified before a causal hypothesis is accepted may be too strict; but it does at least have the merit of clarity. It is easy to tell *whether* a mechanism has been postulated; moreover testing the hypothesis by replicating the *mechanism* may sometimes be a more straightforward (because lab-based or clinical) business than replicating the association itself in a large observational study. Whereas analysing the methodology of a published study in order to form a view as to the security of its results is devilishly difficult. Indeed it may be impossible, depending on the accuracy and completeness of the published methods section.

One solution to this tension, then, would be to require the discovery of mechanisms as an admittedly too strong necessary condition on causal inference. In pragmatic contexts where great harm appears to be a real possibility, but where a mechanism cannot be identified, some other decision-theoretic principle, such as the precautionary principle, might be appealed to. This stubborn mechanistic stance is, however, a last resort, because of the difficulties we have been discussing, and because identifying sound principles for decision-making under uncertainty is itself a difficult task. Even erring on the side of safety is not straightforward. The HRT/heart disease example shows that we can be wrong, not only in our causal inferences, but also in deciding which way it is safe to err.

So far I have been focusing on the view that mechanism discovery is necessary for causal inference. What about the claim that it is sufficient? Here, the mechanistic stance might appear to be on stronger ground: showing *how* A causes B seems to entail that A *does* cause B.

Unfortunately, that does not mean that 'look for mechanisms' is a good methodological principle. It has at least two serious weaknesses. First, it is too vague. Ironically, it does not tell us *how* to look for mechanisms. It states a goal, but gives no indication how to get there. This, I suggest, explains why the mechanistic stance has on occasion led to a bias towards existing knowledge. Mechanisms we already take ourselves to know about satisfy the methodological directive; but the same directive doesn't help us find mechanisms we don't know about.

Second, the search for mechanisms can mislead, because it can allow us to believe we have achieved an understanding of something when we have not. Showing *how* A causes B indeed entails showing *that* A causes B. But giving information about an event's causes is not sufficient to explain that event, nor to allow you to devise an effective intervention. If I explain my late arrival by telling you that I was born, I am citing a cause of my late arrival: but I am probably not providing a good explanation of my late arrival. Similarly, identifying a mechanism by which, say, hydrochloric acid leads to ulceration of the stomach lining is indeed sufficient for showing *that* the presence of acid causes ulceration. But there is an explanation for the presence of excessive acid, in the case of many sufferers, namely the presence of bacteria, *Helicobacter pylori*. Showing that stomach acid causes ulceration by identifying the mechanism is a good method for proving causation; but epidemiologists (like other scientists) are typically interested in more than cataloguing the causes of the phenomena they study. They are interested in explaining them, and intervening to change them. Each of these goals plausibly requires the identification of causes. But not just *any* causes. The mechanistic methodology misleads because it provides a sufficient criterion for causation, but no guidance on whether the 'right' causes have been identified, or what the 'right' causes are.

This philosophical point is also well-illustrated by famous historical episodes. I have already mentioned the discovery of *H pylori*. Thoroughly documenting the mechanism by which hydrochloric acid causes stomach ulceration did not lead to the discovery of *H pylori*. Moreover, the hypothesis that peptic ulcer might in many cases be an infectious disease was initially treated with considerable scepticism – because it was thought that bacteria could not survive in such an acidic environment as the stomach. What led to the discovery of *H pylori* was initially chance observation of unknown bacteria in patients with peptic ulcer, followed up by observational studies, clinical work on the bacteria, and a dramatic piece of self-experimentation (for a summary see Angel 2008, Ch 2).[9] Of course, the discovery of the mechanism by which *H pylori* causes ulcer is also an important feature of this episode. I don't mean to deny that discovering mechanisms is important and useful: only that the directive to do so is not a reliable guide for causal inference in epidemiology. This is not because it does not establish causation, but because it does not identify the *right* causal associations. Identifying causes, *any* causes, is not

[9] One of the discoverers, Barry Marshall, drank a solution containing the bacteria and developed gastritis, then took antibiotics and recovered.

enough; there may be *other* causes – like *H pylori* – which better explain the phenomena in question, or offer readier foci for intervention. *Mere* causal inference is not all it is cracked up to be.

What makes the *H pylori* case such a neat illustration is that acid *does* play a role in ulceration. The bacteria cause excessive acid production, which is what directly causes ulceration. So there *is* a mechanism there. In the cases of Snow and Semmelweis, on the other hand, the mechanisms identified we would now regard as non-existent. So in those historical cases, a defender of the mechanistic stance might argue that, had the *real* mechanism of disease transmission been believed rather than miasma theory, the hypotheses of Snow and Semmelweis would have been better received. That may well be so. But this argument confuses reality with our grasp on it. A sound methodological principle cannot rely on our already knowing what we are trying to find out. In the case of miasma theory, many medical scientists *thought* they knew the mechanism of disease transmission. Partly as a consequence, they failed to properly appreciate the evidence before them. This is not simply a case of scientists being convinced of something and failing to give due weight to disconfirming evidence: it is a case of scientists believing they understand *how* something happens, and rejecting causal hypotheses that appear incompatible with this mechanism.

This problem seems to be at least partly what Doll and Peto have in mind when the advocate the black box stance. They argue that epidemiological evidence can warrant many causal inferences, without the underlying mechanisms being known. This sounds like a sort of call to arms for epidemiologists, a rallying cry for them to have faith in the methods of their discipline, and in particular to pay attention to the causal associations revealed by the evidence, without worrying about how the causal associations might work. Nevertheless, the black box stance does not offer much in the way of positive methodological recommendations. And it does not entirely dispense with the fundamental motivation of the mechanistic stance: that if we want to *really* explain – or control – something, we need to know how it works. In the next section I will propose a reconciliation.

3.4 Contrasts

It may be tempting to see the contrast between the mechanistic stance and the black box stance as a disagreement about the goals of epidemiology. One epidemiologist puts it like this:

> ...the epidemiologist who tries to explain, and if possible eliminate, variations in disease occurrence without much regard for mechanisms, stands in contrast to the laboratory scientist who prefers to disentangle the mechanisms first.
>
> (Vandenbroucke 1988, p. 708)

The mechanistic stance might be seen as a more properly scientific view, interested in deep explanation; while the black box stance might be thought of as a more pragmatic view, interested primarily in designing public health interventions by the most direct inferential route available.

This is a misunderstanding, in my opinion. The strengths and weaknesses of the mechanistic stance apply equally to the goals of explanation and intervention. Let me enumerate the principle strengths and weaknesses of the mechanistic stance, starting with the strengths.

(i) Interpreting causal generalisations in epidemiology as existence claims about underlying mechanisms resolves gives a clear and plausible meaning to those generalisations, and helps us to understand the role of replication in epidemiology.

(ii) How widely we can generalize from a known association seems to be directly linked to how well the underlying mechanism is understood.

(iii) Discovering an underlying mechanism proves the truth of the causal hypothesis in question.

(iv) A mechanistic explanation of a causal association increases our understanding of that association.

(v) Knowledge of the mechanism underlying one causal association gives us, or at least can lead to, knowledge of *other* causal associations.

Each of these is a strength, whether the goal is scientific explanation or public health intervention. (i) is perhaps the most philosophical advantage, but it is surely of some importance to the scientist and the public health policy maker alike to have a good grasp on the nature of the causal generalisations they employ. For the scientist, it yields greater understanding; for the policy-maker, the ability to avoid practical consequences of misunderstanding. (ii) is evidently of central interest to the policy-maker, since how widely the results of a given study apply is central to the question of what interventions it warrants. It is also of evident interest for someone who is interested primarily in explanation, since it bears on how much a given causal hypothesis might explain. (iii) is of interest for both explanation and intervention, since false causal inferences are neither explanatory nor reliable guides for intervention. (iv) is evidently of value for those interested primarily in explanation, but it is also of use to the intervention-focused. This is because when we understand a mechanism, we are often able to identify more than one point at which we might intervene on that mechanism (if we want to prevent it) or more than one point at which it might be vulnerable to breaking down (if we want to protect it). (v) is useful from an explanatory perspective, since knowledge of other associations is a symptom of explanatory power, as well adding to the grand total of our knowledge. And it is evidently useful from the point of view of intervention, because it enables us to avoid unintended consequences.

A similar exercise shows the weaknesses of the mechanistic stance to apply regardless of whether explanation or intervention is the goal. To summarize the weaknesses:

(i) A general causal hypothesis can be warranted before the underlying mechanism is discovered (indeed, a warranted causal hypothesis is a great reason to look for a mechanism).

(ii) The directive to seek mechanisms is non-specific, and does not tell us how to find mechanisms; on some occasions, this appears to have led to a bias in favour of known mechanisms and against causal associations for which no mechanism is yet known.

(iii) A causal hypothesis may be deficient even though the underlying mechanism is understood, because it may lead us to believe that we have obtained a greater understanding of a phenomenon than we really have.

From an intervention-oriented point of view, (i) is a serious drawback of the mechanistic stance, because it shows that the mechanistic stance could lead to serious unnecessary delays on intervention. Less obviously, it is also a drawback from the explanatory point of view, because it leads to the unnecessary rejection of good explanations. From an explanatory point of view, the mechanistic stance is guilty of a why-regress fallacy. The 'why regress' is simply the fact that it is always possible to ask 'Why?', including on occasions when the explanation offered is a good one (Lipton 2004, pp. 21–2). What I call a why-regress fallacy is the refusal to accept an explanation on the grounds that the explanation itself has not been explained. As a general rule, such grounds are fallacious, because explanations can be good as far as they go, without providing the entire causal history of the explanandum. Otherwise, epidemiologists would study the Big Bang. (ii) is clearly a problem for any application of the mechanistic stance on causal inference, regardless of motivation. (iii) applies to intervention- and explanation-oriented foci equally, since it means that we miss out on potentially more fruitful interventions or explanations, respectively.

What, then, is the diagnosis of the tension between mechanistic and black box stances? I suggest it arises from a simple confusion of metaphysics and methodology. On the one hand, it does not follow from the strengths listed that discovering mechanisms is necessary for causal inference, nor that the discovery of mechanisms is a sufficient guide for explanation or intervention. None of these points directly motivate the mechanistic methodological stance on causal inference; what they motivate is *interpreting* causal hypotheses as existence claims about underlying mechanisms, and *seeking* these underlying mechanisms. Neither directive tells us anything about the *method* of causal inference; at most, they tell us about the goal. On the other hand, it does not follow from the weaknesses listed that we should abandon a mechanistic interpretation of causal hypotheses, or that we should abandon the search for mechanisms. It only follows that we should not set the identification of a mechanism as a necessary condition for causal inference; nor confuse the power of a mechanistic explanation for a given causal association with the explanatory power of the association itself, with respect to the goals of our investigative activities.

The advantages of a mechanistic interpretation of causal hypotheses in epidemiology were laid out in Section 3.2. My suggestion is that we endorse this interpretation, but resist the temptation to take the mechanistic methodological stance with respect to causal inferences. The mechanistic metaphysics is good, but its methodology is bad.

The obvious next question is: how should causation be inferred? I do not think there is an easy answer. There are just a bundle of methods, organised around a common theme of identifying patterns of differences and similarities (Mill 1843; Lipton 2004), but increasingly statistically sophisticated (Spirtes, Glymour and Scheines 1993; Pearl 2000), and (it is to be hoped) increasingly reliable. The point of this paper is not to contribute a new method of causal inference, but to identify and debunk a tempting bad method. That method is what I called the mechanistic stance. But at the same time, I hope to have shown that a mechanistic metaphysics for causal generalisations has a great deal to offer epidemiology. There is, therefore, no need to *choose* between the identification of causal associations at the population level, and the identification of underlying mechanisms. On the mechanistic interpretation, causal hypotheses at the population level *are* existence claims about underlying mechanisms. There need be no opposition between epidemiologists conducting observational studies, and those trying to 'disentangle' mechanisms in a laboratory. They are studying the same thing.

It is not necessary to identify the underlying mechanism in order to have *warrant* for a causal hypothesis. But it is necessary in order to *explain* the association. A hypothesis can be perfectly warranted, without being understood. Conversely, a hypothesis may be well (mechanistically) understood, but may itself fail to provide a good explanation of the phenomenon in question for the purposes at hand. To prove a causal hypothesis, it is sufficient to identify an underlying mechanism. But identifying a mechanism is no guarantee of the explanatory power of the hypothesis itself – the explanatory power of the hypothesis that stomach acid causes ulcer, for example. Mechanism is sufficient for causation; but causation is not sufficient for explanation, or for purposes of intervention. Oxygen is a cause of every car crash, and I make that assertion not on the basis of any statistical information, but entirely on the basis of my knowledge of the mechanisms underlying internal combustion engines and human respiration. Yet oxygen offers good explanations of few, if any, car crashes; and controlling the oxygen supply is not a particularly promising avenue for policy makers to pursue.[10]

What I have not done is offer any detailed analysis of the notion of a mechanism. The goal of this paper is to see what the notion can offer causal inference in epidemiology, not to analyse that notion itself. Nevertheless, I would like to finish by sketching a view of the relation between mechanistic explanation and other kinds of causal explanation. I am inclined to see mechanistic explanation as of a kind with causal explanation in general. I take it as widely accepted that causation is not sufficient for explanation, and that for a causal association to explain, it must amount to a difference between fact and foil (Lewis 1986; Lipton 2004, Ch 3). In a public health context, health provides a plausible contrast class, as I have argued elsewhere (Broadbent, 2009b). To give a mechanistic explanation of an association is to tell a story about how events of one type cause events of another, by filling in the intervening steps in the causal chain, and specifying what conditions must hold. It is tempting to see mechanistic explanations as causal explanations, with associations as their explananda. Presumably the relevant contrast class is the failure of that association to hold. Identifying possible failures is not purely a hypothetical exercise: typically there will be plenty of actual failures, since epidemiological generalisations are typically exception-ridden.

If this sketch of a contrastive analysis is approximately correct, it would explain why mechanistic explanations are so useful in epidemiology and public health: because they provide information about what makes the difference between cause-events leading to effect-events, and not doing so. And intervening on links of this sort is the central purpose of public health policy. However, this sketch also illustrates why mechanistic explanation is not necessary for good causal inference: because a causal difference between fact and foil can be identified, even before the causal link between the proposed explanans and the fact is understood. We can know *that* drinking dirty water is the cause of the difference between people with cholera and those without, even if we don't know how drinking dirty water results in cholera. And the contrastive approach shows why mechanistic explanation is not sufficient for effective intervention or explanation. If a causal hypothesis fails to identify a causal difference between good health and ill, then it will offer neither explanation

[10] Sometimes a distinction is drawn between distal and proximate causes of disease; and sometimes biological causes are also distinguished, which may be either distal or proximate. However, drawing and defending such distinctions is not easy, as a large literature in philosophy and also in jurisprudence shows (for references see, respectively: Broadbent 2008; 2009a). A discussion of these distinctions would not be relevant here, since there is no particular reason to hope that the most explanatory cause in a given circumstance will be either proximate, or distal, or biological. Moreover, even if there were some reason for epidemiologists to favour one of these categories, a choice would still have to be made among causes within them.

of nor intervention on that difference; and a mechanistic explanation of the causal hypothesis (e.g. miasma theory) cannot improve matters.

The correct analysis of mechanistic explanation is not, however, the purpose of this paper. To apply the moral reflexively, we can know that mechanistic explanation works, without knowing how it works. I hope to have shown the strengths of a mechanistic interpretation of causal hypotheses in epidemiology, and to have illuminated the pitfalls of a tempting but erroneous companion methodology.

Acknowledgements

I am grateful to Kevin Brosnan, Brendan Clarke, Donald Gillies, Stuart Glennan, Jeremy Howick, the audience at a conference on Causality and Mechanisms held in Kent 2009, and an anonymous referee, for their useful comments and criticisms.

References

Angel, Katherine (2008). Causality and psychosomatic histories in contemporary Anglo-American Biomedicine. Ph D thesis, University of Cambridge.

Broadbent, Alex (2008). The difference between cause and condition. *Proceedings of the Aristotelian Society* 108: 355–364.

Broadbent, Alex (2009a). Fact and law in the causal inquiry. *Legal Theory* 15: 173–181.

Broadbent, Alex (2009b). Causation and models of disease in epidemiology. *Studies in History and Philosophy of Biological and Biomedical Sciences* 40: 302–311.

Brunner, Eric, and Marmot, Michael (2006). Social organization, stress, and health. In *Social Determinants of Health*, ed. Michael Marmot and Richard Wilkinson, pp. 6–30. 2nd edn. Oxford: Oxford University Press.

Cameron, Donald and Jones, Ian G. (1983). John Snow, the Broad Street pump and modern epidemiology. *International Journal of Epidemiology* 12: 393–396.

Carter, K. Codell (1994). *Childbed Fever: A Scientific Biography of Ignaz Semmelweis.* Greenwood Press.

Carter, K. Codell (2003). *The Rise of Causal Concepts of Disease.* Aldershot: Ashgate.

Devereaux, P.J. *et al.* (2008). Effects of extended release metoprolol succinate in patients undergoing non-cardiac surgery (POISE trial): a randomised controlled trial. *The Lancet.* doi:10.1016/S0140-6736(08)60601-7. http://www.thelancet.com/.

Doll, Richard and Peto, Richard (1981). *The Causes of Cancer.* Oxford: Oxford University Press.

Farooqi, Sadaf, and O'Rahilly, Stephen (2006). Genetics of obesity in humans. *Endocrine Reviews* 27, no. 7: 710–718.

Gillies, Donald (2005). Hempelian and Kuhnian approaches in the philosophy of medicine: the Semmelweis case. *Studies in History and Philosophy of Biological and Biomedical Sciences* 36: 159–181.

Glennan, Stuart (2002). Rethinking Mechanistic Explanation. *Philosophy of Science* 69: S342–S353.

Hill, Austin Bradford (1965). The environment and disease: association or causation? *Proceedings of the Royal Society of Medicine* 58: 259–300.

Lewis, David (1973). Causation. *Journal of Philosophy* 70: 556–567.

Lewis, David (1986). Causal explanation. In *Philosophical Papers, Volume II*: 214–241. Oxford: Oxford University Press.

Lewontin, Richard (1974). The analysis of variance and the analysis of causes. *American Journal of Human Genetics* 26: 400–411.

Lipton, Peter (2004). *Inference to the Best Explanation.* 2nd edn. London: Routledge.

Marmot, Michael (2006). Health in an unequal world: Social circumstances, biology, and disease. *Clinical Medicine* 6, no. 6: 559–572.

Mill, John Stuart (1843). *A System of Logic, Ratiocinative and Inductive.* 8th edn. New York: Longman's, Green, and Co.

Pearl, Judea (2000). *Causality: Models, Reasoning and Inference.* Cambridge: Cambridge University Press.

Peto, Richard (1984). The need for ignorance in oncer research. In *The Encyclopedia of Medical Ignorance,* ed. R. Duncan and M. Weston-Sraim, pp. 129–133. Oxford: Perganon Press.

Rothman, Kenneth J., and Greenland, Sander (2005). Causation and causal inference in epidemiology. *American Journal of Public Health (Supplement 1)* 95, no. S1: S144–S150.

Russo, Federica, and Williamson, Jon (2007). Interpreting causality in the health sciences. *International Journal of the Philosophy of Science* 21, no. 2: 157–170.

Rutter, Michael (2007). *Identifying the environmental causes of disease: how should we decide what to believe and when to take action?* The Academy of Medical Sciences.

Schonemann, Peter H. (1997). On models and muddles of heritability. *Genetica* 99: 97–108.

Sesardic, Neven (2005). *Making Sense of Heritability.* Cambridge: Cambridge University Press.

Spirtes, Peter, Glymour, Clark, and Scheines, Richard (1993). *Causation, Prediction, and Search.* New York: Springer-Verlag.

Vandenbroucke, Jan P. (1988). The 'causes of cancer': A miasma theory? *International Journal of Epidemiology* 17, no. 4: 708–709.

Wardle, Jane, Carnell, Susan, Howarth, Claire M.A. and Plomin, Robert (2008). Evidence for a strong genetic influence on childhood adiposity despite the force of the obesogenic environment. *American Journal of Clinical Nutrition* 87: 398–404.

·4·

Causal modelling, mechanism, and probability in epidemiology

Harold Kincaid

Abstract

This chapter looks at interrelated issues concerning causality, mechanisms, and probability with a focus on epidemiology. I argue there is a tendency in epidemiology, one found in other observational sciences I believe, to try to make formal, abstract inference rules do more work than they can. The demand for mechanisms reflects this tendency, because in the abstract it is ambiguous in multiple ways. Using the Pearl directed acyclic framework (DAG), I show how mechanisms in epidemiology can be unnecessary and how they can be either helpful or essential, depending on whether causal relations or causal effect sizes are being examined. Recent work in epidemiology is finding that traditional stratification analysis can be improved by providing explicit DAGs. However, they are not helpful for dealing with moderating variables and other types of complex causality which can be important epidemiology.

Introduction

This chapter focuses on the three subthemes of this volume – causality, mechanism, and probability – largely through the lens of recent causal modelling approaches in epidemiology combined with some general morals from the philosophy of science. The general morals concern tendencies in the sciences to try to make formal methods do more than they can and to down play domain-specific substantive assumptions in scientific inference, a process sometimes called 'black boxing'. The issues about mechanisms that I pursue largely concern claims by epidemiologists (Hafeman & Schwartz, 2009), claims echoed by social scientists (Hedström & Swedberg, 1998; Morgan & Winship, 2007), that various explanations in their fields are inadequate because they lack mechanisms. I use causal modelling results and my philosophy of science morals to help evaluate that criticism and to show some ways in which traditional epidemiology – analytic stratification analysis without use of explicit causal models – sometimes tries to get more out formal inference methods than they can really yield and how some recent uses of causal modelling in epidemiology does the same. One particular instance of this over extension, I argue, comes from appealing to probabilities in epidemiology when they are ungrounded.

Section 4.1 explains the framework and some general morals about mechanism, causation and probability. Section 4.2 looks at the standard epidemiological practice of identifying risk factors by stratification analysis. I argue that traditional epidemiology tries to get by with very little in the way of causality and makes much less use of probability than it advertises. Section 4.3 turns to more recent developments in epidemiology employing Pearle's graphical approach to causality. I discuss both its strengths and limits in understanding the place of mechanism and probability in analysing causality in epidemiology.

4.1 A framework

Some philosophical preliminaries will be useful for framing the kind of issues I want to discuss in epidemiology. There are two truths from the past decades of history and philosophy of science about scientific method and explanation that I would defend: (1) that scientific inference and explanation cannot be adequately captured by an a priori, domain independent logic of science or, put in a positive vein, domain specific, substantive assumptions play key roles in scientific inference; and (2) scientists often believe or act as if they believe that their results are the product of precisely such a logic when they are not.

There are many lines of reasoning to (1). They include:

(a) The argument ad Carnapium given by Quine: if Carnap could not find a successful inductive logic, then there is not one or, more seriously, informative general inductive logics have not been forthcoming.[1] The causal modelling techniques discussed later in the chapter illustrate the situation: there are useful things that can be said using them, but their range of application is restricted because of the strong assumptions and prior background knowledge needed.

(b) The argument from holism: every part of the web of belief is at least indirectly connected to every other, and given a big enough empirical change in one part of the web, we might have to give up those parts of the web that look like logical truths. Thus no inference rule is indefeasible (Quine & Ullian, 1970).

[1] I take it that 'do not violate the probability calculus, first order logic, etc'. is not informative in that it does not say anything about standard methodological disputes, for example on the importance of novel data. I have argued elsewhere that Bayes' theorem is uninformative in this way (Kincaid, 2002).

(c) The argument from underdetermination: there are always possible alternative theories compatible with any given set of data and therefore there cannot be a logic of inference that tells us which is correct, given those data (Longino, 1990).

(d) The argument from conceptual humility: There is no reason to think our concepts, including the concepts of justification, rationality, explanation and other epistemic concepts, have a logic of necessary and sufficient conditions that determine their domain of application (Wilson, 2006).

(e) The argument from history and social studies of science, which is really many related arguments or pieces of evidence about science in practice: Specific instances of inference rules such as parsimony or inference to the best explanation turn out to involve non-logical, domain substantive empirical assumptions (Day & Kincaid, 1994; Sober, 1988). Scientific experiments in our best science seem to be declared decisive by a complex social process that looks contingent (Gallison, 1987). Apparently general scientific inference rules or virtues can conflict and have to be balanced on substantive grounds (Kuhn, 1977). Scientific papers do not reflect the process and uncertainties of the reasoning that went into them. There can be science that is well done by the most obvious standards such as replication, peer review, etc. and yet that involves non-epistemic values (Longino, 1990).

Obviously these arguments are of varying quality, clarity, and upshot. In particular, it is important to note the difference between conclusions about what cannot in principle be done anywhere in science and conclusions about what some specific research programme currently cannot do in practice. Science is a multifaceted process, and pronouncements about it as a whole in fact violate the pragmatist spirit that motivates many of the arguments given above. Of course, rules of inference that approach a logical status are great when you can get them, because if usable by real humans, they guarantee reliability. However, the above arguments are at least some reason to think that we should be alert to the substantive, domain-specific knowledge that is often needed for successful inference.

It is when conclusions are claimed to follow from rules alone, when in truth they do not, that problems can begin. My claim (2) above that scientists make such exaggerated assertions is an empirical one. One line of evidence that will be further supported in this paper comes from the widespread abuse of statistical significance in the social, behavioural and biomedical sciences. Other evidence is found in the studies cited in (e) above. On common-sense grounds it seems clear that scientists have to use methods developed by others without fully understanding how they work because of the scientific division labour. Not understanding the full details of a method can lead to underestimating its limitations. It is the hope for and the difficulties of finding tight inference rules that I shall use in framing my discussion of mechanism, probability, and causation in epidemiology.

Some general morals follow from the above perspective that I will argue are equally applicable to epidemiology. The first is the now perhaps standard truism of Cartwright's (Cartwright, 1994) that if there are no causes in, there will be no causes out. There is no logic that gets you from probabilities or associations alone to causal conclusions. However, at this level of abstraction, the claim is not that interesting (if the premises do not use the predicate 'cause' it is not going to be in a validly drawn conclusion). The interesting question is which causes in are needed to get what causes out. That is an important, non-obvious question of real import to epidemiology and one that will be discussed below.

Another moral concerns the place of mechanisms. Asking that mechanisms be provided is a demand that can mean multiple things resulting in different claims:

1. In the philosophy of science literature (Bechtel, 2008), mechanisms are usually thought of as the component processes realizing some higher level capacities, e.g. the mechanism of memory. I call these 'vertical' mechanisms'. However, many requests for mechanisms are about providing intervening or mediating variables between a putative cause and its effect. Call these 'horizontal' mechanisms. The latter is generally more relevant to epidemiology, particularly to the causal modelling work I shall discuss.
2. We can also want mechanisms for explanation as opposed to confirmation. Those who want mechanisms because they might rule out spurious causation are targeting their role in confirmation and are generally invoking horizontal mechanisms. These are the kinds of concerns that I discuss in epidemiology and causal modelling. The philosophy of science literature on mechanisms largely emphasizes the idea that mechanisms are needed to provide sufficiently deep explanations, working with at least implicitly an ideal of the 'full' explanation. The work in epidemiology that I discuss does not generally come from this motivation.
3. Mechanisms might be important for establishing that there exist causal relations between variables or for establishing causal effect sizes. Their 'importance' might mean knowing the mechanisms is essential or not essential but useful.

Identifying these distinctions helps support my general doubts that there are useful universal methodological rules when it comes to the demand for mechanisms. There are just too many free parameters in that demand for us to take it as binding in the abstract. So, for example, consider the claim that identifying mechanisms is necessary for ruling out spurious causation. If I am worried that the association observed between A and B might not be evidence for causation because they might be the common causes of some third factor, it may not help me to know how A and B's properties are realized, i.e. to know the vertical mechanism. What is needed is not the microdetail but evidence about possible other causes at the level of A and B.

I next want to draw some limited morals about probability from the general framework. We certainly do have clear formal rules for handling probability, and those are universal and a priori if anything is. However, their range of applicability may not be, and we must again be aware of extension beyond reasonable bounds. Arguably the two most basic situations grounding probability judgements come from either (a) there are chance mechanisms, processes, or devices or (b) measurement of consistent degrees of belief. Random sampling from a population, random assignment to treatment or control groups, or random measurement errors are the main instances of the first category and Bayesian updating of the second. We have to assess case by case which of these groundings, if any, is applicable. This may all seem obvious, but we shall see in that epidemiology sometimes uses probability notions where none of these groundings exist.

4.2 Traditional epidemiology

Much of standard epidemiology – what I will call 'traditional epidemiology' in contrast to newer work with explicit causal modelling that I discuss later – consists in identifying risk factors in observational data for disease via stratification analysis to eliminate confounders. I want to sketch some typical work in this vein to serve as the source of my general discussion. My focus is on the epidemiology of leukemia and, more specifically, the roles of benzene exposure and diet in leukemia.

Benzene is a hydrocarbon extracted from petroleum for a variety of uses. The first cases reporting a link between benzene and leukemia date back to the 1920s. It was only in the 1960s that evidence beyond case reports began to appear. Supporting studies are either case-control studies or cohort studies (Glass *et al.* 2006). Case-control studies identify a set of present or past cases of the leukemia and compare benzene exposure in that group to exposure and disese in a control group. Cohort studies follow a group of individuals over time, tracking exposure and disease status. A standard outcome measure is the odds ratio, which is the odds of disease in the exposed divided by the odds of disease in the non-exposed. The data are typically analysed in two ways. In simple stratification analysis correlation coeffecients are calculated within the relevant stratifying subset, for example potential exposures to other carniogens, producing a new estimate of the odds ratio. More complex analysis make use of multiple regression, where other possible 'risk factors' are included and the adjusted odds ratio is reported along with either a signficance level or confidence interval for the association. Benzene, for example, is consistently statistically signficantly associated with chronic lympahtic leukemia but the associations with other leukemias do not usually meet standard significance levels and are described as 'not significantly associated with' these diseases.

Numerous studies have also 'implicated' (a standard wording) diet in cancer. The connection between diet and leukemia, however, is based largely on the one study of Jensen *et al.* (2004). That study identified cases of childhood leukemia in a northern California registry and interviewed all who agreed to participate (83%) about diet of the mother during pregnancy. Controls were selected from the same geographic area based on birth certificates and were matched on variables such as race. Again odds ratios were caculated, multiple regressions run with other risk factors, and associations reported with significance levels. Consumption of fruits and vegetables was inversely associated with acute lymphatic leukemia, the more agressive form of lymphatic leukemia commonly found in children. Benzene exposure was not ascertained.

These kinds of studies are a dominant form of inquiry and research reporting in epidemiology. They predominate in the journals. The main textbooks (Rothman, Greenland & Lash, 2008) consist mostly of discussion of the techniques for doing these kinds of studies. Their general form – (a) reporting multiple regression correlations from (b) observational samples relying on (c) tests of significance and confidence intervals as the acceptance criteria – is also common across the social sciences. Probability, mechanism, and causality get relative little focus in this traditional approach, and the discussions they do get are skewed by the logic of science gloss I mentioned above, or so I want to argue.

A somewhat curious element of this traditional practice is that it is acausal. The Glass *et al.* study is illustrative. The word 'cause' is used only twice, both times in the initial background discussion referring to the work of others. The paper's conclusion is repeatedly stated as establishing an association between low levels of benzene and leukemia. The standard risk factor analysis paper that makes up the majority of the published work in epidemiology shares the same trait: results are always reported as associations, not as causes.

Is this eschewal of causal conclusions merely reasonable humility about the limits of inferring causes from correlations? No doubt it in part is.[2]

[2] A referee wondered whether the reluctance to use causal language was counter-evidence to my thesis that there is a tendency in epidemiology, as in all science, to extend formal methods beyond their reasonable reach. In so far as epidemiology really is motivated by that goal it is indeed and that is a good thing. However, as I point out, despite the admonition to not confuse correlation with causation, epidemiology in the end makes lots of causal claims based on associations. What is exciting about work in causal modelling is that makes such assertions possible in a disciplined way, but also shows when those implicit moves are illegitimate.

However, the roots go considerably deeper than that. Historically, the origins of epidemiology are closely tied to 'positivist' doubts about causation as a legitimate scientific notion. Karl Pearson was extremely influential among the early practitioners of epidemiology. In his *Grammar of Science*, Pearson (1900) argued that the concept of cause had no place in modern science. Causation was too metaphysical a notion; association, on the other hand, could be given full mathematical rigour and should replace causal talk in modern science. So from the start epidemiology was built on correlations taken as ends in themselves.

Moreover, the prejudice against causes is built into the analytic methodology of epidemiology. The ubiquitous stratification techniques actually can be inconsistent with a causal interpretation. Standard practice is to report relative risks or odd ratios after adjustment for any factor that might be thought to change the size of relevant risk. Confounders are often defined in purely statistical terms – confounding of an association between two variables occurs when there is a third factor or variable that, when controlled for, changes the value of the correlation between the two variables at issue. If one were looking for causes, such a procedure would be guaranteed sometimes to produce wrong results. Multiple causal interpretations are possible, as I will discuss in more detail below, when controlling for a third variable changes the correlation between two variables under study. Conditioning on the common effect of two independent variables creates correlations that do not represent causal influence. Conditioning on a moderator variable – one that influences the effect size of another cause (more on this later) – reduces correlation as does conditioning on a common cause or an intermediate cause. So the very procedures that are used by themselves have no consistent causal interpretation. No doubt some or many epidemiologists realize this on some level, and I will report on their attempts to get clear on a better notion of confounding in the next section.

Thus this attempt to stick to associations alone is really not sustainable, for both pragmatic and theoretical reasons. The pragmatic reasons come because epidemiology wants to be relevant to policy – whether it is governments intervening or individuals deciding how to behave – and interventions concern what can be causally influenced. So a typical epidemiological report will describe only associations and provide no explicit causal model, but then conclude with something relevant to policy. The theoretical reason that epidemiology cannot be consistently associationist and non-causal is that associations alone are unacceptably arbitrary. Associations are always associations in a population or sample from a population. If the population is not in some sense a causally homogeneous one, then indefinitely many uninteresting associations of the 'coffee users on Tuesday have less leukemia' sort can be found. The number of associations is restricted only by our ability to imagine possible predicates or categories.

I would use this last point along with several others to argue that the place of probability in actual epidemiological practice in more minimal than might be imaged. My worry about arbitrarily many associations could be put as a variant of Cartwright's dictum 'no causes in, no causes out' that reads 'sometimes, no causes in, no objective probabilities out'. An objective probability, I take it, is one that picks up a real distinction in nature. Cashing out 'real distinction' is of course a matter of controversy, but grounding inferences to other populations seems essential to the notion. Doing that requires us to think there is a causal process behind the probabilities or correlations that we identify. That does not mean that objective probabilities or what philosophers would call 'nomic' generalizations or correlations must always represent a cause – generalizable correlations can result from a causal process involving colliders, for example, which will generate objective non-causal probabilities if the same causal process can be found out of sample. However, I believe that the associations of epidemiology, stripped of any causal basis, may not provide for reliable inferences to populations beyond those where the associations are initially found, because there are a great many accidental associations in any population. Probability talk in such circumstances is misleading, which is precisely a standard critique of probabilistic accounts of causation.

I also am suspicious that the main targets of epidemiological explanation – generic risk claims – are probabilities, though I admit that they may seem to be (see Russo & Williamson, 2007). The kinds of things epidemiologists want to explain are relative risks, odds ratios, and population attributable fractions of disease. These are not measures that vary from zero to one. They are based on frequencies, for sure, and can be converted to percentages. However, though they are loath to say it until that final paragraph with the policy and behaviour implications, what epidemiologists really want these to measure is effect size, a causal notion that need not be cashed out in probability terms. Relative risk can make sense in a single fixed population with deterministic causes that results from no sampling distribution or random assignment. Use of epidemiological information to make risk claims about individuals in the process of diagnosis may well be probability claims, but they are part of clinical medicine, not epidemiology proper.

Not only can measures of relative risk make sense in such populations, these kinds of populations are predominantly what epidemiology deals with. This fact provides another reason to think that probability plays a more limited role in epidemiology than it might seem. Most epidemiological studies do not involve samples picked randomly from a population (and generally randomized experiments are thought not to be part of epidemiology with the exception of clinical epidemiology). Take the work on leukemia cited above. None of it involves random samples except on a few occasions for the control groups, and then the samples come from a 'population' that is in effect a convenience sample. This work shows that diet has some connection to leukemia in those individuals living in northern California willing to participate in a study. No one would pretend these are random samples from anything.

Epidemiologists nonetheless report significance levels taken as the probability of seeing a given result when there is no real correlation. However, we have to ask 'no real correlation where?' Since these probabilities are not explicitly taken to be subjective degrees of belief, their interpretation remains unclear. The closest I can find to a coherent answer in this regard is that the population is some hypothetical super-population from which the population under study is an instantiation and random sample (Morgan & Winship, 2007). However, that framework is to my knowledge notably understudied. Why, for instance do we think that the population is a random sample from the hypothetical population? And why do we want to make inferences about hypothetical populations anyway, unless we are doing so to talk about counterfactual causal possibilities rather than sampling error?

However, there is a natural Bayesian interpretation of the probability claims made by significance tests for these nonrandom samples. I can reasonably ask what probability I should attach to the claim 'if there were some randomizing process such as measurement error that was involved in generating my data, then it is probable/improbable that I should see data like this'. One can then use objective mathematical facts about the hypothetical source or error to assign a conditional probability, which is objective not in the sense of having been generated by a real mechanism but in the sense of following deductively from assumptions.[3] However, traditional analytic epidemiology is decidedly non-Bayesian.

That non-Bayesian commitment is also illustrated by the other ways that probability is minimized in epidemiology, namely, in the studied avoidance of using Bayes' theorem to make sense of results. Like much other biomedical and social research, epidemiologists sometimes use significance levels as straight indicators of probable truth or falsity. This, of course, contravenes Bayes' theorem in that both the prior probability of the hypothesis and the likelihood of the data on the maintained hypothesis – the power – are needed to interpret a significance level. Note that the role attributed to significance levels does not result simply from a reversion to subjective priors, for power calculations are rare as well. Rothman *et al.* (2008) devote three pages to the concept.

These familiar practices seem to me a clear instance of hoping that significance testing can provide a logic of inference without the need for further

[3] A standard way to do this, one not often invoked in epidemiology, is through what is called permutation analysis. For example, suppose one has values for treatment for cases and controls. The mean of observed differences is then compared to the distribution of the mean that is produced if the labels case and control are switched in many different ways.

substantive knowledge. Standard practice portrays itself as rule driven inference when it clearly cannot be, at least where the rules are valid.

Pushing the formulas to do more than they can also shows up when what I have been calling traditional epidemiology does try to talk about causality. One practice that epidemiology shares with the social sciences and other biomedical fields is the use of the R-squared statistic as a measure of how well the given causes account for an outcome. A second practice more specific to epidemiology is reporting what is called the 'population attributable fraction' as a way of measuring how much of the disease burden is due to some risk factor. Both are calculated with the correlational evidence that is common to traditional epidemiology. Not surprising, as purely statistical measures, neither is a reliable guide to causal importance.

Put in the usual regression terms, *R*-squared – 'the explained variance' – is calculated in terms of the predictive errors of the regression. It is the squared ratio of the covariance to the product of the standard deviations:

$$R^2 = \frac{\mathrm{Cov}\,(X, Y)}{\mathrm{StdDev}\,(X) \times \mathrm{StdDev}\,(Y)}.$$

It is common to take a high *R*-squared to mean that the causal factors included in the model capture most of the causal influence.

However, this causal interpretation is more than the formula can warrant. The formula gives a statistical measure of how close the data points are to the regression line estimated from them. In the simple case where X causes Y with no other causes involved and we regress percentage changes of Y on changes in X with measurement error, it is obviously the slope of the regression line – the percentage change in Y that is associated with a percent change in X – that measures the size of the causal effect. The data points may be close to a regression line with a shallow slope and they may be far from one with a steep slope. The *R*-squared statistic is orthogonal to measures of causal influence.

It is not that this confusion has gone unnoticed. It has been. Rather, my point is: the hope for purely formal criteria leads to using formal measures beyond their legitimate domain of application. Warnings about interpretation are ignored.

The 'population attributable fraction' is a statistical measure specific to epidemiology that compares the amount of disease burden in a population exposed to a risk factor to the burden with no exposure. Thus it is:

> $(\mathrm{P}(D)\mathrm{P}(D|E-))/\mathrm{P}(D)$, where P(D) is the (unconditional) probability of disease over a specified time period, and $\mathrm{P}(D|E-)$ is the probability of disease over the same time period conditional on non-exposed status.

There has recently been a large debate between obesity researchers and researchers on other major causes of diseases such as cancer and cardiovascular disease over whose disease contributes most to mortality (Flegal *et al.* 2005). The measure being used in the debate is the population attributable factor.

The population attributable fraction is no better a measure of causal importance than *R*-squared (Levine, 2008). People are subject to overlapping risk factors for the same condition. If we assess them one by one for causal importance in the manner recommended by this formula, the total causal contributions will sum to more than one or, put alternatively, there will be more explained deaths than there actually are.

Let me now turn to the role of mechanisms in traditional epidemiology. Needless to say, if traditional epidemiology avoids causal claims, then it is likely to avoid mechanisms as well. So it does. Here, however, I would argue that it is on better grounds than with its approaches to probability and causation, and indeed for causal reasons that epidemiologists invoke indirectly.

Consider the cases of benzene, diet, and leukemia cited above. I think they nicely illustrate my point (Ok, they were selected to do so) that whether mechanisms are needed for respectable science depends on the context. Currently there is mostly only speculation about the molecular route by which benzene might cause cancer (Atkinson, 2009). Actually the problem is not that routes are hard to picture but that there are too many imagined routes and not much evidence to pick among them. The molecular changes involved in cancer are enormously complex and diverse, so much so that there are reasons to doubt that the cancers form much of a natural kind (Kincaid, 2008). There are many routes to tumorigenic transformations and various metabolites of benzene could be involved in various of those routes. There is no definitive evidence for any of these possibilities as the mechanism by which benzene causes cancer, and most of these possible pathways have only been understood in the last decade.

Yet arguably the evidence was good that benzene causes cancer some time ago. Although the traditional epidemiological reports shy away from causal claims, they do provide evidence that allows for causal interpretation. The cohort and case control studies can mimic, if not ideally realize, the logic of the clinical trial. They can do so by showing that differences in exposure are associated with differences in outcome and then arguing that there is no third factor that might explain the association. The clinical trial logic is precisely designed to establish causality without having to understand the intervening steps or mechanism. Of course, the evidence is fallible and adding in the mechanisms to the story would strengthen the evidence in various ways we will discuss in the next section.

The dietary case tells a different story. The mechanism is not known and there are not many concrete ideas about how diet would influence leukemia. One concrete hypothesis is that foods with DNA topoisomerase II (DNAt2) inhibitor eaten during pregnancy reduce DNA damage. However, in the study described above, when the subset of possible protective dietary factors was restricted to those with DNA topoisomerase II (DNAt2) inhibitor, the inverse correlation with leukemia lost statistical significance. We have considerably less confidence in the plausibility of a mechanism in the case of diet than in the case of benzene.

This weakness in the evidence takes on considerable importance because the correlational evidence for a link between diet and leukemia is much shakier than in the case of benzene. The number of studies is small and the association between diet and leukemia is not always seen. The effect is likely to be small compared to that of benzene and thus proportionately harder to find. The number of possible confounding variables is large and it is hard to make a case that they have all been controlled for. As we noted, the study described here did not control for benzene exposure, certainly a possible confounder. Diet is in much greater need of a mechanism if we are going to label it as a cause of leukemia.

4.3 Causal modelling

I turn in this section to explicit causal modelling efforts in epidemiology, with the focus again on themes related to mechanisms, causality, and probability. I ask to what extent the Pearl (2000) programme can shed light on the need for mechanisms, on the role of probability in identifying causality, and the limitations of the Pearl approach for some epidemiological questions.

A central and intuitive element of the Pearl programme is the directed, acyclic graph (DAG), an instance of which is given in Figure 4.1. The graph is directed in that there are arrows between the variables representing causal relations. A cause is direct if does not go through another variable or 'node' to

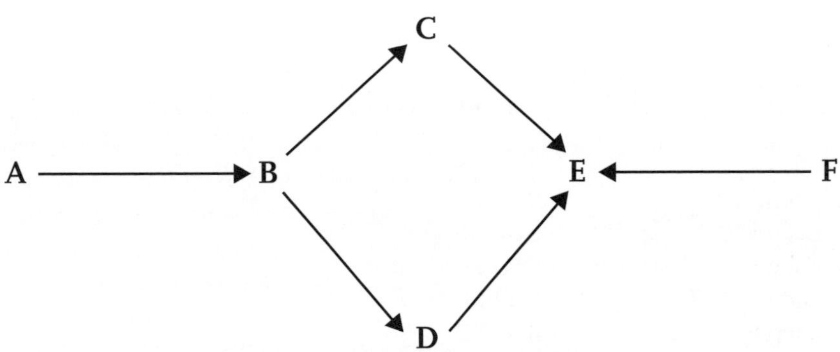

Fig. 4.1 Directed acyclic graph (DAG) with six variables.

exert its influence. The graph is acyclic in that it applies only to causal systems without mutual causation. Thus no arrow points directly to two nodes. All the causes pointing directly or indirectly into a given cause are called its parents or ancestors and all those causes pointed to directly or indirectly are its children and its descendants, respectively. A node with two arrows pointing to it is a collider, a node with arrows pointing to two or more variables is a common cause, and a path between two variables that passes through another node is one that involves in intermediate or mediating cause. A path is open or active, conditional on a set *Z* of other variables, if *Z* contains no common causes or intermediate causes or contains a collider.

Pearl showed how to derive the relevant conditional and marginal dependencies and independencies from causal graph of the sort just described. So, for example, conditioning on a collider or any of its descendants 'opens' the path and thus creates a correlation between the variables leading into it. Conditioning on a non-collider intermediate variable blocks the path and thus removes correlations.

From the rules of *d*-separation the independencies and associations follow as described in Table 4.1. These independencies then can be looked for in the data by asking whether the relevant covariances or correlation coefficients between the variables are zero for the specific independence relations. The corresponding set of dependencies for the causal relations in Figure 4.1 can be tested in the data by looking for the corresponding non-zero covariances or correlation coefficients. These provide direct tests of causality in a disciplined way.

What is the relation of these diagrams to probability claims? Pearl and nearly every other presenter of the DAG framework puts the relations between causal variables in terms of probability distributions. However, DAG models have no inherent connection to probability. It is important to separate the causal model itself from associated statistical issues that result from sampling or measurement uncertainties. If the variables in Figure 4.1 are measured

Table 4.1 Independencies and associations implied by the model of Figure 4.1.

Implied independencies	Marginal associations	Conditional associations
A indep C/B	A and B	C and D/E
A indep D/B	B and D	C and D/F
A indep E/CD	B and C	
A indep F	C, D and E	
B indep E/ACD	E and F	
B indep F/A	B and E	
C indep D/B		
D indep F/B		

with certainty, then what must be observed are functional dependencies and independencies only, not probabilities. Since probabilities are functions, they can express dependencies and independencies. Yet they are not essential to do so. No notion of probability as taking on values between zero and one are necessary. Measurement and sampling issues can be added onto the DAG framework, but the probabilities they require are not part of the basic logic of testing causal claims.

Not only are probabilities unneeded to test DAGs, the methods described above for testing them tell us about the presence or absence of causes, but not about the size of causal effects. Putting the DAG framework in terms of probabilities can make it look like effect sizes are part of the model because the probabilities are often the relative frequencies in the population which then seem like a measure of causal size in the sense of the proportion of As (say, the exposed) are Bs (say, diseased). However, as the graphs make clear, size information is not represented.

It is important to be clear that the testing logic is not about effect size in order to see that estimates of effect size generally presuppose that the maintained causal relations are the true ones. The measure of effect does not test the causal model but rather assumes it. This difference is not often made explicit , though see Morgan and Winship (2007). Sizes of causal effects can be inferred by a variety of methods, though of course sometimes stringent conditions have to be met. Most directly related to the Pearl framework perhaps is the decomposing of causal effect sizes by means of the size of correlation coefficients analysed by the rules of path analysis. So the indirect effect of A on C in Figure 4.1 is the product of the coefficient of the arrow from A to B and the coefficient of B to C. But calculating these effect sizes takes the causal relations themselves between A, B and C as true. This is an instance of Cartwright's dictum 'no causes in, no causes out'. It is likewise a quite concrete sense in which mechanisms are needed.

Traditional epidemiology we saw is skeptical about causation and thus general does not explicitly state the full assumed causal model when providing 'adjusted' odd ratios. Adjusted ratios are contrasted to the 'crude' ratios. Odds ratios, we saw earlier, are measures of effect size. Adjusted ratios reflect effect size after controlling for other risk factors. Using the DAG framework to provide explicit causal models can help show when this traditional practice is on the right track and when it is not (Rothman *et al.* 2008).

Collider bias nicely illustrates why having an explicit causal model is necessary for both inferring causation and causal effect size. Recall that a collider is a node in a causal graph into which two arrows run. If we condition on that node by stratifying on it or including it in a regression, we produce spurious correlation. So, in Figure 4.2, if we condition on C in either diagram, then we have fixed its value and thus the values of A and B, creating a non-causal correlation. This means that we can both make mistakes in causal inference

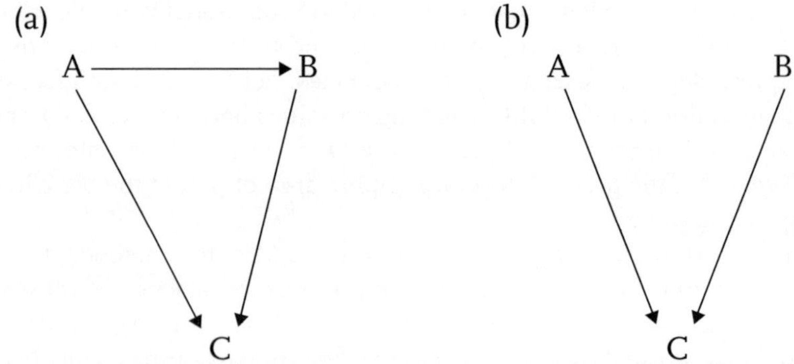

Fig. 4.2 Confounding caused by conditioning on a collider.

and causal size inference. Figure 4.2(b) shows the cause where we create a dependency that is non-causal. Figure 4.2(a) shows the case where there is a causal relation, but by conditioning on C we are making that association look stronger than it is, because we are creating a further association.

Traditional epidemiology works by taking a list of possible causes and confounders and begins an analytic adjustment procedure. The procedure invokes statistical associations without explicit causal models. DAG reasoning shows that this procedure is prone to bias – to inferring causes where they are not and to inferring incorrect estimates of their size (Schisterman, Cole, & Platt, 2009).

These are specific ways that knowing the horizontal mechanism can be necessary for confirming claims about effect size. I want to turn now to the use of mechanisms in identifying not effect sizes but the basic causal relations that are needed to estimate them. Earlier I argued informally that mechanisms are not always needed to infer causation. That can now be seen more formally by thinking about the causal model in Figure 4.1 if my concern is to evaluate the claim that A causes E. Consider the collapsed model in Figure 4.3 where I do not know the mechanism in terms of B, C, D and have no measurement of them. Assume I do have measures of A, B,E, and F. The reduced model AEF nonetheless entails that A and E are independent conditional on B and that A and B, B and E, and E and F are correlated.

If these independencies and dependences are reflected in the relevant zero and non-zero correlations in the data, I have evidence for the claim that A causes E that in no way depends on knowing about the intervening mechanism.

Fig. 4.3 DAG of model without mechanism.

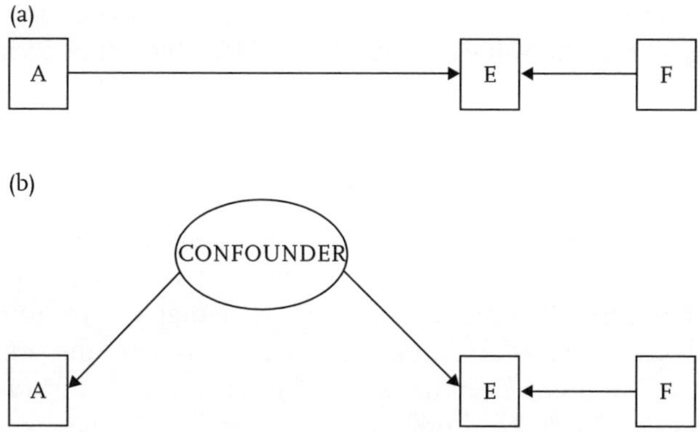

Fig. 4.4 (a) Simple collapsed model of Figure 4.1 and (b) Possible confounder.

However, knowing the intervening process – the horizontal mechanism – can nonetheless increase my confidence that A causes E. To see that imagine that I have evidence only about the simple model in Figure 4.4(a) and I am worried that there is some unknown variable *U* which is confounding my correlation between A and E and so that the real model is as in Figure 4.4(b).

If I now measure B, C, and D and find that I have evidence for the mechanism in Figure 4.1, then for *U* to be a confounder it would also have to account for these further correlations by means of our potential confounder *U*. Obviously spelling out the rest of the mechanism in Figure 4.1 correspondingly would increase our power to reject *U* as a confounder. So while I do not need the mechanism to know that A causes E, having measurements on plausible mechanism makes it easier to do so. I take this to be a core idea behind the claim that mechanisms increase the security of our knowledge.

Mechanisms can also contribute to confirmation by helping provide a more precise understanding of the cause involved. Take our case earlier of leukemia and diet. Eating fruits and vegetables may be a cause of reduced leukemia burden. Identifying a mechanism through which that influence happens would simultaneously help us identify what it was about eating fruits and variables that was beneficial, e.g. antioxidants.

In this case identifying the vertical mechanism is part of identifying the horizontal mechanism. However, the notion of vertical mechanism here is much weaker than the notion described in the recent mechanisms literature. In that literature a mechanism is much more of a system where component parts are identified. The mechanism behind eating fruits and vegetables is presumably just some separable component of them.

I now turn to some shortcomings of the Pearl framework for epidemiology. A standard concern in traditional epidemiology is with moderator variables or what is sometimes called interaction effects. Epidemiologists say that an interaction effect occurs statistically when the correlation between two variables changes with the value of a third variable. The causal idea is that the effect of one factor A on another B is influenced by a third variable C. So rather than A and C being independent causal influences on B, it is their joint effect that does the trick.

Such causal relations are likely to be important in epidemiology. Many diseases are probably the effect of gene–environment interactions where the gene by itself is not enough to cause the disease but only to ground a susceptibility. Disease then occurs when the relevant environmental factors are added. So in the case of depression it is likely that that a genetic background combined with a stressful environment combine to produce depressed individuals, where neither alone would do so. Copy number variants of genes where severity depends on the number of genes while the presence of the disease at all requires an environmental insult are also likely to be a common situation. Similarly, cancers are thought to arise from multiple genetic hits and it is easy to imagine that the number of disease cases could be a function of the level of exposure of multiple agents.

There are in fact two different cases here that we could distinguish, given our earlier distinction between models of causal relations and models of causal effect sizes. Moderators that are involved in the causal relation are instances of Mackie's necessary components of a sufficient complex cause. In the case of A and C jointly causing B, the value of both A and C cannot take on a zero value and have result B. In the effect size instance the measured size of B depends on the size of A and C, which in this case is really the more general connection, with the causal relations issue being an instance where the relevant values of the variables are only zero or greater than zero.

VanderWeele and Robins (2007) claim that all possible types of interaction effects fall into one of four categories according to which of four possible DAGs describe them. Using diabetes (D) as the outcome, they imagine that E is a drug for hypertension in a clinical trial that interacts with a genotype variable X. Mother's genotype C that causes X, proxy variable *R* that is associated with X, and mother's hair colour M that is caused by X as well give them all the possible statistical confounding associations. The four possible interaction situations are illustrated in Figure 4.5, where the parentheses pick out the modifier in terms of statistical interaction.

Unfortunately, these proofs produce much less than they advertise. They provide a classification of all possible statistical interactions between the kind of factors in a traditional epidemiological analysis (confounders, indirect measures, etc.), given a traditional DAG. The problem is that the traditional DAG representation has no place for real causal interaction as opposed to statistical interaction. VanderWeele and Robins have not actually described causal moderation or interaction in the DAG frame work. Their case of drug and genetics influencing hypertension makes each a causal influence

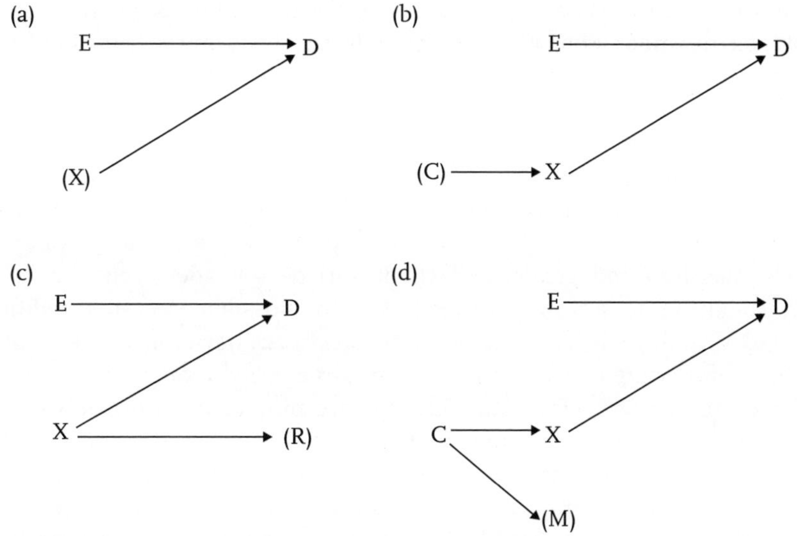

Fig. 4.5 (a) Direct effect modifier X. (b) Indirect effect modifier C. (c) Effect modification by proxy R. (d) Effect modification by common cause M.

on hypertension independently of the other. While the effect size of one on hypertension will depend on the value of the other, the causal relation itself does not. In other words, if either takes on a zero value, the other still has an effect.

Keeping clear on the distinction between establishing a causal relation and an effect size thus pays off here. The depression case mentioned above is not one where there is a gene that causes depression and a social environment that also independently contributes but rather is one where both are required. That kind of situation has no representation in the DAG framework, for it calls for a representation that looks something like Figure 4.6.

Another way to put this point is that causal moderation violates the modularity assumption that a DAG presupposes. The modularity assumption says that the causal relations are such that I could in principle remove an

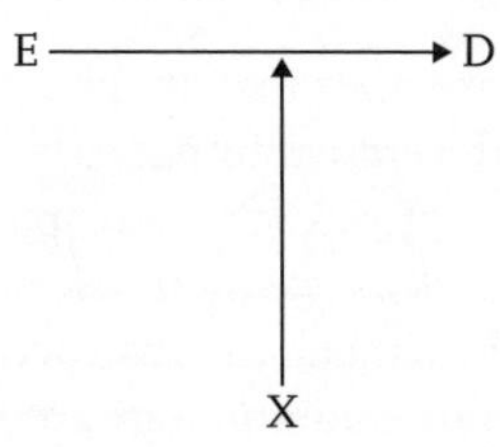

Fig. 4.6 Effect modification.

arrow between two variables while leaving the rest of the causal graph intact. However, the moderation effect shown in Figure 4.6 is precisely the kind of situation that cannot be so far as I can see be represented in the Pearl framework. Epidemiology thus provides fruitful sources for Cartwright's (2007) thesis that modularity can often fail.

When modularity fails, then testing via the expected independencies will not work. The moderator cause will exhibit dependencies, but not the kinds of dependencies that will be screened off in the case of standard causes in a DAG that have independent effects on variables. VanderWeele and Robins have sought to get a formal account of all the possible ways that traditional risk factor analyses might get at moderation effects, however they do so at the expense of missing the substantive issues at stake about causality.

I want to turn finally to some issues concerning estimating effect sizes in epidemiology, when we assume we know the correct set of causal relations. My point will be again that 'knowing the mechanism' means many things – in this case, that measuring effect size requires knowing the functional form connecting changes in causal levels with changes in causal effects. DAG models of causal relations make no assumptions such as linearity, but when effect sizes are measured such assumptions generally become essential.

Typically causal effect sizes are measured in both traditional epidemiology (when it is willing to talk about causes) and in the more sophisticated work based on explicit DAG models by use of regression models. Regression models are best estimates of the average value of the dependent variable based on average values of the dependent variables. These models typically assume that (see Abbott, 2001):

1. Variables are acting on the same timeframes.
2. Causal influences effect the variance of the variables being measured, for example, rather than that causes that effect levels of variables.
3. Causes are symmetric, that is, an increase in the value of variable will increase the value of the variable and a decrease will produce a decrease.
4. There is a causal effect at every value of the variable.
5. The causal effect on one individual unit is independent of the number of individuals affected by that cause.

All of these assumptions are ones that are likely to be violated sometimes by causes in epidemiology. Different carcinogens may act over different periods to produce the same amount of risk. Dose–response relations might be nonmonotonic. Environmental risk assessment takes seriously the idea of threshold effects, and we can imagine mechanisms such as toxin clearing capacity to support them. Social factors in epidemiology are likely to involve cases where the number of individual affected is a factor in the effect on individuals, i.e. where there are scale effects.

These complications show that there can be a significant wedge between what it takes to know that there is a causal relation and what it takes to know its size. Knowing its size may take considerably more information about mechanisms; Pearl has shown that finding causal relations can proceed without presupposing any strong form of assumption about the functional forms of those relations. As I suggested in the first section, different causal claims can make different presuppositions about what the underlying mechanisms might be, some logically stronger than others. Those differences are important in evaluating the need for mechanisms, and suggest that in epidemiology mechanisms are more important in determining the effect size of risk factors as compared to finding risk factors in the first place.

4.4 Conclusion

Epidemiology presents a rich source for looking at the connections between probability, causality, and mechanisms. When pressed, the work discussed here from epidemiology and from the application of DAG style causal modelling relied less on probabilistic notions than might be thought. The kind of horizontal mechanisms emphasized in the philosophy of science literature also played little role. However, mechanisms as intervening variables is much more important, especially when explicit causal modelling is involved. The Pearl type applications in epidemiology provide a nice way to make it clear when and how such mechanisms are needed. Our discussion here shows that considerably more progress has been made in clarifying these issues when it comes to understanding causal relations as opposed to causal effect size. Causal interaction and causality that does not easily fit the standard regression assumptions are much in need of further investigation.

References

Abbott, A. (2001). *Time Matters: On Theory and Method.* Chicago: University of Chicago Press.

Atkinson, T. J. (2009). A review of the role of benzene metabolites and mechanisms in malignant transformation: Summative evidence for a lack of research in nonmyelogenous cancer types. *International Journal of Hygiene and Environmental Health, 212*(1): 1–10.

Bechtel, W. (2008). *Mental Mechanisms: Philosophical Perspectives on Cognitive Neuroscience.* New York: Lawrence Erlbaum Associates.

Cartwright, N. (2007). *Hunting Causes and Using Them.* Cambridge: Cambridge University Press.

Cartwright, N. (1994). *Nature's Capacities and Their Measurement.* Oxford University Press, New York.

Day, T. and Kincaid, H. (1994). Putting inference to the best explanation in its place. *Synthese, 98*(2): 271–295.

Flegal, K. M., Graubard, B. I., Williamson, D. F., and Gail, M. H. (2005). Excess deaths associated with overweight, underweight, and obesity. *Jama*. Am Med Assoc 2005 Apr 20; 293(15): 186, 1–7.

Gallison, P. (1987). *How Experiments End*. Chicago: Chicago University Press.

Glass, D. C., Gray, C. N., Jolley, D. J., Gibbons, C., and Sim, M. R. (2006). *The Story So Far. Annals of the New York Academy of Sciences, 1076* (1 Living in a Chemical World: Framing the Future in Light of the Past), pp. 80–89.

Hafeman, D. M., and Schwartz, S. (2009). Opening the black box: A motivation for the assessment of mediation. *Int J Epidemiol*. 38(3): 838–845.

Hedström, P. and Swedberg, R. (1998). *Social Mechanisms: An Analytical Approach to Social Theory*. Cambridge: Cambridge University Press.

Jensen, C. D., Block, G., Buffler, P., Ma, X., Selvin, S., and Month, S. (2004). Maternal dietary risk factors in childhood acute lymphoblastic leukemia (united states). *Cancer Causes and Control, 15*(6): 559–570.

Kincaid, H. (2002). Scientific Realism and the Empirical Nature of Methodology, In S. Clarke and T. Lyons, *Recent Themes in the philosophy of Science* (Dordrecht: Kluwer, 2002), pp. 39–62.

Kincaid, H. (2008). Do we need theory to study disease? *Perspectives in Biology and Medicine, 51*(3): 367–378.

Kuhn, T. S. (1977). *The Essential Tension: Selected Studies in Scientific Tradition and Change*. Chicago: University of Chicago Press.

Levine, B. J. (2008). The other causality question: Estimating attributable fractions for obesity as a cause of mortality. *International Journal of Obesity*, Aug 2008 Supplement 3, Vol. 32, p. S4–S7, 4p.

Longino, H. E. (1990). *Science as Social Knowledge: Values and Objectivity in Scientific Inquiry*. Princeton: Princeton University Press.

Morgan, S. L., and Winship, C. (2007). *Counterfactuals and Causal Inference: Methods and Principles for Social Research*. Cambridge: Cambridge University Press.

Pearl, J. (2000). *Causality: Models, Reasoning, and Inference*. New York: Cambridge University Press.

Pearson, K. (1900). *The Grammar of Science*. Cambridge: MIT Press.

Quine, W. V., and Ullian, J. S. (1970). *The Web of Belief*. New York: Random House.

Rothman, K. J., Greenland, S., and Lash, T. L. (2008). *Modern Epidemiology*. Philadelphia: Lippincott Williams & Wilkins.

Russo, F. and Williamson, J. (2007). Interpreting probability in causal models for cancer. In F. Russo and J. Williamson (eds), *Causality and Probability in the Sciences*. College Publications, pp. 217–242.

Schisterman, E. F., Cole, S. R., and Platt, R. W. (2009). Overadjustment bias and unnecessary adjustment in epidemiologic studies. *Epidemiology, 20*(4): 488.

Sober, E. (1988). *Reconstructing the Past: Parsimony, Evolution, and Inference*. Cambridge: MIT Press.

VanderWeele, T. J. and Robins, J. M. (2007). Four types of effect modification: A classification based on directed acyclic graphs. *Epidemiology, 18*(5): 561–568.

Wilson, M. (2006). *Wandering Significance: An Essay on Conceptual Behavior*. New York: Oxford University Press.

·5·

The IARC and mechanistic evidence

Bert Leuridan and Erik Weber

Abstract

The International Agency for Research on Cancer (IARC) is an organization which seeks to identify the causes of human cancer. For each agent, such as betel quid or Human Papillomaviruses, they review the available evidence deriving from epidemiological studies, animal experiments and information about mechanisms (and other data). The evidence of the different groups is combined such that an overall assessment of the carcinogenicity of the agent in question is obtained.

In this paper, we critically review the IARC's carcinogenicity evaluations. First we show that serious objections can be raised against their criteria and procedures – more specifically regarding the role of mechanistic knowledge in establishing causal claims. Our arguments are based on the problem of confounders, of the assessment of the temporal stability of carcinogenic relations, and of the extrapolation from animal experiments. Then we address a very important question, viz. how we should treat the carcinogenicity evaluations that were based on the current procedures. After showing that this question is important, we argue that an overall dismissal of the current evaluations would be too radical. Instead, we argue in favour of a stepwise re-evaluation of the current findings.

5.1 Introduction

The IARC, the International Agency for Research on Cancer, is a division of the World Health Organization. In the Preamble to the IARC *Monographs* we read:[1]

Through the *Monographs* programme, IARC seeks to identify the causes of human cancer. (IARC 2006, p. 1)

[1] The Preamble to the IARC *Monographs* can be found at the beginning of each *Monograph* (for the current preamble, see e.g. in IARC 2007, p. 9–31). A slightly different version of the current preamble can be found on the web (IARC 2006). Throughout the history of the IARC, in the past four decades, the procedures and the criteria that are listed in the Preamble have repeatedly been changed (Hopkins 1994, pp. 194–196; Cogliano 2006).

More specifically, the objective of the programme is ...

... to prepare, with the help of international Working Groups of experts, and to publish in the form of *Monographs*, critical reviews and evaluations of evidence on the carcinogenicity of a wide range of human exposures. (IARC 2006, p. 2)

The term 'agent' is used to refer to 'any entity or circumstance that is subject to evaluation in a Monograph' (IARC 2006, p. 2). The exposures or agents include individual chemicals, but also ...

... groups of related chemicals, complex mixtures, occupational exposures, physical and biological agents and lifestyle factors. (IARC 2006, p. 1)

How is the carcinogenic risk of exposures assessed by the IARC?[2] The available evidence consists of epidemiological studies (field experiments with humans), animal experiments and information about mechanisms (and other data). The available studies are first evaluated separately. Then their conclusions are combined per group. Finally, the evidence of the different groups is combined into one final assessment.

In this chapter, we will critically review the IARC's carcinogenicity evaluations. In the first part, we will briefly present their procedures and criteria (Section 5.2) and show that serious objections can be raised against them – more specifically regarding the role of mechanistic knowledge in establishing causal claims (Sections 5.3–5.5). This means we will only focus on the evidential role of mechanisms, not on their possible explanatory roles (cf. Glennan 2002, Bechtel and Abrahamsen 2005). Then we will address a very important question, viz. how we should treat the conclusions of all *Monographs* that were based on the current procedures (Section 5.6). We will show that this question is important (given the possible economic and social consequences of the IARC assessments), but that an overall dismissal of the current evaluations would be too radical. Instead, we argue in favour of a stepwise re-evaluation of the current findings.

5.2 Relevant features of the IARC procedures and criteria

The IARC's evaluation procedure consists of three phases and involves three kinds of studies: epidemiological studies, experimental studies on animals, and mechanistic information and other data. (We will often use the labels 'epidemiological', 'experimental' and 'mechanistic' to refer to these respective

[2] Despite the title of the *Monographs*, the goal of the IARC is to evaluate cancer hazard (i.e. the potential of agents to cause cancer), not cancer risk (i.e. the probability that this potential is realized for a particular section of the population in a defined set of circumstances). (Hopkins 1994, pp. 193–194)

kinds of studies or evidence.) In the first phase, all studies are evaluated separately. In the second phase, assessments are made of the epidemiological, the experimental and the mechanistic group respectively. In the third phase, the evidence of the different groups is combined such that an overall assessment of the carcinogenicity of the agent in question is obtained.

1. Let us first look at the epidemiological studies. For ethical reasons, these studies with humans are almost without exception prospective or retrospective; randomized experiments are very rare. In the first phase, each study is assessed according to three criteria: viz. whether they are plagued by *bias, confounding* or *chance*. Bias is defined as

> ...the operation of factors in study design or execution that lead erroneously to a stronger or weaker association than in fact exists between an agent and disease. (IARC 2006, p. 9)

In order to exclude bias it is required that

> ...the study population, disease (or diseases) and exposure should have been well defined by the authors. Cases of disease in the study population should have been identified in a way that was independent of the exposure of interest, and exposure should have been assessed in a way that was not related to disease status. (IARC 2006, p. 9)

Confounding occurs when

> ...the relationship with disease is made to appear stronger or to appear weaker than it truly is as a result of an association between the apparent causal factor and another factor that is associated with either an increase or decrease in the incidence of the disease. (IARC 2006, p. 9)

In order to rule out confounding it is required that

> ...the authors should have taken into account – in the study design and analysis – other variables that can influence the risk of disease and may have been related to the exposure of interest. Potential confounding by such variables should have been dealt with either in the design of the study, such as by matching, or in the analysis, by statistical adjustment. (IARC 2006, p. 9)

In order to exclude chance the authors must report the basic data on which their conclusions are based, but also their statistical methods:

> Finally, the statistical methods used to obtain estimates of relative risk, absolute rates of cancer, confidence intervals and significance tests, and to adjust for confounding should have been clearly stated by the authors. (IARC 2006, p. 10)

Studies that score badly on these criteria have a low credibility, so their weight in the final evaluation is very low.

After the individual screening, the epidemiological studies are compared with each other. The aim of this second phase is to arrive at one of the following conclusions (IARC 2006, pp. 19–20):

(1) There is *sufficient* epidemiological evidence of carcinogenicity.
(2) There is *limited* epidemiological evidence of carcinogenicity.
(3) The epidemiological evidence of carcinogenicity is *inadequate*.
(4) There is epidemiological evidence suggesting *lack* of carcinogenicity.

Conclusion (1) is drawn if 'a positive relationship has been observed between exposure and cancer in studies in which chance, bias and confounding could be ruled out with reasonable confidence' (IARC 2006, p. 19). Conclusion (2) is drawn if a positive association is observed for which a causal interpretation is credible, but chance, bias or confounding could not be ruled out with reasonable confidence. Conclusion (3) is drawn if there are no studies available, or if the available studies are of insufficient quality or consistency (for the first two conclusions it is required that the positive association occurs in a large majority of the studies). Conclusion (4) is drawn if there are several adequate studies which consistently show no positive association.

2. The experiments with animals are also screened individually in the first phase. One of the considerations is of course whether the animals were allocated randomly to the experimental or the control group: if that condition is not satisfied, the main possible advantage of animal experiments (viz. that they can be randomized trials) is not exploited. Another consideration is whether both male and female animals were used (this prevents a possible *bias*). And of course the data (number of animals studied, number of tumours, length of survival, etc.) should be reported and analysed adequately (elimination of *chance*).

After the individual screening, the results of animal experiments are combined in the second phase. The possible conclusions are (IARC 2006, pp. 20–21):

(1) There is *sufficient* evidence of carcinogenicity in experimental animals.
(2) There is *limited* evidence of carcinogenicity in experimental animals.
(3) The evidence of carcinogenicity in experimental animals is *inadequate*.
(4) There is evidence suggesting *lack* of carcinogenicity in experimental animals.

Conclusion (1) is drawn if there are high quality studies (randomised, elimination of chance) for two or more species, consistently showing an increased incidence of tumours. Increased incidence in a well-conducted study in both sexes of a single species, can also provide sufficient evidence. Conclusion (2) is drawn if the data suggest a carcinogenic effect but are limited for making a definitive evaluation (e.g. because only one sex of a single species is investigated). The criteria for conclusions (3) and (4) are similar to those for epidemiological studies.

3. The mechanistic data[3] include information about toxicokinetics (absorption, distribution, metabolism, and elimination of agents) and mechanisms of carcinogenesis (How does the agent affect the organs, tissues or cells? Does it e.g. lead to genetic mutations?). For carcinogenic effects that have been observed in experimental animals, an evaluation is made of the strength of the evidence that it is due to a particular mechanism. The second-phase categories that are used here are 'weak', 'moderate' and 'strong', but these labels are less clearly defined than those of the epidemiological and experimental studies (IARC 2006, pp. 21–22). For instance, experimental studies which show that suppressing key elements of a mechanism prevents the development of tumours, provide strong evidence for the conclusion that the mechanism operates in the type of experimental animal that is studied. There is also an assessment of how likely it is that a particular mechanism operates in humans. And much attention is paid to the questions whether 'multiple mechanisms might contribute to tumour development, whether different mechanisms might operate in different dose ranges, whether separate mechanisms might operate in humans and experimental animals and whether a unique mechanism might operate in a susceptible group'. (IARC 2006, p. 21)

4. Finally, in the third phase, the three types of evidence are brought together. The agent under investigation is put into one of the following groups (IARC 2006, pp. 22–23):[4]

Group 1: The agent is *carcinogenic to humans.*

Group 2A: The agent is *probably carcinogenic to humans.*

Group 2B: The agent is *possibly carcinogenic to humans.*

Group 3: The agent is *not classifiable as to its carcinogenicity to humans.*

Group 4: The agent is *probably not carcinogenic to humans.*

The Preamble presents a set of rules governing this overall assessment. It is important to note that these are not treated as rigorous rules and that past decisions have resulted in apparent exceptions (as is indicated by formulations starting with 'exceptionally' or 'in some cases'). We will not list all exceptional rules. Thus the following presentation is somewhat simplified.

An agent is placed in Group 1 if there is *sufficient* epidemiological evidence of carcinogenicity. Exceptionally, an agent may also be labelled carcinogenic

[3] Originally, mechanistic data were not taken into account. In 1982, the first kind of mechanistic information (viz. genotoxicity evaluation) was taken into account, and in 1991 several other types of mechanistic data (e.g. concerning gene-expression) were incorporated in the IARC criteria (Hopkins 1994, pp. 194–195). In the current preamble (i.e. IARC 2006), even more weight is attached to mechanistic information (Cogliano 2006).

[4] At this moment, nearly 1,000 agents have been classified. Of these, 107 are in Group 1, 58 are in Group 2A, 249 are in Group 2B, 512 are in Group 3 and, finally, 1 agent (Caprolactam) is in Group 4.

to humans if there is *sufficient* experimental evidence and *strong* mechanistic evidence. But normally, if the epidemiological evidence is less than *sufficient*, it is combined with the evidence from experimental animals and/or with the mechanistic evidence and results in a classification lower than Group 1. For instance, an agent is classified in Group 2A in the following cases:

(a) If there is *limited* epidemiological evidence and *sufficient* experimental evidence.

(b) (In some cases) if there is *inadequate* epidemiological evidence, but *sufficient* experimental evidence and *strong* evidence that the carcinogenesis is mediated by a mechanism that also operates in humans.

(c) (Exceptionally) if there merely is *limited* epidemiological evidence of carcinogenicity in humans.

In condition (b) the mechanistic evidence is used to warrant extrapolation from animals to humans. If this warrant is absent – case (a) – stronger epidemiological evidence is required than in cases where there is such a warrant.

This role of mechanistic evidence, which relates to the extrapolation from animal experiments to humans, can be further clarified by means of the difference between the following rules:

(d) If there is *inadequate* epidemiological evidence and *sufficient* experimental evidence, but *strong* evidence that the mechanism of carcinogenicity in experimental animals *does not* operate in humans, the agent is classified in Group 3.

(e) If there is *inadequate* epidemiological evidence and *sufficient* experimental evidence, but no such negative mechanistic evidence, then the agent is classified in Group 2B.

5.3 Mechanisms and the problem of confounders

1. Biomedical scientists investigating the causes of diseases face a fundamental ethical problem. Randomized experiments with the target population (i.e. humans) provide the most reliable method for establishing causal relations in the biomedical sciences:

> A decisive test of whether smoking causes heart disease, then, would be to take a large sample of human infants randomly selected from the human population, divide them into two equal groups, and force one group to smoke for the rest of their – no doubt abbreviated – lives. (Dupré 1993, pp. 202–203)

However, these randomized experiments are usually impossible for ethical reasons: they may cause physical harm to the experimental subjects, as in

Dupré's example. Biomedical scientists can avoid the unethical experiments by doing merely observational studies on humans (prospective or retrospective designs) and by doing randomised experiments with animals.[5]

From Section 5.2.4 it is clear that the IARC procedures do take into account the role that information about mechanisms can play in extrapolating results from animal experiments to humans. However, mechanisms can have at least two other evidential roles that are neglected in the IARC procedures. The first role relates to the problem of confounders and is discussed in this section. The second relates to extrapolation over time and is discussed in Section 5.4.

2. The problem of confounders originates from the fact that in a prospective or retrospective design the individuals 'put themselves' into the experimental or the control group by the way they act.[6] For instance, in a prospective design set up to investigate the relation between smoking and heart disease, people that decided to smoke end up in the experimental group, non-smokers in the control group. Because of this non-random selection, there may be disturbing factors. For instance, if there are more heart diseases among the smokers, this may be due to the fact that both smoking and heart disease are positively influenced by coffee drinking. Randomized experiments avoid this problem by the random division into experimental and control group.

The standard solution to this problem is 'conditioning on potential confounders'. But this solution has its limitations, as Dan Steel points out with respect to the social sciences:

> I agree that there are cases in which one can draw reasonable conclusions about what causes what without the aid of experiment or substantial knowledge of underlying mechanisms. However, the usefulness of conditioning on potential causes does not undermine the proposal that mechanisms significantly aid causal inference in the social sciences, since social scientists are rarely able to measure all potential common causes. Indeed, the inability to exhaustively consider all potential common causes is a basic element of the problem of confounders, to which mechanisms are being considered as a partial solution. (Steel 2004, p. 63)

This problem is not limited to the social sciences. Potential disturbing factors (confounders) can be eliminated by means of statistical methods on a one-by-one basis. But we can never be sure that no untested variables will ever turn out to be confounders, and we cannot test all possible variables. For instance,

[5] As an anonymous referee rightly pointed out, randomized experimental designs also have other shortcomings than those cited above. For example, they can show us *that* two variables are causally linked but not *how* they are linked. It follows that even where randomized experimental designs are feasible, mechanistic information may still add to our knowledge.

[6] This terminology should not be taken literally. It is not required that individuals are actively responsible for ending up in the experimental or the control group (due to their behaviour). Only where they end up does not depend on any manipulation by the researcher.

we can exclude the possibility that coffee drinking is a common cause in the above example, but we cannot be sure that there is no other variable which causes both smoking and heart disease and is responsible for the correlation. We cannot exclude the possibility that smoking and heart disease have a common cause; we can only test individual variables and exclude them as common causes. More generally, it seems impossible to rule out confounding 'with reasonable confidence' by means of conditioning alone.[7]

How can causal mechanisms help here? Steel (2004) distinguishes two possible roles for mechanisms relating to the problem of confounders. The first possible role is negative: if we don't find a plausible mechanism linking two variables, we can conclude that the correlation between them is spurious (i.e. there is a common cause). Steel argues that this negative role does not work, because we can always find plausible mechanisms. This argument is a bit too strong, however. At least it does not apply straightforwardly to the present context. In biomedical research, it does not suffice to just come up with a plausible mechanism. Mechanistic hypotheses have to be justified empirically. Moreover, the IARC itself sometimes explicitly uses strong evidence suggesting lack of a mechanism (and other relevant data), in tandem with inadequate epidemiological evidence and experimental evidence suggesting lack of carcinogenicity, as a reason to classify an agent in Group 4. (Yet this decision is not taken frivolously.)[8] Hence we suggest that the negative role of mechanisms does bear on carcinogenicity studies (even if, as Steel argues, it does not work in the social sciences).

The second possible role is positive: if we find a mechanism for which we have good evidence of, we can conclude that there is a causal relation between the two correlated variables. In the case of carcinogenesis, the description of the mechanism would contain claims about how the presence of certain chemical substances (e.g. in the blood) leads to the presence of other chemical substances in cells and to changes in properties of cells (e.g. genetic mutation). These processes can be investigated *in vitro*. This is important, because *in vitro* it is possible to do randomized experiments, in which the problem of confounders is unlikely to occur. An ideal mechanistic argument for a claim about carcinogenicity (or other hazard) consists of a chain of lower-level causal

[7] The picture is somewhat more complicated. In the epidemiological literature, two general methods for dealing with the problem of confounders in observational studies are distinguished. 'The first is to consider them in the design of the study by matching on the potential confounder or by restricting the sample to limited levels of the potential confounder. The other method is to evaluate confounding in the analysis by stratification [...] or by using multivariate analysis techniques such as multiple logistic regression.' (Greenberg *et al.* 2004, chapter 10, 'Confounding') (See also the quote from the IARC preamble in Section 5.2.1.) But these complications do not affect the main problem addressed by Dan Steel, a problem which is explicitly recognized in the epidemiological literature: 'Only known confounders can be addressed in observational research.' (Greenberg *et al.* 2004, chapter 10, 'Summary')

[8] See also case (d) in Section 5.2.

claims of which each element is supported by a randomized experiment. Similarly, in the social sciences, an ideal mechanistic argument uses causal claims with respect to the behaviour of individuals which have been tested in randomized trials. Both in the social and in the biomedical sciences, the usefulness of mechanistic evidence 'relies upon causal relationships among components being more directly accessible than those at the macrolevel' (Steel 2008, p. 195).[9]

Looking back at Section 5.2.1 we see that this evidential role of mechanisms is largely neglected in the IARC procedures. It is assumed that one may attempt to exclude the possibility of confounding with reasonable confidence without invoking mechanisms. A sceptic, following Steel's line of reasoning, might argue that confounding can never be excluded with reasonable confidence in this way. So there never is *sufficient* epidemiological evidence for carcinogenicity: that is an empty category. This is an argument for suggesting a change in the IARC procedures. Mechanistic evidence should also be used to better exclude the possibility of confounding in individual epidemiological studies.

5.4 Mechanisms and temporal stability

Consider the following statements, that have an identical logical form:

No gold sphere has a mass greater than 100,000 kg.
No enriched uranium sphere has a mass greater than 100,000 kg.

The second statement is deemed temporally much more stable than the first. The critical mass for enriched uranium is just a few kilograms, so the second statement is not only true at this moment, but will remain true unless some principles that govern the universe change. The truth of the first statement seems to be more contingent (it just happens to be the case that no one did produce such a sphere yet). Examples of even less stable generalizations are 'All screws in Smith's car are rusty' and 'All coins in my pocket are made of copper'.

Likewise, probabilistic causal claims that are true at this moment can also differ with respect to their temporal stability. Consider first an example from

[9] Let us briefly discuss some doubts raised by an anonymous referee who states that mechanistic evidence is as undetermined as any other and that it is trivial to formulate multiple, contradictory plausible mechanisms for any pathogenic process. We already stated that in biomedical research it does not suffice to come up with just a plausible mechanism. It may be trivial to formulate plausible mechanistic hypotheses, but it takes a lot of work to support them empirically. What we need to rule out confounders is a well-justified model of a mechanism, not just a *mechanism sketch*, i.e. a description of a (possible) mechanism containing missing pieces which we do not yet know how to fill in – cf. Machamer *et al.* (2000, p. 18).

the social sciences. In a book on ethical problems in the social sciences, Paul Davidson Reynolds discusses an experiment which investigates the effects of negative income tax (his source is Kershaw 1972):

> The research involved the examination of the effects of different negative income tax plans (direct cash payments) to 'guarantee' a predetermined minimum household income: partial reductions in payments occurred if household earnings increased. The basic question was the extent of labor-force participation of individuals in households with a guaranteed income – i.e. would they work less? The study also estimated the costs of a guaranteed income program if adopted as the major welfare strategy for the nation. The initial study involved 1400 families in five cities in the New Jersey–Pennsylvania area randomly assigned to one of the eight plans (negative income tax schedules) or to a control group (families receiving no guaranteed income). (1982, p. 36)

The eight plans differed in the amount of money that was given if there was no other income, and in the reductions in payment that occurred when there was another income. But in each plan, the reductions were only partial. The aim of the study was to determine whether the advantages of a guaranteed income plan (administrative simplicity, dignity, equity, . . .) were or were not outweighed by a possible disadvantage, viz. reduced labour-force participation.

The experiments reported by Reynolds were performed to evaluate guaranteed income as a nationwide welfare strategy. In the early 1970s (when these experiments were performed) the result was that the effect of guaranteed income on labour-force participation was small. Suppose now that the US Government would have taken this result as a basis for adopting negative income tax as the major welfare strategy. Then 35 years later they might have found out that the effect has changed. Causal relations can become weaker or stronger over the years. For instance, if people become less materialistic, they might attach more value to free time and less to extra consumption, so the effect of a guaranteed income on labour-force participation might increase. Causal relations can even be reversed (from positive to negative, or the other way around).

No matter what the nature of our evidence is (random experiments, prospective or retrospective designs) we face the challenge of extrapolating our results to the future. Without such extrapolation, the results have no policy relevance (see Section 5.6 for a more elaborate discussion of policy relevance). How far the extrapolation must go, depends on what we use the causal relation for. Since no government wants to change its welfare strategy fundamentally too often, extrapolation is required for quite a large period. Regardless of how far one should try to extrapolate, it is clear that extrapolation is impossible without insight into the stability of the underlying social mechanism. Once we know the mechanism, we can investigate how changes at the micro-level (people's beliefs, desires and individual decisions) may affect the macro-level (the relation between negative income tax and labour-force participation). If there is a change at the micro-level that is likely to occur and that has an effect on the causal relation at the macro-level, extrapolation is a risky business. If such changes are unlikely, extrapolation is quite safe.

Let us now go back to the biomedical sciences. There are three ways in which causal generalizations can be unstable in time. First, evolution in the age structure of the population of interest may have an effect on the strength of a causal relation, or even result in new causal relations. Compare a population where everyone dies before the age of 80 with a population in which a substantial part reaches the age of 100 years. It is possible that a compound constitutes a hazard in the second population but not in the first simply because the effects of the compound are very slow and only manifest themselves above the age of 80.

Second, there is risk-instability. Consider the following example. Various risk factors of breast cancer are being explored. In a paper arguing for a causal link between exposure to electromagnetic fields and breast cancer, McElroy *et al.* (2007, p. 266) claim that about half of the variation in breast cancer rate is still unexplained by well explored risk factors such as ionizing radiation, abortion, alcohol consumption, hormone use, etc. That is why they investigate exposure to electromagnetic fields. However, there is also research into the effect of maternal diet on breast cancer. The hypothesis is that maternal diet may increase the risk of breast cancer by inducing changes in the foetus, which alter the susceptibility of the daughter to risk factors that occur later in her life, such as the ones mentioned above (see Hilakivi-Clarke and Clarke 2006, p. 340). This example shows that it is possible that our susceptibility to factors that initiate cancer can vary quite quickly, under the influence of changes in (maternal) diet. This is a typical example of risk-instability. Maternal diet may change quickly in a population. In general, it is possible that a compound does not constitute a hazard at time x (because everyone in the population has a certain property P which neutralizes the effect of the compound) while it does constitute a hazard at time y (because e.g. only half of the population has property P).

A third way in which causal relations can be unstable over time is mechanism instability.[10] Ye *et al.* (2009) study the *evolutionary* mechanism of cancer progression (as opposed to molecular mechanisms). They note that a large number of molecular mechanisms and pathways are known that underlie tumorigenicity. However, no common molecular mechanism underlying all kinds of cancer is known. According to Ye *et al.* genome instability (more specifically, the increased frequency of non-clonal chromosome aberrations) is the common mechanism:

[10] We thank one of the referees for drawing our attention to the difference between risk-instability and mechanism-instability.

> Increasing evidence illustrates that the somatic evolution of cancer is similar to natural evolution with system stability mediated genetic heterogeneity playing a key role [...]. [...] An emerging genome-centric concept on cancer evolution states that overall genome level variation coupled with stochastic gene mutations serve as a driving force of cancer evolution by increasing the cell population diversity [...]. (2009, p. 288)

Genome level variation or instability raises population heterogeneity qua molecular mechanisms,[11] which in turn raises the probability of a specific pathway leading to cancer (this, together with natural selection at the somatic cell level, constitutes the evolutionary mechanism of cancer; 2009, p. 296). It may itself be caused by genetic, metabolic and environmental (cf. the agents reviewed by the IARC elements (2009, p. 295).

Thus the picture is as follows: for some or other reason (e.g. the presence of some carcinogenic agent), genome level instability is induced. This raises the number of potential molecular pathways or mechanisms, some of which may lead to cancer. But these 'mechanisms are constantly changing during cancer evolution' (2009, p. 289), which means that different cells may contain different pathways. Here's the crux: if the hallmark of cancer progression is molecular heterogeneity, what reason would we have to presuppose these mechanisms remain stable over time in the human population?

Mechanism-instability is also present in the case of breast cancer susceptibility we used to illustrate risk-instability (cf. supra). Hilakivi-Clarke and de Assis write that '[a]lterations in the fetal hormonal environment, caused by either maternal diet or exposure to environmental factors with endocrine activities, can modify the epigenome, and these modifications are inherited in somatic daughter cells and maintained throughout life.' (2006, p. 340) The fetal hormonal environment induces changes to the mechanism underlying carcinogenesis.

These examples show that there is yet another reason for changing the IARC procedures. The stability over time of carcinogenicity evaluations should be explicitly addressed and mechanistic evidence should be included as much as possible in this assessment. Moreover, carcinogenicity conclusions that have little stability (or whose stability is unknown) should be re-evaluated more frequently than more stable conclusions.[12]

[11] The population here is composed of cells, not individuals.

[12] In the history of the IARC evaluations, certain agents have been re-evaluated in different *Monographs*. For example, the possible carcinogenicity of tobacco smoking has been evaluated in IARC (1986) and re-evaluated in IARC (1987) and in IARC (2004). Likewise, tobacco habits other than smoking have been evaluated in IARC (1985) and again in IARC (1987) and in IARC (2007). The most recent evaluations (2004, 2007) did not only rely on more recent studies, they also invoked the most recent criteria (cf. footnote 3).

5.5 Extrapolation from animals to humans

The use of mechanistic evidence in the IARC procedures is lacking in still another way. We mentioned that mechanistic data are used to guide the extrapolation of experimental evidence to humans. At the moment, however, this use is open to improvement.

In Section 5.2.4 we discussed the rules governing the attribution of agents to Group 2A. We found that an agent may be placed in this group if there is *limited* epidemiological evidence of carcinogenicity and *sufficient* evidence of carcinogenicity in experimental animals (this was labelled 'case (a)'). In some cases, an agent may also be placed in this group in case there is *inadequate* epidemiological evidence of carcinogenicity, *sufficient* evidence of carcinogenicity in experimental animals and *strong* evidence that the carcinogenesis is mediated by a mechanism that also operates in humans (this was labelled 'case (b)'). The difference between the cases (a) and (b) is remarkable. Once it is acknowledged that there is no *sufficient* epidemiological evidence available (i.e. when it is either *limited* or *inadequate*) and that hence we need to rely on experimental evidence in animals, we should, strictly speaking, be prepared to show that this experimental evidence is relevant to humans.[13] But if mechanistic evidence is needed for extrapolation in case (b), why isn't it needed in case (a)? In short, we recommend that the use of mechanistic evidence in extrapolation is treated as consistently as possible in the IARC procedures. Whatever the experimental evidence in animals may be, its relevance for humans should be made clear. (Note that this role for mechanistic evidence is independent of its possible use to rule out confounders with reasonable confidence.)

5.6 The status of the current IARC conclusions

Let us briefly recapitulate the conclusions of the last three sections. First, we argued that epidemiological studies, given their non-experimental nature, may always fall victim to the problem of confounders. Therefore we suggested a change in the IARC procedures: mechanistic evidence should also be used to better exclude the possibility of confounding in individual epidemiological studies. Secondly, we showed that mechanistic evidence is needed in order to assess the temporal stability of causal claims – even in the biomedical sciences – and we argued that this was yet another reason to change the procedures of the IARC. Finally, we drew attention to the unequal role played by mechanistic evidence regarding the extrapolation of experimental evidence on laboratory animals to human beings.

[13] The need for such a warrant becomes very clear if we realize that different animal species may suggest different causal relations. For instance, aflotoxin B_1 causes liver cancer in rats but not in mice (Steel 2008, p. 82).

These findings and recommendations raise an important question, viz. how should we treat the conclusions of all *Monographs* that were based on the current procedures? Should we dismiss them completely? Our answer to this question consists of three parts. First we further motivate the above question. Prima facie, there are good reasons to dismiss the current evaluations (Section 5.6.1). Then we will analyse the possible consequences of such a decision, leading to the conclusion that an overall dismissal of the current evaluations would be too radical (Section 5.6.2). Finally, we will argue for a stepwise re-evaluation of the current IARC conclusions (Section 5.6.3).

1. Prima facie, there are good reasons to dismiss the current evaluations. We may fear that the current procedures result in *false positives*: some chemical substances, biological agents, ... are declared carcinogenic while they are not. If there can be *sufficient* epidemiological evidence without any mechanistic backing (see Section 5.3), some agents may erroneously end up in Group 1 (see Section 5.2). And if positive experimental evidence on animals may be considered relevant for human beings without any mechanistic warrant, some agents may erroneously end up in Groups 2A or 2B. We may also fear that the carcinogenicity relations that were discovered in the old *Monographs* have changed (cf. the problem of stability over time we discussed in Section 5.4).[14]

False positives would be problematic given that the IARC conclusions serve as the basis for regulation and legislation in large parts of the world. As such, they indirectly have huge financial and economic consequences. (Note that the IARC itself does not directly engage in regulation or legislation, cf. infra.)

> The Monographs are used by national and international authorities to make risk assessments, formulate decisions concerning preventive measures, provide effective cancer control programmes and decide among alternative options for public health decisions. (IARC 2006, p. 3)

For example, one of the tasks of the California Environmental Protection Agency (Cal/EPA) is to 'publish a list of chemicals known to the State of California to cause cancer, birth defects or other reproductive harm'.[15] One of the reasons why a chemical may be listed is that the IARC (or a similar 'authoritative body') has identified it as carcinogenic (Cogliano *et al.* 2004, p. 1269). This list has direct regulatory consequences.

> Proposition 65 imposes certain requirements that apply to chemicals that appear on this list. These requirements are designed to protect California's drinking water sources from contamination by these chemicals, to allow California consumers to make informed choices about the products they purchase, and to enable residents or

[14] The problem of the stability of carcinogenicity claims may also result in false negatives.

[15] Quoted from http://www.calepa.ca.gov/publications/factsheets/1997/prop65fs.htm

workers to take whatever action they deem appropriate to protect themselves from exposures to harmful chemicals.[16]

The legislation of the European Commission provides a second example. For example, in the Commission Directive 2009/2/EC of 15 January 2009 on the classification, packaging and labelling of dangerous substances, it is stated that attention should be paid to the outcome of future discussions within the IARC on the carcinogenicity of nickel substances.[17]

In the past, the use of the IARC's conclusions as a basis for regulation and legislation has been criticized by the industry. For example, twenty years ago Barnard *et al.* (1989, p. 85) condemned the fact that the then IARC procedures 'made no attempt to evaluate whether animal evidence is predictively relevant to human cancer risk'.[18] Furthermore, they regretted that

> [b]ecause of a misunderstanding of the limited scope of the analysis involved, the IARC [...] lists have recently been used as a basis for legislative and regulatory decisions. (Barnard *et al.* 1989, p. 81)

It should be noted, however, that the IARC itself does not take part in regulation or legislation. Quite the reverse, the preamble explicitly states that

> The evaluations of IARC Working Groups are scientific, qualitative judgements on the evidence for or against carcinogenicity provided by the available data. These evaluations represent only one part of the body of information on which public health decisions may be based. Public health options vary from one situation to another and from country to country and relate to many factors, including different socioeconomic and national priorities. Therefore, *no recommendation is given with regard to regulation or legislation, which are the responsibility of individual governments or other international organizations.* (IARC 2006, p. 3, our emphasis)

That the IARC evaluations represent only part of the body of information on which public health decisions are based emerges in two ways. The first way is alluded to in the last quote: one agent (whatever is the group it is attributed to) may be treated differently in different countries. Secondly, the weight of the IARC's verdict on the carcinogenicity to humans of some agent X is not proportional to the issuing regulatory decisions and in many cases a wide range of possible decisions is open for consideration. Hence agents with the same IARC classification may be treated differently in one and the same country. For example, the sale and use of alcoholic beverages, which are carcinogenic to humans (Group 1) is permitted throughout the European Union (of course, in some countries they are more heavily taxed than in

16 Quoted from http://www.calepa.ca.gov/publications/factsheets/1997/prop65fs.htm

17 Commission Directive 2009/2/EC in *Official Journal of the European Union*, 16.1.2009, L 11/7.

18 At the time of publication, the authors were all linked with chemical companies (Barnard with Cleary, Gottlieb, Steen and Hamilton; Moolenaar with Dow Chemical Co.; and Stevenson with Shell Chemical Co.). Barnard and Stevenson were also members of the American Industrial Health Counsil.

others). By contrast, the importation, supply and new use of asbestos (which are also in Group 1) is strictly prohibited throughout the European Union. In both cases (alcoholic beverages and asbestos), there are threats of serious damage or harm to human health (cancer!), and in both cases the same level of scientific certainty is attributed (Group 1). Yet strong precautionary measures are taken in the case of asbestos, but not in the case of alcoholic beverages.

To conclude, given the insufficient use of mechanistic evidence by the IARC, one may fear that the current procedures result in false positives. These may have huge economic and social consequences given that the IARC conclusions serve as the basis for regulation and legislation in large parts of the world (even though the IARC itself does not engage in regulation or legislation). It follows that the above question is important and that prima facie there are good reasons to dismiss the current evaluations.

2. However, a dismissal would be unjustified. In general, rejecting the best available knowledge solely because it is not the best possible knowledge is counterproductive (provided this best available knowledge is reasonably reliable). Scientific knowledge is rarely sought after for intellectual reasons only. It is aspired for its possible use: as a basis for policy. Although the use of the IARC conclusions in policy (regulation and legislation; cancer prevention – IARC 2006, p. 1) provides reasons to deem the flaws in the IARC procedures problematic, we will see that this very same use safeguards them from an overall dismissal. Given what is at stake (viz. the life and the quality of life of thousands of people), we should prefer false positives over total ignorance.

In Sections 5.3–5.5 we showed that the IARC procedures are open to improvement. We did not show that they are completely flawed. Quite the contrary, they incorporate several protocols to provide as sound a scientific basis for evaluation as possible.

Firstly, it is clear from Section 5.2 that the IARC bases its findings on a broad empirical basis, reviewing 'all pertinent epidemiological studies and cancer bioassays in experimental animals' (IARC 2006, p. 3) plus part of the mechanistic and other relevant data on the condition that they are 'published or accepted for publication in the openly available scientific literature' or stem from 'government agency reports that are publicly available' (IARC 2006, p. 4).

Secondly, all participants of the IARC working groups need be qualified and impartial. It is the working groups that are responsible for developing the *Monographs*. Their members are selected by IARC staff together with other experts (IARC 2006, p. 5). The goal of the IARC is to invite the best-qualified experts (Cogliano *et al.* 2004, p. 1273). Study summaries may not be written by or reviewed by someone associated with the study being considered (IARC 2006, p. 6). Potential participants also have to declare, in confidence,

> any interests that could constitute a real, potential or apparent conflict of interest, with respect to his/her involvement in the meeting or work between (a) commercial entities and the participant personally, and (b) commercial entities and the administrative

unit with which the participant has an employment relationship. (quoted in Cogliano *et al.* 2004, p. 1273)

In line with the WHO procedures, an apparent conflict of interest exists when the expert's objectivity could be questioned by others, even if the interest does not necessarily influence the expert (Cogliano *et al.* 2004, p. 1273).[19]

Finally, the working groups strive after consensus evaluation (or otherwise majority vote) and the working group members engage in peer-review.

IARC Working groups strive to achieve a consensus evaluation. Consensus reflects broad agreement among Working Group Members, but not necessarily unanimity. The chair may elect to poll Working Group Members to determine the diversity of scientific opinion on issues where consensus is not readily apparent. (IARC 2006, p. 6)

Together these protocols (broad empirical basis, qualified and impartial experts, and peer review and consensus) ensure that the current IARC conclusions are reasonably reliable for policy. For the sake of prudence, we should not opt for an overall dismissal of the current IARC evaluations and our methodological criticisms should not be used by industrial lobbies to undermine the role of the IARC as providing the scientific basis for regulation and legislation.

3. Instead of dismissal, we would argue for a stepwise re-evaluation. We should stick to the IARC conclusions unless and until they are contradicted by more recent *Monographs* and updates.[20] In this way, we do not lose the pragmatic value of the current body of knowledge. At the same time, we strive for increasingly reliable conclusions.

Here the industry can play a role. It can suggest which agents are eligible for re-evaluation. Yet such a suggestion cannot by itself undermine our confidence in the current findings. Until a re-evaluation is finished and published, the current conclusions remain our best available knowledge and should serve as a basis for policy.

5.7 Conclusions

The procedures of the IARC should be improved by making more appropriate use of mechanistic evidence. It may be feared that the current evaluations result in false positives that can be avoided. In particular we recommend that

[19] It may be the case that the best-qualified experts have real or apparent conflicts of interest and hence may not serve as working group members. In that case they may act as invited specialists. Invited specialists take part in subgroup and plenary discussions but they may not serve as meeting or subgroup chairs, draft text that discusses cancer data or contribute to the evaluations. (Cogliano *et al.* 2004, p. 1273)

[20] In footnote 12 we already mentioned that in the history of the IARC, certain agents have been re-evaluated in different Monographs.

mechanistic evidence is used more consistently with regard to extrapolation of experimental findings on animals to cancer in humans, that it is used to better rule out the possibility of confounding in epidemiological studies and that it is used to assess the temporal stability of carcinogenicity claims.

But from this it does not follow that the current evaluations have to be dismissed – at least not until they are contradicted by more recent *Monographs* or updates. Given what is at stake (viz. the life and the quality of life of thousands of people), we should prefer false positives over total ignorance.

Acknowledgements

We thank two anonymous referees, Leen De Vreese, Isabelle Drouet, Anton Froeyman, Federica Russo and Rafal Urbaniak for their comments on earlier drafts of this paper. We also thank the audiences at the First Biennial Conference of the Society for Philosophy of Science in Practice (Enschede) and at CaPitS2008 (especially Nancy Cartwright, Stathis Psillos and Paolo Vineis) for their comments. The research for this paper was supported by the Fund for Scientific Research – Flanders through project nr. G.0651.07. Bert Leuridan is Postdoctoral Fellow of the Research Foundation – Flanders (FWO).

References

Barnard, R.C., Moolenaar, R.J. and Stevenson, D.E. (1989). IARC and HHS lists of carcinogens: Regulatory use based on misunderstanding of the scope and purpose of the lists, *Regulatory Toxicology and Pharmacology*, 9: 81–97.

Bechtel, W. and Abrahamsen, A. (2005). Explanation: a mechanist alternative, *Studies in History and Philosophy of Biological and Biomedical Sciences* 36: 421–441.

Cogliano, V.J. (2006). Use of carcinogenicity bioassays in the *IARC Monographs, Annals of the New York Academy of Sciences*, 1076: 592–600.

Cogliano, V.J. (2007). The IARC Monographs: a resource for precaution and prevention, *Occupational and Environmental Medicine* 64 (9): 572.

Cogliano, V.J. *et al.* (2004). The science and practice of carcinogen identification and evaluation, *Environmental Health Perspectives* 112, (13): 1269–1274.

Cogliano, V.J. *et al.* (2008). Use of mechanistic data in IARC evaluations, *Environmental and Molecular Mutagenesis* 49: 100–109.

Dupré, J. (1993). *The Disorder of Things*. Cambridge & London: Harvard University Press.

Glennan, S.S. (2002). Rethinking Mechanistic Explanation, *Philosophy of Science* 69, 3: S342–S353.

Greenberg, R.S. *et al.* (2004, e-book). *Medical Epidemiology*, 4th edition. McGraw-Hill Professional, http://www.accessmedicine.com/content.aspx?aID=546168.

Hilakivi-Clarke, L. and de Assis, S. (2006). Fetal origins of breast cancer, *Trends in Endocrinology and Metabolism* 17: 340–348.

Hopkins, J. (1994). The role of cancer mechanism in IARC carcinogen classification, *Food and Chemical Toxicology*, 32, 2: 193–198.

IARC (1985). *Tobacco Habits Other than Smoking; Betel-Quid and Areca-Nut Chewing; and Some Related Nitrosamines. IARC Monographs on the Evaluation of Carcinogenic Risks to Humans*, volume 37, IARC, Lyon.

IARC (1986). *Tobacco Smoking. IARC Monographs on the Evaluation of Carcinogenic Risks to Humans*, volume 38, IARC, Lyon.

IARC (1987). *Overall Evaluations of Carcinogenicity: An Updating of IARC Monographs Volumes 1 to 42. IARC Monographs on the Evaluation of Carcinogenic Risks to Humans*, supplement 7, IARC, Lyon.

IARC (2004). *Tobacco Smoke and Involuntary Smoking. IARC Monographs on the Evaluation of Carcinogenic Risks to Humans*, volume 83, IARC, Lyon.

IARC (2006).*Preamble to the IARC Monographs on the Evaluation of Carcinogenic Risks to Humans*. http://monographs.iarc.fr/ENG/Preamble/CurrentPreamble.pdf

IARC (2007). *Smokeless Tobacco and Some Tobacco-specific N-nitrosamines. IARC Monographs on the Evaluation of Carcinogenic Risks to Humans*, volume 89, IARC, Lyon.

Kershaw, D.N. (1972). A negative income tax experiment, *Scientific American* 227, (4): 19–25.

Machamer, P., Darden, L., and Craver, C.F. (2000). Thinking about mechanisms, *Philosophy of Science*, 67: 1–25.

McElroy J. *et al.* (2007). Occupational exposure to electromagnetic field and breast cancer risk in a large, population-based, case-control study in the United States, *Journal for Occupational and Environmental Medicine* 49: 266–274.

Reynolds, P.D. (1982). *Ethics and Social Science Research*. Englewood Cliffs, New Jersey: Prentice-Hall.

Steel, D. (2004). Social mechanisms and causal inference, *Philosophy of the Social Sciences* 34: 55–78.

Steel, D. (2008). *Across the Boundaries. Extrapolation in Biology and Social Science*. New York: Oxford University Press.

Ye, C.J. *et al.* (2009). Genome based cell population heterogeneity promotes tumorigenicity: The evolutionary mechanism of cancer, *Journal of Cellular Physiology* 219, 288–300.

·6·

The Russo–Williamson thesis and the question of whether smoking causes heart disease

Donald Gillies

Abstract

One of the main problems in establishing causality in medicine is going from a correlation to a causal claim. For example, heavy smoking is strongly correlated with lung cancer, but so is heavy drinking. There is normally held to be a causal link in the former case, but not in the latter. The Russo–Williamson thesis suggests that to establish that A causes B, one needs, in addition to statistical evidence, evidence for the existence of a mechanism connecting A and B. This thesis is examined in the case of the claim that smoking causes heart disease. It is shown that the correlation between smoking and heart disease was established by 1976 before any plausible linking mechanism was known. At that stage, there were doubts about whether a genuine causal connection existed here. Details of the history of research in atherosclerosis from 1979 to the late 1990s are then given, and it is shown that there is now a plausible mechanism connecting smoking and heart disease, and that, correspondingly, most experts now accept that smoking causes heart disease. This historical case study therefore provides support for at least one version of the Russo–Williamson thesis.

6.1 From correlation to causation

One of the most fundamental questions in the analysis of causality is how one can get from correlation to causation. I will illustrate this problem by considering two examples of correlations. The first is generally recognized to have causal implications, while the second most probably does not.

The first of these examples is a very famous one. It is the correlation between smoking and lung cancer. This correlation was found in many separate studies, but I will describe what is perhaps the best known of them. This study was started by Bradford Hill in 1951, not long after he began to suspect that smoking might be a cause of lung cancer. He and Doll wrote at the end of October that year to all the doctors on the British Medical

Register who were believed to be resident in the United Kingdom to ask them if they would participate in a survey concerning smoking. 34,440 agreed to take part and they were then followed for the next 40 years. Their smoking habits were monitored from time to time, and when they died the cause of death was noted. Reports on the results were published occasionally as the study progressed. Here I will quote some of the results to be found in Doll and Peto (1976). Peto had by this time replaced Bradford Hill in handling the survey. The 1976 paper deals with the mortality rates of the male doctors over the 20 years from 1 November 1951 to 31 October 1971. During that time, 10,072 of those who had originally agreed to participate in the survey had died, and 441 of these had died of lung cancer. Doll and Peto calculated the annual death rate for lung cancer per 100,000 men standardized for age. The results in various categories were as follows (1976, p. 1527):

Non-smokers	10
Smokers	104
1–14 gms tobacco per day	52
14–24 gms tobacco per day	106
25 gms tobacco per day or more	224

A cigarette is roughly equivalent to 1 gm of tobacco.

These results do indeed show a striking correlation between smoking and lung cancer. Smokers are on average more than 10 times more likely to die of lung cancer than non-smokers, and this figure rises to more than 22 times for heavy smokers who consume 25 gms or more of tobacco per day. These results are highly significant statistically.

This correlation was accepted at the time by most researchers (if not quite by all!) as establishing a causal link between smoking and lung cancer. Indeed Doll and Peto themselves say explicitly (p. 1535) that the excess mortality from cancer of the lung in cigarette smokers is caused by cigarette smoking.

However, not all correlations are taken as establishing causal links. Freudenheim *et al.* (2005) report a series of studies of alcoholics carried out between 1974 and 1981 which showed that the high consumption of alcohol by these individuals was accompanied by high levels of morbidity and mortality from lung cancer. So we have a strong correlation between heavy drinking and lung cancer. However, few would accept that there is a causal link here. Instead the correlation would be attributed to the confounding factor of smoking. A large percentage of alcoholics are also heavy smokers, and most researchers would judge that it was the smoking of alcoholics rather than their drinking which increased the risk of lung cancer. Admittedly Freudenheim *et al.* (2005) think that heavy drinking might slightly increase the risk of lung cancer, but they add (p. 657): "Residual confounding by smoking may explain part of the observed relation". For the purposes of this chapter, I will assume that the correlation between heavy smoking and lung cancer does show a causal relation, while the correlation between heavy drinking and lung cancer does not indicate any causal relation.

The contrast between these two cases shows why it is important to go beyond correlation and raise the question of causality. In Gillies (2005) I argued that the essential feature of causality is its relation to action. Let us consider a causal law of the form: A causes B. On the basis of this law we can carry out two kinds of actions which I call (2005, p. 827) *productive* and *avoidance*. A productive action is one which tries to produce B by instantiating A, while an avoidance action is one which tries to avoid B by preventing A from occurring. Now in medicine it is, generally speaking, avoidance actions which are relevant. If we accept that smoking causes lung cancer, then not smoking is a good strategy for avoiding lung cancer. However, if the link between heavy drinking and lung cancer is only a correlation with no causal implications, then it is certainly not a good strategy to give up drinking in order to avoid lung cancer. Such a strategy might well have not the slightest effect on the risk of getting lung cancer. This point could be put another way as follows. Causality is a more complicated concept than correlation, and it is more difficult to establish causal links than correlations. However, the extra effort is worthwhile, since, once we have established a causal link, we can easily infer what actions are appropriate for us to take, whereas this cannot be done easily from correlations alone. This shows the importance of the step from correlation to causation. Let us therefore consider the question of what kind of evidence can justify us in taking such a step.

A distinction is often made between *observational* and *experimental* evidence. Observational evidence is obtained by observing some process without intervening in it in any way. Experimental (or interventional) evidence is obtained by producing, in a controlled way by means of an intervention, a process which is then observed. Admittedly this distinction becomes somewhat doubtful in the micro world of quantum mechanics because of Heisenberg's uncertainty principle. In this area any observation turns out to be an intervention which disturbs the process being observed. However, in this chapter, I will be dealing with examples from medicine where the distinction is unproblematic.

As we have seen, causality involves a relation to human action. This suggests the following principle: correlations can be established by observational evidence alone, but to establish a causal claim one needs at least some experimental (or interventional) evidence. This principle can be illustrated by the standard example of a correlation which is not a causal link. Let us consider a particular barometer, and suppose we establish by careful observation that this barometer's reading falling to a low level is strongly correlated with rain occurring. Is this correlation a causal link? Further careful observations will not give us the answer, but we can test out the hypothesis of causality by making an experimental intervention. We add to the barometer a device which enables us to reduce the barometer's reading to a low level by turning a knob. We have only to turn this knob and see whether rain occurs in order to confirm or refute the causal hypothesis.

Let us next generalize from this simple example. Suppose it has been established by observation that A is correlated with B. One way of testing whether A also causes B is to make what could be called a *direct intervention*. This consists in intervening in order to produce A under some controlled conditions and then seeing whether B results. A direct intervention of this sort is perhaps the simplest way of either confirming or refuting a causal claim. Unfortunately, however, it is not always possible.

Let us take our example of whether smoking causes heart disease. In this case a direct intervention might consist of a randomized control trial of the following kind. We take a sample of 30,000 humans chosen at random. We then select at random 15,000 who are forced to become smokers. The remaining 15,000 are forced to become non-smokers. We then follow up these individuals over a period of say 40 years and see what the differential death rates are in the two groups. However, as Pearl remarks (2000, p. 353), controlled experiments of this sort 'are impossible (and now also illegal) to conduct'. The example of smoking causing heart disease is, in this respect, typical of most causal claims in medicine. These characteristically take the form: X causes D, where D is a disease and X is a putative cause of D. In testing such claims it is obviously impossible for ethical, practical, and usually legal reasons, to make the direct intervention of implementing X for a group of humans under controlled conditions and seeing whether D results.

If then direct interventions are ruled out in many medical situations, it is obvious that, if the principle formulated earlier is correct, that we must resort to indirect interventions in order to establish causality. But what form should such indirect interventions take? An answer to this question is provided by the Russo–Williamson thesis to which I now turn.

6.2 The Russo–Williamson thesis

In their 2007 paper, Russo and Williamson write as follows (pp. 158–9):

> ... the health sciences infer causal relations from mixed evidence: on the one hand, mechanisms and theoretical knowledge, and, on the other, statistics and probabilities.... To establish causal claims, scientists need the mutual support of mechanisms and dependencies.... The idea is that probabilistic evidence needs to be accounted for by an underlying mechanism before the causal claim is established ...

I will refer to this suggestion as the *Russo–Williamson thesis* (or RWT for short). The general sense of this thesis is quite clear, but, like most philosophical theses, it can be formulated in a number of slightly different ways. I will now make some comments on the thesis, and then formulate the precise version of the thesis which I will try to defend in the rest of the chapter.

Russo and Williamson speak of *establishing* causal claims. This seems to me a quite appropriate terminology which indeed I have used in Section 6.1. However, it should be taken in a qualified sense. 'Establishing' here does not mean 'establishing with complete certainty or beyond doubt' because scientific claims can never be established in this sense. 'Establishing' should rather be understood in something like the following sense. Suppose a scientific claim has become so well confirmed by the available evidence that it can be accepted for the time being as the basis for action. Then this claim can be said to be established. However, this by no means excludes the possibility that the further advance of science will lead us to modify, or even reject, the claim.

Russo and Williamson speak in one place of 'statistics and probabilities' and in another of 'dependencies'. I will in this confine myself to 'statistics' and 'statistical evidence'. This is connected with another feature of Russo and Williamson's 2007 paper. They limit themselves to the health sciences. Indeed the paper is entitled: 'Interpreting causality in the health sciences'. Now it seems to me that the Russo–Williamson thesis might well be extended to causality in general. However, the health sciences are a particularly good area to study it, at least initially. This is because in this area the two types of evidence which are brought to bear are quite sharply distinguishable. The statistical evidence comes from epidemiological observations of populations, such as the study of doctors and smoking described in Section 6.1. This evidence, however, is usually supplemented by experimental evidence obtained in the laboratory from biochemical investigations of cells and tissues, or from physiological investigations into animals which have been treated in various ways. This enables us to connect the Russo–Williamson thesis to the principle proposed in Section 6.1 concerning observational evidence and experimental (or interventional) evidence. In a typical medical example the epidemiological evidence is observational statistical evidence while the evidence for the existence of a linking mechanism is provided by laboratory experiments.

There is one further distinction which will prove useful in our specific formulation of the Russo–Williamson thesis. This is the distinction between a *plausible* mechanism, and a *confirmed* mechanism. A plausible mechanism is one which is confirmed by our general background knowledge but not necessarily by particular investigations and experiments designed to test it out. This distinction can be illustrated by the example given in Section 6.1 of the difference between the cases of smoking and drinking in relation to lung cancer. Cigarette smoke contains a large number of chemicals of various types and, in the case of those who smoke regularly, some of these are likely to find their way into the lungs. It seems likely that some of these chemicals will damage the tissue of the lungs, and some indeed might be carcinogens which initiate a cancerous tumour. It was part of the background knowledge of 1976 that there existed chemical carcinogens capable of initiating cancerous tumours. Indeed this had been established by experimental evidence. Consequently it was known in 1976 that there was a plausible mechanism linking smoking to lung cancer. By contrast there does not seem to be an obvious mechanism connecting the consumption of alcohol to lung cancer. Of course the body is very complicated, and such a mechanism might turn out to exist after all. However, background knowledge does not make the existence of such a mechanism very plausible. This suggests that we may be more prepared to go from correlation to causation if there exists a plausible mechanism linking the two factors.

In contrast to smoking and lung cancer in 1976, a mechanism may, in some other cases, be uncovered by a lengthy series of experiments, and its correctness confirmed by these results. Here I will speak of a confirmed mechanism. Of course there is really a continuum here which depends on the degree to which the mechanism has been confirmed. However, the rough distinction between plausible and confirmed mechanisms is helpful for clarifying the situation. I will understand confirmed mechanisms to be also plausible, but not vice versa.

Corresponding to this distinction, we can have two forms of the Russo–Williamson thesis – a strong form and a weak form. According to the strong form, a causal link between A and B can only be said to be established if it has been shown that there is a confirmed mechanism linking A and B. For the weak form, it may suffice just to show that there is a plausible mechanism linking A and B. I prefer the weak form of the thesis since it fits better with the classic example of smoking and lung cancer described in Section 6.1. I would argue that Doll and Peto were quite justified in claiming in 1976 to have established that smoking causes lung cancer, even though the mechanism linking the two was, at that stage, only plausible rather than confirmed. So, after these preliminaries, I can formulate the precise form of the Russo–Williamson thesis which I will defend in the rest of the chapter. It may be stated as follows: 'In order to establish that A causes B, observational statistical evidence does not suffice. Such evidence needs to be supplemented by interventional evidence, which can take the form of showing that there is a plausible mechanism linking A to B.' This is a weak version of the RWT, but not a trivial one. It has important implications for artificial intelligence. At the moment, there are a number of research programmes whose aim is to obtain causal relations from observational statistical data automatically using machine learning. If, however, the RWT as just formulated is correct, these programmes cannot succeed. In the example to be considered in

Sections 6.3–6.5, epidemiological data showing a correlation between smoking and heart disease was not regarded as sufficient to establish a causal link. To establish causality, scientists regarded as necessary the additional evidence provided by laboratory investigations which showed that there is a plausible mechanism linking smoking to heart disease.

Against this view, it could be argued that modern machine learning programs do not employ the kind of simple statistical methods employed by Doll and Peto, but analyse the data using a variety of much more sophisticated techniques such as investigating the effect of conditionalizing on a variety of variables. These more sophisticated statistical techniques would, so it might be claimed, be sufficient to establish the causal link between smoking and heart disease without any need for the evidence provided by laboratory experiments. I find such a claim quite unconvincing, but to discuss it further would take me too far from the aim of the present chapter. Let me just say that this example constitutes a challenge to advocates of the automated learning of causal relations from observational statistical data.

From now on I will limit the Russo–Williamson thesis (or RWT) to the specific version just formulated. My aim in the rest of the chapter is to show the RWT is supported by the way that medical research into smoking and heart disease developed in the period from 1976 to the late 1990s.

6.3 Smoking and heart disease. Is there a linking mechanism?

Let us now return to Doll and Peto's paper of 1976. Their survey of mortality among male doctors in the UK did not confine itself to lung cancer, but investigated all the recorded causes of death. In fact the most common cause of death among the doctors was ischaemic heart disease. This accounted for 3191 deaths as compared to 441 from lung cancer. Now death from ischaemic heart disease was also positively correlated with smoking. The annual death rate from ischaemic heart disease per 100,000 men, standardized for age, in various categories was as follows (Doll and Peto 1976, p. 1527):

Non-smokers	413
Smokers	565 (+37%)
1–14 gms tobacco per day	501 (+21%)
14–24 gms tobacco per day	598 (+45%)
25 gms tobacco per day or more	677 (+64%)

A cigarette is roughly equivalent to 1 gm of tobacco.

These figures are less striking than those relating smoking to lung cancer, but they still show a positive correlation which is highly significant statistically. Moreover Doll and Peto say (1976, p. 1534): 'Our data for ischaemic heart disease are similar to those that have been reported in many other studies throughout the world'. They cite in this context six studies carried out between 1965 and 1975.

Doll and Peto are, however, cautious about drawing causal conclusions from this correlation. Ischaemic heart disease is one of the conditions which they list in their table III, and they say (1976, p. 1528):

> Half the conditions in table III were positively related to smoking, some very strongly so, ...To say that these conditions were related to smoking does not necessarily imply that smoking caused... them. The relation may have been secondary in that smoking was associated with some other factor, such as alcohol consumption or a feature of the personality, that caused the disease. Alternatively, smoking habits may have been modified by the disease or the relation may have been an artefact due to misdiagnosis,...

As we have seen, Doll and Peto (1976) do definitely regard the link between smoking and lung cancer as causal in nature. However, as regards ischaemic heart disease, they make the weaker claim that (p. 1535) that the excess mortality from ischaemic heart disease in cigarette smokers is probably wholly or partly attributable to smoking. This caution is very much in agreement with the Russo–Williamson thesis, because the background knowledge of heart disease in 1976 did not support any plausible mechanism linking smoking to heart disease. Let us briefly review some features of this background knowledge.

The vast majority of heart disease is owing to *atherosclerosis*. The word comes from the Greek terms: athere = porridge, and sclerosis = hardening. It refers to the formation of plaques in the arteries. To be scientific then, we should seek a link not between smoking and heart disease, but between smoking and atherosclerosis. This will be our strategy in the rest of the paper.

Let us suppose large atherosclerotic plaques have been formed in the coronary arteries supplying blood to the heart. One of these may break off releasing fatty lipids and other debris into the artery. Alternatively, or in addition, contact with the plaque may cause a blood clot or *thrombus* to form. The effect of this is to cut off temporarily, or severely reduce, the blood flow to the heart, and this in turn can cause damage and/or death (*infarction*) of heart muscle tissue (*myocardium*). This is the mechanism of a heart attack, or myocardial infarction as it is known in the medical world. A similar mechanism is responsible for strokes.

Atherosclerotic plaques in the arteries were investigated by pathologists in the nineteenth century, and in 1910 the German chemist Windaus claimed that such plaques consisted of calcified connective tissue and cholesterol. The importance of cholesterol was reinforced by a paper published by two

Russians (Anitschkov and Chalatov) in 1913. They had succeeded in inducing atherosclerosis in rabbits by feeding them a cholesterol-rich diet. Later investigations showed that cholesterol is carried in the blood in two forms: low-density lipoprotein (LDL) and high-density lipoprotein (HDL). In fact it is only LDL which is responsible for atherosclerosis. HDL inhibits rather than encourages the disease.

So in 1976 it was known that a high concentration of LDL in the blood favoured the development of atherosclerosis. But why should smoking accelerate this process? The background knowledge of the time suggested no plausible mechanism linking smoking and atherosclerosis. However, research into heart disease between 1976 and the late 1990s greatly increased medical knowledge of how and why atherosclerotic plaques form, and this new knowledge does support a plausible mechanism by which smoking may increase the rate of formation of such plaques. I will give a brief account of some of the research into atherosclerosis between 1979 and the late 1990s in the next two sections.

6.4 Research into atherosclerosis 1979–89

To follow the course of this research, two technical terms must first be mastered, namely *macrophages* and *monocytes*. Macrophages (literally 'large eaters') are well-known to the general public. They are white cells which attack and destroy microbes, such as bacteria, which invade the body. Macrophages have *receptors* or *binding sites* by means of which they attach themselves to their prey. This prey is then engulfed, and destroyed. At least the macrophages attempt to destroy their prey. They are not always successful. Monocytes are circulating predecessors of macrophages. If additional macrophages are needed at any point, monocytes arrive and turn into macrophages at the location of the action in the tissue.

Let us now return to the problem of how LDL carried round by the blood stream can be converted into atherosclerotic plaques. A crucial step was taken in elucidating the process was taken by Goldstein *et al.* in their paper published in 1979. In this paper they showed that if LDL is modified by being acetylated, it gets taken up in large quantities by macrophages using a specific binding site or receptor, which later became known as the 'scavenger receptor'. The macrophages which take up the modified LDL become, in their attempts to destroy it, bloated with lipid and resemble the foam cells to be found in the fatty streaks which are the first stage in the formation of atherosclerotic plaques. This was a most suggestive observation, but of course acetylation of LDL in the body seemed an unlikely occurrence. As Goldstein *et al.* say (p. 337):

Although *in vivo* acetylation of plasma LDL seems unlikely at this point, some chemical or physical alteration of LDL occurring in plasma or interstitial fluid may make it susceptible to recognition by the macrophage binding site.

Naturally this remark prompted the search for the appropriate *in vivo* alteration of LDL, and it was duly found in the next few years (Henriksen *et al.* 1981; Steinbrecher *et al.* 1984). The alteration was the oxidation of LDL. This can occur quite naturally in the body. It is induced by incubation of LDL with the endothelial cells which line the arteries. It can also be induced by incubation with smooth muscle cells, or with macrophages (Jürgens *et al.* 1987). Moreover oxidized LDL, just like acetylated LDL, is taken up by the scavenger receptor on macrophages resulting in the formation of foam cells bloated with lipid. Oxidised LDL has two further properties of great interest (Jürgens *et al.* 1987). It inhibits the movement of macrophages, while attracting monocytes.

Putting all these elements together, we arrive at a mechanism by which fatty streaks can form within the artery walls. This is described by Steinberg *et al.* (1989), a paper with the significant title: 'Beyond cholesterol. Modifications of low-density lipoprotein that increase its atherogenicity'. The key point is that LDL, even in large quantities, causes no problems as long as it remains in its natural state. However, if it is oxidized, then trouble starts. Oxidized LDL is attacked by macrophages which bind it with their scavenger receptor, and then attempt to destroy it. The result is the formation of foam cells. If some of these are in the wall of an artery, then, because oxidized LDL inhibits the movement of macrophages, they remain there. Moreover oxidized LDL attracts monocytes, which turn into macrophages and generate more foam cells. Since macrophages also oxidize LDL, a self-reinforcing process can start, resulting in the formation of fatty streaks in the artery walls.

It may seem a little odd that macrophages should try to dispose of oxidized LDL since the results of this attempt are somewhat unfortunate. However it should be remembered that oxidized LDL can be very damaging to the cells of the body, so that its disposal, even at some cost, may be on balance justified. It would obviously, however, be better for LDL not to be oxidized, and in fact there are many devices to protect LDL from oxidation. LDL carries round with it in the blood stream a whole package of antioxidants which protect it against oxidation. The principal component of this package is vitamin E, but the package also contains (Esterbauer *et al.* 1989, p. 256) beta-carotene, lycopine, and retinylstearate. Moreover in the plasma of normal blood there are large quantities of two powerful anti–oxidants – vitamins C and E. These devices have obviously evolved to prevent the oxidation of LDL and the harmful consequences of this oxidation.

So by the end of the 1980s it was established that the oxidation of LDL was an important step in the process which led to atherosclerotic plaques. Now this seems at once to explain why smoking accelerates atherosclerosis. Cigarette smoke gives rise to what are known as reactive oxygen species (or ROS) (Lehr *et al.* 1994, p. 7691). These include superoxide and hydrogen peroxide. Their effect would be to increase the tendency towards oxidation in the body, to introduce what is called 'oxidative stress'. The existence of such oxidative stress in smokers was strongly confirmed in a study by Morrow *et al.* (1995). They introduced a new and superior index of the amount of lipid peroxidation occurring *in vivo*. This was the level of F_2-isoprostanes. Sure enough this level proved to be much higher in smokers than non-smokers. Morrow *et al.* (1995) *et al.* conclude (1995, pp. 1201–2):

> Our finding that the production of F_2-isoprostanes is higher in smokers than in nonsmokers provides compelling evidence that smoking causes oxidative modification of biologic components in humans. This conclusion is greatly strengthened by the finding that levels of F_2-isoprostanes in the smokers fell significantly after two weeks of abstinence from smoking. These results provide a basis for hypotheses that link oxidative damage of critical biomolecules to the pathogenesis of diseases caused by smoking.

At this point it may well look as if we have not just a plausible mechanism, but a confirmed mechanism linking smoking to heart disease. Smoking introduces oxidative stress. This results in more LDL being oxidised, which in turn leads to the development of fatty streaks in the artery walls and atherosclerotic plaques. However, there is a difficulty which shows that this mechanism, however convincing it may appear at first sight, cannot be correct. The difficulty concerns the place at which the oxidation of the LDL takes place. Steinberg *et al.* argue in their 1989 that LDL is not oxidized in the blood stream but within the artery wall. This is what they say (1989, p. 919):

> For several reasons, it seems that the oxidation of LDL probably occurs not in the circulation but within the artery wall. First, even if oxidized LDL were generated in the plasma, it would be swept up within minutes by the liver. Second, oxidation is inhibited by plasma and so probably requires the favorable conditions of a sequestered microenvironoment.

As we pointed out, normal blood plasma contains large quantities of vitamins C and E, and these strongly inhibit the oxidation of LDL. If, however, the blood contains a great deal of LDL, some LDL may diffuse into the artery wall. Here it is no longer protected by the antioxidants of the blood plasma, and so, in the presence of oxidizing agents such as macrophages, its own package of protective antioxidants may be exhausted and it may become oxidized.

Now smoking introduces additional oxidizing agents into the blood stream (the ROS), but these do not penetrate into the artery wall, and so should have no effect on the formation of the atherosclerotic plaques as so far described. So far then the mechanism (if any) by which smoking accelerates the formation of atherosclerotic plaques remains a mystery. However, as I will describe in the next section, further researches in the 1990s shed new light on the question.

6.5 Research into atherosclerosis in the 1990s

In the 1990s, investigations began into another aspect of the process of formation of atherosclerotic plaques. If monocytes adhere to the endothelial cells lining the arteries, they can work their way into the artery wall, turn into macrophages and accelerate any on-going process of plaque formation. Now cigarette smoke consists of 92% gaseous components and 8% particulate constituents. These particulate constituents are known as cigarette smoke condensate. In 1994, Kalra *et al.* discovered that cigarette smoke condensate increases significantly the tendency of monocytes to adhere to the endothelial cells. As they say (Kalra *et al.* 1994, p. 160):

> It thus appears that cigarette smoke particulate constituents potentiate adherence of monocytes to the vascular endothelial cell lining. This presumably is followed by transmigration of adhered monocytes into the subendothelium space to form foam cells and subsequent atherosclerotic lesion formation.

Interestingly Kalra *et al.* elucidate the mechanism which brings about the increased adherence of the monocytes. They describe it as follows (1994, p. 155):

> ...the recruitment of monocytes to the endothelial surface could occur as a result of change in the adhesive properties of the endothelial surface.
>
> The results presented here show that cigarette smoke condensate (CSC), the particulate phase of cigarette smoke, causes an increase in the expression of CD11b on monocytes and ICAM-1, ELAM-1, and VCAM-1 adhesion molecules on endothelial cells with a concomitant increase (70–90%) in the basal adherence of monocytes to cultured endothelial cells.

Daniel Steinberg, who played a key part in the elucidation of the mechanisms described in the previous section, made a useful comment on the situation in his 1995 paper. Here he pointed out that the studies carried out so far had been mainly concerned with the formation of fatty streaks in the artery walls, but that these are only the earliest atherosclerotic lesions. He writes (1995, p. 37):

> How long does it take for a new fatty streak to become a clinically threatening lesion? We cannot be sure, but we know that by age 25 some 20–30% of the aorta is covered by fatty streak lesions and yet myocardial infarction seldom occurs before age 50. If lesion progression is more or less linear as a function of time, we might conclude that it takes 20 years or more for a new fatty streak to become the nidus for coronary thrombosis.

This suggests that more attention should be devoted to the development of atherosclerotic plaques than to their initial stages. One factor to which attention is drawn by Poston and Johnson-Tidey in their 1996 concerns the rate at which monocytes adhere to the endothelium of the artery. They write (1996, p. 75):

"...at 20 or 37 °C, human peripheral blood monocytes...showed selective binding to atherosclerotic plaques, compared with non-atherosclerotic arterial intima...Adhesion occurred to the endothelium of the plaque area...The binding to non-atherosclerotic artery endothelium was much less, and the difference was highly significant...

They go on to suggest that a positive feedback mechanism may develop in which atherosclerotic plaques as they develop attract more and more monocytes which cause them to develop further. As they point out (p. 73) there could be a link here to the question of the effects of smoking on atherosclerosis in the light of some earlier work of Lehr *et al.* (1994), to which we now turn.

Lehr *et al.* investigated the effects of cigarette smoke on hamsters. They showed that cigarette smoke increased the rate of adhesion of leukocytes to arterial endothelium, and that this rate of adhesion was significantly reduced by vitamin C, but not by vitamin E. They argue that the increase in the rate of adhesion was due to the reactive oxygen species (ROS) produced by cigarette smoke (CS). From this they draw the following conclusion (Lehr *et al.* 1994, p. 7691):

The fact that the water-soluble vitamin C, but not the lipid-soluble antioxidants vitamin E and probucol (which contribute little to serum antioxidant activity), afforded protection from CS-induced changes indicates that CS-induced leukocyte adhesion and aggregate formation with platelets involves isolated, direct attacks of aqueous-phase ROS, rather than the sequelae of membrane lipid peroxidation. Like dietary vitamin C, a single intravenous injection of vitamin C just 5 min prior to CS exposure resulted in a similar protection from CS-induced leukocyte/platelet/endothelium interaction, suggesting that vitamin C does not need to be incorporated into cells in order to be effective, but that it merely needs to be circulating in the bloodstream in order to neutralize aqueous phase ROS.

We saw earlier that the oxidation of LDL which is relevant to the formation of atherosclerotic plaques was thought in the 1980s to occur within the artery walls rather than in the blood stream and so was unlikely to be produced by smoking. However, the results of Lehr *et al.* indicated that there was another oxidation process which did occur in the blood stream and which affected atherosclerosis. This was the process which resulted in an increased rate of activation, aggregation, and adhesion of leukocytes to the endothelium of the artery. The process was known to involve oxidation because it was inhibited by antioxidants, and it was known to occur in the bloodstream because it is inhibited by vitamin C but not vitamin E. Lehr *et al.* remark (1994, p. 7692):

> Corroborative evidence can be derived from epidemiological surveys which consistently demonstrate a significant consumption of vitamin C, but not of vitamin E, in the plasma of smokers.

Here then, at last, was a mechanism linking smoking to atherosclerosis. Smoking produced oxidative stress. This increased the adhesion of leukocytes to the endothelium of the artery, which in turn accelerated the formation of atherosclerotic plaques. This mechanism was certainly plausible, and indeed could be regarded as at least partially confirmed. So it was sufficient for the Russo–Williamson thesis in the form in which I have formulated and defended it.

6.6 Conclusions

In this chapter I have formulated a version of the Russo–Williamson thesis according to which observational statistical evidence alone is not sufficient to establish that A causes B. Such evidence can only establish that A and B are correlated. One way of going from correlation to causation is to show that there is a plausible mechanism linking A to B. This version of the thesis was tested out and confirmed by three examples from medicine. The first example is the claim that smoking causes lung cancer. This was taken as established by Doll and Peto in 1976 on the basis of strong observational statistical evidence, which accords with our version of the RWT because there was a plausible mechanism linking smoking and lung cancer. By the early 1980s, a number of observational studies had shown that there was also a strong correlation between drinking heavily and lung cancer. However, this was not taken as showing a causal connection, but rather as being explained by confounding factors such as smoking. This again agrees with our version of the RWT since there was, and is, no plausible mechanism linking heavy drinking to lung cancer. The third example concerned smoking and heart disease. We saw that Doll and Peto established a strong correlation between the two in 1976, and yet were hesitant about inferring a causal connection. They thought that causality was probable but had not been established. This again agrees with our version of the RWT because no plausible mechanism linking smoking and heart disease had at that time been shown to exist. However I went on to show that between 1979 and the late 1990s research into atherosclerosis did bring to light a mechanism linking smoking to an acceleration of the rate of formation of atherosclerotic plaques. This mechanism was at least plausible and perhaps even confirmed. So it justified the increasing acceptance that smoking causes heart disease. However, an account of the research shows that the path to a plausible mechanism here was a winding one. Some earlier results suggested a mechanism which for quite subtle reasons could not be correct, and the linking mechanism which now looks plausible has a rather different character. This is a nice illustration of the difficulties of establishing plausible mechanisms through research in the medical field.

Acknowledgements

An earlier version of this paper was read at an informal workshop on Causality and Linking Mechanisms, held at the University of Kent in Canterbury on 23 July 2008. I would like to thank those present (Lorenzo Casini, Brendan Clarke, Phyllis McKay Illari, Federica Russo, and Jon Williamson) for many helpful comments and a stimulating discussion which, among other things, greatly improved my formulation of the version of the Russo–Williamson thesis which is defended in this chapter. I would also like to thank Brendan Clarke for suggesting the example of the correlation between heavy drinking and lung cancer, and for supplying the reference (Freudenheim *et al.* 2005) in which details about it are to be found. I also received useful comments from Jeremy Howick, John Worrall and an anonymous referee which enabled me to improve several points in the paper. I could never have written my account of research into atherosclerosis from 1979 to the late 1990s without the help of Robin Poston who has himself carried out research in this area for many years. Robin Poston kindly took the trouble to explain the key points of the development to me, and to suggest readings which would make me familiar with the material. He also read carefully an earlier draft of my account of this research and suggested many improvements. For philosophers of science the help of sympathetic research scientists is really essential.

References

Doll, Richard, and Peto, Richard (1976). Mortality in relation to smoking: 20 years' observations on male British doctors, *British Medical Journal*, **2**, pp. 1525–1536.

Esterbauer, Hermann, Striegl, Georg, Puhl, Herbert, Oberreither, Sabine, Rothender, Martina, El-Saadani, Mohammed, and Jürgens, Günther (1989). The role of vitamin E and carotenoids in preventing oxidation of low density lipoproteins, *Annals of the New York Academy of Sciences*, **570**, pp. 254–267.

Freudenheim, Jo L., Ritz, John, Smith-Warner, Stephanie A., Albanes, Demetrius, Bandera, Elisa V., van den Brandt, Piet A., Colditz, Graham, Feskanich, Diane, Goldbohm, R. Alexandra, Harnack, Lisa, Miller, Anthony B., Rimm, Eric, Rohan, Thomas E., Sellers, Thomas A., Virtamo, Jarmo, Willett, Walter C., and Hunter, David J. (2005). Alcohol consumption and risk of lung cancer: A pooled analysis of cohort studies, *American Journal of Clinical Nutrition*, **82**, pp. 657–667.

Gillies, Donald (2005). An action-related theory of causality, *British Journal for the Philosophy of Science*, **56**, pp. 823–842.

Goldstein, Joseph L., Ho, Y.K., Basu, Sandip K., and Brown, Michael S. (1979). Binding site on macrophages that mediates uptake and degradation of acetylated low density lipoprotein, producing massive cholesterol deposition, *Proceedings of the National Academy of Sciences USA*, **76**(1), pp. 333–337.

Heinriksen, Tore, Mahoney, Eileen M., and Steinberg, Daniel (1981). Enhanced macrophage degradation of low density lipoprotein previously incubated with cultured endothelial cells: Recognition by receptors for acetylated low density lipoproteins, *Proceedings of the National Academy of Sciences USA*, **78**(10), pp. 6499–6503.

Jürgens, Günther, Hoff, Henry F., Chisolm, Guy M. III, and Esterbauer, Hermann (1987). Modification of human serum low density lipoprotein by oxidation – Characterization and pathophysiological implications, *Chemistry and Physics of Lipids*, **45**, pp. 315–336.

Kalra, Vijay K., Ying, Yong, Deemer, Kathleen, Natarajan, Rama, Nadler, Jerry L., and Coates, Thomas D. (1994). Mechanism of cigarette smoke condensate induced adhesion of human monocytes to cultured endothelial cells, *Journal of Cellular Physiology*, **160**, pp. 154–162.

Lehr, Hans-Anton, Frei, Balz, and Arfors, Karl-E. (1994). Vitamin C prevents cigarette smoke-induced leukocyte aggregation and adhesion to endothelium *in vivo*, *Proceedings of the National Academy of Sciences USA*, **91**, pp. 7688–7692.

Morrow, Jason D., Frei, Balz, Longmire, Atkinson W., Gaziano, J. Michael, Lynch, Sean M., Shyr, Yu, Strauss, William E., Oates, John A., and Roberts, L. Jackson II (1995). Increase in circulating products of lipid peroxidation (F_2-isoprostanes) in smokers. Smoking as a cause of oxidative damage, *The New England Journal of Medicine*, **332**(18), pp. 1198–1203.

Pearl, Judea (2000). *Causality. Models, Reasoning and Inference*, Cambridge: Cambridge University Press.

Poston, Robin N. and Johnson-Tidey, Ruth R. (1996). Localzied adhesion of monocytes to human atherosclerotic plaques demonstrated *in vitro*. Implications for atherogenesis. *American Journal of Pathology*, **149**(1), pp. 73–80.

Russo, Federica and Williamson, Jon (2007). Interpreting causality in the health sciences, *International Studies in the Philosophy of Science*, **21**(2), pp. 157–170.

Steinberg, Daniel (1995). Clinical trials of antioxidants in atherosclerosis: are we dong the right thing? *The Lancet*, **346**, pp. 36–38.

Steinberg, Daniel, Parthasarathy, Sampath, Carew, Thomas E., Khoo, John C, and Witztum, Joseph L. (1989). Beyond Cholesterol. Modifications of low-density lipoprotein that increase its atherogenicity, *The New England Journal of Medicine*, **320**(14), pp. 915–924.

Steinbrecher, Urs P., Parthasarathy, Sampath, Leake, David S., Witztum, Joseph L., and Steinberg, Daniel (1984). Modification of low density lipoprotein by endothelial cells involves lipid peroxidation and degradation of low density lipoprotein phospholipids, *Proceedings of the National Academy of Sciences USA*, **81**, pp. 3883–3887.

PART III
Psychology

·7·

Causal thinking

David Lagnado

Abstract

How do people acquire and use causal knowledge? This chapter argues that causal learning and reasoning are intertwined, and recruit similar representations and inferential procedures. In contrast to covariation-based approaches to learning, I maintain that people use multiple sources of evidence to discover causal relations, and that the causal representation itself is separate from these informational sources. The key roles of prior knowledge and interventions in learning are also discussed. Finally, I speculate about the role of mental simulation in causal inference. Drawing on parallels with work in the psychology of mechanical reasoning, the notion of a causal mental model is proposed as a viable alternative to reasoning systems based in logic or probability theory alone. The central idea is that when people reason about causal systems they utilize mental models that represent objects, events or states of affairs, and reasoning and inference is carried out by mental simulation of these models.

You arrive at your holiday apartment, welcomed by the local cat and a chorus of crickets outside the window. During the night your sleep is interrupted by intermittent high-pitched squeals. At first you assume it is the cat, but on careful listening the noises sound mechanical rather than animate. You get up and walk around the flat, and notice that the light on the smoke detector is flashing red – suggesting that the battery is running down. You recall a similar problem with the smoke alarm in your own house, and an equally annoying high-pitched squeal as the battery died out. You remove the battery from the fire alarm, and the squeals stop.

Next morning, as you make breakfast, the squeals seem to return. But the smoke detector is lying dismantled on the table. Perhaps the capacitor is still discharging, emitting the occasional squeal? Or a cricket has started mimicking the sound of the dying smoke detector? But then you notice that whenever you turn on the kitchen tap, there is a high-pitched noise that sounds very similar to last night's squeals. You turn on the tap at random moments throughout the day, and it is nearly always followed by a squeal. Problem solved! But maybe the smoke detector, like the local cat, was falsely accused. Perhaps it was the dodgy plumbing all along? Just as you start to re-insert the battery to test out this idea you remember that you are on holiday.

This is an everyday example, but it illustrates several of the key aspects of causal thinking. In this chapter I will use it as a running example to identify some shortcomings with current theorizing in the psychology of causal inference. I will also suggest ways in which our understanding of causal cognition might be improved. In particular, I will argue that causal learning and reasoning are intertwined, that there are multiple sources of evidence for causal relations, and I will speculate about the role of mental simulation in causal inference.

7.1 Interplay of learning and reasoning

At a general level the smoke detector example highlights the interplay between causal learning (typically conceived as the induction of causal relations from patterns of observations) and causal reasoning (drawing inferences on the basis of assumed causal relations) – and the artificiality of separating and studying these activities in isolation, as is common in the psychological literature. Thus, as you try to work out the cause (or causes) of the high-pitched squeals, you engage in a variety of interleaved inferential activities, including hypothesis generation and testing, hypothetical and counterfactual reasoning. For example, your observation that the smoke detector light is flashing red leads you to hypothesize that the battery is running low, and that a low battery is the cause of the noises. You reason (hypothetically) that if the low battery is the cause, then removing the battery altogether should stop the squeals. You confirm this hypothesis by removing the battery and noting that the squeals stop.

Even in this simple learning episode, which is only the start our scenario, you have deftly switched between various forms of inference. You have inferred a putative causal relation from observations, but also engaged in hypothetical reasoning and hypothesis-testing. In mainstream cognitive psychology, however, causal learning and causal reasoning are usually treated as separate areas of research, with different theories and empirical paradigms.

Research in causal learning focuses on the induction of causal relations from data, with little concern for the other reasoning activities that might accompany this induction. A typical experiment presents people with covariation information about potential causes and effects (either in summary format, or presented sequentially) and asks them to assess the strength of the putative causal relation. Applied to our example, this would correspond to showing people a sequence of cases in which the smoke detector battery is either low or high, and the detector either does or does not make a squeal. The central question of interest to experimenters and theorists is how people arrive at a causal estimate from the patterns of covariation, and there is substantial debate about this (Cheng, 1997; Griffiths & Tenenbaum, 2005; Shanks, 2004; Vallee-Tourangeau *et al.* 1998). However, very little is said about what, if any, reasoning occurs in such experiments, or how this reasoning takes place. Indeed theorists of an associative persuasion maintain that people are simply acquiring associations between mental representations of the presented variables (e.g. mentally associating 'low battery' with 'squeals')[1]. But this is only part of the story. There is compelling empirical evidence that in causal learning contexts people are not merely associating the two variables, but are hypothesizing that one *causes* the other (Waldmann, 1996; Waldmann and Holyoak, 1992; see Lagnado *et al.* 2007 for a review), an inference that engenders a range of further inferences (e.g. that if you were to replace the low battery with a new one, then the squeals will stop; and if the battery had been high, then there would have been no squeals). Of course these inferences are fallible. You might have the wrong causal model. But your conjecture that one variable causes another carries with it these additional inferences, in a way that the postulation of a mere association does not. The claim that two variables are associated by itself tells us nothing about what would happen to one variable if we were to change the other.

More generally, an analysis of causal learning that focuses only on the associations that people can learn between variables does not account for a variety of other inferential activities that people engage in, or the wealth of information (beyond covariation data) that they might use to support these inferences.[2]

On the other hand, research in causal reasoning tends to focus on how people make conditional or counterfactual inferences on the basis of presupposed causal relations, with little regard for how these causal models are acquired or generated. For example, one of the dominant accounts of reasoning, mental model theory, argues that causal reasoning involves the construction of possible states of the world and the search for counterexamples (Goldvarg and Johnson-Laird 2001). In particular, the meaning of 'A causes B' is cashed out in terms of the possibilities consistent with 'A materially implies B' and the constraint that A temporally precedes B. However, nothing is said about how people acquire the background causal knowledge that allows them to generate appropriate possibilities, and make appropriate connections between these possibilities (e.g. distinguish between possible states that are causally connected rather than merely correlated).

Other investigations into causal reasoning (e.g. explanation-based reasoning, Hastie & Pennington 2000; causal heuristics, Kahneman & Tversky 1982) also fail to account for how people construct and revise causal models. This would not be a problem if learning and reasoning were separate activities that engaged entirely distinct processes; but these inferential activities are

[1] There are several varieties of associative theory, but this appears to be a shared assumption across most variations (Dickinson, 1980; Hall, 2002; Shanks, 1995).

[2] This is not to undermine the important role that associative learning can play in cognition, but to emphasize that causal thinking will often go beyond the acquisition of associations.

best studied together, as part of a general system of causal inference (see Lagnado 2009). Postulating common representations and mechanisms also explains why both learning and reasoning are subject to similar constraints (e.g. attentional or working memory limitations).

From a formal perspective, causal Bayesian networks (Spirtes, Glymour and Scheines 1993; Pearl, 2000) provide a unified framework for representation, learning and inference (for criticisms of some of the assumptions underlying this framework see Cartwright, 2007, Williamson, 2005). Numerous psychologists have adopted this framework as a model of human inference (Gopnik and Schultz, 2007; Lagnado *et al.* 2007; Sloman and Lagnado, 2005; Sloman *et al.* 2009; Tenenbaum *et al.* 2007). The framework is a great advance insofar as it provides a normative benchmark against which to appraise human inference, and a guide for the construction of descriptive models. But the formalism alone does not mandate any specific psychological account; and, indeed, a psychological theory need not be tied too tightly to the normative theory.

7.2 Multiple sources of evidence for causal beliefs

Psychological research into causal learning has been dominated by the question of how people induce causal relations from patterns of covariation. However, covariation-based learning is only part of the picture, and exclusive focus on this question threatens to distort our understanding of causal cognition.

As well as engaging in several kinds of inference (not just induction), people use various sources of information to infer causal relations (Einhorn and Hogarth, 1986; Lagnado *et al.* 2007; Waldmann *et al.* 2006). These 'cues to causality' include information about temporal order, interventions, spatiotemporal contiguity, similarity, analogy and prior knowledge. The smoke detector scenario in fact illustrates most of these possibilities. For example, the temporal proximity between squeals and tap turns was an important clue to identifying the hot water system as a likely cause (if the temporal interval had been much greater, or more variable, you would have been less likely to associate the two). The repeated interventions on the hot tap, at a random selection of times throughout the morning, helped to rule out possible confounding causes, and establish the hot tap as a cause of the squeals. The analogy of the current situation to a previous encounter with the noises admitted from a smoke detector helped guide the hypothesis formulation and testing. The similarity in sound of the squeals meant that a single cause was sought (e.g. smoke detector or hot water system). The role of prior information was also ubiquitous (see next section).

This is not to deny the role of covariation information (which also plays its part in our scenario), but to emphasize that it is just one cue among many. Indeed covariation information by itself is seldom sufficient for inferring a unique causal model. For example, consider a variation on our story about the mysterious squealing where we ignore the smoke detector and faulty hot water system, and focus on the local cat. Suppose that the only available evidence is your observation of a strong correlation between the appearances of the cat and the sound of the squealing noises. In the absence of any other information (e.g. about temporal order; spatiotemporal contiguity etc.) this evidence does not distinguish between a causal model in which the cat causes the squeals, or a model in which the squeals cause the cat's presence (perhaps it is tracking a squealing mouse or large insect, or even a mate), or a common cause model in which both the cat's presence and the squeals are due to a third factor.

One way to distinguish between these models is to intervene by removing the cat from the premises, and ensuring it does not return. Do the intermittent squeals persist? If so, then the cat is ruled out as a cause. If not, then it is ruled in. Another route is to acquire additional information about the potential cat → squeal relation, perhaps in terms of spatial or temporal information about the cat and the squeals. The critical moral is that the mere observation of a correlation does not provide evidence for a unique causal relation (for more details see Lagnado *et al.* 2007; Sloman, 2005).

Recent psychological studies have shown that people are indeed able to use interventions to learn causal models; and, moreover, to learn models that cannot be learned from covariational information alone (Gopnik *et al.* 2004; Lagnado and Sloman, 2002, 2004, 2006; Meder *et al.* 2008; Steyvers *et al.* 2003; Waldmann & Hagmayer, 2005). Not only do people learn better when they can intervene on a system, but they also make appropriate inferences about the effects of their interventions (Sloman and Lagnado, 2005). These experiments, as well as numerous others conducted by Tenenbaum and Griffiths and colleagues (e.g. Tenenbaum and Griffiths, 2003; Tenenbaum *et al.* 2007), convincingly demonstrate that people do not solely rely on covariational information to induce causal structure. Instead, they make use of a range of sources of information, including a central role for the evidence gathered from interventions.

In many real-world contexts people are provided with a rich variety of information about the causal systems they interact with. The control of objects, tools and simple devices (e.g. pens, scissors, can-openers) are readily learned through a combination of interventions, sensorimotor feedback, and spatiotemporal information. To explore this kind of learning we introduced a novel experimental paradigm in which subjects manipulated on-screen sliders in a real-time learning environment. Their task was to discover the causal connections between these sliders by freely changing the settings of one slider and observing the resultant changes in the other sliders. Subjects excelled at this task, rapidly learning complex causal structures with a minimum of exploration (Lagnado *et al.* 2007; Lagnado & Loventoft-Jessen, in prep.).

This set-up revealed two important points about people's capacity for causal learning. First, it only took a few manipulations of a slider for subjects to leap to a causal conclusion about the link between one slider and another. The causal relation 'popped-out' due to the confluence of various factors: the spatiotemporal similarities in the motions of the sliders that were causally connected; the sensitivity of control that one slider exerted on the other; the opportunity for subjects to intervene when they chose (thus ruling out confounding variables).[3] Second, once subjects had explored the system for several minutes, they were able to construct mental models of the causal connections between sliders, and imagine the effects of interventions that they had not yet taken.

This was shown in a follow-up experiment (Lagnado & Loventoft-Jessen, in prep.), in which subjects made use of 'double interventions' in order to disambiguate between models. For example, if subjects are only able to move one slider at a time, it is impossible to distinguish between a model with a chain (A → B → C) and a similar model with an additional link from A to C. In both cases, changes in A lead to changes in B and C, and changes in B lead to changes in C alone. One way to distinguish these models is to disable B, and then see whether changes in A still lead to changes in C. If they do not, then the true model is the chain. In the experiment, subjects engaged in an initial learning phase where they were restricted to moving one slider at a time. At the end of this phase they were asked which causal model (or models) best explained the observations they had made. In a second test they were asked to choose one disabling intervention (in combination with a single slider move) that would allow them to distinguish between models. Many subjects were able to select the correct disabling intervention, showing that they could mentally represent possible causal models, and imagine the effects of interventions on this model (in particular, what would happen if they disabled one slider, and then moved another).

This experiment supports several of the claims made in this chapter. It shows that people can make use of various sources of information, including interventions and spatiotemporal similarities, to learn causal models. It shows that people can engage in hypothetical reasoning in order to disambiguate complex causal structures, thus confirming the interplay between learning and reasoning. Finally, it anticipates the discussion of mental simulation and causal reasoning presented in later sections.

A central claim in this chapter is that it is mistaken to focus on just one source of information, to the exclusion of other sources. A related mistake would be to conflate one source of evidence for causality (e.g. covariation)

[3] These findings have parallels with Michottean paradigms (Michotte, 1954; for recent discussion see Wagemans *et al.* 2006). However, learning in our experiments was not dependent on spatial contiguity, and the causal mechanisms linking the sliders were invisible.

with the conception of causality that people actually have or ought to have.[4] It seems clear from the psychological literature that people's lay conception of causality is rich and multi-faceted, and not reducible to a purely probabilistic notion.

Given that people use multiple sources of information to infer causal beliefs, the question arises as to how this information is combined. In some contexts this will be relatively trivial, because the different cues will converge on the same causal conclusion. This often occurs when agents act in the natural environment – the information given by interventions tend to be nicely correlated with spatial and temporal information – I swat a fly, and the fly dies at a nearby time and place. However, the correlations between cues are sometimes broken – turning the hot water tap causes a squeal to emanate from pipes above my head. Here I use my hazy knowledge of the hot water system to explain this discrepancy. More problematic are cases where two separate cues point in different directions, for example when the temporal ordering suggests one causal model, but the covariation information suggests a different model (see Lagnado & Sloman 2006, for experiments that explore this kind of situation in the context of the appearance and transmission of computer viruses).

The open question is how people combine these different cues, especially when they suggest different causal conclusions. One general approach is to estimate the reliability of each cue, and combine them relative to this weight. However, this might not reflect what people actually do – certain cues in certain contexts might trump other cues altogether (this is what was suggested in Lagnado & Sloman, 2006, with temporal order trumping covariation information). Another possibility is that just as people tend to entertain or test only one causal hypothesis at a time, they also use cues in a sequential fashion. Lagnado and Sloman suggested that people set up an initial hypothesis or model on the basis of time order alone, and then used covariation information to test this model. This was supported by a later experiment that elicited people's causal models regularly throughout the course of the experiment (but more research is needed).

7.2.1 Prior causal knowledge

In most situations causal learning takes place against the backdrop of extensive prior causal knowledge. This knowledge can be very general – that cats sometimes screech, that water systems malfunction, that batteries run low, or more specific – that smoke detectors are battery operated, that the red light on

[4] It could be argued that such a mistake is sometimes made in philosophical circles – especially when theorists attempt to define causation purely in probabilistic terms (Suppes, 1970). Indeed this mistake is perhaps perpetuated in more recent theories of causality based on causal Bayesian networks. Sources of evidence for causal models, whether from observational or interventional probabilities, should not be taken as definitional of causality.

a smoke detector indicates a low battery, etc. This knowledge includes spatial and temporal information – e.g. concerning the relation between location and sound, and mechanical information of all sorts – e.g. the usual functioning of a smoke detector. Moreover, people do not require detailed (or correct) knowledge about causal systems in order to use this knowledge to acquire new beliefs. Simple beliefs will often suffice to figure out a novel causal relation. For example, one can infer that the low battery is causing the squeals without detailed knowledge about the inner workings of batteries or smoke detectors (although it helps to know that designers might have constructed the detector so that it warns users when it is about to fail).

A clear illustration of the role of prior knowledge is provided by cases of one-trial learning, where people learn (or assume that they learn) a causal relation after exposure to just one exemplar. For example, in our tale of the smoke detector, it took just one test (in which the battery was removed and the squeals stopped) to establish the hypothesis that a low battery was causing the squeals. It might be argued that this test actually involved a couple of observations – e.g. low battery and squeal, no battery and no squeals. But the point is that the number of observations were definitely too low for standard covariation-based learning algorithms to do their work. Most learning algorithms, including those developed within the CBN framework (e.g. Spirtes *et al.* 1993), require repeated observations before a relation can be learned (in the same way that statistical analyses require datasets larger than one). In cases of rapid learning, prior causal knowledge is used in combination with a simple piece of inferential reasoning.

An associative theorist might respond that one-trial learning can be captured in a contingency-based learning rule, so long as the learning rate parameter is high. In other words, a single observation that the squeals stop when the battery is removed provides enough covariation information to support the causal conclusion that the low battery was the cause of the noises. But this move seems unprincipled, in the sense that one would not want to licence single-trial learning in all contexts. Whether or not one makes the leap to a causal conclusion from just one exemplar (or a few) depends heavily on what other prior background knowledge one has. This inductive leap should only be taken when the background is rich and sufficient to ground the inference (e.g. given basic knowledge about how batteries work; how smoke detectors might be designed, etc.). A single-case co-occurrence in a context where there is little prior knowledge to support the inference, or even knowledge that goes against it, is less likely to lead to rapid learning.

Thus any attempt to address one-trial learning by adjusting the learning parameter in an associative mechanism effectively concedes that additional background information is being used in these cases (and is reflected in the adjustment of this parameter). The crucial point is that it is the background knowledge that is modulating the inferences drawn in one-trial cases, not the covariation information. This background knowledge supports additional reasoning about the situation – and this explains our ability to learn causal relations from impoverished data (see also Tenenbaum and Griffiths 2003).

There are numerous routes by which people attain prior knowledge – they might have been taught it, read about it, or possibly acquired it firsthand through their own experiences with the causal system in question. The important point is that it is rare for people to be confronted with a causal inference problem for which they have no relevant prior knowledge. Even infants seem to enter the world with certain prior assumptions that help them acquire more specific causal knowledge (Schlottmann, 2001; Scholl, 2005). Despite its ubiquity, the interaction of prior causal knowledge with novel problem situations, and the ability to construct new causal models from prior assumptions, has not been systematically investigated in mainstream cognitive psychology (but see Ahn and Kalish, 2000, Tenenbaum and Griffiths, 2003, Waldmann, 1996).

7.2.2 Prior assumptions behind interventions

One of the key aspects of causal thinking is that it serves as a guide to action. If done right, it allows us to predict and anticipate the effects of our actions, including those that we have never taken before. Pearl (2000) summarizes this neatly with his claim that causal models are 'oracles for interventions'. The flipside of this is that causal models can often be learned through carrying out appropriate interventions. For instance, when I conjecture that the low battery is causing the squeals, I construct a simple causal model: low battery → squeal. I then reason that according to this model, if I intervene and replace the old battery with a new one, then the squeals will stop. I can then test this prediction by actually replacing the battery, and observing whether or not the squeals stop. Once I have established the correctness of this causal model, I can use it to make predictions on other occasions. Of course I must be aware that the context might change in ways that make this model inappropriate. There is no guarantee that what works on one occasion will work in the future, or generalize to other slightly different circumstances. I will make assumptions that may have better or worse justifications. Thus, I can safely assume that the same smoke detector will work similarly tomorrow (although I can't be sure – perhaps when I replace the battery another component will break), and also assume that the smoke detector next door operates in the same way. But I will be on dangerous ground if I assume that a very different device (e.g. a battery-operated baby doll) will stop squealing once I replace the battery.

This shows that our causal reasoning depends on assumptions, many of them tacit, about the robustness of the causal models we can learn. Indeed a crucial element in our ability to think causally is our ability to gauge when we can generalize and transpose our models to novel environments (cf. Cartwright, 2007; Steele, 2007). This seems to be an unexplored area in cognitive psychology.

7.2.3 Causal Bayesian networks over- and under-estimate human reasoners

This concludes our brief survey of the multiple sources of evidence for causal beliefs (for more details see Einhorn & Hogarth 1986; Lagnado *et al.* 2007). One significant point to emerge from this concerns the applicability of the Causal Bayesian Networks (CBN) framework as a model of human inference. A strong advantage for this framework is that it formalizes the distinction between interventional and observational (correlational) learning, and suggests various algorithms for learning causal models under either regime (given certain crucial assumptions). However, there are reasons to question its wholesale application to everyday human causal inference.

In particular, it appears that the CBN framework both over- and under-estimates the capabilities of human reasoners. It seems to over-estimate people's abilities to perform large-scale computations over large bodies of hypotheses and data. People have limited-capacity working memory, and this serves as a bottleneck for complex computations with many variables and relations (Cowan 2001; Halford *et al.* 2007; Miller, 1956). It is likely that human reasoners adopt strategies to overcome these limitations, such as performing local computations (Fernbach & Sloman, 2009) and chunking information into hierarchically structured representations (Bower, 1970; Lagnado & Harvey, 2008).

On the other hand, current attempts to apply CBN to human inference also seem to underestimate human capabilities. As noted above, there is a wealth of information about causality aside from statistical covariation, including spatiotemporal information, similarity, temporal order etc. People are able to incorporate this information in their search for causal structure, but this is not yet captured in standard causal Bayesian models. This is not to deny the relevance of the CBN framework, or the considerable benefits it brings to the study of causal cognition. But it is a starting point for formulating better psychological theories, not an endpoint.

7.3 Mental models and simulations

So far we have talked about causal inference at a relatively abstract level, without delving into the mechanics of how people actually carry out these inferences. This level of description is appropriate when comparing people's inferences against a normative model of inference, such as that provided by logic, probability theory or causal Bayesian networks. But it tells us little about the psychological processes that underpin these inferences. For example, if someone's inferences correspond to those prescribed by the normative model, there remains the question of how these inferences were actually carried out. There will usually be a variety of possible psychological processes that could have reached the normatively correct conclusions.[5]

It is instructive here to compare with the case of deductive reasoning. Some theorists argue that when people make deductive inferences (e.g. from premises 'If X, then Y' and 'X', infer conclusion 'Y') they apply formal inference schema to syntactically structured mental representations (Inhelder & Piaget 1958; Braine and O'Brien, 1998; Rips, 1983). This 'mental logic' theory is controversial, especially in light of the well-documented failures in people's deductive reasoning, and its sensitivity to both content and context (Wason, 1983; Evans, 2002). One alternative to this position is mental model theory (Johnson-Laird, 1983; 2006). On this theory people evaluate deductive inferences by envisaging and combining possible states of affairs. I will not go into the details of this debate. What is important for current purposes is that we should not simply assume that when people reason causally they use a causal logic that operates on sentential mental representations. But what are the alternatives?

Mental model theory provides an alternative perspective here. As noted above, the theory claims that causal reasoning involves the envisioning and combining of possible states (Goldvarg and Johnson-Laird, 2001). There are various reasons why the theory by itself does not seem satisfactory. Prominent amongst these are the lack of constraints on the possible states implied by causal relations (material implication is too inclusive a relation), the failure to account for people's causal judgments (Sloman, Barbey & Hotaling, 2009) and the difficulty the model has in distinguishing inferences based on observations from those based on interventions (for details see Glymour, 2007; Sloman and Lagnado, 2005).

Despite these shortcomings, mental model theory does contain the seeds of a plausible account. To articulate this, it helps to return to the classic work on mental models by Kenneth Craik:

> If the organism carries a 'small-scale model' of external reality and of its own possible actions within its head, it is able to try out various alternatives, conclude which is the best of them, react to future situations before they arise, utilize the knowledge of past events in dealing with the present and the future, and in every way react in a much fuller, safer, and more competent manner to the emergencies which face it.
>
> (Craik 1952, p. 61).

Craik's suggestion is that people anticipate the effects of their actions by simulating these actions on a mental model of the external world. The key idea is that manipulations of the mental model parallel the correspondent

[5] This does not mean that conformity to the normative model tells us nothing about the nature of the psychological processes. For instance, successful causal inference presumably requires the capability to represent networks of directed relations between variables.

manipulations of the world. In other words, causal inferences are carried out by mental simulations that somehow encapsulate the causal processes in the world. This proposal is innovative, but raises a host of questions, especially with regard to what constitutes a mental model, and how these models are simulated. For example, what aspects are represented, how these are represented, and how a simulation is actually run so that it respects real world processes. In the following I will pursue one possible extension of Craik's original ideas, one most relevant for causal inference.

7.3.1 Mechanical reasoning

Mechanical reasoning furnishes some clear examples of mental simulation (Hegarty, 2004). Consider the following problem: You are presented with a sequence of gears (see Figure 7.1), and told that the leftmost gear is rotated in a clockwise direction. Your task is to predict the direction of movement of the rightmost gear. How do you solve this prediction problem? Psychological studies (e.g. Schwartz & Black 1999) suggest that people engage in mental simulation of the gear system. They tend to mentally rotate the first gear, and imagine how the second gear will rotate in response (it helps if you include some of the gear teeth!). This is continued in a piecemeal fashion until the final gear is rotated. After practice with this way of solving the problem, people often graduate to a more analytic solution, whereby they acquire the rule that adjacent gears will move in opposite directions. But for this task, and a host of other problems (e.g. pulley systems, water in glasses, etc.), people predominantly use mental simulation to answer questions of causal inference.

Hegarty (2004) has extracted several important principles from these studies:

(1) Simulation of complex physical systems is piecemeal rather than holistic, and occurs in the direction of causality (and time). For instance, to solve the gear problem, people simulated the gears sequentially, in a chain, starting from the initial cause and leading onto the final effect. In

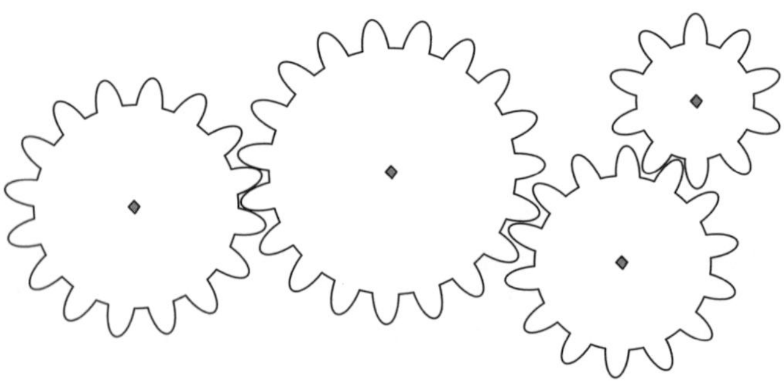

Fig. 7.1 System of interlocking gears.

another example involving pulley problems, Hegarty (1992) found that when people had to infer the movement of a pulley located in the middle of a causal chain, their eye fixations implied that they simulated causally prior pulleys that led to the movement of the pulley in question, but not pulleys that were causally downstream of this intermediate pulley.

These findings suggest that simulation does not operate on a complete or wholesale model of the physical set-up, but proceeds by selectively operating with smaller sub-components (i.e. individual causal links)[6]; it also suggests that simulations are constrained by the limitations of working memory.

(2) Simulation is not solely based on visual information, but can include non-visual information such as force and density. For example, people's mental simulations of the movements of liquids in a container are sensitive to the effects of gravity and the viscosity of the liquid (Schwartz, 1999). This suggests that mental simulation does not simply correspond to the manipulation of visual images, but can incorporate more abstract variables that explain a system's behaviour in the world.

(3) Simulations can include motor representations, especially when people are simulating their own (or other's) actions. Indeed there is now a rich literature on the role of motor representations in thinking (Jeannerod, 2006), and some sophisticated computational models of action that use internal models to predict sensory consequences of both actual and imagined actions (Wolpert, 2007).

(4) People use a variety of strategies to solve mechanical inference problems; these include mental simulation, but also rule-based strategies and analogical reasoning. These strategies are not mutually exclusive, and thus people might use a combination of strategies to solve a problem. As noted above, Schwartz & Black (1999) found that in the gear problems people progressed from using mental simulation to formulating and implementing a simple rule. Schwartz and Black speculate that mental simulation is best suited to novel problem situations, where people cannot draw on a ready-made set of formal rules.

In addition to these principles, Hegarty distinguishes between visual and spatial representation. The former represents visual aspects of things, such as colour and brightness, whereas the latter represents spatial relations and location and movement in space. Hegarty argues that mechanical reasoning predominantly depends on manipulations of spatial rather than visual images.

[6] Applied to the causal learning literature, this fits with suggestions made by Lagnado *et al.* (2007) and recent empirical work by Fernbach & Sloman (2008).

The work on mental simulation in mechanical reasoning has garnered strong empirical support (see Hegarty 2004, Nersessian, 2008). There remain open questions about the exact nature of simulation (e.g. how and in what respects do mental simulations track the real world causal processes), but the descriptive claim that people engage in mental simulation is generally well accepted. It also seems relatively straightforward to apply these ideas to the psychology of causal inference (for a related program see Wolff, 2007). The key idea would be that when people reason about causal systems they utilize mental models that represent objects, events or states of affairs, and reasoning and inference is carried out by mental simulation of these models. Moreover, these mental models admit of multifarious formats ranging from visual or spatial images, sensorimotor models to amodal representations.

7.3.2 Mental simulation of interventions

One of the most basic kinds of simulation involves the predictions of the effects of our own motor actions. In such cases mental simulation is likely to be tied quite closely to sensorimotor representations (e.g. forward models, Jeannerod, 2006, Wolpert, 2007), although these simulations can incorporate more abstract and non-visual elements too (e.g. gravity or friction). There is a natural progression to the simulation of others' actions (Brass and Heyes, 2005), and then to actions that need not involve an agent – e.g. natural causes such as the wind blowing; fire spreading etc. In these contexts the notions of cause and effect become less tied to agency and actions and their immediate effects. Note the parallel in the development of interventionist theories of causation in philosophy. These theories were initially tied to human agency (Collingwood, 1940), but subsequently developed in terms of potential manipulations without anthropomorphic connotations (Woodward, 2003). The evolution of mental causal simulation might have followed a similar course – at first tied to first-person agency and the immediate effects of actions, but progressing to simulations that need not involve agents, and incorporate more abstract causal variables.

One advantage of linking causal inference to simulation is that it can explain a variety of empirical findings which show that inference is enhanced with concrete materials and when spatial imagery/ visualization are supported (Barsalou, 1999; 2003). It also provides a ready explanation for situations where people are misled in their inferences. For instance, when the ease of mental simulation is mistakenly taken as an accurate guide to what actually happened. A compelling example of this is the power of direct witness testimony in legal cases (Heller, 2006). Vivid details given by an eyewitness about how the accused committed the crime greatly aid the jurors in imagining this scenario, and facilitate the move from speculation to conviction. Indeed computer animations that attempt to reconstruct the visual aspects of a crime are increasingly popular in legal cases.

However, as well as showing how people's causal inferences might get distorted, the simulation account can also explain how they can make inferences in accord with normative models of causality. This is because mental simulation, by the very nature of being a simulation of the external causal system, will automatically observe the basic causal 'logic' of that external system. For example, simulating forwards from cause to effects naturally obeys the logic of intervention whereby an intervened-on variable propagates its effects downstream but not upstream (i.e. obeys 'do' surgery). For example, when I imagine myself intervening to turn the leftmost gear in Figure 7.1 I simulate the consequences of this action, such as the turning of the adjacent gear, but I do not typically simulate other possible causes of these effects, such as someone else turning the second gear instead. Predictive inference (inferring effects from causes) is thus relatively easy, because the system's behaviour is simulated forward (in time) from representations of causes to representations of effects. Diagnostic inference (inferring causes from effects) is more complex, because there might be various different possible causes of the effect in question, and hence a need to compare several simulations (for recent work that fits with the differences between diagnostic and predictive inference see Fernbach & Sloman 2009, also see Hagmayer & Waldmann, 2000). More complicated still are situations demanding both predictive and diagnostic inference. In such cases it is likely that people rely on piecemeal simulations of separate links (cf. Hegarty, 2004, Fernbach & Sloman, 2009). Indeed the simulation-based account of causal reasoning makes a range of testable predictions about the kinds of causal inference that people will find easy or hard, and the methods by which they might attempt to solve complex causal inferences.

One important question for this kind of approach is how it might be extended to causal inferences that do not involve directly or easily perceived causal systems? As well as making causal inferences in the physical or mechanical domain, we are able to reason about far more complex systems and about unobserved variables. Just think of the complexity in predicting the behaviour of your favourite soap-opera character. But the same issue arises with simpler cases too. Consider the inference that the low battery in the detector is causing the squeals. Presumably we need not know much about the actual physical processes that make the low battery cause the squeals. So what is involved in the hypothetical inference that removing the battery will stop the squeals? Do we still run simulations in such a situation?

One line of response to this issue is to note that mental simulation is a broad church, and accepts a plurality of possible representational formats – perceptual, motoric, amodal etc. Although its origins might lie in mental models that are closely tied to our immediate experiences with the world (e.g. sensorimotor representations), these can become increasingly more abstract (e.g. spatial and map-like representations, cf. Tolman, 1948; Grush, 2004). Thus, although mental simulation can be accompanied by modal imagery (e.g. when you imagine a diver doing a somersault), visual imagery is not an essential part of simulation. It is possible to engage in mental simulation without explicit visual imagery. And it is also possible to use visual imagery to simulate a very abstract inference (e.g. imagining the economy taking a nose-dive). Indeed in cases of more abstract causal reasoning it is likely that representational schemes and models are created on the fly to solve specific inference problems, with different representations and mappings being used on different occasions. One day I might make predictions about the effects of the credit crunch by using a schematic model of a high-board diver, on another day I might prefer to use a model of a spreading fire, and at another time I might give up on imagery altogether. This highlights the tight coupling between representation and inference, and the flexibility of our representational resources.

The main point is that mental simulations are not restricted to sensorimotor models, but can incorporate a rich array of entities and processes. Indeed even the act of perception involves complex higher-level representations in terms of objects and the interrelations they bear. Thus mental models can be hierarchically structured, with components that can be recombined and reused (and are proposition-like in this respect).

Much of this is speculation ahead of empirical enquiry, but it is notable that all four principles advocated by Hegarty (2004) in the domain of mechanical reasoning seem to apply to the more general context of causal inference. Causal reasoning proceeds piecemeal (Fernbach & Sloman 2009; Lagnado *et al.* 2007; Hagmayer & Waldmann, 2000), it is not tied to visual representations, it takes advantage of motoric information (Jeannerod, 2006; Wolpert, 2007), and seems to admit of a variety of strategies ranging from image-based to abstract amodal simulation. The latter then paves the way for the formulation of general causal rules (see below). Suggestive evidence is also provided by the experiment described in an earlier section of this chapter (Lagnado & Loventoft-Jessen, in prep.). This experiment suggested that people were able to mentally represent possible causal models, and simulate possible interventions on these models. This included the ability to represent two different kinds of operation (disabling one slider and moving another), and draw appropriate inferences from this.

7.3.3 Is causal reasoning more basic than logical reasoning?

At this juncture it is useful to compare the proposed approach with the standard mental model theory advanced by Johnson-Laird and colleagues. The latter theory assumes that when people engage in deductive reasoning they use iconic but amodal representations. The theory then explains people's well-known shortcomings in logical reasoning by adding principles that capture people's failures to represent the full set of possibilities (e.g. focus just on A & B when entertaining 'If A then B'). An alternative approach would be to accept that people can sometimes use amodal representations, but argue that the primary form of inference is causal – via mental simulation of these representations (which might include aspects that are modal), and that logical reasoning piggy-backs on this ability. This would explain many of the shortcomings in logical reasoning (e.g. influence of content and context), and also explain people's superior capability for causal reasoning. A similar argument might be made with respect to recent claims that logical reasoning is subserved by probabilistic reasoning (Oaksford & Chater, 2007). Here again the many shortcomings in people's explicit probabilistic reasoning might be explicable by their use of (causal) mental simulations in these situations (see Krynski & Tenenbaum, 2007, and Kahneman & Tversky, 1982, for related arguments). Of course the details of such an argument need to be spelled out and empirical studies need to be designed in this light. But it seems a suggestive possibility.

Moreover, it yields a simple account for how people can slowly acquire mastery of logical and probabilistic reasoning. They gradually capitalize on their ability to manipulate amodal representations, and integrate this with a more sentence-like symbolic language presented to them while they are learning. This process resembles the observed transition from simulating gears to learning a formal rule (Schwartz & Black, 1999). However, the role of some kind of imagery is probably never lost, and can persist even in rarefied domain of scientific inference (Hadamard 1954).

In short, the speculative claim here is that people have a core capability for *causal* reasoning via mental simulation, and deductive and inductive reasoning builds on this foundation. This shift of perspective might explain the characteristic biases and shortcomings in lay people's logical and probabilistic reasoning.

7.4 Conclusions

I have argued for the interplay of causal learning and reasoning, the multiplicity of sources of evidence for causal relations, and the role of mental simulation in causal inference. These three strands are themselves intertwined. Learning and reasoning utilize the same kinds of representations and mechanisms: they both rely on mental models, and in both cases inference depends on the simulation of these models. The fact that these mental models admit of multifarious formats (e.g. spatial, perceptual, sensorimotor) reflects the rich causal information available from the world and our interactions with it. Nevertheless, our ability to construct evermore abstract amodal representations, catalysed by the invention of external representational forms such as diagrams and language, enables us to draw causal inferences that take us beyond the surface of our perceptions.

References

Ahn, W. & Kalish, C. (2000). The role of covariation vs. mechanism information in causal attribution. In R. Wilson, & F. Keil (eds.) *Cognition and Explanation*, Cambridge, MA: MIT Press.

Barsalou, L. (1999). Perceptual symbol systems. *Behavioral and Brain Sciences, 22*, 577–660.

Barsalou, L.W., Simmons, W. K., Barbey, A. K. & Wilson, C. D. (2003). Grounding conceptual knowledge in modality-specific systems. *Trends in Cognitive Science.*, 7, 84–91.

Braine, M. & O'Brien, D. (1998). *Mental Logic*. Mahwah, NJ: Lawrence Erlbaum Associates.

Brass, M. & Heyes, C.M. (2005). Imitation: Is cognitive neuroscience solving the correspondence problem? *Trends in Cognitive Science*, 9, 489–495.

Bower, G.H. (1970). Organizational factors in memory. *Cognitive Psychology*, 1, 18–46.

Cartwright, N. (2007). *Hunting Causes and Using Them: Approaches in Philosophy and Economics*, Cambridge: Cambridge University Press.

Cheng, P.W. (1997). From covariation to causation: A causal power theory. *Psychological Review, 104*, 367–405.

Collingwood, R. (1940). *An Essay on Metaphysics*. Oxford: Clarendon Press.

Cowan, N. (2001). The magical number 4 in short-term memory: A reconsideration of mental storage capacity. *Behavioral and Brain Sciences*, 24, 87–185.

Craik, K. (1952). *The Nature of Explanation. Cambridge University Press, Cambridge, UK.*

Einhorn, H. J. & Hogarth, R. M. (1986). Judging probable cause. *Psychological Bulletin*, 99, 3–19.

Evans, J. (2002). Logic and human reasoning: An assessment of the deduction paradigm. *Psychological Bulletin*, 128, 978–96.

Fernbach, P.M. & Sloman, S.A. (2009). Causal learning with local computations. *Journal of Experimental Psychology: Learning, Memory, and Cognition. 35 (3), 678–693.*

Glymour, C. (2007). Statistical jokes and social effects: Intervention and invariance in causal relations. In Gopnik, A., & Schultz, L. (eds.), *Causal learning: Psychology, Philosophy, and Computation*, pp. 294–300. Oxford: Oxford University Press.

Goldvarg, Y. Johnson-Laird, P.N. (2001). Naïve causality: A mental model theory of causal meaning and reasoning. *Cognitive Science*, 25, 565–610.

Gopnik, A. & Schultz, L. (2007). *Causal Learning: Psychology, Philosophy, and Computation*. Oxford: Oxford University Press.

Gopnik, A., Glymour, C., Sobel, D. M., Schulz, L. E., Kushnir, T., & Danks, D. (2004). A theory of causal learning in children: Causal maps and Bayes nets. *Psychological Review*, 111, 1–31.

Griffiths, T.L. & Tenenbaum, J.B. (2005). Structure and strength in causal induction. *Cognitive Psychology*, 51, 334–384.

Grush, R. (2004). The emulation theory of representation: Motor control, imagery, and perception. *Behavioral and Brain Sciences, 27*, 377–396.

Hadamard, J. (1954). *The Psychology of Invention in the Mathematical Field*. New York: Dover.

Hagmayer, Y., & Waldmann, M.R. (2000). Simulating causal models: The way to structural sensitivity. In L. Gleitman & A. Joshi (Eds.), *Proceedings of the Twenty-second Annual Conference of the Cognitive Science Society*. Mahwah, NJ: Erlbaum.

Hall, G. (2002). Associative structures in Pavlovian and instrumental conditioning. In C.R. Gallistel (ed.), *Stevens' Handbook of Experimental Psychology*, 3rd edition, Vol. 3, pp. 1–45. New York: John Wiley.

Hastie, R. & Pennington, N. (2000). Explanation-based decision making. In T. Connolly, H. R. Arkes and K. R. Hammond (eds): *Judgment and Decision Making: An Interdisciplinary Reader* (2nd ed.). pp. 212–28. Cambridge University Press.

Halford, G.S., Cowan, N., & Andrews, G. (2007). Separating cognitive capacity from knowledge: A new hypothesis. *Trends in Cognitive Sciences*, 11 (6), 236–242.

Hegarty, M. (2004). Mechanical reasoning as mental simulation. *Trends in Cognitive Science*, 8, 280–285.

Hegarty, M. (1992). Mental animation: Inferring motion from static diagrams of mechanical systems. *Journal of Experimental Psychology: Learning, Memory and Cognition*, 18, 1084–1102.

Heller, K. J. (2006). The cognitive psychology of circumstantial evidence. *Michigan Law Review*, 105, 241–305.

Inhelder, B. and Piaget, J. (1958). *The Growth of Logical Thinking from Childhood to Adolescence*. New York: Basic Books.

Jeannerod, M. (2006). *Motor Cognition: What Actions Tell the Self.* Oxford: Oxford University Press.

Johnson-Laird, P.N. (1983). *Mental Models*. Cambridge: Cambridge University Press.

Johnson-Laird, P.N. (2006). *How We Reason*. Oxford: Oxford University Press.

Kahneman, D. & Tversky, A. (1982). The simulation heuristic. In D. Kahneman, P. Slovic & A. Tversky (eds.), *Judgment under Uncertainty: Heuristics and Biases* (pp. 201–208). New York: Cambridge University Press.

Krynski, T.R. & Tenenbaum, J.B. (2007). The role of causality in judgment under uncertainty. *Journal of Experimental Psychology: General*, 136, 430–450.

Lagnado, D.A. (2009). A causal framework for learning and reasoning. *Behavioral and Brain Sciences*, 32, (2), 211–212.

Lagnado, D.A. (2010). Thinking about evidence. To appear in Dawid, P, Twining, W., Vasilaki, M. eds. *Evidence, Inference and Enquiry*. British Academy/OUP. (In Press).

Lagnado, D.A. & Sloman, S.A. (2004). The advantage of timely intervention. *Journal of Experimental Psychology: Learning, Memory & Cognition*, 30, 856–876.

Lagnado, D.A. & Sloman, S.A. (2002). Learning causal structure. In W. Gray & C. D. Schunn (Eds.), *Proceedings of the Twenty-Fourth Annual Conference of the Cognitive Science Society*, Erlbaum.

Lagnado, D.A., Waldmann, M.R., Hagmayer, Y., & Sloman, S.A. (2007). Beyond covariation: Cues to causal structure. In Gopnik, A., & Schultz, L. (eds.), *Causal learning: Psychology, Philosophy, and Computation*, pp. 154–172. Oxford: Oxford University Press.

Lagnado, D.A. & Harvey, N. (2008). The impact of discredited evidence. *Psychonomic Bulletin and Review*, 15, 1166–1173.

Lagnado, D.A. & Sloman, S. A. (2006). Time as a guide to cause. *Journal of Experimental Psychology: Learning, Memory & Cognition*, 32, 451–460.

Lagnado, D. A. & Loventoft-Jessen, J. (in prep.). Learning causal models through multiple interventions.

Meder, B., Hagmayer, Y., & Waldmann, M. R. (2008). Inferring interventional predictions from observational learning data. *Psychonomic Bulletin & Review*, 15, 75–80.

Michotte, A. (1954/1963). *The Perception of Causality*. London: Methuen.

Miller, G. A. (1956). The magical number seven, plus or minus two: Some limits on our capacity for processing information. *Psychological Review*, 63, 81–97.

Nersessian, N. J. (2008). *Creating Scientific Concepts*. Cambridge, MA: MIT Press.

Oaksford, M. & Chater, N. (2007). *Bayesian Rationality: The Probabilistic Approach to Human Reasoning*. Oxford: Oxford University Press.

Pearl, J. (2000). *Causality*. Cambridge: Cambridge University Press.

Rips, L. J. (1983). Cognitive processes in prepositional reasoning. *Psychological Review*, 90, 38–71.

Shanks, D. R. (2004). Judging covariation and causation. In D. Koehler & N. Harvey (Eds.), *Blackwell Handbook of Judgment and Decision Making*. Oxford: Blackwell.

Shanks, D. R. (1995). *The Psychology of Associative Learning*. Cambridge: Cambridge University Press.

Schlottmann, A. (2001). Perception versus knowledge of cause-and-effect in children: When seeing is believing. *Current Directions in Psychological Science*, 10, 111–115.

Scholl, B. J. (2005). Innateness and (Bayesian) visual perception: Reconciling nativism and development. In P. Carruthers, S. Laurence, & S. Stich (Eds.), *The Innate Mind: Structure and Contents* pp. 34–52. Oxford: Oxford University Press.

Schwartz, D.L. (1999). Physical imagery: Kinematic vs. dynamic models. *Cognitive Psychology*. 38, 433–464.

Schwartz, D.L. & Black, T. (1999). Inferences through imagined actions: Knowing by simulated doing. *Journal of Experimental Psychology: Learning, Memory & Cognition*, 25, 116–136.

Shanks, D. R. & Dickinson, A. (1987). Associative accounts of causality judgment. In G. H. Bower (Ed.), *The Psychology of Learning and Motivation: Advances in Research and Theory* Vol. 21, pp. 229–261. San Diego, CA: Academic Press.

Sloman, S. A. (2005). *Causal Models; How People Think About the World and its Alternatives*. New York: Oxford University Press.

Sloman, S. A., Barbey, A. K. & Hotaling, J. (2009). A causal model theory of the meaning of cause, enable, and prevent. *Cognitive Science*, 33, 21–50.

Sloman, S. A. & Lagnado, D. A. (2005). Do we do? *Cognitive Science*,. 29, 5–39.

Spirtes, P., Glymour, C., & Schienes, R. (1993). *Causation, Prediction, and Search*. New York: Springer-Verlag.

Steel, D. (2008). *Across the Boundaries: Extrapolation in Biology and Social Science*. New York: Oxford University Press.

Steyvers, M., Tenenbaum, J. B., Wagenmakers, E. J., & Blum, B. (2003). Inferring causal networks from observations and interventions. *Cognitive Science, 27*, 453–489.

Suppes, P. (1970). *A Probabilistic Theory of Causality*. Amsterdam: North-Holland.

Tenenbaum, J.B., & Griffiths, T.L. (2003). Theory-based causal inference. *Advances in Neural Information Processing Systems 15*, pp. 35–42. Cambridge, MA: MIT Press.

Tolman, E.C. (1948). Cognitive maps in rats and men. *Psychological Review*, 55, 189–208.

Wagemans, J., van Lier, R. & Scholl, B.J. (2006). Introduction to Michotte's heritage in perception and cognition research. *Acta Psychologica*, 123, 1–19.

Waldmann, M.R. (1996). Knowledge-based causal induction. In D. R. Shanks, K. J. Holyoak, & D. L. Medin (Eds.), *The Psychology of Learning and Motivation* Vol. 34, pp. 47–88. San Diego, CA: Academic Press.

Waldmann, M.R. & Holyoak, K.J. (1992). Predictive and diagnostic learning within causal models: Asymmetries in cue competition. *Journal of Experimental Psychology, General*, 121(2), 222–236.

Waldmann, M.R., & Hagmayer, Y. (2005). Seeing versus doing: Two modes of accessing causal knowledge. *Journal of Experimental Psychology: Learning, Motivation, and Cognition*, 31, 216–227.

Waldmann, M.R., Hagmayer, Y, & Blaisdell, A.P. (2006). Beyond the information given: Causal models in learning and reasoning. *Current Directions in Psychological Science*, 15 (6), 307–311.

Wason, P.C. (1983). Realism and rationality in the selection task. In J. St. B. T. Evans (ed.), *Thinking and Reasoning: Psychological Approaches*. London: Routledge & Kegan.

Williamson, J. (2005). *Bayesian Nets and Causality: Philosophical and Computational Foundations*. Oxford: Oxford University Press.

Wolff, P. (2007). Representing causation. *Journal of Experimental Psychology: General*, 136, 82–111.

Wolpert, D.M. (2007). Probabilistic models in human sensorimotor control. *Human Movement Science*, 26, 511–524.

Woodward, J. (2003). *Making Things Happen. A Theory of Causal Explanation*. Oxford: Oxford University Press.

·8·

When and how do people reason about unobserved causes?

Benjamin Rottman, Woo-kyoung Ahn, and Christian Luhmann

Abstract

Assumptions and beliefs about unobserved causes are critical for inferring causal relationships from observed correlations. For example, an unobserved factor can influence two observed variables, creating a spurious relationship. Or an observed cause may interact with unobserved factors to produce an effect, in which case the contingency between the observed cause and effect cannot be taken at face value to infer causality. We review evidence that three types of situations lead people to infer unobserved causes: after observing single events that occur in the absence of any precipitating causal event, after observing a systematic pattern among events that cannot be explained by observed causes, and after observing a previously stable causal relationship change. In all three scenarios people make sophisticated inferences about unobserved causes to explain the observed data. We also discuss working memory as a requirement for reasoning about unobserved causes and briefly discuss implications for models of human causal reasoning.

An observed correlation between two events does not imply a direct causal relationship between them. One reason that is particularly important in developing theories of human causal learning is that unobserved or unattended cause(s) may account for all or part of the observed correlations.

For instance, an article published in *Nature* reported that young children who sleep with a nightlight are much more likely to develop myopia later in life (Quinn, Shin, Maguire, & Stone, 1999). This was interpreted as implying a causal relationship by the popular press. For instance, CNN reported, even low levels of light can penetrate the eyelids during sleep, keeping the eyes working when they should be at rest. Taking precautions during infancy, when eyes are developing at a rapid pace, may ward off vision trouble later in life (CNN, May 13, 1999). A later study, however, suggested that a common cause is responsible for this correlation; myopic parents are more likely to leave a light on for children, and myopic parents are more likely to have myopic children (Gwiazda, Ong, Held, & Thorn, 2000).

While the above example illustrates how a positive correlation between two variables does not imply that one causes the other, the opposite can happen as well; we observe no correlation between two events when, in fact, there is a causal relationship between them. For example, a recent study demonstrated that pollution and daily temperature range are positively associated in the summer, but negatively associated in the winter (Gong, Guo, & Ho, 2006). Prior to learning that season plays a causal role, it would appear as if there is no relationship between pollution and temperature range because there is no correlation, even though there is an important relationship. Restated, there was a period of time during which an unknown variable (season) obscured the causal relationship between two observed variables, and the researchers had to learn about this interaction.

Considering these examples, it should be obvious that assumptions and beliefs about unobserved causes are vital in inferring causal relationships from observed correlations. In some sense, it is remarkable that we can make any valid causal inferences from observed correlations alone. There can be any number of unobserved causes at play, and people cannot possibly reason through all possible combinations whenever they make causal inferences.

This paper examines laypeoples' inferences about unobserved causes. We will first elaborate on the problems involving unobserved causes. Then, we will argue that people actually perform fairly sophisticated reasoning about unobserved causes, and that such reasoning is engaged due to a certain set of assumptions that they hold about the world. We also review psychological studies supporting our argument.

Problems with reasoning about unobserved causes

Consider a simple causal reasoning scenario involving a light switch and a light. Suppose you go into a room for the first time, and you observe the light (i.e. on or off; 1 or 0, respectively in Table 8.1 under Light) across eight consecutive trials when the switch is up or down (1 or 0 respectively in Table 8.1 under Switch). One possible causal interpretation is that there is no causal relationship between the observed switch and the light, and there is an unobserved switch that is *entirely responsible* for the light's behaviour (see Table 8.1, 'Entirely Responsible' column).

But, there are many other equally plausible possibilities in which the observed switch is causally responsible for the effect in combination with another unobserved switch (see Table 8.1). One is that the observed switch interacts with another switch through a *biconditional* interaction such that the light turns on only when the two switches are either both up or both down. Yet another possibility is that there are two unobserved switches, and at least *two out of three* of these switches must be up to make the light turn on. Depending on whether one believes in the biconditional interaction or two out of three unobserved switches, one's future intervention to make the light go on would change (e.g. if it is two unobserved switches, keeping the switch up would maximize the time the light is on, but for a biconditional

Table 8.1 Light switch example

Trials	Observed events		Different types of possible unobserved switches			
	Switch	Light	Entirely Responsible	Biconditional	2 Out of 3	Always Present
1	0	0	0	1	00	1
2	1	0	0	0	00	1
3	1	1	1	1	10	1
4	0	1	1	0	11	1
5	0	0	0	1	10	1
6	1	0	0	0	00	1
7	1	1	1	1	11	1
8	0	1	1	0	11	1

case, flipping the switch whenever the light goes off would likely maximize the time the light is on).

The point of this example is to illustrate that there are so many possible ways that unobserved causes could interact that it would be impossible for people to consider all of these configurations. Does this mean that people do not spontaneously reason about unobserved causes? The answer must be no, given the obvious fact that people do make causal inferences based on correlations, and they must make (or act as if they make) some assumptions about unobserved causes in order to do so. (For instance, inferring that X causes Y based on a positive correlation between X and Y requires assuming that there is no unobserved, confounding variable.) The important question, then, is what assumptions and inferences people make about unobserved causes, and what triggers inferences about unobserved causes given that people cannot always consider all possible unobserved causes? The current chapter reviews studies from our labs that provide some answers to these questions.

In the following sections, we first briefly review how existing models of causal learning handle unobserved causes. Then we argue that people hold assumptions that trigger *specific* inferences about unobserved causes. We claim that people believe that (i) an event must be caused by another event (causal determinism), (ii) any systematic pattern or regularity among events must be causally determined, and (iii) causal relations stay stable across different times and contexts. When causal determinism is violated, when a systematic pattern is not explained by observed causes, or when causal relations are not stable, we argue that people infer an unobserved cause to explain the apparent violation of the assumption. Then we present experimental results suggesting that people do spontaneously make such inferences about unobserved causes, and describe how such inferences further influence the causal inferences people draw from observed correlations. Finally, we will discuss one cognitive requirement for reasoning about unobserved causes.

8.1 Unobserved causes in models of human causal learning

Many models have been developed to explain how people learn the causal strength of a particular cause and effect relationship. Luhmann and Ahn (2007) and Hagmayer and Waldmann (2007) have provided detailed reviews of how these models handle unobserved causes. Here we provide a brief summary.

One class of models makes no assumptions about unobserved causes, and thus makes no inferences about unobserved causes. For example, ΔP (Jenkins & Ward, 1965), an associative measure, estimates causal strength as the difference in probability of the effect (E) being present when the cause (C) is present vs. absent: $P(E|C) - P(E| \sim C)$. Though ΔP is a very intuitive way of calculating the influence of C on E, it runs into a critical problem; people are more sensitive to certain types of evidence such as when both C and E are present and are less sensitive to other types of evidence such as when both are absent. Many subsequent descriptive models have tried to capture this phenomenon by differentially weighting the evidence (e.g. Arkes & Harkness, 1983; Downing, Steinberg, & Ross, 1985; Einhorn & Hogarth, 1986; Nisbett & Ross, 1980; Schustack & Sternberg, 1981; Shaklee and Tucker, 1980). However, these approaches did not provide a theoretical explanation for the phenomenon.

Cheng (1997), Novick and Cheng (2004), see also Pearl (2000), provided a parsimonious theoretical explanation for this phenomenon by appealing to unobserved causes. Cheng argued that differential weighting of evidence is a normative result of accounting for ceiling effects, when an unobserved cause frequently produces the effect (see Section 8.4.1 for a discussion). However, Cheng's model requires a number of assumptions. Specifically, for a generative observed cause, unobserved causes are assumed to interact in a noisy-or fashion with observed causes, to be generative, not inhibitory, and to be independent from observed causes. These very strict assumptions limit the applicability of the model and it is not entirely clear whether people actually make these assumptions (Luhmann & Ahn, 2007; Hagmayer & Waldmann, 2007; White, 2005, 2009).

A very different approach to unobserved causes makes the straightforward assumption that all unobserved causes, taken as a whole, are *present* across all learning trials. For example, the Rescorla–Wagner model (Rescorla & Wagner, 1972; Dickinson, 1984) includes a background context node that can be viewed as an aggregation of all unobserved causes. When an effect occurs without the observed cause, this node gains associative strength, which can be used as an estimate of the causal strength of an unobserved cause. However, it is easy to see that the consequence of this assumption would be quite unsatisfactory for reasoners. For example, in Table 8.1, last column, an unobserved cause present on every trial would be completely unable to explain the light's behaviour; neither the observed switch nor the unobserved cause correlates with the status of the light and thus the only unsatisfactory conclusion is that the light was acting randomly without any cause.

Some recent models have attempted to explain peoples' sophisticated reasoning about unobserved causes including (i) inferring whether an unobserved cause is present or absent on a particular trial, and (ii) inferring the causal strength of an unobserved cause. For example, if an effect is observed without an observed cause, one would likely infer that an unobserved cause is responsible. Furthermore, given that an observed cause is present, one would more likely infer that an unobserved cause is also present if the effect is present rather than absent (see Hagmayer & Waldmann, 2007, for a detailed explanation of these examples). One new model, BUCKLE (Luhmann & Ahn, 2007), has been developed specifically for these types of inferences. BUCKLE is explained in the Section 8.2.

In sum, few models have been developed to account for reasoning about unobserved causes, though there have been some recent attempts to explain how people learn about the presence and causal strength of an unobserved cause. In the next section, we provide further evidence of reasoning about unobserved causes that a more comprehensive model should account for.

8.2 Causal determinism about individual events

One of the more primitive assumptions that lay reasoners appear to make is causal determinism, that every event has a cause and that events cannot occur in the absence of any precipitating causal event.[1] This assumption of causal determinism is captured in the causal principle from ancient philosophy; 'nothing happens without a cause' ('*nihil fit sine causa*' Audi, 1995). For someone who believes in causal determinism, events with no apparent cause should suggest the existence of hidden causes. Much of the empirical work suggesting that people believe in causal determinism has investigated children's beliefs about agency and magic, which is reviewed below.

8.2.1 Children

Children's beliefs about agency

A major question in developmental psychology pertains to children's beliefs about agency, the idea that there are entities with free will (e.g. humans and animals) that are primary sources of causal influence. For example, the motion of animate agents may be assumed to be generated internally and to not require further explanation (e.g. Wegner, 2002; Leising, Wong, Waldmann, &

[1] Here we distinguish causal determinism, the idea that every event must have a cause, from deterministic causality, the belief that an effect must be present if its cause is present (Goldvarg & Johnson-Laird, 1994; Koslowski, 1996; Luhmann & Ahn, 2005).

Blaisdell, 2008). In contrast, the motion of non-agents (e.g. billiard balls) must be explained by referring to external causal forces. When an inanimate object assumed not to have self-agency appears to move on its own (e.g. a baseball moving like a bird rather than in an arc), this violation of determinism should be surprising.

Saxe, Tenenbaum, & Carey (2005), see also Saxe, Tzelnic, & Carey (2007), tested this reasoning with infants in the following way. They had infants repeatedly observe a beanbag (a non-agent) flying through the air from one side of a small stage to the other. (Studies with infants often use a 'habituation' phase, during which the infant becomes accustomed to seeing the same event and stops paying attention to the event. In a later phase, if infants show increased interest, this is taken to imply 'surprise.') After an infant was habituated to this event, he/she was presented with this same event (the beanbag flying across the stage) followed by a human hand entering from one side of the stage, either the side from which the beanbag was launched, or the opposite side. The infants were more 'surprised' (spent more time looking at the hand) when the hand entered from the opposite side of the stage as the beanbag. The reasoning is that when the hand entered from the same side as the beanbag, the infants could reason backwards that the hand had been behind the stage all along and could have thrown the beanbag. However, when the hand entered from the opposite side as the beanbag, there is no cause of the beanbag's motion – the beanbag cannot propel itself and the hand was on the opposite side. Critically, this result disappeared entirely when the beanbag was replaced with a puppet (an agent) that appeared to propel itself across the stage without requiring another agent. In sum, the infants were 'surprised' only when an inanimate object appeared to move itself, a violation of causal determinism, but they were not surprised when an animate object, assumed to have self-agency, moved itself.

Children's beliefs about magic

Other developmental work has examined circumstances that evoke magical explanations from children (see Woolley, 1997 for a review). For example, Phelps & Woolley (1994) presented children (ages 4–8) with several real-world objects and asked them about their operation. Children were shown two objects that were, unbeknownst to the children, magnets of opposing polarity. The children were first asked to make a prediction (e.g. whether one object could move the other without touching it) and then to provide an explanation once a surprising event occurred (e.g. after one object pushed the other without touching it). This study revealed that if a child could not explain the event with a physical explanation, he/she tended to appeal to magic or 'tricks' (both of which refer to hidden causes). Thus, children's reliance on magic for explanation, an inference to the ultimate hidden cause, appears to be strongly driven by events that violate causal determinism.

8.2.2 Adults

Luhmann & Ahn (2007) have recently conducted a series of experiments to explore beliefs about causal determinism in adults. The study was designed to investigate whether adults make significant inferences to hidden causes and what, if any, influence on behavior such simple inferences might have. The study used a typical causal learning task in which subjects were asked to learn about a pair of potential causes (gray and white buttons) and their influence on a single effect (light turning on). Participants observed the presence/absence of the different events in a trial-by-trial manner. On each trial, learners observed the presence/absence of one cause (whether the gray button in Figure 8.1 was pressed or not) and the presence/absence of the effect (whether the light was on/off). Unlike typical causal learning experiments, one of the two causes in our study was 'hidden' from subjects (the white button in Figure 8.1). No information was ever provided to the participants about the presence/absence of the second, hidden cause. After they completed the trial sequence, participants evaluated the strength of the causal relationship between the observed cause and the effect and the strength of the relationship between the hidden cause and the effect.

To determine whether learners made notable inferences about the hidden cause, we manipulated whether or not the different sequences included trials that violated causal determinism; trials in which the effect was present but the observed cause was absent (the light was on, but the button was not pressed; Figure 8.1). We call these trials unexplained effects, as effects are present in the absence of any observed causes. As shown in Table 8.2, the 'unnecessary' and 'zero' conditions included unexplained effects, whereas the 'perfect' and 'insufficient' conditions did not. Our results demonstrated that sequences that included unexplained effects, or violations of causal determinism, led subjects to believe that the hidden cause was a stronger (generative) cause than sequences that did not include unexplained effects (last column of Table 8.2).

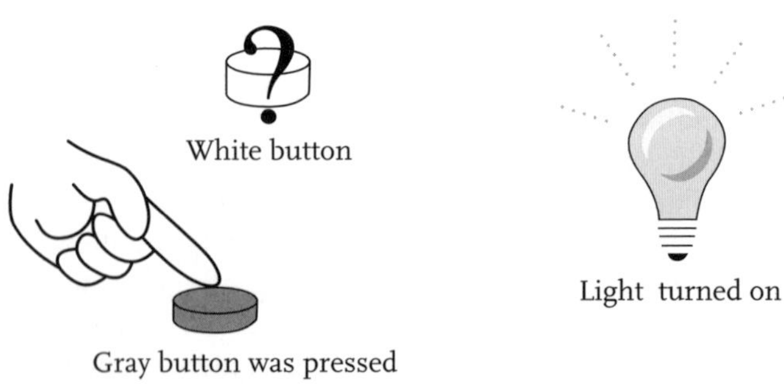

Fig. 8.1 Sample unexplained effect trial used in Luhmann & Ahn (2007).

Table 8.2 Summary of conditions and mean probability and causal strength estimates of unobserved cause in Luhmann & Ahn (2007, Experiment 3).

Unexplained Effects	Conditions	Observed frequencies of four trial types	Average trial-by-trial likelihood judgments of hidden cause being present (10) vs. absent (0) on each trial type	Average Causal Strength Estimates of Hidden Cause
Present	Unnecessary	*E* ~*E* *O* 7 0 ~*O* **7** 7	*E* ~*E* *O* 5.8 - 5.8 ~*O* **7.5** 2.2 4.8 6.7 2.2	70.8 (6.44)
	Zero	*E* ~*E* *O* 7 7 ~*O* **7** 7	*E* ~*E* *O* 5.4 3.4 4.4 ~*O* **7.8** 1.8 4.8 6.6 2.6	74.4 (5.35)
Absent	Perfect	*E* ~*E* *O* 7 0 ~*O* 0 7	*E* ~*E* *O* 4.1 - 4.1 ~*O* - 3.2 3.2 4.1 3.2	32.7 (6.83)
	Insufficient	*E* ~*E* *O* 7 7 ~*O* 0 7	*E* ~*E* *O* 5.3 3.8 4.5 ~*O* - 2.7 2.7 5.3 3.2	50.0 (7.56)

Note: *O* is the observed cause, and *E* is the effect. ~ represents the absence of an event. Unexplained effect (~*OE*) trials are shown in bold. Standard errors are in parentheses.

Because violations of causal determinism were prima facie evidence for the operation of an unobserved, generative cause, we suggested that subjects were using these specific occasions as the basis for their causal strength judgments of the hidden cause.

To validate our explanation, we asked learners on each trial to judge how likely the hidden cause was present using a scale that ranged from 0 (definitely absent) to 10 (definitely present, see the fourth column, Table 8.2). These probability judgments allowed us to directly measure learners' beliefs about the hidden cause on all four types of trials. As expected, learners believed that the hidden cause was likely present when causal determinism was violated (~*O E* trials in bold in the fourth column in Table 8.2). In fact, learners believed that the hidden cause was more likely to be present on these occasions than on any other type of trial. Thus, similar to infants, violations of causal determinism lead adults to infer hidden causes.

The finding that people infer unobserved causes during unexplained effects may seem fairly intuitive. However, this experiment allowed us to uncover additional, potentially less intuitive, and more sophisticated inferences about hidden causes.

One demonstration of sophisticated reasoning about unobserved causes is that participants' real-time judgments about the presence/absence of the unobserved cause explain their judgments of the causal strength of the unobserved cause. To demonstrate this, we computed ΔP, a measure of co-variation, between the unobserved cause and the effect (i.e. $P(E|U)-P(E|\sim U)$) based on participants' average probability judgments of the presence of the unobserved cause shown in Table 8.2 (converted to probabilities that ranged from 0 to 1); they were 0.40, 0.40, 0.10, and 0.19, respectively, for the four conditions in Table 8.2.[2] Impressively, these virtual covariations correlate with the subsequent judgments of the causal strength of the unobserved cause (shown in the last column of Table 8.2). It should be stressed that these were 'virtual' covariation in the sense that it only existed in the heads of the learners. No actual covariation existed because one of the two potentially covarying events was hidden. Furthermore, subjects were never asked to estimate the overall covariations between these two events; we computed them over participants' probability judgments of the presence of the unobserved cause. That is, when the virtual ΔP between the hidden cause and the effect was higher, subjects' causal strength estimate of the hidden cause was higher, and vice versa, as if the 'virtual' covariation we computed had been directly observed by learners.

Even more subtly, each subject's idiosyncratic beliefs about hidden cause–effect covariation could be used to estimate his/her own inferences of the causal strength of the unobserved cause. Some subjects believed in strong covariation between the hidden cause and the effect. Other subjects' probability judgments showed weaker covariation. Remarkably, the individual differences in this virtual covariation measure (i.e. ΔP between the unobserved cause and the effect) significantly predicted subjects' subsequent judgments of the causal strength of the hidden cause in each of the four conditions: $rs = 0.48$, 0.43, 0.59, and 0.52, respectively (all $ps < 0.05$, Luhmann & Ahn 2007, unpublished analyses). Those subjects whose inferences implied strong hidden cause–effect covariation judged the causal relationship to be stronger than those subjects whose inferences implied weak covariation. This pattern of beliefs suggests particularly elaborate reasoning about hidden causes.

These data could be also used to evaluate some of the theoretical claims about hidden causes. For example, as mentioned above, prominent theories of causal inference (e.g. Cheng, 1997) require that hidden causes occur independently of observed causes; that is, the likelihood of a hidden cause, U, in the presence of an observed cause, O, is the same as the likelihood of U in the absence of O, $P(U|O) = P(U|\sim O)$. In contrast, according to subjects' probability judgments shown in Table 8.2, this requirement was violated in the majority of situations we tested. The hidden cause was judged to be

[2] To compute this measure of ΔP, the quantities $P(E|U)$ and $P(E|\sim U)$ were derived by applying Bayes rule to the quantities $P(U|E)$ and $P(U|\sim E)$ which were computed using learners' trial-by-trial likelihood judgments. In this way, the resulting ΔP measures the extent to which each subject believed the unobserved cause covaried with the effect.

more likely when the observed cause was present and less likely when the observed cause was absent (i.e. $P(U|O) > P(U|{\sim}O)$), as illustrated by the marginal means of O and $\sim O$ in the fourth column of Table 8.2. Nonetheless, subjects were uniformly willing to estimate the strength of both the hidden and observed cause. This suggests that people might not believe that independence of hidden causes is a requirement for valid causal inference (Luhmann & Ahn, 2005; see also Hagmayer & Waldmann, 2007).

These data also provide insight into the conditions under which people infer unobserved causes to be generative or inhibitory. In the previous studies, the unobserved cause was always judged to be generative. However, Schulz and Sommerville (2006) demonstrated that four-year-olds sometimes infer preventative hidden causes. In their study, children were presented with a cause that produced an effect four times. They then observed eight trials when the cause unreliably produced the effect (sometimes the effect was present when the cause was present, sometimes the effect was absent when the cause was present). Finally, the children were shown a button box that the experimenter had hidden during the cause–effect sequence. When asked to prevent the effect, children pressed the previously hidden button, indicating that they thought it was preventative. To summarize, Schulz and Sommerville found that instances when a cause is present but the effect is absent ($O{\sim}E$ observations) lead children to infer an inhibitory cause, but Luhmann and Ahn found that adults inferred a generative cause.

Luhmann and Ahn (2007) reasoned that $O{\sim}E$ observations could be interpreted in multiple ways. For instance, $O{\sim}E$ may occur (a) because an unobserved cause prevented the effect from happening or (b) because the observed cause is not entirely sufficient to bring about the effect. Thus, if a learner believes that the observed cause is weak, then the learner does not have to infer that the unobserved cause is inhibitory in order to account for $O{\sim}E$ observations. Indeed, in conditions with $O{\sim}E$ observations in the experiment described above (e.g. the Insufficient and Zero conditions), learners believed that the observed cause was relatively weak and the hidden cause was relatively strong and generative (e.g. note the relatively high causal judgment of the unobserved cause in the Insufficient cause). However, if people already believe that the observed cause is strongly generative, they might infer an unobserved inhibitory cause to explain $O{\sim}E$ evidence.

To reconcile our findings with those of Schulz and Sommerville (2006), we designed an experiment that provided pre-training to learners. This pre-training was designed to convince learners that the observed cause was, on its own, a sufficient cause of the effect. Once this pre-training was complete, we then presented the same Insufficient condition we used in the experiment described above. In light of the pre-training, learners' judgments indicated that they believed that the hidden cause was preventative. This result suggests an important difference between $O{\sim}E$ and ${\sim}OE$. Observations of ${\sim}OE$ are violations of causal determinism and require inferring a hidden generative cause. In contrast, observations of *O∼E* are somewhat more ambiguous. If learners entertain the possibility that the observed cause has no causal influence at all, or if they allow for the possibility that the observed cause produces its effect unreliably, then there is no need to appeal to hidden causes at all. Alternatively, if learners believe that the observed cause reliably produces its effect (e.g. the children in Schulz and Sommerville (2006) experiment or the adults after our pre-training), then observations of *O∼E* suggest the operation of hidden, preventative causes.

8.2.3 BUCKLE: A model of unobserved cause learning

Because existing theories of causal inference were unable to account for these results, Luhmann and Ahn (2007) proposed an alternative account, instantiated as a computational model called BUCKLE. The basic operation of BUCKLE involves (1) making inferences about the presence or absence of hidden causes (via Bayesian inference) and (2) then adjusting beliefs about the causal strength of all causes (both hidden and observed). These two steps are performed on each trial.

In Step 1, BUCKLE makes use of four pieces of information to estimate the probability that *U* is present: whether *O* and *E* are present or absent and the causal strengths of *O* and *U* that were calculated on the previous trial. Suppose that the learner observes an *OE* trial. If the learner believes that *O* is strong, then there is little reason to posit that *U* is present because *O* could have produced *E*. The stronger the causal strength of *O*, the less likely that *U* is to be present. However, if *O* is weak, then *U* is needed to explain the presence of *E*. In fact, if *O* has zero causal strength, then *U must* be present, otherwise there is no way to explain the presence of *E*. Finally, suppose that the learner observes an *∼OE* trial. Like the previous case, *U* also must be present because *O* is not present to produce *E*. These last two examples show how BUCKLE embodies the assumption of causal determinism; if *O* is unable to produce *E* because it is absent or has zero causal strength, then *U* must have been present and must have caused it. This sort of reasoning drives BUCKLE's inference about the presence of *U* on a given trial.

In Step 2, BUCKLE updates its estimate of the causal strengths of *O* and *U*. To do so, BUCKLE first predicts whether *E* should be present or not based on its current causal strengths of *O* and *U*, knowledge about the presence/absence of *O*, and its inference about the presence of *U* (from Step 1). Intuitively, this prediction is made the following way. Suppose we know that *O* is absent. Then, the only way *E* could be present is by *U*. Thus, the probability that *E* would be present is the causal strength of *U* multiplied by the probability that *U* is present. Alternatively, if *O* is present, the probability of *E* being present is increased if *O* is strong, if *U* is strong, and if *U* is likely present. Once BUCKLE has made its prediction about *E*, BUCKLE calculates the difference between this prediction and knowledge about the actual presence/absence of E. This step is very similar to the Rescorla and Wagner (1972) model. If BUCKLE under-predicts E, then the causal strengths of the present causes are increased. If BUCKLE over-predicts E, they are decreased.

BUCKLE is capable of accounting for the patterns of inferences described above, both the trial-by-trial judgments of the probability of an unobserved cause being present and the causal strength judgments. Additionally, BUCKLE explains how inferences that people make about unobserved causes interact with their inferences about observed causes such as order effects (Luhmann & Ahn, 2007).

In summary, these studies have shown that when an event occurs that cannot be explained by an observed cause, people infer an unobserved cause to explain the event. These studies have so far focused on inferences that people make about single events. In the next section, we will discuss inferences people make about unobserved causes from patterns of events.

8.3 Causal determinism about systematic patterns among events

We argue that when people perceive a pattern in a sequence of events, they are reluctant to treat it as purely accidental, and instead they infer that the pattern was planned or produced through a causal mechanism. A classic example involves the pattern of bombs dropped on London by the Nazis during World War II (Gilovich, 1991, pp. 19–21; Hastie & Dawes, 2001, pp. 160–161). Even though the locations of the bombings have since been shown to be statistically random, many British citizens thought they saw clusters of bombings, and consequently inferred that German bombers deliberately avoided locations where German spies lived, creating the perceived clustering. This is a perfect example of how people infer an unobserved cause to explain an observed pattern (even though the pattern is statistically absent).

Perhaps the most basic types of patterns from which people infer a causal mechanism are those studied in introductory statistics courses: differences between the mean scores of two groups and correlations between two variables. There exists extensive literature about causal learning of this sort (e.g. Cheng, 1997; Jenkins & Ward, 1965). Here we focus on other patterns from which people are likely to infer an unobserved cause.

8.3.1 Autocorrelation

One type of pattern that has been investigated in previous literature is autocorrelation, when a previous event is statistically correlated with a future event. We will first discuss two types of irrational beliefs about autocorrelation that people have been shown to endorse and later discuss how beliefs about the underlying unobserved causal mechanisms moderate these fallacies.

One famous example of irrational belief in 'negative autocorrelation' is called the gambler's fallacy. Specifically, gamblers often believe that if they lost on the previous gamble (e.g. roulette bet), they are 'due for a win' on the next gamble. That is, people sometimes believe that the previous event negatively predicts the next event. A similar phenomenon occurs with other processes that are expected to be random. For example, people think that having six boys in a row (BBBBBB) is less likely than a specific intermixed sequence (e.g. GBBGBG) even though both are equally likely. As famously stated in a Dear Abby column, a woman who just had her eighth girl in a row claimed that 'this one was supposed to have been a boy' as if the previous births negatively influenced the chances of the future birth (DEAR ABBY column, reprinted in Hastie & Dawes, 2001, p. 159).

Consider another fallacy termed 'hot-hand.' This fallacy is named after the belief that basketball players go through hot streaks of many baskets in a row and cold streaks of many misses in a row. In fact, statistical analyses have not found even a single basketball player whose streaks deviate from chance; however, hot-hand has been found in other sports (Adams, 1995; Dorsey-Palmateer & Smith, 2004; Gilden & Wilson, 1995; 1996; Smith, 2003).

What mediates whether people believe in hot-hand or gambler's fallacy? Restated, for a given random series of events, why do people sometimes infer streaks (positive autocorrelation) and other times infer alternation (negative autocorrelation)? Burns and Corpus (2004) proposed that inferring positive vs. negative autocorrelation depends upon the causal mechanism that people believe to have generated the data. Specifically, people believe that some causal mechanisms have 'momentum' and cause streaks, whereas mechanisms that are 'random' do not produce streaks. Burns and Corpus presented participants with scenarios intended to imply either a random mechanism (e.g. roulette wheel) or a mechanism with momentum (e.g. basketball-shooting under competition). Participants were told that in a series of 100 events, the frequencies of the two outcomes had been equal, but that the event ended with a streak of one outcome. In the scenarios that participants believed to be produced by a random mechanism, participants were more likely to predict that the next event would break the streak (i.e. negative autocorrelation), whereas in the conditions that participants believed to have been produced by a non-random mechanism with momentum, participants were more likely to believe that the next event would continue the streak.

Ayton and Fischer (2004) conducted an experiment demonstrating the reverse effect; participants observed a binary sequence of events and were then asked to choose whether they thought the sequence was produced by either a mechanism meant to imply randomness (e.g. roulette wheel) or by a mechanism meant to imply momentum (e.g. basketball shooting). After observing positive/negative autocorrelation within the sequence of events, people tended to infer non-random/random mechanisms, respectively. In sum, these studies show that people infer different unobserved mechanisms based on different observed patterns.

These studies have compared two generating procedures, random and streaky mechanisms. One interesting possibility is that people may also infer a mechanism that frequently alternates. For example, it seems likely that most people do not go to the same restaurant twice in a row – after going to a restaurant once, one would probably switch to a different restaurant for diversity. In a series of data that is strongly alternating, it seems likely that people would infer a third type of mechanism that produces alternating sequences.

8.3.2 Tolerance and sensitization

A recent study by Rottman and Ahn (2009) demonstrates that people infer a causal mechanism given other kinds of patterns: tolerance and sensitization. An example of a *tolerance* scenario is tolerance to coffee. The first time a person drinks one cup of coffee he/she may feel very awake. However, after repeatedly drinking one cup of coffee, he/she becomes tolerant and one cup of coffee has little effect. The person may then drink two cups of coffee and initially feel very alert, but after repeatedly drinking two cups of coffee, again becomes tolerant. In sum, tolerance involves a *decreasing* effect over time when the cause is held constant. *Sensitization* is essentially the opposite of tolerance; sensitization involves an *increasing* effect over time when the cause is held constant. For example, many antidepressants require repeated exposure for full effectiveness. Two pills of antidepressant may initially have no effect, but after repeated exposure, two pills may be sufficient to make a person very happy. If the person cuts down to one pill of antidepressant, the decrease may initially result in a decrease in happiness, but if the person becomes sensitized to the reduced amount of antidepressant, over time, one pill may become sufficient.

To determine whether people are sensitive to these tolerance/sensitization schemata, Rottman and Ahn showed participants scenarios in which machines were tested 14 times in a row for their emissions (e.g. noise, light, heat, or smell). The input to the machines was a lever that could be set to three positions, analogous to the number of cups of coffee or number of pills of antidepressant for the scenarios described above. In one set of 'ordered' conditions, the emissions increased (sensitization) or decreased (tolerance) over repeated use. In another set of 'unordered' conditions, there was no temporal pattern so the data looked random. Figure 8.2 depicts the ordered and unordered versions of both the tolerance and sensitization conditions.

After observing these trials in each condition, participants were asked to 'rate how confident you are that the lever has the capacity to affect' the

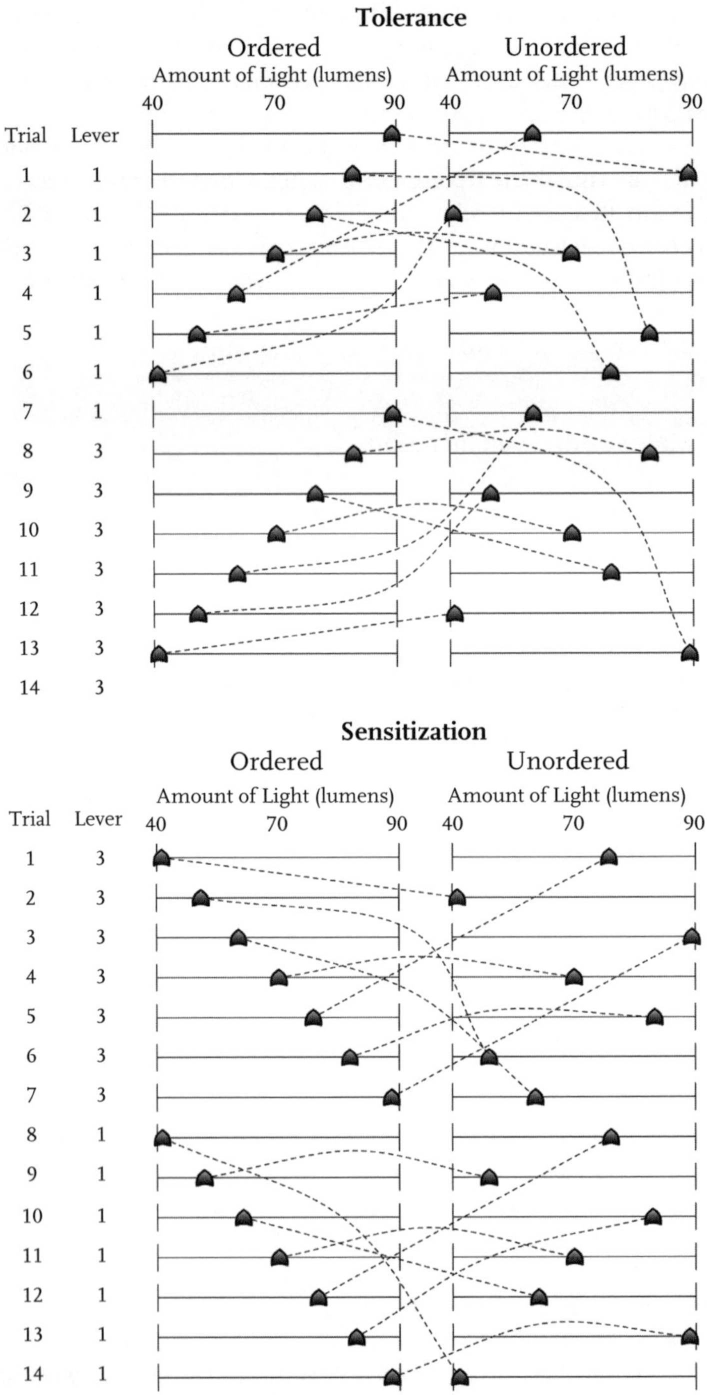

Fig. 8.2 Tolerance and sensitization, ordered and unordered conditions.

Note: Dashed lines (not shown to participants) illustrate how the same data were reordered from the ordered to unordered conditions.

emission (e.g. noise, light, heat, or smell). Participants gave considerably higher ratings in the ordered tolerance and sensitization conditions than in the unordered conditions. What is particularly interesting about the results is that people in both the ordered and unordered conditions saw identical data in terms of the simple correlation between the lever and the emission (i.e. not considering the temporal dimension). This is the type of information that has been used as the basis of causal inferences by most traditional causal induction models, and thus, they would have predicted no difference between the two conditions. Furthermore, the overall correlation was zero. Despite this, people in the ordered condition were moderately confident that the lever had causal efficacy.

One way to explain this finding appeals to unobserved causes. For example, in the ordered-tolerance condition, participants likely inferred a process that occurs within each individual machine such that a latent inhibitory variable increases over time. This is why the emissions decrease over time. In the ordered-sensitization condition, participants likely inferred an unobserved inhibitory cause that decreases over time, explaining why the emissions increase. It seems unlikely that people would think that time itself directly influences the emissions, however, over time, the machine may become 'worn in' and produce less emission. In this case, the variable responsible for 'wearing in' would be the unobserved variable that inhibits the emissions and is correlated with time.

In this account, people would use the temporal pattern to infer an unobserved cause, and the combination of the lever and this unobserved cause completely explains the emissions. This explains why participants judged that the lever influences the emissions. After all, in the ordered conditions, the emission is statistically dependent upon the lever once time or the unobserved cause is taken into account. However, when there is no temporal pattern as in the unordered condition, there is no reason to infer an unobserved cause that changes with time. Consequently people have no way to make sense of the influence of the lever and judge it not causally efficacious.

It is not difficult to find real-world examples of this reasoning. Caffeine is an adenosine antagonist; caffeine inhibits sleep through blocking adenosine, which promotes sleep. However, over repeated caffeine exposure, the number of adenosine receptors increases, making caffeine less effective at blocking them. The number of adenosine receptors is thus an unobserved cause that changes over time within an individual person.

The above situations depict repeated treatments on the same machine. To further test whether people understand the tolerance/sensitization scenarios, we created another set of scenarios in which the increasing/decreasing patterns occur to many different machines. Going back to the coffee scenario, one person's coffee drinking can influence the effectiveness of coffee for that same person at a later time. However, one person's coffee drinking should not influence the effectiveness of coffee for a different person at a later time – tolerance to coffee must happen within one entity. If people only apply the tolerance/sensitization schemata for one-entity scenarios, then they should give higher causal efficacy ratings for the lever in one-entity scenarios compared to many-entity scenarios that depict the exact same input/output data patterns. In a second experiment designed to test this one-entity vs. many-entity distinction, we found that people were more confident in the causal efficacy of the lever in the one-entity than many-entities conditions both for sensitization and tolerance.

This experiment further clarifies the inferences about the unobserved variable. In this experiment, the data patterns in the one-entity and many-entity conditions were identical for the lever, emissions, and temporal order. However, it is only in the one-entity condition that one can plausibly infer a latent process; an unobserved cause within each individual machine changes and affects the emissions even though the observed cause's strength remains constant. As previously explained, if people infer an unobserved cause in the one-entity scenario, the combination of this unobserved variable and the lever completely accounts for the pattern of emissions, which explains why people rated the lever to be efficacious. However, it would be too bizarre to infer a latent process occurring within each individual getting transferred to the person who happens to drink coffee next or the machine that happens to be tested next. If people do not infer an unobserved cause in the many-entities scenario, the pattern of data between the lever and emission does not make sense; after all, as is also true in the one-entity scenario, there is no simple correlation between the lever and emission. This explains why participants gave lower ratings for the causal efficacy of the lever in the many-entities condition. In sum, this study suggests that inferences about an unobserved cause that changes over time within one entity influenced inferences about the relationship between the observed cause and effect.

8.3.3 Developmental origins of beliefs about order

The previous studies have focused on cases when adults perceive a pattern and infer an unobserved cause to explain the pattern. Some previous studies have plotted the development of children's beliefs and inferences about causal mechanisms. Friedman (2001) found that four-year-old children believe that it is plausible for animate agents to create an ordered pattern from randomness but less plausible for non-animate causes (e.g. the wind) to do the same. That is, even without seeing a particular cause occur, four-year-olds infer that one type of unobserved cause is more plausible than another.

Newman, Keil, Kuhlmeier, and Wynn (2010) found similar results among even 12-month-old children. They created scenarios in which the infants initially saw either an ordered or unordered pile of blocks. Then an opaque barrier occluded the blocks and either a rolling ball or an animate agent (a self-propelled circular face) moved behind the occluder, presumably coming in contact with the blocks. Finally, the occluder was removed displaying either ordered and unordered blocks. The infants were more surprised (looked longer) when the ball appeared to create order from disorder than disorder from order, but they looked equally long at the two conditions for the animate agent. In sum, from a fairly early age, children understand that only animate mechanisms can create ordered patterns and infer an unobserved agent to explain an observed pattern.

8.3.4 Other types of patterns and discussion

The tolerance/sensitization experiments described above made us aware that there might be other types of patterns that people may use to infer unobserved causes. When making the unordered tolerance/sensitization conditions (see Figure 8.2), we tried to make the temporal patterns look as random as possible. However, despite our best efforts, in informal discussions after the experiment we discovered that some participants still saw patterns in the data. (See Hastie & Dawes, 2001, p. 355, for an example of the many possible patterns one might infer from a series of six sequential coin flips.) Participants saw increasing or decreasing patterns within subsets of data and interpreted them as meaningful trends (e.g. an increasing pattern in Trials 10–14 in the tolerance-unordered condition in Figure 8.2 despite Trials 9 and 13). Thus, tolerance/sensitization may potentially be triggered for noisier data than what we presented to participants or subsets of data. Some participants also saw alternating patterns (e.g. Trials 1–5 in the tolerance-unordered condition in Figure 8.2) of the form we proposed at the end of Section 8.3.1. Another type of pattern people would likely infer in other situations is a periodic or sinusoidal pattern. A sinusoidal pattern is similar to positive autocorrelation in that the previous trial predicts the next trial, but different in that the period of repetition may be constant which is not necessarily the case for autocorrelation. In all of these scenarios, we believe that people would likely attribute an observed pattern to an unobserved cause, which could further influence their judgments about observed causes. (But see the last section of this chapter for a discussion of boundary conditions.)

There are a number of important future directions of this research. First, it would be useful to determine whether people have a limited set of schemata or patterns they primarily search for when learning new causal relationships. A limited set or taxonomy of plausible schemata could reduce the complexity of causal learning given that there are infinite numbers of possible patterns caused by unobserved variables. Exploring the diversity of causal schemata may help us better understand the limits of causal learning as well as how people make generalizations from schemata they know and learn new schemata.

There might also be important individual differences in inferring unobserved causes due to pattern detection. Certain people, for example, paranoid people, may have a higher likelihood of seeing a pattern where none exists and attributing the pattern to an unobserved cause. Prior experience with certain types of mechanisms or schemata may also make a person more likely to infer a particular type of mechanism.

Finally, though some of these phenomena (e.g. hot-hand fallacy) have been studied extensively, they are not usually considered the domain of causal reasoning, but rather decision making. It would be useful to integrate research about systematic patterns with more traditional causal learning paradigms that have focused on single events (i.e. the previous section). After all, when observing a new set of data, people are sensitive to both single events and patterns of events, and a general model of causal learning should incorporate both.

8.4 Beliefs in stability of causal relations

The previous two sections have suggested that people infer an unobserved cause to understand unexplained events and systematic patterns of events. In this section, we suggest that people also infer an unobserved cause if they notice that the relationship between an observed cause and effect changes. For example if you know that a medicine has a particular side effect for most people, but find a group of people who do not develop the side effect, it would make sense to infer an unobserved cause to explain the difference (e.g. the group has an unusual gene). Restated, it seems likely that people will infer an unobserved cause when a causal relationship is not stable. We will discuss two types of stability of causal relationships: stability across different samples and stability over time.

8.4.1 Causal power–stability across samples

Cheng (Power PC; Cheng, 1997) proposed that when people judge whether X causes Y, people intuitively estimate *causal power*, the 'probability with which [X] influences [Y]' (Buehner, Cheng, & Clifford, 2003). Consider the following scenario: you are testing the side-effects of a new drug and discover that when given to 100 people without headaches, 50 of these people develop a headache. Suppose you gave the drug to 100 people, 50 of whom already have a headache. How many out of these 100 would have a headache after taking the drug? According to Power PC theory 75 people would have a headache. In the first situation, the medicine caused 50% of the people to get a headache. In the second scenario, 50 people already have a headache, and the medicine will cause 50% of the remaining people to get a headache. The base rate percent of people who already have a headache may vary from situation to situation, but Cheng argues that the percent of people who do not already have a headache and will get a headache should be constant across scenarios.

Before moving on, it is useful to understand why Cheng makes this argument, though some readers may prefer to skip ahead to the results of her experiment. One easy way to calculate causal strength is simply to subtract the probability of an effect (E) occurring in the presence of an observed cause (O) minus the probability of the effect occurring in the absence of the cause: $P(E|O) - P(E|{\sim}O)$ (i.e. ΔP measure mentioned earlier). However, this calculation has the problem that it is influenced by ceiling effects. When the base rate of the effect occurring without the cause is greater than zero, the causal strength cannot be the maximum 1 even if the effect always occurs in the presence of the cause. To get around this problem, Cheng uses a number of assumptions to calculate causal power.[3] First, she assumes that an effect can occur for two reasons: if the observed cause produces the effect or if an unobserved cause (U) produces the effect (or they can both produce it simultaneously). This assumption is very useful because it implies that if E occurs in the absence of O then U must be responsible. Thus the probability of the effect occurring in the absence of the cause, $P(E|{\sim}O)$ is an estimate of the frequency that the unobserved was responsible for the effect. Second, she assumes that the observed cause occurs independently from unobserved causes, $P(O|U) = P(O|{\sim}U)$, and that O and U influence E independently. These assumptions are also very useful because we now assume that $P(E|{\sim}O)$ is an estimate of how frequently the unobserved cause produces the effect in general, even when O is present. To determine causal power, ΔP is divided by $1 - P(E|{\sim}O)$ which effectively normalizes it on the probability that the unobserved cause produced the effect, resulting in the increase in probability of the effect due to the observed cause regardless of unobserved causes.

$$\text{Causal power} = \frac{P(E|O) - P(E|{\sim}O)}{1 - P(E|{\sim}O)}.$$

The causal power of a particular cause/effect relationship is thus supposed to be the same in samples regardless of unobserved causes (the base rate of E). If different causal powers are observed in different samples, one likely explanation is that the assumptions about the unobserved causes are violated, and that the apparent relationship between the observed cause and effect is partially due to unobserved causes.

Liljeholm and Cheng (2007) tested whether people believe that the causal power of a specific cause is stable across situations. They created two conditions, each of which had three scenarios like the headache scenario. In one condition, the three scenarios had the same causal power but different base rates of headache. In a second condition, the base rate of headache was the same (zero people initially had a headache) but the causal power was different

[3] See Cheng (1997) for the complete, formal treatment. Note that we use different notation for consistency within this chapter.

across the three scenarios. After observing the data for the three scenarios in each condition, participants answered whether they thought the medicine interacts with some unobserved factor across the experiments or whether the medicine has the same influence across the three scenarios. Whereas only one third of participants thought the medicine interacted with an unobserved factor in the causal power constant condition, 86% thought that the medicine interacted with an unobserved factor in the condition in which causal power varied.

In sum, causal power seems to be one way in which people expect causal relations to be stable across scenarios. When it is not stable, people infer an unobserved cause that interacts with the observed cause and is responsible for the discrepant causal power estimates. The type of stability of causal relations discussed here relates to stability across different contexts that are distinguished for learners. That is, Liljeholm and Cheng (2007) presented participants with three scenarios each framed as an individual study with different hypothetical patients. In the next studies, participants learned about changing causal relations on their own.

8.4.2 Grouping effects – stability over time

One of the challenging aspects of causal learning is that there are infinitely many possible interacting factors and these interacting factors change over time and context. Sometimes we have a priori beliefs about possible interacting factors; however, often we do not know about interacting factors, or whether interacting factors have changed. In this case, we may learn about interacting factors by noticing a difference in a causal relationship over time. When we observe a change in a causal relationship over time, we will likely conclude that an unobserved interacting factor changed. One critical assumption for making such an inference is that unobserved factors are stable for long enough periods of time that we can notice the difference.

Consider the following double-switch scenario that was briefly discussed at the beginning of this chapter listed in Table 8.1 as *biconditional*. Some lights are connected to two switches (e.g. often at opposite ends of a hallway). There are two important characterizations of this scenario. First, whenever one switch is flipped (assuming that the other light switch is not flipped at exactly the same time), the state of the light will change. Second, neither of the two switches has an 'on' or 'off' position – there is not necessarily any correlation between the position of a given switch and the state of the light. Figure 8.3 provides a wiring diagram of a double light switch. The light will be on whenever the switches make a complete circuit (if both switches are up or if both switches are down).

Suppose you enter a room for the first time and discover that when you flip a switch up, a light goes on, and when you flip it down, the light goes off (grey cells, Steps 1–4 in Table 8.3). If you assume that other potential causes of the

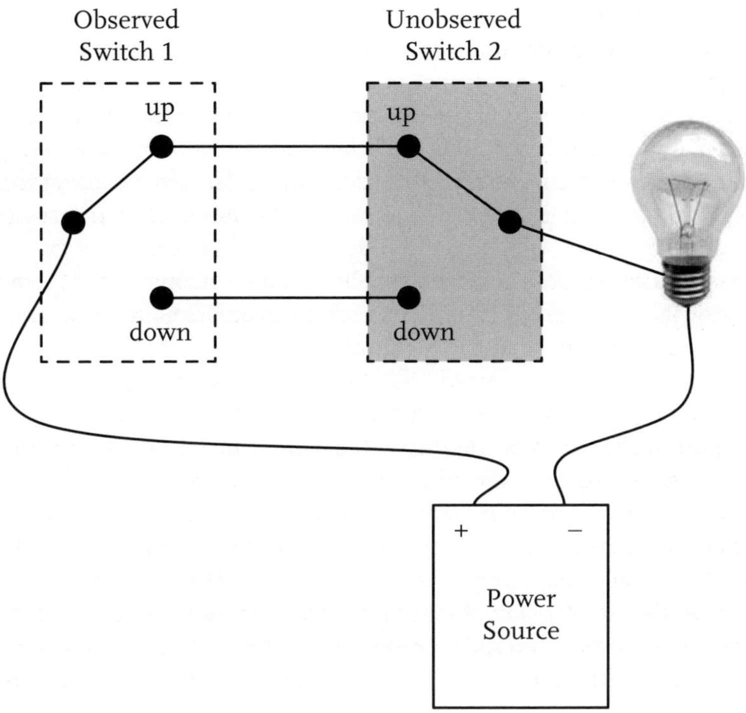

Fig. 8.3 Wiring diagram for double light switch scenario.

Table 8.3 Double light switch, grouped scenario.

Steps	Switch	Light	U	I	Inferences:
1	0	0	1	1	
2	1	1	1	1	Switch influenced light.
3	0	0	1	1	
4	1	1	1	1	Unobserved factor changed.
5	1	0	0	0	
6	0	1	0	0	
7	1	0	0	0	Switch influenced light.
8	0	1	0	0	

Note: For Tables 8.3 and 8.4, 'U' represents an unobserved interacting factor and 'I' represents a factor learners are likely to infer. 0 represents down for the switch and off for the light, and 1 represents up for the switch and on for the light.

light are fairly stable (and do not happen to change at the same moment you flipped your switch), you would infer that the switch influences the light. Later (Steps 4–5), the light turns off without anyone touching the switch (perhaps your daughter flipped the other switch unknown to you; U in Table 8.3). Afterwards, when the switch is down, the light is on, and off when up (Steps 5–8). From this scenario, you might be very confident that your switch influences the light; there were two long periods when the status of the switch correlated with the status of the light. Additionally, because the light mysteriously turned off, you might infer an unobserved factor (I in Table 8.3) that interacts with your switch, explaining the overall zero contingency between the switch and light.

However, inferring the observed switch to be efficacious depends upon the stability of the unobserved cause. For example, consider the same data from Table 8.3, rearranged as in Table 8.4. Initially, the switch is down and the light is off (Step 1). In Step 2 the switch is flipped up, but the light still stays off. In order to believe that the switch is causally efficacious, one must infer that at the moment the switch was flipped, an unobserved factor coincidentally changed and counteracted the effect of the observed switch, as specified under column 'U' (unobserved interacting factor). Then, in Step 3, the light turns on without flipping the switch, and so on. Thus, for the situation shown in Table 8.4, it would be extremely difficult to infer the switch to be causally efficacious: The switch cannot be the sole cause of the light because there is zero contingency with the light. Furthermore, it would be difficult to infer it as part of an interaction because doing so would require inferring an unobserved factor operating as specified under column 'U,' which is counterintuitive; the unobserved interacting factor is highly unstable and exceedingly complicated to track. Instead, the simplest account (intuitively) would be to infer an unobserved factor that is *entirely* responsible for turning the light on and off. Such a factor would be perfectly correlated with the light, as specified under column 'I' (inferred factor). If a learner inferred 'I', he/she would likely infer that the switch is not causally responsible for the light at all.

These two examples were meant to demonstrate that if an unobserved cause is relatively stable for periods of time with a few salient different periods as in

Table 8.4 Double light switch, ungrouped scenario.

Steps	Switch	Light	~~U~~	I	Inferences:
1	0	0	~~1~~	0	Switch did not influence light.
2	1	0	~~0~~	0	Unobserved factor changed.
3	1	1	~~1~~	1	Switch did not influence light.
4	0	1	~~0~~	1	Unobserved factor changed.
5	0	0	~~1~~	0	Switch did not influence light.
6	1	0	~~0~~	0	Unobserved factor changed.
7	1	1	~~1~~	1	Switch did not influence light.
8	0	1	~~0~~	1	

Table 8.3, a learner is likely to infer that an interaction is taking place with an unobserved cause. However, if the scenario is very unstable, as in Table 8.4, then the learner is less likely to infer an interaction with an unobserved cause. Instead, they would likely infer that the unobserved cause is not responsible for the effect at all.

To investigate these inferences, we gave participants a cover-story about machines that produce blocks of various shapes (e.g. square or triangle), and asked participants to determine if the position of a lever on the machines affects the shape of the blocks. Participants then observed videos of 20 trials in a continuous temporal sequence; from trial to trial, the lever sometimes switched between the left and right position and the shape of block (e.g. square or triangle) sometimes changed. In all conditions, the lever was statistically uncorrelated with the shape of the blocks.

Rottman and Ahn (in prep. see Rottman & Ahn 2009, for partial results) manipulated two aspects of the scenarios. First, we manipulated the grouping of the trials similar to that shown in Tables 8.3 and 8.4. In the 'grouped' condition, there were relatively stable periods of time when one shape was associated with one position of the lever, and other periods when the association flipped. In the 'ungrouped' condition, these two different associations were more intermixed so that there were no discernable stable periods. If grouping allows people to infer an unobserved cause that is stable for periods of time and then switches, people should infer an interacting unobserved cause more in the grouped than ungrouped condition.

Another manipulation of the study was whether the scenario involved only one machine changing over time or different machines. In the one-machine condition, all 20 trials occurred with one machine. That is, in the one-machine conditions, the lever on the machine was sometimes flipped back and forth between left and right, and the shape of the block produced by the machine sometimes changed over 20 trials. In the many-machines conditions, 20 different machines were observed once each: the lever of each machine was set either to the left or the right, and the machine produced either a square or triangle. Even though the many-machine conditions were identical to the one-machine conditions in every other way, we reasoned that participants would not make different inferences about the unobserved cause between the grouped vs. ungrouped conditions. Because each machine is different, we reasoned that participants would not make use of the temporal grouping information to infer a stable unobserved cause; after all, participants had no reason why they were presented with the machines in the particular order. If the temporal stability information was not deemed important and people did not use it to infer stability, there should not be any difference between the grouped and ungrouped conditions for many-machines. Such a finding would suggest a caveat to the assumption of stability: people only distinguish between stable and unstable scenarios for

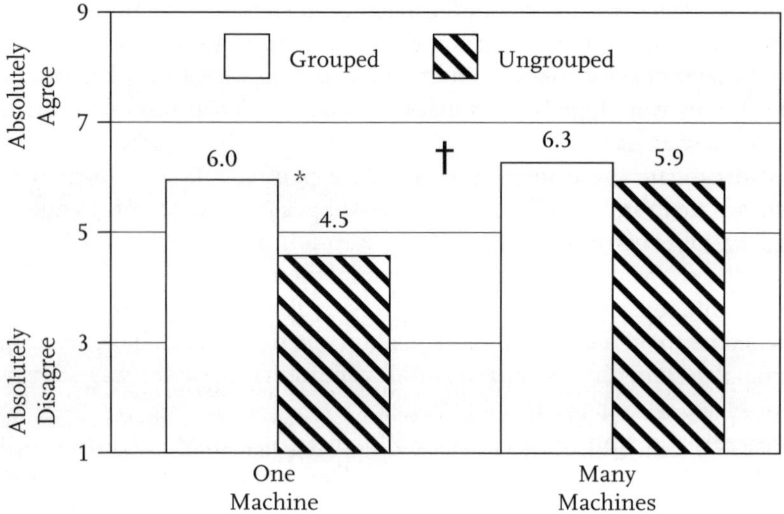

Fig. 8.4 Combination of lever and unobserved factor influenced shape.

Note: * paired t-test, $p < 0.05$, † interaction in repeated-measures ANOVA, $p < 0.05$.

inferring interactions with unobserved causes when time is a meaningful variable.

After observing each scenario, participants rated their agreement with whether 'A combination of the lever and some other factor influenced the shape of the blocks' from 1 ('Absolutely Disagree') to 9 ('Absolutely Agree') (see Figure 8.4). As expected, in the one-machine condition, participants inferred an interaction with an unobserved cause more in the grouped than ungrouped condition. However, in the many-machine condition, when time was not a meaningful factor, there was no difference between the grouped and ungrouped conditions in participants' inferences about an interaction with an unobserved cause.[4] In other words, only when time is a meaningful variable (i.e. the one-machine condition), do people use temporal stability to infer an interaction with an unobserved cause. When time is not meaningful (i.e. the many-machine condition), there is no difference between grouped and ungrouped conditions.

Inferring that there is an unobserved interacting factor has further implications for peoples' views of the observed cause. Specifically, Rottman &

[4] As seen in Figure 8.4, participants were more likely to think that there was an interaction with an unobserved cause in the many-machines condition than in the one-machine condition. The reason for this finding is likely because they thought the different machines interact differently with the switch – the different machines is the second interacting factor.

Ahn (2009; Experiment 1) demonstrated that the more grouped the scenario, the higher the causal strength ratings that participants gave it. Across these two experiments, when a scenario is grouped, people are able to infer the interacting unobserved cause and still believe that the observed cause influences the effect even though there is no correlation between the two. However, when the scenario is ungrouped, people are less likely to infer an unobserved factor and more likely to infer that the observed cause is not related to the effect (after all, there is zero correlation).

One of the important implications of this study is that people spontaneously distinguish between scenarios with stable unobserved causes and scenarios with unstable unobserved causes, even though both can appear to have zero correlation between the cause and effect. Consider the study briefly discussed in the introduction that investigated the role of pollution (observed cause) on daily temperature range (effect; Gong, Guo, & Ho, 2006). The researchers found that pollution *decreases* daily temperature range during the winter, but pollution *increases* daily temperature range during the summer. Summarized, the season flips the direction of the influence of pollution on diurnal temperature range. This example makes an important point: the researchers did not know about the interaction with season a priori. At some point, they must have noticed that the relationship between pollution and weather is flipped depending on the season, and if they had overlooked this important factor, the relationship between pollution and diurnal temperature range would have been obscured and might appear to not exist. By observing a stream of data and noticing periods of stability, the researchers could uncover that an unpredicted variable (season) plays an important interacting role.

In summary, these studies suggest that people believe causal relationships to be fairly stable across contexts, and if they notice a difference across samples or times, they posit an unobserved factor to explain the difference. Presumably people would also infer systematic patterns to be stable across different contexts, and would likewise posit an unobserved factor if they notice a difference. For example, if a cause exhibited tolerance for one sample and sensitization for another, or positive autocorrelation for one sample and negative autocorrelation for another, people would likely infer an unobserved cause to explain the difference. Additionally, if a cause appeared to switch from a noisy-or to a biconditional functional relationship, people would also infer that an unobserved factor changed. In this way, the inferences we have discussed about single events, patterns of events, and relationships between causes/effects can be viewed in a hierarchy. If an unexplained change occurs anywhere along the hierarchy from the lowest level (single events), middle level (patterns of events) or the highest level (relationships between causes/effects), people will infer that an unobserved cause is responsible.

8.5 Working memory – A requirement for reasoning about unobserved causes

We have now discussed a number of situations when people infer unobserved causes. However, reasoning about unobserved causes is also cognitively challenging. As we have already explained, there are many possible unobserved and unattended causes, and many ways in which those causes can interact with observed causes. In this section, we propose that reasoning about unobserved causes requires considerable working memory capacity. We will now review a particular phenomenon, recency/primacy effects, in which beliefs about unobserved causes play a central role. Then we will demonstrate how working memory mediates this phenomenon.

Suppose you initially observe a set of data, mostly showing positive covariation between two events, followed by data mostly showing negative covariation between the same events. For instance, for the first half of the baseball season, you notice that your favorite baseball team was more likely to win when you were wearing your 'lucky' socks, but for the second half of the baseball season, you notice that your team was more likely to lose when you were wearing your 'lucky' socks. Would you consider your socks to be still lucky? There are many possible strategies a reasonable learner could take to answer this question. One could average across all of the available data, concluding that the socks have nothing to do with winning. Alternatively, one could give more weight on the most recent data, concluding that wearing those socks actually hurts performance. Or one can give more weight on initial data, concluding that wearing those socks improves performance. It is difficult to tell which one of these three is the most rational strategy, and in fact, the experimental results using this paradigm show that people demonstrate all three strategies (Dennis & Ahn 2001; Glautier, 2008; Lòpez, Shanks, Almaraz, & Fernàndez, 1998; Marsh & Ahn, 2006; Shanks, Lòpez, Darby, & Dickinson, 1996).

In a recent study, Luhmann and Ahn (in press) found empirical evidence that the conflicting findings of primacy/recency effects can be explained by learners' beliefs about unobserved causes. Specifically, people who reason more about unobserved causes tend to show primacy, whereas people who reason less about unobserved causes tend to show recency effects in their causal strength judgments. Learners were presented with sequences of covariation information involving medications and potential side effects. Sequences always used the same set of observations, but were constructed to present the majority of positive evidence first, followed by the majority of negative evidence (Positive–Negative) or vice versa (Negative–Positive). Sometimes during the learning sequence, participants were asked to explain why the effect did or did not occur, and at the end of the sequence, learners made causal strength judgments.

Some subjects were particularly likely to explain the outcome of a specific trial by appealing to unobserved causes. For example, consider a learner in the Positive–Negative condition who observed the first half of positive evidence, and then observes some contradictory negative evidence (i.e. a trial when the cause occurs but the effect does not). Participants were then prompted to choose one explanation for why this happened: '[the cause] prevented [the effect]', 'it is pure coincidence that' [the effect] did not occur after [the cause]', or 'for some reason, [the cause] failed to cause [the effect].' If the participant appeals to an unobserved cause, he/she would choose the third option, which subtly references alternative causal influences (i.e. 'for some reason'). The unobserved cause could have overridden the observed cause and prevented the effect, allowing participants to continue to believe that the observed cause was generative. In fact, participants who choose this option gave higher causal strength ratings in the Positive–Negative condition. (These participants exhibited a primacy effect because their higher causal strength ratings reflect the initial positive contingency.) These results suggest that unexpected covariation information elicits reasoning about unobserved, alternative causes in some learners. Such reasoning tends to 'excuse' the new, contradictory information and leave the prior causal beliefs relatively untouched.

However, other learners appealed to unobserved causes less and instead used this same conflicting information to directly modify their causal beliefs. For example, in the Positive–Negative case, upon encountering the negative evidence, a learner could take this evidence at face value and modify the initial belief that the cause generates the effect to conclude that the cause is not related to the effect or even prevented the effect. Such learners subsequently gave lower causal strength estimates (a recency effect).

So far, these results suggest that primacy/recency effects in causal judgments are related to whether people appeal to unobserved causes. However, why do some people appeal to unobserved causes more than others? We hypothesize that one of the reasons is the ease with which learners are able to reason about unobserved causes that produces these different learning strategies. Marsh and Ahn (2006) demonstrated that learners with higher verbal working memory capacity were more likely to show a primacy effect. Thus, we reasoned that working memory may facilitate reasoning about unobserved causes, which we know influences primacy/recency in causal strength judgments. To test this hypothesis, Luhmann and Ahn (in press) created two situations that experimentally manipulated the ease of reasoning about unobserved causes.

First, Luhmann and Ahn (in press) increased the cognitive load during the task, which we predicted would decrease the ease of reasoning about unobserved causes and produce a recency effect. The learners performed the same task explained above while simultaneously performing a difficult secondary task (counting backwards by 3s) This manipulation impaired reasoning about unobserved causes. Participants took individual trials at face value and modified existing hypotheses. For example, in the Positive–Negative condition, when faced with the negative evidence, participants simply said that the cause inhibits the effect, presumably because it would be too taxing on working memory to postulate unobserved causes. Furthermore, participants' overall causal strength judgments showed a recency effect.

Second, we made it easier for learners to reason about unobserved causes by simply making the unobserved causes observed. During the second, contradictory half of the event sequence, learners were told that an alternative cause was present. Note that they were not told that the effect occurred because of this alternative cause (that is, there still is an ambiguity as to what was the true cause of the effect). Yet, this manipulation increased participants' reasoning about unobserved causes and they interpreted information that contradicted their prior causal beliefs as "something" going wrong, leaving their beliefs relatively untouched. Furthermore, their overall causal strength judgments showed a primacy effect.

In sum, these studies demonstrate that working memory moderates inferences about unobserved causes for primacy/recency effects. Based on these results, it is plausible that working memory would moderate other inferences about unobserved causes, such as those in the double light switch scenario or perhaps those discussed in the section on Power PC. Given that reasoning about unobserved causes (e.g. potential confounds; see Cheng, 1997) is necessary for normative causal inference, future research on the limits and conditions under which people reason about unobserved causes is particularly important.

8.6 Conclusions

In this chapter, we demonstrated that people make a number of sophisticated inferences about unobserved causes. First, people infer an unobserved cause when a single unexplained event happens: children appeal to magic when they don't have a physical explanation for why one object would move without another object touching it (Phelps & Woolley, 1994), and adults infer that a hidden button was pressed when a light bulb illuminates without an observed button being pressed (Luhmann & Ahn, 2007). Second, people infer an unobserved factor to explain patterns of events that cannot be explained by the observed causes, such as a series of 10 coin flips all landing on heads (Ayton & Fischer, 2004), or a person becoming tolerant to caffeine (Rottman & Ahn, in press). (In both these scenarios, the mechanism that produced the pattern is not observed.) Third, people infer an unobserved cause to explain changes in the relationship between a cause and effect, for example, if a cause sometimes generates the effect and sometimes inhibits the effect (Rottman & Ahn, 2009).

Though we have not focused much on causal learning models, these experiments suggest some important implications for the development of future models. Most existing models of causal learning have focused upon observed causes and make fairly simple assumptions about unobserved causes. For example, existing models assume that unobserved causes are always present (Rescorla & Wagner, 1972) or are not confounded with observed causes (e.g. Cheng, 1997; see Luhmann & Ahn, 2007, for a discussion). However, people do not seem to make these assumptions and instead make dynamic inferences about unobserved causes; people do not believe that unobserved causes are constant and make sophisticated inferences about the presence and changes in unobserved causes.

Many of the phenomena discussed in the current chapter involve scenarios that unfold over time (i.e. autocorrelation, tolerance/sensitization, the double light switch scenario, and primacy/recency effects), yet most existing models do not capture time sufficiently. Many influential models have been designed primarily to capture causal phenomena that do not occur over time, and thus they aggregate across all trials (e.g. Cheng, 1997; Jenkins & Ward, 1965; Griffiths & Tenenbaum, 2005). As already discussed, influential animal-learning models that do model learning over time (e.g. Rescorla & Wagner, 1972) often make overly-simplistic assumptions about unobserved causes. Bayesian inference has become a particularly common way to model causal learning, and it proved very useful in BUCKLE. However, as noted by Danks (2007), 'causal Bayes nets do not currently provide good models of continuous time phenomena, though continuous time Bayes nets are the subject of ongoing research (Nodelman 2002, 2003)'. We believe that capturing phenomena that unfold over time should be an important aspect of future models.

We have also demonstrated that when working memory is taxed, people have difficulty reasoning about unobserved causes (Luhmann & Ahn, in press) Yet, despite the additional cognitive challenge of reasoning about unobserved causes, we believe that it occurs as a normal part of causal learning. Many of the previously mentioned studies have demonstrated that people spontaneously reason about unobserved causes. Furthermore, we have described multiple phenomena in which reasoning about unobserved causes influences the inferences people make about observed causes, the more typically studied form of causal learning. For example, when people infer that an unobserved cause flips the relationship between the observed cause and effect (generative vs. preventative), since the interaction explains why there may be zero overall correlation, people infer that the observed cause still influences the effect (Rottman & Ahn, 2009).

One general way to summarize these findings is that people are always on the lookout for unexplained data and consequently infer unobserved causes to make sense of the relationship between observed causes and effects. Any attempt to explain causal learning based on overly simplistic and static assumptions about unobserved causes will not be able to account for the dynamic interplay between observed and unobserved causes demonstrated in these studies. Perhaps reasoning about unobserved causes should be viewed as a fundamental feature of 'causal' reasoning.

Acknowledgements

This research was supported by an NSF Graduate Research Fellowship (Rottman) and NIMH Grant MH57737 (Ahn). The authors thank two anonymous reviewers for useful suggestions.

References

Adams, R. M. (1995). Momentum in the performance of professional tournament pocket billiards players. *International Journal of Sport Psychology, 26*, 580–587.

Arkes, H. R. & Harkness, A. R. (1983). Estimates of contingency between two dichotomous variables. *Journal of Experimental Psychology: General, 112*, 117–135.

Audi, R. (1995). *The Cambridge Dictionary of Philosophy*. Cambridge, MA: Cambridge University Press.

Ayton, P. & Fischer, I. (2004). The hot hand fallacy and the gambler's fallacy: Two faces of subjective randomness? *Memory & Cognition, 32*, 1369–78.

Burns, B. D. & Corpus, B. (2004). Randomness and inductions from streaks: 'Gambler's fallacy' versus 'hot hand'. *Psychonomic Bulletin & Review, 11*, 179–84.

Cheng, P. W. (1997). From covariation to causation: A causal power theory. *Psychological Review, 104*, 367–405.

CNN (1999, May 13). Night-light may lead to nearsightedness. Retrieved June 25, 2009, from http://www.cnn.com/HEALTH/9905/12/children.lights/index.html

Danks, D. (2007). Causal learning from observations and manipulations. In M. C. Lovett & P. Shah (Eds.), *Thinking with Data* pp. 359–388. New York: Lawrence Erlbaum Associates

Dennis, M. J. & Ahn, W. K. (2001). Primacy in causal strength judgments: The effect of initial evidence for generative versus inhibitory relationships. *Memory & Cognition, 29*, 152–64.

Dickinson, A., Shanks, D., & Evenden, J. (1984). Judgment of act-outcome contingency: The role of selective attribution. *Quarterly Journal of Experimental Psychology, 36A*, 29–50.

Dorsey-Palmateer, R. & Smith, G. (2004). Bowlers' hot hands. *The American Statistician, 58*, 38–45.

Downing, C. J., Steinberg, R. J., & Ross, B. H. (1985). Multicausal inference: Evaluation of evidence in causally complex situations. *Journal of Experimental Psychology: General, 114*, 239–263.

Einhom, H. J. & Hogarth, R. M. (1986). Judging probable cause. *Psychological Bulletin, 99*, 3–19.

Friedman, W. J. (2001). The development of an intuitive understanding of entropy. *Child Development, 72*, 460–73.

Gilden, D. L. & Wilson, S. G. (1995). On the nature of streaks in signal-detection. *Cognitive Psychology, 28*, 17–64.

Gilden, D. L. & Wilson, S. G. (1996). Streaks in skilled performance. *Psychonomic Bulletin & Review, 2*, 260–265.

Gilovich, T. (1991). *How We Know What Isn't So: The Fallibility of Human Reason in Everyday Life*. New York: Free Press.

Glautier, S. (2008). Recency and primacy in causal judgments: Effects of probe question and context switch on latent inhibition and extinction. *Memory & Cognition, 36*, 1087–93.

Goldvarg, E. & Johnson-Laird, P. N. (2001). Naive causality: A mental model theory of causal meaning and reasoning. *Cognitive Science, 25*, 565–610.

Gong, D., Guo, D., & Ho, C. (2006). Weekend effect in diurnal temperature range in China: Opposite signals between winter and summer. *Journal of Geophysical Research, 111* D18113, doi:10. 1029/2006JD007068.

Griffiths, T. L. & Tenenbaum, J. B. (2005). Structure and strength in causal induction. *Cognitive Psychology, 51*, 334–384.

Gwiazda, J., Ong, E., Held, R., & Thorn, F. (2000). Vision: Myopia and ambient night-time lighting, *Nature, 404*, 144.

Hagmayer, Y. & Waldmann, M. R. (2007). Inferences about unobserved causes in human contingency learning. *Quarterly Journal of Experimental Psychology (2006), 60*, 330–55.

Hastie, R. & Dawes, R. M. (2001). *Rational Choice in an Uncertain World*. Thousand Oaks, CA: Sage.

Jenkins, H. M. & Ward, W. C. (1965). Judgment of contingency between responses and outcomes. *Psychological Monographs: General and Applied, 79*, 1–17.

Koslowski, B. (1996). *Theory and Evidence: The Development of Scientific reasoning*. Cambridge, MA: MIT Press.

Leising, K. J., Wong, J., Waldmann, M. R., & Blaisdell, A. P. (2008). The special status of actions in causal reasoning in rats. *Journal of Experimental Psychology: General, 127*, 514–527.

Liljeholm, M. & Cheng, P. W. (2007). When is a cause the 'same'? Coherent generalization across contexts. *Psychological Science, 18*, 1014–21.

Lòpez, F. J., Shanks, D. R., Almaraz, A., & Fernàndez, P. (1998). Effects of trial order on contingency judgments: A comparison of associative and probabilistic contrast accounts. *Journal of Experimental Psychology: Learning, Memory, and Cognition, 24*, 672–694.

Luhmann, C. C. & Ahn, W. (in press). Order Effects during Learning: Expectations and Interpretations, *Journal of Experimental Psychology: Learning, Memory, and Cognition*

Luhmann, C. C. & Ahn, W. K. (2005). The meaning and computation of causal power: Comment on Cheng (1997) and Novick and Cheng (2004). *Psychological Review, 112*, 685–93.

Luhmann, C. C. & Ahn, W. K. (2007). BUCKLE: A model of unobserved cause learning. *Psychological Review, 114*, 657–77.

Marsh, J. K. & Ahn, W. K. (2006). Order effects in contingency learning: The role of task complexity. *Memory & Cognition, 34*, 568–76.

Newman, G. E., Keil, F. C., Kuhlmeier, V., & Wynn, K. (2010) Sensitivity to design: Early understandings of the link between agents and order. *Proceedings of the National Academy of Sciences.*

Nisbett, R. E. & Ross, L. (1980). *Human Inference: Strategies and Short-comings of Social Judgment.* Englewood Cliffs, NJ: Prentice-Hall.

Nodelman, U., Shelton, C. R., & Koller, D. (2002). Continuous time Bayesian networks. In *Proceedings of the 18th international conference on uncertainty in artificial intelligence,* pp. 378–387.

Nodelman, U., Shelton, C. R., & Koller, D. (2003). Learning continuous time Bayesian networks. In *Proceedings of the 19th international conference on uncertainty in artificial intelligence,* pp. 451–458.

Novick, L. R. & Cheng, P. W. (2004). Assessing interactive causal influence. *Psychological Review, 111*(2), 455–85.

Pearl, J. (2000). *Causality: Models, Reasoning, and Inference.* Cambridge: Cambridge University Press.

Phelps K. E. & Woolley, J. D. (1994). The form and function of children's magical beliefs. *Developmental Psychology, 30,* 385–394.

Quinn, G., Shin, C. , Maguire, M. & Stone, R. (1999). Myopia and ambient lighting at night, *Nature,* 399, 113–114.

Rescorla, R. A. & Wagner, A. R. (1972). A theory of Pavlovian conditioning: Variations in the effectiveness of reinforcement and nonreinforcement. In A. H. Black, & W. F. Prokasy (eds.), *Classical Conditioning II: Current Research and Theory.* New York: Appelton-Century-Crofts.

Rottman, B. M. & Ahn, W. (2009). Causal inference when observed and unobserved causes interact. *Proceedings of the 31st Annual Conference of the Cognitive Science Society.*

Rottman, B. M. & Ahn, W. K. (2009). Causal learning about tolerance and sensitization. *Psychonomic Bulletin and Review,* 16(6), 1043–9.

Rottman, B. M. & Ahn, W. (in prep.) Effect of Grouping of Evidence types on Learning about interactions between observed and unobserved causes.

Saxe, R., Tenenbaum, J. B., & Carey, S. (2005). Secret agents: Inferences about hidden causes by 10- and 12-month-old infants. *Psychological Science, 16,* 995–1001.

Saxe, R., Tzelnic, T., & Carey, S. (2007). Knowing who dunnit: Infants identify the causal agent in an unseen causal interaction. *Developmental Psychology, 43,* 149–58.

Schustack, M. W. & Sternberg, R. J. (1981). Evaluation of evidence in causal inference. *Journal of Experimental Psychology: General, 110,* 101–120.

Shaklee, H., & Tucker, D. (1980). A rule analysis of judgments of covariation between events. *Memory and Cognition, 8,* 459-467.

Shanks, D. R., Lòpez, F. J., Darby, R. J., & Dickinson, A. (1996). Distinguishing associative and probabilistic contrast theories of human contingency judgment. In D. R. Shanks, K. J. Holyoak, & D. L. Medin (eds.), *Causal Learning: The Psychology of Learning and Motivation.* pp. 265–311. San Diego, CA: Academic Press.

Smith, G. (2003). Horseshoe pitchers' hot hands. *Psychonomic Bulletin & Review, 10,* 753–758.

Wegner, D. M. (2002). *The Illusion of Conscious Will.* Cambridge, MA: MIT Press.

White, P. A. (2005). The power PC theory and causal powers; Reply to Cheng (1997) and Novick and Cheng (2004). *Psychological Review, 112,* 675–682.
White, P. A. (2009). Not by contingency: Some arguments about the fundamentals of human causal learning. *Thinking & Reasoning, 15,* 129–166.
Woolley, J. D. (1997). Thinking about fantasy: Are children fundamentally different thinkers and believers from adults? *Child Development, 68,* 991–1011.

·9·

Counterfactual and generative accounts of causal attribution

Clare R. Walsh and Steven A. Sloman

Abstract

Causal attribution is central to people's ability to understand and make sense of the world. It is necessary to explain why events occurred, to predict the consequences of actions as well as other events, and to assign credit and blame. The question we aim to address in this chapter is how people make these attributions. Our interest is in the explanation of single events, for example, how someone explains what caused this particular window to break, rather than general occurrences, such as what kinds of things usually cause windows to break. This question has been studied for millennia and yet the answer remains elusive. We will review some of the existing research and offer a bit of new evidence about how people make causal attributions. Our review will focus on two major types of theory: counterfactual and generative theories.

9.1 Counterfactual theories of causal attribution

Counterfactual theories have their origins in the work of Hume (1748/1988). On these accounts, a cause is an event that makes a difference to another event. This definition does not rely on any reference to the mechanism or process connecting the two events. Because causation cannot be directly perceived, people use other cues. For instance, they may judge that C caused E if events similar to E tend to follow events similar to C (Hume, 1748/1988). Hume's work established a strong tradition in psychological theorizing about causation that takes co-variation to be the major determinant of causal beliefs (e.g. Cheng and Novick, 1990; Kelley, 1973).

Another prong in the Humean analysis of causation led to the development in philosophy of counterfactual theories (Lewis, 1973). These theories share with co-variation theories a reliance on difference-making rather than any reference to the mechanism mediating cause and effect. Although co-variation may provide a cue to causation, it cannot distinguish causation from mere correlation. Two events (e.g. poor vegetation and a dry riverbed) may co-vary not because one causes the other but rather because both share a common cause (e.g. a shortage of rain). Counterfactuals enable us to separate correlation from causation. According to the counterfactual view of causation, C caused E provided that if C hadn't occurred, then E wouldn't have occurred (Lewis, 1973). Lewis proposed that we evaluate counterfactuals with reference to the closest possible world. If we assume C and E both occurred and we imagine the closest possible world to the actual world in which C did not occur, C will be judged to have caused E provided that E also would not have occurred in this counterfactual world. We may judge that the lack of rain caused the dry riverbed because in the closest world in which it did rain, the riverbed wouldn't be dry.

In practice, when people mentally simulate alternative scenarios, they may not always generate ones that are minimally different from the actual world (Walsh & Johnson-Laird, 2009) and in fact calculating a complete ordering of possible worlds in terms of closeness may exceed cognitive limitations. An alternative possibility is that people generate counterfactuals about specific situations (worlds) by building models based on their existing knowledge of the causal relations that generally hold between events of the types that occur in that situation (Halpern & Pearl, 2001; Hitchcock, 2001; Pearl, 2000; Woodward, 2003). For example, they may generate a causal model showing that, in general, rain tends to cause rivers to flow and vegetation to grow. Each event in the model can have different values. For example, rain could have the value 'present' or 'absent'. When we change the value of an event then the values of any consequences of that event in the model can also change. The updated model can then be used to evaluate a counterfactual world. For example, if we change the value of rain to absent, then its consequences will also change so that we can infer that if there hadn't been rain, then the river wouldn't flow and vegetation wouldn't grow. In contrast the values of any antecedents in a model will remain unchanged. For instance, if we change the value of a river from flowing to dried up, then we don't change the values of rain and therefore we don't infer that a dry riverbed caused a shortage of rain. In sum then, on this view the knowledge of the general causal relations between events can be used to make causal attributions for a particular event.

9.1.1 Empirical evidence

Research on the role of counterfactuals in judging causation has addressed three main questions. First, it has examined whether a change to the possible counterfactual alternatives to an action leads to different judgments of the causal role of that action. It has also examined whether the availability of counterfactual alternatives to an event influences the extent to which the event is judged to have caused the outcome and, finally, whether the availability of counterfactual alternatives to an event influences the likelihood the event will be selected as 'the cause' of an ensuing event from a set of necessary conditions.

Do people consider counterfactual alternatives in their causal ratings?
One way to examine whether people use counterfactuals to make causal judgments is by comparing their responses to scenarios in which the facts are held constant and the counterfactual alternatives are manipulated. Wells & Gavanski, (1989) did this by presenting participants with a scenario based on a true story about a taxi driver who refused to take a fare from a paraplegic couple because he thought the taxi would be too crowded with them and their wheelchairs. Instead the couple take their own specially-equipped car. Unfortunately a bridge on their journey collapses shortly before they arrive and in the dark they drive off the bridge and drown. The question asked was the extent to which the taxi driver caused and was responsible for the outcome. In one version of the story the taxi driver made it safely across the bridge before it collapsed whereas in the other he did not. And as counterfactual theory predicts, participants judged him to be more causal and responsible in the former case where a change to his actions could have prevented the outcome (Wells & Gavanski, 1989). Similarly it has been shown that when people are presented with a scenario where the counterfactual outcome is omitted, causal ratings tend to be intermediate, that is, they are lower than when it is known that a change to the event would undo the outcome and higher than when it is known that a change to the event would leave the outcome unchanged (McCloy & Byrne, 2002). In these cases, causal ratings were sensitive to counterfactual alternatives.

Does the availability of counterfactuals influence causal judgments?
The studies described so far suggest that counterfactuals play an important role in judging the extent to which an event caused an outcome. Extensive evidence suggests that some counterfactuals are generated more readily than others. In this section, we examine whether this differential availability may affect causal judgments.

Despite the fact that in any situation there are an infinite number of counterfactuals, it turns out that there are considerable regularities in the kinds of alternatives that people imagine most readily (Kahneman & Tversky, 1982a). Taking Hofstadter's (1979) example, imagine running into a swarm of bees while driving down a country road. Some alternatives are likely to come to mind very readily, for example, if only I had closed my window. In contrast, others seem very unlikely, for example, if only those bees were dollar bills. Psychological studies indicate that people are more likely to mentally simulate counterfactual alternatives to some events than others. They imagine alternatives to exceptional events more than normal ones (Kahneman & Miller, 1986), actions more than inactions (e.g. Byrne & McEleney, 2000; Gilovich & Medvec, 1994; Kahneman & Tversky, 1982b), and controllable more than uncontrollable events (Girotto, Legrenzi, & Rizzo, 1991; McCloy & Byrne, 2000). The mutability of an event is also influenced by its relation to other events that happened. People tend to imagine alternatives to the first event in a causal chain (Wells, Taylor, & Turtle, 1987) but to the last event in a temporal sequence of events (Miller, 1990; Walsh & Byrne, 2004).

These findings lead to the hypothesis that events that have available counterfactual alternatives tend to be perceived as more causal. There is some evidence to support this. People more readily imagine counterfactual alternatives to actions than failures to act perhaps because inaction is the norm (Kahneman & Miller, 1986) or because it matches the pre-action state which people have already represented in mind (Byrne & McEleney, 2000). They also tend to perceive actions as more causal than failures to act even when both lead to the same outcome. Spranca, Minsk & Baron (1991) collected causal ratings for the actions of a tennis player John. The night before a match, his competitor Ivan is about to choose between two food dressings at dinner, one of which John knows contains an ingredient that Ivan is allergic to Ivan takes the wrong one and gets food poisoning. People judged John's actions as more causal when he recommended the wrong dressing to Ivan than when he failed to warn Ivan after he chose it. Other studies have used more indirect measures such as blame, victim compensation and perpetrator punishment for crimes committed. These also suggest that the availability of a counterfactual alternative that would undo the outcome leads to more extreme judgments. For example, when people are asked to recommend a punishment to a burglar who robs a home during the family's three month summer vacation, they recommend a harsher punishment when the burglary takes place the night before the family returns than when it takes places in the middle of the vacation. In the former case, it's easier to imagine that the family might have been there thus foiling the burglar (Macrae, Milne & Griffiths, 1993). Similarly, people tend to recommend harsher punishment to a mugger who attacks someone when they are on an unusual than a usual route (Macrae *et al.* 1993; see also Roese & Olson, 1996). The results suggest that counterfactuals are used to make these judgments of blame and punishment, and that people use the counterfactual that comes most readily to mind, perhaps because people have difficulty juggling multiple possibilities (Byrne, 2005). However, given that the measures in these latter studies were indirect, we can't be sure that they reflect effects of counterfactual availability on causation.

A number of studies have attempted to obtain a more direct measure of the impact of counterfactual availability on causation. These studies either provide participants with an explicit counterfactual alternative or get them to generate one and then examine its influence on causal judgments. Using response latency measures, Roese & Olson (1997) found that judging a counterfactual conditional, i.e. whether a change to event C would have led to a change in outcome E, increased the speed of response to a subsequent corresponding causal statement, whether C caused E. In contrast however, making a causal judgment first had no effect on the speed of judging counterfactual conditionals. The first result suggests that people do benefit from having an available counterfactual when making a causal judgment. However, the fact that the priming advantage is only in one direction suggests that causal judgments don't always involve the simulation of a counterfactual alternative. If they did, we would expect to find that judging causation should facilitate subsequent counterfactual judgments.

A similar pattern tends to be found in studies that examine whether making a counterfactual alternative available influences causal ratings. For instance in one study (Spellman & Kincannon, 2001), participants were presented with details of a court case about a drunk driver who hit another car and caused an accident. The victim refused a blood transfusion due to religious beliefs and died. Participants then heard either the prosecution argument about how a change to the defendant's actions could have avoided the outcome or the defence argument describing a number of alternative counterfactual possibilities such as a case where the victim accepted treatment or the defendant hit someone else and as a result no-one died. Participants who heard the prosecution's argument rated the defendant as more causal than those who heard the defence's argument (Spellman & Kincannon, 2001).

The same result occurs when individuals generate their own counterfactuals. For instance, after reading a date rape story, individuals who thought about how a change to the victim's behaviour could have changed the outcome tended to judge the victim as more at fault than individuals who thought about how a change to the victim's behaviour would have lead to the same outcome (Branscombe *et al.* 1996, see also McCloy & Byrne, 2002; Nario-Redmond, 1996 for similar effects). However, like the latency study, rating the extent to which a particular event C caused E generally has no influence on the likelihood that individuals will subsequently select that event as a means to undo the outcome (Wells & Gavanski, 1989). Increasing the salience of a counterfactual alternative influences causal judgments, suggesting once again that individuals don't generate all the relevant counterfactuals in rating how causal an event is. Instead, they may use only the most relevant ones. As a consequence, causal ratings tend to be higher for events that have readily available alternatives or when the relevant counterfactual is made explicit.

Does the availability of a counterfactual increase the likelihood that the corresponding actual event will be selected as 'the cause' of an outcome?

Given any outcome, there are usually multiple factors that were necessary to bring it about. But people usually pick out one event as 'the cause'. Another issue that researchers have therefore investigated is whether the availability of a counterfactual about a particular event increases the likelihood that that event will be chosen as the cause of the outcome. In one study, participants read a scenario about a man who takes a flight despite his wife's fear of flying and the fact that normally they never fly. His wife considers pleading with him to take the train instead, knowing that he would do it if she asked, but decides against it. Unfortunately, there is an engine failure and the plane crashes killing everyone on board. When participants were asked to imagine how the outcome could have been avoided they tended to focus on Mrs. Wallace's failure to encourage her husband to take the train. However when they are asked what caused the outcome they selected the engine failure (Mandel & Lehman, 1996; see also N'gbala & Branscombe, 1995). The results suggest that although the availability of a counterfactual alternative to a particular event may increase the perceived causal role of that event, counterfactual availability does not influence the likelihood that an event will be selected as 'the cause' from a set of necessary conditions.

Summary

Psychological studies have provided evidence that people use counterfactuals when attributing causation. When the facts are held constant, people tend to judge an event to be more causal if a change to that event would produce a change to the outcome. In addition, people spontaneously generate counterfactuals for some events more than others and those events with readily available counterfactual alternatives tend to elicit stronger ratings for judgments associated with causation, such as, blame, guilt, and victim compensation. Finally, although counterfactuals may be used to decide the extent to which a particular event caused an outcome, the availability of counterfactual alternatives does not seem to influence judgments about which one of a set of events should be judged as 'the cause'. People tend to select different events when asked to imagine what caused an outcome and how it could have happened differently.

9.2 Generative theories of causation

An alternative view to counterfactual theories is that a cause is something that generates an outcome. This idea can be traced back to Kant (1781) and can be found in current philosophical (Dowe, 2000; Salmon, 1984) and psychological theories (Schultz, 1982). According to this view, causation involves a mechanism or causal process that links cause and effect. There is a transmission along a causal pathway and an exchange of some conserved quantity such as energy or momentum (Dowe, 2000; Salmon, 1997). For example, the reason that we believe that throwing the stone caused the window to break is that a force is transferred from the stone to the window.

9.2.1 Empirical evidence

One striking piece of evidence that causation is not always inferred from repeated associations came from the work of Michotte (1946/1963). He demonstrated that in some cases, causation is directly perceived. In his demonstrations, Michotte used images, such as moving circles to represent billiard balls. In the classic example, one circle moves across the screen, it stops when it comes in contact with another circle and at that moment the second circle moves off in the same direction. In this demonstration, people generally go beyond these images when describing what they see and infer that the first circle hit the second circle and set it in motion. Michotte showed that at least in some cases, causation can be directly perceived and hence that it can be inferred from a single observation. A number of other methods have been used to examine the role of mechanisms in causal attribution.

What kinds of information do people search for when attributing causation?
One way to examine whether people rely on mechanisms to decide whether one event caused another is to examine what information they search for in trying to discover the cause of an outcome. Ahn *et al.* (1995) presented participants with events such as 'John had a car accident on Route 7 yesterday' and allowed them to ask questions in order to establish the cause. If people rely on information about the regular co-occurrence of cause and effect, they might ask questions about events at other times and places. For example, they could ask whether John often has accidents, whether there are frequently accidents on this road, or whether there were many accidents in other places yesterday. In contrast, if people rely on an understanding of the generative processes to establish the cause of an event, then they may search information about the underlying mechanism such as whether there was a fault with the car or whether there was bad weather that caused hazardous driving conditions. The results showed that the majority of questions sought out mechanisms for how the event occurred. In attributing causation, people aim to discover not what has happened on previous occasions but rather how the event came about on this occasion.

Do people use mechanistic knowledge when attributing causation?
A number of studies have examined whether children and adults use knowledge of the kinds of mechanisms that can mediate a cause–effect relation to attribute causation. In one such study (Bullock, Gelman, & Baillargeon, 1982), individuals were presented with a box with two windows and a jack-in-the-box at the end. A ball rolled along one window whereas a light passed simultaneously down the other. When they reached the end, a jack popped out of the box. Both adults and young children judged that the ball caused the jack to pop out presumably because they knew that a ball can produce movement of an object through physical contact. When the box with the windows was disconnected from the jack-in-the-box so that there was a gap between them, people tended to switch their judgments. It was no longer possible for the ball to hit the jack. In contrast, electrical mechanisms are often hidden and appear

to occur at a distance. Only 3 year olds continued to attribute causation to the ball. Adults and older children (4 and 5 years of age) attributed causation to the light in this case, showing that when they make causal judgments they take into account the kinds of mechanisms that can mediate between a cause and effect.

When mechanistic and alternative cues are available, which do people rely on to attribute causation?

In order to examine which cues people draw on most in attributing causation, a number of studies have presented individuals with an outcome preceded by multiple antecedents each of which is supported by a different cue to causation. For instance, Ahn *et al.* (1995) gave participants descriptions of events such as 'Kim had a traffic accident last night' and two possible factors, one supported by mechanism information, e.g. 'Kim is near-sighted and tends not to wear her glasses when driving' and one supported by co-variation information, e.g. 'Traffic accidents were more likely last night than on other nights'. Participants gave higher ratings of causal strength to factors that were supported by mechanism information. Schultz (1982) used a similar method to see which cues young children prefer to use. He showed the children events such as a spotlight appearing on the wall and their task was to decide which of two lamps caused the light. One of the lamps actually generated the spotlight whereas the other did not but was supported by other causally relevant cues. For instance in one condition, when the green lamp generated the spotlight, switching on the yellow lamp (but not the green one) was temporally contiguous with the appearance of the spotlight. In another condition, the yellow lamp was supported by a spatial contiguity cue (it actually touched the spotlight). In a final condition, switching on the yellow lamp only co-varied with the appearance of the spotlight on a number of earlier trials. Children ranging in age from 3 to 13 predominantly used generative transmission to attribute causation. The only exception was that 3 year old children preferred to use spatial contiguity over generative transmission.

Summary

Evidence generally supports the view that people use generative transmission to attribute causation. They readily perceive causation when two objects interact. When attempting to discover a cause, they are more likely to search for information about possible mechanisms than for covariation information. They are sensitive to what mechanisms could possibly produce an effect. For example, they are more likely to attribute causation to an electrical mechanism than a direct physical force when there is a spatial gap between the cause and effect. And when forced to choose between events that are supported by different causal cues, they tend to attribute causation to the event that generated the outcome.

9.3 Challenges for counterfactual and generative theories of causation

9.3.1 Pre-emption

Situations that involve over-determination or pre-emption provide a challenge for counterfactual theories in their original form. In these cases, there is more than one cause which is sufficient for an outcome hence undoing either one will not undo the outcome. In the classic example, two individuals, Billy and Suzy, throw rocks at a bottle. Suzy throws seconds before Billy so it is her rock that hits the bottle. But if she hadn't thrown, Billy's rock would have hit the bottle and so the outcome would be unchanged. These cases have received a lot of attention from counterfactual theorists and two recent modifications to the theory have been proposed. Lewis's (2000) reformulation solves the problem of pre-emption by proposing that the degree of causal influence reflects the extent to which *whether, when and how* one event happens depends on whether, when and how another event happens. The modified theory states that C causally influences E if an alteration to C would have led to an alteration of E. For instance, if Suzy's throw had been harder, then the shattering would have been different whereas a change to Billy's throw would make no difference.

Another solution was proposed by Halpern & Pearl (2001; see also Hitchcock, 2001). According to this account, people mentally run a counterfactual simulation while holding certain events constant. For instance, if we hold constant the fact that Billy's rock didn't hit the bottle, then if Suzy hadn't thrown the bottle wouldn't have broken.

Pre-emption tends to be unproblematic for generative theories of causation. According to these accounts causation is attributed to the event that passes a conserved quantity along a path that culminates in the outcome. In the example above, Suzy will be judged to have caused the outcome because it is the force of her rock interacting with the bottle that produced the shattering.

9.3.2 The attribution of prevention

Although the literature on causation is quite extensive, much less attention has been paid to how people attribute prevention. Counterfactual theories of causation can easily be extended to deal with cases of prevention. To prevent something may mean the same as to cause it not to occur (Collins, 2000). Lewis (2000) treats double prevention as a case of causation. In other words, if A prevents B and B prevents C, then we should say that A causes C. According to this view, we can say that P prevents E provided that if P hadn't occurred, then E would have. For instance, if Billy hadn't caught the ball, then the window would have broken.

Prevention is a bigger challenge for generative theories because in general prevention involves the absence of an outcome. Absences are difficult for generative accounts because nothing is transferred from the preventer to the target outcome. Dowe (2000) developed a generative theory of prevention that defines prevention as qualitatively different from causation. He proposed that A prevented B if and only if A occurred and B did not, there was a causal interaction between A and another process X, and if A hadn't occurred then X would have caused B. For example, Jack prevented the child from being hit by a car if and only if (i) he grabbed the child and she wasn't hit by the car and (ii) if he hadn't grabbed her, she would have been hit by the car. Unlike causation which relies on purely generative processes, Dowe (2000) classes prevention as a form of quasi-causation because of its reliance on non-generative processes. Similarly, Wolff, Barbey & Hauskneckt (2010) define prevention as a removal of a force that had a tendency to produce an outcome. There are two things to note about these accounts. First, these definitions include a counterfactual element. Hence, they need to deal with any challenges faced by the counterfactual theory such as cases of pre-emption. Second, the theories assume that the meanings of causation and prevention need not be symmetrical. In other words, *to prevent E* means something different from *to cause not E*. Below, we describe a study that we ran to test whether this is the case.

9.4 Studies of causation and prevention

9.4.1 Experiment 1: Do people attribute causation and prevention in the same way?

In order to provide an initial test of whether people attribute causation and prevention in the same way, we generated two scenarios based on classic examples of late pre-emption (cf. Hall & Paul, 2003). Although studies have been carried out to examine how people attribute causation when there are two sufficient causes (Mandel, 2003; Spellman & Kincannon, 2001), we know of no empirical studies that have tested how people respond to cases of pre-emptive prevention.

Consider the following two scenarios, one of which involves pre-emptive causation:

There is a bottle on the wall. Billy and Suzy are standing close by with stones and each one throws a stone at the bottle. Their throws are perfectly on target. Billy happens to throw first and his reaches the bottle before Suzy's. The bottle breaks.

and one of which had a similar structure to the first except that the links were preventive:

There is a bottle on the wall. Frank and Jane are standing close by. While they are there someone else aims to throw a ball at the bottle. The aim is perfectly on target. Frank and Jane both step in front of the bottle. Frank happens to step in front of Jane and catches the ball. The bottle doesn't break.

The 'causation scenario' included an actual mechanism going from cause to effect, that is, a stone is thrown, it hits a bottle and the bottle breaks. It was followed by two questions which asked whether each actor caused the outcome:

Did Billy cause the bottle to break?
Did Suzy cause the bottle to break?

In contrast, in the 'prevention scenario' the actions involved an interruption to a causal mechanism; that is, Frank stops the ball and as a result the ball doesn't hit the bottle. Again, the scenario was followed by two questions, this time based on prevention:

Did Frank prevent the bottle from breaking?
Did Jane prevent the bottle from breaking?

The counterfactual theories developed by Lewis, (2000) and Halpern & Pearl (2001) were designed to ascribe causation to Billy and not Suzy. On the assumption that 'A prevents B' means the same as 'A causes not B', the theories make the parallel prediction for prevention, ascribing prevention to Frank and not Jane.

According to generative mechanism theories, people should also ascribe causation to Billy but not Suzy. In this case, there is a clear mechanism linking the action (Billy's throw) to the outcome (the bottle breaking). On Dowe's (2000) account, people should also ascribe prevention to Frank but not Jane, although the reason is different. In this case, the action (Frank's catch) interacts with the marble and it is natural to infer that if Frank had not caught the marble, it would have hit the bottle and broken it. Hence in this study, both theories make the same predictions. Our aim was to provide an initial test of whether people's attributions are consistent with these predictions.

One hundred individuals read the scenarios described above and responded to the questions by answering 'yes' or 'no'. As Table 9.1 shows, the results for the causation scenario strongly corroborate the predictions of the mechanism and counterfactual views. A large majority (90%) attributed causation to Billy but not Suzy.

The results for the prevention scenario are much less clear cut. The majority of participants also attributed prevention to the first actor only, i.e. Frank but not Jane (60%). However, the percentage of participants who made an attribution to the first actor only was significantly lower in the prevention scenario than in the causation scenario (McNemar Test, $p < 0.001$). For the prevention

Table 9.1 The percentage of 'yes' responses to the four questions in Experiment 1.

	Causation scenario	Prevention scenario
First actor total *(Billy / Frank)*	95	83
First actor only	*90*	*60*
Both actors	*5*	*23*
Second actor total *(Suzy / Jane)*	6	29
Second actor only	*1*	*6*
Both actors	*5*	*23*
Neither	4	11

scenario, a large minority (23%) ascribed prevention to both actors. This result is not predicted by Dowe's mechanism theory or by recent counterfactual theories which would predict that only the first actor prevented the outcome. One explanation for this result is that some people may understand prevention not just as an actual interruption to a causal mechanism (Frank actually caught the ball) but rather as a potential interruption to that mechanism (Jane would have caught the ball). The ball would have transmitted a force breaking the bottle if Frank and Jane had not intervened to interrupt that process. People often talk about prevention in the sense of having the potential to block some event even if that event does not occur (e.g. the lock is preventing the bike from being stolen). The result also suggests that although people tend to select one event as being 'the cause' of an outcome, they may be less likely to do so for prevention.

9.4.2 Experiment 2: Comparing causation and prevention when a mechanism is unblocked

Our first study suggests that people may think differently about cases of causation and prevention. However, the scenarios differed in two ways. First, in one case we asked questions about causation whereas in the other we asked about prevention. Second, the causation scenario involved a continuous mechanism linking cause and effect whereas the prevention scenario involved an event that interrupted a mechanism. In order to test whether the different results really depend on a difference in the meaning of causation and prevention, we carried out a study in which we collected measures of both using the same scenario.

Our second objective was to pit counterfactual against generative theories of causation, hence we constructed a scenario for which the two theories make different predictions. We examined attributions to actions that unblocked a causal pathway. There was no causal process linking the actions to the outcome, but they were nevertheless responsible for a change to the outcome:

There is a coin wobbling on edge at the end of the table. It is about to fall over and land on tails. There is a book directly in front of the coin. Max and Anne are standing close by. While they are there someone else rolls a marble toward the coin. The roll is perfectly on target and in the absence of the book it will hit the coin, knock it over and the coin will land on heads. Max and Anne both reach out to lift the book. Max happens to reach in front of Anne and he lifts the book. The marble hits the coin, and the coin falls over and lands on heads.

After reading the scenario, participants answered the following four questions.

Did Max cause the coin to land on heads?
Did Anne cause the coin to land on heads?
Did Max prevent the coin from landing on tails?
Did Anne prevent the coin from landing on tails?

If attributions of cause and prevent depend on a change to the outcome and not on a causal mechanism, then we expect people to attribute both to the first actor (Max) in this scenario. In contrast, if a mechanism from cause to effect is important in attributing causation, then we expect that people should not attribute causation to Max. However, they should judge that he prevented the outcome. Consistent with Dowe's (2000) definition of prevention, Max interacted with the book and if he hadn't, then the coin would have landed on tails.

Sixty-eight participants read the scenario and responded to each of the four questions by answering 'yes' or 'no'. The results are set out in Table 9.2. Our first aim was to compare attributions of causation and prevention. As Dowe (2000) predicted, Max was judged to have prevented the outcome (53%) more often than to have caused it (38%). Nonetheless, this time a significant minority (41%) attributed prevention to neither actor. This result contrasts with the previous experiment, where the largest minority attributed prevention to both actors. The main difference between the two studies is that in the previous study the prevention scenario ended with no change of state, the bottle remained unbroken, whereas in this experiment the scenarios end with an outcome different from the original state, i.e. the coin was spinning but in the end lands on heads. In this case, the outcome may be attributed to a different mechanism, namely the spinning of the coin, and hence the actors may be perceived to have played a lesser role and are thus less likely to be assigned any kind of causal role. This latter finding is once again inconsistent with both Dowe's theory and counterfactual theories that have been adapted to deal with preemption (Halpern & Pearl, 2000; Hitchcock, 2001; Lewis, 2000). Also, in contrast to causal judgments, when people make prevention judgments they are less likely to distinguish between the preempting cause and the preempted back-up.

Our second objective was to examine whether people attributed causation to Max, as counterfactual theories predict, or not, as generative theories predict. As Table 9.2 shows, the majority of participants did not attribute causation to

Table 9.2 The percentage of 'yes' responses to the eight questions in Experiment 2.

	Mechanism unblocked	
	Cause question	Prevent question
First actor total *(Max)*	38	53
First actor only	*34*	*47*
Both actors	*4*	*6*
Second actor total *(Suzy / Jane)*	4	12
Second actor only	*0*	*6*
Both actors	*4*	*6*
Neither	62	41

him (62%). The result supports the predictions made by generative theories. In contrast, the result is inconsistent with counterfactual theories (Halpern & Pearl, 2000; Hitchcock, 2001; Lewis, 2000) because a change to Max's behaviour would have led to a change in the outcome. When there is a change to the default outcome, in the absence of a mechanism, people are less likely to make a causal inference. In other studies using the same coin toss scenario, we have shown that when a mechanism is present (i.e. a rolling marble changes the way that the coin falls), then the majority of people attribute causation to it (Walsh & Sloman, in press).

9.5 Conclusions

There is evidence to suggest that both counterfactuals and mechanisms can influence causal judgments. Counterfactual theory is supported by the finding that when a change to an event leads to a change in the outcome, people rate it as more causal than when a change to the event would not undo the outcome. Nonetheless, people generate counterfactuals for some events more readily than others and as a consequence these events tend to be assigned a greater causal role.

There is also evidence to support generative theories. When trying to discover the cause of an event, people tend to spontaneously search for mechanistic information. In addition, when co-variation and mechanistic information are pitted against one another, people prefer to rely on mechanistic information than co-variation.

In most situations, counterfactuals and causal processes point to the same causal judgment. One case where they make different predictions is when an event blocks or unblocks a mechanism. For example, the scenario used in our second experiment described the removal of a book that is blocking a causal pathway between a marble and a coin and as a consequence, the coin falls on heads instead of tails. Counterfactual theory suggests that the action caused the outcome. However, there is no causal process linking the removal of the book and the coin's fall. Hence, generative theory predicts that the action is not causal. The results of our study show that the majority of participants make judgments consistent with generative theory. At least in some cases, people rely on an analysis of mechanisms more readily than on counterfactuals. Of course in many real life situations, people do not know how the mechanisms work (Rozenblit & Keil, 2002). However, it is likely that people need only believe that a mechanism is possible (Ahn and Kalish, 2000; Walsh & Sloman, in press). Our study suggests that once it is clear that there is no mechanism linking an event to an outcome, people tend not to attribute causation to that event, even if it changes the outcome.

The results of our studies include some novel findings regarding judgments of prevention. First, compared to causation, there is less consensus regarding whether or not an event prevented an outcome (see also Collins, 2000) and so it is unsurprising that no theory can explain all the results. Second, our results show that the meanings of causation and prevention are not symmetrical. To prevent something does not always mean the same thing as to cause it not to occur. When an action unblocks a causal pathway, individuals are more likely to judge the actor to have prevented the outcome than to have caused it. The result is predicted by Dowe's generative theory of prevention. Finally, people are less likely to distinguish between preempting and preempted events when making judgments about prevention than about causation.

Causation focuses people's attention on the facts. It asks people to make judgments about how the actual outcome came about. Hence, most people rely on an analysis of the mechanisms that brought that event about. This may also explain why people often rely on counterfactuals. Counterfactual possibilities are also determined by the mechanisms governing the actual event; the counterfactual arises when we imagine the same mechanisms to have different antecedent conditions. As a result, counterfactuals can inform us about which mechanisms are at play. People seem to be more likely to rely on a counterfactual if it is made explicit or if it comes to mind readily. Prevention, on the other hand, focuses people's attention on the counterfactual alternatives. It asks people to make a judgment about an event that might have happened. Prevention questions therefore force people to think about how things might have happened differently. One reason that judgments of prevention are less clear cut may be that people consider different possibilities and these give rise to different judgments. In addition, if people consider multiple possibilities, it may explain why they are less likely to pick out a single event as having prevented an outcome.

Causal judgments are ubiquitous in our everyday thinking as well as in domains ranging from science to the law. We suggest some steps toward the development of an understanding of this fundamental process. Future theories may benefit from considering how people make judgments of prevention as well as causation.

Acknowledgements

This work was supported by NSF Award 0518147 to Steven Sloman.

References

Ahn, W. & Kalish, C. (2000). The role of covariation vs. mechanism information in causal attribution. In R. Wilson & F. Keil (eds.) *Cognition and Explanation*, Cambridge, MA: MIT Press.

Ahn, W., Kalish, C.W., Medin, D.L., & Gelman, S.A. (1995). The role of covariation versus mechanism information in causal attribution. *Cognition, 54*, 299–352.

Branscombe, N.R., Owen, S., Garstka, T. & Coleman, J. (1996). Rape and accident counterfactuals: Who might have done otherwise and would it have changed the outcome? *Journal of Applied Social Psychology, 26*, 1042–1067.

Byrne, R.M.J. (2005). *The Rational Imagination: How People Create Alternatives to Reality*. Cambridge, MA: MIT Press.

Byrne, R.M.J. & McEleney, A. (2000). Counterfactual thinking about actions and failures to act. *Journal of Experimental Psychology: Learning, Memory, and Cognition, 26*, 1318–1331.

Bullock, M., Gelman, R., & Baillargeon, R. (1982). The development of causal reasoning. In W.J. Friedman (ed.), *The Developmental Psychology of Time*, pp. 209–254. New York: Academic Press.

Cheng, R.W. & Novick, L.R. (1990). A probabilistic contrast model of causal induction. *Journal of Personality and Social Psychology, 58*, 545–567.

Collins, J. (2000). Preemptive Prevention. *Journal of Philosophy, 97*, 223–34

Dowe, P. (2000). *Physical Causation*. New York: Cambridge University Press.

Gilovich, T. & Medvec, V.H. (1994). The temporal pattern to the experience of regret. *Journal of Personality and Social Psychology, 67*, 357–365.

Girotto, V., Legrenzi, P., & Rizzo, A. (1991). Event controllability in counterfactual thinking. *Acta Psychologica, 78*, 111–133.

Hall, N. & Paul, L.A. (2003). Causation and preemption. In P. Clark & K. Hawley (eds.), *Philosophy of Science Today*, pp. 100–130. Oxford: Oxford University Press.

Halpern, J.Y. & Pearl J. (2001). Causes and explanations: A structural-model approach – Part I: Causes. In *Proceedings of the Seventeenth Conference on Uncertainty in Artificial Intelligence*, pp. 194–202. San Francisco, CA: Morgan Kaufmann.

Hitchcock, C. (2001). The intransitivity of causation revealed in equations and graphs. *Journal of Philosophy, 98*, 273–299.

Hume, D. (1988). *An Enquiry Concerning Human Understanding*. A. Flew (ed.) La Salle, IL: Open Court. (Originally published 1748.)

Hofstadter, D.R. (1979). *Gödel, Escher, Bach: An Eternal Golden Braid.* New York: Basic Books

Kahneman, D. & Miller, D.T. (1986). Norm theory: Comparing reality to its alternatives. *Psychological Review, 93,* 136–153.

Kahneman, D. & Tversky, A. (1982a). The simulation heuristic. In D. Kahneman, P. Slovic, & A. Tversky (eds.), *Judgement under Uncertainty: Heuristics and Biases,* pp. 201–208. New York: Cambridge University Press.

Kahneman, D. & Tversky, A. (1982b). The psychology of preferences. *Scientific American, 246,* 160–173.

Kant, I. (1781). *The Critique of Pure Reason.* 1985 Web version by Palgrave Macmillan.

Kelley, H.H. (1973). The processes of causal attribution. *American Psychologist, 28,* 107–128.

Lewis, D. (1973). *Counterfactuals.* Oxford: Blackwell.

Lewis, D. (2000). Causation as influence. In J. Collins, N. Hall, and L.A. Paul (eds.), *Causation and Counterfactuals.* Cambridge: MIT Press.

Macrae, C.N., Milne, A.B., & Griffiths, R.J. (1993). Counterfactual thinking and the perception of criminal behaviour. *British Journal of Psychology, 84,* 221–226.

Mandel, D.R. (2003). Judgment dissociation theory: An analysis of differences in causal, counterfactual, and covariational reasoning. *Journal of Experimental Psychology: General, 132,* 419–434.

Mandel, D.R. & Lehman, D.R. (1996). Counterfactual thinking and ascriptions of cause and preventability. *Journal of Personality and Social Psychology, 70,* 450–463.

McCloy, R. & Byrne, R.M.J. (2000). Counterfactual thinking about controllable events.*Memory and Cognition, 28,* 1071–1078.

McCloy, R. & Byrne, R.M.J. (2002). Semifactual 'even if' thinking. *Thinking & Reasoning, 8,* 41–67.

Michotte, A. (1946). *The Perception of Causality* (1963 translation by T. & E. Miles). London: Methuen.

Nario-Redmond, M.R. & Branscombe, N.R. (1995). It could have been better or it might have been worse. Implications for blame assignment in rape cases. *Basic and Applied Social Psychology, 18,* 347–366.

N'gbala, A. & Branscombe, N.R. (1995). Mental simulation and causal attribution: When simulating an event does not affect fault assignment. *Journal of Experimental Social Psychology, 31,* 139–162.

Pearl, J. (2000). *Causality: Models, Reasoning and Inference.* Cambridge: Cambridge University Press.

Roese, N.J. & Olson, J.M. (1996). Counterfactuals, causal attributions, and the hindsight bias: A conceptual integration.*Journal of Experimental Social Psychology, 32,* 197–227.

Roese, N. J., & Olson, J.M. (1997). Counterfactual thinking: The intersection of affect and function. *Advances in Experimental Social Psychology, 29,* 1–59.

Rozenblit, L.R. & Keil, F.C. (2002). The misunderstood limits of folk science: an illusion of explanatory depth. *Cognitive Science, 26,* 521–562.

Salmon, W. (1984). *Scientific Explanation and the Causal Structure of the World.* Princeton: Princeton University Press.

Salmon, W. (1997). Causality and explanation: A reply to two critiques. *Philosophy of Science, 64,* 461–477.

Shultz, T.R. (1982). Rules of causal attribution. *Monographs of the Society for Research in Child Development, 47,* 1–51.

Spellman, B. A. & Kincannon, A. (2001). The relation between counterfactual ('but for') and causal reasoning: Experimental findings and implications for jurors' decisions. *Law and Contemporary Problems: Causation in Law and Science, 64,* 241–264.

Spranca, M., Minsk, E., & Baron, J. (1991). Omission and commission in judgment and choice. *Journal of Experimental Social Psychology, 27,* 76–105.

Walsh, C.R. & Byrne, R.M.J. (2004). Counterfactual thinking and the temporal order effect.*Memory & Cognition, 32,* 369–378.

Walsh C.R. & Johnson-Laird, P.N. (2009). A change of mind. *Memory & Cognition, 37,* 624–631.

Walsh, C.R. & Sloman, S.A. (in press). The meaning of cause and prevent: The role of causal mechanism. *Mind and Language.*

Wells, G.L., & Gavanski, I. (1989). Mental simulation of causality. *Journal of Personality and Social Psychology, 56,* 161–169.

Wells, G.L., Taylor, B.R. & Turtle, J.W. (1987). The undoing of scenarios.*Journal of Personality and Social Psychology, 53,* 421–430.

Wolff, P. Barbey, A., & Hauskneckt, M. (2010). For want of a nail: How absences cause events. *Journal of Experimental psychology:* General, 139(2): 191–221.

Woodward, J. (2003). *Making Things Happen: A Theory of Causal Explanation.* New York: Oxford University Press.

·10·

The autonomy of psychology in the age of neuroscience

Ken Aizawa and Carl Gillett

Abstract

The implications of multiple realization for scientific methodology have recently been hotly debated. For example, neuroscientists have discovered distinct realizations for what appears to be a single psychological property and some philosophers have recently maintained that in such cases scientists will always abandon commitment to the single, multiply realized psychological property in favour of two, or more, uniquely realized psychological properties. In this chapter, we explore such methodological claims by building on the dimensioned theory of realization and a companion theory of multiple realization. Using concrete cases, we show that such an 'eliminate-and-split' methodology is not always the case in actual practice. Furthermore, we also establish that whether scientists postulate unique or multiple realizations is not determined by the neuroscience alone, but only in concert with the psychological theory under examination. Thus, in a sense we articulate, in the splitting or non-splitting of properties, psychology enjoys a kind of autonomy from neuroscience.

Suppose that scientists discover a high level property G that is *prima facie* multiply realized by two sets of lower level properties, $F_1, F_2, \ldots, F_n$, and $F^*_1, F^*_2, \ldots, F^*_m$. One response would be to take this situation at face value and conclude that G is in fact so multiply realized. A second response, however, would be to eliminate the property G and instead hypothesize subtypes of G, G_1 and G_2, and say that G_1 is uniquely realized by $F_1, F_2, \ldots, F_n$, and that G_2 is uniquely realized by $F^*_1, F^*_2, \ldots, F^*_m$. This second response would eliminate a multiply realized property in favour of two uniquely realized properties.[1]

Clearly these are two logically possible responses to this type of situation, so when faced with it how do scientists respond in real cases? This is a matter of providing a descriptively adequate account of actual scientific practice. In support of the view that scientists opt for the 'eliminate-and-split' strategy, one might propose that it is illustrated by the way scientists responded in the

[1] A third possible scientific strategy would be to keep G and add subtypes G_1 and G_2. This strategy would leave G to be multiply realized, which would make it useless for blocking multiple realization.

case of memory. Once upon a time, it was thought that there existed a single kind of memory. With the advance of science, however, it was discovered that it is possible to perform certain sorts of brain lesions that would lead to the selective loss of certain memory functions, while certain other sorts of brain lesions would lead to selective loss of certain other memory functions. These neurobiological dissociation experiments, one might say, support the view that, instead of a single overarching type of memory, there are distinct subtypes of memory, procedural memory and declarative memory. Thus, generalizing from this example, it might be suggested that the eliminate-and-split strategy is always the approach of scientists in such cases.

We believe that this argument is based upon serious oversimplifications. To begin with, note that there is the assumption that scientists treat all discoveries about differences in realizers in the same way. We contend, however, that actual practice is far more complicated than this. For one thing, realistic biological and psychological cases typically have a greater diversity of lower realizing properties than is commonly appreciated. Consequently, discoveries about differences in lower level realizers might be expected to interact in a variety of different ways with the higher level realized properties. Once we take this last possibility seriously, we contend that one finds that in actual scientific practice not all discoveries about differences in realizing properties influence higher level theory in the same way. In particular, scientists do not uniformly adopt the eliminate-and-split strategy. As we show by reference to actual examples, discoveries about different lower level realizers *are* handled in different ways depending upon the nature of the higher level theory.

By considering actual cases, we show that finding variations in some lower level realizers, say, $F_1, F_2, \ldots, F_i$, sometimes leads scientists to posit differences in the higher level realized property G, thus following the eliminate-and-split strategy. But in other cases even though scientists find variation in other realizers, say $F_j, \ldots, F_n$, such differences do not lead the scientists to posit differences in higher level properties. To speak somewhat metaphorically, we might label the former sort of realizers 'parallel realizers', since findings about differences in the lower level realizers give rise to scientists positing parallel differences in our theories about the higher level realized property. We might then label the latter sort of realizers 'orthogonal realizers' because differences among them do not lead researchers to change their theories about the higher level property. The idea behind the name for these realizers is, therefore, that differences among them are, in some sense, orthogonal to the higher level account. Such examples show that scientists do not simply follow an eliminate-and-split strategy. Perhaps more importantly, these cases show that, although psychology takes account of neuroscience, the details of *how* it does this are determined by the needs of psychological theorizing in partnership with lower level theories.

Our cases also reveal that even the distinction between cases involving parallel and orthogonal realizers fails to do justice to all the nuances of actual scientific practice. For we show that in some cases in biology and psychology, discoveries about differences in lower level realizers lead scientists to posit what they term 'individual differences' in the *same* higher level property of subjects. These examples indicate that even the distinction between orthogonal and parallel realizers needs to be amended in still further ways. On the one hand, the realizers that give rise to individual difference are not orthogonal realizers, since discoveries about variations in them does lead to changes in our higher level theories about the realized properties. But, on the other hand, the realizers that give rise to individual differences affect higher level theories about realized properties in a manner distinct from the eliminate-and-split strategy. For scientists continue to posit the same higher level property, though distinguishing variations within it. We therefore also distinguish between two kinds of parallel realizers, 'strong' and 'weak', in the following ways. We have *strong* parallel realizers in cases where differences in these realizers prompt scientists to eliminate the original realized property and posit distinct realized properties for the different realizers. Thus strong parallel realizers underpin the eliminate-and-split approach. However, we also sometimes have *weak* parallel realizers in examples where the variations in realizers do prompt revisions in our higher level accounts about the realized property, but where scientists posit individual differences within the same realized property.

Even when variations in realizers prompt changes in our higher level theories, we show that such revisions do not always follow the eliminate-and-split model. Once again, we also show that the nature of the higher level theory plays a key role in whether scientists take parallel realizers to be weak or strong. Thus the autonomy of psychology in the age of neuroscience is, in part, a kind of methodological, rather than ontological autonomy. Psychological theory shapes how psychology accommodates the discovery of differences in neuroscientific realizers in partnership with lower level theories, rather than the lower level theories simply necessarily dictating changes through their discoveries.

Throughout the chapter we focus directly upon the case of properties and their instances, but our work has obvious consequences for causal processes and our theorizing about them. Like most other writers in the metaphysics of science we endorse the causal theory of properties and take properties to be individuated by their contributions of powers.[2] Putting things crudely, processes in the sciences are grounded by the manifestation of the powers contributed to individuals by such properties and we therefore plausibly have

[2] We thus endorse a weakened version of the theory defended by Shoemaker (1980) under which in the actual world all instances of a property contribute the same powers under the same conditions.

different kinds of process where we have different properties and powers. Consequently, competing views over the implications of discoveries about multiple realization for the diversity of higher level properties also have implications for the diversity of higher level processes, too. For example, the eliminate-and-split strategy entails that we increase the kinds of higher level process we accept when we discover cases of multiple realization, since it claims we increase the number of higher level properties we posit. Although we do not explicitly focus on the implications for processes, our critical work therefore also shows that such claims about higher level processes are also too simple and too quick because they fail to reflect the nuances of actual practices.

To articulate and defend these views, in Section 10.1 we briefly review the dimensioned view of realization and a theory of multiple realization that naturally and elegantly accompanies it. The remainder of the chapter then draws attention to the ways in which current scientific research treats the properties of the eye that realize normal human colour vision. This research is extremely useful for the study of realization and multiple realization, since scientists have a relatively firm grasp of the natures of both realizer and realized properties at multiple levels. Section 10.2 reviews some of the basic features of colour processing in the eye. This sets the stage for Section 10.3 where we consider three examples from the sciences that illuminate the various ways in which discoveries concerning lower level realizers do, or do not, influence the properties postulated in higher levels. Finally, Section 10.4 examines the realization of procedural and declarative memory as analysed by Craver (2004) and shows how the morals developed in the preceding sections bear on this example.

10.1 The dimensioned view of realization and a theory of multiple realization

As the dimensioned view of realization and its companion theory of multiple realization have been described and defended in detail in numerous other publications, only a brief review of them will be presented here.[3] The core idea of the dimensioned view of realization is that, typically, many lower level property instances will together realize an instance of a higher level property. The official statement of the view is that

[3] The dimensioned view is introduced and defended by Gillett (2002, 2003). It is combined with a theory of multiple realization and applied to various neuroscientific and psychological examples in Aizawa and Gillett (2009a, 2009b, Unpublished). Those who reject our theories of realization and multiple realization might read what follows more restrictively as simply articulating what we take to be some of the implications of this combination of views.

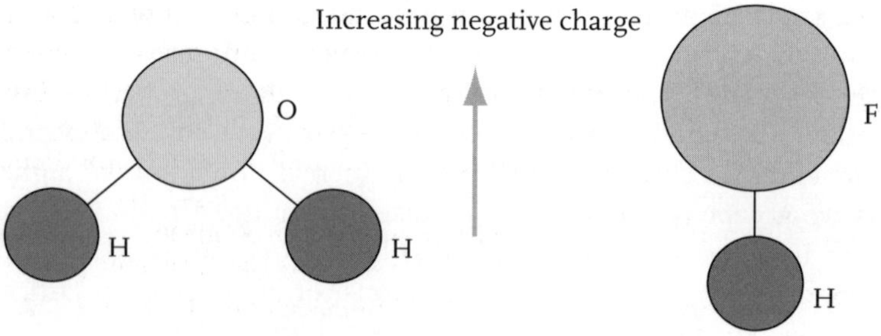

Fig. 10.1 The polarity of water and hydrogen fluoride.

> Property/relation instance(s) F_1–F_n realize an instance of a property G, in an individual *s* under conditions \$, *if and only if*, under \$, F_1–F_n together contribute powers, to *s* or *s*'s part(s)/constituent(s), in virtue of which *s* has powers that are individuative of an instance of G, but not vice versa.

So, to take a very simple example from chemistry, let *s* be an individual water molecule with the property G of being polar, i.e. more negatively charged in one direction than in another. (See Figure 10.1.) What makes a water molecule polar? It has to do with the greater electronegativity of oxygen versus hydrogen along with the angle of the bond between the two hydrogen atoms and the oxygen atom. The two instances of the hydrogen's electronegativity of 2.2 on the Pauling scale, the one instance of the oxygen's electronegativity of 3.44 on the Pauling scale, and the bond angle of 105° between the two hydrogen bonds leads electrons to cluster nearer the oxygen atom, hence for the 'oxygen side' of the molecule to be more negative where the 'hydrogen side' of the molecule is more positive. These facts can be inserted in the schema above in the obvious way.[4]

The core idea of multiple realization is that one must have instances of one set of properties F_1–F_n that realizes an instance of G and another set of instances of distinct properties F_1^*–F_m^* that realizes another instance of G and that these properties are not identical.[5] Things are not *that* simple, however, since one does not count the realization of, say, pain at the neuronal level and at the biochemical level as multiple realizations of pain. One must add that

[4] It is sometimes held against the dimensioned view that it appeals to property instances, rather than simply properties, and that it is overly technical on this score. In the example of polarity, however, we can see quite clearly how one really needs to appeal to the number of instances of the property of having an electronegativity of 2.2 – rather than merely to the property of having an electronegativity of 2.2 – in order to explain the realization of the polarity of a water molecule.

[5] Note that we focus throughout on the multiple realization of properties through the differential realization of their instances. However, we should note that a single instance in a certain individual may also be multiply realized over time. Having noted this possibility we leave it to one side in order to focus on the more usual case of the multiple realization of properties.

the two distinct realizers that multiply realize G must be at the same level. The official formulation of multiple realization is, therefore, that

> A property G is multiply realized *if and only if* (i) under condition \$, an individual *s* has an instance of property G in virtue of the powers contributed by instances of properties/relations F_1–F_n to *s*, or *s*'s constituents, but not vice versa; (ii) under condition \$* (which may or may not be identical to \$), an individual *s** (which may or may not be identical to *s*) has an instance of property G in virtue of the powers contributed by instances of properties/relations F_1^*–F_m^* of *s** or *s**'s constituents, but not vice versa; (iii) F_1–$F_n \neq F_1^*$–F_m^* and (iv), under conditions \$ and \$*, F_1–F_n of *s* and F_1^*–F_m^* of *s** are at the same scientific level of properties.

To continue with the example of polarity, we can explain how it is multiply realized. A water molecule has this property in virtue of the electronegativity of the hydrogen and oxygen atoms and the angle at which they are bonded. A hydrogen fluoride molecule, however, is polar in virtue of the hydrogen's electronegativity, fluorine's electronegativity, and the bond between them. (See Figure 10.1.)

Of the many important features of the dimensioned view, the one that will be most important for the present discussion of multiple realization and the possible elimination and subtyping of properties is the fact that there are typically many distinct lower level realizers F_1–F_n for a single higher level property instance G. Once we begin to examine actual scientific cases with this in mind we recognize the possibility of different ways in which higher level theory can handle discoveries about different lower level realizers. Sometimes different sets of lower level realizers may still result in the very same higher level property. Other times, different sets of lower level realizers may prompt recognition of individual differences across instances of the same higher level realized property. While still other lower level differences may be such that they force us to say that these realizers actually result in different realized properties. What we will see is that one subset of realizers, F_1–F_g will be handled one way, another subset F_h–F_j will be handled in another, and still another subset, F_k–F_n will be handled in yet another, depending upon features of the higher level theory. Sometimes differences in realizers together result in instances of the same realized property – perhaps with individual differences across these instances – and sometimes they together result in instances of distinct realized properties.

10.2 The mechanisms of colour vision in the eye

The mechanisms of colour vision are realized in many regions of the body and the central nervous system, including the eye, the lateral geniculate nucleus, areas V1, V2, V4, and likely very many more. Our present philosophical

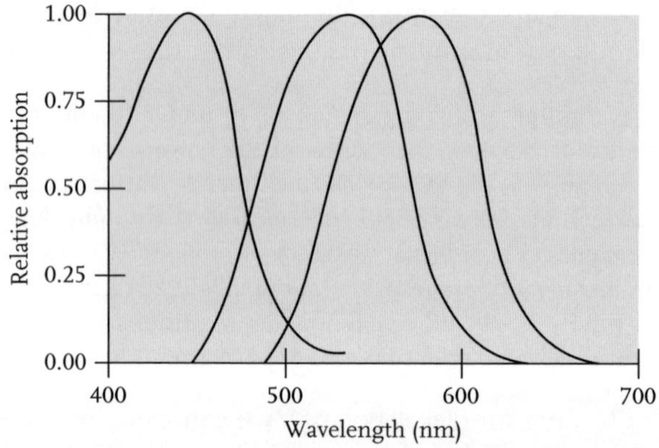

Fig. 10.2 Photopigment absorption curves.

concerns will, however, be best served by limiting our attention to the mechanisms within the eye. Insofar as there is realization and multiple realization of colour vision by the apparatus of the eye, there will be at least that much realization and multiple realization in the entirety of the visual system.

If we begin at the level of the entire eye, we can say that the visual system begins to interact with light as soon as photons enter the cornea. Since the cornea, aqueous humor, lens, and so forth, are not perfectly transparent, these components influence the retina's response to incoming light. Moreover, since they do not absorb all wavelengths of light equally, they change the spectral distribution of incoming light, hence the colour that a person perceives. The pre-receptoral components of the eye that absorb most of the incoming light are the lens and the macular region of the eye (which contains the vast majority of the colour processing cones of the eye). What will matter for us is the fact that an eye's response to light depends on three distinct components: the lens, the macula, and the photoreceptors.

If we descend to the level of the retina, we naturally turn our attention to the colour photoreceptors, the cones. On the standard trichromatic theory of human colour vision, the ability to perceive colour is based on making comparisons of signals from three distinct types of cones – red, green, and blue – each sensitive to a slightly different range of the visible spectrum of electromagnetic radiation.[6] (See Figure 10.2.) It is well known that abnormalities in the cones can lead to abnormalities in colour perception. Protanopes lack red

[6] The terminology for describing these cones is not consistent across the disciplines that study them. In psychology and psychophysics, one is more likely to find the cones described as L-cones, M-cones, and S-cones or long-wavelength-sensitive (LWS), medium-wavelength-sensitive (MWS), and short-wavelength-sensitive (SWS), where biochemical studies of the opsins often use red, green, and blue. Nothing, as far as we can tell, depends on our choice of terminology.

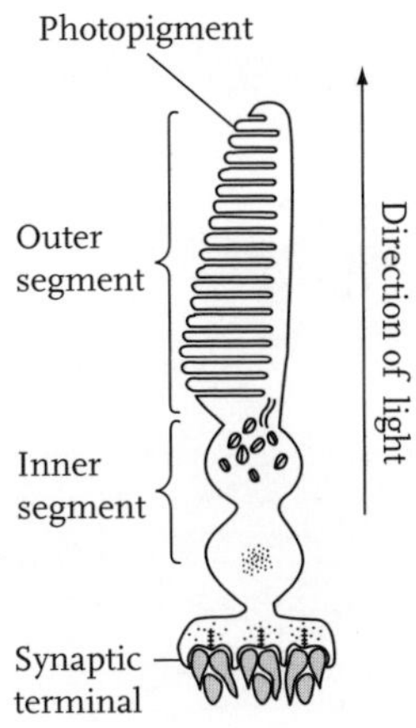

Fig. 10.3 Human cone.

cones and deuteranopes lack green cones. They, therefore, perceive the visible spectrum of light differently than do those with normal colour vision.

Moving to the cellular level, we discover that each cone has photopigment molecules embedded in the membrane of its outer segment. (See Figure 10.4.) Each cone's photosensitivity is determined by three principal factors: the length of the cone's outer segment, the concentration of the photopigment in the outer segment, and the sensitivity of the individual photopigment molecules. These first two features involve relatively pedestrian physics, but the final one concerns the biochemistry of photopigments, a topic of significant interest in the sciences of colour vision.

At the biochemical level, a given photopigment molecule consists of a protein component – a red, green, or blue cone opsin – and a non-protein component – an 11-*cis*-retinal chromophore. The chromophore component of a photopigment is responsible for the actual process of photon capture and is the same in all photopigments, where the opsin component modulates the frequencies of light to which the chromophore is sensitive. Differences in the amino acid sequences of the normal red, green, and blue cone opsins, thus give rise to the differences in light sensitivity of the complete photopigment molecules.

As our final bit of scientific information on human colour vision, we note that the photopigments are only one component in the biochemical cascade that links photon capture to neuronal signaling. (See Figure 10.3.) Upon absorption of a photon, a single photopigment molecule will change conformation from 11-*cis*- retinal to all-*trans*-retinal. After this conformational change, the retinal chromophore is released and the opsin molecule is activated. The activated opsin binds to a single G protein molecule located on the inner surface of the cell membrane. This G protein molecule, in turn,

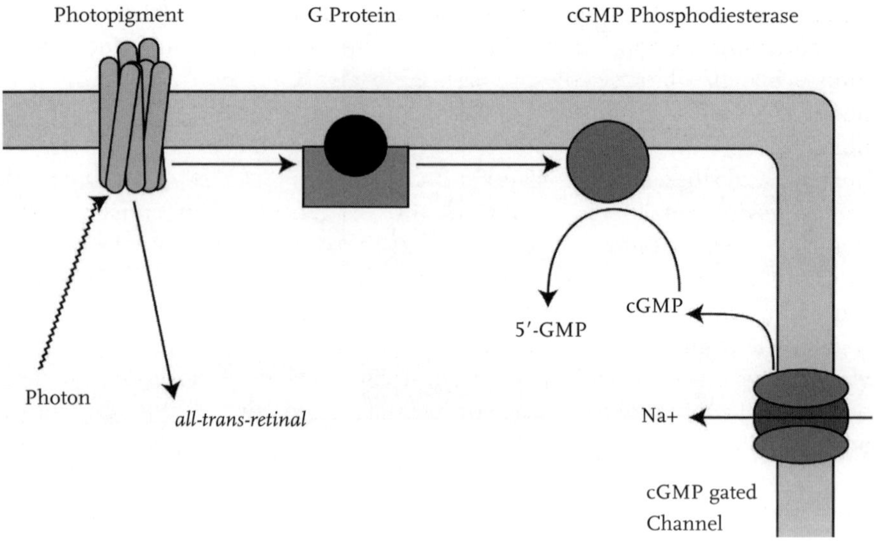

Fig. 10.4 The Phototransduction Biochemical Cascade.

activates a molecule of an enzyme, cGMP phosphodiesterase, which breaks down many molecules of cGMP to 5′-GMP. When the intracellular cGMP concentration subsequently decreases, cGMP molecules are removed from cGMP-gated Na^+ channels, leading to the closure of the channels. Closing of the channels blocks the influx of Na^+ into the cell. In concert, vast numbers of photopigment molecules, G protein molecules, ion channels, and Na^+ ions go through this process leading to the hyperpolarization of the cell. This hyperpolarization propagates from the outer segment of the cone to the synaptic contact, where it reduces the rate of release of the neurotransmitter glutamate. This reduction in neurotransmitter release is the cone's signal that the cell has been illuminated.

10.3 The multiple realization of normal colour vision

Our central concern in this chapter is to explore the ways in which the discovery of differences in lower level realizers influences how scientists handle higher level properties. As a first philosophical concern, it is important to clarify what is at issue. The debate between the splitting versus the non-splitting strategy is not a debate about the descriptive powers of natural language. When the opponent of multiple realization observes that, faced with a possible case of multiple realization of a property G, one might recognize, say, two properties G_1 and G_2, and say that G_1 is uniquely realized by F_1,

$F_2, \ldots, F_n$, and that G_2 is uniquely realized by $F_1^*, F_2^*, \ldots, F_m^*$, the claim is not merely one about what one or another natural language allows a scientist to express. It is uncontroversial to claim that in English we can speak of, say, being a green cone that is realized by cone opsin A and being a green cone that is realized by cone opsin B. Scientists can certainly use English to discriminate between properties that are realized in one way and properties that are realized in another. Such linguistic facts are no more interesting than the linguistic fact that scientists can speak of the property of being a green cone that realizes trichromatic vision and the property of being a green cone that realizes dichromatic vision. That is, the linguistic ability to individuate properties by reference to what realizes them is no more interesting than the ability to individuate properties by reference to what they realize. The matter of splitting versus not splitting properties is not one of linguistic usage. Instead, it is a question about the ontology scientists advance; it is about what properties scientists postulate in higher level theories in the face of discoveries at lower levels. Of course, the distinction between what is a linguistic matter and what is an ontological or theoretical matter is not perfectly clear, but such a distinction appears to be necessary if there is to be a substantive issue.

A second philosophical concern is clarity about what is meant by 'normal colour vision'. In the literature being examined here, a person is said to have normal colour vision if that person makes normal colour discriminations. Such normalcy does not include other features of colour vision, such as rapidity of response, luminance sensitivity, etc. One could perhaps define a more robust, multidimensional concept of normalcy or perhaps find another conception of normalcy in other segments of the vision science literature, but such conceptions would not be the one in play in the research being reviewed here. We can tell that the concept of normalcy being invoked here depends only on how one makes colour discriminations, since this is the only type of test used to screen for normal colour vision. Thus, in reading through the description of methods, one might find that subjects were screened for normal colour vision using the Ishihara test. This very simple test involves 24 plates consisting of a circular field of dots of various sizes and colours. Normal trichromats easily recognize a numeral in the pattern of dots on each plate, where those having one or another colour deficiency will not recognize a numeral on one or more plates. Part of what makes this test so popular is how easily and quickly normalcy can be determined.[7]

[7] One can be worried about what sort of normativity there might be in the concept of 'normal colour vision', but much of this worry might be avoided by simply changing the higher level property that is invoked. So, for example, all of the arguments that are developed here would go through essentially unchanged even if we invoked other higher level properties, such as being an anomalous trichromat, being a dichromat, being a deuteranope, being a protanope, or being a tritanope. The property of having normal colour vision is more useful than these others for two reasons. First, the property of having normal colour vision is easily described operationally as in

10.3.1 Normal colour vision and photopigment diversity

With philosophical preliminaries out of the way, we can now relate the science to the metaphysics of realization and multiple realization. At first glance, one might think that the theory of colour vision would strongly support the splitting of higher level properties. A number of studies have documented the existence of polymorphisms in the green and red photopigments.[8] For the red photopigment, it has been estimated that roughly 44% of the population has an amino acid chain, often designated Red (ala^{180}), that has an alanine at position 180, where about 56% of the population has an amino acid chain, often designated Red (ser^{180}), with a serine at position 180. For the green photopigment, it has been estimated that roughly 94% of the population has an amino acid chain, often designated Green (ala^{180}), that has an alanine at position 180, where about 6% of the population has an amino acid chain, often designated Green (ser^{180}), with a serine at position 180.[9] These different amino acid chains contribute slightly different absorption spectra, which are properties that they contribute to the realization of normal human colour vision. For example, Merbs & Nathans (1992) report that the wavelength of maximum absorption, λ_{max}, for Red (ala^{180}) is 552.4 nm and that the Red (ser^{180})$\lambda_{max} = 556.7$.[10] Thus, one might expect that the property of having normal colour vision would be subtyped.[11] The subtyping strategy proposes that vision scientists will postulate four types of colour vision corresponding to the four combinations of photopigments:

Normal colour vision with Red (ala^{180}), Green (ala^{180}),
Normal colour vision with Red (ala^{180}), Green (ser^{180}),
Normal colour vision with Red (ser^{180}), Green (ala^{180}),
Normal colour vision with Red (ser^{180}), Green (ser^{180}).[12]

Vision scientists could therefore describe an instance or instances of normal colour vision as being realized by one or another of these properties. However,

the body of the text above. Second, the literature on this property is more extensive than that on the other properties.

8 See (Neitz & Neitz, 1998; Sjoberg, 1998; Winderickx *et al.* 1992).

9 This composite data is assembled in Sharpe, Stockman, Jägle, & Nathans (1999).

10 Using different techniques, Sharpe *et al.* (1998) report that Red (ala^{180})$\lambda_{max} = 557.9$ and that the Red (ser^{180})$\lambda_{max} = 560.3$, where with still different techniques, Asenjo, Rim, & Oprian (1994) report that Red (ala^{180})$\lambda_{max} = 557.9$ and that the Red (ser^{180})$\lambda_{max} = 560.3$.

11 In fact, in a paper to be discussed below, the authors actually appear to broach the possibility of subtyping normal colour perceivers on the basis of differences they find in the subjects' cone photopigments: 'The ratio of two Rayleigh-type matches is a rapid and convenient measurement for assessing the L-cone λ_{max} in the eye of an individual observer and therefore may be useful for classifying normal trichromats into phenotypic sub-types' (He & Shevell, 1994, p. 367). We shall return to this claim later.

12 The subtyping strategy does not specify how the higher level psychological properties will be named or described; it only asserts that there will be some form of subtyping. Here we subtype the properties by reference to the molecules involved.

Table 10.1

Pigment	Mean λ_{max}	SD
Green	529.7	2.0
R2G3	529.5	2.6
R3G4 (Ala^{180})	529.0	1.0
R3G4 (Ser^{180})	533.3	1.0
R4G5 (Ala^{180})	531.6	1.8
R4G5 (Ser^{180})	536.0	1.4
Red (Ala^{180})	552.4	1.1
Red (Ser^{180})	556.7	2.1
G2R3 (Ala^{180})	549.6	0.9
G2R3 (Ser^{180})	553.0	1.4
G3R4	548.8	1.3
G4R5	544.8	1.8

the fact is that in these actual examples researchers have not abandoned the unitary property of having normal colour vision in favor of a set of four higher level properties.

The common red and green photopigment polymorphisms are only the tip of the diversity iceberg. There are, in fact, a relatively large number of distinct red and green photopigments whose absorption spectra have been determined by a variety of methods. Just to give a hint of this diversity, we report, in Table 10.1, data from Merbs and Nathans (1992).

Despite this well-known diversity in the red and green cone opsins and the well-known differences in their absorption spectra, vision scientists have not abandoned the category of normal colour vision. Nor have they introduced an elaborate and systematic taxonomy of dozens of subtypes of normal colour vision as suggested by the property splitting strategy.

Actual practice with regard to normal colour vision does not follow the property splitting strategy. Instead, vision scientists appear to accept, or at least tolerate, the existence of non-identical realizers of the higher level property of normal colour vision.[13] This is not to say, however, that scientists simply dismiss differences in lower level realizers as irrelevant to the higher level theory or properties. There is not *that* kind of autonomy of psychology. Scientists often study differences in lower level realizers as a means of explaining what they refer to as individual differences, differences from one human to the next. In vision science, a common approach to studying individual differences among normal colour perceivers begins by creating a pool of normal subjects

[13] To repeat a point made in an earlier footnote, there is nothing special in this regard concerning normal human colour vision. The argument applies just as well to the property of being a tritanope. It applies only slightly less well to being a protanope or being a deuteranope, since by definition these deficiencies mean a lack of red or green cones.

by selecting only those who make correct classifications on all the Ishihara plates. Once the pool of normals is assembled, a more sensitive test, such as Rayleigh matching, is used to measure subjects' ability to make finer colour discriminations. In a Rayleigh match, subjects might be shown a target hemifield of 589 nm light, then asked to adjust the amounts of 545 nm and 670 nm light displayed in a second test hemifield so as to have the two hemifields match.

He and Shevell (1994) report some results that are especially pertinent and illuminating. They develop a variant of the Rayleigh match test, a dual-Rayleigh match, which essentially involves subjects making one match using 545 nm and 670 nm light and another using 545 nm and 620 nm light. They argue that this dual match procedure enables them to locate the source of individual differences in the photopigments. Thus, the title of their paper is 'Individual differences in cone photopigments of normal trichromats measured by dual rayleigh-type colour matches'. In other words, even among those individuals who have the property of having normal colour vision, there are variations in colour matches that arise from differences in photopigment absorption spectra. The introduction of He and Shevell's paper emphasizes the same point:

> The color matches of normal trichromatic observers show substantial and reliable individual differences. This implies the population of normal trichromats is not homogeneous, an observation that leads to the question of how one normal trichromat differs from another. In general, the physiological mechanisms that contribute to color-matching differences among normal observers may be classified as either pre-receptoral or receptoral. Pre-receptoral spectrally selective filtering can significantly affect color matches and therefore can cause individual differences. The influence of pre-receptoral filtering, however, can be eliminated with well-known techniques,... This implies that individual differences among normal trichromats are due in part to receptoral variation (He & Shevell, 1994, p. 367)

He and Shevell clearly recognize the impact of lower level realizers on higher level behaviour, but they do not deny the existence of normal colour vision and they do not subtype normal colour vision by means of receptoral differences. That is, they do not entertain the sorts of proposals one finds in the memory literature where, as an apparent result of discoveries about differences in realizers, psychologists deny the existence of a unitary kind of memory in favour of subtypes of memory, such as long-term memory and short-term memory or procedural memory and declarative memory. We, thus, have cases that do not follow the eliminate-and-split strategy. Moreover, we see that an appeal to individual differences is a feature of actual scientific practice not recognized in our simple distinction between splitting and non-splitting strategies. Finally, we also appear to have some measure of autonomy of psychology from any putative dictates of lower level science.

10.3.2 Normal colour vision and pre-receptoral properties

In the abstract to their paper, He and Shevell conclude with a comment that offers some comfort to the advocates of the property splitting strategy. They actually broach the possibility of subtyping normal colour perceivers on the basis of differences they find in the subjects' cone photopigments: 'The ratio of two Rayleigh-type matches is a rapid and convenient measurement for assessing the L-cone [i.e. red cone] λ_{max} in the eye of an individual observer and therefore may be useful for classifying normal trichromats into phenotypic sub-types' (He & Shevell, 1994, p. 367). So, scientists are at least willing to entertain the possibility of applying the property splitting strategy. The point to be made through our additional examples, however, is to indicate that the property splitting strategy is not likely to be invoked as uniformly and systematically as might be suggested by the simple formal schema with which this chapter began.

Return now to He and Shevell's distinction between the two sources of individual differences: pre-receptoral and receptoral. He and Shevell are willing to entertain the possibility of subtyping normal colour vision along the lines of differences in photopigments, that is, based on differences among certain realizers of normal colour vision. They do not, however, entertain the possibility of subtyping normal colour vision along the lines of differences in the pre-receptoral properties. They do not entertain the possibility of subtyping normal colour vision by means of combinations of differences in lens optical density and macular pigment optical density even when they explicitly note the effect these pre-receptoral features have on colour discriminations. Nor does such a taxonomy seem to appear in the vision science literature. As with differences in photopigment sensitivity, differences in the optical absorption properties of the lens and macula are treated as sources of individual variation among normal colour perceivers.

10.3.3 Normal colour vision and the phototransduction biochemical cascade

To this point, we have claimed that there are two kinds of counterexamples to the property splitting strategy in vision science. One is based on the properties of photoreceptors; the other is based on the properties of the lens and macula. The most interesting counterexamples, however, involve the properties of the elements in the biochemical cascade. Recall that, within a single cone, there are multitudes of molecules and ions of many types involved in the biochemical cascade that leads from photon capture to a change in neurotransmitter release. There are the cone opsin molecules, the G proteins, the cGMP phosphodiesterase molecules, the cGMP molecules, the phospholipid molecules of the cell membrane, the sodium ions, the potassium ions, the ion channel components, and so forth. Each of these molecules and ions has one or more properties that it contributes to phototransduction, hence to normal colour vision. Of course, each of these molecules and ions will have properties that are irrelevant to normal colour vision, so that those properties will not be among the realizers. But, each molecule and ion will still have relevant properties spelled out in standard accounts of phototransduction.

Set aside the ions, the water, and cGMP and focus on the proteins in the biochemical cascade. Suppose that each of the proteins admits of mutations that only slightly alter the functionality of the protein in the cascade.[14] That is, just as there are variations in the amino acid sequences of the opsins, suppose that there variations in the amino acid sequences of the G proteins, the cGMP phosphodiesterases, and the monomeric components of the cGMP-gated Na^{+} channels. If one reflects on the combinatorics of just these proteins, one finds that the number of types of normal human colour vision that one would have to postulate would increase dramatically. If we bear this in mind, we can see how impractical it would be to develop a theory of colour vision that hypothesizes a distinct type corresponding to each distinct set of lower level realizers. We do not have a comprehensive account of theoretical virtues in the higher sciences, since this is monumental task. None the less, it is plausible that one does not want a theory of human color discrimination abilities that tracks literally *all* the *bona fide* different realizers of normal colour vision given their vast numbers, since this would, to take just one example, mean that we can formulate few if any generalizations across subjects.

The problem here is not merely that the combinatorics of subtyping colour vision by way of its many lower level property instances is cumbersome. It is also that using *all* of the lower level realizer properties to individuate higher level properties leaves us without higher level theories that can track important regularities or generalizations at the higher level. Think of the properties of the G proteins, such as the rate at which they are activated by the retinal-free membrane-bound opsin or the rate at which they activate cGMP phosphodiesterase. The many different values of these properties realize, in part, the colour processing properties of humans. Nevertheless, these properties, and their differences, are not the kinds of properties scientists want in their theory of human colour discriminations. Because these properties do not bring about changes in the color discriminations humans make, they are what we are calling 'orthogonal' realizers of colour discriminations.

[14] Here it would be convenient to be able to cite some studies that document the variability in the proteins, but such studies are hard to come by, if they even exist yet. Thus, rather than direct measurements of variability in the G proteins, the cGMP phosphodiesterases, etc., one must settle for considerations of the general nature of proteins. These are likely to be variable due to the supposed underlying genetic variability, which is essential for evolution by natural selection. The lack of direct evidence might, thus, be taken to make this illustration more speculative than the preceding two.

To this point we have mentioned a number of times the idea that scientists do not treat all discoveries about differences in realizers equally. Now we are in a position to elaborate on this point by connecting our initial taxonomy of types of realizer to the cases we have examined. Some of the lower level properties that realize normal human colour vision are such that we discover small differences in their natures so that we are forced to posit 'parallel' variations in colour discrimination capacities. The latter are what we earlier termed 'parallel' realizers. And with such realizers discovering the differences amongst them entails our accepting variations in the colour discriminations persons can make. Differences in the light absorbing properties of the lens, macula, and photopigments are thus parallel realizers of colour discriminations. The lower level differences along the 'dimension' of light absorption lead to parallel higher level differences along the 'dimension' of colour discrimination.

It is these parallel realizers that have the most 'intuitive' appeal as a basis for adopting the subtyping and hence eliminate-and-split strategy, but, as we have seen, even in these cases this appeal is, at least at times, limited only to recognizing individual differences within a broader category. As a result, we can now see why a further distinction needs to be made amongst parallel realizers. Where differences amongst realizers leads only to scientists positing individual differences in the same higher level property, then we have what we termed 'weak' parallel realizers; when such discoveries lead scientists to posit two higher level properties, following the eliminate-and-split approach, then we have 'strong' parallel realizers.

As our last case highlighted, as well as these strong and weak parallel realizers, there are also cases of orthogonal realizers. Discoveries of differences in these orthogonal realizers are such that differences in the properties they contribute to normal colour vision – differences in such things as the activation rates and reaction rates – do not make a difference to colour discriminations. Thus, they do not lead scientists to posit different higher level realized properties. Differences along the 'reaction rate dimension' are orthogonal to differences in the colour discrimination, so differences in orthogonal realizers do not provide even a *prima facie* basis for invoking the property-elimination-and-subtyping strategy.

It is important to forestall some misguided objections to the parallel-orthogonal distinction among realizers. So let us emphasize, first, that both parallel and orthogonal realizers are in fact realizers. Both types of lower level realizers stand in the kind of non-causal, non-logical determination relation we take to be definitive of causal-mechanical realization. The point about orthogonal realizers is not that they have no higher level consequences. They have to have such consequences in order to be realizers at all. The point is, instead, that orthogonal realizers do not have higher level consequences of a particular sort, higher level consequences along a particular dimension relevant to the higher level property. Second, it is also important to note that being a parallel or orthogonal realizer is relative to both the higher level and lower level properties in question. Finally, one should not suppose that orthogonal realizers are not to be construed as realizations whose variations have relatively little impact on higher level properties. Instead, the idea is that the discovered variations in the orthogonal realizers lead to *no variation* in the higher level realized properties. In contrast, weak or strong parallel realizers are such that the discovered variations do lead to some variation in the higher realized properties.

10.4 Some broader philosophical context

Our chapter began with a succinct question about the nature of scientific theorizing. How do scientists accommodate findings about differences in lower level realizers in their higher level theories? We believe that the question, and our answer to it, should be of interest to philosophers of science who wish to understand the nature of scientific practice. That is one motivation for our project, but another stems from the fact that other philosophers have already touched on this question and given an answer that differs from our own. These philosophers have reasoned, in one way or another, that differences in lower level realizers will always lead to higher level differences that block multiple realization.[15] Rather than attempting to track all the argumentative paths that have been taken, we will select one that fits most closely with the framework we have established here, namely, Carl Craver's treatment of 'dissociable realization' (Craver, 2004). In fact a feature of Craver's analysis is that it also illustrates one of the general morals of our analysis – namely, that whether a higher level property is split or not, depends, at least in part, on the needs of good theory at the higher level.

Craver's project is to explain the reasoning underlying dissociation experiments in which brain lesions can impair one form of memory, such as declarative memory, while preserving another form of memory, such as procedural memory, thereby supporting the view that there is no such thing as memory *simpliciter*, but instead two distinct subtypes of memory, namely, declarative and procedural memory. At the heart of Craver's analysis is a principle of No Dissociable Realization NDR. What we want to show is that, upon clarification (NDR), becomes a principle that endorses the properties splitting strategy that we have argued is not uniformly adopted in science. Formulated in terms of properties, it is the following:

[15] Here we have in mind Shagrir (1998); Craver (2004); Couch (2005); Shapiro (2008), and Polger (2008).

(*NDR**) Instances of a property have one and only one realizer. If there are two distinct realizers for a putative instance of a property, then there are really two properties, one for each realizer. (Cf. Craver, 2004, p. 962).[16]

The first thing we need to do is to refine Craver's analysis to remove an ambiguity in the notion of 'distinct realizers'. Consider two water molecules. Both of these molecules are polar, so both have oxygen and hydrogen atoms with properties that together realize the property of being polar. Here we should say that the properties of the water's constituent oxygen and hydrogen atoms provide what we might call *numerically distinct realizations* of the property of being polar. However, they do not provide what we might call *property distinct realizations* of the property of being polar. It is property distinct realizations that are implicitly taken to be involved in multiple realization. So, it is because a water molecule is polar in virtue of having two instances of hydrogen's electronegativity and one instance of oxygen's electronegativity (among other properties), where a hydrogen fluoride molecule is polar in virtue of having one instance of hydrogen's electronegativity and one instance of flourine's electronegativity (among other properties), that a water molecule and a hydrogen fluoride molecule give us multiple realization.[17]

So, how should we interpret the phrase 'distinct realization' in *NDR**? Let us consider the options. First, suppose we have a numerically distinct interpretation:

(*NDR***) Instances of a property have one and only one numerically distinct realizer. If there are two numerically distinct realizers for a putative instance of a property, then there are really two properties, one for each numerically distinct realizer. (Cf. Craver, 2004, p. 962).

This, however, cannot be the correct principle. What it says, in essence, is that there cannot be a single property of being a kidney. If there are two numerically distinct realizers for the property of being a kidney, say, the left kidney and the right kidney, then there are really two properties – such as the property of being the left kidney and the property of being the right kidney – one for each numerically distinct realizer. Craver, however, rightly rejects this proposal (Craver, 2004, p. 967). Presumably, scientists do not introduce the

[16] In a late section of his paper, Craver proposes that arguments involving dissociable realization only work for properties and activities. This is why we skip over the NDR formulation in terms of natural kinds directly to *NDR** formulated in terms of properties. This does not distort Craver's views.

[17] Some might prefer to mark the distinction we have in mind here by saying that two individual water molecules provide two tokens of the same type of realization of polarity, where a water molecule and a hydrogen fluoride molecule provide two tokens of two distinct types of realization of polarity.

subtypes of left kidney and right kidney, since this would tend to obscure scientific generalizations concerning kidneys. So, consider the property distinct interpretation:

> (NDR^{***}) Instances of a property have one and only one property realizer. If there are two property distinct realizers for a putative instance of a property, then there are really two properties, one for each property distinct realizer. (Cf. Craver, 2004, p. 962)

When framed in this manner, we can see that the principle looks to be a statement of the necessary property splitting strategy. However, we have now seen this approach faces problems. Among the oversimplifications inherent in this position is the tacit presupposition that, when we discover variations in realizers, the higher level theory has no role to play in deciding whether or not such differences at the lower level do, or do not, necessitate positing new properties at the higher level.

In fact, once we consider the role of higher level theory, we can return to our opening example of memory and explain how it does not, after all, lend support to our simplistic version of the property splitting strategy. Lesion studies by themselves do not distinguish between property distinct and numerically distinct realizations. Remove a bit of tissue X_1 from location L_1 and a bit of tissue X_2 from location L_2 and let these distinct lesions have behavioural consequences. This alone does not tell scientists whether X_1 and X_2 have distinct neuroscientific or psychological properties. X_1 and X_2 might be the left and right instances of a common structure, such as the left and right eye, the left and right kidney, or perhaps the left and right halves of area V1. In such a case one might have merely numerically distinct, rather than property distinct realizations of a neuroscientific property.

Second, even if scientists were to have evidence that X_1 and X_2 are property distinct realizations, that would not tell them whether they are property distinct realizations of distinct higher level properties or property distinct realizations of the same higher level property. To put the matter in another way, given that declarative memory and procedural memory have property distinct realizations, it is not merely the distinctness of the lower level realizing properties that motivates them to split memory into declarative and procedural forms. Distinct sets of lower level properties can either give us two property distinct realizations of a single higher level kind (hence multiple realization) or two property distinct realizations of two higher level kinds (hence unique realizations). But, as we argued in the preceding sections, scientists facing a choice between these two options do not simply look to lower level realizers to make this decision. Instead, they look to principles of good higher level theory construction in making this choice. Higher, psychological level differences between procedural memory and declarative memory contribute to the splitting of properties; not mere differences in lower level realizers.

Craver, in fact, implicitly recognizes this when he mentions a number of the psychological level differences. He writes,

> Declarative memories are triggered by the presentation of facts or the occurrence of events in the life of the person, and they play important roles in, for example, conversation, autobiography, or the simple act of reminiscing. Nondeclarative forms of memory (like procedural memory, iconic memory, priming, etc.) have their own unique triggering conditions (procedural memories are acquired by doing things, iconic memory by visual impressions, etc.) and play different roles in the life of the organism. These differences are reflected in the different kinds of stimuli used to produce and evoke memories of the different types (Craver, 2004, p. 966).

So, by our lights, the case of memory does not provide an illustration of the view that scientists subtype a higher level property when they find that it has distinct lower level realizers – thus taking the findings of the lower level on their own to determine decisions about which properties to posit at the higher level. Instead, we can now see that the case of memory supports our view that the decision whether to subtype properties at the higher level, or not, is driven, at least to some degree, by considerations of what makes for the better higher level theory and hence by higher level theory in *partnership* with the lower level accounts.

10.5 Conclusion

Neuroscientists and psychologists, at least at times, choose not to eliminate and subtype higher level properties when faced with the discovery of differences in lower level realizers. They have not postulated a myriad of distinct types of normal human colour vision each one of which corresponds to a distinct set of realizers. Neuroscientific and psychological theorizing does not hew to an extreme view according to which higher level taxonomy is always a slave to lower level taxonomy. But scientific practice also does not embrace the other polar extreme according to which neuroscience and psychology simply ignore differences in lower level realizers. We have shown that scientific theorizing does not necessarily adopt either extreme and we have described, at least in outline, how it actually proceeds in certain cases.

Sometimes scientists acknowledge the effects of lower level realizers by using them to explain individual differences at higher levels of analysis. To show this, we noted that biochemical differences in photopigments explain individual differences in subtle colour discrimination tasks, such as Rayleigh matching, even among individuals who are classified as colour normal by coarser tests, such as the Ishihara test. We also noted that differences in light absorption by the lens and macula are also used to explain individual differences among colour normals. The property of colour normalcy is, thus, retained by vision scientists, despite individual differences within that type. In other words, colour normalcy is retained in the face of the discovery even of parallel realizers and we have thus seen that we need to accept there are both weak, as well as strong, parallel realizers.

In addition to cases involving parallel realizers, we also saw there are cases where differences in realizers are acknowledged, but where this does not lead scientists to posit individual differences in a specific higher level property.[18] These are cases of orthogonal realizers. Our illustration of this approach was the apparent role of the properties of the components of the phototransduction biochemical cascade in normal colour vision. This result will be surprising to the philosophers who have reasoned that the discovery of distinct sets of realizers should always lead to the subtyping of higher level properties, hence that all distinct sets of realizers constitute strong parallel realizers. With orthogonal realizers, differences in realizers result in no difference at the higher level relevant to the higher level properties under discussion. Thus differences in the properties of the components of the phototransduction biochemical cascade do not lead to differences in colour discriminations. In orthogonal realizers, one has discoveries about differences in lower level realizer properties that scientists find no interest in incorporating into their higher level theories.

What do the foregoing observations about a segment of vision science, if we assume them to be descriptively accurate, tell us about neuroscience and psychology? Why do neuroscientists and vision scientists reason as they do? As we have seen, the short answer is that there is some measure of autonomy of psychology even in the age of neuroscience. Lower level sciences can have closer or more distant relations to higher level sciences, as revealed by parallel and orthogonal realizers, but exactly how lower level science influences higher level science is determined, at least in part, by the needs of higher level science. Higher level science is not a mere repository of lower level differences, but a body of theoretical knowledge in its own right and thus a partner with lower level science in our ongoing project of investigating the world around us.

References

Aizawa, K. & Gillett, C. (2009a). Levels, individual variation and massive multiple realization in neurobiology. In J. Bickle (ed.), *The Oxford Handbook of Philosophy and Neuroscience*, pp. 539–581. Oxford: Oxford University Press.

[18] Another sort of case we have not discussed here is when one gets multiple realization of a higher level property G by having the differences between F_1, F_2, ..., F_n, and F_1^*, F_2^*, ..., F_m^* 'cancel each other out'. To take a suggestive example, one might get multiple realization of a given stroke volume of an automobile engine's cylinder by distinct combinations of stroke length and cylinder area.

Aizawa, K. & Gillett, C. (2009b). The (multiple) realization of psychological and other properties in the sciences. *Mind & Language, 24*(2), 181–208.

Aizawa, K. & Gillett, C. (Unpublished). Multiple realization and methodology in neuroscience and psychology.

Asenjo, A., Rim, J., & Oprian, D. (1994). Molecular determinants of human red/green color discrimination. *Neuron, 12*(5), 1131–1138.

Couch, M. (2005). Functional properties and convergence in biology. *Philosophy of Science, 72*(5), 1041–1051.

Craver, C. (2004). Dissociable realization and kind splitting. *Philosophy of Science, 71*(5), 960–971.

Gillett, C. (2002). The dimensions of realization: A critique of the Standard view. *Analysis, 62*(276), 316–323.

Gillett, C. (2003). The metaphysics of realization, multiple realizability, and the special sciences. *The Journal of philosophy*, vol. 100, 591–603.

He, J.C. & Shevell, S.K. (1994). Individual differences in cone photopigments of normal trichromats measured by dual Rayleigh-type color matches. *Vision research(Oxford), 34*(3), 367–376.

Merbs, S. & Nathans, J. (1992). Absorption spectra of the hybrid pigments responsible for anomalous color vision. *Science, 258*(5081), 464–466.

Polger, T. W. (2008). Evaluating the evidence for multiple realization. *Synthese, 167*(3), 457–472.

Shagrir, O. (1998). Multiple realization, computation and the taxonomy of psychological states. *Synthese, 114*, 445–461.

Shapiro, L.A. (2008). How to test for multiple realization. *Philosophy of Science, 75*(5), 514–525.

Sharpe, L., Stockman, A., Jagle, H., Knau, H., Klausen, G., Reitner, A., *et al.* (1998). Red, green, and red-green hybrid pigments in the human retina: Correlations between deduced protein sequences and psychophysically measured spectral sensitivities. *Journal of Neuroscience, 18*(23), 10053–10069.

Sharpe, L., Stockman, A., Jägle, H., & Nathans, J. (1999). Opsin genes, cone photopigments, color vision, and color blindness. In K.R. Genenfurtner & L.T. Sharpe (Eds), *Color vision: From genes to perception*, 3–51. Newyork: walkr de Gruyter.

Shoemaker, S. (1980). Causality and properties. In Van Inwagen (ed.) *Time and Cause.* Dordrecht: Reidel.

·11·

Turing machines and causal mechanisms in cognitive science

Otto Lappi and Anna-Mari Rusanen

Abstract

A body of recent literature has proposed that explanation in neurosciences, including cognitive neuroscience, is mechanistic. It has also been argued that the mechanistic model could be extended to cover explanations in computer sciences and cognitive sciences. Mechanistic explanation as standardly conceived is a form of *causal* explanation, and it requires that the explanatory mechanisms are *concrete, implemented mechanisms*. However, 'computing mechanisms' can mean two things. On the one hand, it can refer to concrete – causal – computing mechanisms, such as brains (*ex hypothesi*) or man-made computers, etc. On the other hand, it can also refer to abstract computing mechanisms such as abstract Turing machines. Therefore, the notion of computation can be used in cognitive science in at least two ways. Since there are computational explanations, in which Turing machines are considered as abstract mechanisms, the current formulation of mechanistic explanation does not cover those explanations.

11.1 Introduction

Turing machines are simple computational entities which were originally used to define the class of computational tasks that may be carried out by mechanical means. In this chapter we look at the explanatory roles the Turing machine[1] plays in cognitive science. With respect to Turing machines' explanatory role, we assume that (a) rational processes may be defined in such a way (viz. as computational functions) that they can be carried out by mechanical systems, and (b) the Turing machine specifies a means whereby one can show that a mechanistic system can be designed to perform rational processes.

[1] There are different ways to present Turing machines (single tape, multiple tapes, etc.) which are basically notational variants for different purposes. An intuitive introduction to the concept of Turing machines and algorithmic information (discussed in Section 11.4) is given in the appendix.

A body of recent literature has proposed that the explanation in neurosciences, including cognitive neuroscience, is mechanistic (Bechtel & Richardson, 1993; Bechtel & Abrahamsen, 2005; Craver, 2005; Craver & Darden, 2005; Machamer, Darden & Craver, 2000). Some of these philosophers have argued that the mechanistic model could be extended to cover explanations in psychology and computer sciences as well (for instance, Bechtel & Abrahamsen, 2005; Wright & Bechtel, 2007; Bechtel, 2008; Piccinini, 2006b, 2007, 2008). This chapter concerns the question: how well does this account of explanation fit explanations based on Turing machines.

In what follows we will argue that although the mechanistic organization of Turing machines may be important for their explanatory usefulness, not all computational explanations in cognitive science that make use of Turing machines conform to the account of mechanistic explanation put forward by these mechanistic philosophers. The reason, in a nutshell, is the abstract nature of Turing machines. Mechanistic explanation as standardly conceived is a form of *causal* explanation, and it requires that the explanatory mechanisms are *concrete, implemented mechanisms.* However, 'computing mechanisms' can mean two things. On the one hand, it can refer to concrete – causal – computing mechanisms, such as brains (*ex hypothesi*) or man-made computers, etc. On the other hand, it can also refer to abstract computing mechanisms such as abstract Turing machines. Therefore, the notion of computation can be used in cognitive science in at least two ways. Since there are computational explanations, in which Turing machines are considered as abstract mechanisms, the current formulation of mechanistic explanation does not cover those explanations.

11.2 Computation and cognition

One foundational task of cognitive science is to define and to explain the information processing capacities of natural systems, e.g. human brains, and provide a scientific account of how a cognitive system produces the adaptive and systematic 'intelligent' behaviour that it does. In this enterprise, Turing machines are theoretically significant because they specify a design whereby a mechanical system can be perform complex information processing tasks (which would seem to require 'intelligence'), viz. symbolic computation (Turing, 1936, 1950). In cognitive science this idea was used as a basis for the hypothesis that cognitive processes (complex perceptual processes, problem solving, inference, etc.) can be considered as computable functions, and that cognition is a species of computation.

From this perspective, Turing's work on the concept of computability provides not only the theoretical foundation on which theoretical computer science was built, but also forms part of the background of cognitive science, and has greatly influenced thinking about cognition. For instance, the dominant paradigm in cognitive science from the 1950's until the early 1980's, the so-called classical paradigm, viewed the mind as a symbol system (Miller, Galanter & Pribram, 1960; Newell & Simon, 1972; Newell, 1980; Fodor, 1975; Fodor, 1983, 1994; Fodor and Pylyshyn, 1988; Pinker, 1994). The idea is that mechanical or biological systems can perform cognitive tasks because they are *computers*, i.e. mechanisms capable of representing information in some appropriate medium, and performing computations on them (operations that some Turing machine could be designed for).[2]

In this classical paradigm, one might view Turing machines as the basis for implementable cognitive architectures. In such an approach, explanations based on Turing machines would be interpreted to involve not only theories of representations and information processing *tasks*, but also an account of *how* a cognitive system has, and exercises, a certain cognitive capacity. The behaviour of a Turing machine would be interpreted to represent literally the steps a system goes trough as it goes through from an input to output. However, few if any supporters of the classical paradigm believe Turing machines are mechanical models of the mind/brain in this literal sense. Thus, the Turing machine is significant not because it offers a plausible mechanistic model of cognitive operations – it does not – but because it offers a more abstract way to explain certain features of cognitive systems within a computational framework.

To anticipate the main conclusion, we claim it is useful to consider this theoretical role of Turing machines in terms of Marr's (1982) three levels of explanation (computation, representation and implementation, discussed in Section 11.4): abstract computational entities such as the Turing machine can be used at Marr's computational level to specify the information processing tasks as mappings, functions from one kind of information to another. These abstract entities are thus used in *a computational theory of competence* for a specific cognitive capacity – vision, language, belief revision, decision making,

[2] When combined with the view of cognition as information processing, Turing machines also specify a specific kind of information processing architecture (the so-called classical architecture; Fodor & Pylyshyn, 1988, etc.). A variety of approaches have challenged the 'classical paradigm' for a variety of reasons. Some forms of connectionism, dynamical systems theory, and some advocates of embodied cognition among others do not view cognition in terms of operations on representations (e.g. Van Gelder, 1995). But many advocates of more modern computational architectures still view cognition as computation – just not classical computation. For instance, Smolensky once wrote that connectionist models might 'challenge the strong construal of Church's Thesis as the claim that the class of well-defined computation is exhausted by those of Turing Machines' (Smolensky, 1988, p. 3). We are concerned with representational/computational explanations in cognitive science generally, and the Turing machine's role therein. Therefore, our analyses are meant to apply to both classical modern computationalist theories, but not non-represenational theories.

etc.[3] The Turing machine figures in such explanations,[4] and this form of explanation is not mechanistic in the standard sense. It is special to cognitive sciences. This form of explanation has no direct analogy in the non-cognitive and non-representational branches of biosciences and the neurosciences, for which the current mechanistic model of explanation was mainly developed. The upshot is that the explanatory relevance of Turing machines in cognitive science is not based on a 1:1 correspondence between the 'behaviour' of the constituent parts of the Turing machine (machine table, tape, symbols) and the behaviour of a system of interest, such as the human mind-brain.[5]

11.3 Mechanistic explanation in cognitive science

Historically, one point of contention in the foundations of cognitive science has been the tension between general laws and specific models. To what extent are cognitive phenomena subject to general, universal, law-like regularities, such as those observed in many aspects of the physical world? German psychophysicists studied empirically the existence of law-like dependencies in the process of sensory transduction. In the domain of learning theory, behaviourists explored the possibility of formulating species-universal laws of learning (conditioning) as relationships between stimuli and responses. More recently, computational research[6] into inductive procedures has been characterized as a search for rational 'univeral laws or generalization' for cognitive science (Anderson, 1991; Shepard, 1987; Chater & Vitanyi, 2003).

[3] The term competence level theory comes from Chomsky's work. Marr explicitly relates the two when he writes: 'Chomsky's (1965) theory of transformational grammar is a true computational theory in the sense defined earlier. It is concerned solely with specifying what the decomposition of an English sentence should be, and not at all with how that decomposition should be achieved' (1982, p. 28). We use the term 'theory of competence' (for a cognitive capacity) synonymously with 'computational level theory' (for that capacity).

[4] Note that we are *not* claiming that all explanations involve Turing machines, let alone that *all* computational explanation in cognitive science is 'Marrian'.

[5] The Turing machine can, of course, be used for theoretical purposes other than explanation. One can view it, for instance, as a conceptual tool (constraint) on theory formation, and one could argue that this is not explaining, and constraints on theory formulation carry no explanatory weight, since explanation requires a description of real mechanisms in the world. We would rather not take a strong stance on the matter. We simply note that Turing Machines are involved in computational explanations, and leave open whether or not they should be considered metaphysically 'real'.

[6] By computational research we mean *all* research that views cognition in terms of complex information processing tasks, i.e. which studies cognition in terms of complex internal representations (cf. footnote 2), and explains cognitive phenomena at least in part at from the point of view of Marr's (1982) computational level. The implied contrast is thus *not* that between 'computational modeling' and 'mathematical modeling', or between analytic treatment and numerical simulation, for instance.

This attitude towards explanation is possibly a result from a comparison with the physical sciences, where explanation is not considered to be achieved by modeling the behaviour of the constituent parts of a system in mechanistic terms, but only when mechanisms are subsumed under general laws.

However, the cognitive revolution in the 1960s moved the strategic focus from the universal laws to specific algorithmic models of specific phenomena. Rather than adapting the received explanatory strategy of the hard natural sciences, it became natural to think of cognitive science as 'reverse engineering', and view minds/brains as collections of highly complex computational devices. The guiding heuristics of research were borrowed partially from computer sciences and combined with the explanatory purposes of psychological sciences. Computer scientists design specific solutions to specific problems instead of searching for universal laws. This attitude was applied to cognitive science: the primary approach is to decompose complex information processing tasks such as problem solving (Newell, 1980) or object vision (Marr, 1982) into simpler information processing tasks, and to seek mechanical biological realizations for the most basic tasks.

Also philosophers of science have suggested alternatives to the received, nomological view of scientific explanation. These 'mechanistic' philosophers propose that explanation of the behaviour and capacities of complex systems (such as those found in the biological and cognitive sciences) does not typically involve laws, but specific models of particular mechanisms.[7] These philosophers have focused more on biology rather than physics as the prototypical 'modern' science. Philosophers such as William Bechtel, Stuart Glennan, Lindley Darden and Carl Craver have offered an account of explanation in which a phenomenon is explained by describing a mechanism that produces the phenomenon. Originally this account was put forward as an account of explanation in biochemistry and molecular genetics, and some aspects of neuroscience, but it was soon proposed to be extended to explanations in cognitive sciences and cognitive psychology (Bechtel & Richardson, 1993; Machamer, Darden & Craver, 2000; Bechtel & Abrahamsen, 2005; Glennan, 2005; Craver, 2005; Craver & Darden, 2005; Wright & Bechtel, 2007; Bechtel, 2008). The account spread to consciousness studies

[7] As such, the idea that the explanation of a phenomenon involves the mechanisms responsible for that phenomenon is nothing new. The early atomists, such as Democritus, explained things by referring to the features of observable objects in terms of the shape and motion of their constituents. In the seventeenth century Descartes applied his mechanistic perspective to the physical world, which included the living organisms, animal behaviour and that part of human behaviour that was not guided by the immaterial mind (which Descartes called reflexes, and considered to be shared with animals). In this sense Descartes extended the mechanistic explanation into the domains of biology, and partially the domain of psychology. However, since Descartes and his contemporaries did not have the conception of rational behaviour (thought and speech) being carried out by *mechanical* means, i.e. for instance, by Turing machines, they were thought to lie beyond the capacity of any mechanism.

as well (Revonsuo, 2006), and simultaneously it was proposed how computational explanations in computer sciences can take the form of a mechanistic explanation (Piccinini, 2004, 2006a, 2006b, 2007).

11.3.1 Mechanistic explanation of cognitive phenomena

What is distinctive about mechanistic explanations is their appeal to the components of a system and their causal interactions. According to this account, to explain a phenomenon is to give an account of how a causal mechanism, a hierarchical system composed of component parts and their properties, gives rise to, sustains, or produces the phenomenon. Each component is able to perform (causally) some operation and interact (causally) with other parts of the mechanism so that the coordinated operation of the parts is what constitutes the systemic activity of the mechanism.

Constructing a mechanistic model involves mapping elements of a mechanistic model to the system of interest, so that the elements of the model correspond to identifiable constituent parts with the appropriate organization and causal powers to sustain that organization. A mechanistic explanation requires a realistic and causal interpretation of the implementation of the mechanism. To see this more clearly, think of a clockwork model of the solar system (Ptolemaic or Copernican). This model presents a mechanism that allows one to predict the observable behaviour of the planets.[8] The clockwork clearly does not *explain* the causes of the regularities in a mechanistic way, since there are really no gears and differentials in space that are responsible for the behavioural regularities (i.e. the epicyclic geocentric orbits or the heliocentric orbits of the two systems, respectively). Indeed, we know now that there are no such gears, but instead the regularity is traced back to universal laws (Newton's laws) governing the motions of the planets, resulting in trajectories that unfold 'as if' run by the mechanism. A genuine mechanistic explanation requires that the trajectories should not merely conform to the constraints imposed by the mechanism, but that the (parts of) the mechanism should actually causally produce these trajectories.

While mechanists vary slightly in their precise definitions of a mechanism, there seems to be a clear consensus in the literature that mechanisms are *causally* responsible for phenomena, and that mechanistic explanation is therefore a form of *causal* explanation. For instance, Glennan defines a mechanism underlying a behaviour as a complex system which *produces* that behaviour by of the interaction of a number of parts according to direct *causal* laws (Glennan, 1996, p. 52). The term 'causal' distinguishes an actual cause from simple correlations, and the 'direct, causal laws' attempt to capture the idea that one part in the mechanism must be the immediate actor on the

[8] It can be interpreted merely as a heuristic device that is able to save the phenomena without explaining them (Duhem, 1969).

next part (Glennan, 1996, p. 55). Subsequently, Glennan has characterized the 'laws' as 'direct, invariant, change-relating generalizations' instead of using the philosophically loaded concept 'law' (Glennan, 2002, p. 345, fn.1). In this formulation a mechanism for a behaviour 'is a complex system that produces that behavior by the interaction of a number of parts, where the interactions between parts can be characterized by direct, invariant, change-relating generalizations' (Glennan, 2002, p. 344). Bechtel and Abrahamsen define a mechanism as 'a structure *performing* a function in virtue of its component parts, component operations, and their organization,' and add that 'The orchestrated functioning of the mechanism is *responsible* for one or more phenomena' (Bechtel & Abrahamsen, 2005, p. 423, emphasis added). In Craver's characterization, mechanisms are collections of entities and activities, which are organized in the *production* of regular changes from start or set up conditions to finish or termination conditions (Machamer, Darden & Craver, 2000; Craver, 2001; Craver, 2007). In this account, a mechanism is a structure performing a function, given initial and boundary conditions, in virtue of its component parts, component operations performed by those parts, and the organization of the parts into a functional whole (the 'system'). For example, the heart's function, the behaviour to be explained, is to maintain blood pressure and circulation. This is achieved by an intricate system of contractile fibres and a neural mechanism, whereby the fibres synchronize their contractions.

The explanandum of a mechanistic explanation is thus some behaviour of the system or some output generated by the mechanism (Machamer, Darden & Craver, 2000; Craver, 2005, 2006). The explanans of a phenomenon is either the model of the mechanism (the epistemic version) that describes the *causal* factors or the causal mechanism (the ontic version) responsible for carrying out the component operations that produce or sustain the phenomenon (Bechtel & Abrahamsen, 2005; Craver, 2006).[9] In addition, Bechtel and Richardson identified two heuristic assumptions, 'decomposition' and 'localization' (Bechtel & Richardson, 1993). Decomposition as an explanatory strategy consists of the assumption that the overall activity results from the execution of the component operations, and localization as a scientific strategy rests on the assumption that these tasks are performed by particular kinds of components into which we can identify and analyse the system.

Now, if we apply the mechanistic account of explanation to cognitive phenomena, then to explain a cognitive phenomenon is to explain how a certain cognitive mechanism gives rise to, sustains, or produces that phenomenon.

[9] Mechanists disagree on whether explanation should be considered in 'epistemic' or 'ontic' terms. The distinction derives originally from Salmon (1984). The epistemic view is that a model of a mechanism gains its explanatory status based on its being part of an argument, while the ontic conception says that a model explains if it represents an actual mechanism in nature (see Bechtel & Abrahamsen, 2005; Wright, 2006; Bechtel, 2008).

The explananda would be systems of cognitive operations, and the behaviours would be complex systems performing these operations. If we apply the heuristic strategies of decomposition and localization, the mechanistic explanation of a complex cognitive operation would present a cognitive mechanism that is able to carry out the complex cognitive operation by breaking it up into simpler operations which – ultimately – are made up from primitive operations carried out directly by the hardware, explaining how the hardware sustains or implements cognition. This has proved to be a very promising strategy in many areas of cognitive sciences: It underlies the classical symbolic paradigm in artificial intelligence, and cognitive psychology as well as many connectionist models. This strategy is also implicit in the epistemic goal of cognitive neuroscience, viz. localizing the neural basis of cognitive phenomena and the lower level physiological neural mechanisms.

However, we argue that as a species of abstract mechanisms Turing machines are used in computational explanations in a manner that does not conform to this strategy of explanation. The problem is not that one could not decompose Turing machines into parts, nor that Turing machines would lack the characterizing properties of mechanisms. As we illustrate below, Turing machines are mechanisms. However, they are not concrete causal mechanisms. And in the standard account of mechanistic explanations, the mechanisms involved are always causal mechanisms. A mechanism is not merely a collection of its parts; it also includes the way the parts *interact* with each other, spatially, temporally and causally.

11.3.2 Abstract and concrete mechanisms

If Turing machines are not concrete mechanisms, what kinds of mechanisms are they? A mechanism must be something more than just an aggregation of its parts. In standard mechanistic accounts this 'something more' is the causal organization of the mechanism. For example, in Craver's characterization, mechanisms are collections of entities and activities, which are organized in the production of regular changes from start or set up conditions to finish or termination conditions (Machamer, Darden & Craver, 2000; Craver, 2001, 2007). In Craver's terminology this organization, which distinguishes mechanisms from mere aggregates or events is the *active organization* of a mechanism (Craver, 2001, 2007). Craver illustrates this feature of mechanistic organization by comparing the active organization of mechanisms with the conditions of aggregation developed by William Wimsatt:[10] Suppose that a property or activity (ψ) of the whole (S) is explained by the properties or activities $\{x_1, \ldots, x_n\}$ of its parts $\{X_1, \ldots, X_m\}$. The ψ- property of S is an aggregate of the x-properties of X's when:

[10] Originally from Wimsatt (1997), but we use Craver's modified version (Craver, 2007, p. 135).

(W1) ψ is invariant under the rearrangement and intersubstitution of Xs;

(W2) ψ remains qualitatively similar (if quantative, differing only in value) with the addition and subtraction of Xs;

(W3) ψ remains invariant under the disaggregation and reaggregation of Xs; and

(W4) There are no cooperative or inhibitory interactions between Xs that are relevant to ψ.

Each of these criteria (W1–W4) points to the absence of organization among the parts in mere aggregates. This interconnectedness is the essence of a mechanism; the components of mechanisms, in contrast to those of mere aggregates, are in various *dependecy relations* to each other. (A model of the mechanism specifies these dependencies.) A part cannot be freely intersubstituted with another (W1), because other parts depend on the characteristics of that part. The parts cannot be removed or multiplied without changes elsewhere (W2). Changing the relationships between interdependent components would break down the organization of a mechanism (W3), and in the mechanisms there are systematically different kinds of dependencies (inhibitory and cooperative interactions) between the components (W4).

Turing machines fulfil criteria W1–W4, and hence they qualify as mechanism. A Turing machine clearly is not a mere aggregate. In its canonical representation a Turing machine consists of a list of instructions or rules that can be represented as the machine's machine table. The inputs and outputs are represented as strings of symbols on a tape. The system is clearly organized in a way that is *not* invariant with respect to rearrangement or intersubstitution of symbols on the tape, or cells in the machine table, as changing the structural arrangement of the components (symbols and machine table entries) can dramatically change behaviour (the function computed). Further, there is interaction between the components of the machine in that operations depend on the location and 'movements' of the scanning head. Thus Turing machines are mechanisms.

When the notion of computing mechanism is interpreted as a *concrete* mechanism, symbols are understood as states, and symbol structures are spatiotemporal arrangements of these symbols (Piccinini, 2007). Under this interpretation computation is a *process* unfolding in real time. However, if we allow that the conceptual glue that binds a mechanism together need not be causation, and the relations among the parts of a mechanism need not be spatiotemporal, we can apply Wimsatt's criteria to *abstract Turing machine, i.e. abstract mechanisms* as well.

Computation is mathematically defined in terms of symbol structures (e.g. strings of letters produced from a finite alphabet), and instructions that generate new symbol structures from old ones in accordance with a recursive function. A computation is a proof-like sequence of symbol structures related to each other by operations specified by the instructions. None of this involves the notion of physical causation. Neither do any of the criteria for mechanistic organization W1–W4 strictly imply causal organization. We conclude that there is no good reason to question the idea that an abstract computing mechanism is a coherent one.

Thus the term 'computing mechanism' can refer to two things. On the one hand, it can refer to concrete computing mechanisms such as (*ex hypothesi*) brains, computers and so on. On the other hand, it can refer to abstract computing mechanisms such as abstract Turing machines. In a similar way that a concrete Turing machine is more than just an aggregate of its parts and has a mechanistic organization, an abstract Turing machine is more than an aggregate of its parts and has a 'mechanistic' organization as well.

Turing machines and their computations are abstract in a very strong sense: Turing machines can be defined to have properties that are not and cannot be implemented in any real computational system (infinite memory, unlimited processing time and other idealized properties), but the most important thing is that they do not operate in real space and time. Their theoretical use does not require one to commit to a literal localization of the parts, or a decomposition of the mind/brain into a tape and a read/write head, for example.

This creates a problem for those who would subsume Turing machines under the banner of the current mechanistic explanation. The current mechanistic account of explanation, as it stands, talks exclusively about *concrete* mechanisms: they require the mechanisms in explanations to be *implemented*. Mechanisms are anchored in their components, and those components occupy space and they act in real time (Craver, 2007). In short, mechanistic explanation requires a realistic causal interpretation of the implementation of the mechanism whereas Turing machines are abstract.

11.4 Computational explanation based on abstract mechanisms

A useful way to look at the basic distinction between concrete and abstract computational mechanisms is to view them operating at the different levels of explanation: algorithmic (performance) and computational (competence). It is important to understand that these theories describe cognitive organization at a different level of abstraction. The theory of abstract computational mechanisms lies at a higher level of abstraction than specific, concrete mechanisms.

Whereas the abstract computational level specifies the information represented and operated on, the level of representations and algorithms describes the syntactic or formal means by which the information is explicitly represented and operated on. Using abstract mechanisms at the computational level does not commit one to the same assumptions regarding the internal causal organization of an organism, as assuming concrete mechanisms at the algorithmic level would.

This distinction has not always been fully appreciated. However, the computational level is indispensable in computational explanations in the neurocognitive sciences, and abstract Turing machines are part of the story. They are used in theories that characterize representational requirements and constraints. For instance, the recent rational and probabilistic approach to cognition includes theories that are formulated at this level of abstraction (for example, Chater, Tenenbaum & Yuille, 2006). The role of this computational level is captured nicely in the following quote from Anderson (1991b, p. 471):

> A rational theory [...] provides an explanation at a level of abstraction above specific mechanistic proposals. [...] One might take the view [...] that we do not need a mechanistic theory, that a rational theory offers a more appropriate explanatory level for behavioral data. This creates an unnecessary dichotomy between alternative levels of explanation, however. It is more reasonable to adopt Marr's view that a rational theory (which he called 'the computational level') helps define the issues in developing a mechanistic theory (which he called the level of 'algorithm and representation'). In particular, a rational theory provides a precise characterization and justification the mechanistic theory should achieve

As an example – chosen here on the basis of involving Turing machines – Chater (1996) represents perception as the computational task of encoding perceptual stimuli into the *simplest* possible representation. Here simplicity is defined in terms of algorithmic complexity (Chaitin, 1977; Kolmogorov, 1968; Solomonoff, 1964a, 1964b). This is a theory at the computational level, because it defines the *computational task* of perception as producing the simplest possible encoding of the stimulus. This is a *law of perceptual organization*, and the Turing machine is part of this law, because it is part of the definition of complexity and simplicity: the complexity of an encoding is defined as the length of the shortest Turing machine that will produce that encoding. No mechanistic hypotheses of the causal agents, Turing machines or otherwise, are put forward. Generally, it is *not* the case that the encoding would actually be literally *produced* by the shortest (or indeed any) Turing machine. The abstract computational mechanisms are here used to specify the information processing *tasks*, mappings from one kind of information to another. What the abstract computational theory allows one to do is to specify the information processing task precisely and unambiguously. This level constitutes *a theory of competence* for a specific cognitive capacity – here it is perception, but analogous theories can be formed for language, reasoning, decision making, etc.

The computational theory can be used to characterize the information constraints that define the cognitive task which the brain of the organism faces. In that sense the computational level is perhaps rather more like the *laws of motion* governing the motion of the planet, rather than mechanically causing them. When the system is working appropriately its behaviour is *governed* by the computational principles, though the principles themselves need not be written down anywhere and consulted or explicitly represented as constituent elements with causal power to *drive* the behaviour of the system. Descriptions of singular *performances* are given at the algorithmic or implementation levels. In the case of neurocognitive sciences these performances would be at the functional (symbolic or connectionist) or neurological (systemic, cellular, molecular) levels.[11] These levels give a description of the concrete mechanisms that *fulfil* the tasks described at the computational level. They explain *how* one ends up from one representation (which makes explicit a piece of input-information) to another (for output-information), such that this derivation respects constraints already defined at the level of computation.

What is the explanatory role of the abstract, Marrian, computational level? This computational theory answers 'what' questions and 'why' questions. These are questions such as 'what is the goal of the computation?' (e.g. to add natural numbers) and 'why is it appropriate?' (Given the representational convention of the arabic numerals, the algorithm is faithful to the rules governing the operation of addition, which is defined at the level of computation.) The algorithmic level answers 'how' questions ('How are the number representations – numerals – meant to be manipulated?') by describing formal means of explicitly representing the information and syntactically manipulating these representations.

With a computational theory one can explain many things about the concrete mechanisms. For example, if one considers, *why* this pattern of synaptic changes is such-and-such, one can answer *because they serve to store the value of x needed in order to compute y*. Or, *why* is the wiring in this type of ganglion cell such-and-such? *Because the wiring computes, or approximates a computation of, some variable x*. In other words we are able to explain many phenomena about the lower levels by their representational character, and the appropriateness of the mechanism for the computational task. As Marr stressed, although the concrete mechanisms are perhaps more readily investigated, since they have causal powers to affect our measuring equipment, after all, the computational level is an equally important level from an information-processing theory point of view (Marr, 1982).

[11] These are the levels that Marr calls the levels of algorithms and representation, and implementation, but often algorithmic level descriptions are called 'computational', which can cause some confusion.

11.5 Conclusions

Turing machines are often understood to be abstract entities. As such, they are not part of the causal order of things. They do not run in real time. No one has ever built nor started an abstract Turing machine. Despite this, they offer not only the theoretical foundation for computer science, but also are a part of many theories and explanations in cognitive science.

In this article we have emphasized the role of Turing machines in computational explanations, and considered whether the standard account of mechanistic explanations is appropriate for describing these explanations. We have argued that although Turing machines are mechanisms, computational explanations in cognitive science making use of Turing machines do not conform to the standard account of mechanistic explanation. This is because mechanistic explanations, as traditionally conceived, require a realistic and causal interpretation of the mechanism, whereas abstract computing mechanisms, such as Turing machines, are not concrete causal mechanisms.

Acknowledgements

We thank Jani Raerinne and two anonymous reviewers for their comments and suggestions.

References

Anderson, J.R. (1991b). Is human cognition adaptive? *Behavioral and Brain Sciences*, 14: 471–457.

Bechtel, W. & Abrahamsen, A. (2005). Explanation: A mechanistic alternative. *Studies in the History and Philosophy of Biological and Biomedical Sciences*, 36: 421–441.

Bechtel, W. & Richardson, R. (1993). *Discovering Complexity, Decomposition and Localization as Strategies in Scientific Research*. New Jersey: Princeton University Press.

Bechtel, W. (2008). *Mental Mechanisms: Philosophical Perspectives on Cognitive Neuroscience*. London: Routledge.

Chaitin, G. (1977). Algorithmic information theory. *IBM Journal of Research and Development*, 21: 350–359.

Chater, N. (1996). Reconciling simplicity and likelihood principles in perceptual organization. *Psychological Review*, 103(3): 566–581.

Chater, N., Tenenbaum, J.B., & Yuille, A. (2006). Probabilistic models of cognition: Conceptual foundations. *Trends in Cognitive Sciences*, 10(7): 287–291.

Chater, N. & Vitanyi, P.M.B. (2003). The generalized universal law of generalization. *Journal of Mathematical Psychology*, 47: 346–369.

Chomsky, N. (1965). *Aspects of the Theory of Syntax*. Cambridge, MA: MIT Press.

Churchland, P. & Sejnowski, T. (1992). *The Computational Brain*. Cambridge, MA: MIT Press.

Craver, C.F. (2001). Role functions, mechanisms and hierarchy. *Philosophy of Science*, 68: 53–74.

Craver, C.F. (2005). Beyond reductionism: mechanisms, multifield integration and the unity of neuroscience. *Studies in the History and Philosophy of Biological and Biomedical Sciences*, 36: 373–395.

Craver, C.F. (2006). When mechanistic models explain. *Synthese*, 153: 355–376.

Craver, C.F. (2007). *Explaining the Brain: What a Science of the Mind-Brain Could Be*. New York: Oxford University Press.

Duhem, P. (1969). *To Save the Phenomena. An Essay on the Idea of Physical Theory from Plato to Galileo*. Chicago, IL: University of Chicago Press.

Fodor, J.A & Pylyshyn, Z. (1988). Connectionism and cognitive architecture. A critical analysis. *Cognition*, 28: 7–30.

Glennan, S. (1996). Mechanisms and the nature of causation. *Erkenntnis*, 44: 49–71.

Glennan, S. (2002). Rethinking mechanistic explanation. *Philosophy of Science* 69: 342–353.

Glennan, S. (2005). Modeling mechanisms. *Studies in the History and Philosophy of Biomedical Sciences*, 36: 443–464.

Grush, R. (2001). The semantic challenge to computational neuroscience. In P. K. Machamer, R. Grush & P. McLaughlin (eds.), *Theory and Method in the Neurosciences*, pp. 155–172. Pittsburgh: University of Pittsburgh Press.

Kolmogorov, A.N. (1968). Logical basis for information theory and probability theory. *IEEE Transactions on Information Theory*, 14(5): 662–664.

Machamer, P.K., Darden, L., & Craver, C.F. (2000). Thinking about mechanisms. *Philosophy of Science*, 67: 1–25.

Marr, D. (1982). *Vision: A Computational Investigation into the Human Representation of Visual Information*. San Francisco: W.H. Freeman.

Miller, G.A., Galanter, E., & Pribram, K.H. (1960). *Plans and the Structure of Behavior*. New York: Holt, Rinehart & Winston.

Newell, A. (1980). Physical symbol systems. *Cognitive Science*, 4(2): 135–183.

Newell, A. & Simon, H. (1972). *Human Problem-Solving*. Englewood Cliffs, NJ: Prentice Hall.

Piccinini, G. (2004a). Functionalism, computationalism and mental contents. *Canadian Journal of Philosophy, 34*: 375–410.

Piccinini, G. (2004b). Functionalism, computationalism, and mental states. *Studies in the History and Philosophy of Science, 35*: 811–833.

Piccinini, G. (2006a). Computational explanation and mechanistic explanation of mind. In M. DeCaro, F. Ferretti & M. Marraffa (eds.), *Cartographies of the Mind: The Interface Between Philosophy and Cognitive Science*. Dordrecht: Kluwer.

Piccinini, G. (2006b). Computational explanation in neuroscience. *Synthese*, 153(3): 343–353.

Piccinini, G. (2007). Computing mechanisms. *Philosophy of Science*, 74: 501–526.

Piccinini, G. (2008). Some neural networks compute, others don't. *Neural Networks*, 21.2–3: 311–321.

Revonsuo A. (2006). *Inner Presence. Consciousness as a Biological Phenomenon*. Cambridge, MA: MIT Press.

Salmon, W.C. (1984). *Scientific Explanation and the Causal Structure of the World.* Princeton: Princeton University Press.

Shepard, R.N. (1987). Toward a universal law of generalization for psychological science. *Science,* 237(4820), 1317–1323.

Solomonoff, R.J. (1964a). A formal theory of inductive inference. Part I. *Information and Control,* 7(1): 1–22.

Solomonoff, R.J. (1964b). A formal theory of inductive inference. Part II. *Information and Control,* 7(2): 224–254.

Turing, A. (1936). On computable numbers, with an application to the entscheidungsproblem. *Proceedings of the London Mathematical Society,* ser. 2. vol. 42: 230–265.

Turing, A.M. (1950). Computing machinery and intelligence. *Mind,* 59(236): 433–460.

Gelder, T.V. (1995). What might cognition be, if not computation? *The Journal of Philosophy,* 92(7): 345–381.

Wimsatt, W.C. (1997). Aggregativity: Reductive heuristics for finding emergence. *Philosophy of Science* 64(4): 372–84.

Wright, C. & Bechtel, W. (2007). Mechanisms and psychological explanation. In P. Thagard (ed.), *Philosophy of Psychology and Cognitive Science,* pp. 31–77. Amsterdam: Elsevier.

Appendix: Turing machines and algorithmic complexity

An algorithm is a list of instructions that can be followed mechanically (without rational understanding), and which, when followed, will compute a function. It will take arguments of the function as input, and produce values as output. Turing's (1936) idea was to analyse the concept of algorithmic information processing into its essential details. Hence, the idea of the Turing machine. We use the *definite singular* for the common *idea* behind the various simple abstract devices covering the various classes of Turing machines, and the plural when we are talking about different machines for computing specific functions.

Here is one way to present the idea of the Turing machine: A Turing machine M has a *tape,* an indefinitely extendable list of cells. Symbols drawn from a *finite alphabet* A may be written in the cells. Without loss of generality, we can assume the symbols to be drawn from the binary alphabet consisting of $A = \{1, 0\}$ (as symbols from any finite alphabet may be recoded into finite binary strings). The machine has a *finite control,* a read/write head, which can be in one *internal state* $q_1, q_2, \ldots, q_n$ out of a finite number n of internal states. The head *scans* the tape one cell at a time, writes a symbol from the alphabet on the tape (possibly the one already on the tape, effecting no change), and moves one step to the left or to the right or stays put. (Again, we could give the head the capability of scanning, reading and writing more symbols at a time, but this would not make a difference.) The operation of the head is based on a finite list of *rules,* which determines for each ordered pair of scanned symbol and internal state the operation to be performed: erasing or replacing the symbol (or leaving it alone), changing its internal state (or remaining in the same state), and possibly moving to the left or right (or staying put). The rules can be represented as a *machine table* with two columns for the symbols, and n rows, one for each of the n possible internal states. The entries in the cells of the table will contain the operation to be carried out when scanning a particular symbol in a particular state.

A Turing machine will accept certain *input* strings on the tape $x_1, x_2, \ldots,$ which are finite digital objects, and, determined by its list of rules, operate on them. If the machine halts, then it is said to have produced an *output*, which is the string of symbols on the tape. Different machines (with different control rules) will compute different mathematical functions (mappings from input strings to output strings).

There are special kinds of Turing machines, called *universal machines* $U_1, U_2, \ldots$ which accept as input encoded representations of any Turing machine (fully specified by their *finite* control), and inputs the encoded Turing machine would accept, producing as output the very output that the encoded machine would produce. The encoding of a Turing machine on the tape of the universal machine is called a *program*. The symbols corresponding to the input for the simulated machine is called *data*. The universal machine carries out on the data the operations given in the algorithmic instructions in the program.

This way a universal machine U can compute any function that there exists some Turing machine to compute. According to the *Church–Turing thesis* the class of functions which can be computed 'mechanically' is the class of functions that can be computed by some Turing machine and therefore any universal machine U can compute any and all mechanically computable functions, by simulating the Turing machine M_k appropriate for computing the function f_k.

We can now define the *algorithmic complexity* of a (digital) object x as the *length* (in binary digits) of the shortest program that will generate that object as its output (with empty input). A universal machine U accepts as inputs programs corresponding to the (denumerable) set of all Turing machines. For each of those programs, either x is their output or it isn't. Since the set of Turing machines is countable, we can enumerate the programs, ordering them by size, and select the smallest program. (Assuming we can determine for each Turing machine whether or not it will produce x and halt; unfortunately, this halting problem was shown by Turing to be non-computable, and therefore the property of algorithmic complexity is itself non-computable.)

There are different universal machines which will have slightly different encodings for the Turing machines, and thus the length of the program for a given Turing machine will vary slightly depending on our choice of universal machine. However, a central result in algorithmic complexity theory, the *invariance theorem* establishes equivalence, up to c, a small additive constant. That is, for all universal machines U_p, U_r, $K_p(x) \leq K_r(x) + c$, where $K_p(x)$ is the length of the shortest program for universal machine p that will output x, $K_r(x)$ is the length of the shortest program for universal machine p that will output x, and c is a small (positive or negative) constant. Therefore, the complexity $K(x)$ of x is intrisic to x and not dependent on our choice of unversal machine.

·12·

Real causes and ideal manipulations: Pearl's theory of causal inference from the point of view of psychological research methods

Keith A. Markus

Abstract

Pearl's work on causation has helped focus new attention on the nature of causal reasoning and causal inference in behavioural science. Pearl takes an axiomatic approach, presenting axioms as first principles, but these may be better understood as boundary conditions for the application of the theory. Pearl adopts a non-eliminative but instrumental approach to causation which creates some tension with the tradition of ruling out rival hypotheses in the behavioural sciences. Finally, much causal reasoning in the behavioural sciences involves reasoning across possible worlds that differ in their causal structure, which becomes awkward within the basic architecture of Pearl's system. A neighbourhood semantics approach could represent this type of reasoning more naturally. Consideration of these issues may be helpful both to behavioural scientists working to incorporate Pearl's work and also to those working outside the behavioural sciences attempting to explain causal reasoning within those sciences.

Following a resurgence of interest in causal inference (Hoover, 2004; Russo and Williamson, 2007), Pearl's (2000[1]) book has helped rejuvenate methodological interest in causal inference in psychology and the behavioural sciences. The present chapter explores three issues relating Pearl's theory to causal reasoning in these sciences. The axioms of Pearl's theory can be read two ways: as universal principles or as assumptions that set the boundaries of application. Whereas Pearl (2009) seems to favour the former reading, the first section of this chapter agues in favour of the second reading. One can sustain the former reading with respect to the narrow issue of the formal account of causal models and the axioms. However, as soon as one broadens one's view to

[1] The present chapter was written in June 2009 without the benefit of Pearl's (2009) second edition. The second edition was incorporated during revisions undertaken in November of the same year.

consider how well these together characterize the domain of causal reasoning from models, outside of the formal representation, only the second reading applies.

The second section focuses on Pearl's causal non-eliminative instrumentalism (causation is needed but in the head) and its lack of fit with the tradition of evaluating rival explanatory hypotheses (Cook and Campbell 1979; Shadish, Cook and Campbell, 2002) in behavioural science research. A detailed discussion of Pearl's proposals regarding causal inference falls beyond the scope of the present chapter. The goal is merely to examine prevailing practices of causal inference in the behavioural sciences as a backdrop to Pearl's theory of reasoning from causal models as it relates to behavioural science research.

The final section argues that the nature of counterfactual reasoning in the behavioural sciences motivates allowing causal structure to vary across possible worlds, in addition to value assignments. Expanding the model accomplishes this better than approaches that work within the model because the expanded model maintains a clearer distinction between variables and causal functions. The chapter may be useful both to researchers interested in Pearl's work and to those interested in studying causal reasoning in psychology and behavioural science.

12.1 Interpreting Pearl's axioms

> We present three properties of counterfactuals – composition, effectiveness, and reversibility – that hold for all causal models (Pearl, 2009, p. 228).

> With all due respect to multiculturalism, all approaches to causation are variants or abstractions of the structural theory presented in this book (Chapter 7) (Pearl, 2009, p. 353).

This section begins by reviewing the structural account of causal models and the axioms that characterize them presented by Pearl (2009, Chapter 7). Some consideration is given to the proper interpretation of these components of Pearl's account. Then, several counterexamples are considered, including Pearl's theory itself. When considered in relation to the domain of causal reasoning with causal models, and not just the particular type of model stipulated early on, it seems clear that Pearl's theory covers only a restricted domain. As such, the axioms that characterize the theory serve to delimit the domain of applicability rather than providing universal truths that apply to all causal models.

The discussion that follows will take for granted several fundamental concepts. Pearl's theory assumes a context in which the cognitive agent (a researcher, say) already has causal knowledge and seeks to reason from that knowledge (this discussion does not directly address the theory of causal induction). This knowledge involves a universe of one or more individual property bearers, and sets of mutually exclusive and exhaustive non-relational properties organized as variables over those individuals. Although Pearl's theory does not seek to provide a reductive account of causation (Woodward, 2003a, 2003b), it assumes that causal reasoning involves determinative relationships between the values of these variables for a given unit or set of interconnected units. For example, one variable might code the position of a light switch as on or off and another variable might code the state of a light as on or off. The switch and the light constitute distinct units, but a causal relation holds between the two variables if the position of the light switch causally determines the status (on-ness or off-ness) of the light (clearly a non-reductive theory) possibly given some boundary conditions (such as the presence of electricity in the line and a working light bulb) which may be included in the causal model or simply assumed constant.

The causal relation is further assumed to be non-extensional as is now commonly accepted (Davidson, 2001; Mellor, 1995). In other words, a claim that the light switch position causes the light to go on or off does not simply assert that all actual instances of appropriately paired lights and switches conform to a general law that rules out disconcordant pairings of switch states and light states such as lights shining on a switch in the off position. Such purely extensional assertions lack the deductive strength to support the kind of reasoning that the theory seeks to provide an account of. Instead, causal assertions carry further content with implications constraining the structure of non-actualized possible states. The causal assertion regarding the light and switch implies not only that if the switch is in the on position the light is shining but further that had the switch been placed in the on position (although it was not) the light would have shone (although it did not). As such, causal claims support counterfactual inferences (Collins, Hall and Paul, 2004). Rubin and others (Rubin, 1974; Holland, 1986) have characterized these counterfactuals in terms of potential responses. The light has shining as its potential response to the placement of the switch in the on position and not shining as its potential response to the opposite. The light and switch can only actualize one of these potential responses at any given time, but the causal claim entails both even for lights that come in and out of existence without ever having been turned on. This approach to causation is sometimes referred to as a black-box approach because it merely maps the nomothetic causal relationships between variables without offering an account of the process that maintains these relationships. Nonetheless, it is commonly assumed that whatever makes counterfactual assertions true or false must exist extensionally in the actual state of affairs, and the underlying processes offer a likely candidate for truth-makers of black-box assertions about counterfactual states of affairs and the causal assertions that entail them (Markus, in press).

The current discussion further assumes that Pearl's (2009) theory of causal induction (Chapter 2) describes inferences to the kind of causation described in Pearl's theory of causal inference (Chapter 7). A reviewer suggested an alternative reading in which these two aspects of Pearl's book are unrelated to one another, making the above stated assumption potentially misleading to the reader. For reasons given immediately below, I retain the above assumption but hope that clearly acknowledging it here eliminates any potential to mislead. The reviewer is certainly correct in pointing to a substantial body of literature in cognitive psychology that attempts to work out a computational theory of inference from probabilistic dependence to causal beliefs of the form *A* causes *B* without attempting to spell out the content of such beliefs in much detail (Cheng, 1997; Goodman, 1983). Indeed, operationally defining causal belief as assent to statements of the above form might make the reference of such statements inscrutable (Quine, 1960). My attempts to construct a satisfactory reading of Pearl's (2009) book along these lines have only further convinced me that the integrations of inference to causal beliefs with inferences from causal beliefs constitutes an integral element of Pearl's approach and a strength of that approach, but others may succeed where I have failed. As a practical matter, such alternative approaches fall beyond the scope of the present chapter but the reader should not take this to suggest that cognitive psychologists have not developed associative learning theories of causation or that one cannot consider reading Pearl's work in this tradition.

12.1.1 Pearl's theory of causal inference from causal models

The axioms presented below animate a general theory of causal models. The theory defines a causal model, M, as an ordered set of the form $\langle U, V, F\rangle$. U contains variables that have no causes within the model whereas V contains variables that have causes within the model. F contains a set of functions to each variable in V from the union of U and V omitting the member of V for which the function gives the causes from among the causes of itself. The potential response $Y_x(u)$, then, corresponds to the solution for Y of the set of functions F substituting in the value x of X fixed by intervention (Pearl, 2000, pp. 203–204) and the values u of U.[2] As discussed in more detail in a later section, the term 'intervention' may carry some unwanted connotations in the present context. The variables are elements of a mental model (possibly represented externally by symbolic means) and interventions are therefore mental operations on the mental representations. The interventions do not refer to actions on the systems represented by the causal models. One can answer counterfactual questions about how one's life might change if one won a lottery through interventions on one's mental model even though one

[2] Note that in Pearl's notation u refers to fixed values of background variables whereas in Rubin's notation u refers to the unit that has those values.

cannot intervene in the world to make it transpire that one wins the lottery without violating the constitutive conditions of what makes it a (fair) lottery.

To switch to a more behavioural example, consider the Müller–Lyer illusion (Judd, 1905). As shown in Figure 12.1, line *A* appears longer than line *B*. The lines have equal length, and the different orientations of the arrow heads causally influence the perception of line length. The experimental phenomenon does not uniquely determine the representation as a causal model, as Pearl recognizes. As one approach, one could code the position of the arrow heads as one variable (X) and the perceptual judgment as another variable (Y). For convenience, one could code both variables to have the values '*A*' and '*B*' such that $x =$ '*A*' means that line *A* has the outward pointing arrowheads and line *B* inward and $y =$ '*A*' means that line *A* is judged longer than line *B* (mutatis mutandis for $x =$ '*B*' and $y =$ '*B*'). The causal relationship between variable X and variable Y would then be the identity function such that $X = Y$. The potential response $Y_x(u)$ then represents the perceptual judgment of relative length that would result for the observer if the arrowheads were fixed the value x of X (e.g. $Y_A(u) =$ '*A*' for $X =$ '*A*'). Alternatively, one could select lines as units, code their arrowhead position and judged length as variables (W and X), and then code their judged relative-length compared to a line of equal length with outward-facing arrows (say) as '*same*', '*shorter*', or '*longer*' (Y) in which case one could model a causal chain such that W causes X and X causes Y. Additional models can be generated from alternative representations with their convenience waning as the choice of units and variables departs further and further from common-sense natural-language descriptions of the situation.

Pearl (2009, Chapter 7; 2000; Galles and Pearl, 1997, 1998) presents three axioms to characterize causal counterfactual assertions of the form $Y_x(u) = y$ in the context of such models. Y and X refer to variables for which an individual takes determinate values x and y. Each of these may be multivariate, consisting of a set of univariate variables. The assertion, then, reads that intervening to set X to x has the result of causing Y to take the value y assuming a fixed value for u. Roughly, the first axiom asserts that a statement

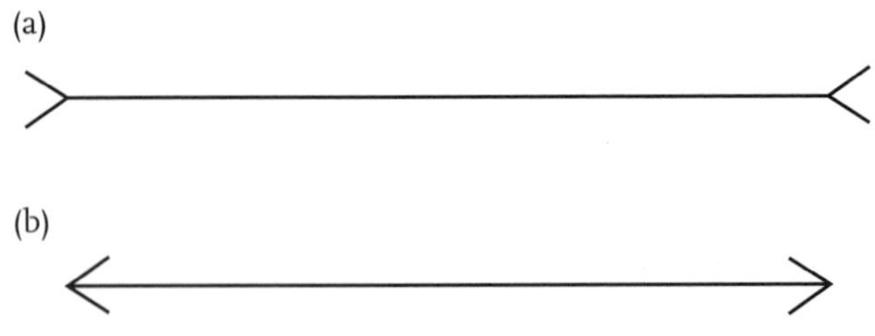

Fig. 12.1 The Müller–Lyer illusion.

of this form remains invariant to manipulations that fix other variables in the system to the values that they would have without the intervention. The second axiom asserts that an intervention to fix the value of a variable for a given unit will result in the variable taking on that value for that unit. The third axiom offers a patch for systems with causal loops and adds no additional content to causal systems without loops. In the potential response notation explained above, these axioms take the following form where '$\supset$' represents the material conditional ('If *A* then *B*' in the very weak sense of 'not [*A* and not *B*]' for which Pearl uses '$\Rightarrow$'), the equal signs denote nothing more than material numeric equality (i.e. a purely extensional relation that tells us nothing of what is possible or necessary), '$\exists$' represents existential quantification ('For some *x*'), '$\forall$' represents universal quantification ('For all *x*', which Pearl does not formalize), '|' represents a material condition on the quantification (e.g. '$(\forall x|x = y)(P)$' is equivalent to '$(\forall x)((x = y) \supset P)$'), '$\wedge$' indicates logical conjunction ('and'), and $S(.)$ indicates that the variable inside the parenthesis is univariate (i.e. scalar). For simplicity, the scope of the material conditional exceeds that of equality or other logical connectives such that '$A = B \supset C \wedge D$' means the same as '$(A = B) \supset (C \wedge D)$'. Finally, to help make them stand out, I enclose the formula to which the quantifiers apply in square brackets ('[...]') instead of parentheses.

Composition: $(\forall u|u \in U)(\forall\langle W, X, Y\rangle)(\forall x|x \in X)(\forall w|w \in W)[W_x(u) = w \supset Y_{xw}(u) = Y_x(u)]$.

Effectiveness: $(\forall\langle W, X\rangle)(\forall x|x \in X)(\forall w|w \in W)[X_{xw}(u) = x]$

Reversibility: $(\forall\langle Y, W\rangle|S(Y) \wedge S(W))(\forall X)(\forall x|x \in X)(\forall w|w \in W)(\forall y|y \in Y)[(Y_{xw}(u) = y) \wedge (W_{xy}(u) = w) \supset Y_x(u) = y]$

I have here chosen to render Pearl's axioms as quantifying over ordered sets of variables (e.g. $\forall\langle W, X, Y\rangle$) rather than sequentially quantifying over each variable (e.g. $(\forall W)(\forall X)(\forall Y)$) because Pearl's verbal statement does not clearly distinguish these and the former seems more general. The question of which way to interpret Pearl's verbal statements opens the door to a thicket of thorny issues which could easily occupy a separate discussion (Hintikka, 1996). Nonetheless, I do not believe that the arguments contained within the present chapter depend on this issue in any important way, particularly where only universal quantifiers are used. I have also made explicit quantification over *w*, *x*, and *y*, which denote values of the quantified variables *W*, *X*, and *Y*. I interpret Pearl to have intended an implicit universal quantification over these as well, presumably restricted to the range of possible values of the corresponding variable.

Returning to the Müller–Lyer illusion, assume that the arrowhead position of line *A* is coded as *W*, the arrowhead position of line *B* is coded as *X*, and the judgment of relative length is coded as *Y* with the perceiver taken as the unit (W and X representing conditions presented to the perceiver and thus a perceiver property and not just a property of two distinct lines). Composition implies that if W = '*inward*' results when X is manipulated (in the model) to X = '*outward*', then the potential response to the manipulation of W = '*inward*' and X = '*outward*' is identical to the potential response to the manipulation of X = '*outward*' without manipulating W in its current state of W = '*inward*'. In both cases, the potential response equals the judgment that line A is longer than line B as illustrated in Figure 12.1. Readers who feel tempted to over-interpret the 'if-then' relation as anything more than the material conditional may find it helpful to translate 'If A then B' to its logical equivalent 'Not (A and Not B).' Applied to the above, this states that in no instance does '$W_{X=\text{'outward'}}(u)$ = '*inward*' hold true but '$Y_{X=\text{'outward'}, W=\text{'inward'}}(u) = Y_{X=\text{'outward'}}(u)$' fail to hold true. Effectiveness implies that if one (mentally) manipulates X to '*outward*' in conjunction with manipulating W to any value one likes, the result will be that X ='*outward*'. Somewhat less intuitively, reversibility implies the following. If manipulating X = '*inward*' and W = '*outward*' yields the potential response that line A is judged longer and (as a distinct and independent counterfactual) manipulating X = '*inward*' and Y = '*A longer than B*' yields the potential response that W = 'outward', then manipulating only X = '*inward*' itself yields the potential response that Y = '*A longer than B*'. The first conjunct of the antecedent phrase (left-hand side) assures that under the condition that W = '*outward*', X = '*inward*' yields the expected result. The second conjunct assures that left to its own devices W ='*outward*' (in this example because X and Y have no causal impact on W and the potential response on W remains constant under any intervention on X and Y). These together assure the expected potential response to an intervention on X alone (right-hand side or consequent).

To these three basic axioms, Pearl (2000) adds two supplementary axioms. The first states that a potential response on a variable exists for every possible intervention. The second states that no more than one such potential response exists. The prime is used to distinguish distinct logical variables representing not-necessarily-distinct but determinate values of the corresponding upper-case causal variable (e.g. x and x' represent distinct but possibly equal values of X).

Existence: $(\forall u|u \in U)(\forall y|y \in Y)(\exists x|x \in X)[X_y(u) = x]$

Uniqueness: $(\forall u|u \in U)(\forall X|S(X))(\forall x|x \in X)(\forall x'|x' \in X)(\forall y|y \in Y)[(X_y(u) = x) \wedge (X_y(u) = x') \supset x = x']$

Thus, some potential response perceptual judgment (Y) exists for any manipulation on the arrowhead directions (W and X). No manipulation on W and X exists that fails to produce a potential response. Moreover, a manipulation to any combination of values for X and W yields no more than one value as the potential response on Y. In every case, the potential response takes a value from '*A longer*', '*B longer*' and '*same length*' and in every case the potential response has only one of these values.

12.1.2 Counterexamples to Pearl's axioms

Pearl (2000; Galles and Pearl, 1997, 1998) provided proofs that the proposed axioms correctly characterize the proposed causal model structure. However, the book does not devote space to considering how well either of these characterize the informal domain of discourse and reasoning with which the terminology used to present the model and axioms connects them. This distinction appears central to evaluating the interpretation of the axioms. If one follows Pearl's lead and focuses narrowly on the correspondence of the axioms to the specific model structure proposed in the book, then Pearl appears well founded in presenting these as universal truths about such models. In contrast, if one focuses more broadly on the correspondence of the axioms and model structure to the domain of causal reasoning with causal models as practiced in behavioural science research, then this interpretation does not hold up. Examples exist that seem like unambiguous examples of causal reasoning from causal models but which do not conform to the model structure or axioms presented within Pearl's theory. In this context, then, the axioms seem much better interpreted as demarcating the range of applicability of the theory. The theory only applies to instances of causal reasoning from models that conform to the axioms.

Efficacy versus effectiveness in program evaluation

One example of a form of causal reasoning that does not conform to Pearl's axioms involves the distinction between efficacy and effectiveness in the field of program evaluation. Efficacy refers to the causal effect of a program, such as a policy, treatment, or service provision, that the program would produce under ideal conditions. One can think of efficacy as the optimum possible causal effect. In contrast, effectiveness refers to the causal effect of a program that the program would produce under realistic conditions. Given that realistic conditions cannot exceed ideal conditions, otherwise the realistic conditions would constitute the ideal conditions, effectiveness generally remains less than or equal to efficacy.

One can easily assimilate these concepts to Pearl's theory by incorporating a moderator variable (discussed in more detail in the final section). At the coarsest level of representation, one might introduce a dichotomous variable coded '1' for ideal conditions and '0' for realistic conditions, and then make the effect of the program conditional on the value of the conditions variable: lower for conditions = 0 and higher for conditions = 1. However, such an assimilation does violence to the texture of the initial construction of ideal versus realistic conditions, and here lies the interest to the present concern.

If a program evaluation researcher had in mind a few key variables that need to have the correct values in order for the program to work – for example, children need to have a few basic needs met before they can benefit from education – then the above strategy might capture this very successfully (but see the third section below). However, in such an instance, the researcher is very unlikely to couch the issue in terms of ideal versus realistic conditions. He or she would much more likely list the handful of conditions for the program to produce its optimal effect. Talk of ideal versus realistic conditions involves a more diffuse cognitive representation. Even if one assumes the number of factors potentially influencing the effect of the program to be finite, talk in terms of conditions suggest a very large set of such variables, more than can be effectively enumerated and represented in a mental model. Certainly it might involve more effects than can be statistically estimated given available sample sizes. As a result, an early stage of any modelling endeavour involves the selection of a core set of variables for inclusion in the model, relegating the rest to background conditions. Talk of efficacy versus effectiveness most naturally describes causal reasoning in which causal effects in the model depend upon unmodelled (and probably unmodelable) background conditions. As such, the equations in the model cannot provide point estimates of potential responses because the parameters in the equations governing causal effects vary over unmodelled background conditions rather than having fixed values. This feature introduces an inherent vagueness of reference outside the range of what the type of models assumed in Pearl's theory can represent (Keefe & Smith, 1997; Williamson, 1994). Nonetheless, talk of efficacy and effectiveness proves extremely useful for program evaluation research design, marking important absolute and relative distinctions between what alternative research designs estimate and the interpretations that various research results support.

To make the above argument concrete, consider the following example. I know that my vacuum cleaner does not work as well when it becomes too full. The phase 'too full', however, remains vague in the sense that it does not refer to a specific degree of fullness. There is no point at which the addition of one more grain of dust would cross the semantic threshold from not-too-full to too full, or vice versa for the removal of single grain. As a result, I cannot represent this causal knowledge properly using variables and functions because I cannot specify the function that relates fullness to suction in anything more than a vague way. Another way of looking at this is that vague predicates violate the uniqueness axiom described above, an axiom that Pearl correctly describes as implicit in his definition of a causal model. Of course, I could carry out the tedious empirical task of adding dust one grain at a time and estimating the function, but the point is that I don't need to. I make very useful practical inferences and decisions based on this semantically vague representation on a regular basis and have no need for a more precise representation for any purposes outside of making my inferences conform to Pearl's theory of inference from non-vague representations. In the previous paragraphs, I mean to suggest that 'ideal conditions' serves as a useful but vague predicate in the context of program evaluation in precisely the same way as does 'too full' in the context of vacuum cleaner maintenance. More broadly, one might consider any causal reasoning from fuzzy logic models (McNeill and Freiberger, 1993) as causal reasoning that requires an extension of Pearl's theory.

Non-modular representations of causal systems

A second example involves the assumption of modularity required by models of the form assumed in Pearl's theory (Cartwright, 2007; Pearl, in press). Cartwright's criticisms of modularity assumptions may not apply directly to Pearl's theory for reasons developed in some detail in the next section. Nonetheless, the kind of causal systems that she considers as counterexamples give every appearance of providing examples of the sorts of systems that one can model and reason causally about. Indeed, it seems unlikely that readers could comprehend her examples were this not the case. Such reasoning provides a further example of causal inference from models that do not conform to Pearl's axioms.

Modularity involves the ability to independently manipulate each cause in a causal system. Cartwright gives the example of a toaster in which the lever causes both the rack to move up or down and also causes the electrical circuit to open or close activating the heating mechanism. It is not possible in this set-up to manipulate the rack level independently of the circuit or vice versa – without interfering with the normal operation of the machine that sustains the causal relationships such as by breaking the weld between the rack and the level, or bypassing the circuit. As noted above, the issue of manipulation is not of interest at present. One could envision imaginary interventions and quite possibly draw correct conclusions about toasters from these. What is interesting is that the natural description of Cartwright's toaster violates Pearl's effectiveness axiom. If one takes $x =$ *lever down*, $w =$ *rack up*, and $y =$ *circuit closed* for u representing any further assumed conditions, then $Y_x(u) = y$ but $Y_{xw}(u)$ seems indeterminate, perhaps violating the existence or uniqueness axiom as well. By definition, it seems impossible to have x and w without breaking the toaster and the causal system only applies to a toaster in normal working order. Nonetheless, it seems entirely possible to communicate about and reason about the causal mechanism of the toaster using a non-modular representation despite the fact that it violates the stipulated criteria for causal models. Pearl's theory cannot account for such causal reasoning. It can replace it with alternatives, but cannot explain it as it stands.

Composition, causes and effects

The composition axiom serves the important purpose in Pearl's theory of imposing a common assumption for many types of causation, the assumption that causes and effects must constitute distinct events. Thus composition incorporates into Pearl's theory the assumption that the causal relation has antireflexivity as a property, that nothing can cause itself. To see this, consider x = *switch* on, w = *switch on or battery disconnected*, y = *light on*. These value assignments satisfy the antecedent: $W_x(u) = w$, if switch-on then switch-on-or-battery-disconnected. However, if in the consequent, $Y_{xw}(u) = Y_x(u)$, we satisfy w by disconnecting the battery, then $Y_{xw}(u)$ = light off $\neq Y_x(u)$ = light on. What goes wrong in this example involves the fact that the variables X (switch on?) and W (switch on or battery disconnected?) do not have the property of representing distinct events. Instead X constitutes a constituent of W. Thus, one can interpret the composition axiom as ensuring, *inter alia*, that the intended sorts of variables *compose* the model. Of course, such non-distinct events can create trouble for other axioms as well. For example, let $\sim w$ = not w, entailing both that the switch is not on and the battery is connected. If one assigns $X_{x\sim w} = x$ then one violates the effectiveness axiom with respect to $\sim w$ because this entails that $W_{x\sim w} = w$ by substitution, if one assigns $X_{x\sim w} = \sim x$ (switch off) then one violates the same axiom with respect to x. If one accepts neither or both, one violates either existence or uniqueness.

Should one take this as a universal first principle that nothing worth calling causation can hold between non-distinct events and that nothing worth calling causal reasoning can proceed on the basis of such events? Consider three horn players playing a musical game in which the saxophone player plays a melody and the trumpet and trombone players harmonize with two constraints: The trio must voice a triad (three notes separated by continuous major or minor thirds) and the trio cannot play a C major triad. Now suppose that the saxophone player voices an E and this cannot change because it is the melody note. The trombone player plays a C and the trumpet player plays an A, forming an A minor triad (C–E–A). Had the trumpet player instead played a G, this would have violated the C-major rule (C–E–G). In this circumstance, it seems reasonable to say that the trumpet player playing an A caused the trio to voice an A-minor instead of a C-major triad (because the other instruments voiced notes common to both).

The most natural way to encode this into a model might be to define the triad variable over a domain that excludes C-major. Recall that Pearl defines probability in terms of events, and then defines variables as encoding events, in the sense of property bearers taking on certain properties, over restricted domains of possible values. One would not want to meddle with this restriction because it plays an essential role in allowing Pearl's system to handle causal reasoning over restricted domains, such as when scientists do not extrapolate beyond the range of their data. However, in this case, it would violate existence because the omission of the C-major triad from possible values of the variable would mean that the potential response value did not exist when the variables for each player are manipulated to form C–E–G. To avoid this problem, we might admit C–E–G (C-major) as a possible value of the variable encoding what triad the trio plays. However, now we violate composition, as above, because jointly manipulating the individual horns to, say, C–E–A, and the trio to C-major (C–E–G) cannot yield both due to the fact that these are not distinct events. Nonetheless, it seems quite plausible to reason causally about this musical game. Indeed, if the trumpet player voiced a G, we might imagine the trombone player dropping the C down to a B to form an E-minor triad (B–E–G). The cognitive processing involved in this game play seems well described in causal terms: If I continue to play C and the trumpet player voices a G over the E melody note, then that will cause us to play C major, we don't want that, so if I play a B instead of a C, that will cause us to play E minor, etc. Other plausible examples of causal reasoning about non-distinct events involving parts and wholes include a holistic state of clinical depression caused by a collocation of specific negative cognitions about self and others; global judgments of legal guilt caused by specific cognitions about facts of the case; satisfaction caused by cognitions of amount wanted, amount had, and importance of the object of satisfaction; and the overall success of an interview caused by the success of each question–answer exchange (Markus, 2008a).

Causal reasoning without rigid designators

The existence and uniqueness axioms also serve the purpose of introducing a constraint on the universe of property bearers about which causal reasoning takes place and also the domain of values of variables about which causal reasoning takes place. Imagine a photocopy machine such that the original is placed on the glass, the button is pushed, and the copy comes out onto the tray. If the button had not been pushed the copy would not be in the tray, but one hesitates to assert that the copy has the property of not being in the tray because the copy does not exist and thus has no properties in this possible world. However, if the copy does not exist, then no value of the variable exists that the copy has as a property, and thus the example violates the existence axiom. Conversely, if the button had been pressed twice *a* copy would be in the tray because *both* copies would be in the tray. Would *the* copy be the top copy or the bottom copy in the tray? In this case, the copy in the actual world corresponds to two distinct entities in the hypothetical possible world, and these two entities may differ in their properties. If 'the copy' then referred at once to each copy, this would violate the uniqueness axiom. Similar exceptions occur if the domain of variable values varies across possible worlds. If the button were never pressed, which would be the top copy in the tray? One could say that the top copy variable ranges over all possible copies, but has no value in this instance. Alternatively, one could say that the domain of values of the variable reduces to the null set in this instance, in which case the variable has no possible values that it could take. Either way, one violates the existence axiom.

Pearl's system thus seems focused on causal reasoning in which designators of property bearers and properties have the property of rigidity: They refer to one and the same entity in all possible worlds (Kripke, 1980). Traditionally, this assumption takes the form of assuming that if it is possible that an individual exists with a given property then an individual exists that possibly has this property (Barcan Marcus, 1993). The Existence axiom nearly states this directly for domains of variables if one recognizes possibility in the potential response operator. This seems implicit in the way that models fix variable domains. The lack of explicit notation for property bearers makes it more complex to derive the assumption with respect to property bearers (individuals) but one can take it as implicit in the definition of a model, as Pearl seems to do. Nonetheless, it seems clear that people can and do reason causally about situations where the assumption does not hold. Complete and consistent systems of modal logic exist which do not have this assumption, commonly referred to as the Barcan formula, as a theorem (Barcan Marcus, 1993; Hughes & Cresswell, 1996). No obvious reason presents itself to deny that the content of such reasoning can involve causation.

Structural explanations in cognitive psychology

A fourth type of example involves structural explanation as commonly formulated in cognitive psychology. One example would be the explanation of memory retention in terms of positing separate mechanisms for short-term and long-term memory (Miller, 1956; Peterson & Peterson, 1959). Another example would be the explanation of semantic memory response times in terms of the structure that organizes semantic memory (Collins and Quillian, 1969). However, perhaps the most illuminating example in the present context is Pearl's theory itself. Pearl seeks to explain causal reasoning in terms of a certain structural account of the representation of causal knowledge in causal models, and operations on those models to make inferences from the knowledge. Pearl (2009) makes no attempt to present his theory in the form of a causal model, and seemingly for good reason. Any attempt to do so would distort the theory and fail to capture its explanatory power. Pearl's theory explains how certain types of causal reasoning can take place given a certain structure of the cognitive representations of causal information. Simply spelling out the causal dependencies between inputs and outputs would not shed the same light on the cognitive structure that sustains these causal dependencies. The strength of structural explanations seems to come from their ability to show that a set of nomothetic relationships such as those presented in a causal model follow naturally from a certain structural organization. The structural hypothesis explains the nomothetic relationships that a causal model only describes. Of course, one could code various structures into a causal variable and various causal models into an effect variable, but the substance of the structural explanation would still reside in the complex coding of the variables whereas the causal model would only encode a relatively trivial portion of the structural explanation. Nonetheless, such structural explanations seem to provide potent bases for causal reasoning.

12.1.3 Section summary

The previous subsections have considered five instances of causal reasoning from bases that seem like plausible candidates for models but which do not conform to the axioms of Pearl's theory. The general picture seems clear: Pearl's theory does not admit as a causal model anything that cannot be represented in the form $M = \langle U, V, F \rangle$ or support causal reasoning that does not conform to his axioms. Nonetheless, much causal explanatory talk in the behavioural sciences seems not to fit this prescribed structure. Either Pearl's theory fails to account for a large swath of causal reasoning, or it rules out a large swath of seemingly causal reasoning as something other than causal reasoning by stipulation.

Despite being called axioms, the axioms are not presented as self-evident first principles. Instead, the definition of a causal model is presented as primitive, and the axioms are shown to be implied by the model. In following this procedure, however, it is important to recognize that showing that the axioms characterize the proposed model does not show that either fully characterizes the domain of discourse or reasoning that they are put forward as an account of. In this case, it seems clear that a number of examples present themselves as putative examples of causal reasoning in the behavioural sciences that cannot be adequately represented by the kind of causal model to which Pearl restricts his theory. As such, the axioms serve not as universal truths about causal reasoning but rather as expressions of the limits to the range of application of the theory. Of course, none of this detracts from the fact that the axioms characterize the formal model presented in the theory. In this narrower context, one can interpret them as universal truths. It is just that they hold universally for a restricted universe that does not seem to capture the full range of causal reasoning from models found in the behavioural sciences.

One might note that Pearl's (2009) presentation waivers to some degree between a descriptive and prescriptive account of causal reasoning. The prescriptive reading would excuse the theory from covering all cases of causal reasoning. However, this seems like a poor strategy for interpreting Pearl's theory. There seems to be little basis to suppose that the examples of causal reasoning that fail to conform to the axioms represent substandard or inferior forms of causal reasoning. Moreover, even if all behavioural science were to strive toward causal models that conform to the axioms, it seems clear that researchers cannot defer all causal reasoning until after this goal is achieved. They need to reason causally from the models that they have as the research proceeds, and these may not conform to Pearl's axioms.

Finally, it is worth noting that one can play the converse of the game played above: One can search for examples of non-causal reasoning characterized by Pearl's axioms. However, such examples do not carry much import for the interpretation of Pearl's theory because he offers a theory that covers both veridical causal reasoning and mistaken causal reasoning. So long as the reasoning itself follows the prescribed pattern, the model used for reasoning can entirely mischaracterize the phenomenon to which the causal reasoner applies it.

12.2 Causal eliminativism, causal instrumentalism and causal realism

Symbolic modularity does not assume physical modularity. Surgery [on causal graphs] is a symbolic operation which makes no claims about the physical means available to the experimenter . . . (Pearl 2009, p. 364).

[W]e are dealing here with symbolic, not physical, manipulations. Our task is to formulate a meaningful mathematical definition of 'the causal effect of one variable on another' in a symbolic system called a 'model.' This permits us to manipulate symbols at will, while ignoring the technical feasibility of these manipulations. Implementation considerations need not enter the discussion of definition. (Pearl, 2009, p. 375).

One key dimension on which theories of causation differ involves the degree of literalism that they ascribe to causal assertions (Markus, 2004). One can describe a view that characterizes causation as a property ascribed instrumentally within a mental representation as an instrumentalist view. On such a view, causation exists only within the cognitive apparatus rather than in the objects or processes represented by that apparatus. In short, causation resides in the head. In contrast, one can describe a view that characterizes causation as a property ascribed literally to the things and processes cognized as they exist independent of that cognition as a realist view. Such a view can allow for variation in how different theories or models represent the same things and processes, but not so much variation as allowed by an instrumentalist view. Any reasonably accurate description would need to represent the same causal relationships in some form, allowing that the specific form might vary. To take Hillary Putnam's phrase out of context, no matter how you cut it, causation is not in the head. The details of both such views remain works in progress (Leplin, 1984) but the broad characterizations are nonetheless useful. Although the book involves some important ambiguities, Pearl (2000) appears to present an instrumentalist view of causation.

How does such a reading make sense of Pearl's professed realism about causation?[3] I concede that there may not be a possible reading that makes a consistent whole out of every stray claim found in Pearl's expansive book. However, I read such claims as Kantian in flavour. Pearl makes these claims in response to a pervasive skepticism about causation and efforts to explain it away by reducing it to extensional phenomena such as the relative frequency notion of probability. In the relevant passages, Pearl asserts an ontology that affirms the existence of something in the world that we cannot fully grasp in purely extensional, probabilistic terms, but can fully grasp in causal terms. Causal representations constitute our means of representing and reasoning about this property of things-in-themselves. An alien intelligence based on different principles might represent and reason about the same property by different means. Causation is our means. Thus asserting realism about causation, for Pearl, comes down to shorthand for asserting the reality of the property we grasp through causal representations, although causality itself remains internal to our representational system and not an inherent property of what we represent by that means. Contrasting Pearl's theory with a robust realism that posits causation as an inherent property of real-world systems in no way diminishes the further contrast between Pearl's theory and attempts to eliminate the notion of causation by reducing it to probability or something else describable in entirely extensional language. Pearl's is a non-eliminativist instrumentalism.

One way to appreciate this is to consider the importance of an instrumental interpretation to rendering plausible the axioms discussed in the previous section. If one adopts a realist notion of causation, the axioms seem much less plausible. Under a realist interpretation, the axioms would refer not to mental manipulations of a mental model but rather to actual manipulations of the system under study. This seems to be the kind of intervention discussed by Cartwright (2007). The efficacy/effectiveness distinction discussed above provides one example of where a realist interpretation of Pearl's axioms seems out of step with the discourse of behavioural science researchers. The effectiveness axiom states that a manipulation intended to produce the value x of variable X will succeed, but the efficacy/effectiveness distinction rests precisely on the assumption that this will not always hold true. Instead, the difference between efficacy and effectiveness refers to the gap between the intended intervention and the imperfect intervention as implemented under realistic conditions. Program evaluation researchers routinely assume

[3] This question was raised independently by an anonymous reviewer of the present chapter and also by Denny Borsboom in response to Markus (2004). For example Pearl (2009, p. xv–xvi) contrasts his earlier view that causation 'simply provides useful ways of abbreviating and organizing intricate patterns of probabilistic relationships' with his current view that takes 'causal relationships to be the fundamental building blocks both of physical reality and of human understanding of that reality'.

that one can expect a gap between the intended intervention and what one actually succeeds in bringing about. For example, it is easier to imagine the full implementation of a new curriculum than to actually complete such an implementation. Likewise, it is easier to imagine a course of treatment than to attain full compliance with that course of treatment.

Another example involves a contrast between the kind of interventions consistent with the effectiveness axiom and the kind of interventions discussed by Bollen (1989) in the context of empirical studies using structural equation models. The effectiveness axiom describes what might be called a poke-hold intervention. One fixes $X = x$, and holds X at that value as the causal effects of the intervention play out. In contrast, Bollen describes what one might call poke-release interventions. One sets $X = x$, but then releases X so that consequences of the intervention can change the resulting value of X as the effects work their way through the system.[4] As such, poke-release interventions violate the axioms. When considering an instrumental view where interventions are mental operations, poke-hold interventions seem most natural because the purpose is to manipulate the model to determine the predicted result. When considering a realist view, however, poke-release interventions make sense because the goal is to learn the ultimate result of a temporally bounded operation. One takes some course of action, and then allows it to have its effects after the completion of the action. Poke-hold interventions have some applications under a realist interpretation involving manipulations of actual causal systems, but assuming them as universally the only applicable kind of intervention has very little plausibility under such an interpretation.

To take an example involving a different axiom, consider how the Composition axiom applies to real-world interventions. Individuals living in the United States saved 1.59% of their personal income in 2008 (Bureau of Economic Analysis, 2009). Recall that the composition axiom asserts that the potential response associated with an intervention on one variable does not change if an additional variable in set by intervention to the value that it would have as a result of the intervention on the first variable. Imagine, then, a bill passing through the United States congress which implements an economic incentive (X) to save a larger proportion of one's income (Y). Imagine that this incentive all by itself would produce a potential response of, say, 3% savings. Further assume that the intervention would produce a potential response for individual outlays as a proportion of disposable income (post-tax) of 96.59% (W, down from 98.19% in 2008). Now, in contrast, imagine the same bill with

[4] Pearl (2009, p. 164) describes the total effect associated with poke-release interventions as having 'no operational significance worthy of the phrase 'effect of X' '. A reviewer points out that one could represent poke-release interventions using dynamic time-indexed models within Pearl's model structure. However, the fact that Pearl does not routinely do this, or define total effects in terms of poke-release interventions still supports the general claim that ideal interventions offer a more plausible reading of his theory.

an additional component legislating that the average personal outlays must not exceed 96.59% of disposable income. The composition axiom requires that both bills have the same effect on savings. However, it seems entirely plausible that the impact of the second bill might differ due to the shock value of the added component, political reaction to it, or the impact that this has on the causal mechanisms that link the incentive to savings. Applied to real-world interventions rather than mental operations, this axiom assures that interventions are transparent in the sense of that they have no side effects. This is entirely plausible for mental interventions on mental models because one can easily imagine the intervention that way. However, in real-world interventions it seems unlikely that legislating an upper limit on the proportion of disposable income that can be spent would have no impact even if the rate legislated matched the potential response produced by the incentive without legislating the maximum outlay rate. As such the axiom loses some of its plausibility under a realist interpretation of this particular model.[5]

12.2.1 Can one have Pearl's cake and idealize it too?

A reviewer thoughtfully asked whether one could interpret Pearl's theory as referring to symbolic manipulations in some contexts and real manipulations in others. I do not think so, but I do think that it proves very helpful to spell out why not. Indeed, I take this as the crux of my argument. The approach to semantics adopted by Pearl assumes that what an assertion is about makes the assertion true or false, and what makes it true or false is what it is about. To make the argument concrete, imagine that one were interested in the philosophy of spoons and presented a semantics for 'x is a spoon' that relied only on idealized geometric representations that real spoons can only approximate. The truth or falsity of spoon assertions would thereby turn only on properties of the representation, not the actual objects in the world that one might choose to speak of as spoons. Because these actual objects play no role in fixing the truth of spoon assertions, the spoon assertions do not refer to them. Nonetheless, idealized spoon models might still play a useful role in reasoning about spooning and spoon collecting with the recognition that the spoon-ness rest in our cognitive representation of these objects rather than the objects themselves. Analogously, in the face of criticism that real-world manipulations do not approximate his axioms, Pearl consistently responds that the manipulations that fix the semantics of causal assertions involve only symbolic manipulations of the representation, not real-world manipulations. It follows that these symbolic manipulations and their results wholly determine the truth of causal assertions, not the behaviour of real-world systems under

[5] A reviewer correctly points out that the axiom might hold for a more complex model. However, Pearl's theory does not limit itself to good models. It seeks to present a theory of causal inference from both good and bad models alike.

manipulation, and that they assertions therefore do not refer to these real-world systems. My basic argument, then, come down to this: One cannot have it both ways. If causal assertions refer to causation as an inherent property of the world, then one cannot brush away the kinds of concerns about real-world manipulations that others have raised (Cartwright, 2007; Woodward, 2003a; 2003b questioned some imaginary manipulations). On the other hand, if the semantics of causal assertions rests only on manipulations of the representations themselves and not real-world systems, then causation resides in the means of representation of real-world systems rather than inherently in the real-world systems themselves. Causal assertions may refer to real objects when these objects pay into their truth conditions, but the causal element of these assertions does not itself refer to an inherently causal element of the objects.

12.2.2 Causal instrumentalism and the praxis of psychological research

One can always argue prescriptively that if behavioural scientists do not currently adopt an instrumentalist view of causation, they nonetheless should. As noted, Pearl's exposition sometimes waivers between description and prescription, sometimes suggesting that the theory is hardwired into the human cognitive system, sometimes presenting a prescriptive methodology for researchers. However, it remains useful to note areas of ill fit between Pearl's (2000) theory taken as a descriptive theory and how behavioural scientists think about causation.

This instrumentalism about causation contrasts with attempts to develop a methodology to measure causal effects as they exist in the world within the context of Rubin's causal model (RCM; Rubin, 1974). Pearl (2000) discusses Rubin's theory of estimating causal effects and somewhat awkwardly relates his theory to RCM. Nonetheless, one fundamental difference between the RCM perspective and that of Pearl involves the admissibility of causes that one cannot actually manipulate. For example, RCM considers demographic characteristics like gender and ethnicity non-causes because one cannot randomly assign experimental participants to these factors. Woodward (2003) has suggested that such cases might be explained away in terms of inadequately precisely specified counterfactuals. I am not aware that anyone working in the RCM tradition has responded to this suggestion, but I would anticipate a response along the following lines. The correct counterfactuals are ones that involve changes from the actual course of events that do not precede the experiment by any great length of time such that the participants can still be reasonably understood as the same people with minor changes in their recent life history. To go back before their birth and alter their genes to change their gender or ethnicity would precipitate a sufficiently great change in the course of their life events that they would no longer be recognizably the same person.

In short, RCM involves a form of modest essentialism in which too great a change in someone's life history precludes trans-world identity. There may be a possible world in which a female participant's parents had a son with an otherwise genetically identical make-up, but there cannot be a possible world in which that same female participant was born male. This stands in stark contrast to Pearl's (2000) view where any manipulation is possible because manipulations refer to mental operations operating on the mental representation. The above difference gets at a deeper difference between RCM and Pearl's approach. RCM begins with the assumption that causal effects exist independently of our causal models, and thus can be measured through experiments and other worldly operations.

The instrumentalism in Pearl's approach can be neatly distinguished from the causal realism inherent in much behavioural research by an assertion regarding the uniqueness of adequate causal descriptions. As a first pass, one might formulate such an assertion as follows.

(1*) If model m_1 is an adequate causal description of system s and m_2 is an adequate causal description of s, then $m_1 = m_2$.

This will not do because causal realism accepts adequate partial descriptions that are not identical but do not contradict one another. E.g. '*C* causes *M* and *M* causes *E*' allows both '*C* causes *E*' and '*M* causes *E*' as partial descriptions.

(1) If m_1 and m_2 jointly entail a contradictory statement about s and neither individually entails such a contradiction, then at least one of the two models is not an adequate causal description of system s.

Causal realist approaches will generally endorse some form of Assertion 1 whereas instrumentalist approaches will not. Two causal representations may both be useful even if they present contradictory causal representations of the same system. In contrast, if causal processes are inherent to the system itself, then only one of two contradictory causal descriptions can hold true of the system. Thus, if one seeks to characterize the meaning of 'cause' in terms of what one can infer from a causal assertion, then Assertion 1 supports further inferences that the same causal assertion would not support if interpreted without Assertion 1. Specifically, a causal assertion supports the denial of all contradictory causal assertions under a realist reading whereas an instrumentalist reading merely needs to avoid mixing alternative causal assertions but need not assume that only one can hold.

Assertion 1 plays a central role in the Campbellian tradition of causal inference (Cook and Campbell, 1979; Shadish, Cook and Campbell, 2003). Here the emphasis rests with identifying plausible rival hypotheses and designing data collection in a way that allows the researcher to rule out such rival hypotheses. Any approach that rejects Assertion 1 runs counter to this tradition because it admits contradictory causal explanations as complementary rather than characterizing them as rival hypotheses. The above does not imply that instrumentalist approaches view all causal models as equally good, but it does show that they are importantly slower to deem models as rival hypotheses.

Pearl's (2009) tendency to assume a set of variables as cognitive givens draws attention away from the kinds of situations that most illuminate the above difference. Often times, disputes between different groups of researchers investigating a common topic turn on the selection of the set of variables used to describe the phenomenon of interest. Three examples will serve to illustrate. The first is a simple contrived example involving electrical storms. The second provides a realistic example from the psychology literature involving research on gender roles and cognitive schema. The final example offers an idealized but plausible example involving different levels of explanation and somewhat less clear-cut interpretation.

As a simple example, consider causal models of electrical storms. From a realist perspective, placing causation in the storm process, the correct causal model might have the storm causing electrical discharges which in turn cause both lightning and thunder. A simpler model might have the storm cause the lightening and the lightning cause the thunder. From a realist perspective, either lightning causes the thunder or it does not, thus the two models contradict one another and only one can hold true (holding the references of the terms constant). From an instrumental perspective, however, causation is just a tool of cognitive organization for interacting with the environment. Each of these models might have interactive utility in different contexts of action, so there is no problem adopting each of them for use in different contexts. Because lightening and thunder tend to come together, either model is likely to serve as a useful tool for things like anticipating thunder.

Bem (1975) hypothesized that androgynous individuals – those high in both masculine and feminine characteristics – are more able to adapt to different situations by behaving in ways associated with either masculine or feminine gender roles because they are less likely to process information using cognitive schema associated with those gender roles (Bem 1979, 1981a, 1981b). Bem developed the Bem Sex Roles Inventory to validate the construct of androgyny (Bem 1974). The inventory differed from other measures at the time in having separate scales for masculinity and femininity. Bem defined androgynous individuals as those high in both masculinity and femininity as compared to those high in one or the other (those low in both, she labelled undifferentiated but did not include in the causal hypothesis; Bem, 1981b).

A literal translation of Bem's hypothesis would involve a causal model in which the causal variable distinguishes androgyny from gender differentiation (high on one but not both). This variable would then cause individuals to make greater use of gender role cognitive schema, and thus behave less flexibly across social situations – forming a chain model with three variables. In contrast, one could eliminate the androgyny variable altogether and make masculinity and femininity the causes with an interactive function requiring high values of each to produce the effect. On a realist interpretation, either androgyny causes flexibility or the interaction between masculinity and femininity causes flexibility – the two accounts compete as rival hypotheses. From an instrumental perspective, they each may provide an adequate guide to action and one might thus adopt one or the other as a matter of convenience. The body of accepted psychological theory can consistently include both theories on an instrumental reading of causation but not a realist reading.

This example illustrates a very general phenomenon in causal modeling. Androgyny equals (Masculinity + Femininity)/2 and traditional bipolar Masculinity/Femininity scales equal +/−(Masculinity − Femininity)/2 depending on the direction of the scale. This is a simple example of a 45 degree rotation of the data within the geometric space defined by the two variables (Steiger and Schönemann, 1978). One could just as easily define the space in terms of androgyny and the traditional bipolar scale and rotate it to derive the Masculinity and Femininity variables. Of course, there is nothing unique about a 45 degree rotation, one could create alternative pairs of variables with alternative rotations of the initial two variables. Adding more causes expands the dimensionality of the space and thus the alternative rotations. Moreover, as Pearl (2000) notes, this rotational indeterminacy applies to all variables in a model, not just causes. If causal relationships exist in the world prior to description, and they hold between determinate sets of variables essentially, then a realist account of causation needs to select one such rotation as the correct set of variables in order to identify the correct causal relationship. An instrumental view of causation more readily allows for indifference between causal relationships formulated between alternative rotations of the variable set. The Campbellian emphasis on ruling out rival hypotheses fits more comfortably with the former, and Pearl's theory fits more comfortably with the latter.

A third example involves different levels of description, and the problems of reductionism. Garfinkel (1981) offers the example of explaining the size of the rabbit population in a particular region in terms of the fox population. One could form a model out of a set of variables describing individual rabbits and foxes providing a causal explanation of how placing more foxes in the model leads to fewer living rabbits. However, such a model would be enormously complex. One could produce a more parsimonious model by considering potential responses at the population level (g) rather than the individual level (i). Whereas X_i might represent rabbit i being caught by a fox, $X_g{}'$ might represent an overall decline in the rabbit population (g). A number of other factors might impact on $X_g{}'$ aside from the size of the fox population, but the model would remain much simpler than one specified at the individual level (i).

One way to construe the relationship between these two levels is to incorporate them into a two-level model in which increasing the fox population, W_g', increasing the occurrence of hungry foxes in proximity to rabbits, W_i, which increases the number of caught rabbits, X_i, which in turn decreases the rabbit population, X_g'. On this model, a causal chain links W_g' to X_g' and thus the model allows the simplification that W_g' causes X_g'. In contrast, however, one could construe the relationship between W and W' as parallel to that between X and X', such that more individual foxes at the individual level (the proximity clause more or less falls out but could be represented as a separate variable if desired) causes the larger population at the population level just as fewer rabbits at the individual level causes the smaller population at the population level. In this case, the model is not consistent with the claim that W_g' causes X_g' but instead, the relationship is merely an artifact of the causal relationship between W_i and X_i at the individual level. This latter view might be more consistent with conceptualizing the cross-level relationships in terms of part-whole relationships (Markus, 2008a). Alternatively, one might take a more reductionist view, the view that Garfinkle argued against, that the causal relationship at the population level reduces to the causal relationship at the individual level, in which case the seeming rival hypotheses might dissolve could one successfully address the difficulties with such a reduction. In this case, then, the best interpretation seems less clear, but the possibility of conflicting causal theories on a realist reading of causation arise under at least some interpretations.

The point here is not to argue for or against causal instrumentalism or causal realism. It seems possible that one or the other offers a better overall strategy for behavioral science but it also seems possible that sometimes one offers a better strategy and sometimes the other, which would lead to a form of causal pluralism (Cartwright, 2007; Markus, 2004, 2008a). The present discussion does not seek to adjudicate that issue. Instead, the present discussion seeks only to develop the more modest point that Pearl's causal instrumentalism contrasts with certain aspects of prevailing theory and practice of causal inference in behavioural science research. As such, the former may not provide a complete descriptive account of the latter.

12.3 Counterfactuals and causal inference in the behavioural sciences

Pearl (2009) presented three types of inferences ('queries' in the jargon of artificial intelligence, as in a query to a database) that his theory is designed to explain: *Predictions* involve inferring the value of some variable from the values of others. *Interventions* involve inferences about what would result from an intervention that changes the values of one or more variables (and fixes them to those values). *Counterfactuals* involve inferences about the values of variables had other variables taken different values given the actual values of the variables in the system. (Note that counterfactual queries in Pearl's jargon differ from both counterfactual assertions and simple counterfactual conditionals – they are doubly conditional.)

While this is certainly an impressive and ambitious enough list, it omits an important type of inference that plays an important role in applied behavioural science. Addressing this omission, in turn, places stress on the basic architecture of Pearl's theory. At first blush, each of the inferences addressed by Pearl's (2009) theory involves only changes in the values of variables for specified individuals (i.e. property bearers such as people or sprinklers). All assume a fixed causal structure (although some are answered by blocking various causal effects in the system). In contrast, behavioural scientists often wish to make inferences about interventions that aim not to change the value of variables for individuals but rather to modify the causal system for them (Markus, 2008b). For example: If children are informed about the dangers of smoking, will advertisements have a weaker causal effect on smoking? If students have opportunities to ask questions in smaller groups, will lectures to larger groups have a larger effect on learning? If individuals who lose their jobs receive job training, will the downturn in the economy have less of an effect on levels of unemployment?

Formally, of course, one can incorporate at least some such inferences into Pearl's system using the same methods commonly applied to statistical models by behavioural science researchers. For example, if the effect of class size depends upon opportunities to ask questions in small groups, one can model learning (L) as a function of the product of class size (C) and small group (S) variables: $L = \beta_0 + \beta_1 C + \beta_2 S + \beta_3 CS + e$, where the βs represent causal effects and e represents residual variance in L. Rearranging the equation shows that S becomes part of the effect coefficient for C:$L = \beta_0 + (\beta_1 + \beta_3 S)C + \beta_2 S + e$. Thus the causal effect of C on L depends upon the value of S. Similarly, including an S^2 term makes it possible for the causal effect of S to depend upon the level of S in a similar manner. Clearly, Pearl's theory can handle these sorts of models, but they strain the semantics of the proposed model structure, $M = \langle U, V, F \rangle$. The model structure separates variables (U and V) from functions (F) with the intent of manipulating the variables to counterfactual values while holding the functions constant. Parameterizing changes in causal structure in terms of values of variables that serve as part of the causal functions between variables blurs the semantic division between these two sets of elements of the model structure and makes their dependencies less than transparent.

One can also handle such questions using the method of augmented directed acyclic graphs described by Pearl (2009, pp. 70–72). To use this approach, one explicitly defines new variables with functions as values.

For example, instead of encoding learning as an interactive function of class size and small groups, as above, one would instead encode a variable giving the function linking learning (L) to class size (C) as a new variable, F'. Next one encodes F' as a function of small group opportunities (S) and other background variables, if any. This can be a very effective method for encoding systems where one variable affects the causal structure relating other variables to one another. However, the approach clearly stresses the basic architecture separating variables from functions to the point that the distinction collapses. The set F no longer contains all of the functions and the variables in V no longer exclude functions, blurring the basic structure $M = \langle U, V, F \rangle$.

Also, these approaches create an unnecessary tension between the semantic portion of the theory, the *Do Calculus*, used to answer the kinds of questions outlined above (Pearl 2009, Chapter 3) and the inductive part of the theory (Chapter 2). The inductive part of the theory rests on a central assumption called the *stability* (or the *faithfulness assumption*) which requires that causes never exactly cancel out. For example, clocks run at the same speed at the poles and equator because the effect of the Earth's spin exactly cancels out the effect of increased diameter, violating the stability assumption. In a model with an interaction, a variable can have zero effect for certain values of a mediating variable. For example, cigarette advertisements might have no effect on children saturated with material means to desensitize them to such advertisements; or, more blatantly, type size will have no effect on reading speed when any light source is absent. This situation can violate the stability assumption,[6] creating a tension between these two parts of the theory.[7]

Although he does not formalize it this way, one can understand Pearl's theory as a form of neighbourhood semantics (Hughes & Cresswell, 1996, Chapter 12). The Do calculus selects certain possible worlds within the model structure, and then evaluates the truth of propositions within the selected neighborhoods of worlds. The functions in F place one set of restrictions on possible worlds. For example, assume that smaller class sizes (C) improve learning (L) but that smaller class sizes also direct spending away from other areas (Sp) which then works against improved learning (Figure 12.2). Deleting the equation for C and fixing $C = c$ further restricts the neighbourhood of relevant worlds. Finally, following through the implied direct and indirect effects on L identifies a world (or possibly set of worlds) corresponding to the unique potential response $L_c(u)$.

[6] Specifically, Pearl's (2009, p. 48) stability assumption requires that modifying the parameter values in the model not introduce or remove probabilistic independence within the model. The interaction effect can violate this principle if the data happens to represent a set of parameter values in which the moderator cancels the effect of the causal variable on the outcome, such as the lights being off for the reading speed experiment.

[7] Recall here the earlier caveat regarding the assumption that these are two parts of the same theory.

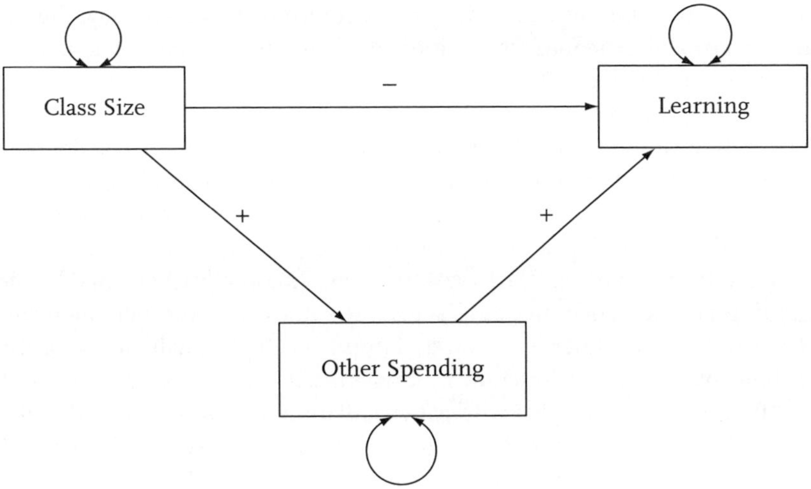

Fig. 12.2 Causal model for class size, other spending, and learning.

One could broaden the model structure to $M^* = \langle W, R, U, V, F\rangle$ where W denotes a set of possible worlds, R a neighbourhood relation, and $\langle U, V, F\rangle$ Pearl's original model structure, where R is chosen to map worlds onto sets of functions in F such that neighbourhoods correspond to worlds sharing the same causal structure. This set up assumes that all worlds share the same individuals and variables over those individuals, although further generalization is possible (Barcan Marcus, 1992). This expanded structure would more naturally allow for inferences regarding changes in F. The relevant counterfactuals would then involve a shift from one neighbourhood to another, and the evaluation would then follow the same procedure for Pearl's Do Calculus within the counterfactual neighbourhood rather than the original, which is to say, with the counterfactual causal structure. This approach does not require blurring the lines between variables and causal functions. Stability would naturally hold within neighbourhoods if not across the board.

Thinking about Pearl's theory in this way helps bring out some interesting aspects of existential quantification, asserting that such and such exists, with respect to Pearl's approach. Consider a special substitution instance of the existence axiom: $(\forall x|x \in X)(\exists x|x \in X)(X_x(u) = x)$. By substituting x for y, one obtains the rather unusual situation of applying both a quantitative and existential quantifier to the same variable, which seems to entail that every value of X exists. Clearly, however, Pearl did not intend that his system entail that every value of a variable is simultaneously realized in the actual world. Instead, asserting the existence of a value of a variable seems to assert a form of abstract mathematical existence equivalent to asserting that an abstract value is available to be realized, independent of whether that value obtains or not. This understanding of existence differs from, say, the usual interpretation of existence for an individual, requiring that the actual world contain that individual. This difference may reflect a fundamental difference between the ontology of variables and that of individual property bearers: Apparently, different types of things exist differently. The above considerations also illustrate the critical role played in Pearl's system of restricting the range of variable values through the definition of variables in models, only in this way does the system block the existence of every imaginable value.

Finally, this window on Pearl's system suggests some further considerations regarding the existence, in some sense, of values in some neighbourhoods but not others. Consider a three-way lamp such that installation of a three-way bulb produces the function linking switch settings to luminosity $f_1 = \{1 = \text{off}, 2 = \text{dim}, 3 = \text{bright}\}$ whereas installation of a two-way bulb produces the function $f_2 = \{1 = \text{off}; 2, 3 = \text{bright}\}$. Thus, possible worlds divide into two neighbourhoods on the basis of these two functions. The assertion that 'dim' exists as a value holds true in any possible world, including f_2 worlds, as Pearl construes things, because 'dim' is possible in some possible worlds, namely f_1 worlds. However, in some contexts, it does seem that it might be useful to distinguish 'dim' as existing in f_1 worlds in a way that it does not exist in f_2 worlds where one can only bring it about by changing light bulbs and thus shifting to the f_1 neighbourhood of worlds. The proposed expanded model structure makes that easier to accomplish.

12.4 Conclusions

The present chapter had three goals. The first section argued that one can best interpret Pearl's axioms as delimiters of the range of applicability of his theory rather than as universal truths about causal models. This conclusion turns on the distinction between causal models as defined by the stipulated model structure and the domain of naturally occurring discourse in which people reason causally from models that may or may not conform to Pearl's stipulated structure. Although the axioms correspond to the more limited domain, they appear to fall short of fully characterizing the broader domain of discourse. The second section argued that the plausibility of the axioms rests on a (non-eliminitivist) instrumentalist view of causation as a characteristic of the cognitive mechanism rather than something existing independently in the world. This aspect of Pearl's theory creates a tension with the practice of treating conflicting causal accounts as rival hypotheses, at most one of which can prove acceptable. The third section briefly suggested a form of inference that plays a crucial role in behavioural science research but that proves awkward for Pearl's theory. An extension was suggested that makes such inferences less awkward.

The overarching goal was to explore Pearl's theory from the perspective of psychological or behavioural science research methodology. Researchers and methodologists are still struggling to digest Pearl's theory nearly ten years after his book attracted widespread attention – much the way it took time for philosophers to digest Lewis's (1986) theory. My own view is that in large part the basic assumptions and assertions of the theory have yet to sink in. It is my hope that the present chapter will serve the double purpose of clarifying some key aspects of Pearl's theory for researchers and methodologists while also offering a useful perspective on how the theory lines up with the ways that psychologists and other behavioural scientists think about causation for those studying such reasoning from outside of these sciences.

Acknowledgments

I wish to thank the editors for encouraging this chapter after I missed the initial call for papers, and two anonymous reviewers for detailed and constructive reviews from which this chapter benefitted greatly.

References

Barcan Marcus, R. (1993). *Modalities: Philosophical Essays*. New York: Oxford University Press.

Bollen, K. A. (1989). *Structural Equations with Latent Variables*. New York: John Wiley.

Bureau of Economic Analysis (2009). Comparison of personal saving in the NIPAs with personal saving in the FFAs. CSV file downloaded 27 June 2009 from http://www.bea.gov/national/nipaweb/Nipa-Frb.asp

Cartwright, N. (2007). *Hunting Causes and Using Them: Approaches in Philosophy and economics*. Cambridge: Cambridge University Press.

Collins, A. M. & Quillian, M. R. (1969). Retrieval time from semantic memory. *Journal of Verbal Learning & Verbal Behavior, 8*, 240–247.

Collins, J., Hall, N., & Paul, L. A. (2004). *Causation and Counterfactuals*. Cambridge, MA: MIT Press.

Cook, T. D. & Campbell, D. T. (1979). *Quasi-Experimentation: Design & Analysis Issues for Field Settings*. Boston: Houghton-Mifflin.

Davidson, D. (2001). *Essays on Actions and Events*. Oxford: Oxford University Press.

Galles, D. & Pearl, J. (1997). Axioms of causal relevance. *Artificial Intelligence, 97*, 9–43.

Galles, D. & Pearl, J. (1998). An axiomatic characterization of causal counterfactuals. *Foundations of Science, 3*, 151–182.

Garfinkel, A. (1981). *Forms of Explanation: Rethinking the Questions in Social Theory*. New Haven, CT: Yale University Press.

Goodman, N. (1983). *Fact, Fiction and Forecast* (4th edn). Cambridge, MA: Harvard University Press.

Hintikka, J. (1996). *The Principles of Mathematics Revisited*. Cambridge: Cambridge University Press.

Holland, P. W. (1986). Statistics and causal inference. *Journal of the American Statistical Association, 81,* 945–970.

Hoover, K. D. (2004). Lost causes. *Journal of the History of Economic Thought,* 26, 149–164.

Hughes, G. E. & Cresswell, M. J. (1996). *A New Introduction to Modal Logic*. London: Routledge.

Judd, C. H. (1905). The Muller–Lyer illusion. *Psychological Monographs, 7,* 55–81.

Keefe, R. & Smith, P. (1997). *Vagueness: A Reader*. Cambridge, MA: MIT Press.

Kripke, S. A. (1980). *Naming and Necessity*. Cambridge, MA: Harvard University Press.

Leplin, J. (1984). *Scientific Realism*. Berkeley: University of California Press.

Lewis, D. (1986). *On the Plurality of Worlds*. Malden, MA: Blackwell.

Markus, K. A. (2004). Varieties of causal modeling: How optimal research design varies by explanatory strategy. In K. van Montfort, J.Oud & A. Satorra (eds.), *Recent Developments on Structural Equation Models: Theory and Applications*, pp. 175–196. Dordrecht: Kluwer Academic Publishers.

Markus, K. A. (2008a). Hypothesis formulation, model interpretation, and model equivalence: Implications of a mereological causal interpretation of structural equation models. *Multivariate Behavioral Research, 43,* 177–209.

Markus, K. A. (2008b). Constructs, concepts and the worlds of possibility: Connecting the measurement, manipulation, and meaning of variables. *Measurement, 6,* 54–77.

Markus, K. A. (2010). Structural equations and causal explanations: Some challenges for causal SEM. *Structural Equation Modeling*, 17, 654–676.

McNeill, D. & Freiberger, P. (1993). *Fuzzy Logic: The Revolutionary Computer Technology that is Changing our World*. New York: Simon and Schuster.

Mellor, D. H. (1995). *The Facts of Causation*. London: Routledge.

Miller, G. (1956). The magical number seven, plus or minus two: Some limits on our capacity for processing information. *Psychological Review, 63,* 81–97.

Pearl, J. (2000). *Causality: Models, Reasoning, and Inference*. Cambridge: Cambridge University Press.

Pearl, J. (2009). *Causality: Models, Reasoning, and Inference* (2nd edn). Cambridge: Cambridge University Press.

Pearl, J. (2010). Review of N. Cartwright 'Hunting causes and using them'. Approaches in philosophy and economics. *Economics and Philosophy*.

Peterson, L. R. & Peterson, M. J. (1959). Short-term retention of individual verbal items. *Journal of Experimental Psychology, 58,* 193–198.

Quine, W. V. O. (1960). *Word and Object*. Cambridge, MA: MIT Press.

Rubin, D. B. (1974). Estimating causal effects of treatments in randomized and nonrandomized studies. *Journal of educational psychology, 66,* 688–701.

Russo, F. & Williamson, J. (2007). *Causality and Probability in the Sciences*. London: College Publications.

Shadish, W. R., Cook, T. D., & Campbell, D. T. (2002). *Experimental and Quasi-Experimental Designs for Generalized Causal Inference*. Boston: Houghton-Mifflin.

Steiger, J. H. & Schönemann, P. H. (1978). A history of factor indeterminacy. In S. Shye (ed.), *Theory Construction and Data Analysis in the Behavioral Sciences*, pp. 136–178. San Francisco: Jossey-Bass.

Williamson, T. (1994). *Vagueness*. London: Routledge.

Woodward, J. (2003a). *Making Things Happen: A Theory of Causal Explanation*. Oxford: Oxford University Press.

Woodward, J. (2003b). Critical notice: *Causality* by Judea Pearl. *Economics and Philosophy, 19*, 321–340.

PART IV
Social sciences

·13·

Causal mechanisms in the social realm

Daniel Little

Abstract

The essay considers the specific characteristics of causal relations among social structures, processes, and activities. Against the Humean idea that causal relations are defined by facts about regular succession, the essay argues that the notion of a causal mechanism is fundamental. Causal realism asserts that causal connections between events and conditions are real and are conveyed by the powers and properties of entities. It is therefore necessary to consider the ontology of a given realm in order to be able to identify how mechanisms work in this realm. In the social realm causal mechanisms are constituted by the purposive actions of agents within constraints. Examples of social mechanisms are considered at several levels of detail, and more extended treatments are offered for transportation, violent crime, epidemiological processes, and system safety as examples of social domains where we can analyse underlying social mechanisms in order to understand the outcomes. The view de-emphasizes the feasibility of strong predictions in the social sciences; even when we have good reason to expect that a given set of social mechanisms are at work, it is often impossible to aggregate their interactions with confidence.

13.1 Introduction

To explain an outcome is to demonstrate what conditions combined to bring it about – what caused the outcome in the circumstances, or caused it to be more likely to occur. The most fundamental aspect of an explanation is a hypothesis about what caused the circumstance we want to explain. So social explanation requires that we provide accounts of the social causes of social outcomes. Why are there disparities in health outcomes between white and black populations in the United States? Why are rates of violent crime so different across the cities of the world? What accounts for the pattern of the spread of disease in the nineteenth and twentieth centuries? Why did labour mobilization on the docks take such different courses on the east and west coasts? In each case we want to provide an explanation that identifies the causes of these various social outcomes.

What is a cause? Generally speaking, a cause is a condition that either necessitates or renders more probable its effect, in a given environment of conditions. Often a cause is also necessary for the production of its effect – 'if C had not occurred, E would not have occurred'. The probabilistic version is analogous: 'If C had not occurred, the likelihood of E would have been lower'. The exception to this feature of causation is the rare set of cases where an outcome is 'overdetermined' – that is, cases in which there are multiple factors present, each of which would bring about the outcome in isolation. (J. L. Mackie's work is fundamental in defining what we mean by 'necessary and sufficient causal conditions' (Mackie 1974), and Wesley Salmon explores the intricacies of probabilistic causation in much greater detail (Salmon 1984).)

We can also ask the question, what is a *social* cause? It is not the case that all social outcomes are the result of social causes; for example, the depopulation of New Orleans in 2005 was caused by a natural fact, Hurricane Katrina. *Social* causes involve the actions of individuals within the context of social institutions and the actions of others. Actions are purposive performances by individual agents within social and natural constraints; institutions are sets of rules embodied in the beliefs, values, and behaviours of groups of individuals. We can now limit the concept of a social cause in this way: A social cause is a circumstance that ineliminably involves the actions and institutions of purposive agents. The circumstance may also involve natural and environmental factors; but in order for the circumstance to count as a social cause, there must be components of the event or process that involve agency within institutional settings. The breaking of the levees in New Orleans was the result of circumstances that included both natural and social components – the occurrence of the storm but also the institutional breakdown of the Army Corps of Engineers that had led to the poor condition of the levees in 2005. The former event is a natural occurrence while the latter event is a social cause of the physical failure of the levees.

This account of social causation depends upon something that Hume abhorred: the idea of necessity connecting cause to effect. For natural causes we have a suitable candidate in the form of physical mechanisms governed by the laws of nature. When a neutron of such-and-so energy strikes a U-235 atom, the atom breaks into two smaller atoms and three neutrons; the outcome follows from the laws of nature governing neutrons and nuclei. However, there are no 'laws of society' that function ontologically like laws of nature (Little 1993). A society is not a law-governed system. So how can there be 'social necessity'? Fortunately, there is an alternative to law-based necessity, in the form of a *social mechanism*. This chapter will explore the concept of a causal social mechanism in some detail. It will examine the social ontology that provides the best understanding of the 'substratum' of social causation – the idea of methodological localism. And it will review some clear examples of mechanisms that have been identified in various areas of the social sciences to illustrate how causal reasoning seems to work in several areas of social investigation.

13.2 Social mechanisms

The central tenet of causal realism is a thesis about the reality of causal mechanisms or causal powers. Causal realists maintain that we can only assert that there is a causal relationship between X and Y if we can offer a credible hypothesis about the sort of underlying mechanism that might connect X to the occurrence of Y. The sociologist Mats Ekström puts the view this way: 'the essence of causal analysis is . . . the elucidation of the processes that generate the objects, events, and actions we seek to explain' (Ekstrom 1992, p. 115). Authors who have urged the centrality of causal mechanisms for both explanatory and ontological purposes include Nancy Cartwright (Cartwright 1989), Jon Elster (Elster 2007), Rom Harré (Harre and Madden 1975), and Wesley Salmon (Salmon 1984). (Hedstrom and Swedberg's collection on mechanisms in the social sciences is a key source on this topic (Hedström and Swedberg 1998).)

Nancy Cartwright is one of the most original voices within contemporary philosophy of science. Cartwright places real causal mechanisms at the centre of her account of scientific knowledge. As she and John Dupré put the point, 'things and events have causal capacities: in virtue of the properties they possess, they have the power to bring about other events or states' (Dupré and Cartwright 1988). Cartwright argues, for the natural sciences, that the concept of a real causal connection among a set of events is more fundamental than the concept of a law of nature. And most fundamentally, she argues that identifying causal relations requires substantive theories of the causal powers (capacities, in her language) that govern the entities in question. Causal relations cannot be directly inferred from facts about association among variables. As she puts the point, 'No reduction of generic causation to regularities is possible' (Cartwright 1989, p. 90). The importance of this idea for sociological research is profound; it confirms the notion shared by many researchers that attribution of social causation depends unavoidably on the formulation of good, middle-level theories about the real causal properties of various social forces and entities – the social mechanisms that convey social causation. The account of social mechanisms to be sketched here concurs with Cartwright's approach in the important sense that it places priority on the causal powers of the substrate in which a causal process is thought to be located.

Aage Sørensen summarizes a causal realist position for sociology in these terms: 'Sociological ideas are best reintroduced into quantitative sociological research by focusing on specifying the mechanisms by which change is brought about in social processes' (Sørensen 1998, p. 264). He argues that sociology requires better integration of theory and evidence. Central to an adequate explanatory theory, however, is the specification of the mechanism that is hypothesized to underlie a given set of observations. 'Developing theoretical ideas about social processes is to specify some concept of what brings about a certain outcome – a change in political regimes, a new job, an increase in corporate performance.... The development of the conceptualization of change amounts to proposing a mechanism for a social process' (pp. 239–240). Sørensen makes the critical point that one cannot select a statistical model for analysis of a set of data without first asking the question, what in the nature of the mechanisms we wish to postulate to link the influences of some variables with others? It is necessary to have a hypothesis of the mechanisms that link the variables before we can arrive at a justified estimate of the relative importance of the causal variables in bringing about the outcome. What Sørensen's account does not provide is an analysis of the nature of the mechanisms that need to be identified in providing a social explanation; so the account provided below, analysing causal mechanisms in terms of features of structured agency, is a natural complement to his view.

A particularly important recent effort to make use of causal mechanisms as a foundation for social research is found within the literature on social contention – the occurrence of medium- and large-scale episodes of contention in a variety of social settings. Charles Tilly, Doug McAdam, and Sidney Tarrow have applied framework of causal mechanisms with a great deal of rigour in *Dynamics of Contention* (McAdam, Tarrow, and Tilly 2001) and a volume of associated research. They provide a simple definition of mechanisms: 'a delimited class of events that alter relations among specified sets of elements in identical or closely similar ways over a variety of situations' (p. 24). And processes are concatenations of mechanisms: 'regular sequences of such mechanisms that produce similar (generally more complex and contingent) transformations of these elements' (p. 24). 'We employ mechanisms and processes as our workhorses of explanation, episodes as our workhorses of description. We therefore make a bet on how the social world works: that big structures and sequences never repeat themselves, but result from differing combinations and sequences of mechanisms with very general scope' (p. 30). They summarize their theoretical ambitions concisely: 'Our aim is not to construct general models of revolution, democratization, or social movements, much less of all political contention whenever and wherever it occurs. On the contrary, we aim to identify crucial causal mechanisms that recur in a wide variety of contention, but produce different aggregate outcomes depending on the initial conditions, combinations, and sequences in which they occur' (p. 37).

The account of social causation offered below wholeheartedly endorses the idea that social explanations need to proceed on the basis of an analysis of underlying social mechanisms. But it is important not to reify the concept of a mechanism into a rigid component of higher-level social processes. And in fact, one can justly be skeptical about the discreteness and elementality of the items McAdam, Tarrow and Tilly offer as examples of social mechanisms. Take brokerage as a mechanism of social contention – isn't this really an umbrella term that encompasses a number of different kinds of negotiation and alliance-formation? So brokerage is not analogous to 'expansion of ice during freezing' – a clear example of a physical causal mechanism that is homogeneous across physical settings. Brokerage is rather a 'family-resemblance' term that captures a number of different instances of collective behaviour and agency.

If we find this line of thought somewhat persuasive, it suggests that we need to locate the causal connectedness among social settings at an even deeper micro-level. It is the situation of 'agents with interests, identities, networks, allies, and repertoires' that constitutes the causal nexus of social causation on contention – not a set of frozen mid-level groups of behaviours such as brokerage or radicalization. Instead, these mid-level concepts are descriptive terms that allow us to single out some broadly similar components of social contention. This is the reason for the emphasis that I will lay on locating the 'substrate' of social causation in structured purposive action. The level at which we find real causal connections in the social world is the level of the socially situated and socially constituted individual in interaction with other individuals.

13.3 What is a causal mechanism?

What is a causal mechanism? Consider this formulation:

> A causal mechanism is (i) a particular configuration of conditions and processes that (ii) always or normally leads from one set of conditions to an outcome (iii) through the properties and powers of the events and entities in the domain of concern.[1]

Mechanisms bring about specific effects based on the properties of the substrate of processes and events in this domain. For example, 'over-grazing of the commons' is a mechanism of resource depletion in the context of a non-regulated community of users (Hardin 1968). We can reconstruct precisely why this would be true for rationally self-interested actors in the presence of a public good: rational agents use more of the 'free' public resource to increase their own private consumption, and this behaviour aggregates to over-use of the public resource. This is how we specify condition (iii) for the overgrazing mechanism. Further, it is the case that, whenever the conditions of the mechanism are satisfied, the result regularly ensues; in any case where the dominant motive for agents is rational self-interest, we can expect that a common resource will be over-used.

So we do not need to postulate 'laws of society' in order to see how social causation might work. Instead, we can directly identify the features

[1] This is an extension of the formulation offered by Little (1991, p. 15).

of purposive action within given structures that make the mechanism work. Human actions and refrainings are the 'stuff' of social causation, and features of human agency underwrite the 'necessity' of social mechanisms. So we can properly understand a claim for social causation along these lines: 'C causes E' means 'there is a set of causal mechanisms working through features of structured agency that convey circumstances including C to circumstances including E'. It follows from this analysis that mechanisms implicate regularities. But these regularities are low-level and may not be observable in macro-level social behaviour (for example, because of the mixing of several causal processes and the possibility of countervailing mechanisms in play). So they do not serve to play the role of a set of governing laws of society, analogous to laws of nature.

The discovery of social mechanisms often requires the formulation of mid-level theories and models of these mechanisms and processes – for example, the theory of free-riding. These theories and models are 'theories of the middle range' in much the sense that Robert Merton meant to convey when he introduced the term (Merton 1963): accounts of the real social processes that take place above the level of isolated individual action but below the level of full theories of whole social systems. Marx's theory of capitalism illustrates the latter; Jevons's theory of the individual consumer as a utility maximizer illustrates the former. Coase's theory of transaction costs is a good example of a mid-level theory (Coase 1988): general enough to apply across a wide range of institutional settings, but restricted enough in its claim of comprehensiveness to admit of careful empirical investigation. Significantly, the theory of transaction costs has spawned major new developments in the new institutionalism in sociology.

So this provides an answer to the fundamental question: explaining a social outcome or pattern involves providing an account of the social-causal mechanisms that typically bring it about, or brought it about in specific circumstances. This position may be described as 'causal realism', and it serves both as an ontological thesis and as a guide to social-science methodology. (An important advocate for a realist interpretation of science is Roy Bhaskar's *A Realist Theory of Science* (Bhaskar 1975).) But what is the nature of the substrate of social causation? What do social mechanisms consist of? What makes them operate in the patterned and regular ways that we hypothesize for them?

13.4 Methodological localism

Crucial to a valid understanding of social mechanisms is a general answer to the question, what ontological features of the social world facilitate and empower the causal connection from one end of a social mechanism to another? What stands in the place of 'causal powers of natural entities' in the social world?

The answer is a fairly straightforward one: it is facts about intentional agents, socially situated in embodied social relations, that constitute the motive power of social causation. This corresponds to the ontology of 'methodological localism' I have developed elsewhere (Little 2006). So let us address this question directly. How do social causes work?

Some social theorists have treated social constructs as unified macro-entities with their own causal powers. Structuralist theories maintain things like 'capitalism causes people to value consuming more than family time' or 'democracy causes social cohesion'. Likewise, some theorists have held that moral systems and cultures cause distinctive patterns of behaviour – 'Confucian societies produce cohesive families'. Each of these claims places a large social entity in the role of a causal factor.

Is this a coherent way of talking? Can large structures and value systems exercise causal influence? The problem here is that statements like these look a lot like 'action at a distance'. We are led to ask: *How* do capitalism, democracy, or Confucianism influence social outcomes? In other words, we want to know something about the lower-level mechanisms through which large social factors impact upon behaviour, thereby producing a change in social outcomes. We want to know quite a bit about the 'microfoundations' of social causation (Little 1994).

One point seems obvious – and yet it is often overlooked or denied. Social behaviours are carried out by individuals, and individuals are influenced only by factors that directly impinge upon them (currently or in the past). Consider a particular voter's process of deciding to support particular candidate. This person experienced a particular history of personality formation – a particular family, a specific city, a work history, an education. So the person's current political identity and values are the product of a sequence of direct influences. And at the current moment, this socially-constructed person is now exposed to another set of direct influences about the election contest – newspapers, internet, co-workers' comments, attendance at political events, etc. In other words, his or her current political judgments and preferences are caused or influenced by a past and current set of experiences and contexts. This story brings in social factors at every stage – the family was Catholic, the city was Chicago, the work was a UAW-organized factory. So the individual is socially influenced and formed at every stage. But here is the important point: every bit of that social influence is mediated by locally experienced actions and behaviours of other socially formed individuals. 'Catholicism', 'Chicago culture', and 'union movement' have no independent reality over and above the behaviours and actions of people who embody those social labels.

The approach taken here to social causation insists on the centrality of concrete social mechanisms embedded in the actions of social actors. This perspective is sometimes called methodological individualism. I prefer to call it methodological localism (Little 2006). This is the view that the foundation for the explanation of social action and outcome is the local, socially located and socially constructed individual person. The individual is socially constructed, in that her modes of behaviour, thought, and reasoning are created through a specific set of prior social interactions. And her actions are socially situated, in the sense that they are responsive to the institutional setting in which she chooses to act. Purposive individuals, embodied with powers and constraints, pursue their goals in specific institutional settings; and regularities of social outcome often result.

Methodological localism affirms that there are large social structures and facts that influence social outcomes. But it insists that these structures are only possible insofar as they are embodied in the actions and states of socially constructed individuals. The 'molecule' of all social life is the socially constructed and socially situated individual, who lives, acts, and develops within a set of local social relationships, institutions, norms, and rules.

So here we have a fairly comprehensive basis for a theory of social mechanisms. There is such a thing as social causation. Institutions, structures, demographic features, and widespread social arrangements have specific causal effects on the societies in which they exist. The mechanisms that convey these causal powers exist in a social ontology of socially situated and socially constituted individuals, acting and refraining in response to their own motivations and beliefs and the rules, conventions, and constraints that exist around them. And the most basic challenge of social inquiry and social explanation is to discover some of the specific arrangements and features of mentality that aggregate to bring about surprising social outcomes.

13.5 Varieties of causal mechanisms

The general nature of the mechanisms that underlie sociological causation has been very much the subject of debate. Two broad approaches may be identified: agent-based perspectives and social-influence theories. The former follow the strategy of aggregating the results of individual-level choices into macro-level outcomes; the latter attempt to identify the factors that work behind the backs of agents to influence their choices. Thomas Schelling's apt title *Micromotives and Macrobehavior* captures the logic of the former approach, and his work profoundly illustrates the sometimes highly unpredictable results of the interactions of locally rational behaviour (Schelling 1978). Jon Elster has also shed light on the ways in which the tools of rational choice theory support the construction of large-scale sociological explanations (Elster 1989). The second approach, the social-influence approach, attempts to identify socially salient influences such as race, gender, educational status, and to provide detailed accounts of how these factors influence or constrain individual trajectories – thereby affecting sociological outcomes. These should not be understood as being contradictory approaches; rather, they each direct explanatory inquiry at different parts of the same nexus of socially situated agency. The first set of approaches pays primary attention to the motives and reasonings of agents within a given set of constraints; while the second set gives more attention to the broad social factors that influence individual agency.

How do social mechanisms work? The basics are fairly clear: individuals have goals, values, and beliefs, they exist within social and natural constraints, and their actions *bring about* a variety of social outcomes. But how do features of 'agents within structures' bring about social outcomes? We can give a somewhat more detailed analysis of some of the ways that social facts might cause other social facts by surveying a wide sample of causal explanations from the social science literature. This approach leads to an open-ended list of kinds of social mechanisms.

1. Rational-intentional mechanisms. Why do empires establish a policy of rotating senior military officials? Because emperors want to avoid the creation of warlords.
2. Imitation mechanisms. Why did the no-huddle offense become so common in the National Football League in the 1980s? Because it was successful for a few teams, and others copied the offense in the hope that they too would win more games.
3. Conspiracy mechanisms (covert strategems of the powerful). Why did the United States move away from passenger railroads as the primary form of intercity transportation? Because powerful actors took political actions to assure that private automobiles would be encouraged as the primary form of transport.
4. Aggregative mechanisms (aggregate consequences of individual-level strategies). Why does technological innovation occur continuously within a market-based society? Because each firm is constantly looking for lower-cost and higher-value-added methods of manufacturing, and these individual efforts aggregate to an industry trend towards innovations in products and technologies.
5. Mentality mechanisms (behaviour is changed by changing beliefs and attitudes). Why were so many Quaker men conscientious objectors at great personal cost during World War II? Because their religious beliefs categorically rejected the violence in war and they refused to participate in this immoral activity.
6. Social network mechanisms (information and norms proliferate through concrete sets of social relationships among individuals). Why was the Soviet military system less adaptive in combat than the Israeli

military system? Because information flow among officers and troops was more rapid and more bidirectional in the latter than the former.

7. Evolutionary mechanisms. Why does the level of firm efficiency tend to rise over time? Because the net efficiency of a firm is the product of many small factors. These small factors sometimes change, with an effect on the efficiency of the firm. Low efficiency firms tend ultimately to lose market share and decline into bankruptcy. Surviving firms will have features that produce higher efficiency.
8. Filtering mechanisms. Why are passengers on commercial aircraft better educated than the general population? Because most airline passengers are business travellers, and high-level and mid-level business employees tend to have a higher level of education than the general population.
9. Critical mass mechanisms. A new social networking site experiences slow growth for the first eighteen months of operation until it reaches *N* users; it then takes off with rapid growth for the next eighteen months. We attempt to explain this change by arguing that *N* is a critical mass of users, stimulating much more rapid growth in the future.
10. Path-dependency mechanisms. Why do we still use the very inefficient QWERTY keyboard arrangement that was devised in 1874? Because this arrangement, designed to keep typists from typing faster than the mechanical keyboard would permit, was so deeply embodied in the typing skills of a large population and the existing typewriter inventory by 1940 that no other keyboard arrangement could be introduced without incurring massive marketing and training costs.

This is not an exhaustive list of types of social causation, and there is some overlap among these types. The first four examples fall roughly into the broad category of agent-centred explanations; the next three examples illustrate the social-influence model; and the final three examples illustrate 'system-level' features of the environment of social change (selective filtering of events, the mathematics of critical mass, and the momentum of prior social choices). There are no doubt another dozen examples of explanatory schemata that could be adduced as well. What this list illustrates, however, is that there are a variety of ways, both direct and indirect, through which social causation can be conveyed from one set of social facts to another. They all involve the same basic ontology of social causation – agents acting within structures leading to social outcomes – but the nature of the pathway from cause to effect is different in the various types.

Emphasis on causal mechanisms for adequate social explanation has several beneficial effects on sociological method. It takes us away from easy reliance on uncritical statistical models. But it also may take us away from excessive emphasis on large-scale classification of events into revolutions, democracies, or religions, and toward more specific analysis of the processes and features that serve to discriminate among instances of large social categories. Charles Tilly emphasizes this point in his arguments for causal narratives in comparative sociology (Tilly 1995). He writes, 'I am arguing that regularities in political life are very broad, indeed transhistorical, but do not operate in the form of recurrent structures and processes at a large scale. They consist of recurrent causes which in different circumstances and sequences compound into highly variable but nonetheless explicable effects' (Tilly 1995, p. 1601).

We do a poor job of understanding industrial strikes if we simply collect a thousand instances and perform statistical analysis on the features we've measured against the outcome variables. We do a much better job of understanding them if we put together a set of theories about the features of structure and agency through which a strike emerges and through which individuals make decisions about participation – the mechanisms that commonly arise in the social processes of industrial contention. Analysis of the common 'agent/structure' factors that are relevant to mobilization will permit us to understand individual instances of mobilization, explain the soft regularities that we discover, and account for the negative instances as well.

13.5.1 Examples of social mechanisms

Are there any credible social mechanisms? In fact, a cursory survey of comparative sociology, political science, and the new institutionalism provides a very large body of explanations that identify common social mechanisms – for example, 'collective action problems often cause strikes to fail', 'increasing demand for a good causes prices to rise for the good in a competitive market', 'transportation systems cause shifts of social activity and habitation'. Here is a small sample of familiar social mechanisms that have been invoked to explain important social outcomes across a range of social settings.

- public goods problems (Hardin 1982)
- political entrepreneurship (Bates 1981)
- principal-agent problems (Ensminger 1992)
- features of ethnic or religious group mobilization (Hardin 1995)
- market mechanisms and failures (Akerlof 1970)
- rent-seeking behaviour (Seligson and Passé-Smith 1993)
- mechanisms of corruption (Klitgaard 1988)
- the social psychology of race (Steele and Aronson 1995)
- the moral emotions of family and kinship (Hareven 2000)

- the dynamics of a transport network (Vance 1986)
- the 'moral economy' of the crowd (Thompson 1971)
- the communications characteristics of medium-size social networks (Latour 2005)
- the psychology and circumstances of solidarity (Taylor 1982).

These are all mechanisms that work at the level of socially situated actors. They characterize one or more of the features of agency and structure. We understand how they work; individuals with specified motivational and cognitive characteristics, placed within the context of the social settings identified by the mechanisms, will behave in ways that bring about the outcome. And they are abstract enough that they can be identified in a wide range of settings: the feudal manor, the collective farm, the Wall Street law firm. In fact, we might say that the most fundamental value of theories in the social sciences is the formulation of models of mechanisms at this level, providing a toolkit for social explanation (Elster 2007).

It is important to observe that social mechanisms rarely work in isolation; so their operation usually cannot be observed in a pure state. Take the mechanism of 'collective action failures in the presence of public goods'. (Russell Hardin's analysis of collective action problems is highly useful; (Hardin 1982).) Here the heart of the mechanism is the analytical point that rationally self-interested decision-makers will take account of private goods but not public goods; so they will tend to avoid investments in activities that produce public goods. They will tend to become 'free riders' or 'easy riders' (Popkin 1981). The social regularity that corresponds to this mechanism is a 'soft' generalization – that situations that involve a strong component of collective opportunities for creating public goods will tend to demonstrate low contribution levels from members of affected groups. So public radio fundraising will receive contributions only from a small minority of listeners; boycotts and strikes will be difficult to maintain over time; lakes and estuaries will tend to be over-fished. And in fact, these regularities can be identified in a range of historical and social settings.

However, the 'free rider' mechanism is only one of several that influence collective action. There are other social mechanisms that have the effect of enhancing collective action rather than undermining it (Ostrom 1990). For example, the presence of competent organizations makes a big difference in eliciting voluntary contributions to public goods; the fact that many decision-makers appear to be 'conditional altruists' rather than 'rationally self-interested maximizers' makes a difference; and the fact that people can be mobilized to exercise sanctions against free riders affects the level of contribution to public goods. (If your neighbours complain bitterly about your smoky fireplace, you may be incentivized to purchase a cleaner-burning wood or coal.) The result is that the free-rider mechanism rarely operates by itself – so the expected regularities may be diminished or even extinguished.

Let us look at a few examples of causal explanation of social outcomes in greater detail.

13.5.2 Transportation as a social mechanism

Transportation systems function to move people, goods, and ideas. Rail systems, road networks, airline systems, and water transport provide links between places that permit more reliable and low-cost movement of people and goods from point to point than previously available. The history of transportation is simultaneously a history of technology change, population movement, colonialism, economic growth, business development, and the spread of disease.

Transportation systems are particularly interesting when we consider their capacity for conveying social causation. Consider these examples of causal relations mediated by transportation systems:

- Extension of a rail network stimulates the growth of new towns, villages and cities in North America in the 1880s.
- Establishment of a direct air travel link between A and B causes the more rapid spread of disease between these locations.
- Breakdown of the administration of the rail system leads to logistics bottlenecks and military defeat of the French army in the Franco-Prussian War (Howard 1961).
- Regular river travel throughout the Canton Delta in China leads to the rapid spread of revolutionary ideas during the Republican Revolution, as travellers and merchants move easily from place to place (Hsieh 1978).
- Commodity price correlation increases between Chicago and New York as a result of regular and cheap rail transport and communication between these two cities (Cronon 1991).
- New business institutions (grain futures markets and grain elevators) are created to take advantage of cheap regular rail transport (Cronon 1991).

A rail system provides convenient transportation among a number of places, while providing no service at all between other pairs of locations. So a rail system certainly has direct effects on social behaviour; it structures the activities of the several million residents of a major city by making some places of residence, work, shopping, and entertainment substantially more accessible than other places. And there are a number of other social characteristics that are structured by a commuter rail system as a consequence: for example, patterns of class stratification of neighbourhoods, patterns of diffusion of infectious disease, patterns of ethnic habitation around the city, patterns of diffusion of social styles and dialect, and so on (Warner 1969). In brief, a rail system has definite social effects. It creates opportunities and constraints that affect the ways in which individuals arrange their lives and plan their daily activities. And other forms of social behaviour and activity are conveyed through the conduits established by the transport system.

Moreover, a rail system is a physical network that has an embodied geometry and spatiality on the ground. Through social investments over decades or more, tracks, stations, power lines, people movers, and fuel depots have been created as physical infrastructure for the transportation network. Lines cross at junctions, creating the topology of a network of travel; and the characteristics of travel are themselves elements of the workings of the network – for example, the rate of speed feasible on various lines determines the volume of throughput of passengers through the system. And neighbourhoods and hotels agglomerate around important hubs within the system.

In addition to this physical infrastructure, there is a personnel and management infrastructure associated with a rail system as well: a small army of skilled workers who maintain trains, sell tickets, schedule trains, repair tracks, and myriad other complex tasks that must be accomplished in order for the rail system to carry out its function of efficiently and promptly providing transportation. This human organization is surely a 'social structure', with some level of internal corrective mechanisms that maintain the quality of human effort, react to emergencies, and accomplish the business functions of the rail system. This structure exists in the form of training procedures, operating manuals, and processes of supervision that maintain the coordination needed among ticket agents in stations, repairmen in the field, track inspectors, engineers, and countless other railroad workers. And this structure is fairly resilient in the face of change of personnel; it is a bureaucratized structure that makes provision for the replacement of individuals in all locations within the organization over time.

So a rail network has structural and causal characteristics at multiple levels. The physical network itself has structural characteristics (nodes, rates of travel, volume of flow of passengers and freight). This can be represented statically by the network of tracks and intersections that exist; dynamically, we can imagine a 'live' map of the system representing the coordinated surging of multiple trains throughout the system, throughout the course of the day. The railroad organization has a bureaucratic structure – represented abstractly by the organizational chart of the company, but embodied in the internal processes of training, supervision, and recruitment that manage the activities of thousands of employees. And the social and technical ensemble that these constitute in turn creates an important structure within the social landscape, in that these physical and human structures determine the opportunities and constraints that exist for individuals to pursue their goals and purposes.

The central point here is that transportation is a robust family of causal mechanisms that mediate many important social processes and outcomes. And its causal effectiveness is fairly transparent: new transportation opportunities create new options for social actors, who take advantage of these opportunities in choosing a place to live and work, in pursuing political goals, in moving armies, and in generating income. So transportation is a causal mechanism whose microfoundations are especially visible, and whose causal consequences are often very large.

13.5.3 What is to explain about violent crime?

Let us turn next to a sociological research problem of great longevity: the explanation of variations in the crime rates of different places. Every city has a crime rate – the incidence of murders, assaults, car thefts, or burglaries per 100,000 residents. And there are very significant differences across cities and countries with respect to the incidence of violent crime. In particular, some cities in the world experience extremely high levels in the incidence of violent crimes (for example, Johannesburg, South Africa). Compare these national statistics describing 'murders per 100,000 population': England/Wales, 1; USA, 6; Brazil and Russia, 21; Columbia, 58; South Africa, 59. Why are there such major variations across countries with respect to the murder rate? Why are some cities and countries so much more violent than others? What are the most important factors that cause a community to have a high (or low) rate of violent crime? Why do individuals in some societies have a higher propensity for violent crime than in others?

The study of the causes of violence and crime has been a part of Western sociology since the beginning. A variety of causes have been suggested: the incidence of absolute poverty, the extent of inequalities, the phenomenon of relative deprivation, the drug trade, the breakdown of traditional community and family values, and the effects of racism, to name a few.

Notice that we can put the causal question in several ways:

- What causes variation across communities with respect to crime rates?
- What factors increase the likelihood of a particular individual becoming a violent criminal?
- What social factors cause an increase or reduction in the crime rate?

That is, we can ask about explaining variation across cases; we can ask about explaining particular individuals' behaviour; we can ask about 'inducing' and 'inhibiting' causes of changes in the crime rate; and there are other causal questions as well.

Consider this small set of possible causal mechanisms that might influence violent behaviour:

- rational incentives (risk and gain calculation)
- material circumstances (unemployment, education)
- community cohesiveness
- a broad sense of injustice (exploitation, unfair exclusion)
- moral and religious values
- alienation and disaffection
- racial or ethnic polarization
- imitation of the behaviour of others
- organizations (gangs, youth groups, social networks)
- laws and policing.

Now we might try to construct an agent-centred theory of violent crime based on these sorts of factors conjoined with a description of the social environment at the time – and then predict variation across time and place as a behavioral result of the incorporation of these factors influencing action.

A very different approach – and one that is probably closer to the quantitative-methodology mainstream in sociology today – is to assemble a set of cases (a list of United States cities, for example); measure a number of variables for each city (unemployment rate, index of social capital, level of education, degree of neighbourhood segregation, presence/absence of mass transportation, ...); and then test the degree of correlation that one or more of these variables has with the observed variation in the crime rate. Do variations in the incidence of violent crime correlate with levels of unemployment? Then unemployment is a plausible causal factor in determining the level of the crime rate. Does some measure of social capital correlate negatively with variations in the crime rate? Then the social cohesion that is hypothetically linked to higher scores for social capital in a community is a negative causal factor in variations in the crime rate. And so forth.

Both these approaches are compatible with the research methodology associated with trying to identify causal mechanisms. If it is in fact true that a young person's disposition to engaging in violent crime is decreased if he/she is a member of a church – then we ought to find at the macro-level that a higher index of church membership will be associated with a lower crime rate. However, given the multiplicity of causal factors that are likely to be at work, the purely statistical approach is unlikely to yield satisfactory results. Reciprocally, if we discover a positive correlation between 'degree of neighbourhood segregation' and the crime rate – then we need to be able to disaggregate the story and discover the individual-level mechanisms through which segregation increases individuals' propensity for violent crime.

So explaining differences in crime rates across places requires that we provide hypotheses about how various social factors might influence criminal behaviour; which is simply another way of saying that these explanations require hypotheses about the underlying social mechanisms.

13.5.4 Race and asthma

How can a group characteristic be a causal factor in enhancing some other group characteristic? What kinds of social mechanisms might convey the group characteristic onto its effect on the population?

Suppose the facts are these: that African-Americans have a higher probability of developing asthma, even controlling for income levels, education levels, age, and urban-suburban residence. And suppose that the researcher summarizes his/her findings by saying that 'being African-American causes the individual to have a higher risk of developing asthma'. How are we supposed to interpret this claim?

My preferred interpretation of statements like these is to hypothesize a causal mechanism, presently unknown, that influences African-American people differentially and produces a higher incidence of asthma. Here are a few possibilities for a mechanism that might have this effect:

- African-Americans as a population have a lower level of access to quality healthcare and are more likely to be uninsured. Asthma is a disease that is best treated on the basis of early diagnosis. Therefore African-Americans are more likely to suffer from undiagnosed and worsening asthma.
- Asthma is an inner-city disease. It is stimulated by air pollution. African-Americans are more likely to live in inner-city environments because of the workings of residential segregation. So race causes exposure which in turn causes a higher incidence of the disease.
- There might be an unidentified gene that is more frequent in people with African ancestry than non-African ancestry and that makes one more susceptible to asthma.
- There might be a nutritional component to the onset of asthma, and it could be that cultural differences between the two communities lead the African-American population to have higher levels of exposure to the nutritional cause of the disease.

And of course we could proliferate possible mechanisms.

In each case the logic of the account is similar. We proceed by hypothesizing a factor or combination of factors that increase the likelihood of developing asthma; and then we try to determine whether this collateral factor is more common in the African-American population. Some of these stories would amount to spurious correlations, while others would constitute stories in which the fact of race (as opposed to a factor with which race is accidentally correlated) plays an essential role in the causal story. (Reduced access to healthcare and inner city air pollution fall in this category, since it is institutionalized racial segregation that causes the higher-than-normal frequency of urban residence for African-Americans.)

Race is not itself a causal mechanism, but rather a social factor that plays into a variety of mechanisms. So for example, race is associated with differential health outcomes and incarceration rates. The task for the sociologist is to discover some of the mechanisms or pathways through which one's racial status exercises influence on his/her health or incarceration outcomes. Here is one particular pathway:

> racial status {**affects**} residential status in inner city or suburb {**affects**} exposure to airborne pollutants {**affects**} likelihood of developing respiratory disease.

Essentially the mechanism linking racial status to health outcomes in this story is the package of concrete social processes through which segregation is maintained: mortgage and insurance redlining practices, neighborhood resistance to new families of another race, real estate steering practices, and discrimination in rental housing, for example.

So this is a potential interpretation of the causal meaning of a statement like 'race causes an increased risk of X'. It provides a schematic explanatory statement along these lines: racial status causes propensity to asthma through mechanism M. What would be most perplexing to the quantitative researcher is if there were multiple sets of causal mechanisms, each independent of the others and each creating a race-specific difference in incidence of the disease. For example, it might be that both exposure to air pollution and lack of health insurance lead to a higher incidence of the disease; and further, it might be that inner-city residents do in fact have adequate healthcare but exposure to inner-city pollution; while suburban African-Americans might have less healthcare and limited exposure to air pollution. In this set of facts, both African-American populations would display higher-than-normal incidence, but for different and unrelated reasons. We explore this possibility in the following section.

13.6 Is sociology analogous to epidemiology?

Sociology attempts, among other things, to establish causal connections between large social factors (race, socio-economic status, residential status) and social outcomes of interest (rates of delinquency). This sounds quite a bit like the reasoning done by epidemiologists to assign 'risk factors' to the occurrence of a disease based on individual and environmental characteristics – for example, 'people exposed to high levels of radiation such as survivors of nuclear reactor accidents are at an increased risk for developing non-Hodgkin's lymphoma'. The epidemiologist works with a large data set of persons with a given disease and then tries to discover factors X, Y, Z whose statistical distribution differs for the disease population relative to the general population. Is quantitative sociology analogous in any way to the use of large disease data sets to attempt to identify risk factors? In other words, is there a useful analogy between sociology and epidemiology? Are there similarities in the forms of causal reasoning that are to be found in the two areas of research?

Suppose that the divorce rate for all American men is 30%. Suppose the rate for New York City males with income greater than $200,000 is 60%. We might want to draw the inference that something about being a high-income male resident of New York causes a higher risk of divorce for these persons. And we might want to justify this inference by noticing that it is similar to a parallel statistical finding relating smoking to lung cancer: the probability of acquiring lung cancer is significantly higher for smokers than non-smokers. So sociology is similar to epidemiology. Certain factors can be demonstrated to cause an elevated risk of a certain kind of outcome. There are 'risk factors' for social outcomes such as divorce, delinquency, or drug use.

Is this a valid analogy? I think it is not. Epidemiological reasoning depends upon one additional step: a background set of assumptions about the ontology and etiology of disease, specifying the mechanisms and causal environment of disease. A given disease is a specific physiological condition within a complex system of cells and biochemical processes. We may assume that each of these physiological abnormalities is caused by some specific combination of external and internal factors through specific causal mechanisms. The causal pathways of normal physiological functioning are discrete and well-defined, and so are the mechanisms that cause disruption of these normal causal pathways. Within the framework of these guiding assumptions, the task of the statistics of epidemiology is to help sort out which factors are causally associated with the disease. The key, though, is that we can be confident that there is a small number of discrete causal mechanisms that link a real causal factor to the disease.

The case is quite different in the social world. Social processes are not similar to physiological processes, and social outcomes are not similar to diseases. In each case the failure of parallel derives from the fact that there are not unique and physiologically specific causal systems at work. Cellular reproduction has a specific biochemistry. Malignancy is a specific deviation from these cellular processes. And specific physical circumstances cause these deviations. But becoming a criminal is a much more fluid and multiform process.

The problem is that a social outcome like 'violent society' is not the result of a homogeneous social process across multiple social settings. Rather, it is a heterogeneous mix of social developments and events; and these components are different in different times and places. And outcomes that might be considered the social equivalent of disease – a rising murder rate, for example – are also composites of many distinct social happenings and processes. So social systems and outcomes lack the simple, discrete causal uniformity that is a crucial underpinning of epidemiological reasoning.

This is not to say that there are not underlying causal mechanisms whose workings bring about a sharp increase in, say, the population's murder rate. Rather, it is to say that there are numerous, heterogeneous and cross-cutting such mechanisms. So the resultant social outcome is simply the contingent residue of the multiple middle-level processes that were in play in the relevant time period. And the discovery that 'X, Y, Z factors are correlated with a rise in the incidence of O' isn't causally irrelevant. But the effects of these factors must be understood as working through their influence on the many mid-level causal mechanisms.

In particular, several basic propositions of epidemiological reasoning cannot be affirmed in the case of social causation:

(a) If there is a causal mechanism linking X to Y, then there will be a statistical association between X and Y. (And the contrapositive: If there is no statistical association between X and Y, then there is unlikely to be a causal mechanism leading from X to Y.)

(b) If there is a statistical association between X and Y, then there is likely to be a single causal mechanism leading from X to Y.

Neither of these propositions is routinely true in the case of social outcomes. Statement (a) is untrue, because there may *ceteris paribus* conditions and alternative causal pathways that result in a lack of correlation between X and Y – in spite of the real causal mechanism linking them. And statement (b) is untrue, because the observed correlation between X and Y may be simply the aggregate effect of a large number of separate causal processes involving X and Y. So quantitative sociological reasoning is not analogous to epidemiological reasoning, for this reason: there is a substantially greater possibility of multiple causal pathways and conditions in the case of the social world, leading to the result that discovery of gross correlations between factors is unlikely to correspond to unique causal mechanisms and pathways leading to the observed outcome.

13.7 Conclusion

We have argued for several key points concerning social causation. First, there is such a thing as social causation. Causal realism is a defensible position when it comes to the social world: there are real social relations among social factors (structures, institutions, groups, norms, and salient social characteristics like race or gender). We can give a rigorous interpretation to claims like 'racial discrimination causes health disparities in the United States'.

Second, we have argued in support of the idea that causal relations depend on the existence of real social-causal mechanisms linking cause to effect. Discovery of correlations among factors does not constitute the whole meaning of a causal statement. Rather, it is necessary to have a theory of the mechanisms and processes that give rise to the correlation. Moreover, it is defensible to attribute a causal relation to a pair of factors even in the absence of a correlation between them, if we can provide evidence supporting the claim that there are specific mechanisms connecting them. So mechanisms are more fundamental than regularities.

Third, we have tried to make good on a key intellectual obligation that goes along with postulating real social mechanisms: to provide an account of the ontology or substrate within which these mechanisms operate. This we have attempted to provide through the theory of methodological localism – the idea that the causal nexus of the social world is the behaviours of socially situated and socially constructed individuals. To put the claim in its extreme form, every social mechanism derives from facts about institutional context, the features of the social construction and development of individuals, and the factors governing purposive agency in specific sorts of settings. And different research programs target different aspects of this nexus.

And finally, we have looked at a few typical forms of sociological reasoning in detail, in order to see how the postulation and discovery of social mechanisms play into mainstream sociological research. Properly understood, there is no contradiction between the effort to use quantitative tools to chart the empirical outlines of a complex social reality, and the use of theory, comparison, case studies, process-tracing, and other research approaches aimed at uncovering the salient social mechanisms that hold this empirical reality together.

References

Akerlof, George. 1970. The market for 'lemons': Quality, uncertainty and the market mechanism. *Quarterly Journal of Economics* 84: 488–500.

Bates, Robert H. 1981. *Markets and states in tropical Africa: The Political Basis of Agricultural Policies*, California Series on Social Choice and Political Economy. Berkeley: University of California Press.

Bhaskar, Roy. 1975. *A Realist Theory of Science*. Leeds: Leeds Books.

Cartwright, Nancy. 1989. *Nature's Capacities and their Measurement*. Oxford: Oxford University Press.

Coase, R. H. 1988. *The Firm, the Market, and the Law*. Chicago: University of Chicago Press.

Cronon, William. 1991. *Nature's Metropolis: Chicago and the Great West*. New York: W. W. Norton.

Dupré, John, and Nancy Cartwright. 1988. Probability and causality: Why Hume and indeterminism don't mix. *Nous* 22: 521–536.

Ekstrom, Mats. 1992. Causal explanation of social action: The contribution of Max Weber and of critical realism to a generative view of causal explanation in the social sciences. *Acta Sociologica* 35 (2): 107 (16).

Elster, Jon. 1989. *The Cement of Society: A Study of Social Order*. Cambridge: Cambridge University Press.

Elster, Jon. 2007. *Explaining Social Behavior: More Nuts and Bolts for the Social Sciences*. Cambridge: Cambridge University Press.

Ensminger, Jean. 1992. *Making a market: The Institutional Transformation of an African Society*, The Political Economy of Institutions and Decisions. Cambridge: Cambridge University Press.

Hardin, Garrett. 1968. The tragedy of the commons. *Science* 162: 1243–48.

Hardin, Russell. 1982. *Collective Action*. Baltimore: The Johns Hopkins University Press.

Hardin, Russell. 1995. *One for All: The Logic of Group Conflict*. Princeton: Princeton University Press.

Hareven, Tamara K. 2000. *Families, History, and Social Change: Life Course and Cross-Cultural Perspectives*. Boulder: Westview Press.

Harré, Rom, and Edward H. Madden. 1975. *Causal Powers: A Theory of Natural Necessity*. Oxford: Basil Blackwell.

Hedström, Peter, and Richard Swedberg (eds.) 1998. *Social Mechanisms: An Analytical Approach to Social Theory*, Studies in Rationality and Social Change. Cambridge: Cambridge University Press.

Howard, Michael Eliot. 1961. *The Franco-Prussian War; The German Invasion of France*. London: R. Hart-Davis.

Hsieh, Winston. 1978. Peasant insurrection and the marketing hierarchy in the Canton Delta, 1911–12: in Wolf, Arthur P., ed. 1978. *Studies in Chinese Society*. Stanford: Stanford University Press.

Klitgaard, Robert E. 1988. *Controlling Corruption*. Berkeley: University of California Press.

Latour, Bruno. 2005. *Reassembling the Social: An Introduction to Actor-Network-Theory*, Clarendon Lectures in Management Studies. Oxford: Oxford University Press.

Little, Daniel. 1991. *Varieties of Social Explanation: An Introduction to the Philosophy of Social Science*. Boulder: Westview Press.

Little, Daniel. 1993. On the scope and limits of generalizations in the social sciences. *Synthese* 97: 183–207.

Little, Daniel. 1994. Microfoundations of Marxism. In *Readings in the Philosophy of Social Science*, edited by M. Martin and L. McIntyre. Cambridge: MIT Press.

Little, Daniel. 2006. Levels of the Social. In *Handbook for Philosophy of Anthropology and Sociology*, edited by S. Turner and M. Risjord. Amsterdam: Elsevier Publishing.

Mackie, J. L. 1974. *The Cement of the Universe; A Study of Causation*. Oxford: Clarendon Press.

McAdam, Doug, Sidney G. Tarrow, and Charles Tilly. 2001. *Dynamics of Contention*, Cambridge Studies in Contentious Politics. New York: Cambridge University Press.

Merton, Robert K. 1963. *Social Theory and Social Structure*. New York: Free Press.

Ostrom, Elinor. 1990. *Governing the Commons: The Evolution of Institutions for Collective Action*. Cambridge: Cambridge University Press.

Popkin, Samuel L. 1988. Public choice and rural development – free riders, lemons, and institutional design. In Bates, Robert H., ed. 1988. *Toward a Political Economy of Development*. A Rational choice Perspective Berkeley. University of California Press.

Salmon, Wesley C. 1984. *Scientific Explanation and the Causal Structure of the World*. Princeton: Princeton University Press.

Schelling, Thomas C. 1978. *Micromotives and Macrobehavior*. New York: Norton.

Seligson, Mitchell A., and John T. Passé-Smith (eds.) 1993. *Development and Underdevelopment: The Political Economy of Inequality*. Boulder: L. Rienner Publishers.

Sørensen, Aage B. 1998. Theoretical mechanisms and the empirical study of social processes. In *Social Mechanisms: An Analytical Approach to Social Theory*, edited by P. Hedström and R. Swedberg. Cambridge, U.K.; New York: Cambridge University Press.

Steele, Claude M., and Joshua Aronson. 1995. Stereotype threat and the intellectual test performance of African Americans. *Journal of Personality and Social Psychology* 69 (5): 797–811.

Taylor, Michael. 1982. *Community, Anarchy and Liberty*. Cambridge: Cambridge University Press.

Thompson, E. P. 1971. The moral economy of the English crowd in the eighteenth century. *Past and Present* 50: 71–136.

Tilly, Charles. 1995. To explain political processes. *American Journal of Sociology*. 100: 1594–610

Vance, James E. 1986. *Capturing the Horizon: The Historical Geography of Transportation Since the Transportation Revolution of the Sixteenth Century*. New York: Harper & Row.

Warner, Sam Bass. 1969. *Streetcar Suburbs: The Process of Growth in Boston, 1870–1900.*, Publications of the Joint Center for Urban Studies. New York: Atheneum.

·14·

Getting past Hume in the philosophy of social science

Ruth Groff

Abstract

A realist, powers-based metaphysics is very much on the table in contemporary metaphysics, and is beginning to take hold in philosophy of mind and philosophy of science. On this picture, causality is (roughly) a matter of the powers that things have to effect change(s) in other things. The realist view is at odds with every version of Humeanism, according to all of which causation is not, in the end, about the exercise of powers, but rather, in one way or another, about regular sequences. My chapter has two parts. In the first part I consider how it is that analytic philosophers of social science have been able thus far to side-step the critique of Humeanism. In the second part, I consider how analytic philosophy of social science might look different, were Humeanism no longer to be its tacit metaphysics.

Such is the influence of custom, that, where it is strongest, it not only covers our natural ignorance, but even conceals itself, and seems not to take place, merely because it is found in the highest degree.[1]–Hume

14.1 Introduction

As anyone who has ever been to a lively meeting governed by *Robert's Rules of Order* knows, one way to block discussion of a difficult issue is on procedural grounds. I don't mean to question the need for sound procedures (or, for that matter, to invoke the idea of intentional strategy) – only to point to a familiar instance of a shift from substance to method. I'd like to suggest that something similar, albeit with an added twist, happens systematically in analytic philosophy of social science. The analogue of the blocked substantive proposal is direct argument about ontology. In mainstream analytic philosophy of social science, ontological discussion is effectively cut short by turning metaphysical questions into questions about norms regarding explanation – a conceptual sleight of hand that has been dubbed 'the epistemic fallacy' by

[1] David Hume, *Enquiries Concerning Human Understanding and Concerning the Principles of Morals, 3rd edition* (New York: Clarendon Press, 2000), Section IV, Part I, pp. 28–29.

Roy Bhaskar.[2] The twist is that, in contrast to the lively meeting governed by *Robert's Rules*, it is in the nature of the case here that the procedures upon which discussion is re-focused will themselves be 'partisan', since alternate methodologies carry with them alternate ontological commitments.[3] Thus it is not exactly that questions of substance are simply tabled (thereby, of course, tacitly preserving the metaphysical status quo), but rather that ontological controversies are either transposed into a kind of debate-by-proxy at the level of methodology, or, in the absence of such debate, resolved unilaterally by methodological fiat.

The metaphysical issue that I want to examine in light of this dynamic is causality. For the first time in a long while, there is again a divide amongst metaphysicians between those who are realists about causality and those who are not. To be a realist about causality is to think that, in one way or another, causality involves a display of powers, by the kinds of things that bear or are them inherently, so as to effect changes in other things. To hold this view is to think, *contra* Hume and ultimately Kant alike, that there is indeed such a thing as natural necessity. In the contemporary literature, Brian Ellis, Stephen Mumford, Alexander Bird and Nancy Cartwright, for example, are well-known proponents of this view – preceded in the recent period by Rom Harré and E. H. Madden, George Molnar and Roy Bhaskar. The position, broadly Aristotelian in many if not all of its iterations, is sometimes referred to as dispositional realism. Anti-realism about causality, meanwhile, derives in its fully developed empiricist form from Hume, via Mill. Mackie and Lewis are Humeans often cited by contemporary analytic philosophers and methodologists of social science. On the anti-realist view, dominant within the discipline of philosophy throughout the modern period, there are no such metaphysically necessary connections. There is simply the fact that there are sequences of event that are constantly conjoined – or, for some, conjoined with a high degree of probability.[4] From this perspective, standard until recently, causality has nothing to do with the exercise of things' inherent powers. Indeed, at the metaphysical core of the anti-realist position is precisely a disavowal of such dispositional properties.

[2] Roy Bhaskar, *A Realist Theory of Science* (Sussex: Harvester Press Limited, 1975), p. 16: 'Empirical realism is underpinned by a metaphysical dogma, which I call the epistemic fallacy, that statements about being can always be transposed into statements about our knowledge of being'; and *passim*.

[3] This is a point well explored by Charles Taylor. See, e.g. 'Value neutrality in political science' and 'Interpretations and the sciences of man', reprinted in Taylor, *Philosophy and the Human Sciences: Philosophical Papers 2* (Cambridge: Cambridge University Press, 1985).

[4] Kant is an anti-realist about causality on this definition. There are necessary connections, for Kant, but the necessity in question derives from the synthetic operation of reason, not from the objects of empirical experience.

Humeanism, as it is often called, is now under significant strain in contemporary metaphysics. By contrast, it remains the predominant view in analytic philosophy and methodology of social science, where it has been challenged in a sustained fashion only by critical realists, who have remained marginal within the analytic literature.[5] The situation is a curious one, however. For one thing, I would venture to suggest that even in the positivist stronghold of the US, most philosophers and methodologists of social science would balk at the idea that in aligning themselves with Hume, they have forsworn belief in something like regular old causality. Only Hume himself, one is tempted to say – and perhaps Kant – was prepared to acknowledge the metaphysical import of the anti-realist stance. Moreover, the Humean position tends to be advanced only implicitly. Despite the prevalence of the view, therefore, one can't help but wonder what the outcome would actually be, were there to be a floor-fight on the question, i.e. a substantive debate within analytic philosophy and methodology of social science on the merits of Humean anti-realism about causality versus the merits of a powers-based, realist account of causality.

It's the very lack of such an airing that I want to address. Despite the significance of the issue for the subject area, Humean-derived anti-realism about causality has not been a topic of overt discussion within mainstream analytic work in the philosophy and methodology of social science. And this, as I say, is odd. As a practical matter, it may simply be that everyone who has been to graduate school (especially in the US, and especially in analytic philosophy) has been trained to be an anti-realist about causality, and that no one especially cares to revisit the point – excepting, of late, those who specialize in metaphysics. But there's more to it than this. Even allowing for broad but 'soft' support for Humeanism, it remains the case that the underlying ontological question is more difficult to get at, conceptually, than it ought to be. How *is* it, one wonders, that such a fundamental commitment – one pertaining to an issue that goes to the heart of the concerns of the specialty – has been sheltered from serious critique?

The answer, I shall argue, has to do with the epistemic fallacy. Operationally, that is, if one were to try to force a floor-fight on the question of whether or not causality is more or less what Hume said it is, one would do well to understand the peculiar way in which the focus on explanation functions, in this area of specialization, to advance the Humean position while at the same time displacing attention from metaphysics onto applied epistemology, such that metaphysical considerations appear to be extraneous.

[5] Beginning with Roy Bhaskar, *The Possibility of Naturalism: A Philosophical Critique of the Contemporary Human Sciences*, originally published in 1979. Other critics of Humean methods in social science who were themselves influenced by Aristotle and/or by Marx – e.g. Adorno, Horkheimer, Marcuse, Taylor, MacIntyre – either were not realists about causality, or were so only ambiguously.

14.2 From causality to explanation

One reason why it may seem plausible to regard questions about explanation and questions about causality as interchangeable, akin to fractions versus percentages as designators of proportion, is that intuitively it would seem that to offer an explanation of something is to say why it happened, and to mean by that 'what caused it'. It seems hardly a step at all, when one thinks of it this way, to go from being able to specify philosophically what the features of a proper explanation are to being able to specify what causality is. Yet, no sooner than is explanation rather than causality established as the governing category, it turns out that 'to explain' can be (and often is) taken to mean any number of things other than 'to say what caused it to happen'; for example, to explain may be thought to mean 'to render intelligible', 'to predict', 'to state the function of', etc.[6] Therefore, although each of these alternate explanation-forms can be fruitfully mined for its own implicit metaphysics, let me make it clear that I want to focus on specifically *causal* explanation.

Arguably, the single most influential model of causal explanation within twentieth century analytic philosophy and methodology of social science was Hempel's – still relevant, if no longer orthodoxy, after more than half a century. As is well-known, on the Hempelian covering law model, an explanation consists of a true universal statement expressing a law-like regularity, plus a statement of antecedent conditions relative to the phenomenon which is to be explained, such that a description of the phenomenon follows deductively from the conjunction of the major and the minor premise.[7] Thus, to explain q is to show that its following upon p empirically is an instance of a law to the effect that where there's a p, there will subsequently be a q.[8] As Hempel and Oppenheim put it in 1948, 'the question "*Why* does the phenomenon happen?" is construed as meaning "according to what general laws, and by virtue of what antecedent conditions does the phenomenon occur?"[9]

This is not the place to rehearse decades of commentary on Hempel, or to try to re-invent any philosophical wheels. For the purposes of the present discussion, however, I do want to stress that what is striking about explanations that conform the covering law model is that they do not actually tell us

[6] I am reminded of this by Julian Reiss, 'Do we need mechanisms in the social sciences?' *Philosophy of the Social Sciences*, Volume 37, Number 2, June 2007.

[7] Carl G. Hempel and Paul Oppenheim, 'Studies in the logic of explanation', *Philosophy of Science*, Volume 15, Number 2, April 1948.

[8] By choosing the letters p and q here to refer to events, I do not mean to conflate events and propositions. I use them as a literary trope, to underscore Hempel and Oppenheim's own effort to reduce metaphysical necessity to logical necessity. Nor should it matter for the present discussion that in some places the metaphor stands in for kinds of event, in others for particular events. Thanks to Irem Kurtsal Steen for raising these issues.

[9] *Ibid.*, p. 136.

what produces q. Is it p? If so, if it's something about p itself that produces q, we would expect an explanation to identify the relevant property of p that brings about the effect q. On the covering law model, however, we don't find out anything about p, other than that it seems always to be followed by q, and that a particular p occurred prior to a particular q. If it's *not* p, though, that produces q, then what is it? Is it the law itself, the fact that 'In all cases, if p, then q'? This seems less plausible. For one thing, it's not clear what it would mean, exactly, to say that the fact that the sequence 'p, q' is a regular one is itself what produces q's; such a claim seems a befuddled, question-begging response to the causal question. But even if we do assume, for the sake of argument, that the regularity of the sequence 'p, q' is the kind of thing that can produce q's, then – as above – we would expect an explanation in which it was the posited cause to consist of an account of the dispositional property borne by the sequence. Instead we have only a dogmatic assertion of order.

What, then, *does* do the producing, on the covering law model of causal explanation? Hempel and Oppenheim themselves fudged the issue, writing:

> [i]f E describes a particular event, then the antecedent circumstances described in sentences C1, C2, ..., Ck may be said jointly to 'cause' that event, in the sense that there are certain empirical regularities, expressed by the laws L1, L2, ..., Lr, which imply that whenever conditions of the kind indicated by C1, C2, ..., Ck occur, an event of the kind described in E will take place.[10]

A passage such as this invites one to think that it's the antecedent conditions that have the causal power after all, that it's p. But upon reflection it's not really so. The antecedent conditions only do the causing if we put the concept of cause inside quotation marks, and stipulate that what we mean by it is 'be a minor premise in a syllogism' – one in which the regularity in question is assumed as the major premise.[11] Clearly, in fact – and un-problematically in keeping with Hume – nothing does any actual causing, from this perspective. As an explanation-form, the covering law model presupposes a regularity-based metaphysics, devoid of necessary connections grounded in the powers of things.

Strictly speaking, we might want to label the stance Kantian, rather than empiricist. To put it in the language of the *Prolegomena*, the statement of law that forms the major premise represents what Kant calls there a 'judgment of experience', rather than a 'judgment of perception'.[12] The complication, though, is that the Hempelian empiricist, in virtue of his or her empiricism, in fact has legitimate recourse only to the latter. One could respond that the contemporary Humean has access to genuine statements of law via the device

[10] *Ibid.*, p. 139.

[11] Again, I put it this way to make the point; I do not mean myself to conflate events and propositions.

[12] Immanuel Kant (trans. Paul Carus), *Prolegomena To Any Future Metaphysics That Can Qualify As A Science*, (La Salle, Illinois: Open Court Publishing Company, 1988), Sec. 18–20; pp. 54–60.

of imagined possible worlds – though unlike a Kantian, the Humean has no theoretical apparatus with which to translate what remains for him or her mere logical necessity into something with even the minimal metaphysical traction of Kant's transcendental necessity. However, even if one grants this for the sake of argument, little is gained ontologically by moving from Hume to Kant. While the Kantian picture includes lawful order as a feature of experience, it no more allows for causality construed as the display of powers or dispositional properties than does Humeanism. In this respect – decisive, for the present discussion – transcendental idealism is simply a very sophisticated form of anti-realism about causality.[13]

That a Humean metaphysics is built into the covering law model is an obvious if important point, given Hempel's attachment to positivism. Less obvious is the way in which, in general, advancing anti-realism about causality via the category of explanation works to place the issue outside of the standard parameters of debate in this area of philosophy. At one level, such an elision is possible because those who are involved in the discussion can legitimately claim that they are not talking about causality. Patently, they're concerned with epistemological and methodological problems, not with metaphysical ones. Professional specialization functions to normalize this distinction. Solving problems that have to do with explanation is what epistemologists, philosophers of science and methodologists do; determining what causality is is what metaphysicians do. Paradoxically, though, the elision is also sustained precisely because it would seem strange to say, in response to someone who'd just provided a causal account of *q*, 'Yes, I see that you have given us a good explanation, but tell me – what actually caused *q*?' Here the logic is reversed: the idea is not that explanation and causality are *separate* concerns, but rather that if the explanatory issues have been resolved, so too, thereby, must have been the causal ones. The combination of the hypostatized division between epistemology and metaphysics, on the one hand, and a causal explanation-form that stands in for causal reasoning while precluding reference to actual productive causes, on the other – the combination of these two factors results in a situation in which an effort to take on anti-realism about causality shows up as a kind of sub-disciplinary category mistake.

[13] Cf. Bhaskar, *A Realist Theory of Science*; Theodor Adorno (trans. Rodney Livingstone) *Kant's Critique of Pure Reason* (Stanford: Stanford University Press, 2001), esp. Lecture 13; Ruth Groff, *Critical Realism, Post-positivism and the Possibility of Knowledge* (London:Routledge 2004), chapter 2. Kant argues in the *Critique of Pure Reason* that causality is a category of the understanding. The synthetic operation of reason supplies the transcendental fact of causal order; we can therefore be certain that our experience of phenomenal objects will always be so ordered. If one is persuaded by the argument, it shores up the Humean account epistemologically. But it doesn't change the ontology, as evidenced by Kant's insistence that he is a transcendental idealist and an empirical realist, but not a transcendental realist. Bhaskar in his early work referred to his own realist, powers-based view of causality as transcendental realism, which he contrasted with both empirical realism and transcendental idealism.

If the covering law model remains a paradigmatic form of explanation in positivist social science and analytic philosophy and methodology of social science, its hold has loosened. The approach currently favoured by quantitative researchers allows for *q* – to retain the metaphor – now conceived as a dependent variable, to be explained via a statistically significant correlation with *p* (the independent variable[s]). Once identified, the correlation between *p* and *q* is deemed causal, if it is, via reference to a theory that accounts for it, which theory need not necessarily include law-statements. This approach is so widespread that I'll call it simply the generic variable analysis model. For qualitative researchers, the case study takes the place of variable analysis. Given that the sample size is smaller in a case study (sometimes as small as 1), patterns that emerge tend not to be conceived in terms of statistical regularity.[14] Here too, however, the idea is that identified patterns may be deemed causal only insofar as they can be explained.

The standard quantitative and qualitative models differ metaphysically from the covering law model in that, in principle at least, the theories that explain the identified correlations or patterns may do so via reference to a causal bearer of some kind, i.e. to something that actually has the power to bring about an effect. As Doug Porpora has argued, proponents of a powers-based metaphysics are perfectly entitled to run regressions, even if many don't.[15] My hunch in this regard is that, in practice, the majority of those who do invoke real causal powers (a) engage in qualitative rather than quantitative research; (b) are trained in sociology and political economy rather than in economics or political science (in political science, more will be found in comparative politics than in American politics); and (c) are likely to be based outside the US. Certainly critical realism, the school most associated with this approach, is well-established in the UK while remaining virtually unknown in the US.

Still, while it is indeed possible to combine standard quantitative and qualitative explanation-forms with a commitment to real causal powers, anti-Humean realism about causality is not, in fact, the norm within positivist social science or analytic philosophy or methodology of social science. Rather, Hume prevails – albeit implicitly. The question, then, is this: how is Humeanism tacitly advanced by the standard quantitative and qualitative approaches – as well as by meta-level reflection upon these models by analytic philosophers of social science and mainstream methodologists?

Let's begin with variable analysis. Note that one can imagine a version of this model in which statistically significant correlations simply stand on

14 See James Mahoney and Gary Goertz, 'A tale of two cultures: Contrasting quantitative and qualitative research,' *Political Analysis* (2006), 14: 227–249, for a useful comparison of the structure of quantitative and qualitative research.

15 Douglas V. Porpora, 'Do realists run regressions?' in Jose Lopez and Garry Potter (eds.), *After Postmodernism: An Introduction to Critical Realism* (New York: The Athlone Press, 2001).

their own: if *p* and *q* are determined to be conjoined a statistically significant percentage of the time, then we may conclude that *p* 'causes' *q* – by which we just mean, precisely, that *p* and *q* are highly likely to be conjoined. As Gudmund Hernes observes (himself advancing a anti-realist model of explanation via mechanisms, which I shall address below), 'It is interesting to note that much that goes under the name of "causal analysis" stops at [this] point (e.g. when it [has] established a solid correlation between some independent and dependent variables).'[16] In defining the variable analysis approach in the way that I initially did, however, I wanted to acknowledge that many quantitative researchers would argue that beginning and ending with regularity does not allow one to distinguish between those correlations that are spurious and those that are genuinely causal.[17] Thus there is the added stipulation that in addition to identifying a correlation between dependent and independent variables, one must be able to explain a correlation in order to infer that it is indeed causal. The model in question is both widespread and basic enough that where one is most likely to find it articulated is in methods textbooks. For example, the author of *Social Research Methods: Qualitative and Quantitative Approaches* writes, 'Condition 4: Theory – Finally, even when you have established nonspurious, consistent, strong covariation, as well as a logical time sequence for two or more variables, you need a *theory* that *explains* the association.'[18]

The way that Humeanism is advanced through this model is that the causal dynamics identified by the explanatory theory will turn out to be themselves conceived along Humean lines. Thus, in the quantitative case, the correlation between dependent variable *q* and independent variable *p* will be deemed non-spurious via a causal theory to the effect that '*q* depends upon *p* because of *x*', say, but – albeit now at the level of the authorizing theory – the concept of 'cause' will be once again understood in an anti-realist manner, in terms of regularity. There is room for variation in this regard: to cause may be thought to mean 'to always come first, under specified conditions' or 'to be a necessary and sufficient antecedent of', or 'to support subjunctive conditional statements that are true in all possible worlds',[19] or (for those attracted to

16 Gudmund Hernes, 'Real virtuality,' in Peter Hedstrom and Richard Swedberg (eds.), *Social Mechanisms: An Analytical Approach to Social Theory* (Cambridge: Cambridge University Press, 1998), p. 76, n.1.

17 Of course, the very recognition of such a distinction is already a contradiction, for the Humean, who denies that causality involves metaphysically necessary connections.

18 H. Russell Bernard, *Social Research Methods: Qualitative and Quantitative Approaches* (Thousand Oaks, CA: Sage Publications, 2000), p. 55. Russell uses the term 'mechanism' and the term 'theory' interchangeably, but it is clear that correlation must be accompanied by an explanation, even if it is unclear what the ontological status of a mechanism is, for Bernard.

19 Inclusion of the Lewisian option is informed by Henry E. Brady, 'Models of causal inference: Going beyond the Neyman–Rubin–Holland theory,' conference paper, Annual Meeting of the Political Methodology Group (University of Washington, Seattle, WA, July 2002), as paraphrased

the idea of an infinite regress) 'to be a statistically significant independent variable, explicable via a theory'. But the options are only alternate versions of the principle of constant conjunction: unless one is a realist, a Humean conception of causality will be built into the explanatory theory that, for the Humean, is itself charged with sustaining causal inference.

The qualitative model is no different in this regard. As Mahoney and Goertz observe, standard qualitative explanations more often presuppose the (Mackie-style) view that causality should be understood in terms of necessary and sufficient conditions, rather than presupposing what they call the 'correlational' view, associated with quantitative research. But the conceptual logic that I have identified above is unchanged by this, as the former is simply a variation on the Humean theme. There is nothing to alter the underlying metaphysics – only an ungrounded insistence (explicit, in this version) on the empirical necessity of the order that, on the Humean view, just is causality. In this respect, the Mackie approach is tantamount to Kant without a transcendental argument. The standard qualitative model too, then, builds Humean anti-realism about causality into the very explanatory theory that is charged with grounding causal inference in relation to given findings. In short, in both the quantitative and the qualitative models, metaphysical reliance upon Humean constant conjunction is simply pushed out a frame, from the level of data to the level of modality-conferring theory.

What is especially significant about all of this for my own argument is not so much that Humeanism continues to be the default ontology of especially American, often positivist, social science; but rather that it can be combined with the idea that it is not – i.e. with the idea that competing versions of regularity theory somehow differ in a deep way, or that it is possible to remain neutral on what causality is, whilst engaging in causal explanation. I don't want to be misunderstood, in saying this. The proponent of a standard approach to quantitative or qualitative research is indeed entitled to regard him or herself as rejecting a direct, unadorned regularity theory of causation, such as is embedded in the covering law model. The variable analysis model, as I've been calling it, differs from the Hempelian model not just in that the law requirement has been loosened, but also in that the restriction on what counts as a causal connection has, at least ostensibly, been tightened: correlations alone, even apparently constant ones, do not add up to causation without a supplementary 'theory' to provide the missing necessity that we associate with causality. And the same is so of the patterns and/or conditions of possibility that figure in the qualitative, case-study version of the model. But, as I've argued, Humean anti-realism about causality carries the day all the same. Regularity is simply relocated, as it were, onto the authorizing theory,

by David Dahua Yang, 'Empirical social inquiry and models of causal inference,' *The New England Journal of Political Science*, (2) 1: 51–88, 2006, and as expressed in the original.

along with a proviso that it may be expressed in terms of INUS conditions or the logic of possible worlds, rather than in terms of correlations. At the level of data, causality now shows up as 'when a regular pattern can be explained', a view no less odd than is that associated with the covering law model. The upshot philosophically is that the standard quantitative and qualitative models seem to save the phenomenon of productive causation from Hempel, while at the same time implicitly affirming the Humean view that there is in reality no such thing.

An implicit commitment to Humeanism combined with pseudo-argument over causality can be seen even more clearly in the literature related to causal pluralism. One line of discussion in this area works by reinforcing the notion that alternate versions of Humeanism are importantly different metaphysically. Mahoney and Goertz, for example, in their otherwise-lovely lexicon for translating across the quantitative-qualitative divide, observe that – as previously noted – one camp affirms a correlational or probabilistic conception of causality while the other is concerned with necessary and sufficient conditions. This is presented as a genuine dichotomy, with no indication of its relative superficiality[20] (surprising, given that Mahoney elsewhere criticizes those who reduce mechanisms to covariance[21]). Similarly, though the treatment is more nuanced, in 'Models of causal inference' Henry Brady offers a four-category typology of definitions of causality: regularity, counterfactual, manipulation, mechanisms or capacities.[22] Brady correctly, in my view, regards Mackie's INUS account as a version of regularity theory. But although he acknowledges that Lewis (with whom he associates the counterfactual position) makes reference to Hume, he presents the device of possible worlds as a significant metaphysical alternative to Humeanism.

A second type of intervention, consistent with the first, is to propose that the (supposed) differences between competing definitions of causality should be tolerated, set aside or seen to be ultimately reconcilable: in the end, everyone should simply continue on with the business of doing social science. Mahoney and Goertz, Brady, David Yang (following Brady), John Gerring and Jeroen Van Bouwel and Erik Weber all make arguments of this kind.[23] To recall my opening analogy, this corresponds to that point in the meeting when someone invariably says 'We've talked this to death. Can't we just move on?' Of course, 'just moving on', amounts to whatever was being talked about

[20] Mahoney and Goertz, *op. cit.*

[21] James Mahoney, 'Beyond correlational analysis: Recent innovations in theory and method,' *Sociological Forum*, 2001, Vol. 16, No. 3.

[22] Brady, *op. cit.*

[23] Mahoney and Goertz, *op. cit.*; Brady, Yang *op. cit*; John Gerring, 'Causation: A unified framework for the social sciences,' *Journal of Theoretical Politics*, 2005, 17(2): 163–198; Jeroen Van Bouwel and Erik Weber, 'De-ontologizing the debate on social explanations: A pragmatic approach based on epistemic interests,' *Human Studies*, 2008, 31: 423–442.

being dropped in favour of a return to the status quo – in this instance a ubiquitous, implicit attachment to Humeanism. The version of the move that is of most interest philosophically, it seems to me, is the pragmatist one. John Gerring, for example, defends a probabilistic account of causality – causality is the increased likelihood that a given thing will happen – on the grounds that it captures what practicing social scientists all already believe, and so is a definition upon which all or most can agree.[24] I say that this version is the most interesting because it raises to the level of principle the deflection of genuine ontological debate.

The most recent major development in the mainstream literature on explanation in social science is talk of 'mechanisms'. Proponents of the mechanisms model contend that a proper explanation should tell us how p's causing q actually happens – to carry through with my original metaphor. Jon Elster, for example, one of the authors most widely associated with this approach, holds that, as explanatory devices, laws alone are too general; the picture that they yield is not 'fine-grained' enough.[25] Daniel Little goes further, suggesting that explanation via reference to regularity lacks causal force entirely, is purely descriptive. Neither correlations nor necessary and sufficient antecedent conditions themselves tell a causal story, Little claims.[26] Certainly, a model of causal explanation that requires the researcher to identify actual causal mechanisms would seem to pose a significant alternative both to the deductions of the covering law model and the regularity-plus-theory of the standard generic variable analysis and case study models. But upon closer examination, the mainstream mechanisms model is more of the same, metaphysically.

Elster offers this definition of a mechanism: 'Roughly speaking', he says in *More Nuts and Bolts*, 'mechanisms are *frequently occurring and easily recognizable causal patterns that are triggered under generally unknown conditions or with indeterminate consequences*. They allow us to explain, but not to predict.'[27] He points to folk proverbs as proto-typical statements of mechanisms, e.g. 'absence makes the heart grow fonder'.[28] Though not all mechanisms have been or can be articulated as proverbs, the reference to proverbs highlights, for Elster, the sense in which mechanisms are not as universal as laws, as well as the fact that mechanisms are not a basis for prediction. Proverbs, he notes, often come in contradictory pairs: absence makes the heart grow fonder, but so too is it that 'out of sight, out of mind'.[29] Moreover, Elster emphasizes, mechanisms as an explanatory category allow one to jump in in the middle of a story, as it were, since it is part of their definition that their 'triggering' conditions are

[24] Gerring, *op. cit.*

[25] Jon Elster, *Explaining Social Behavior: More Nuts and Bolts for the Social Sciences* (New York: Cambridge University Press, 2007), pp. 24, 33.

[26] Daniel Little, *Varieties of Social Explanation* (Boulder: Westview Press, 1991), chapter 2.

[27] Elster, *op. cit.*, p. 36.

[28] *Ibid.*, p. 37.

[29] *Ibid.*, pp. 37–38.

generally unknown. (A mechanism whose triggering conditions *are* known, Elster says, may be replaced with a statement of law, albeit what he calls a 'weak' law.[30]) Elster makes much of all of this, and of the epistemic gain, to use Charles Taylor's term, associated with moving from laws to mechanisms. But what does it add up to metaphysically? Humean constant conjunction. Local patterns, in the case of mechanisms; universal regularity, in the case of the laws that govern triggering conditions.

Elster is what I will call an anti-realist about mechanisms in that his so-called mechanisms are in fact regular (if not constant) conjunctions. Another way to be an anti-realist about mechanisms is to regard them as nothing other than theoretical constructs. For example, the following are stated definitions of 'mechanism' drawn from a number of articles in Peter Hedstrom and Richard Swedberg's well-known edited collection *Social Mechanisms: An Analytical Approach to Social Theory*, to which Elster was also a contributor. Thomas Schelling says,

> I propose – and I believe I am paraphrasing Hedstrom and Swedberg in their introductory essay – that a social mechanism is a plausible hypothesis, or set of hypotheses, that could be the explanation of some social phenomenon... Alternately, a social mechanism is an *interpretation*... of a model.[31]

Gudmund Hernes holds that a mechanism is 'an abstract, dynamic logic by which social scientists render understandable the reality they depict'.[32] Diego Gambetta adds, 'I take "mechanisms" to be hypothetical causal models that make sense of *individual* behavior.'[33] Arthur Stinchcombe references James Coleman, as does Hernes: 'I have defined mechanisms before... as bits of "sometime true theory" (the phrase is due to James S. Coleman 1964, pp. 516–19).'[34] While these authors maintain anti-realism about causality by equating mechanisms with theories (which is not to say that ideas cannot have real causal powers, a view defended by some critical realists) rather than with regularities, others in the literature sound more like Elster.[35]

Daniel Little does, for example – although I suspect that he doesn't quite believe what he says in *Varieties of Social Explanation*. Echoing the title of Elster's original work on the topic, *Nuts and Bolts for the Social Sciences*, Little

[30] *Ibid.*, p. 44.

[31] Thomas C. Schelling, 'Social mechanisms and social dynamics,' in Hedstrom and Swedberg (eds.), *op. cit.*, p. 33.

[32] Hernes, *op. cit.*, p. 74.

[33] Diego Gambetta, 'Concatenations of mechanisms,' in Hedstrom and Swedberg, (eds.) *op. cit.*, p. 102.

[34] Arthur L. Stinchcombe, 'Monopolistic competition as a mechanism: Corporations, universities, and nation-states in competitive fields,' in Hedstrom and Swedberg (eds.), *op. cit.*, p. 267.

[35] For excellent overviews of the treatment of mechanisms in the analytic literature, see Renate Mayntz, 'Mechanisms in the analysis of social macro-phenomena,' *Philosophy of the Social Sciences*, 2004, Vol. 34, No. 2: 237–259; and Mahoney, 'Beyond correlational analysis: Recent innovations in theory and method,' *op. cit.*

starts off his own book talking about nuts and bolts literally. Here's the lengthy passage in which he introduces the concept of a causal mechanism:

I contend that the central idea in causal explanation is that of a causal mechanism leading from C to E, so let us begin with that notion. A bolt is left loose on an automobile wheel; after being driven several hundred miles the wheel works loose and falls off. The cause of the accident was the loose bolt, but to establish this finding we must reconstruct the events that conveyed the car from its loose-bolt state to its missing-wheel state. The account might go along these lines: The vibration of the moving wheel caused the loose bolt to fall off completely. This left the wheel less securely attached, leading to increased vibration. The increased vibration caused the remaining bolts to loosen and detach. Once the bolts were completely gone the wheel was released and the accident occurred. Here we have a relatively simple causal story that involves a number of steps, and at each step our task is to show how the state of the system at that point, in the conditions then current, leads to the new state of the system.[36]

The example also recalls the Hempel and Oppenheim piece discussed above, in which the authors contrast a properly scientific explanation with the sense in which 'it may be explained that a car turned over on the road "because" one of its tires blew out while the car was traveling at high speed'.[37] Unlike Little, Hempel and Oppenheim judge the latter to be unscientific. But what does it turn out to mean, for Little, to say that the loose bolt caused the accident? Just this: that it figures into a sequence of events characterized by regularity. Thus he can write, 'a causal mechanism, then, is a series of events governed by lawlike regularities that lead from the explanans to the explicandum'.[38] Or, more formally, defining the what he calls the causal mechanism thesis, CM: C is a cause of E = df there is a series of events Ci leading from C to E, and the transition from each Ci to Ci+1 is governed by one or more laws Li.[39]

Once again it's worth pausing to ask what's going on. If the covering law and regularity-plus-explanation models were metaphysically peculiar, the causal mechanisms model has a downright *Alice in Wonderland* quality to it. Here we find not just a simple, unapologetic appeal to constant conjunction, or even an effort to cash out causality in epistemic terms, as 'when a correlation or discernable pattern can be explained', but something stranger yet, viz., so-called causal mechanisms that are precisely not mechanisms at all, but simply regularities or conceptual posits. As with the other models, nothing on the mainstream mechanisms model is in a position of actually *doing* anything, in the sense of actively producing an effect. Thus here too, with an extra bit of ironic panache, the explanation-form functions as a delivery mechanism (no pun intended) for a Humean metaphysics. Finally, what makes the whole

36 Little, *op. cit.*, p. 15.
37 Hempel and Oppenheim, *op. cit.*, pp. 148–149.
38 Little, *op. cit.*, p. 15.
39 *Ibid.*, p. 14.

discussion truly surreal is that, despite the anti-realism about causality built into the model, its proponents have been criticized for being overly concerned with the causal question of why things happen.[40]

14.3 From explanation back to causality

I hope to have shown that anti-realism about causality is the implicit metaphysics of the leading models of explanation in analytic philosophy and methodology of social science, and that at the same time the form of the mainstream discourse has largely shielded this same metaphysics from contestation. Having brought the default commitment to Humeanism to light, I want to consider now what some of the salient issues might be if one were to extend the current *non-Humean* work in metaphysics to this area of philosophy. The latest thinking in metaphysics, as noted at the outset, has at its core the idea of powers or dispositional properties – capacities to effect change that things, from a dispositional realist perspective, have or bear intrinsically, as non-contingent features of what they are. Causality – again - from this perspective, consists of the display or exercise of such powers.[41] How a thing behaves relative to other things is thought to be a function of its dispositional properties; which dispositional properties a thing has is thought to be connected to what it is. Necessity inheres in this, the actual world; it is a matter of metaphysics, not logic.

Three issues that would be raised for analytic philosophers of social science were dispositional realism to become the default position, and/or that appear in a different light when viewed from a *non-Humean* perspective are: (a) the question of what kinds of phenomena may be said to be bearers of causal powers; (b) the existence of emergent powers; and (c) the principle of methodological individualism. I shall address these points in turn. In so doing, I am not undertaking to defend the *non-Humean* approach, but rather, as I've just said, to illustrate how a shift in underlying metaphysics could be expected to alter the terms of the discussion. Before I do this, however, let me make two observations. First, it is important to note that dispositional realists also employ the term causal mechanism. Those more accustomed to thinking along Humean lines should be careful to appreciate that what realists about causality mean by the term is entirely different from what anti-realists mean by it, or don't mean. For Humeans, as we've seen, mechanisms are patterns of events or heuristic devices. For dispositional realists, by contrast, mechanisms are real processes, involving the exercise of things' causal powers. Second, it

[40] Reiss, *op. cit.*

[41] For recent statements of the position see, e.g. Brian Ellis, *Scientific Essentialism* (Cambridge: Cambridge University Press, 2001); Stephen Mumford, *Dispositions* (Oxford: Oxford University Press, 2003); Anjan Chakravartty, *A Metaphysics for Scientific Realism: Knowing the Unobservable* (Cambridge: Cambridge University Press, 2007).

should also be noted that while those on the cutting edge of work on powers in metaphysics are only now beginning to turn their attention to social phenomena, the contemporary critique of Humeanism was arguably launched by the publication in 1975 of *Causal Powers*, by Harré and Madden and *A Realist Theory of Science* (followed quickly by *The Possibility of Naturalism*), by Roy Bhaskar, thinkers who were quick to apply the metaphysics in question to the philosophy of social science. Bhaskar's work, in particular, spawned the school of thought known as critical realism, within which there is now a 30 year history of scholarly debate from a *non-Humean*, powers-based perspective.[42] This literature is an important resource upon which analytic philosophers newly concerned with powers and social phenomena can and should draw.

Once one is thinking of causality in terms of powers rather than in terms of regularity, a question that comes immediately to the fore is 'What kinds of things *have* causal powers?' With respect to the social sciences, the contentious issue will be whether or not it is only individual persons who do. What about sociological 'things', such as collectivities, institutions, structures, etc.? Might they be bearers of causal powers as well? Needless to say, this is an issue that has been much discussed by longer-standing proponents of a powers-based approach, as well as by structuralists of various persuasion. At the moment, though, I am less interested in the answer than I am in the question. Specifically, notice straight away that once one has set aside a commitment to Humeanism, one may no longer understand oneself to have resolved the ontological concern by turning it into an epistemological one. In this case, for instance, one will not be able to ask (instead) if statements of regularity concerning institutions or structures can themselves be explained, or if the existence of institutions or structures appears to be a sufficient antecedent condition of certain sorts of outcome. Rather, one will be forced to engage with the question of whether or not institutions and/or structures can *do* things, in the way(s) in which, on the *non-Humean* view, things that are causes can. Again: not, e.g. 'Can they be said to come first in a non-spurious correlation, or to be a necessary and/or sufficient condition of something?' but rather, 'Are they the kind of object that can produce an effect?'

'Do sociological objects have causal powers, and if so in what sense?' is not a question that a Humean philosopher of social science need ask. It may be tempting to think that this is once again a matter of specialization: the question is one for a metaphysician, not for a philosopher of social science. And there is, as before, some truth to this. But the philosophical reason why the issue is not on the table (as opposed to the matter of how the areas of

[42] In addition to the *Journal of Critical Realism*, see the *Journal for the Theory of Social Behavior*. Routledge publishes critical realist titles under two different series. Representative authors of well-known works on powers-based philosophy of social science include: Roy Bhaskar, Margaret Archer, Andrew Sayer, Doug Porpora, Tony Lawson, Steve Fleetwood, amongst others. Much Marxist philosophy of social science is also implicitly powers-based.

specialization fall out within the profession) – the philosophical reason is that Humeans don't believe that *anything* can be said to have productive causal powers. Were the dominant metaphysics one that did admit of powers – or if it comes to be so once again – there is every reason to think that properly conceptualizing the powers of sociological entities would be a central task of philosophers of social science.

From a powers-perspective, the question of whether or not sociological phenomena can be bearers of causal powers leads to the closely related issue of emergence. Are there such things as emergent powers, powers that social objects have that individuals or aggregates of individuals do not have? This question is very close to the first one, but whereas before we wanted to know if social objects are the type of entity that can produce effects, here we want to know something about powers themselves – specifically, whether or not those that exist at one level must be equivalent, ontologically, to those that exist at a 'lower' level.

Here too, I want to focus on the question rather than on the answer. The point is this: as with the question of whether or not sociological entities may be causal bearers, the question of whether or not there are emergent properties is normally reformulated by Humeans such that it becomes an epistemological rather than metaphysical question. Thus, in philosophy of mind as well as in philosophy of social science, the question of emergent properties is routinely posed (and pursued) not as 'Do properties come into being at one level that do not exist at the lower level?' but instead as 'Must our explanations at the higher level include concepts that are not reducible to those that are relevant at the lower level?' One might think that a positive answer to the latter question would suggest a positive answer to the former, such that the issue is merely a matter of having cast an ontological commitment to emergent properties in epistemic terms. But this is not how it goes. Standard non-reductive physicalism, for instance, is precisely the view that while the answer to the epistemological question is yes, the answer to the ontological one is no. Mental properties can't be *explained* without reference to higher-level concepts, but it is only physical properties that may be said to *exist*.

A particularly striking version of the transposition of ontology into epistemology in relation to emergent properties is in fact to be found in Hempel and Oppenheim's 1948 article cited above. There, Hempel and Oppenheim first define emergent properties in epistemic terms, as properties that cannot be 'inferred' from 'information' about 'constituent parts'.[43] Next they assert that such 'mysterious' unexplainables must not be thought to be real. Supposedly emergent properties are, using my terminology rather than theirs, simply reifications – illusory phenomena produced by sloppy thinking about the contingent character of prediction vis-à-vis available theoretical resources.

[43] Hempel and Oppenheim, *op. cit.* pp. 148–150.

There are no 'non-reducible' properties, they maintain; there are only properties the existence and nature of which we are not yet able to predict from existing micro-level theory. And non-controversially, such limitations present no special cause for concern. 'The observations presented in the preceding discussion', they conclude,

> strip the idea of emergence of these unfounded connotations: emergence of a characteristic is not an ontological trait inherent in some phenomena; rather, it is indicative of the scope of our knowledge at a given time; thus it has no absolute, but a relative character; and what is emergent with respect to the theories available today may lose its emergent status tomorrow.[44]

Once more, as with the earlier treatment of causality, what we find is an ontological position – viz. the denial of emergent properties – being advanced via an epistemic argument – viz. that the partiality of our current knowledge should not be thought to be a cause for undue concern. A *non-Humean* approach does not commit one to affirming the existence of emergent properties. It does, however, make the reminder that we don't yet know all there is to be known about the world seem a less than compelling reason to reject them. As Timothy O'Connor has observed, 'I'm inclined to think that any tendency to suppose that the emergence of macrodeterminative properties in material substances is strictly inconceivable must be diagnosed as an instance of the withering effect on one's imagination that results from long-standing captivation by a certain picture of the world.'[45]

Finally, I would like to say a word about the principle of methodological individualism. The first thing to note is that, as with the debate over emergent properties, the debate over methodological individualism, such as it is, is the form that the relevant would-be ontological debate takes within the Humean context. Rather than asking if holistic, or macro-level sociological objects exist, the question is posed methodologically: 'Must the unit of analysis in social scientific explanations be the individual?' As it happens, more often than not the answer to the reformulated question is yes; adequate explanations in the social sciences must indeed be at the micro-level. Elster, for example, simply treats this as a given. 'In principle', he writes,

> explanations in the social sciences should refer only to individuals and their actions. In practice, social scientists often refer to supraindividual entities such as families, firms, or nations, either as a *harmless shorthand* or as a *second-best approach* forced upon them by lack of data or of fine-grained theories.[46]

[44] *Ibid.*, pp. 150–151.

[45] Timothy O'Connor, 'Agent causation,' in Gary Wilson (ed.), *Free Will, Second Edition* (New York: Oxford University Press, 2003), p. 263.

[46] Elster, *op. cit.*, p. 13.

Daniel Little's epistemic response is more refined than Elster's is here, but the underlying ontology is the same.

It's worth looking more closely at Little's version of the transposition that I've identified. Little distinguishes between three theses which, he argues, jointly constitute the principle of methodological individualism. The first he calls the ontological thesis: 'all social entities are reducible without remainder to logical compounds of individuals'.[47] Second is the meaning thesis, the claim that concepts that refer to social entities 'must be *definable* in terms of concepts that refer only to individuals and their relations and behavior'.[48] Third is the explanation thesis: 'all social facts and regularities must ultimately be explicable in terms of facts about individuals'.[49]

Little says that the ontological thesis is 'manifestly true', on the grounds that social objects do not exist independently of individuals.[50] Even Durkheim, he claims, argued only for non-reducible norms and meanings, not for non-reducible entities, e.g. institutions, societies.[51] (Or, in Durkheim's case, a collective consciousness, the existence of which Durkheim clearly *did* defend.)[52] The meaning thesis, by contrast, Little regards as false. Although non-reducible social objects do not in his view exist (as per the ontological thesis), we cannot do without concepts that refer to them. And these concepts are not equivalent in meaning to concepts referring to individuals. Note that the implications of accepting the ontological thesis while rejecting the meaning thesis are ignored. The problem that's glossed over is this: if the ontological thesis is true, then the objects to which higher-level concepts refer do not actually exist. But if the objects to which such concepts refer to do not exist, then it is not at all clear why it should be necessary, as Little argues it is, to retain said concepts in order to make sense of non-pathological, everyday individual behaviour – behaviour that (the love of literature aside) is presumably not necessarily and systematically oriented toward imaginary, non-existent objects. Finally, on the explanation thesis, Little equivocates. On the one hand, he argues that there should be at least type-identity between social and individual-level explanations; on the other hand, he concludes that as long as one can identify empirically verifiable regularities, one has a viable explanation – at whatever level.[53] I have set out Little's position in some detail because, notwithstanding the obvious care he has taken not to prejudge

[47] Little, *op. cit.*, 183.

[48] *Ibid.*, p. 184.

[49] *Ibid.*, p. 186.

[50] *Ibid.*, p. 184.

[51] *Ibid.*

[52] See, especially, *The Division of Labor in Society*, in addition to other works; Durkheim's commitment to the existence of non-reducible sociological entities is beyond question it seems to me.

[53] Little, *op. cit.*, pp. 186–189.

the issue, it provides another example of the way in which (a) metaphysical commitments are transposed, in the Humean literature, into epistemic ones (e.g. in this case, social entities deemed not to exist reappear as concepts necessary for the explanation of everyday actions of individuals); and (b) the prevailing metaphysical position is thereby affirmed. In this case, the ontology that is advanced involves atomism as well as anti-realism about causality.

I do not mean to suggest that dispositional realism entails commitment to holism, either ontological or methodological, any more than it entails a commitment to emergent properties. With respect to holism, the deciding ontological consideration for the realist about causality may well be precisely whether or not a purported object can be shown to exhibit non-reducible causal powers. If there is evidence for the existence of such powers in a given case, then there will be an argument to be made for the existence of the entity that bears them. And if an entity exists and has causal powers, then there will be reason to think that it ought to be allowed to figure in explanations of effects of the type that it has the power to produce. From a dispositional realist perspective, the main point is that none of this should be settled dogmatically or via appeal to epistemic rather than ontological argument. However, it does seem likely that insofar as the hold of methodological individualism is part and parcel with that of Humeanism, its soundness may become less self-evident as the challenges to the metaphysics continue to mount. More broadly, there is reason to hope that these same challenges will also call into question the routine translation of ontological questions into epistemological ones. After all, the phenomenalist notion that objects are, in the end, nothing other than the experiences of subjects, is itself the metaphysical end-game of empiricism.

14.4 A word on meta-theory

By way of conclusion, let me emphasize that my primary aim in the foregoing analysis has not been to offer a positive argument for realism about causality. Nor have I sought to flesh out a powers-based model of social scientific inquiry. In both cases, I have merely flagged relevant literature. The argument that I have been concerned to make is, instead, meta-theoretical. It may be worth asking what the importance is, of such a contribution. Recalling Kuhn's various uses of the term paradigm – from an over-arching theoretical framework to a 'paradigmatic' puzzle or example – one response might be to say that the significance of meta-theoretical analysis is that it advances philosophical reflection in ways that, by loose analogy to what Kuhn called normal science, 'normal philosophy' often does not. Crucially, meta-theoretical discussion renders explicit the foundational assumptions of a prevailing paradigm. On Kuhn's model, those assumptions are eventually called into question in the course of normal science, via the persistence of anomolies – or, in philosophy, irresolvable problems or worries. We might say that meta-theoretical analysis, by contrast, allows one to engage in revolutionary science in an on-going way, even if only to remind oneself that the dominant scheme, viable or not, has its own internal logic. As a matter of intellectual habit, this more comprehensive style of thinking is arguably not as widely cultivated in analytic philosophy as it might be; more common is the idea that one is confronted with discrete problems, which one solves as best one can, with the requirement only that the totality of one's various theoretical commitments not be self-contradictory. Meta-theoretical argument highlights that there is a there, there, with respect to the prevailing approach, which in turn can promote a kind of conceptual bilingualism, or even multilingualism. Kuhn may have been right about the limits of inter-paradigmatic debate. But he may not have been, and/or philosophy may be different from natural science in this regard. Whatever else is so, fluency across paradigms cannot but facilitate serious and productive philosophical exchange. In the present case, my hope is that by delineating the ways in which the Humean view of causality has been tacitly fortified within analytic philosophy and methodology of social science, I will have helped to clear the ground for a meaningful debate on the topic of causal powers to be had in this area.

References

Adorno, Theodor W. (trans., Rodney Livingstone), *Kant's Critique of Pure Reason* (Stanford: Stanford University Press, 2001).

Bernard, H. Russell, *Social Research Methods: Qualitative and Quantitative Approaches* (Thousand Oaks, CA: Sage Publications, 2000).

Bhaskar, Roy, *A Realist Theory of Science* (Sussex: Harvester Press, 1975).

Bhaskar, Roy, *The Possibility of Naturalism: A Philosophical Critique of the Contemporary Human Sciences, Third Edition* (New York: Routledge, 1998).

Brady, Henry E., Models of causal inference: Going beyond the Neyman–Rubin–Holland theory, conference paper, Annual Meeting of the Political Methodology Group, University of Washington, Seattle, WA, July 2002.

Chakravartty, Anjan, *A Metaphysics for Scientific Realism: Knowing the Unobservable* (Cambridge: Cambridge University Press, 2007).

Ellis, Brian, *Scientific Essentialism* (Cambridge: Cambridge University Press, 2001).

Elster, Jon, *Explaining Social Behavior: More Nuts and Bolts for the Social Sciences* (New York: Cambridge University Press, 2007).

Durkheim, Emile, *The Division of Labor in Society* (New York: The Free Press, 1964).

Gambetta, Diego, Concatenations of mechanisms, in P. Hedstrom and R. Swedberg, (eds.) *Social Mechanisms: An Analytical Approach to Social Theory* (Cambridge: Cambridge University Press, 1998).

Gerring, John, Causation: A unified framework for the social sciences, *Journal of Theoretical Politics*, 2005, Vol. 17, No. 2: 163–198.

Groff, Ruth, *Critical Realism, Post-positivism and the Possibility of Knowledge* (London: Routledge, 2004).

Harré, Rom and Madden, E. H., *Causal Powers: A Theory of Natural Necessity* (Totowa, NY: Rowman and Littlefield, 1975).

Hempel, Carl and Oppenheim, Paul, Studies in the logic of explanation, *Philosophy of Science*, 1948, Vol. 15, No. 2: 135–175.

Hernes, Gudmund, Real virtuality, in P. Hedstrom and R. Swedberg (eds.), *Social Mechanisms: An Analytical Approach to Social Theory* (Cambridge: Cambridge University Press, 1998).

Hume, David, *Enquiries Concerning Human Understanding and Concerning the Principles of Morals, 3rd edition* (New York: Calendon Press, 2000).

Kant, Immanuel (trans. Paul Carus) *Prolegomena To Any Future Metaphysics That Can Qualify As A Science* (La Salle, Illinois: Open Court Publishing Company, 1988).

Kant, Immanuel (trans. and ed. Paul Guyer and Allen W. Wood), *Critique of Pure Reason* (New York: Cambridge University Press, 1997).

King, Gary, Keohane, Robert O., and Verba, Sidney, *Designing Social Inquiry: Scientific Inference in Qualitative Research* (Princeton: Princeton University Press, 1994).

Little, Daniel, *Varieties of Social Explanation* (Boulder: Westview Press, 1991).

Mayntz, Renate, Mechanisms in the analysis of social macro-phenomena, *Philosophy of the Social Sciences*, 2004, Vol. 34, No. 2: 237–259.

Mahoney, James, Beyond correlational analysis: Recent innovations in theory and method, *Sociological Forum*, 2001, Vol. 16, No. 3: 575–593.

Mahoney, James and Goertz, Gary, A tale of two cultures: Contrasting quantitative and qualitative research, *Political Analysis*, 2006, 14: 227–249.

Mumford, Stephen, *Dispositions* (Oxford: Oxford University Press, 2003).

O'Connor, Timothy, Agent causation, in Gary Wilson (ed.), *Free Will, Second Edition* (New York: Oxford University Press, 2003).

Psillos, Stathis, *Causation and Explanation* (Ithaca: McGill-Queen's University Press, 2002).

Porpora, Douglas V., Do realists run regressions? in J. Lopez and G. Potter (eds.), *After Postmodernism: An Introduction to Critical Realism* (New York: The Athlone Press, 2001).

Reiss, Julian, Do we need mechanisms in the social sciences? *Philosophy of the Social Sciences*, June 2007, Vol. 37, No. 2: 163–184.

Schelling, Thomas C., Social mechanisms and social dynamics, in P. Hedstrom and R. Swedberg (eds.), *Social Mechanisms: An Analytical Approach to Social Theory* (Cambridge: Cambridge University Press, 1998).

Stinchcombe, Arthur L., Monopolistic competition as a mechanism: Corporations, universities, and nation-states in competitive fields, in P. Hedstrom and R. Swedberg (eds.), *Social Mechanisms: An Analytical Approach to Social Theory* (Cambridge: Cambridge University Press, 1998).

Taylor, Charles, *Philosophy and the Human Sciences: Philosophical Papers 2* (Cambridge: Cambridge University Press, 1985).

Van Bouwel, Jeroen and Weber, Erik, De-ontologizing the debate on social explanations: A pragmatic approach based on epistemic interests, *Human Studies*, 2008, 31: 423–442.

Yang, David Dahua, Empirical social inquiry and models of causal inference, *The New England Journal of Political Science*, 2006, Vol. 2, No. 1: 51–88.

·15·

Causal explanation: Recursive decompositions and mechanisms

Michel Mouchart and Federica Russo

Abstract

This chapter deals with causal explanation in quantitative-oriented social sciences. In the framework of statistical modelling, we first develop a formal structural modelling approach which is meant to shape causal explanation. Recursive decomposition and exogeneity are given a major role for explaining social phenomena. Then, based on the main features of structural models, the recursive decomposition is interpreted as a mechanism and exogenous variables as causal factors. Arguments from statistical methodology are first offered and then submitted to critical evaluation.

15.1 The quest for causal explanations in the social sciences

Emile Durkheim (1960) had the ambitious goal to explain suicide as a social phenomenon. Accordingly, in his masterpiece *Le Suicide*, he looked for the social causes of suicide. Durkheim's interest in the determinants of suicide was motivated by the observation of a great variability in the suicide rate. This variability appeared to be quite irrelevant across time within the same population, but was instead considerable across different societies. By examining how the suicide ratio varied across societies, Durkheim aimed to detect the social factors this variation depended on and thus to explain why, for instance, societies with a more integrated family structure had lower suicide rates. More recently, the demographer John Caldwell proposed a model to explain child survival in developing countries. Notably, Caldwell (1979) investigated maternal education as a major causal factor. He observed that maternal education alone could account for more variance than all other relevant socio-economic factors altogether (e.g. mother's place of residence, husband's occupation and education, type of marriage, etc.), and therefore *this* factor deserved special attention. López-Ríos *et al.* (1992), to give another example, were interested in explaining a significant lower mortality rate in Spain in the 1980s after socio-economic policies in the 1970s were carried out. In particular, they were interested in assessing the causal effect of factors such as economic and social development on the one hand, and of the use of sanitary infrastructure on the other hand.

What these examples from various areas in the social sciences have in common is that they all seek to provide an explanation of a phenomenon of interest – more specifically, they seek to provide a *causal* explanation. A causal explanation is provided, intuitively, once the factors or causes that bring about the phenomenon are identified. This, however, is still too loose a characterization. In this chapter we investigate how causal explanations are built in the social sciences.

We proceed as follows. Section 15.2 presents the structure of the explanation that a statistician provides for a phenomenon of interest. Three aspects are highlighted. First, an explanation is incomplete, or partial, because it is based on a stochastic representation of the world where the stochastic component stands for what is *not* explained. Second, an explanation is given by decomposing a complex causal mechanism into a sequence of 'simpler' explanatory mechanisms. Third, the explanation is causal, that is we identify cause–effect relations through the condition of exogeneity. Section 15.3 then addresses the question of the interpretation of the recursive decomposition and of why it carries explanatory power. We argue that a recursive decomposition is to be interpreted as a causal mechanism and that what allows the causal interpretation is exogeneity. The goal is not to provide (yet another) definition of the concept of mechanism, but rather to clarify what it is meant when social scientists interpret a component, or factor, of a recursive decomposition as a 'mechanism'.

15.2 The structure of the statistician's explanation

15.2.1 Explanation in a stochastic environment

To explain a social phenomenon, the statistician is usually provided with a data set containing observations coming, for instance, from a survey or a census. The statistician's explanation of a phenomenon related to a data set is based on a statistical model, which is basically a set of probability distributions on an observation space, namely:

$$\mathcal{M} = \{\mathbf{S},\ P^{\theta} \quad \theta \in \Theta\} \tag{15.1}$$

where $\mathbf{S}$ is the sample, or observation, space and for each $\theta \in \Theta$, P^{θ} is a probability distribution on $\mathbf{S}$. Thus, θ characterizes a particular distribution P^{θ} and is called a parameter. The statistician analyses data *as if* it had been generated by one of these distributions; at this stage, the distributions in (15.1) have only a representational role without structural or causal implications – a topic to be considered later in this section. Thus a statistical model can be thought of as a set of plausible hypotheses to uncover the so-called data generating process.

Earlier papers (Mouchart *et al.* 2009; Russo *et al.* 2008) noticed that a statistical model can be seen as a stochastic representation of the phenomenon of interest. This is due to partial and incomplete knowledge of a phenomenon that leads the statistician to model the phenomenon stochastically. This stochastic representation is the cornerstone of the statistician's explanation: the interpretation of parameter θ provides the explanation of the phenomenon whereas the stochastic component of the model stands for the unexplained part of the phenomenon. Here, 'hazard' (stochastic, random) is used as opposed to 'explainable'. We use it as an epistemic concept independent of the metaphysical issue of whether the real world is deterministic or indeterministic.

Let us illustrate with a very simple example. Suppose we weigh an object n times with imperfect scales. We accordingly observe $X = (X_1, X_2, \ldots, X_n)$. A simple statistical model might be $X_i \sim \text{ind.}\, N(\mu, \sigma^2)$, i.e. each X_i is assumed to be identically independently distributed (iid) as a normal distribution with mean μ and variance σ^2. The statistician's explanation then runs as follows: because of the imperfection of the scales, the measurements X_i are interpreted as a realization of a probability distribution and by interpreting the parameter $\theta = (\mu, \sigma^2) \in \mathbb{R} \times \mathbb{R}_+$ the statistician explains that each measurement X_i is relative to a same 'true' weight μ and that the error distribution is characterized by a variance σ^2. The statistician is able to account for the fact that the measurements X_i tend to cluster around the same value μ but not for the fact that the measurements X_i are at a distance from that value. In this sense the statistician's explanation is partial.

15.2.2 The multivariate aspect of social phenomena

In social contexts, interest is usually given to the multivariate aspect of a phenomenon of interest. The statistician who analyses a social phenomenon considers a vector of variables selected on the basis of background knowledge and of the available data; this is far from being an easy task but a thorough discussion of the criteria to select the variables of interest falls beyond the scope of this chapter. This multivariate aspect makes the task of the statistician's explanation far more complex. In the sequel, we develop three steps that should be distinguished: (i) marginal-conditional decomposition, (ii) structural modelling, and (iii) recursive decomposition.

Marginal-conditional decomposition

Explaining a complex multivariate phenomenon is typically operated by decomposing it into a sequence of simpler 'pieces'. The natural decomposition of a multivariate probability distribution is obtained by a marginal-conditional decomposition, i.e. in the bivariate case $X = (Y, Z)$:

$$p_X = p_Z \, p_{Y|Z}. \tag{15.2}$$

Here the bivariate process generating $X = (Y, Z)$ is decomposed into two univariate processes: a marginal process generating Z and a conditional process generating Y given Z ('generating' in the sense suggested in Section 15.2.1). To illustrate, consider the following example. Suppose that the statistician observes data on the price, Y, and the quantity, Z, of fish transacted upon the fishermen's return from the sea. A simple model might decompose this bivariate process, generating $X = (Y, Z)$, into a marginal process generating Z and deemed to represent the good or bad fortune of the fishing activity, and a conditional process, generating $(Y \mid Z)$, representing the operation of the auctioning process. Let us now assume, for the sake of simplicity, that each distribution is normal. Thus the left-hand side of (15.2) would be written:

$$\begin{pmatrix} Y \\ Z \end{pmatrix} \sim \mathcal{N}\left[\begin{pmatrix} \mu_Y \\ \mu_Z \end{pmatrix}, \begin{pmatrix} \sigma_Y^2 & \sigma_{YZ} \\ \sigma_{YZ} & \sigma_Z^2 \end{pmatrix}\right] \tag{15.3}$$

whereas the right-hand side would be written

$$Z \sim \mathcal{N}(\mu_Z, \sigma_Z^2) \qquad (Y \mid Z) \sim \mathcal{N}(\alpha + \beta Z, \sigma_{Y|Z}^2). \tag{15.4}$$

Although models (15.3) and (15.4) are mathematically equivalent, model (15.4) is explanatory because the parameters (μ_Z, σ_Z^2) and $(\alpha, \beta, \sigma_{Y|Z}^2)$ characterize the marginal process representing the fishing activity and the auctioning process, respectively, at variance from the parameters of (15.3) that only characterize the random variability of the data. The statistician's explanation essentially lies in the interpretation of the parameters that characterize the distributions, e.g. if $\beta < 0$, the conditional process generating $(Y|Z)$ explains that, when Z increases, Y tends to decrease on conditional average. Similarly, the converse decomposition $p_X = p_Y \, p_{Z|Y}$ would not provide a convincing explanation from an economic point of view.

Structural modelling

Alone, the marginal-conditional decomposition is not enough for explanation because, in the example of the bivariate case, it is mathematically arbitrary to choose $p_X = p_Z \, p_{Y|Z}$ or $p_X = p_Y \, p_{Z|Y}$. Thus, an explanation also requires the statistical model to uncover the *structure* of the data generating process. As long as such a structure is latent, i.e. not directly observable, the statistician will make systematic use of two ingredients: background knowledge and invariance.

Background knowledge

Broadly speaking, background knowledge, or field knowledge, stands for the whole body of knowledge we have of a given field, and may incorporate different aspects such as: knowledge of the socio-demo-political context, knowledge of the same/similar phenomena relative to the same population at different times or to other populations, results of analyses performed with different methods, knowledge of related fields such as biomedicine in epidemiology, etc. Sometimes such knowledge may be gathered into 'well established' theories. In the fish market example given above, background knowledge supports the decomposition $p_X = p_Z\, p_{Y|Z}$ rather than $p_X = p_Y\, p_{Z|Y}$ because the first one allows us to associate a contextually interpretable univariate subprocess to each component of the relevant decomposition: this was precisely the rationale behind (15.4).

Invariance

We mentioned earlier that an observation of the data set is interpreted as *a* realization of one of the distributions constituting the statistical model. However, a model that specifies a different process for each observation of the data set would be rather useless for explanation as it would not be *structural*. Thus a major aim of structural modelling is to distinguish incidental from structural components of a data generating process. This means that a structural model should display an adequate level of stability or invariance under a suitable class of modifications of the environment and/or of interventions. Invariance is a condition of stability of the marginal-conditional structure of the model and of the characteristics (parameters) of the distribution; parameter stability is indeed an important object of statistical testing. Moreover, the stability of the marginal-conditional decomposition may also be tested by evaluating the parametric stability of alternative decompositions. The specification of the invariance property of a structural model is also a basic ingredient of the definition of the population of interest or the population of reference. In the fish market example, it would be important to test whether the auctioning process has the same characteristics in different seasons and/or different harbours. It should be stressed that invariance and stability properties are requirements *complementary* to congruence with background knowledge, that does not necessarily imply stability or invariance. For this reason invariance and stability are the object of (statistical) testing. In particular, intervention may raise difficulties in the stability of the system, a problem already pointed out by Lucas (1976).

Recursive decomposition

Let us now consider the general case where a vector of variables X is decomposed into g components, namely $X = (X_1, X_2, \cdots, X_g)$ (with g typically much larger than 2), and suppose that the components of X have been ordered in such a way that in the complete marginal-conditional decomposition

$$\begin{aligned} p_X(x \mid \omega) &= p_{X_g|X_1,X_2,\cdots,X_{g-1}}(x_g \mid x_1, x_2, \cdots, x_{g-1}, \theta_{g|1,\cdots,g-1}) \\ &\cdot p_{X_{g-1}|X_1,X_2,\cdots,X_{g-2}}(x_{g-1} \mid x_1, x_2, \cdots, x_{g-2}, \theta_{g-1|1,\cdots,g-2}) \cdots, \\ &\cdot p_{X_j|X_1,X_2,\cdots,X_{j-1}}(x_j \mid x_1, x_2, \cdots, x_{j-1}, \theta_{j|1,\cdots,j-1}) \cdots, p_{X_1}(x_1 \mid \theta_1) \end{aligned} \tag{15.5}$$

each component of the right-hand side may be considered, in a first step, as a structural component with mutually independent parameters, i.e. (in a sampling theory framework):

$$\omega = (\theta_{g|1,\cdots,g-1}, \theta_{g-1|1,\cdots,g-2}, \cdots, \theta_1) \in \Theta_{g|1,\cdots,g-1} \times \Theta_{g-1|1,\cdots,g-2}, \cdots \times \Theta_1 \tag{15.6}$$

Under property (15.6) the conditioning variables $X_1, \ldots, X_{j-1}$ of each factor of (15.5), $p_{X_j|X_1,X_2,\cdots,X_{j-1}}$, are called *exogenous for* the parameter of the corresponding conditional distribution, $\theta_{j|1,\cdots,j-1}$. That is to say, the inference on the parameter $\theta_{j|1,\cdots,j-1}$ of the conditional distribution should not depend on the specification of the data generating process of the conditioning variables $X_1, X_2, \cdots, X_{j-1}$. Therefore, exogeneity is a condition that allows the separation of inferences, notably of the inferences on $\theta_{j|1,\cdots,j-1}$ characterizing the conditional distribution and on $\theta_{1,\cdots,j-1}$ characterizing the marginal distribution of the conditioning variables. More explicitly, in a likelihood approach, the separation of inference means that any inference concerning (any function of) $\theta_{1,\cdots,j-1}$ would be based on a likelihood function derived from the marginal distributions $p_{X_1,X_2,\cdots,X_{j-1}}$, independently of the specification of the conditional distribution $p_{X_j|X_1,X_2,\cdots,X_{j-1}}$, whereas any inference concerning (any function of) $\theta_{j|1,\cdots,j-1}$ would be based on a likelihood function derived from the conditional distributions $p_{X_j|X_1,X_2,\cdots,X_{j-1}}$ independently of the specification of the marginal distribution $p_{X_1,X_2,\cdots,X_{j-1}}$. It follows that such a separation of inference takes advantage of more parsimonius modelling and eventually enjoys more robust properties. Later, in Section 15.3.3, we stress other important features of exogeneity.

Remark

The concept of exogeneity has a long history in econometrics. The works of the Cowles Commission in the late 1940s and the early 1950s have been pathbreaking and are still influential nowadays; in particular, Koopmans (1950) puts emphasis on exogeneity in dynamic models. Barndorff-Nielsen (1978) is significant in the development of conditions for separation of inference. Florens and Mouchart (1980, 1985) and Florens *et al.* (1980) bridge the work of Koopmans (1950) and of Barndorff-Nielsen (1978), and provide a coherent account of exogeneity integrating the separation of inference in dynamic and non-dynamic models. Engle *et al.* (1983) present a list of different concepts of the econometric literature and display their connections with exogeneity through the introduction of supplementary conditions. Florens and Mouchart (1985) not only provide a basic concept of exogeneity, but also make the concept explicit in different levels of model specification, namely, global, initial, and sequential, before combining those concepts of exogeneity with non-causality. This analysis is further developed by Florens *et al.* (1993). ■

Equations (15.5) and (15.6) characterize a *completely recursive system*. A recursive decomposition is not complete when, in equation (15.5), some components are random vectors rather than random variables. This typically happens when we cannot order some of the variables due to a lack of knowledge about their causal or temporal priority. In such a case, in the factorization (15.5) there is (at least) one factor which is a distribution of a *vector* of variables, say X_j, conditional on the antecedent ones $(X_1, \cdots, X_{j-1})$. In other words, the conditional process generating $(X_j \mid X_1, \cdots, X_{j-1})$ is *not* decomposed into a sequence of univariate conditional processes, hence the recursive decomposition is not complete. This situation is met under the heading of 'simultaneity' in the econometric literature.

The case of simultaneity is an interesting and quite disputed issue. A classroom example is provided by a simple two-equation market model: supply and demand. When it is known, and specified, that one side is price-setter and the other side is price-taker, a complete recursive decomposition may be operated and the econometric model completely explains the mechanism generating the observed price and quantity and the market equilibrium process is completely understood. However, when a standard competitive approach is adopted, the market-clearing process becomes a black box generating the equilibrium price and quantity, and the econometric explanation has a different nature. Indeed, the supply and demand equation now represents notional concepts, rather than statistical entities such as marginal and conditional distributions. These concepts are of the same nature as the counterfactuals used in a large portion of the literature on causation. In such a case the identification of the parameters requires identifying restrictions that are not empirically testable, but only supported by contextual knowledge and/or economic theory. In other words, the explanation provided by the econometric model is of a speculative nature and the recursive decomposition among the endogenous variables is not operating.

Let us emphasize that a recursive decomposition is essentially an ordering of the variables in such a way that each factor of the right-hand side of (15.5) is structurally valid. Once the number of components p increases, background knowledge, possibly substantiated by statistical tests, typically provides a simplification of the factors in the form of conditional independence properties. More specifically, it is often the case that the distribution of $(X_j \mid X_1, \cdots, X_{j-1})$ is known not to depend on some of the conditioning variables. Thus there is a subset of variables $\mathcal{I}_j \subset \{X_1, \cdots, X_{j-1}\}$ actually relevant for the conditional process generating $X_j \mid X_1, \cdots, X_{j-1}$ as defined by the property

$$X_j \perp\!\!\!\perp X_1, \cdots, X_{j-1} \mid \mathcal{I}_j \tag{15.7}$$

implying that the factor $p_{X_j|X_1,X_2,\cdots,X_{j-1}}$ in (15.5) is actually simplified into $p_{X_j|\mathcal{I}_j}$ and $\mathcal{I}_j$ may be called the *relevant information of the j-th factor*. Once $\mathcal{I}_j$ has been specified for each factor, (15.5) is condensed into

$$p_{X_1,X_2,\cdots,X_g} = \prod_{1\leq j\leq g} p_{X_j|\mathcal{I}_j}. \quad (15.8)$$

This form is accordingly called a *condensed recursive decomposition*. As argued by Mouchart *et al.* (2009), *causes* may then be viewed as exogenous variables in the condensed recursive decomposition, or, alternatively, as the relevant information of a structurally valid conditional distribution.

Readers familiar with the literature on graph models may recognize that a directed acyclic graph (DAG) is a graphical representation of a condensed recursive decomposition and that the causal structure is depicted by the set of ancestors. Also, in the literature on graph models, the concepts of completeness and of recursivity are not identical to those developed in the statistical tradition and discussed in this chapter. This is due to the fact that in the statistical tradition these concepts relate to multivariate distributions; not all structures of probabilistic independences can be represented by graph models. For binary variables, a simple example is the case of a trivariate distribution with pairwise independence but not complete mutual independence. In simple cases, however, the two families of concepts coincide.

Summarizing

The statistician's explanation is *partial*, because it is based on a stochastic representation of a phenomenon of interest, and *structural*, because it is based on a recursive decomposition that seeks to decompose a vector of variables into structurally valid components.

15.2.3 Difficulties

As a matter of fact, several problems hinder explanation from being a simple task. Let us focus on three of them: (i) partial observability, (ii) time delay, dynamic structure, and feedback effects, and (iii) causal chain.

Partial observability

Many models in the social sciences involve latent, i.e. non-observable, variables. Some of these variables could possibly be observed but are non-observable for practical, legal, or ethical reasons whereas other variables are genuinely non-observable because they correspond to theoretical concepts partially observed through indicators, or proxy variables, and are used in framing theoretical models. The statistical model, because it bears on observable variables only, is accordingly constructed in two steps: a first step in the form of a structural model involving both observable and latent variables, and a second step, the implied statistical model (also called the operational model), obtained by integrating out the non-observable variables.

Statistical models of that type are known in statistical methods as 'mixture models', because a marginal distribution may be viewed as a mixture of conditional distributions, and are characterized by severe difficulties. First, the distributions have a complex analytical structure, making the inference process often cumbersome. Second, the integration of non-observable variables requires the introduction of specific supplementary assumptions often impossible to be controlled or to be statistically tested, and justified only by field knowledge. Third, the marginalized distributions no longer represent a data generating process supported by arguments that it is structural, and their parameters may no longer be given a simple structural interpretation. Moreover, integrating the unobservable explanatory variables typically jeopardizes the exogeneity of the remaining variables; this problem has been explicitly worked out for a simple trivariate case by Mouchart *et al.* (2009). The econometric literature on heterogeneity, i.e. on unobservable explanatory variables, often suggests the recovery of loss of exogeneity by introducing further *ad hoc* assumptions, such as the independence between the heterogenous factors and the observable explanatory variables, even though such assumptions may be contextually doubtful and empirically not testable (see also Mouchart *et al.* (2009) for a deeper discussion of those supplementary assumptions).

Time delay, dynamic structure, feedback effects

In many cases, a reasonable specification of the structure of the data generating process requires the introduction of time delays in order to take into account dynamic features of the phenomenon of interest. This makes the observations a sequence of data that are not independent. In particular, the effects of a cause require some delay before being operational and feedback effects often take place through adjustments of individual behaviours. These facts generate further difficulties.

Firstly, the specification of dynamic models is substantially more demanding than the specification of models with mutually independent observations.

Secondly, the time frequency of data is often not high enough for identifying the shortest-term features. In other words, the available data operate a time-aggregation and therefore should be viewed as partial observability of the dynamic structure of the data generating process. In line with the above-mentioned difficulties due to partial observability, econometricians, already in the early 1950s (Wold and Jureen 1953; Bentzel and Hansen 1955), have argued that time-aggregation is the main cause of simultaneity because otherwise an econometric model should become completely recursive.

Thirdly, even without time-aggregation, the presence of feedback effects requires a substantially more complex analysis of exogeneity and causality. Moreover, different levels of model specification should be distinguished, namely (i) a global one, modelling at once all available data, (ii) an initial one, modelling each data conditionally on initial conditions, and (iii) a sequential one, modelling items of data sequentially conditionally on their relative history. Florens and Mouchart (1985) provide an integrated approach to this topic.

Causal chain

The recursive decomposition, be it complete or not, along with the causal interpretation of exogenous factors, makes manifest that variables in $\mathcal{I}_j$ are (direct) causes of X_j and, similarly, other variables are causing X_{j-1}, accordingly producing a chain of causes within the data $X = (X_1, \ldots, X_g)$. Two issues should be made explicit.

First, the 'natural' state of a social phenomenon is not 'one cause–one effect' but rather 'multiple causes–multiple effects'. This leads us to identify not only direct but also *indirect* causes. Thus the crucial aspect of the framework presented here is to provide an *ordered structure* in a 'systemic' approach. In other words, it is not enough to say that 'everything depends on everything': a structure ought to be elaborated in order to explain and shed light within an otherwise black box. But there is no free lunch. The cost to be paid is to learn how to manage a complex causal structure. In the social sciences this issue is crucial, in particular, for policy purposes.

Second, the causal chain constructed within a given data set $X = (X_1, \ldots, X_g)$ is essentially truncated. This means that once the data set has been ordered in such a way that X_g is explained by $X_1, \ldots, X_{g-1}, \ldots, X_j$ explained by $X_1, \ldots, X_{j-1}$, etc., the statistician is still left with explaining X_1. One might argue that either there is no plausible explanation for X_1 or, in the causal chain, X_1 is far enough from the variables $X_j, X_{j+1}, \ldots$ to be explained, so that the indirect effect of the explanatory variables of X_1 could be neglected. However, it should be stressed that, although necessary from an operational viewpoint, the explanation provided by means of the statistical model may fail to be robust with respect to the truncation. The social scientist should be particularly aware of that difficulty, and only field knowledge can be of help at this stage. Nevertheless, the fact that appealing to field knowledge helps in those cases does not introduce a vicious circle nor does it make knowledge to be gained through structural modelling radically different from field knowledge. Rather, this reflects the idea that structural modelling (i) does establish knowledge to be used in other studies, but (ii) does not establish immutable and eternal 'causal truths'. Structural modelling is a dynamic process in which field knowledge and new knowledge constantly interplay. How precisely they interplay is, however, the object of another paper.

15.3 Explanatory mechanisms

15.3.1 Interpreting recursive decompositions

So far we have argued that the statistician's attempt to explain a given phenomenon of interest involves two aspects. First, a genuinely partial explanation by incorporating in the statistical model a stochastic component deemed to represent what is not explained. Second, a recursive decomposition over an ordered vector of variables deemed to disentangle a multivariate, i.e. complex, phenomenon into a sequence of univariate (conditional) processes.

It is worth emphasizing that a recursive decomposition is made of two ingredients. First, a sequential marginal-conditional decomposition of a multivariate distribution. This is a standard operation in the calculus of probability that can be operated over an arbitrary order of a vector of variables. Second, a specific order is selected by requiring structural validity of each component of the decomposition. Such validity requires a close congruence with background knowledge, along with a condition of stability/invariance, which implicitly defines the population of reference. These two elements allow us to introduce a *specific* concept of mechanism that fits structural models. Because this concept is not meant to be a general one and consequently may not attract unanimous consensus, comparison with other alternative approaches may be useful.

Thus, the guiding questions of this section are the following: How do we interpret the recursive decomposition? Why is it explanatory? In a nutshell, recursive decompositions are to be interpreted in terms of *mechanisms*, and we discuss below why mechanisms thus conceived *explain*.

15.3.2 Recursive decompositions and mechanisms

In interpreting the recursive decomposition in terms of 'mechanisms' we distinguish between 'global mechanisms' and 'sub-mechanisms'. The *whole* recursive decomposition characterizes a *global* mechanism, whereas *each* conditional distribution within the recursive decomposition characterizes an (autonomous) *sub*-mechanism within the global one. In this context, decomposing a global mechanism into a sequence of (autonomous) sub-mechanisms is tantamount to disentangling the action of each component in a sequence of the sub-mechanisms operating in a global mechanism. In other words, the explanatory power of a mechanism is operationalized, in structural models, through the recursive decomposition. For instance, in the fish market example, the market, call it the global mechanism, generates a bivariate distribution of the price (Y) and quantity (Z); the econometric explanation consists in distinguishing a supply process represented by the marginal distribution of Z, a demand process represented by the conditional distribution of $(Y|Z)$, and a market equilibrium process based on the quantity-taking behaviour of the demand side.

Thus a recursive decomposition carries explanatory power insofar as it disentangles a global mechanism into sub-mechanisms in the above sense. But what are the specific features of a mechanism in this context of structural models?

Stochastic mechanisms

This concept of mechanism is a stochastic one: a mechanism is *not* deterministic but it rather singles out a stable/invariant and contextually meaningful aspects of the phenomenon of interest (see below). The fact that in social contexts causal explanation is essentially mechanistic does not imply that it also is *mechanicistic*, in the sense that it essentially requires physical deterministic mechanisms in order to explain social phenomena.

Stable mechanisms

Because a mechanism is meant to identify a stable/invariant, and therefore repeatable, aspect of the phenomenon being modelled, identifying a mechanism means separating incidental from structural features of the data generating process. By so doing, the statistician is also able to distinguish spurious from causal correlations.

Mixed mechanisms

In social contexts, mechanisms are not necessarily 'physical', that is made of physical processes or physical entities interacting in one way or another. This is so for several reasons, three of which are:

1. In statistical models used in the social sciences mechanisms try to depict the working forces, i.e. the motivation or rationale for evolving, characterized by variables that possibly lack 'physical' (or even manifest/observable) counterparts. Many social, demographic, or economic variables are conceptual constructs introduced to shape a 'theory', the development of which leads to building measurement devices by means of a number of relevant indicators that are distinct from the definition of the concept. For instance, 'socio-economic status' might be measured from income and level of education, but these indicators are not meant to provide a unique definition of the concept.
2. Many social scientists are interested in mechanisms where very different types of variables interplay. For instance, health economics or some branches of epidemiology are interested in how economic variables influence health variables and vice versa. In this case, although some variables might have a 'physical' counterpart (e.g. baby's weight at birth), not all of them will (e.g. socio-economic status). Consequently, we need a characterization of mechanism broad enough to include both 'physical' and 'non-physical' components. That is to say, we have to model 'mixed' mechanisms.
3. Also, health variables do not influence economic variables (or vice versa) *as such*, but through indirect paths involving intermediate causal variables. Those indirect paths need (or may need) to be specified in order to *explain* the phenomenon of interest. In Caldwell's model mentioned

> in Section 15.1, maternal education does not influence child mortality *as such*, and in fact a major improvement of Caldwell's framework was provided by Mosley and Chen (1984), who developed an analytical framework explaining the indirect paths through which a social variable such as maternal education can have a causal impact on a health variables such as child survival.

For details on mixed mechanisms and, more generally, on modelling mechanisms in causal modelling, see Russo (2009, ch. 6).

Alternative views

The very notion of mechanism is currently a matter of vivid debate (Little, 1991, 1998; Machamer *et al.* 2000; Woodward, 2002, 2003; Bunge, 2004; Psillos, 2004; Bechtel and Abrahamsen, 2005; Reiss, 2007; Craver, 2007). Alternative accounts may feed debate about the *concept* or the *role* of mechanisms in the social sciences. For expository purposes, we selected only two, notably Little (1991, 1998) and Craver (2007).

Little (1991) defends the idea that causal analysis in the social sciences is legitimate but that it depends upon identifying social mechanisms. Little goes as far as saying that such social mechanisms work through the actions of individuals – a position also known as methodological individualism. To discuss the plausibility of a microfoundation approach (see for instance Little, 1991, 1998) would lead us too far away from the main track. Yet, Little's characterization of mechanisms will help us in clarifying our claim that the recursive decomposition represents a social mechanism. According to Little (1991, p. 15) a causal mechanism is a series of events governed by lawlike regularities that lead from the explanans to the explanandum. Mechanisms, within a microfoundational perspective, are grounded in meaningful and intentional behaviour of individuals. The sort of things having causal properties are, for instance, the actions of individuals and groups. One might disagree with Little about the soundness of a microfoundation approach, or about the use of *lawlike* regularities. Notably, Hoover (2001) stresses the causal import of the structural approach in econometrics arguing for a reality of macroeconomic structures that does not boil down to the reality of microeconomic relations and holds the view that mechanisms and causal structures may substitute for laws and do not necessarily need to be supported by appeals to laws.

Yet, notwithstanding divergences on those issues, we surely agree with Little on his account of *statistical analysis* as a form of causal reasoning in social research (Little, 1991, ch. 8). Statistical explanation, in his view, has to be accompanied by a causal story indicating the mechanisms. The identification of the mechanism involves (lawlike) statistical regularities, but is not the end of the explanation; it is only the first step in establishing causal relations. So statistical tools serve to uncover the patterns present in the empirical phenomenon, i.e. the data set. It is in this sense that Little's and our views are close to each other.

Craver (2007) explores the notion and the explanatory power of mechanisms in the domain of neurosciences. At the beginning of his book, Craver (2007, p. 5) discusses the example of how a neuron releases neurotransmitters and concludes:

> This is a mechanism in the sense that it is a set of entities and activities organized such that they exhibit the phenomenon to be explained.

To our understanding, this "skeletal description", as Craver calls it, is broad enough to account for mechanisms in various domains. Should you take the entities to be neurons and the activities neurotransmitter release, the above skeletal description will well fit neuromechanisms. Should you take entities to be socio-demo-economic variables, and activities to be their influence on other socio-demo-economic variables, the above skeletal description will fit equally well social mechanisms. The degree of 'physical' reality one wishes to give to entities and activities may lead to different accounts – notably, to a different ontological commitment to the existence – of mechanisms. In the social sciences, we do not need to endorse the view that elements and relations should always have *physical* counterparts – see the discussion of mixed mechanisms above.

15.3.3 Mechanisms, explanation, causality

Mechanisms and explanation

More on partial explanation

In Section 15.3.1 we have recalled the partial nature of statistical explanation – the stochastic component delineates the frontier between what we explain and what we do not explain. But there is another sense in which explanation is partial.

In case the statistician can operate a *complete* recursive decomposition, the explanation is complete in the sense that each sub-mechanism is identified and thus the global mechanism is fully disentangled. However, in case the statistician is unable – for a whole variety of different reasons, e.g. missing data or insufficient background knowledge – to operate a complete recursive decomposition, the explanation itself is partial. In such a case the conditional distribution bears simultaneously on several variables that are therefore not explained individually.

More on explanatory power

However, whether complete or incomplete, the recursive decomposition – the mechanism – provides an explanation of the phenomenon of interest. In other words, mechanisms, we claim, carry explanatory power. The question is to understand *why* it is so. The answer to this question resides in considering the whole modelling procedure as explanatory on the one hand, and in understanding the explanatory import of exogeneity on the other hand. Simply put, when a conditional distribution is a component of the recursive decomposition, the conditioning variables are exogenous and can be interpreted as causal factors. In the next section, we discuss this idea more thoroughly. A related issue concerns the possibility to simulate. The recursive decomposition explains because the distribution generating the data can be simulated in a contextually meaningful way. In this sense to explain also means 'being able to reproduce'.

Causal factors and exogeneity

Why do we interpret *exogenous* variables as *causal* factors? This is so for three different but related reasons.

First, the *whole modelling procedure* is explanatory. The goal of structural modelling is to characterize clearly identified and interpretable mechanisms. We mentioned in Section 15.2 that the choice of the marginal-conditional decomposition may be arbitrary; this is the reason why we need background knowledge and invariance: to make a selection among the various possible decompositions. In other words, the marginal-conditional decomposition *alone* does not provide a (causal) explanation of a given phenomenon, but the whole modelling procedure does. Indeed, building a structural model is made of a progressive procedure, three steps of which may be identified. In the first stage we select the variables of interest, that is the elements of the mechanism, out of background knowledge. In the second stage we build the statistical model, in particular, we operate a recursive decomposition over the initial joint probability distribution. Finally, in the third stage we confirm or disconfirm the hypothesized mechanism by confrontation with empirical evidence and background knowledge. Briefly put, tests on the recursive decomposition concern, on the one hand, its invariance or stability and, on the other hand, whether it is congruent with background knowledge. Adequacy of the model is also tested by measuring goodness of fit and the amount of variability the model can account for. This is a very concise presentation of the hypothetico-deductive methodology of causal models. For a thorough discussion of hypothetico-deductivism in causal modelling see Little (1998, p. 9), Cartwright (2007, ch. 2), and Russo (2009, ch. 3).

Such a stepwise methodology provides a *causal* explanation because it aims to provide an understanding of a given phenomenon by showing the causal sub-mechanisms that underlie it. Of course, the question arises as to what guarantees the *causal* interpretation of the relations or the mechanisms established in those models. This is the relation between exogeneity and causality, that is why we interpret exogenous factors as *causal* factors.

Second, before discussing in more detail the relation between exogeneity and causality, another issue is worth pointing out. Borrowing Cartwright's adage, *no causes in, no causes out.* According to the hypothetico-deductive methodology used in structural models, the first stage, that is the hypothesis formulation stage, exactly concerns a *causal* hypothesis. Therefore what we will (dis)confirm exactly is a *causal* structure. Causal relations are not inferred from mere correlations, taken out from a 'magical' statistical hat. Structural models, unlike simple associational or descriptive models, have a rich and sophisticated apparatus of assumptions that underwrite their causal interpretations (see for instance Russo (2009, ch. 3) who divides them into three categories: statistical, extra-statistical, and causal). Also, specific tests – notably, invariance and exogeneity tests – allow us to causally interpret the relations showed in the recursive decomposition and therefore the exogenous variables as causes.

Third, here comes the thorny issue: the relation between exogeneity and causality. Identifying causal factors with exogenous variables is based on the following considerations:

- Because causality is a latent concept, causal inference can only be of the type 'to the best of our knowledge'; causal relations pertain to the *interpretation* of a model (i.e. a *representation* of the data generating process) and are, therefore, relative to a model rather than a sole characteristic of the available data. Differently put, structural modelling is not a hunt for the 'true' model nor a device that enables us to discover the 'true' causal relations. Structural modelling is a progressive path toward making intelligible the observed phenomena while adjusting the window of observation to pre-specified targets.
- Exogeneity is a condition of separation of inference. As mentioned earlier, the (partial) explanation of the statistician is cast in the framework of a statistical model, in terms of parameters that characterize the distributions of interest (see the Section 15.2.2). Thus the exogeneity condition (15.6) does not only allows us to separate the inferences on $\theta_{j|1,\cdots,j-1}$ and on $\theta_{1,\cdots,j-1}$, but it also allows us to distinguish the process generating the causes, characterized by $\theta_{1,\cdots,j-1}$, and the process generating the effect, characterised by $\theta_{j|1,\cdots,j-1}$. Separating causes from effects mirrors the asymmetry of causation. This last point makes clearer and more precise the older expression 'exogenous means generated outside the model'.

Needless to say, this is not to suggest that structural models provide immutable and eternal causal explanations of social phenomena. As mentioned above, explanation is intrinsically relative and partial, that is relative to the specific conceptual framework and dependent on available empirical and theoretical information. This means that nothing prevents future explanations to discard previous ones. Furthermore, nothing prevents different social scientists to provide different explanations of the same phenomenon: a causal explanation crucially depends on background knowledge which also includes the social scientist's personal or political beliefs. Finally, such causal explanations involve an implicit stopping rule in order to avoid an otherwise ad infinitum chain of 'explaining the explanatory'.

15.3.4 Evaluation/flexibility of explanation

Before closing this section on explanatory mechanisms, two features of causal explanation are worth mentioning. The first concerns the evaluation of explanations and the second their flexibility.

We can evaluate explanations by considering three interrelated aspects: (i) statistical, (ii) epistemic, and (iii) ontological adequacy. We can give (i) a *statistical* evaluation by measuring, with the coefficient of determination, how much variability is accounted for, and by measuring the goodness of fit. We can also give (ii) an *epistemic* evaluation, by asking whether results are coherent with background knowledge. (iii) An *ontological* evaluation is also possible: if ontological homogeneity between the variables acting in the mechanism is lacking (for instance if the mixed mechanism includes both economic and health variables), it may be desirable to identify and justify indirect paths from the causes to the effect. Causal explanations will then be good or bad depending on how well they meet statistical, epistemic, and ontological requirements.

Such explanations also exhibit high flexibility. The first aspect of flexibility is concerned with 'mixed mechanisms': as discussed earlier, we do not need to stick to a *physical* concept of mechanism. The second aspect relates to the available information we base the explanation upon. The example on bargaining powers and market segmentation presented below shows that even if available data and background knowledge do not allow us to fully explain the phenomenon, *some* explanation is possible. The third aspect is that such explanations allow a *va et vient* between established theories and establishing theories. Established scientific theories are (and ought to be) used to formulate the causal hypothesis and to evaluate the plausibility of results on theoretical grounds. But causal models also participate in establishing new theories by generalizing results of single studies. This reflects the idea that science is not monolithic, discovering immutable and eternal truths. If the model fits the data, the components of the recursive decomposition are structurally stable and congruent with background knowledge, then we can say, to the best our knowledge, that we hit upon a mechanism that explains a given social phenomenon. But what if one of these conditions fails? A negative result may trigger further research by improving the modelling strategies or by collecting new data, thus leading to new discoveries that may question background knowledge.

We now conclude by briefly presenting two examples. In the first one, researchers successfully provided a causal explanation by disentangling the mechanism in a recursive decomposition on five variables. In the second one, researchers did not succeed to fully explain the phenomenon by providing the marginal recursive decompositions due to a lack of data and background knowledge.

Health systems and mortality in Spain

López-Ríos *et al.* (1992) were interested in regional mortality in Spain. Spain met deep socio-economic changes in the mid-1970s, and consequently policy in that period tried to intervene on improving the social and economic situation. This led to a lower mortality rate at the time of the study. This background supported the choice of distinguishing the supply and demand of medical care, unlike the majority of similar studies. In fact, previous studies in demography and medical geography examined the incidence of the health system on regional mortality coming to the conclusion that regional differences in mortality could not possibly be explained by regional differences in the health system. López-Ríos *et al.* (1992), instead, hypothesized that regional mortality is influenced by the health system which was in turn influenced by social and economic development. The vector of variables (economic development, social development, sanitary infrastructure, use of the medical care system, age structure, mortality) was decomposed into basically two sub-mechanisms. In the first, 'economic development' was the exogenous variable influencing mortality through 'social development' and 'sanitary infrastructure'; in the second, 'age structure' was the exogenous variable influencing mortality through 'use of the medical care system'.

Bargaining powers and market segmentation in freight transport in Belgium

Mouchart and Vandresse (2007) analysed multivariate data obtained by face-to-face interviews with companies using the services of freight transport in Belgium. This data provided information, for each contract of a sample, on several characteristics of the contract, such as the price, the distance, the speed of delivery, etc. From the interviews it was clear that each contract was the result of a bargain between the service user and the service provider. For instance, a requirement of quick delivery could be priced at a higher tariff than a slow delivery, depending on the availability of the provider. However, there wasn't available data on every step of the bargaining process: only the final result was known. Moreover, no economic theory, no game-theoretic strategy, nor any field knowledge was available for substantiating any possible recursive decomposition. Consequently, a recursive decomposition of the data generating process was not possible and Mouchart and Vandresse (2007) could only provide an analysis of the joint distribution of all the available data without incorporating any exogeneity assumption. This analysis nevertheless provided some explanation of the global functioning of the freight transport market in terms of imperfection of the competition and of the bargaining power of the actors, but did not provide an explanation about the data generating process of each variable separately.

15.4 Concluding remarks

Quite uncontroversially, explanation belongs to the tasks of the social sciences. What is more controversial, however, are the features and characteristics of explanations in social contexts. This chapter tackled this issue by analysing structural modelling.

We highlighted three features. (i) Explanation is incomplete, or partial, in the sense that it is based on a stochastic representation of the world where the stochastic component stands for what is not explained. (ii) An explanation is given by decomposing a complex causal mechanism into a sequence of 'simpler' explanatory mechanisms. In a nutshell, explaining a complex social phenomenon involves two ingredients. First, we operate a recursive decomposition on a multivariate distribution that represents the phenomenon of interest; the whole recursive decomposition is interpreted as the 'global mechanism'. Second, we consider each component of the recursive decomposition as an 'autonomous' sub-mechanism within the global mechanism insofar as it is composed of a univariate conditional distribution. Decomposing a global mechanism into a sequence of (autonomous) sub-mechanisms is tantamount to disentangling the action of each component into a sequence of the sub-mechanisms operating in a global mechanism. In other words, the explanatory power of a mechanism is operationalized, in structural models, through the recursive decomposition. (iii) Explanation is causal, that is we identify cause–effect relations through the condition of exogeneity. We defined exogeneity as a condition of separability of inferences on the parameters of the marginal-conditional distribution, which allows us to identify the variables that play a causal role in the mechanism.

We emphasized that providing a complete explanation of a complex phenomenon is not always (and often not) possible and that incomplete recursive decompositions have to be accommodated in the toolkit of the social scientist.

Rather than proposing new definitions of key concepts used in structural modelling, we offered a reassessment of the literature connecting the practice of (structural) statistical modelling to the concepts of explanation and mechanism. Notably, we aimed to clarify how social scientists explain social phenomena by building structural models and what it means to interpret a recursive marginal-conditional decomposition as a mechanism.

Acknowledgements

This chapter was originally prepared for presentation at the conference 'CAPITS 2008', University of Kent 10–12 September 2008. We thank the conference participants, Guillaume Wunsch and two referees for extremely helpful and stimulating comments. F. Russo acknowledges financial support from the FRS-FNRS (Belgium). M. Mouchart acknowledges financial support from the IAP Research network grant nr P6/03 of the Belgian Government (Belgian Science Policy).

References

Barndorff-Nielsen, O. (1978). *Information and Exponential Families in Statistical Theory*. John Wiley & Sons., New York.

Bechtel, W. and Abrahamsen, A. (2005). Explanation: A mechanist alternative. *Studies in the History and Philosophy of the Biological and Biomedical Sciences*, 36: 421–441.

Bentzel, R. and Hansen, B. (1954–1955). On recursiveness and interdependency in economic models. *Reviews in Economic Studies*, 22: 153–158.

Bunge, M. (2004). How does it work? The search for explanatory mechanisms. *Philosophy of the Social Sciences*, 34:182–210.

Caldwell, J. (1979). Education as a factor in mortality decline: An examination of nigerian data. *Population Studies*, 33(3): 395–413.

Cartwright, N. (2007). *Hunting Causes and Using Them: Approaches in Philosophy and Economics*. Cambridge University Press, Cambridge.

Craver, C. F. (2007). *Explaining the Brain. Mechanisms and the Mosaic of Neuroscience*. Oxford University Press, Oxford.

Durkheim, E. (1897/1960). *Le Suicide*. Presses Universitaires de France.

Engle, R., Hendry, D., and Richard, J.-F. (1983). Exogeneity. *Econometrica*, 51(2): 277–304.

Florens, J.-P. and Mouchart, M. (1980). Initial and sequential reduction of bayesian experiments. *CORE Discussion Paper 8015*.

Florens, J.-P. and Mouchart, M. (1985). Conditioning in dynamic models. *Journal of Time Series Analysis*, 53(1): 15–35.

Florens, J.-P., Mouchart, M., and Rolin, J.-M. (1980). Réductions dans les expériences bayésiennes séquentielles, paper presented at the colloque processus aléatoires et problèmes de prévision, held in Brusells 24–25 April 1980. *Cahiers du Centre d'Etudes de Recherche Opérationnelle*, 23(3-4): 353–362.

Florens, J.-P., Mouchart, M., and Rolin, J.-M. (1993). Noncausality and marginalization of Markov processes. *Econometric Theory*, 9: 241–262.

Hoover, K. (2001). *Causality in Macroeconomics*. Cambridge University Press, New York.

Koopmans, T. C. (1950). When is an equation system complete for statistical purposes? In Koopmans, T., editor, *Statistical Inference in Dynamic Economic Models*, volume on the Cowles Commission. Monograph 10 of *Statistical Inference in Dynamic Economic Models*. John Wiley & Sons., New York.

Little, D. (1991). *Varieties of Social Explanation: An Introduction to the Philosophy of Social Science*. Westview Press, Boulder.

Little, D. (1998). *Microfoundations, Method and Causation. On the Philosophy of the Social Sciences*. Transaction Publishers, New Brunswick, N.J.

López-Ríos, O., Mompart, A., and Wunsch, G. (1992). Système de soins et mortalité régionale: une analyse causale. *European Journal of Population*, 8(4): 363–379.

Lucas, R. (1976). Econometric policy evaluation. In Brunner, K. and Meltzer, A., editors, *The Phillips Curve and Labor Markets*, volume 1 of *Carnegie–Rochester Conference Series on Public Policy*, 161–168. North-Holland, Amsterdam.

Machamer, P., Darden, L., and Craver, C. (2000). Thinking about mechanisms. *Philosophy of Science*, 67: 1–25.

Mosley, W. and Chen, L. (1984). An analytical framework for the study of child survival in developing countries. *Population and Development Review*, 10, Supplement: Child survival: Strategies for research: 25–45.

Mouchart, M., Russo, F., and Wunsch, G. (2009). Structural modelling, exogeneity and causality. In H. Engelhardt, H-P Kohler, A. P., editor, *Causal Analysis in Population Studies: Concepts, Methods, Applications*, chapter 4, 59–82. Springer, Dordrecht.

Mouchart, M. and Vandresse, M. (2007). Bargaining powers and market segmentation in freight transport. *Journal of Applied Econometrics*, 22: 1295–1313.

Psillos, S. (2004). A glimpse of the secret connexion: Harmonising mechanisms with counterfactuals. *Perspectives on Science*, 12(3): 288–319.

Reiss, J. (2007). Do we need mechanisms in the social sciences? *Philosophy of the social sciences*, 37(2): 163–184.

Russo, F. (2009). *Causality and Causal Modelling in the Social Sciences. Measuring Variations*. Methodos Series. Springer, New York.

Russo, F., Wunsch, G., and Mouchart, M. (2008). Potential outcomes, counterfactuals, and structural modelling. Causal approaches in the social sciences. DP 0826, Institut de Statistique, UCL.

Wold, H. and Jureen, L. (1953). *Demand Analysis*. John Wiley & Sons, New York.

Woodward, J. (2002). What is a mechanism? A counterfactual account. *Philosophy of Science*, 69: S366–S377.

Woodward, J. (2003). *Making Things Happen: A Theory of Causal Explanation*. Oxford University Press, Oxford.

·16·

Counterfactuals and causal structure

Kevin D. Hoover

Abstract

The structural account of causation derives *inter alia* from Herbert Simon's work on causal order and was developed in Hoover's *Causality in Macroeconomics* and earlier articles. The structural account easily connects to, enriches, and illuminates graphical or Bayes net approaches to causal representation and is able to handle modular, non-modular, linear, and nonlinear causal systems. The representation is used to illuminate the mutual relationship between causal structure and counterfactuals, particularly addressing the role of counterfactuals in Woodward's manipulationist account of causation and Cartwright's attack on 'impostor counterfactuals'.

16.1 Introduction

Causality is closely related to the analysis of counterfactuals. Hume, who is often seen as having depreciated the status of causal relations, stressed their importance in political and economic contexts:

> it is of consequence to know the principle whence any phenomenon arises, and to distinguish between a cause and a concomitant effect. Besides that the speculation is curious, it may frequently be of use in the conduct of public affairs. At least, it must be owned, that nothing can be of more use than to improve, by practice, the method of reasoning on these subjects, which of all others are the most important; though they are commonly treated in the loosest and most careless manners. (Hume 1754, p. 304)[1]

The value of causal reasoning is, in part, diagnostic and retrospective: *why did X happen?* Such a backward looking question calls for a counterfactual inquiry: *if Y had not happened, would X have happened?* Causal reasoning is also prospective and related to planning: *if Y were implemented, would X happen?*

A central question addresses the relationship between causes and counterfactuals. David Lewis (1973), for example, defines causes reductively in

[1] Nor is this view of the importance and utility of causal knowledge limited to Hume's economic and political writings – see Hume (1739, pp. 73, 89; 1777, p. 76).

terms of counterfactuals that are given an independent account. James Woodward (2003) also defines causes counterfactually, albeit non-reductively. Woodward's account has become increasingly popular among philosophers of science, although it is not universally accepted. Nancy Cartwright (2007) attacks it, partly over the relationship of causes to counterfactuals. She objects both to defining cause in terms of counterfactual manipulations and to the subsequent use of causal knowledge so defined to evaluate counterfactuals for policy. Cartwright doubts that Woodward's criterion for cause is generally applicable, and she regards the counterfactuals supported by the supposed causal knowledge as irrelevant 'imposters' (Cartwright 2007, esp. ch. 16).

Woodward's account draws substantially on the graph-theoretic analyses of Peter Spirtes, Clark Glymour and Richard Scheines (2001) and Judea Pearl (2000), in which causes are conceived as holding among variables that are connected through functional, but asymmetrical, relations. The graphs in these accounts are maps of the asymmetric flow of causal influence. A main purpose of the current chapter is to suggest that Woodward's version of the graph-theoretic approach implies an unnecessarily impoverished representation of causal relations and that these representations, in turn, lead him to attribute too great a role for counterfactual manipulability in defining cause and to support a too highly constrained account of the structure of the relationships among causes and effects, laying his account open to many of Cartwright's criticisms. I offer an alternative account, the *structural account*, built on work that long predates Woodward's book, but which is less well known to philosophers (Hoover 1990, 1994, 2001). The structural account bears a close family resemblance to Woodward's manipulation account. Yet, there are key differences that provide a richer set of resources, which are adequate to deal with Cartwright's objections and to provide a basis for understanding the connection of counterfactuals to causality.

To avoid confusion, it is worth noting that the account proposed here does not fundamentally conflict with the general approach of modeling causal relationships graphically, developed especially by Pearl (2000) and Spirtes, Glymour and Scheines (2001), and used by Woodward. Rather it clarifies the relationship between graphical representations and systems of equations in a manner that both enriches the graphical approach and demonstrates the fundamental kinship of the two approaches.

16.2 Woodward's manipulation account

While Woodward's account of causation relies on a counterfactual analysis, it is substantially different from the influential counterfactual account due to David Lewis (1973, 1979), which relies on a possible-worlds analysis.

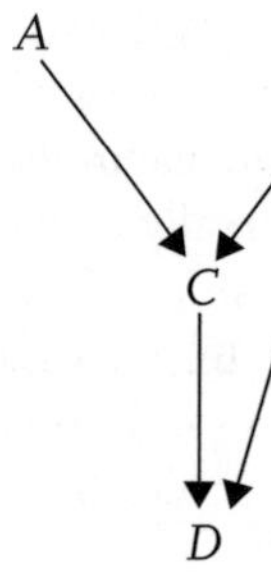

Fig. 16.1

Woodward and I agree that Lewis unnecessarily privileges the notion of non-causal, universal laws (Hoover 2001, ch. 4, sections 4.2–4.4; Woodward 2003, p. 16, ch. 6). Furthermore, the notion of a metric for nearness of possible worlds is fundamentally vague (Woodward 2003, p. 138). Two analysts are vastly more likely to agree on a causal claim than on the truth or falsity of the counterfactual that is supposed to underwrite it or the nearness of the possible worlds that are supposed to decide the truth value of that counterfactual.

In contrast to Lewis, who sees causal relations as connecting token events, Woodward follows Spirtes (2001) and Pearl (2000) as seeing causal relations as connecting variables. Fundamentally, then, Woodward's account is one of type-causation. Token-causation is analysed through assessing the cases in which variables take particular values.

Causal relations among variables are represented both graphically and functionally. Thus in Figure 16.1, A and B cause C; and C and B (directly as well as indirectly), in turn, cause D. These relationships may be made more quantitatively precise by specifying the functional connections among the variables. For example, Figure 16.1 might be the graph of a system of equations:

$$C = \alpha_{CA} A + \alpha_{CB} B, \tag{16.1}$$

$$D = \alpha_{DC} C + \alpha_{DB} B, \tag{16.2}$$

where the α_{ij}, $i, j = A, B, C, D$ are the coefficients that measure the strength of the causal connection between variable j and variable i.[2]

Neither the graphs nor the equations are dispensable, unless we abandon the symmetry of the equal sign and rule out functionally equivalent sets of equations as causally adequate. This can be done implicitly by adopting a rule: *effects on the left; causes on the right*. While Woodward does not adopt such a convention, both Cartwright (2007, p. 13) and Hoover (2001, p. 40) explicitly convert the symmetrical equal sign into an assignment operator: '$c^{=}$' for Cartwright; '$\Leftarrow$' for Hoover. Thus, rewriting (16.1) and (16.2) as

[2] Woodward restricts attention to *acyclical* or *recursive* systems in which there is no mutual causation and no causal chains that turn back on themselves.

$$C \Leftarrow \alpha_{CA} A + \alpha_{CB} B, \qquad (16.1')$$

$$D \Leftarrow \alpha_{DC} C + \alpha_{DB} B, \qquad (16.2')$$

in effect combines Figure 16.1 with equations (16.1′) and (16.2′).

Causal arrows indicate direct causes, a key concept in Woodward's account:

(**DC**) A necessary and sufficient condition for *X* to be a direct cause of *Y* with respect to some variable set **V** is that there be a possible intervention on *X* that will change *Y* (or the probability distribution of *Y*) when all other variables in **V** besides *X* and *Y* are held fixed at some value by interventions. (Woodward 2003, p. 55)

Woodward (2003, p. 98) *intervention variable* (*I*) for a variable *X* with respect to a variable *Y* is defined according to four criteria:

1. *I* causes *X*;
2. *I* acts as a switch so that when it takes the right values it can eliminate the effect of all other variables in determining *X*;
3. any causal path from *I* to *Y* goes through *X*;
4. *I* is independent of any variable *Z* that causes *Y* otherwise than through *X*.

An *intervention* is defined as a token realization of an intervention variable that is an actual cause of the value of *X*.

Counterfactuals play a role at two key points in Woodward's definition of direct cause. First, the definition relies on counterfactuals in that it is enough that the contemplated interventions are possible; he does not require them to be actual. The exact modality captured in a *possible* intervention (equivalently, possible manipulation) is an open question. On the one hand, Woodward rejects as potential causes variables for which we have no notion of manipulation, as well as variables, such as race, sex, or species, for which a change would threaten the fundamental identity of the subject to which the variable is attached (Woodward 2003, p. 113; cf. Hoover 2009a). On the other hand, Woodward rejects the notions that manipulations must be the result of human agency (naturally occurring 'interventions' will suffice) or that they are necessarily practically possible (the moon causes the tides, but how can we practically manipulate the moon in the right sort of way?) (Woodward 2003, p. 113).

The second point at which counterfactuals play an essential role is in the notion that the causal relationship is to be evaluated in isolation by holding other variables fixed. There may, in fact, be no way actually to achieve such holding fixed and, like Lewis, Woodward is willing to countenance the semantic device of 'small miracles' to achieve the necessary isolation (Lewis 1973, p. 560; Woodward 2003, pp. 132, 136). And like Lewis, Woodward evaluates the counterfactual manipulation not in the actual world or, more accurately perhaps, in the actual causal graph, but in one that is different, though derived from it.

The semantic content of the assertion of Figure 16.1 that *C* is a direct cause of *D* is captured in the counterfactual experiment of manipulating some of its variables. Following Pearl, Woodward suggests that we consider an intervention that sets variables other than *C* and *D* to token values – in effect, 'breaking' (or 'wiping out') the causal connections between variables wherever needed to achieve this. Thus Figure 16.1 would be replaced by Figure 16.2 in which the lower-case letters indicate token values for the correlative upper-case variables and in which the causal arrows into *C* are removed. *C* causes *D*, then, if a change in *C*, say, from *c* to *c'* results in a change in *D*, say, from *d* to *d'*. The truth of this counterfactual justifies the direction of the causal arrow from *C* to *D* in Figure 16.1.

Although establishing direct cause relies on the evaluation of a counterfactual, Woodward's account, unlike Lewis's, is not reductive. Manipulation is an admittedly causal relationship. Rather than explaining causation in terms of some more basic notion, Woodward explains the causal structure of one part of a network of variables in terms of the causal structure of other parts. While such a non-reductive account may be metaphysically unsatisfying to those unwilling to take causation as a primitive, it is very much in keeping with Cartwright's (1989, ch. 2) slogan, 'no causes in, no causes out', and provides a framework for a causal epistemology, which explains its appeal to philosophers of science.

How one is to evaluate the truth value of counterfactuals remains an issue. Woodward rejects Lewis's appeal to universal natural laws. In the end, he grounds the evaluation of counterfactuals in the empirical fact of invariance. The invariant connection of the manipulated cause to the effect, under the conditions set out in the definition of direct cause (**DC**) is relied upon to translate the causal map given in the graphs and their associated functions into more complex counterfactual assessments that constitute Hume's useful causal knowledge. Invariance, in Woodward's view, is not absolute but admits of degrees. A relationship may be invariant to some sorts of interventions and not to others (Woodward 2003, ch. 6, section 6.4). And, in general, Woodward

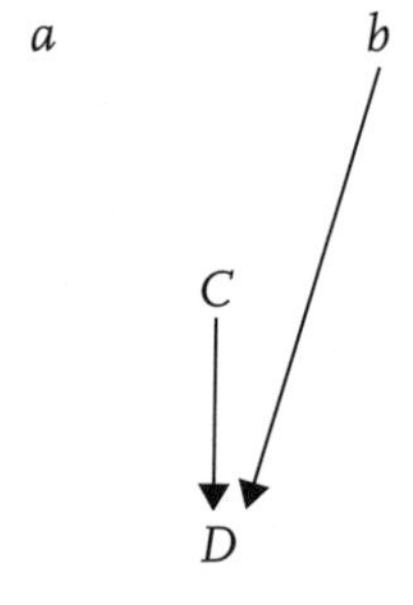

Fig. 16.2

stresses that causal knowledge and the assessments of counterfactuals are deeply contextual and that causal explanation is contrastive.

Woodward's strategy of defining direct cause through a process of counterfactual manipulation and then reconstructing causal networks out of the pieces requires, he believes, that causal relationships possess a kind of autonomy that he calls *modularity*:

> a system of equations will be modular if it is possible to disrupt or replace (the relationships represented by) any one of the equations in the system by means of an intervention on (the magnitude corresponding to) the dependent variable in that equation, without disrupting any of the other equations. (Woodward 2003, p. 48)

The actual systems that, for example, a scientist works with may or not be modular; nonetheless, Woodward maintains

> that when causal relationships are correctly and fully represented by systems of equations, each equation will correspond to a distinct causal mechanism and that the equation system will be modular. (Woodward 2003, p. 49)

Modularity, and whether it is essential to causal relationships, is a major point of dispute between Woodward and Cartwright (see Cartwright 2007, chs. 7, 8; Hausman and Woodward 1999, 2004; cf. Hoover 2009a).

16.3 The structural account

While it is an alternative to Woodward's manipulation account of causation, the structural account bears a family resemblance to it and to the related graph-theoretic analyses of Spirtes *et al.* and Pearl. Its pedigree, however, can be traced back principally to J.L. Mackie's (1980, ch. 3) INUS analysis of causation and to Simon's (1953) analysis of causal order in econometrics (see Hoover 2001, ch. 2). Our present focus is on Simon.

16.3.1 Simon on causal order

Simon (1953) proposes a syntax for representing causal relationships that suits the structural account very well. Consider the representation of a causal structure in a system of equations, such as (16.1) and (16.2) with the addition of

$$A = \alpha_A, \tag{16.3}$$

and

$$B = \alpha_B. \tag{16.4}$$

These equations, in which A and B are set equal to parameters, complete the system of equations, so that once the parameters have been assigned values, the system can be solved. Call the system (16.1)–(16.4) S. In S, we can solve for the value of A from equation (16.3) alone without knowledge of the parameters of the other equations, and we can solve for the value of B from equation (16.4) alone. Each is a *minimal complete subsystem* of S; call them S_A and S_B. Similarly equations (16.1)–(16.3) form a complete subsystem (S_C) in which we can solve for A, B, and C. It is minimal for C, though not for A and B. Equations (16.1)–(16.4) form a complete subsystem ($S_D = S$) that is minimal for D. For Simon, causal order is about the hierarchical relationships of minimal complete subsystems. A and B cause C because S_A and S_B are subsystems of S_C (written $S_A \subset S_C$ and $S_B \subset S_C$). C causes D because $S_C \subset S_D$. A is an *indirect cause* of D because A causes C and C causes D *and* knowing the value of C allows us to dispense with knowledge of the parameters that determine A in solving for D. B is both a direct and indirect cause of D because, as with A, there is a chain of causation running through C; but, in contrast to A, knowledge of C is not enough to allow us to dispense with knowledge of the parameters of B in solving for D.

Simon recognizes that his syntactic approach is inadequate on its own because structures of equations can be written in equivalent forms that syntactically yield different systems of minimally complete subsystems and, therefore, different causal orderings. For example, let system S' consist of equations (16.2), (16.4) and

$$A = \beta_A + \beta_{AB} B + \beta_{AC} C, \tag{16.5}$$

where

$$\beta_A = \frac{\alpha_A}{1-\alpha_{CA}}, \quad \beta_{AB} = \frac{\alpha_{CB}}{1-\alpha_{CA}}, \quad \text{and} \quad \beta_{AC} = -\frac{1}{1-\alpha_{CA}};$$

and

$$C = \beta_C + \beta_{CB} B, \tag{16.6}$$

where $\beta_C = \alpha_A \alpha_{CA}$ and $\beta_{CB} = \alpha_{CB}$.

By construction, systems S and S' have identical solutions; yet by Simon's syntactic criteria the causal ordering of S' is represented by Figure 16.3 – substantially different from the causal ordering of S in Figure 16.1.

Simon's solution to this problem of observational equivalence is to provide a semantic account of causal order (Simon 1953, pp. 24–26; 1955, p. 194). Parameters are not, on this view, fixed constants, but precisely the things that are altered by interventions or manipulations. If the α-parameterization of S were the true one, then any one of its parameters can be set to a new value without affecting any of the other parameters in the system. However, the β-parameters of S' must change in order to maintain the common solution. It works both ways, if the β-parameterization of S' were the true one, then the

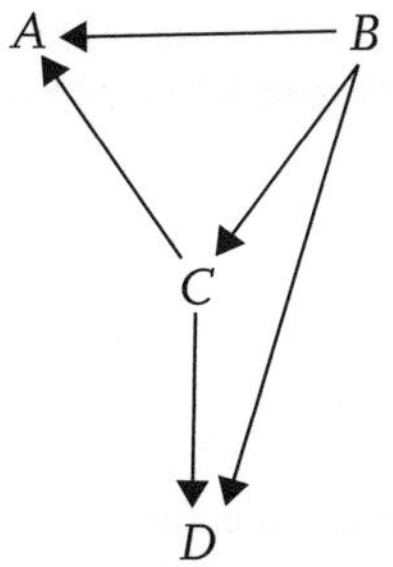

Fig. 16.3

α-parameters of *S* would have to change in the face of a change in one of the β-parameters.

The true causal order, then, is one that allows mutually unconstrained interventions among its parameters or, to put it another way, any change to the variables of the causal system leaves the parameterization and, therefore, the functional form of the remaining causal relations invariant. Invariance of the functional forms in the face of specific interventions is, on this view, the hallmark of a true causal representation; while failure of invariance is a key symptom of causal misrepresentation.

We can restate Simon's semantics by defining a parameter to be one of a set of variation-free variables that represent the scope for interventions in the causal system, where *variation-free* means that the choice of any particular value for a variable does not constrain the admissible choices of values for other variables in the set.

Cartwright objects to this characterization of a parameterization as a set of variation-free variables that govern the values of variables that are constrained by the causal structure: 'this is not generally the distinction intended between the parameters and the variables' (Cartwright 2007, p. 241). But I submit that defining parameter in this way is consistent with ordinary usage. The *Oxford American Dictionary* defines parameter as 'a variable quantity or quality that restricts or gives particular form to the thing it characterizes.' Simon treats parameters as capable of taking different values and uses the parameterization to define the causal order – the form of the causal order of a system of variables.

Cartwright also suggests that this interpretation of Simon's characterization of causal order is incorrect (Cartwright 2007, ch. 13, 14). The best rejoinder is to quote Simon at length:

The causal relationships have operational meaning, then, to the extent that particular alterations or 'interventions' in the structure can be associated with specific complete subsets of equations. We can picture the situation, perhaps somewhat metaphorically, as follows. We suppose a group of persons whom we shall call 'experimenters'. If we like, we may consider 'nature' to be a member of the group. The experimenters, severally or separately, are able to choose the nonzero elements of the coefficient matrix of a linear structure, but they may not replace zero elements by nonzero elements or vice versa (i.e. they are restricted to a specified linear model). We may say that they *control directly* the values of the nonzero coefficients. Once the matrix is specified, the values of the *n* variables in the *n* linear equations of the structure are uniquely determined. Hence, the experimenters *control indirectly* the values of these variables. The causal ordering specifies which variables will be affected by intervention at a particular point (a complete subset) of the structure. (Simon 1953, p. 26)

What Simon refers to as 'coefficients' subject to direct control – that is, able to be freely chosen by the 'experimenters' – is exactly what we call parameters.[3] And a specific change in a parameter is the manner in which an intervention (or Woodward's manipulation) is implemented. We can think of a causal system as a machine whose various operating characteristics are the variables which are controlled indirectly by selecting the settings for various switches and dials.

Significantly, Woodward (2003, p. 96) uses the analogy of switches as a means of explaining the breaking or wiping out of causal arrows involved in intervention. The operation of a switch is not analogous to the wiping out of a causal relationship. (This is perhaps more obvious with respect to dials that allow the setting of a continuously variable quantity. Not for the first time causal analysis is misled by philosophers' penchant for 0/1 or on/off variation.)[4] Flipping a switch does not break a causal system; it operates it. Interventions that change the values of parameters maintain the topology of the causal relationships among variables, calling for variables to take different values but not altering the causal graph itself.

While Simon's conception of causal order in one sense rejects Woodward's approach (causal order is not best understood through the comparison of a causal system to a topologically different system), in another sense it generalizes it. For the parameters represent the scope of possible interventions in a causal system, so that the connection between interventions and outcomes for variables is clear. Simon's conception shifts the focus away from specific token interventions to parameters that can take a variety of values. These are types,

[3] In light of the fact that Cartwright sees 'direct control' as highly restrictive notion and the possibility that her view arises partly from an assumption that direct control requires human agency, it is worth reiterating that Simon counts 'nature' as among the 'experimenters' who can exercise direct control (Cartwright 2007, p. 252, fn. 27; also p. 205).

[4] And, indeed, Simon is not free of the potential confusion; for he refers to the elimination of a causal linkage as the setting of a coefficient to zero; but there is a difference between a parameter having no value and having a range of admissible values which happens to include zero. Causal analysts frequently – and no doubt inadvertently – equivocate on the meaning of zero, failing to distinguish these two cases. See Hoover (2001, p. 45, esp. fn. 13). In the cited passage, Woodward contrasts switches and dials: switches break causal connections, while dials modulate the strength of causal connections. The structural account sees this as a distinction without a difference.

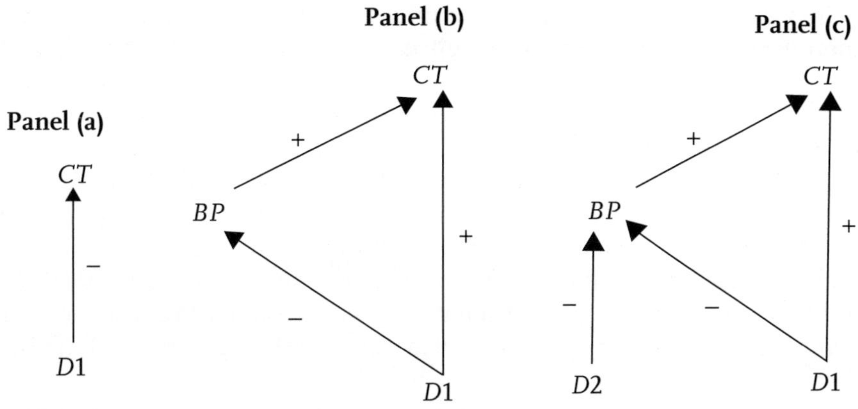

Fig. 16.4

which can instantiate a variety of token manipulations. The causal structure is defined entirely at the type level.

The structural account takes the minimal causal connection between two variables as primitive, offering no deeper account. The nature even of such a primitive causal connection must be understood counterfactually. If a cause has a certain effect in the right circumstances, then we cannot sensibly assert that it has that effect when those circumstances are not actual but not when they are actual. If diamonds scratch glass, the property cannot hold only when a diamond is not actually used to scratch glass. This is the sense in which causal relationships are naturally connected to invariance and it captures the meaning of what it is for a cause to be necessary in the circumstances for an effect.

Primitive causal connections reflect the *natures* or *capacities* (to use Cartwright's 1989 preferred term) of the causes. And capacities must be understood as dispositional, subject to a counterfactual analysis (see Mackie 1973, ch. 4). Cartwright analyses capacities as dispositions that are carried from context to context; yet they do not have to express themselves in every context (Cartwright 1989, pp. 3, 146–147, 191, *passim*). It is no failure of an account in terms of capacities that contextual details matter substantially in whether capacities are actualized. And a capacity account in no way presupposes modularity, which is a sort of independence from context.

While direct causal connections are primitive, they are also relative to the representation or model and may or may not be brute facts. For example, we might imagine that a drug (*D*1) is found experimentally to reduce coronary thrombosis (*CT*). The relationship may be modelled as in Figure 16.4a. (Plus or minus signs next to causal arrows indicate whether the causal influence promotes or inhibits the effect.)

It is not inconsistent with such a model that further research supports a more complex model (Figure 16.4b) in which *D*1 causes *CT* directly and, in fact, promotes coronary thrombosis, while it also reduces blood pressure (*BP*), while high blood pressure promotes thrombosis; the net affect being to reduce thrombosis. Sometimes a model such as Figure 16.4b is taken to imply that the model in Figure 16.4a is defective. A better interpretation is that the models operate in different contexts. If there were no known means of intervening independently on blood pressure or in such a way as would meliorate the adverse direct effect of *D*1, then the model in Figure 16.4a would be a perfectly fine model and a guide to clinical practice. The model in Figure 16.4b would simply explain the mechanism through which *D*1 inhibits thrombosis.

An advantage of the model in Figure 16.4b, however, is that, in articulating the mechanism of operation, it may suggest paths toward better outcomes. For example, knowing that the positive effect of *D*1 operates through *BP* suggests seeking another drug (say, *D*2) that would reduce blood pressure with no direct effect on coronary thrombosis (Figure 16.4c). If research successfully produced such a drug, a better clinical practice might be to administer *D*2 and omit *D*1. There is a sense in which the causal arrow in Figure 16.4a captures a fact that is primitive relative to the model, but which is not brute, in that it has a more complex explanation in a finer grained model. There is no guarantee that primitive causes can be explained through further refinements, though that is often the object of research.

16.3.2 Modularity and difference-making

The structural account agrees with Woodward (2003, e.g., p. 80) that causes are difference makers, yet it marks the difference that they make relative to an intact causal structure, not some related, but different, structure constructed by manipulations that, in effect, break the system. The difference between our approach and Woodward's becomes important with respect to modularity. Where Woodward sees modularity as a fundamental element of a well-articulated causal system, the structural approach does not require modularity at all – a distinct advantage, since many intuitively causal systems are decidedly nonmodular.

Different notions of intervention also distinguish the structural account from Woodward's manipulation account. A parameter can be thought of as a causal variable, so there is no fundamental difference with Woodward's criterion 1 for an intervention variable, cited in Section 16.2: *I* causes *X*. A significant difference arises with Woodward's criterion 2: *I* acts as a switch so that when it takes the right values it can eliminate the effect of all other variables in determining *X*. As already noted, while parameters may well act as switches or dials (i.e. instruments of continuous variation rather than simply on or off), they need not shut off the effects of other causes. The critical feature is not the breaking or wiping out but the accounting for the effects of other causes. Nor does the structural account accept criterion 3: any causal path from I to Y goes through X. This criterion is closely related to modularity, and the structural account does not require modularity. A parameter may have a direct effect on Y as well as an indirect effect on Y though X without undermining a clear causal ordering of X and Y. The case in the next subsection illustrates exactly this situation. The structural account suggests a different interpretation of criterion 3. When it is fulfilled for parameters (I), then we find ourselves in a particularly fortunate position to infer causal direction through an intervention. We should not, however, confuse the epistemic issue of how and when causes are inferable from data with the conceptual issue of what it means to be a cause or with the question of how to represent causal order. The structural account does not accept Woodward's criterion 4: I is independent of any variable Z that causes Y otherwise than through X. If I is interpreted as a parameter as we have defined it, then it is required to be independent only of other parameters and not of all other variables. This is the requirement that parameters be variation-free.

An intervention for the structural account is a token realization of a parameter in the same way that an intervention for Woodward is a token realization of an intervention variable. But unlike Woodward, the type-relations among the variables and the parameters fully determine the causal order without reference to a particular token intervention or manipulation. In relying on token interventions in the definition of direct cause, Woodward again seems to confuse the causal relationship with a strategy that supports the inference of causal relationship.

We can illustrate the issue with a typical macroeconomic model.[5] The demand for real money balances $(m - p)$ is given by

$$m_t - p_t = \delta - \alpha({}_t p^e_{t+1} - p_t) + \nu_t, \tag{16.7}$$

where the subscripts t indicate time periods; m is the logarithm of money; p, the logarithm of the price level; ${}_t p^e_{t+1}$, the expectation at time t of the price level at time $t + 1$; ν, an independent random error; and α and δ are parameters. The central bank's money supply rule is

$$m_{t+1} = \lambda + m_t + \varepsilon_t, \tag{16.8}$$

where λ is a parameter that governs the growth rate of money and ε is an independent random error. Expectations are formed rationally:

$${}_t p^e_{t+1} = E(p_{t+1} | \Xi_t), \tag{16.9}$$

which says that the expectation of the price level is the mathematical expectation of actual prices conditional on all the information available at time t, including the model itself Ξ_t. Janssen (1993, pp. 137–139) argues persuasively

[5] This model is adapted from Hamilton (1995, pp. 326–332).

that 'rational expectations' are not expectations at all, but a consistency criterion or solution concept analogous to the condition that markets clear – that is, that prices are assumed to take the values at which supply equals demand (see also Hoover 2009b). On that assumption, ${}_tp^e_{t+1}$, is not a proper variable – or at least not a causally efficacious one – but an instrument for imposing a certain nonlinear restriction on the parameters of the model.

On that interpretation, the causally relevant solution to the model is given by (16.9) and

$$p_t = m_t - \delta + \alpha\lambda + \nu_t{}^{6}. \tag{16.10}$$

The model is nonlinear in parameters, and it is not modular. The appearance of the multiplicative coefficient $\alpha\lambda$ in (16.11) results from the imposition of rational expectations – λ appears only because it is a parameter of the money-supply rule (16.9). Thus, it is impossible to perform Woodward's manipulation test of whether money causes prices, which calls for setting m_t in (16.9) to a fixed value come what may and, in effect, wiping out any causal arrows, say, from m_t or ν_t to m_{t+1}; for that would remove the basis for the parameterization of (16.11). The causal structure reflected in (16.11) cannot survive such a breaking of the system.

Nevertheless, applying the definition of direct cause from the structural account tells us unequivocally that m_t causes p_t. Rather than calling for miraculous token manipulations, direct cause is implied by the constraints on the variables determined by the parameterization.

The model illustrates another important point. Consider a change in the central bank's money-supply rule – for example, an increase in the growth rate of money indicated by a higher value for λ. As equation (16.11) indicates such an intervention would alter the coefficient $\alpha\lambda$ in (16.11). As a result a statistical test of the relationship of money to prices based on (16.11) would be non-invariant. Woodward and others have treated invariance under manipulation of causes as the hallmark of a causal relationship, and they would, therefore, be tempted to reject the causal status of (16.11) (Woodward 2003, pp. 15–16, ch. 6). What the example actually shows is that a more subtle approach to invariance is necessary.

Consider an intervention that changes prices through some other instrument than money; for example, consider an intervention that changes δ, then naturally (16.11) is non-invariant. But (16.9) is invariant. And this is a general rule in non-modular systems with one-way causes: the causal structure that determines a cause of an effect is invariant to interventions that alter the effect through some other mechanism than the cause in question; while the causal structure that connects a cause to an effect is *not* generally invariant to interventions that alter the cause of the effect in question (Hoover 2001, ch. 8; Cartwright 2007, pp. 99, 105). In fact, the differential invariance is

[6] See Hoover (2001, pp. 64–65) for the derivation of the solution to a slightly more general version of this model.

diagnostic of causal direction in non-modular systems; and invariance in both directions is a test of modularity (see Hoover 2001, ch. 8, section 8.1). The more subtle analysis of invariance nonetheless supports Woodward's view that one can alter the effect by manipulating the cause; one cannot alter the cause by manipulating the effect.

16.3.3 Counterfactual analysis

Lewis explains causal relations by an appeal to counterfactuals and evaluates the truth of counterfactuals through a comparison of possible worlds. Mackie (1973, ch. 3) rejects the notion that counterfactuals *per se* have a truth value. He interprets them as enthymemetic or disguised arguments, which can be evaluated for validity once their structures are articulated and, additionally, for soundness once the truth of their premises is established or, at least, accepted 'for the sake of argument'.

The structural account of causation rejects Lewis's account of the counterfactual basis for causal relationships, but is compatible with Mackie's account. A causal model can be interpreted as a map of possible worlds. Unlike the possible worlds in Lewis's account of counterfactuals, causal models are precise about what changes are possible and what implications they have for the variables in the model. Consequently, they provide an instrument for the construction and articulation of the kind of arguments that Mackie sees as underwriting counterfactuals, and they avoid the hopeless ambiguities of Lewis's metric for the closeness of possible worlds.

Naturally, we cannot avoid the question of the adequacy of causal models as representations of the real world or the need to choose among competing causal models. But these are the ordinary epistemological problems faced in scientific inference. They may be practically difficult and subject to other philosophical reservations; but, once we are satisfied that they have been dealt with adequately, counterfactual analysis itself is not additionally problematic.[7]

16.4 Counterfactuals and policy analysis

16.4.1 Impostor counterfactuals

In her recent book, Cartwright (2007, ch. 16) makes a case against the typical uses of counterfactual analysis in economics. Her theme is reflected in her title, *Hunting Causes and Using Them*; she suggests that the techniques appropriate to hunting causes are not the ones appropriate to using them in counterfactual analysis. The title, however, masks different levels of issues.

[7] Spirtes (2001) provides an extensive account of the using of information encoded in conditional independence relationships of variables as a means of establishing facts relevant to causal inference. Hoover (2001, chs. 8–10) provides both a methodological account and case studies of causal inference based on interventions of a type that induces structural change in causal systems.

It seems off the mark. In the well known story, the recipe for jugged hare (usually wrongly attributed to the famous cookbooks of Mrs. Beeton or Mrs. Glasse) begins 'first, catch your hare...' In an obvious sense, one surely needs to hunt causes (that is, to establish the existence of causal connections empirically) before one can use them for anything.

Cartwright's real concern is with what she calls 'imposter counterfactuals' – that is, cases in which the counterfactual that received empirical warrant is not the one that would appropriately warrant a counterfactual policy analysis. One of Cartwright's objections is simply an extension of her skepticism of modularity. Treated as an empirical strategy, Woodward's manipulation test should reveal the causal relationship if the causal structure is, in fact, modular. Cartwright does not deny that, but stigmatizes such systems as 'epistemically convenient', suggesting that such convenience is necessarily rare.

She also objects to the breaking or wiping out of causal connections as part of Woodward's and Pearl's approach to evaluating a causal relationship on the ground that it is the intact system, not the broken system, that it is needed to evaluate policy counterfactuals. The difficulty is that when implementing a policy, we may in fact not be able to manipulate a cause independently of other causes, so that the counterfactual that Woodward or Pearl seeks to evaluate is simply not a counterfactual that the policy analyst can rely on.

This objection connects to the distinction that Cartwright draws between implementation-neutral and implementation-specific counterfactuals. An *implementation-neutral* counterfactual is one that implies the same effect no matter what means are used to bring about the causal antecedent. An *implementation-specific* counterfactual is one in which the effect is sensitive to the manner in which the causal antecedent is brought about. For example, in the causal structure connecting high blood pressure to coronary thrombosis in Figure 16.4c, the counterfactual question, 'how much would a reduction in blood pressure reduce thrombosis?,' is not well posed because the counterfactual is not implementation-neutral. A reduction of blood pressure to a particular level using drug $D2$ will be more effective than one using drug $D1$, since $D1$ has a direct promoting affect on thrombosis independent of its indirect inhibiting effect operating through its effect on blood pressure.

The structural account of causation suggests that implementation-specific counterfactuals are the rule, in large measure because causal complexity and failures of modularity are the rule. What Cartwright calls implementation-neutral policies are merely policies that benefit from the special features of some causal structures that render them *robust* to the mode of implementation. In such structures, a variety of modes of determining a cause produce the same effect. Such robustness is practically useful in many cases and may be sought for that reason, but there is no reason to connect it to the existence or non-existence of a causal relationship in general. In fact, not infrequently a lack of such robustness is desirable. Pushing up on the plastic tab causes the cap to the medicine bottle to come off, but *only when* the plastic tab is aligned with the arrow on the bottle. The lack of robustness contributes to child safety.

Cartwright (2007, p. 254) assumes that we should prefer implementation-neutral policies. Yet, such a preference is not obvious. And, even if we did prefer them, we would need a more detailed, accurate causal representation to be sure that the policies were *in fact* implementation-neutral. For example, a simple model of the relationship of blood pressure to coronary thrombosis, $BP \rightarrow CT$, might appear to be implementation-neutral; but, if Figure 16.4.c, truly represents the causal structure, whether a fall in blood pressure is brought about by drug $D1$ or $D2$ matters, the policy is implementation-specific, and we have made a mistake.

16.4.2 An illustration from monetary economics

We can illustrate some of the key issues and how the structural account deals with them using the model in equations (16.8)–(16.11).[8] A monetary regime is defined by the parameterization of the central bank's money supply rule (16.9), so that any change in the parameter λ represents a new regime. Imagine that the model (16.8)–(16.11) is, in fact, a true representation of the economy; it is, what econometricians sometimes call the *data-generating process*. Equations (16.9) and (16.11), then describe the actual dynamic process of governing the evolution of money and prices. Of course, economists do not know the data-generating process a priori. A central problem for econometrics is *identification*: how can we recover the parameters of the data-generating process (or of some, close enough approximation) based on observations of the variables (here, of m and p)?

Many macroeconomists estimate so-called *structural vector autoregressions*. Most of the details are not important here, but a few are worth noting. The structural vector autoregression technique gives up on learning about all of the parameters of the dynamic process, focusing instead only on those that relate the contemporaneous values of variables to each other, letting the relationships of lagged to contemporaneous variables be summarized by coefficients that may themselves be difficult-to-disentangle functions of the parameters of the data-generating process. Estimates of these contemporaneous parameters are obtained under a maintained assumption about the causal order of the variables. For example, if we assume (correctly) that in the data-generating process to hand, m_t causes p_t, we would be able to recover good estimates of the true parameters. But most economists make the necessary assumptions

[8] Equations (16.8)–(16.11) represent a simplified version of an actual econometric model used to conduct counterfactual analyses of US monetary policy (see Hoover and Jordá 2001).

about causal ordering on the basis of a priori guesswork, so that considerable doubt hangs about their estimates. This is an area in which the graph-theoretic (or Bayes net) inferential techniques pioneered by Spirtes *et al.* (2001) and first applied to the problem of structural vector autoregressions by Swanson and Granger (1997) have considerable power.[9]

Typically, once an economist has estimated a structural vector autoregression, it is used to evaluate particular counterfactual questions: for example, treating (16.9) and (16.11) as the structural vector autoregression, we might ask, 'what would be the path of prices (p) if money (m) were increased for a single period?' Such a one-period increase is referred to as a 'monetary shock' (or 'impulse') and the path of prices is referred to as an 'impulse-response function'. The shock is typically administered by setting the error term (here ε_t) to a positive value for a single period.

The effect of the shock is nonetheless the same as Woodward's or Pearl's experiments forcing a variable to take a particular value come what may. Typically, the implied breaking of causal relationships is restricted to the current period, so that future values of the money are not fixed but allowed to develop in line with the causal structure. However, one could in principle offer a series of shocks that had the effect of fixing money at every future time period. That this is not often done is partly pragmatic: the own dynamics of the shocked variable are independently interesting, so economists prefer not to suppress them.

It is also partly the result of the *Lucas critique*, the fact that in a model with rational expectations, the dynamics are not invariant with respect to changes in the policy rule.[10] The Lucas critique is exemplified in the point previously made that the coefficient $\alpha\lambda$ shifts with changes in the monetary-policy rule (the setting of the monetary growth rate, λ), so that to know the path of prices (p), we need to know not only the value of m_t but also how m_t was brought about – that is, the value of λ. Impulse–response analysis is sometimes thought to circumvent the Lucas critique. The idea is that a shock to the random-error term in (16.9) leaves the parameters untouched and, therefore, does not induce any failure of invariance in (16.9) or (16.11).

Impulse–response analysis provides a good example of what Cartwright criticizes as impostor counterfactuals. It is used to say something about the effects of monetary policy – for example, what would happen if the Federal Reserve raised the money supply by 1 percent? – and it tries to answer that question by treating the Federal Reserve's action as a shock to a stable system. The impulse–response function *does* answer a well-posed counterfactual ques-

[9] See Demiralp and Hoover (2003), Hoover (2005), Demiralp, Hoover, and Perez (2008), and Hoover, Demiralp, and Perez (2009) for further development, evaluation, and applications of these techniques to economic problems.

[10] Lucas (1976); Hoover (1988, ch. 8, section 8.3; 2001, ch. 7, section 7.4).

tion, but not the one for which we want an answer. The question it actually poses is, what if the money supply were to rise unexpectedly and arbitrarily, say, by 1 percent above its dynamic path and then fall back the next period by the same amount? It truly asks what would happen if there were a shock to the system. But monetary policy is not delivered by shocks. The increases in the money stock that the Federal Reserve typically delivers are reactions to economic conditions aimed at desired goals for economic variables. Furthermore, while the impulse–response function can trace out the effects of a random shock, it cannot trace out the effect of a change in systematic policy, since to deliver a series of shocks in one direction, for example, in order to force money to evolve along a desired path represents a violation of the randomness assumptions that govern the underlying representation of the error term (16.9). It is not that such systematic policy could not be analysed; it is that it cannot be analysed while assuming that the causal structure of the model, which includes the model of the random errors, is constant and – at one and the same time – changes.

In a discussion of the Great Depression, the economist Christopher Sims (1999) recognizes that random shocks to error terms could not adequately address the counterfactual question, what if monetary policy in the 1930s had adopted the rules that characterized it in the 1990s? He obtained econometric estimates of a model more complex than, but similar to, (16.8)–(16.11) for both the 1930s and the 1990s. To evaluate the counterfactual, he, in effect, replaced the monetary-policy rule (16.9) for the 1930s by the one that characterized the 1990s. For our purposes, think of this as simply changing λ to a new value – while holding the estimated value of $\alpha\lambda$ and the other parameters of (16.11) constant at values appropriate to the 1930s. The counterfactual is evaluated by setting m and p in the first period to the values that they actually took at the onset of the Great Depression and then feeding the random error terms from the original estimates into the model with the alternative monetary-policy rule. (Notice that if this procedure were undertaken with the original monetary-policy rule, it would necessarily have simply generated the actual path for money and prices over the Great Depression.) With such counterfactual estimates, Sims felt free to discuss whether modern central bankers would have produced better outcomes.

The ordinary impulse–response analysis was a true impostor counterfactual, in that it answered a counterfactual question, but the wrong one. In contrast, Sims's counterfactual experiment is simply incoherent. If, as he holds, the existence of rational expectations subjects the model to the Lucas critique, then one cannot simply substitute one monetary-policy rule for another. The incoherence is displayed in simultaneously assuming that λ may change while α and $\alpha\lambda$ remain constant. The difficulty is the non-modularity of the causal structure. Were the causal structure modular, as it might be if rational expectations were not an element of the data-generating process, then Sims's method would be coherent.

Hoover and Jordá (2001) offer a different counterfactual analysis. Rather than replacing the monetary rule of the 1930s by that of the 1990s, they simply transfer the entire model of the 1990s to the 1930s by setting the initial values of the 1990s model to their values at the onset of the Great Depression and then feeding the estimated random error terms from the 1930s model into the 1990s model. This procedure makes sense if the elements of the model other than the particular shocks and the monetary-policy rule have not changed between the two time periods. The counterfactual question that it answers is well-posed and not an impostor. Essentially, the procedure is that same as if we had changed λ in the estimated 1930s model and, unlike Sims, allowed $\alpha\lambda$ to take a new value, holding α constant. The central message of Hoover and Jordá's approach is that one should not ignore non-modularity but account for it. Accounting for it raises difficult, but not necessarily insuperable, inferential problems, which – fortunately – are not our direct concern here.

16.4.3 Internal and external validity

Holding α constant is the emblem in our expository model for the constancy of the rest of the structure of the model between the two periods. The assumption that we are justified in doing so is by no means automatic and raises the classic question of *internal* versus *external validity*. The issue is whether a causal relationship (indeed, empirical relationships of other kinds as well) undercovered in a specific context can be transferred and assumed to hold in other contexts. Superficially, it might appear to recapitulate the distinction between implementation-specific and implementation-neutral counterfactuals. The implementation dichotomy comes down, first, to whether or not an empirically warranted causal structure supports the counterfactual; second, to whether the detail in the representation of that structure is fine enough to distinguish between alternative policy implementations; and, finally, to whether the target effects are, in fact, robust to different implementations. In contrast, external validity comes down, first, to whether there is homogeneity in the background conditions between implementations in the two situations; and, second, to the domain of possible interventions.[11]

[11] Hoover (2009a) frames the notion of background conditions in a manner consistent with Woodward's (2003, pp. 145–146) emphasis on contrastive focus using John Anderson's (1938/1962) notion of a causal field. The *causal field* is the set of standing conditions that, while they may themselves be causes, do not change relative in relation to our particular causal interests and, so, define the boundary conditions for a particular causal relationship. Causal relations may be evaluated differently in different causal fields. As a results, causal relationships may be represented or modeled in a variety of (ultimately non-contradictory) ways depending on our differing pragmatic aims.

In the case of the monetary-policy counterfactual, there is no implementation-neutral counterfactual possible: to know what effect a change in the money stock causes we must know how it comes about – an increase in m_t from a shock to ε_t has a different effect than one from a rise in λ. Yet, Hoover and Jordá's counterfactual analysis trades on external validity. In order that the 1990s may speak to the Great Depression, they assume that the actual history of the variables and the causal topology are the same in both periods. And they assume that the parameterization, except for the parameterization of the monetary-policy rule itself (the value of λ) is also the same.

Hoover and Jordá could easily be wrong: the conditions that underwrite external validity may fail. But that is an issue on which empirical evidence can be brought to bear. It is not a special problem for causal analysis but a more general problem for the import of empirical results derived in one set of circumstances (say, in a particular experiment) for other sets of circumstances. Our current concern is not with the problem of external validity but with the problem of using causes in situations in which the external validity of the causal model is not in question.

A good deal of Cartwright's skepticism about causal knowledge in economics and – one presumes – in other fields is apparently generated by a lack of sufficient respect for her own distinction between hunting and using. She writes as if the manner in which causes are hunted limits their possible uses. The structural account, however, clarifies that there are a variety of things we typically need to know to have a useful representation of causal structure. We need to know the causal topology – essentially the pattern of arrows connecting variables or, equivalently, the parameterization (for example, that the parameter space includes α and λ and not, say, $\theta = \alpha\lambda$). We also need to know the functional interrelationship of the parameters, including the manner of potential nonmodularity. And we need to know, for any real-world counterfactuals, the actual values of the parameters. The structural account tells us both what we need to know and where to slot such knowledge as we have obtained into the representation of causal structure.

Cartwright (2007, p. 9, *passim*) characterizes her position as 'causal pluralism.' The structural account aims at a high enough level of generality that any coherent approaches are nested within it. Yet, it is compatible with substantial methodological pluralism: different methods may supply different elements of the knowledge needed to fill in the causal structure. For example, Bayes net methods, as discussed earlier, are helpful in mapping the causal topology. But there are situations – well known to their advocates – in which they are not discriminating, and they do not directly address parameter values. Hoover (2001, ch. 8) offers methods that use patterns of invariance and noninvariance across regime changes that can sometimes resolve the equivalent causal topologies allowed by Bayes net methods. Hoover and Jordá (2001) demonstrate in an enriched version of the model (16.8)–(16.11) that knowledge of interventions in the monetary-policy rule (not even the fine details, but simply the timing of when they occur) may allow us to recover the functional relationships among parameters of nonmodular causal systems. Their approach is an empirical analogue to Woodward's use of manipulations in the evaluation of causal counterfactuals, although it does not involve breaking of causal arrows, but in the manner of Section 16.3 above, considers manipulations in a conserved causal topology.

These empirical methods, at various points, all involve untested assumptions – a number of which have been mentioned already with respect to Hoover and Jordá's counterfactual experiments. But then so does all empirical investigation. These assumptions may not be tested in a particular study, but they are not necessarily untestable or, at least, not necessarily beyond empirically based criticism. They are not, however, all jointly testable at the same time. Only a thoroughly destructive skeptic would be unwilling to make *some* assumptions that seem reasonable and reliable until there exist reasons to doubt their truth more compelling than the mere possibility that they could be false.

16.4.4 Epistemic opportunism

While Cartwright takes the fact that some methods work well only for modular, 'epistemically convenient' situations to be a significant drawback, from a practitioner's point of view it is surprising, but welcome, how often the real world seems to be convenient enough to make empirical progress with relatively simple methods. Cartwright is certainly correct that modularity is not a general property of causation, but it is common enough – to a reasonable approximation – that methods that require it are often practically effective. And where modularity fails, there are other methods, such as the methods based on invariance testing advocated by Hoover (2001) and methods built on similar principles used by Hoover and Jordá (2001). Some nuts have not yet been cracked; some perhaps never will be. Rather than decrying methods that require 'epistemic convenience' generally, it would be better to embrace *epistemic opportunism*: articulate causal models by any means necessary. The structural account gives us a systematic way to interpret what appropriate methods have accomplished.

Acknowledgements

I thank Richard Scheines, Julian Reiss, Clark Glymour, and two anonymous referees for comments on an earlier draft. This paper was written with the partial support of the U.S. National Science Foundation (grant no. NSF SES-1026983).

References

Anderson, John (1938). 'The problem of causality,' reprinted in *Studies in Empirical Philosophy*. Sydney: Angus Robertson, 1962.

Cartwright, Nancy (1989). *Nature's Capacities and Their Measurement*. Oxford: Clarendon Press.

Cartwright, Nancy (2007). *Hunting Causes and Using Them: Approaches in Philosophy and Economics*. Cambridge: Cambridge University Press.

Demiralp, Selva and Kevin D. Hoover (2003). 'Searching for the causal structure of a vector autoregression,' *Oxford Bulletin of Economics and Statistics* 65(supplement), 745–767.

Demiralp, Selva, Kevin D. Hoover and Stephen J. Perez (2008). 'A bootstrap method for identifying and evaluating a structural vector autoregression,' *Oxford Bulletin of Economics and Statistics* 70(4), 509–533.

Hamilton, James D. (1995). 'Rational expectations and the econometric consequences of changes in regime,' in Kevin D. Hoover, editor. *Macroeconometrics: Developments, Tensions, and Prospects*. Boston: Kluwer, pp. 325–245.

Hausman, Daniel M. and James Woodward (1999). 'Independence, invariance, and the causal Markov condition,' *British Journal for the Philosophy of Science*, 50(4), pp. 521–83.

Hausman, Daniel M. and James Woodward (2004). 'Modularity and the causal Markov condition: A restatement,' *British Journal for the Philosophy of Science*, 55, pp. 147–161.

Hoover, Kevin D. (1988). *The New Classical Macroeconomics: A Sceptical Inquiry*. Oxford: Blackwell.

Hoover, Kevin D. (1990). 'The logic of causal inference: Econometrics and the conditional analysis of causality,' *Economics and Philosophy* 6(2), 207–234.

Hoover, Kevin D. (1994). 'Econometrics as observation: The Lucas critique and the nature of econometric inference,' *Journal of Economic Methodology* 1(1), 65–80.

Hoover, Kevin D. (2001). *Causality in Macroeconomics*. Cambridge: Cambridge University Press.

Hoover, Kevin D. (2005). 'Automatic inference of the contemporaneous causal order of a system of equations,' *Econometric Theory*, 21(1), 69–77.

Hoover, Kevin D. (2009a). 'Identity, structure, and causation,' unpublished typescript, Duke University, downloadable from: http://www.econ.duke.edu/~kdh9/research.html.

Hoover, Kevin D. (2009b). 'Microfoundational programs,' unpublished typescript, Duke University, downloadable from: http://www.econ.duke.edu/~kdh9/research.html.

Hoover, Kevin D., Selva Demiralp, and Stephen J. Perez (2009). 'Empirical identification of the vector autoregression: The cause and effects of US. M2,' in Jennifer L. Castle and Neil N. Shephard, editors, *The Methodology and Practice of Econometrics: A Festschrift in Honour of David F. Hendry*. Oxford: Oxford University Press, 2009, pp. 37–58.

Hoover, Kevin D. and Oscar Jordá (2001). 'Measuring systematic monetary policy,' *Federal Reserve Bank of St. Louis Review*, 83(4), 113–137.

Hume, David (1739). *A Treatise of Human Nature*. Page numbers refer to the edition edited by L.A. Selby-Bigge. Oxford: Clarendon Press, 1888.

Hume, David (1754). 'Of Interest,' in *Essays: Moral, Political, and Literary*. Page references to the edition edited by Eugene F. Miller. Indianapolis: Liberty *Classics*, 1885.

Hume, David (1777). *An Enquiry Concerning Human Understanding*. Page numbers refer to L.A. Selby-Bigge, editor. *Enquiries Concerning Human Understanding and Concerning the Principles of Morals*, 2nd edition. Oxford: Clarendon Press, 1902.

Janssen, Maarten C.W. (1993). *Microfoundations: A Critical Inquiry*. London: Routledge.

Lewis, David (1973). 'Causation,' *Journal of Philosophy* 70, 556–567.

Lewis, David (1979). 'Counterfactual dependence and time's arrow,' *Noûs* 13(4), 455–476.

Lucas, Robert E., Jr. (1976). 'Econometric policy evaluation: A critique,' reprinted in Lucas, *Studies in Business Cycle Theory*. Oxford: Blackwell, pp. 104–130.

Mackie, John L. (1973). *Truth, Probability, and Paradox*. Oxford: Clarendon Press.

Mackie, John L. (1980). *The Cement of the Universe: A Study in Causation*, 2nd edition. Oxford: Clarendon Press.

Pearl, Judea (2000). *Causality: Models, Reasoning, and Inference*. Cambridge: Cambridge University Press.

Reichenbach, H. (1956). *The Direction of Time*, Berkeley: University of California Press.

Simon, Herbert A. (1953). 'Causal order and identifiability,' in Hood and Koopmans, editors. *Studies in Econometric Method*. New York: Wiley, chapter 3. Page numbers refer to the reprint in Simon (1957) *Models of Man*. New York: Wiley, pp. 10–36.

Simon, Herbert A. (1955). 'Causality and econometrics: Comment,' *Econometrica* 23(2), 193–195.

Sims, Christopher A. (1999). 'The role of interest rate policy in the generation and propagation of business cycles: What has changed since the '30s?' in Jeffrey C. Fuhrer and Scott Schuh, editors., *Beyond Shocks: What Causes Business Cycles*. Federal Reserve Bank of Boston Conference Series, No. 42. Boston: Federal Reserve Bank of Boston, pp. 121–60.

Spirtes, Peter, Clark Glymour, and Richard Scheines (2001). *Causation, Prediction, and Search*, 2nd edition. Cambridge, MA: MIT Press.

Swanson, Norman R. and Clive W.J. Granger (1997). 'Impulse–response functions based on a causal approach to residual orthogonalization in vector autoregressions,' *Journal of the American Statistical Association* 92(437), 357–67.

Woodward, James B. (2003). *Making Things Happen: A Theory of Causal Explanation*. Oxford: Oxford University Press.

·17·

The error term and its interpretation in structural models in econometrics

Damien Fennell

Abstract

This chapter explores what the error term represents in structural models in econometrics and the assumptions about the error terms that are used for successful statistical and causal inference. The error term is of particular interest because it acts as a cover-all term for parts of the system that are not fully known about and not explicitly modelled. The chapter attempts to bring some of the key assumptions imposed on the error term for different purposes (statistical and causal inference) and to ask to what extent the conditions imposed on the error term can be empirically tested in some way.

17.1 Introduction

Structural econometrics attempts the extremely difficult task of making causal inferences from non-experimental data. Its core approach, which in its modern form dates from the ground-breaking paper of Trygve Haavelmo (1944), is to postulate a statistical model that carries structural (or causal) content. The model may be postulated from theory, from observations and from other background knowledge. One then uses sample data to test its observable implications (to check it is adequacy) and to infer remaining unknown features (for example, to estimate parameters if the model is parametric).[1]

In a highly general form, we can denote a structural model in the following way. Denote the variables of interest to the econometrician as a vector of random variables Z, some of whose components (though not necessarily all) are observable. The probabilistic part of model postulates conditions on the joint probability distribution of Z. Structural content can be introduced in several ways. For example, some can be introduced with a partition of Z into *exogenous* variables, X, and *endogenous* variables, Y. Though not all concepts

[1] In practice the process of model selection and inference of parameters will tend be interlinked, for example, inferring certain parameter values (e.g. zero's) may lead one to simplify the model.

Table 17.1 A general characterization of a structural model

Probabilistic assumptions	$Z \sim D(Z)$ where the joint distribution $D(Z)$ is assumed to have certain properties (e.g. independence among certain variables, a certain distributional form, etc.)
Structural assumptions	$Z = (YX)$, where Y is a vector of endogenous variables, X is a vector of exogenous random variables (where the exogeneity concept has structural not just probabilistic content) Structural content assigned to probabilistic relations (e.g. conditional independencies). $G(Z, U) = 0$, i.e. a set of functional relations (with structural interpretation) that hold among the components of Z, where U denotes an unobserveable vector of 'error' terms.

of exogeneity are structural,[2] in structural models the exogeneity assumption often assumes (in some form) that endogenous variables are causally determined by other variables in the model while the exogenous variables are not. Further structural content can be introduced by interpreting conditional probabilistic independencies among the variables as indicative of causal relations[3] or assumed functional relations among the variables, where these functional relations carry some structural interpretation (as would typically be the case when the functional relations are derived from economic theory). Importantly, functional relations are also a way that error terms are introduced to models, since it is highly unlikely that our knowledge of functional relations will be so powerful as to permit us to claim that exact deterministic relations hold among (independently defined) random variables. Thus the functional relations will typically explicitly represent some omitted content using a vector of error terms, U, to sustain the deterministic relation postulated among the other random variables. In summary, the general structural model can be presented as shown in Table 17.1.

There is a lot of work in econometrics considering the problems of statistical inference – and to a lesser extent causal inference – for general models specified in a way similar to that above. For just two examples, see Hendry (1995) and Spanos (1999). As much of this work is highly technical, I believe there is space for a work that attempts to keep things simple, yet which nevertheless gives a taste of some of the difficult issues facing econometricians in causal inference. In this paper, therefore, I try to raise some of the important issues by looking at the error terms in simple, textbook models. Specifically, the chapter looks at the simplest linear models with errors-in-the-equations and, in keeping with their continued use in econometrics, it looks at simultaneous equations models. Of course, most structural equations models

[2] See, for example, weak exogeneity in Engle *et al.* (1983).

[3] This assumes a bridge principle from conditional independencies to causal relations. This is a key element in theories of probabilistic causality, and has in the last twenty years been developed to a highly sophisticated degree in Causal Bayes Net methods. See, for example, the faithfulness condition in Pearl (2000).

in econometrics are more complicated than this, and as a result it could be argued that what follows is of little relevance. This I think is too uncharitable, as any structural model which postulates deterministic functional relations among observable random variables will include error terms to represent omitted content. Structural equation models of this sort are widely used in econometrics and the philosophical and methodological issues raised here are relevant *mutatis mutandis* to these more complex models.

Finally, it should be emphasized that what follows is not intended to present a way of doing econometric modelling. I am not claiming – though I believe that claims along these lines can be reasonably made – that looking at the error term is the right or best way to build or select an econometric model. To do this would require a more general and technical approach. Instead, the aim of this paper is to be expository, to highlight the kind of issues one ought to be aware of when doing econometrics or when trying to understand some of the philosophical challenges to doing structural modelling in econometrics. It does this by considering what the error term represents in very simple structural models in econometrics and explores some of assumptions about the error terms that are used for successful statistical and causal inference. Ultimately, this is important for understanding the scope of econometric models. The error term is also of particular interest because it acts as a cover-all term for parts of the system that are not fully known about and not explicitly modelled. Therefore, there is always a danger that the error term hides important information which should have been modelled, and which may render the model inadequate in certain ways. That said, as the general model above shows, the error term is merely one part of a more general model, and tests on the error term are ultimately tests of the general model proposed. Thus, the more important point, emphasised by Spanos and others, is that one should test the general model assumed against observation.

The structure of the chapter is as follows. The first part presents a simple econometric model, like that found in introductory textbooks in econometrics, and sets out some of the assumptions imposed on the error term for successful statistical inference, in particular, those well-known conditions required for the ordinary least squares (OLS) method of estimation to yield 'good' (consistent, unbiased, etc.) estimates. These assumptions are well-known and widely discussed. The chapter then asks which of these assumptions can be tested from observations on the residuals[4] for the estimated model. Given the central role of the identification problem for simultaneous models in econometrics, the second part of the chapter investigates what conditions may be imposed on the error term in order to have an identifiable model.

[4] In the chapter, I use 'residuals' to denote the sample of the error terms (assuming the model to hold) and 'error terms' to denote the population random variable in the model of which (provided the model were true) the residuals would be a sample.

Again, as in the previous section, the chapter considers to what extent these conditions imposed on the error term can be tested. The third part of the chapter briefly presents a causal interpretation of the simultaneous equations model based on Herbert Simon (1953). Under this reading the error term is seen to denote the net impact of causal factors not explicitly modelled in a mechanism. In this section, the chapter also presents a restriction on error terms which is necessary for causal inference. To finish, the chapter concludes with an overview of conditions imposed on error term in (simple) econometric structural models. It notes that one assumption in particular, the orthogonality assumption, that the error terms are uncorrelated with the explanatory variables in a model, plays an important role for statistical and causal inference and for securing identifiability.

17.2 The role of the error term in estimation

To provide a concrete focus, consider the following simple 'textbook' simultaneous equations supply and demand model. This is a simultaneous equation model and as such represents the equilibrium relations between price and quantity as determined by underlying dynamic supply and demand mechanisms. For the purposes of this discussion, I assume that the equations have been chosen by appeal to theory and knowledge of the market being modelled.[5]

$$q = \alpha p + \beta i + u_1 \tag{17.1}$$

$$q = \gamma p + \delta c + u_2. \tag{17.2}$$

In this model, q denotes the equilibrium quantity of a good transacted, p denotes the equilibrium price for the good, i denotes consumer income, c denotes production costs, u_1 and u_2 are the error terms that denote factors not explicitly modelled in the supply and demand equations. In this simple model, assume that q, p, i and c are observable. The error terms denote omitted factors; the error terms are unobservable. In this model, q and p are determined in terms of i, c and the error terms. The population parameters α, β, γ and δ are unknown. The functional form is assumed to hold and i, c and the error terms are assumed to follow a particular joint probability distribution.

To relate this model to the general structural model above, in this model we have $Z = (XY)$ where $X = (ic)$ and $Y = (q\ p)$ and $U = (u_1 u_2)$. The functional relations (17.1) and (17.2) are the two equations of $G(Z, U) = 0$. The probabilistic assumptions on Z, follow from the assumptions on the

[5] This glosses over a difficult part of model building, particularly in this case where one is trying to sustain claims that the equilibrium relation can be represented by two static equations as is done here.

distributions of X (i and c), those on U and the assumed relationships of (17.1) and (17.2). The model is structural in virtue of the assumption that income and costs are not determined by the equilibrium quantity and price (the exogeneity assumption) and the assumption the two functional relations (17.1) and (17.2) hold in virtue of the demand and supply mechanisms that generate the equilibrium relations.

For the purposes that follows, I assume – perhaps artificially – that the model has been selected in a reasonable way (either from observation or by good background knowledge). At this stage then the econometrician's inferential problem is to infer the parameter values from sample observations of q, p, i and c.

The simplest estimation method for estimating linear equations in econometrics is the ordinary least squares (OLS) method. This approach picks estimates for the parameters that minimize the sum of the square deviations of the estimated values for the left-hand variable from the observed values for that variable. Provided certain assumptions are met, OLS provides consistent, unbiased and efficient estimators.[6] Of these assumptions, some directly involve the error terms. These are: (i) errors have a constant variance (are 'homoscedastic'); (ii) errors are normally distributed; (iii) errors are uncorrelated with the right-hand variables (orthogonality assumption). So in the example above OLS cannot be applied because p, being determined in the model, is unlikely to be orthogonal to the error term in either equation.[7] Therefore, to estimate this model one first solves for the reduced form equations (the solutions for p and q):

$$p = (\delta c - \beta i + u_2 - u_1)/(\alpha - \gamma) = \alpha' c + \beta' i + v_1 \quad (17.3)$$

$$q = \alpha(\delta c - \beta i + u_2 - u_1)/(\alpha - \gamma) + \beta i + u_1 = \gamma' c + \delta' i + v_2 \quad (17.4)$$

where

$$\alpha' = \delta/(\alpha - \gamma)\ \beta' = -\beta/(\alpha - \gamma) v_1 = (u_2 - u_1)/(\alpha - \gamma)$$
$$\gamma' = \alpha\delta/(\alpha - \gamma)\ \delta' = \beta - \alpha\beta/(\alpha - \gamma) v_2 = u_1 + \alpha(u_2 - u_1)/(\alpha - \gamma). \quad (*)$$

Now, if u_1, u_2 are both uncorrelated with c and i, have mean zero, are normally distributed and have constant variance then it follows that v_1 and v_2 meet all of these assumptions also. Then OLS can be applied to the reduced form equations to yield good (consistent, unbiased) estimates for parameters α', β', γ' and δ'. Then consistent estimates for α, β, γ and δ can be obtained by using formulae (∗) above.[8] This method of estimating parameters for simultaneous equation models is called 'indirect least squares' (ILS).

[6] See Gujarati (1995, chap. 3).

[7] If OLS were used here then the estimates would be biased and inconsistent.

[8] This assumes that one can solve for the original, structural parameters in terms of the reduced form parameters. If this is possible and if the reduced form parameters are identifiable, then the model is identifiable. The example chosen here is identifiable so this is not a problem

What is important to note here is that although a different estimation method has been used for the simultaneous model (ILS rather than OLS) similar assumptions have been imposed on the error terms, *u*, as would have been if OLS were a feasible estimation technique. Therefore, the assumptions on the error terms that ensure OLS yield good estimates in the non-simultaneous equations model, are also assumptions on the error terms that ensure ILS yields consistent[9] estimates in the simultaneous equations model.[10] Of course, the desirable properties of ILS (and OLS) estimators depends on these assumptions being met. I now consider these assumptions, their significance, and how they might be tested empirically.

The first assumption is that the error terms have constant variance. If this assumption is not met then OLS estimates are no longer efficient, though they remain unbiased and consistent.[11] There is a generalization of OLS, called 'generalized least squares' which may – provided there is information about the changing variance of the error – be used to provide efficient estimates. The second assumption is that the error terms are normally distributed. Interestingly, some desirable properties of OLS, such as consistency and unbiasedness hold independently of this assumption. Nevertheless, there are important advantages to the normality assumption, since it provides the basis for the distributions for a whole host of important test-statistics. If the normality assumption is not met then the distributions of the test statistics and of the estimates will almost certainly differ.[12] This is a practical problem, however, and in principle if non-normal distributions were specified for error terms it would be possible to numerically construct new test statistics and new distributions for the estimates. In conclusion, though these two assumptions on the error terms are important, their failure does not jeopardise the most desirable properties of the OLS estimates. Though this sounds promising, however, one should worry whether one can infer that it is *these* assumptions that have failed. I discuss this below.

The third and last assumption is the orthogonality assumption, that the right-hand variables are uncorrelated with the error term. If this assumption

here. In the next section, identifiability and the conditions it may impose on the error term are considered in more detail.

[9] Note that ILS estimates need not be unbiased nor efficient.

[10] That said, there is an important difference, nevertheless. In OLS the orthogonality assumption between the right-hand variables and the error need only hold between right-hand variables in one equation and the error term in that equation. In contrast, in the ILS case since right-hand variables from other equations may also appear on the right-hand side of the reduced form equations, we have made the stronger assumption that each error term from each structural equation is orthogonal to every variable not determined in the model.

[11] See Gujarati (1995, chap. 11) for a more detailed discussion of the consequences of non-constant variances of the error terms.

[12] Though if samples are large, a central limit theorem can be used to show that distributions of error terms will be approximately normal, see Gujarati (1995, p. 316–317).

is not met then the OLS estimates of the slope coefficient of an explanatory variable will be biased and inconsistent. Therefore, it is a key assumption that the error term must meet in order for the OLS estimates to be acceptable.

Having set out the significance of these three assumptions on the error terms for OLS estimation, I now consider to what extent these assumptions can be empirically tested. When attempting to empirically investigate error terms, one place to look is at the residuals for an estimated model, that is, at the differences between the estimated value for the left-hand variable and the actual value it takes. If all of the assumptions of the model are met then the set of residuals are a sample for the error terms . In this way, the model makes predictions about the likely samples of residuals. Investigating the sample residuals can then give useful information about the error terms and whether the assumptions about them hold. In the case of normality, if one could be sure that all the other assumptions were met, then one could infer from deviations from the normal distribution to likely failure of the normality assumption of the errors. Likewise with the constant variance assumption, if there were signs of changing variance, there one would – provided one were sure all the other assumptions of the model were met – have reason to suspect that this assumption for the errors had failed. However, there is a key problem here in that one does not typically know that all the other assumptions hold. Thus when one has a sample that would be highly unlikely under the assumed model (cf. a low p-value for a null hypothesis) then one has reason to suspect that at least one of the assumptions of the model is false, but one cannot infer which has failed. This is a form of the Duhem–Quine problem (see Ariew 2007) that one cannot infer from a false implication of a hypothesis which assumption(s) of the hypothesis fail. This problem is one reason why writers in econometric methodology, like Spanos (1999, p. 739), stress the importance of model specification which tests the model as a whole. If one has an incorrectly specified model, then the assumptions of ones statistical tests will probably not hold and the tests will be unreliable guides to inference.

The orthogonality assumption is also difficult to test. This can be seen by considering the simplest model of all, a regression model with only one explanatory variable, x.

$$y = \alpha x + u. \tag{17.5}$$

In this case, the sample correlation between residuals and the sample values of x will be zero by definition of the OLS estimate for α. Hence, regardless of the sample, in this simple regression model, the residuals are always uncorrelated with the right-hand variable. Therefore, in this case there is no way to test from residuals whether or not the assumption that x is uncorrelated with the error term in the model is met. In regression models with more than one right-hand variable, the more general result is that the residuals are uncorrelated with the sum of the products of the right-hand variables with the OLS estimates of their slope coefficients. Therefore, in these cases the residuals may be correlated in the sample with the one or more of the right-hand variables. Whether or not, however, such correlations can be used to make inferences about the covariance between the error term and the right-hand variables also depends on whether there is any covariance between the right-hand variables.

Econometricians have developed methods for dealing with this problem. They have developed tests for whether variables are suitably orthogonal to the error and methods for consistent estimation where an explanatory variable is not orthogonal to the error term (instrumental variable estimation). Crucially, these methods tend to work by augmenting the model in some way so that the variable whose orthogonality with the error is suspect, is itself modelled in terms of other variables. For instance, in the case where an explanatory variable is correlated with the error term, the instrumental variables method attempts to find an additional 'instrumental' variable which is correlated with the explanatory variable but not the error term, which can be used in place of the non-orthogonal variable for estimation. In short, testing for orthogonality failures between regressors and error terms is difficult, and generally requires using methods more sophisticated than the mere analysis of residuals. Importantly, it appears to necessitate a specification testing, that is, trying to find out if some important variables have been omitted in the model, variables whose explicit inclusion could overcome a failure of orthogonality.

17.3 The error term and identifiability

Identifiability is an important condition for performing statistical and causal inference in econometrics. If a model has unknowns that cannot be inferred uniquely from observation, then (that part of the model) is said to be unidentifiable. The classic, historical example of non-identifiability in economics is that of measuring supply and demand curves from observed market movements.[13] The problem is that observations of price and quantity transacted in a market are the result of both supply and demand mechanisms acting together. Therefore, the identification problem in this case is how to attribute any observed shifts in observed price and quantity of goods sold to supply and/or demand changes. The solution to the problem is to introduce some additional background 'a priori'[14] constraints (using background knowledge) to further limit the number of possible models that are consistent with observation.

[13] See Morgan (1990) for a historical account of the development of ideas in relation to identification in econometrics.

[14] The term 'a priori' is typically used to describe the knowledge used to secure identifiability. This is meant as knowledge prior to that provided by the observations used to parameterize the model. It does not mean that this knowledge is not in itself empirical.

The simple example presented above of a simultaneous equation model is identifiable because one can solve uniquely for the structural parameters from the reduced form parameters (the coefficients in equations (17.3) and (17.4) above) which can be estimated from observation. In this example, identifiability follows from the form of the equations (17.1) and (17.2) which have sufficiently few unknown parameters so that their values can be solved for from the estimates of coefficients in (17.3) and (17.4). Here identifiability is being secured by the a priori *exclusion* of variables from equations (17.1) and (17.2). This method of ensuring identifiability in simultaneous equation models is generalized in the well-known *Rank Condition* for identification.[15] This is a condition on the matrix of parameters in the model which if and only if met ensures it can be solved for from the reduced form equations (which can be estimated using OLS). This condition is necessary and sufficient for identifiability by using exclusions of variables from equations. What is important about identifiability by exclusion here is that it does *not* impose any conditions on the error terms in the model. All that matters for identifiability secured in this way is that there be sufficiently many exclusions of variables from the equations in the model.

This may seem to suggest that there is no interesting connection between identifiability and error terms. However, this is incorrect, since identifiability can also be secured by imposing constraints on the covariance matrix of the error terms in a model. A well-known example is that of the general (non-simultaneous) recursive model:

$$\begin{aligned} x_1 &= u_1 \\ x_2 &= \alpha_{21}x_1 + u_2 \\ \vdots \quad & \qquad \vdots \qquad\quad \vdots \\ x_n &= \alpha_{n1}x_1 \ldots + \alpha_{nn-1}x_{n-1} + u_n \end{aligned}$$

In this model only the first equation is identifiable by exclusions.[16] The other equations do not meet the rank condition and are thus not identifiable without some further constraint. The natural additional constraint to impose here is to assume that the error terms in the equations are orthogonal to each other (i.e. have a diagonal covariance matrix) with which the model becomes fully identifiable. So in this example identifiability of the model depends on an additional assumption that the error terms are orthogonal. Unlike the previous simultaneous equations example, identifiability here rests in part on the error terms meeting an orthogonality condition.[17]

[15] For more on the rank condition see Fisher (1966, p. 39–41), Gujarati (1995, p. 657–669) and Maddala 2001, p. 348–352).

[16] Though this isn't particularly useful since there is no parameter to estimate in the first equation!

[17] See Fisher (1966, chap. 4) for detailed discussion of identification conditions using both exclusions and covariance matrix (of the errors) constraints.

Interestingly, if one uses OLS to estimate the coefficients in the equations in this model then the required orthogonality assumption for OLS implies given the functional form of the model that the errors are orthogonal to one another. Therefore, using OLS to estimate this model implicitly assumes the orthogonality assumption for the error terms, and this renders the model identifiable. Unfortunately, however, this implicit orthogonality assumption is not testable from residuals in this case. This is because, as in the case of the two variable regression model above (17.5), the residuals that are generated by using an OLS estimation technique will be uncorrelated *by construction* with the right-hand variables, and thus, given the assumed functional form, will be uncorrelated by construction with one another. Therefore, regardless of the data, the residuals that result from OLS will be mutually uncorrelated. Therefore, analysing the residuals will not give any indication as to the correctness or otherwise of the assumed orthogonality of the error terms in the model. Justification of the implicit orthogonality assumption which ensures identifiability must be provided in some other way.

This type of problem, that constraints used for identification are not directly testable from the observations used to parameterise the model, is common. In fact, it is unsurprising since constraints used to secure identifiability are provided to supplement the insufficient power of the observations for determining a unique model as the most empirically adequate (i.e. solve the identification problem). It is only where there is a surplus of identifying constraints, that is, where not all the identifiability constraints are required for inferring a unique model that the observations used to pick the unique model can be also used to test surplus identifying constraints. In such a situation, the model is said to be *overidentified*. The example just given is not overidentified, it is *just identified*, that is, the observations and the orthogonal error terms are together just sufficient to pick out one unique model. Therefore, the chosen model is tailor-made to have uncorrelated residuals, since without this assumption, there would not be a unique model that fits observation. In short, with the exception of over-identified models, testing identifiability restrictions requires some observations or information in addition to the observations used to parameterise the model. This is the case of the orthogonal errors assumption used to identify the recursive model presented here.

17.4 The causal interpretation of the error term and its role in causal inference

So far, the chapter has ignored the causal aspect of structural models. Yet what is distinctive about structural models, in contrast to forecasting models, is that they are supposed to be – when successfully supported by observation – informative about the impact of interventions in the economy. As such, they carry causal content about the structure of the economy. Therefore, structural models do not model mere functional relations supported by correlations, their functional relations have causal content which support counterfactuals about what would happen under certain changes or interventions.

This suggests an important question: just what is the causal content attributed to structural models in econometrics? And, from the more restricted perspective of this paper, what does this imply with respect to the interpretation of the error term? What does the error term represent causally in structural equation models in econometrics? And finally, what constraints are imposed on the error term for successful causal inference? In order to begin to answer these, I first present a simple causal semantics, developed by Herbert Simon (1953) especially for the kind of simultaneous (and non-simultaneous) equations models looked at in this chapter.

In Simon (1953) a formal definition of casual order for structural equations models is presented. To obtain the causal order, one first distinguishes between two types of variables in the model, endogenous and exogenous.[18] The endogenous variables are those that are determined by the model (for example q and p in the simultaneous equations example above), while the exogenous variables (for example income, i, and cost of production, c) and the error terms have values that are taken as given, from outside the model. One then solves for the endogenous variables one-by-one using the fewest equations required to solve for them; this stipulates an order for the solution of the endogenous variables. Any variable used to solve for and solved for prior to another variable causally precedes it. One variable directly causally precedes another if it causally precedes it and if it appears in the same equation as the other variable. The resulting ordering among the variables is the causal order.[19]

Consider the earlier supply and demand example, where one categorizes q and p as endogenous and i and c as exogenous.

$$q = \alpha p + \beta i + u_1 \qquad \ldots \qquad \textit{demand}$$
$$q = \gamma p + \delta c + u_2 \qquad \ldots \qquad \textit{supply}.$$

Here one constructs the causal order by noting that q and p can only be solved for together in terms of i and c and the error terms, using both equations. Moreover, since both i and c appear in an equation with q and p, both are direct causes of p and q. Also, since p and q are both determined together (in the same minimal set of equations) they are 'co-determined'. Thus, the

[18] The account given here is based on the more detailed analysis of Fennell (2005, chap. 2). Note that this version deviates slightly from that of Simon (1953). However, the differences are not significant here.

[19] Note that Simon's causal order depends on the functional form of the equations since the order of solution by which the causal order is defined depends on the equations in which the variables appear.

causal order can be represented by (where the arrows denote direct causal precedence):

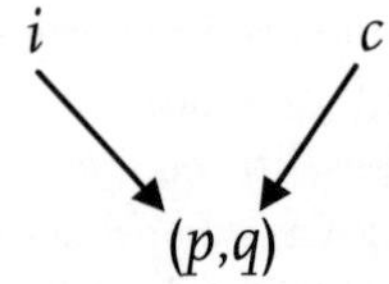

Simon's causal order yields an intuitive result for the example since it makes explicit that income and production costs are direct causes of equilibrium price and quantity, while equilibrium price and quantity are co-determined, just what one would expect for an equilibrium model of supply and demand.

Although Simon's causal order helps to make explicit the content of the functional form of structural equations, it is limited progress because it is merely a formal relation among the variables defined from the functional form of the equations. Despite its 'causal' label, as it stands it says nothing about the content of 'cause'. Luckily, Simon helps by briefly discussing how variables and equations should be interpreted. Simon states that the exogenous variables should be taken to denote factors that are directly controllable by an 'experimenter' or 'nature', and endogenous variables taken to denote factors that are indirectly controllable. Equations are taken to denote mechanisms and error terms are taken to denote the joint role of omitted directly controllable factors in a mechanism. The core idea is that the experimenter or nature has hypothetical[20] direct access to the directly controllable factors and is free to change them. Changing these then has an impact on the other indirectly controllable factors in virtue of the mechanisms that connect the indirectly controllable factors to the directly controllable factors. Under this interpretation, the causal order arises from the joint action of mechanisms, and maps out how changes in a factor will 'in general' lead to changes in other factors. It sets out that changing a cause 'in general' changes its effects, whereas it is possible to change an effect without changing one of its causes.[21] It is this

[20] It is important that the direct control here is hypothetical. Directly controllable factors need not in fact be directly controlled by some *actual* experimenter. This is why, I believe, Simon permited 'interventions' by nature. The point is rather that the causal relations are such that if the 'direct controllable' factor were intervened upon surgically – by an agent or by nature – then they would change directly in virtue of these interventions and the other indirectly controllable factors would change as a result. I take the idea here to be similar to Woodward's more developed (2003) analysis of the causal relation in terms of hypothetical interventions.

[21] Though this gives us some idea of how to causally interpret the structural equations models, there is much which is not discussed by Simon. For example, it is also important in his interpretation that mechanisms be invariant to changes brought about by the experimenter or nature. Otherwise, the equations expressing the mechanisms could be completely changed upon intervention and the equations would tell us nothing about what happens to the indirectly controllable factors as a result of changes to the directly controllable factors. Also, it is important that the

related series of possible changes under direct changes that is represented by the formal ordering relation.

With some causal semantics in place, the causal interpretation error term can be investigated in some more detail. According to the brief discussion above, the error term denotes the net impact of factors in a mechanism, those not explicitly modelled.[22] Yet what does it mean? Can any variable be omitted from an equation and simply 'brought into' an error term? The answer is quite simply 'no', as the following example illustrates.

Suppose one starts with the earlier simultaneous equation example.

$$q = \alpha p + \beta i + u_1 \ \ldots \ \textit{demand}$$
$$q = \gamma p + \delta c + u_2 \ \ldots \ \textit{supply}.$$

Now imagine that one were to 'omit' price from the demand equation, by bringing it into the error term (let $u_1' = u_1 + \alpha p$), to obtain a new first equation.

$$q = \beta i + u_1' \qquad \ldots \ \textit{new demand}$$
$$q = \gamma p + \delta c + u_2 \ \ldots \ \textit{supply}.$$

For these modified equations, the causal order obtained following Simon's method is:

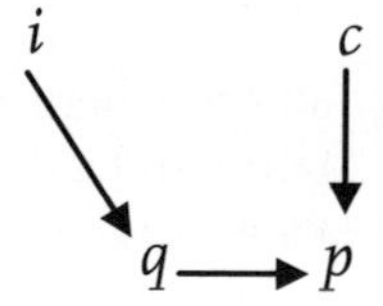

This causal order is different from that of the original system, even though the first model was assumed in constructing the second. By omitting price from the demand equation and bring it, as an omitted factor, into the error term changes the causal meaning of the model. Most strikingly perhaps, there is no longer an equilibrium relation between price and quantity but instead price is a direct cause of quantity transacted. This raises a worry, that the error terms in a model should not include as omitted factors, factors like p because

directly controllable factors be independent of each other in the sense that the experimenter must be 'free' to change them. Fleshing out these issues is an important step – one which I do not attempt here – in setting out a clear interpretation of the structural models. In addition, I do not argue for Simon's semantics over other possibilities, though this too is an important work to do.

22 This type of interpretation of the error term is widespread. For example, according to Kevin Hoover 'error terms might be thought to represent those INUS conditions that, though they help to determine the effects and are not constant, are not explicitly measured or modelled' (2001, p. 50). While Herbert Simon states that "'error terms"... measure the net effects of all other variables (not introduced explicitly) upon the system' (1954, p. 40). Nancy Cartwright (1989, p. 29) states that the error terms are 'supposed to represent the unknown or unobservable factors that may have an effect'.

if they do then the apparent causal semantics of the model may misrepresent the underlying system.

Therefore, the causal interpretation of the error term as the joint impact of factors that are simply 'omitted', i.e. not explicitly modelled is too weak since it does not rule out cases like that just presented. To clarify the causal interpretation of the error term, one could perform an analysis to attempt to find the weakest interpretation of the error term where the causal ordering relations among the explicitly modelled variables remain unchanged (if one were to bring out our bring in factors from the error term).[23] However, for my purposes here, I will simply assume that the factors omitted in the error terms are such that if they were to be introduced explicitly into the model they would be denoted by exogenous variables. This requires that factors omitted from the model, whose net impact is represented in the error terms, be causally prior to the factors whose causes are being modelled by the equations. Though this is stronger than necessary, it is intuitive and avoids the difficulty presented above.

To finish this section, I now consider briefly a key constraint that may be necessary for the error term to meet to use the model for causal inference. To keep the discussion simple, I look only at the simplest model.

$$y = \alpha x + u.$$

Interpreting this model using Simon, where x is exogenous and y endogenous, amounts to reading the right hand variable, x, as a direct cause of y, and u denoting the net impact of a set of omitted direct causes of y. Here the aim is (as in the problem of statistical inference) to infer the unknown value of α given observations of y and x. In this case though, since the problem is one of causal inference, I consider a simple experiment as an ideal way of inferring α.

The obvious experiment that comes to mind is to vary x, to see by how much y changes as a result. This sounds straightforward, one changes x, y changes and one calculates α as follows

$$\alpha = \Delta y / \Delta x.$$

Everything seems straightforward. However there is a concern since u is unobservable: how does one know that u has not also changed in changing x? Suppose that u does change so that there is hidden in the change in y a change in u, that is, the change in y is incorrectly measured by

$$\Delta y_{false} = \Delta y + \Delta u.$$

And thus that α is falsely measured as

$$\alpha_{false} = \Delta y_{false} / \Delta x = \Delta y / \Delta x + \Delta u / \Delta x = \alpha + \Delta u / \Delta x.$$

23 See Fennell (2005, chap. 3) for an analysis of this kind.

Therefore, in order for the experiment to give the correct measurement for α, one needs either to know that u has not also changed or to know by how much it has changed. Since u is unobservable this cannot be known by observation. This leaves as the only option to know – in virtue of the knowledge of how the change was brought about - that in changing x, u has not also been unwittingly changed. Intuitively, this requires that it is known that whatever cause(s) of x which are used to change x, they are not causes of any of factors hidden in u. This is to require that x have what Cartwright (1989, chap. 1) calls an 'open back path' with respect to y, that is, a cause which only causes y via x. The open back path provides a channel by which x could be varied to measure its impact on y. Such an open back path provides a 'clean' way to intervene in x for the purposes of causal inference.[24]

More generally, the example above shows a need to constrain the error term in the equation in a non-simultaneous structural equation model as follows. It requires that the each right-hand variable have a cause that causes y but not via any factor hidden in the error term. This imposes a limit on the common causes the factors in the error term can have with those factors explicitly modelled.

To finish, consider briefly the testability of the assumptions brought to light in this section. Given these assumptions directly involve the factors omitted in the error term, testing these empirically seems impossible without information about what is hidden in the error term. But given the error term is unobserveable, this places the modeller in a difficult situation: how to know that some important factor has not been left out from the model undermining desired inferences in some way. It also shows that there will always be element of faith in the assumptions about the error term.

17.5 Conclusion: Many different error term assumptions? Or a few in many guises?

This chapter has attempted to draw out what the error term represents in structural models and some of the conditions it has imposed upon it for inferential purposes. In the analysis of statistical inference (the OLS method of estimation), it was assumed that the error was normally distributed, had constant variance and was orthogonal to the explanatory variables. In the discussion of identifiability, it was shown that though identifiability can be achieved purely by exclusions of variables from the equations, but that this is not always the case, and that constraints on the covariance of the errors,

[24] Similar conditions to the open back path requirement appear widely in the literature. For instance, James Woodward incorporates a similar condition into his definition of an intervention variable, see Woodward (2003, p. 98).

such as mutual orthogonality, are also used to achieve identifiability. Finally, the paper briefly presented a causal interpretation of the error term, as the net impact of omitted causal factors from a mechanism, and showed that for causal inference purposes, it is important that there be a cause of any explicitly modelled causal factor that does not cause the effect of interest through a factor hidden in the error term.

Though this analysis seems to yield a large number of conditions the error term must meet, it is important not to assume that these conditions are unconnected. In particular, there is a strong connection between the orthogonality assumption, which is central for estimation, and the open-back-path requirement observed for causal inference. This can also be seen by adopting a principle which licenses a move from correlations to causes, for instance, Reichenbach's principle of the common cause that probabilistic dependencies imply a causal connection or common cause(s).[25] This principle implies that if the orthogonality assumption between the error and an explanatory variable fails, then either there is a factor in the error that causes the factor denoted by the variable, vice versa, or that there is a common cause of an explicitly modelled factor and a factor hidden in the error term. In the first two cases, no open back path is possible, while in third the common causal factor is itself not an open back path. Therefore, given Reichenbach's principle, a failure of orthogonality suggests a non open back path is being varied, the situation which frustrated causal inference. Therefore, there appears to be an intimate connection between the orthogonality requirement and the open-back-path requirement. This intimate connection is also visible in the instrumental variables method for overcoming a failure of orthogonality, in which a variable(s) is found which is correlated with the variable that fails orthogonality but uncorrelated with the error term. Interpreting these correlations using Reichenbach's principle, the instrumental variable is a search for some causal structure by which variable (which failed orthogonality with the error) can be varied, without varying the error term. In short, it is a search for an open back path.[26]

In any event, the point here isn't to explore the important connections between the conditions on the error term required for causal inference and those for statistical inference, but rather to show that such connections exist and are fundamental. This, of course, should not be surprising since ultimately the problems of statistical and causal inference overlap greatly. After all, the estimation methods of structural methods aim to measure strengths of causal connection.

[25] There are related principles such as the Causal Markov condition which also allow one to make inferences from correlations to causes. See Spirtes, Glymour and Scheines (1993) for more details.

[26] See Reiss (2003) for a causal discussion of instrumental variables.

The second point in highlighting the connection between the orthogonality and the open-back-path condition is to highlight the centrality of this kind of assumption for inference. As seen above, their failure frustrates inference, and also their testing is not a straightforward matter of analysing residuals. Therefore, this condition, and more specifically the relationship between what is hidden in the error term and what is explicitly modelled deserves careful scrutiny. In the econometrics literature, this condition is typically discussed under the term 'exogeneity' of the explanatory variables in a model, though there are many different definitions of exogeneity and disputes over which is correct for which purposes.[27] The analysis of this chapter suggests, seen from the orthogonality assumption and its causal cousin the open-back-path requirement, is that such exogeneity assumptions can play different roles (estimation, causal inference) when viewed from different perspectives.

Acknowledgements

This research was supported by the AHRC 'Contingency and Dissent in Science' project at the CPNSS, London School of Economics. I am very grateful for their support. I am also very grateful to participants of the ERROR conference, June 2006,Virginia Tech, Blacksburg, Virginia for helpful feedback. Finally, I would like to thank Michel Mouchart and an anonymous referee for very helpful comments.

References

Ariew, R. (2007). 'Pierre Duhem', *The Stanford Encyclopedia of Philosophy*, E. N. Zalta (ed.), URL = <http://plato.stanford.edu/entries/duhem/>

Cartwright, N. (1989). *Nature's Capacities and their Measurement*, Oxford, Clarendon press.

Engle, R. F., D. F. Hendry and J.-F. Richard (1983). Exogeneity, *Econometrica*, 51, 277–304.

Fennell, D. (2005). *A Philosophical Analysis of Causality in Econometrics*, PhD thesis, University of London.

Fisher, F. (1966). *The Identification Problem in Econometrics*, Huntington: Krieger Publishing Company.

Gujarati, D. (1995). *Basic Econometrics*, 3rd edition, New York: McGraw-Hill.

Haavelmo, T. (1944). 'The Probability Approach to Econometrics', *Econometrica*, 12, suppl., 1–115.

Hendry, D. (1995). *Dynamic Econometrics*, Oxford University Press: Oxford.

[27] See Engle *et al.* (1983).

Hoover, K. (2001). *Causality in Macroeconomics*, Cambridge: Cambridge University Press.

Maddala, G. S. (2001). *Introduction to Econometrics*, 3rd edition, New York: John Wiley and Sons.

Morgan, M. (1990). *The History of Econometric Ideas*, Cambridge: Cambridge University Press.

Reiss, J. (2003). Practice ahead of theory: Instrumental variables, natural experiments and inductivism in econometrics. *Causality: metaphysics and methods technical reports*, CTR 12/03, Centre for the Philosophy of the Natural and Social Sciences, London School of Economics.

Simon, H. (1953). Causal ordering and identifiability, reprinted in H. Simon, *Models of Man*, New York: John Wiley and Sons.

Simon, H. (1954). Spurious causation: A causal interpretation, reprinted in H. Simon *Models of Man*, New York: John Wiley and sons.

Spanos, A. (1999). *Probability Theory and Statistical Inference*, Cambridge University Press: Cambridge.

Spirtes, P., C. Glymour and R. Scheines (1993). *Causation, Prediction and Search*, New York: Springer-Verlag.

Woodward, J. (2003). *Making Things Happen: A Theory of Causal Explanation*, New York: Oxford University Press.

·18·

A comprehensive causality test based on the singular spectrum analysis

Hossein Hassani, Anatoly Zhigljavsky, Kerry Patterson, and Abdol S. Soofi

Abstract

In this chapter, we consider the concept of causal relationship between two time series based on the singular spectrum analysis. We introduce several criteria which characterize this causality. The criteria are based on the forecasting accuracy and the predictability of the direction of change. The performance of the proposed tests is examined using different real time series.

18.1 Introduction

A question that frequently arises in time series analysis is whether one economic variable can help in predicting another economic variable. One way to address this question was proposed by Granger (1969). Granger (1969) formalized a causality concept as follows: process X does not cause process Y if (and only if) the capability to predict the series Y based on the histories of all observables is unaffected by the omission of X's history (see also Granger 1980). Testing causality, in the Granger sense, involves using F-tests to test whether lagged information on one variable, say X, provides any statistically significant information about another variable, say Y, in the presence of lagged Y. If not, then 'Y does not Granger-cause X'.

Criteria for Granger causality typically have been realized in the framework of multivariate Gaussian statistics via vector autoregressive (VAR) models. It is worth mentioning that the linear Granger causality is not causality in a broader sense of the word. It just considers linear prediction and time-lagged dependence between two time series. The definition of Granger causality does not mention anything about possible instantaneous correlation between two series X_T and Y_T. (If the innovation to X_T and the innovation to Y_T are correlated then it is sometimes called instantaneous causality.) It is not rare when instantaneous correlation between two time series can be easily revealed, but since the causality can go either way, one usually does not test for instantaneous correlation. In this chapter, several of our causality tests incorporate testing for the instantaneous causality. One more drawback of the Granger causality test is the dependence on the right choice of the conditioning set. In reality one can never be sure that the conditioning set selected is large enough (in short macro-economic series one is forced to choose a low dimension for the VAR model). Moreover, there are special problems with testing for Granger causality in co-integrated relations (see Toda and Phillips 1991).

The original notion of Granger causality was formulated in terms of linear regression, but there are some nonlinear extensions in the literature (see, for example, Chu *et al.* 2004). Hiemstra and Jones (1994) also propose a non-parametric test which seems to be most used test in testing nonlinear causality. However, this method also has several drawbacks: (i) the test is not consistent, at least against a specific class of alternatives (Diks and Panchenko 2005), (ii) there are restrictive assumptions in this approach (Bosq 1998) and (iii) the test can severely over-reject the null hypothesis of non-causality (Diks and Panchenko 2006).

It is also important to note that Granger causality attempts to capture an important aspect of causality, but it is not meant to capture all. A method based on the information theory has realized a more general Granger causality measure that accommodates in principle arbitrary statistical processes (Diks and DeGoede 2001). Su and White (2008) propose a non-parametric test of conditional independence based on the weighted Hellinger distance between the two conditional densities. There are also a number of alternative methods, but they are rarely used.

We overcome many of these difficulties by implementing a different technique for capturing the causality; this technique uses the singular spectrum analysis (SSA) technique; a non-parametric technique that works with arbitrary statistical processes, whether linear or nonlinear, stationary or non-stationary, Gaussian or non-Gaussian.

The general aim of this study is to assess the degree of association between two arbitrary time series (these associations are often called causal relationships as they might be caused by the genuine causality) based on the observation of these time series. We develop new tests and criteria which will be based on the forecasting accuracy and predictability of the direction of change of the SSA algorithms.

The structure of the chapter is as follows. Section 18.2 briefly describes the SSA technique. The proposed criteria and statistical tests are considered in Section 18.3. Empirical results are presented in Section 18.4. Conclusions are given in Section 18.5. Appendix contains some necessary technical details about SSA.

18.2 Singular spectral analysis

A thorough description of the theoretical and practical foundations of the SSA technique (with many examples) can be found in Golyandina *et al.* (2001)

and Danilov and Zhigljavsky (1997). There are many papers where SSA has been applied to real-life time series. In particular, the performance of the SSA technique has been compared with other techniques for forecasting economics time series (Hassani 2007; Hassani *et al.* 2009a–c, Hassani *et al.* 2010, Patterson *et al.* 2010; and see also Hassani 2009d for a new SSA-based algorithm and its application for forecasting).

Consider the real-valued nonzero time series $Y_T = (y_1, \dots, y_T)$ of sufficient length T. The main purpose of SSA is to decompose the original series into a sum of series, so that each component in this sum can be identified as either a trend, periodic or quasi-periodic component (perhaps, amplitude-modulated), or noise. This is followed by a reconstruction the original series.

The state of a process at time t is considered to capture the relevant information of the process up to time t. Moreover, it is the state of a process that is to be predicted. Assume that the process is governed by some linear recurrent formula (LRF), then having the LRF and embedding theory, forecasting the process at time t may be regarded as forecasting the state vector. According to the SSA terminology, the problem of forecasting a new vector requires (a) a window of some suitable length and (b) the number of eigenvalues.

The SSA technique consists of two complementary stages: decomposition and reconstruction, both of which include two separate steps. At the first stage we decompose the series and at the second stage we reconstruct the original series and use the reconstructed series (which is without noise) for forecasting new data points. Below we provide a brief discussion on the methodology of the SSA technique (for more description of the SSA algorithm, forecasting procedure and parameter estimation, see Appendix A).

18.2.1 A short description of the Basic SSA

We consider a time series $Y_T = (y_1, \dots, y_T)$. Fix L $(L \leq T/2)$, the window length, and let $K = T - L + 1$.

Step 1 (*Computing the trajectory matrix*): this transfers a one-dimensional time series $Y_T = (y_1, \dots, y_T)$ into the multi-dimensional series $X_1, \dots, X_K$ with vectors $X_i = (y_i, \dots, y_{i+L-1})' \in \mathbf{R}^L$, where $K = T - L + 1$. Vectors X_i are called *L-lagged vectors* (or, simply, *lagged vectors*). The single parameter of the embedding is the *window length* L, an integer such that $2 \leq L \leq T$. The result of this step is the trajectory matrix $\mathbf{X} = [X_1, \dots, X_K] = (x_{ij})_{i,j=1}^{L,K}$.

Step 2 (*Constructing a matrix for applying SVD*): compute the matrix $\mathbf{XX}^T$.

Step 3 (*SVD of the matrix* $\mathbf{XX}^T$): compute the eigenvalues and eigenvectors of the matrix $\mathbf{XX}^T$ and represent it in the form $\mathbf{XX}^T = P\Lambda P^T$. Here $\Lambda = diag(\lambda_1, \dots, \lambda_L)$ is the diagonal matrix of eigenvalues of $\mathbf{XX}^T$ ordered so that $\lambda_1 \geq \lambda_2 \geq \dots \geq \lambda_L \geq 0$ and $P = (P_1, P_2, \dots, P_L)$ is the corresponding orthogonal matrix of eigen–vectors of $\mathbf{XX}^T$.

Step 4 (*Selection of eigenvectors*): select a group of l $(1 \leq l \leq L)$ eigenvectors $P_{i_1}, P_{i_2}, \ldots, P_{i_l}$.

The grouping step corresponds to splitting the elementary matrices $\mathbf{X}_i$ into several groups and summing the matrices within each group. Let $I = \{i_1, \ldots, i_l\}$ be a group of indices $i_1, \ldots, i_l$. Then the matrix $\mathbf{X}_I$ corresponding to the group I is defined as $\mathbf{X}_I = \mathbf{X}_{i_1} + \cdots + \mathbf{X}_{i_l}$.

Step 5 (*Reconstruction of the one-dimensional series*): compute the matrix $\tilde{\mathbf{X}} = ||\tilde{x}_{i,j}|| = \sum_{k=1}^{l} P_{i_k} P_{i_k}^T \mathbf{X}$ as an approximation to $\mathbf{X}$. Transition to the one–dimensional series can now be achieved by averaging over the diagonals of the matrix $\tilde{\mathbf{X}}$.

18.2.2 Multivariate singular spectrum analysis: MSSA

Multivariate (or multichannel) SSA is an extension of the standard SSA to the case of multivariate time series (see e.g. Broomhead and King 1986). It can be described as follows. Assume we have two time series $X_T = x_1, \ldots, x_T$ and $Y_T = y_1, \ldots, y_T$ simultaneously (a bivariate approach), and let L be window length. Using embedding terminology, we can define the trajectory matrices $\mathbf{M}_X$ and $\mathbf{M}_Y$ of the one-dimensional time series X_T and Y_T, respectively. The trajectory matrix $\mathbf{M}$ can then be defined as

$$\mathbf{M} = \begin{pmatrix} \mathbf{M}_X & \mathbf{M}_Y \end{pmatrix}. \tag{18.1}$$

Note also that the matrix $\mathbf{M}$ can be represented as follows:

$$\mathbf{M} = \begin{pmatrix} \mathbf{M}_X \\ \mathbf{M}_Y \end{pmatrix}. \tag{18.2}$$

The other stages of the Basic Multivariate SSA (or MSSA) procedure are identical to the Basic SSA. The generalization to the case of several series is straightforward. There are numerous examples of successful application of the multivariate SSA (see, for example, Plaut and Vautard, 1994; Danilov and Zhigljavsky, 1997).

18.3 Causality criteria

18.3.1 Forecasting accuracy based criterion

The first criterion we use here is based on the out-of-sample forecasting, which is very common in the framework of Granger causality. The question behind Granger causality is whether forecasts of one variable can be improved using the history of another variable. Here, we compare the forecasted value obtained using the univariate procedure, SSA, and also the multivariate one, MSSA. We then compare the forecasted values with the actual values to evaluate the forecasting error. If the forecasting error using MSSA is significantly smaller than the forecasting error of the univariate SSA, we then conclude that there is a casual relationship between these series.

Let us consider in more detail the procedure of constructing a vector of forecasting error for an out-of-sample test. In the first step we divide the series $X_T = (x_1, \dots, x_T)$ into two separate subseries X_R and X_F: $X_T = (X_R, X_F)$ where $X_R = (x_1, \dots, x_R)$, and $X_F = (x_{R+1}, \dots, x_T)$. The subseries X_R is used in reconstruction step to provide the noise free series $\tilde{X}_R$. The noise free series $\tilde{X}_R$ is then used for forecasting the subseries X_F using either the recurrent or vector forecasting algorithm, see Appendix A. The subseries X_F will be forecasted using the recursive h-step ahead forecast with SSA and MSSA. The forecasted points $\hat{X}_F = (\hat{x}_{R+1}, \dots, \hat{x}_T)$ are then used for computing the forecasting error, and the vector $(x_{R+2}, \dots, x_T)$ is forecasted using the new subseries $(x_1, \dots, x_{R+1})$. This procedure is continued recursively up to the end of series, yielding the series of h-step-ahead forecasts for univariate and multivariate algorithms. Therefore, the vector of h-step-ahead forecasts obtained can be used in examining the association (or order h) between the two series. Let us now consider a formal procedure of constructing a criterion of SSA causality of order h between two arbitrary time series.

Criterion

Let $X_T = (x_1, \dots, x_T)$ and $Y_T = (y_1, \dots, y_T)$ denote two different time series of length T. Set window lengths L_x and L_y for the series X_T and Y_T, respectively. Here, for simplicity assume $L_x = L_y$. Using the embedding terminology, we construct trajectory matrices $\mathbf{X} = [X_1, \dots, X_K]$ and $\mathbf{Y} = [Y_1, \dots, Y_K]$ for the series X_T and Y_T.

Consider an arbitrary loss function $\mathcal{L}$. In econometrics, the loss function $\mathcal{L}$ is usually selected so that it minimizes the mean square error of the forecast. Let us first assume that the aim is to forecast the series X_T. Thus, the aim is to minimize $\mathcal{L}(X_{K+H_x} - \hat{X}_{K+H_x})$, where the vector $\hat{X}_{K+H_x}$ is an estimate, obtained using a forecasting algorithm, of the vector X_{K+H_x} of the trajectory matrices $\mathbf{X}$. Note that, for example, when $H_x = 1$, $\hat{X}_{K+1}$ is an estimate of the vector $X_{K+1} = (x_{T+1}, \dots, x_{T+h})$ where h varies between 1 and L. In a vector form, this means that an estimate of X_{K+1} can be obtained using the trajectory matrix $\mathbf{X}$ consisting of vectors $[X_1, \dots, X_K]$. The vector X_{K+H_x} can be forecasted using either univariate SSA or MSSA. Let us first consider the univariate approach. Define

$$\Delta_{X_{K+H_x}} \equiv \mathcal{L}(X_{K+H_x} - \hat{X}_{K+H_x}), \tag{18.3}$$

where $\hat{X}_{K+H_x}$ is obtained using univariate SSA; that is, the estimate $\hat{X}_{K+H_x}$ is obtained only from the vectors $[X_1, \dots, X_K]$.

Let $X_T = (x_1, \dots, x_T)$ and $Y_{T+d} = (y_1, \dots, y_{T+d})$ denote two different time series to be considered simultaneously and consider the same window length L for both series. Now, we forecast $x_{T+1}, \dots, x_{T+h}$ using the information provided by the series Y_{T+d} and X_T. Next, compute the following statistic:

$$\Delta_{X_{K+H_x}|Y_{K+H_y}} \equiv \mathcal{L}(X_{K+H_x} - \tilde{X}_{K+H_x}). \tag{18.4}$$

where $\tilde{X}_{K+H_x}$ is an estimate of X_{K+H_x} obtained using multivariate SSA. This means that we simultaneously use vectors $[X_1, \ldots, X_K]$ and $\left[Y_1, \ldots, Y_{K+H_y}\right]$ in forecasting vector X_{K+H_x}. Now, define the criterion:

$$F_{X|Y}^{(h,d)} = \frac{\Delta_{X_{K+H_x}|Y_{K+H_y}}}{\Delta_{X_{K+H_x}}} \tag{18.5}$$

corresponding to the h step ahead forecast of the series X_T in the presence of the series Y_{T+d}; here d shows the lagged difference between series X_T and Y_{T+d}, respectively. Note that d is any given integer (even negative). For example, $F_{X|Y}^{(h,0)}$ indicates that we use the same series length in h step ahead forecasting series X; we use the series X_T and Y_T simultaneously. $F_{X|Y}^{(h,0)}$ can be considered as a common multivariate forecasting system for time series with the same series length. The criterion $F_{X|Y}^{(h,0)}$ can then be used in evaluating two instantaneous causality. Similarly, $F_{X|Y}^{(h,1)}$ indicates that there is an additional information for series Y and that this information is one step ahead of the information for the series X; we use the series X_T and Y_{T+1} simultaneously.

If $F_{X|Y}^{(h,d)}$ is small, then having information obtained from the series Y helps us to have a better forecast of the series X. This means there is a relationship between series X and Y of order h according to this criterion. In fact, this measure of association shows how much more information about the future values of series X is contained in the bivariate time series (X, Y) than in the series X alone. If $F_{X|Y}^{(h,d)}$ is very small, then the predictions using the multivariate version are much more accurate than the predictions by the univariate SSA. If $F_{X|Y}^{(h,d)} < 1$, then we conclude that the information provided by the series Y can be regarded as useful or *supportive* for forecasting the series X. Alternatively, if the values of $F_{X|Y}^{(h,d)} \geq 1$, then either there is no detectable association between X and Y or the performance of the univariate version is better than the multivariate version (this may happen, for example, when the series Y has structural breaks which may misdirect the forecasts of X).

To asses which series is more *supportive* in forecasting, we need to consider another criterion. We obtain $F_{Y|X}^{(h,d)}$ in a similar manner. Now, these measures tell us whether using extra information about time series Y_{T+d} (or X_{T+d}) supports X_T (or Y_T) in h-step forecasting. If $F_{Y|X}^{(h,d)} < F_{X|Y}^{(h,d)}$, we then conclude that X is more *supportive* than Y, and if $F_{X|Y}^{(h,d)} < F_{Y|X}^{(h,d)}$, we then conclude that Y is more *supportive* than X.

Let us now consider a definition for a feedback system according to the above criteria. If $F_{Y|X}^{(h,d)} < 1$ and $F_{X|Y}^{(h,d)} < 1$, we then conclude that there is a feedback system between series X and Y. We shall call it F-feedback (forecasting feedback) which means that using a multivariate system improves the forecasting for both series. For a F-feedback system, X and Y are mutually supportive.

Statistical test

To check if the discrepancy between the two forecasting procedures are statistically significant we may apply the Diebold and Mariano (1995) test statistic, with the corrections suggested by Harvey *et al.* (1997). The quality of a forecast is to be judged on some specified function $\mathcal{L}$ as a loss function of the forecast error. Then, the null hypothesis of equality of expected forecast performance is $E(D_t) = 0$, where $D_t = (D_{X_{K+H_x}|Y_{K+H_y}} - D_{X_{K+H_x}})$ and $D_{X_{K+H_x}|Y_{K+H_y}}$ and $D_{X_{K+H_x}}$ are the vectors of the forecast errors obtained with the univariate and multivariate approaches, respectively. In our case, $\mathcal{L}$ is the quadratic loss function. The modified Diebold and Mariano statistic for a h step ahead forecast and the number of n forecasted points is

$$S = \bar{D}\sqrt{\frac{n + 1 - 2h + h(h-1)/n}{n\,\widehat{\text{var}}(\bar{D})}}$$

where $\bar{D}$ is the sample mean of the vector D_t and $\widehat{\text{var}}(\bar{D})$ is, asymptotically $n^{-1}\left(\widehat{\gamma_0} + 2\sum_{k=1}^{h-1}\widehat{\gamma_k}\right)$, where $\widehat{\gamma_k}$ is the k-th autocovariance of D_t and can be estimated by $n^{-1}\sum_{t=k+1}^{n}(D_t - \bar{D})(D_{t-k} - \bar{D})$. The S statistic has an asymptotic standard normal distribution under the null hypothesis and its correction for a finite samples follows the Student's t distribution with $n-1$ degrees of freedom.

18.3.2 Direction of change based criterion

Ash *et al.* (1997) argue that for some purposes, it may be more harmful to make a smaller prediction error yet fail in predicting the direction of change, than to make a larger directionally correct error. Clements and Smith (1999) discuss that the value of a model's forecasts may be better measured by the direction of change. Heravi *et al.* (2004) argue that the direction of change forecasts are particularly important in economics for capturing the business cycle movement relating to expansion versus contraction phases of the cycle. Thus as another measure of forecasting performance, we also compute the percentage of forecasts that correctly predict the direction of change.

Criterion

The direction of change criterion shows the proportion of forecasts that correctly predict the direction of the series movement. For the forecasts obtained using only X_T (univariate case), let Z_{X_i} take the value 1 if the forecast observations correctly predicts the direction of change and 0 otherwise. Then $\bar{Z}_X = \sum_{i=1}^{n} Z_{Xi}/n$ shows the proportion of forecasts that correctly predict the direction of the series movement (in forecasting n data points). The Moivre–Laplace central limit theorem implies that, for large samples, the test statistic $2(\bar{Z}_X - 0.5)N^{1/2}$ is approximately distributed as standard normal. When $\bar{Z}_X$ is significantly larger than 0.5, then the forecast is said to have the ability to predict the direction of change. Alternatively, if $\bar{Z}_X$ is significantly smaller than 0.5, the forecast tends to give the wrong direction of change.

For the multivariate case, let $Z_{X|Y,i}$ takes a value 1 if the forecast series correctly predicts the direction of change of the series X having information about the series Y and 0 otherwise. Then, we define the following criterion:

$$D_{X|Y}^{(h,d)} = \frac{\bar{Z}_X}{\bar{Z}_{X|Y}} \tag{18.6}$$

where h and d have the same interpretation as for $F_{X|Y}^{(h,d)}$. The criterion $D_{X|Y}^{(h,d)}$ characterizes the improvement we are getting from the information contained in Y_{T+h} (or X_{T+h}) for forecasting the direction of change in the h step ahead forecast.

If $D_{X|Y}^{(h,d)} < 1$, then having information about the series Y helps us to have a better prediction of the direction of change for the series X. This means that there is an association between the series X and Y with respect to this criterion. This criterion informs us how much more information we have in the bivariate time series relative to the information contained in the univariate time series alone with respect to the prediction of the direction of change. Alternatively, if $D_{X|Y}^{(h,d)} > 1$, then the univariate SSA is better than the multivariate version.

To find out which series is more supportive in predicting the direction of change, we consider the following criterion. We compute $D_{Y|X}^{(h,d)}$ in a similar manner. Now, if $D_{Y|X}^{(h,d)} < D_{X|Y}^{(h,d)}$, then we conclude that that X is more supportive (with respect to predicting the direction) to Y than Y to X.

Similar to the consideration of the forecasting accuracy criteria, we can define a feedback system based on the criteria characterizing the predictability of the direction of change. Let us introduce a definition for a feedback system according to $D_{X|Y}^{(h,d)}$ and $D_{Y|X}^{(h,d)}$. If $D_{Y|X}^{(h,d)} < 1$ and $D_{X|Y}^{(h,d)} < 1$, we conclude that there is a feedback system between the series X and Y for prediction of the direction of change. We shall call this type of feedback D-feedback. The existence of a D-feedback in a system yields that the series in the system help each other to capture the direction of the series movement with higher accuracy.

Statistical test

Let us describe a statistical test for the criterion $D_{X|Y}^{(h,d)}$. As in the comparison of two proportions, when we test the hypothesis about the difference between two proportions, first we need to know whether the two proportions are dependent. The test is different depending on whether the proportions are

Table 18.1 An arrangement of Z_X and $Z_{X|Y}$ in forecasting n future points of the series X.

$Z_{X\|Y}$	Z_X	Number
1	1	a
1	0	b
0	1	c
0	0	d
Total		$n = a + b + c + d$

independent or dependent. In our case, obviously, Z_X and $Z_{X|Y}$ are dependent. We therefore consider this dependence in the following procedure. Let us consider the test statistic for the difference between Z_X and $Z_{X|Y}$. Assume that Z_X and $Z_{X|Y}$, in forecasting n future points of the series X, are arranged as Table 18.1.

Then the estimated proportion using the multivariate system is $P_{X|Y} = (a + b)/n$, and the estimated proportion using the univariate version is $P_X = (a + c)/n$. The difference between the two estimated proportions is

$$\pi = P_{X|Y} - P_X = \frac{a+b}{n} - \frac{a+c}{n} = \frac{b-c}{n}. \tag{18.7}$$

Since the two population probabilities are dependent, we cannot use the same approach for estimating the standard error of the difference that is used for independent case. The formula for the estimated standard error for the dependent case was given by Fleiss (1981):

$$\hat{\text{SE}}(\pi) = \frac{1}{n}\sqrt{(b+c) - \frac{(b-c)^2}{n}}. \tag{18.8}$$

Let us consider the related test for the difference between two dependent proportions, then the null and alternative hypotheses are:

$$\begin{aligned} H_0 &: \pi_d = \Delta_0 \\ H_a &: \pi_d \neq \Delta_0. \end{aligned} \tag{18.9}$$

The test statistic, assuming that the sample size is large enough for the normal approximation to the binomial to be appropriate, is:

$$T_{\pi_d} = \frac{\pi - \Delta_0 - 1/n}{\hat{\text{SE}}(\pi)} \tag{18.10}$$

where $1/n$ is the continuity correction. In our case $\Delta_0 = 0$. The test statistic then becomes:

$$T_{\pi_d} = \frac{(b-c)/n - 1/n}{1/n\sqrt{(b+c) - (b-c)^2/n}} = \frac{b-c-1}{\sqrt{(b+c) - (b-c)^2/n}}. \tag{18.11}$$

The test is valid when the average of the discordant cell frequencies, $(b+c)/2$, is equal or more than 5. However, then it is less than 5, a binomial test can be used. Note that under the null hypothesis of no difference between samples Z_X and $Z_{X|Y}$, T_{π_d} is asymptotically distributed as standard normal.

18.3.3 Comparison with Granger causality test

Linear Granger causality test

Let X_T and Y_T be two stationary time series. To test for Granger causality we compare the full and the restricted model. The full model is given by

$$x_t = \phi_0 + \phi_1 x_{t-1} + \cdots + \phi_p x_{t-p} + \psi_1 y_{t-1} + \cdots + \psi_p y_{t-p} + \varepsilon_{t_{x|y}} \qquad (18.12)$$

where $\{\varepsilon_{t_{x|y}}\}$ is an iid sequence with zero mean and variance $\sigma_{x|y}$, ϕ_i and ψ_i are model parameters. The null hypothesis stating that Y_T does not Granger cause X_T is:

$$H_0 = \psi_{L+1} = \psi_2 = \cdots = \psi_p = 0. \qquad (18.13)$$

If the null hypothesis holds, the full model (18.12) is reduced to the restricted model as follows:

$$x_t = \phi_0 + \phi_1 x_{t-1} + \cdots + \phi_p x_{t-L+1} + \varepsilon_{t_x} \qquad (18.14)$$

where ε_{t_x} is iid sequence with zero mean and variance σ_x. The forecasting results obtained by the restricted model (18.14) are compared to those obtained using the full model (18.12) to test for Granger causality. We then apply the F-test (or some other similar test) to obtain a p-value for whether the full model results are better than the restricted model results. If the full model provides a better forecast, according to the standard loss functions, we then conclude that Y_T Granger causes X_T. Thus, Y_T would Granger cause X_T if Y_T occurs before and contains information useful in forecasting X_T. As the formula of Granger causality shows, the test, in fact, is a mathematical formulation based on the linear regression modelling of two time series. Therefore, the above formulation of Granger causality can only give information about linear features of the series.

Let us now compare the similarity and dissimilarity of the proposed algorithm which is based on the SSA forecasting algorithm with the Granger causality procedure. As mentioned in the description of the SSA forecasting algorithm, the last component X_L of any vector $X = (x_1, \ldots, x_L)^T \in \mathfrak{L}_r$ is a linear combination of the first $L-1$ components $(x_1, \ldots, x_{L-1})$ such that:

$$x_L = a_1 x_{L-1} + \cdots + a_{L-1} x_1.$$

where $A = (a_1, \ldots, a_{L-1})$ can be estimated using equation (18.24) of Appendix A. Thus, the univariate version of SSA is given by

$$x_t = a_1 x_{t-1} + \cdots + a_{L-1} x_{t-L+1}. \qquad (18.15)$$

As can be seen from (18.15), a univariate SSA forecasting formula is similar to the restricted model. However, the procedure of parameter estimation in the SSA technique and the Granger model are quite different. Both are linear combinations of previous observations, and from this point of view both are similar. The multivariate version of SSA is a system in which X_T and Y_T are considered simultaneously to estimate vectors A and B as follows. The multivariate forecasting system is:

$$\begin{pmatrix} x_t \\ y_t \end{pmatrix} = \begin{pmatrix} \alpha_1 x_{t-1} + \cdots + \alpha_{L-1} x_{t-L+1} \\ \beta_1 y_{t-1} + \cdots + \beta_{L-1} y_{t-L+1} \end{pmatrix} \tag{18.16}$$

where the vectors $A = (\alpha_1, \ldots, \alpha_{L-1})$ and $B = (\beta_1, \ldots, \beta_{L-1})$ are estimated using the multivariate system. As equation (18.16) shows, the multivariate SSA is not similar to the Granger full model. An obvious discrepancy is that we use the value of the series Y in parameter estimation and also in forecasting series X in the Granger based test, while we use the information provided in the subspaces generated by Y in multivariate SSA and not the observed values. More specifically, the Granger causality test uses a linear combination of the values of both series X and Y in the full model, whereas multivariate SSA uses the information provided by X and Y in construction of the subspace and not the observations themselves.

Nonlinear Granger causality test

It is worth mentioning that the simultaneous reconstruction of the trajectory matrices $\mathbf{X}$ and $\mathbf{Y}$ in the MSSA technique is also used in testing for Granger causality between two nonlinear time series. Let us consider the concept of nonlinear Granger causality in more detail. Let $\mathbf{Z} = [\mathbf{X}, \mathbf{Y}]$ be the joint trajectory matrix with lagged difference zero (same value of K in the trajectory matrix $\mathbf{X}$ and $\mathbf{Y}$). In the joint phase space consider a small neighbourhood of any vector. The dynamics of this neighbourhood can be described via a linear approximation and a linear autoregressive model can be used to predict the dynamics within the neighbourhood. Assume that the vectors of prediction errors are given by $\mathbf{e}_{X|Y}$ and $\mathbf{e}_{Y|X}$. The reconstruction and the fitting procedure are now employed for the individual time series X_T and Y_T in the same neighbourhood and the vector of prediction errors $\mathbf{e}_X$ and $\mathbf{e}_Y$ are then computed. Now, we compute the following criteria

$$\frac{\mathrm{Var}(\mathbf{e}_{X|Y})}{\mathrm{Var}(\mathbf{e}_X)}, \quad \frac{\mathrm{Var}(\mathbf{e}_{Y|X})}{\mathrm{Var}(\mathbf{e}_Y)}. \tag{18.17}$$

The above procedure is then repeated for various regions on the attractor, each column of trajectory matrices $\mathbf{X}$ and $\mathbf{Y}$, and the average of the above criteria are used. The above criteria, clearly, can be considered as a function of neighbourhood size. If the ratios are smaller than 1, we then conclude that there is a nonlinear Granger causal relation between two series. The similarity of nonlinear Granger causality test with SSA causality test is only in the construction of the trajectory matrices **X** and **Y** using embedding terminology, which is only the first step of SSA. Otherwise, the Granger nonlinear test is different from the test considered here. Moreover, the major drawback of the standard nonlinear analysis is that it requires a long time series, while the SSA technique works well for short and long time series (see, for example, Hassani and Zhigljavsky 2009a).

Further discussion of the difference between Granger causality and the SSA-based techniques

One of the main drawbacks of the Granger causality is that we need to assume that the model is fixed (we then just test for significance of some parameters in the model); model can be (and usually is) wrong. The test statistics used for testing the Granger causality are not comprehensive. In the certain case of the linear model, testing for Granger causality consists in the repeated use of the standard F-test which is sensitive to various deviations from the model, and the Granger causality is only associated with the lag difference between the two series.

In our approach, the model of dependence (or causality) is not fixed a priori; instead, this is built into the process of analysis. The models we build are non-parametric and are very broad (in particular, causality is not necessarily associated with a lag) and flexible.

The tests for Granger causality consider the past information of other series in forecasting the series. For example, in the linear Granger causality test, we use the series X up to time t and the series Y up to time $t - d$; and the series Y_{T-d} is used in forecasting series X_T. Whereas in the proposed test here, the series Y_{T+d} is employed in forecasting series X_T.

Furthermore, the tests for Granger causality are based on the forecasting accuracy. In this chapter, we have also introduced another criterion for capturing causality which is based on the predictability of the direction of change. As we mentioned above, it may be more harmful to make a smaller prediction error yet fail in predicting the direction of change, than to make a larger directionally correct error (Ash *et al.* 1997).

The definition of Granger causality does not mention anything about possible instantaneous correlation between two series X_T and Y_T, where the criteria introduced enable an interpretation of an instantaneous causality. In fact, the proposed test is not restricted to the lagged difference between two series. It works equally well when there is no lagged difference between series.

Furthermore, real world time series are typically noisy (e.g. financial time series), non-stationary, and can have small length. It is well known that the existence of a significant noise level reduces the efficiency of the tests (linear and nonlinear) for capturing the amount of dependence between two financial series (see, for example, Hassani *et al.* 2010).

There are mainly two different approaches to examine causality between two time series. According to the first one, that is utilized in current methods, the criteria of capturing causality is computed directly from the noisy time series. Therefore, we ignore the existence of the noise, which can lead to misleading interpretations of causal effects. In our approach, the noisy time series is filtered in order to reduce the noise level and then we calculate the criteria. It is commonly accepted that the second approach is more effective than the first one if we are dealing with the series with high noise level (Soofi and Cao 2002).

18.4 Empirical results

18.4.1 Exchange rate

Given the high correlation between the UK (pound/dollar) and EU (euro/dollar) exchange rates, Hassani *et al.* (2009c) used a two-variable vector autoregressive (VAR) model and SSA (univariate and multivariate model) in exchange rate predicting. This approach to prediction is called a-theoretical, since there is no theoretical justifications in asserting that one exchange rate is a predictor of another one. They showed that VAR model is not a good choice in predicting exchange rate series, while SSA (specifically multivariate version) decisively outperforms the VAR model. They also found that the exchange rate series has a unit root, which implies the series is non-stationary.

Moreover, using Johansen maximum-likelihood method, they also found that the exchange rates are cointegrated, and the Granger causality test showed that the UK/dollar rate does Granger cause the EU/dollar exchange rate series and vice versa.

Next we consider testing for causality between the two exchange rate series using the criteria we have introduced in previous section. First, we consider univariate SSA to forecast one step ahead of the UK and EU exchange rate series, and then compare the MSSA and SSA forecasting results to find $F^{(1,0)}_{UK|EU}$ and $F^{(1,1)}_{UK|EU}$. In this particular example, examining $F^{(h,0)}_{UK|EU}$ also shows whether exchange rate series is martingale or not.

To find the vector of forecasting errors, we forecast all observations of the UK and EU series from 1 May 2009 to 26 June 2009. Figure 18.1 shows these series over the period 3 Jan 2000 to 26 June 2009, in these prediction exercises. Each of these series contain 2452 points. It is very clear that the UK and EU series are highly correlated (indeed, the nonlinear correlation coefficient between UK and EU series is about 0.75). It must be mentioned that this correlation only shows the relationship between the main trends of the series. One source of the relation between the UK and EU exchange rate series is obvious as the two series are each a ratio of US series.

Fig. 18.1 The exchange rate series UK (thick line) and EU (thin line) exchange rate series over the period 3 Jan 2000 to 26 June 2009.

We perform one-step ahead forecasting based on the most up-to-date information available at the time of the forecast. Note that we first use SSA in prediction of a single series, e.g. in prediction of the UK series without using euro series. Next, we use both series simultaneously, e.g. we use the EU time series in forecasting the UK series and vice versa. We shall refer to the results of this step $F^{(1,0)}_{UK|EU}$ and $F^{(1,0)}_{EU|UK}$. We also use one-step ahead information of EU time series as additional information in forecasting UK series and vice versa. We shall call this version of results $F^{(1,1)}_{UK|EU}$ and $F^{(1,1)}_{EU|UK}$. Note that we select window length 3 for both single and multivariate SSA in forecasting exchange rate series. The symbol $*$ indicates the significant results on the 1% level.

It can be observed from Table 18.2 that the difference between the MSSA predictions and SSA is significant with respect to $F^{(h,d)}_{UK|EU}$ and $D^{(h,d)}_{UK|EU}$. The results confirm with that we have improved both accuracy and direction of change of the forecasting results. For example, in forecasting one step ahead for the EU series and $d = 0$, compared to the univariate case, we have improved the accuracy and the direction of change of the forecasting results up to 19% and 8% (column 3 of Table 18.2), respectively. Similarly for the UK exchange rate series with zero lagged difference, MSSA enable

Table 18.2 The value of $F_{UK|EU}$, $D_{UK|EU}$, $F_{EU|UK}$ and $D_{EU|UK}$ in forecasting one-step ahead of the UK and EU exchange rate series for $d = 0$ and 1.

$F^{(1,0)}_{UK\|EU}$	$D^{(1,0)}_{UK\|EU}$	$F^{(1,1)}_{UK\|EU}$	$D^{(1,1)}_{UK\|EU}$	$F^{(1,0)}_{EU\|UK}$	$D^{(1,0)}_{EU\|UK}$	$F^{(1,1)}_{EU\|UK}$	$D^{(1,1)}_{EU\|UK}$
0.94	0.92	0.63*	0.88	0.81*	0.92	0.45*	0.84

an improvement in forecasting accuracy and prediction of the direction of change up to 6% and 8% (with respect to $F^{(1,0)}_{UK|EU}$ and $D^{(1,0)}_{UK|EU}$), respectively. Thus, using the information of the UK and EU exchange rate (with zero lagged difference) enable an improvement the results.

The results obtained so far can be considered as zero-lag correlation between two exchange rate series or multivariate version of the SSA with zero lagged difference. These results can be considered as an evidence that there is the SSA causal relationship between the UK and EU exchange rate of order zero. It should be noted that the SSA causality of zero order confirms that there exists instantaneous causality. The SSA causality of zero order, instantaneous causality, suggests that there might be SSA causal relationship of higher order.

To examine this, next we consider MSSA with one more additional observation for one series. For example, we use the UK exchange rate series up to time t, and the EU exchange rate series up to time $t + 1$ in forecasting one step ahead of the UK exchange rate series to obtain $F^{(1,1)}_{UK|EU}$. In fact, there is one lagged difference between two series in one step ahead forecasting. We use a similar procedure in forecasting the EU series. We expect this additional information gives better results in both forecasting accuracy and the direction of change prediction.

Tt can be observed from columns $F^{(1,1)}_{UK|EU}$, $D^{(1,1)}_{UK|EU}$, $F^{(1,1)}_{EU|UK}$ and $D^{(1,1)}_{EU|UK}$, thus the errors for the MSSA forecast and direction of change, with only one additional observation, are much smaller than those obtained using univariate version. These results are also better than the results obtained using the multivariate approach with zero lag difference. This is not surprising though as the additional data used for forecast is highly correlated with the values we are forecasting. As the results show, the accuracy performance of MSSA has been significantly increased. This means using only one additional observation enable an improvement in forecasting accuracy up to 37% and 55% relative to the univariate version for the UK and EU series (according to $F^{(1,1)}_{UK|EU}$ and $F^{(1,1)}_{EU|UK}$), respectively. Similarly for the direction of change, using only one additional observation enable an improvement in predicting the direction of change up to 12% and 16% (with respect to $D^{(1,1)}_{UK|EU}$ and $D^{(1,1)}_{EU|UK}$).

These results imply that the exchange rate time series are not martingales with respect to all available information available at the currency exchange markets. In fact, the results confirm that the series are SSA causal of order 1. Moreover, $F^{(1,1)}_{EU|UK} > F^{(1,1)}_{UK|EU}$ indicates that, in forecasting this period of the series, the UK exchange rate series is more supportive than the EU series. Furthermore, $F^{(1,0)}_{EU|UK} > F^{(1,0)}_{UK|EU}$ is other evidence for this. This means for this particular example, the SSA casual of order zero then consequences SSA casual of order one as well. However, $D^{(0,1)}_{EU|UK} = D^{(0,1)}_{UK|EU}$ and the discrepancy between $D^{(1,1)}_{EU|UK}$ and $D^{(1,1)}_{UK|EU}$ is not substantially indicating that neither is more directive.

Finally, the results of Table 18.2 strongly confirm that there exists F-feedback and D-feedback between the UK and EU exchange rate series. This means, considering both the UK and EU exchange rate series simultaneously, with and without one additional observation, will improve both the accuracy of forecasting and predictability of the direction of change.

18.4.2 Index of industrial production series

As the second example, we consider the index of industrial production (IIP) series. The IIP series is a key indicator of the state of the UK's industrial base and regarded as a leading indicator of the general state of the economy. The IIP series is published on a monthly basis by the Office for National Statistics (ONS). The index is first released as a provisional estimate and then revised each month to incorporate the information that was not available at the time of the preliminary release. A number of studies have been concerned with the size and nature of revisions to important economic time series. Patterson and Heravi (1991a, b, 1992) have extensively analysed the key national income and expenditure time series. There are many other studies for modelling and forecasting of data revision. For example, Patterson (1995a, b) have used state space approach in forecasting the final vintage of the IIP series and real personal disposable income. For more information about the data revision see Patterson and Heravi (1992, 1994, 1995c).

The overall data period for the study includes 423 monthly observations for 1972:1 to 2007:3 on 12 vintages of data seasonally adjusted IIP. The first vintage, which is published one month after the latest month of published data, refers to the first publication in the monthly *Digest of Statistics*. The second vintage refers to the next published figure and so on. For this study we take the 12th vintage as the final vintage (m), then having 12 vintages of data on the same variables.

Let y_t^v be the vth vintage ($v = 1, \cdots, m$) of the data on variable y for the period t, where $v = 1$ indicates the initially published data and $v = m$ the finally published data. (In practice, m may be taken to indicate the conditionally final vintage.) Here $m = 12$. The structure of the data which is published by *Monthly Digest of Statistics* (MDS) is as follow:

$$\begin{pmatrix} y_1^1 & y_1^2 & y_1^3 & \cdots & y_1^m \\ \vdots & \vdots & \vdots & \ddots & \vdots \\ y_{t-m}^1 & y_{t-m}^2 & y_{t-m}^3 & \cdots & y_{t-m}^m \\ \vdots & \vdots & \vdots & \ddots & \\ y_{t-2}^1 & y_{t-2}^2 & y_{t-2}^3 & & \\ y_{t-1}^1 & y_{t-1}^2 & & & \\ y_t^1 & & & & \end{pmatrix}. \tag{18.18}$$

Thus, publication from a particular issue of MDS traces back a diagonal of this data matrix which is a composite of data of different vintages. We expect that there is a SSA causal relationship between preliminary vintage (vth vintage) and final vintage (mth vintage). To answer this, we need to forecast h step ahead ($h = 1, \dots, 11$) of the final vintage, $v = m$, giving the information at time t. The forecast could be obtained using classical univariate time series methods. However, the forecasts are not optimal since other information (vintages) available at time t are not used. For example, in forecasting y^m_{t-m+1} we also have available information of y^v_{t-m+1} for $v = 1, \dots, m-1$, each of which could itself be regarded as a forecast of y^m_{t-m+1}. This matter motivates us to use a multivariate method for forecasting h step ahead of y^m_t. For example, to obtain the final vintage value at time t, y^m_t, we can use the information for the first vintage data $y^1_1, \dots, y^1_t$ and the final vintage data $y^m_1, \dots, y^m_{t-m}$. If the results of h step ahead forecast MSSA are better than SSA, e.g. $F^{(h,m-i)}_{v^m|v^i} < 1$ and $D^{(h,m-i)}_{v^m|v^i} < 1$, we then conclude that there is a SSA causal relationship of order h between i^{th} vintage and final vintage. To asses this, SSA and MSSA models were estimated using data to the end of 2000 and post-sample forecasts are then computed for 64 observations of 2001:1–2006:3. Thus, we have 64 one step ahead post sample forecast errors, at horizon $h = 1$. The number of forecast errors available decreases as the forecast horizon increases, so that at horizons of $h = 2, 3, \dots, 12$ the number of forecast errors are $63, 62, \dots, 52$, respectively. The value of $F^{(h,m-i)}_{v^m|v^i}$ and $D^{(h,m-i)}_{v^m|v^i}$ ($i = 1, \dots, 11$) for each vintage and relative to single SSA are given in Table 18.3. The two parameters L (window length) and r (number of eigenvalues) chosen in the decomposition and reconstruction are also presented in the table.

From Table 18.3, observe that there are gains to using MSSA throughout the revision process, these being between 87% and 67% for vintage up to

Table 18.3 The value of $F^{(h,m-i)}_{v^m|v^i}$ and $D^{(h,m-i)}_{v^m|v^i}$ in forecasting of ith vintage of the index of industrial production series.

ith Vintage	L	r	$F^{(h,m-i)}_{v^m\|v^i}$	$D^{(h,m-i)}_{v^m\|v^i}$
1	13	5	0.22*	0.45*
2	12	5	0.24*	0.47*
3	11	5	0.27*	0.48*
4	10	5	0.31*	0.50*
5	9	5	0.33*	0.55*
6	8	4	0.36*	0.61*
7	7	4	0.39*	0.65*
8	6	3	0.41*	0.70*
9	5	3	0.45*	0.73*
10	4	3	0.49*	0.77*
11	3	2	0.55*	0.82

$v = 5$, reducing to 50% or slightly less for latter vintages (according to the column labeled $F^{(h,m-i)}_{v^m|v^i}$). This is because, as the structure of the data matrix (18.18) shows, even one observation is very important in forecasting a new vector of the data matrix (18.18). All results are statistically significant at the 1% significant level.

For the direction of change results, for each preliminary vintage v, we compare the true direction of $y^m_t - y^m_{t+v-12}$ with the direction of vintage v estimate $y^v_t - y^m_{t+v-12}$ and the SSA estimate $\hat{y}_t - y^m_{t+v-12}$. Table 18.3 provides the percentage of forecasts that correctly predict the direction of change for each vintage. As the results show the percentage of correct signs produced by MSSA are significantly higher than those given by SSA, these being between 55% and 45% for vintage up to $v = 5$, reducing to 18% for latter vintages (according to the column labelled $D^{(h,m-i)}_{v^m|v^i}$).

Thus, these results, without exception, confirm that there exists a SSA causal relationship between each vintage and the final vintage. In fact the results strongly indicate that there is SSA causality between ith vintage and final vintage is of order $m - i$. It should be noted that here i is equal to h step ahead forecast which is the time lag difference between ith vintage and final vintage. Here, as the results show, SSA causality holds for lower lag order such as in the case of the exchange rate series. This confirms that SSA causality of order $m - i$ has consequences for other orders of causality. Note that here the problem of interest is one-side causality as the final vintage is forecasted.

The results of Granger causality tests, also showed that there is a Granger causal relationship between these series. This is not surprising as each column of the data matrix is a revised version of the previous column and therefore they are high correlated. Also, it should be noted that the results of VAR model in forecasting these series are worse than the MSSA results.

18.5 Conclusion

In this chapter, we developed a new approach in testing for causality between two arbitrary univariate time series. We introduced a family of causality tests which are based on the singular spectrum analysis (SSA) analysis. The SSA technique accommodates, in principle, arbitrary processes, including linear, nonlinear, stationary, non-stationary, Gaussian, and non-Gaussian. Accordingly, we believe our approach to be superior to the traditional criteria used in Granger causality tests, criteria that are based on autoregressive integrated moving average (p, d, q) or multivariate vector autoregressive (VAR) representation of the data; the models that impose restrictive assumptions on the time series under investigation.

Several test statistics and criteria are introduced in testing for casuality. The criteria are based on the idea of minimizing a loss function, forecasting accuracy and predictability of the direction of change. We use the univariate SSA and multivariate SSA in forecasting the value of the series and also prediction of the direction.

The performance of the proposed test was examined using the euro/dollar and the pound/dollar daily exchange rates as well as the index of industrial production (IIP) series for the United Kingdom. It has been shown here that the euro/dollar rate causes the pound/dollar rate and vice versa. Moreover, it has been documented that, without exception, there exists a SSA causal relationship between each vintage and final vintage of the IIP data.

References

Ash, J. C. K., Smyth, D. J., and Heravi, S. (1997). The accuracy of OECD forecasts for Japan. *Pacific Economic Review*, **2** (1), pp. 25–44.

Bosq, D. (1998). *Nonparametric Statistics for Stochastic Processes* (2nd edn), Springer, New York.

Broomhead, D. S. and King, G. P. (1986). Extracting qualitative dynamics from experimental data, *Physica D*, **20**, pp. 217–236.

Chu, T., Danks, D. and Glymour, C. (2004). *Data Driven Methods for Nonlinear Granger Causality: Climate Teleconnection Mechanisms*, 2004.

Clements, M. P. and Smith, J. (1999). A Monte Carlo investigation of forecasting performance of emprical SETAR model. *Journal of Applied Econometrics*, **14**, 123–141.

Danilov, D. and Zhigljavsky, A (eds). (1997). *Principal Components of Time Series: The 'Caterpillar' Method*, University of St. Petersburg St. Petersburg (in Russian).

Diebold F. X. and Mariano R. S. (1995). Comparing predictive accuracy, *Journal of Business and Economic Statistics*, **13**(3), pp. 253–63.

Diks, C. and DeGoede, J. (2001). A general nonparametric bootstrap test for Granger causality. In: H.W. broer, B. Krauskopf, G. Vegter (eds.), *Global Analysis of Dynamical Systems* Institute of Physics Publishing, Bristol, UK, pp. 391–403.

Diks, C. and Panchenko, V. (2005). A note on the Hiemstra–Jones test for Granger non-causality. *Studies in Nonlinear Dynamics and Econometrics*, **9**, art. 4.

Diks, C. and Panchenko, V. (2006). A new statistic and practical guidelines for non-parametric Granger causality testing, *Journal of Economic Dynamics and Control*, **30**, pp. 1647–1669.

Fleiss, J. L. (1981). *Statistical Methods for Rates and Proportions* (2nd edn) John Wiley, New York.

Golyandina, N., Nekrutkin, V., and Zhigljavsky, A. (2001). *Analysis of Time Series Structure: SSA and related techniques*, Chapman & Hall/CRC, New York.

Granger, C. W. J. (1969). Investigating causal relations by econometric models and cross-spectral methods, *Econometrica*, **37** (3), pp. 424–438.

Granger, C. W. J. (1980). Testing for causality: A personal viewpoint, *Journal of Economic Dynamics and Control*, **2**, pp. 329–352.

Harvey, D. I., Leybourne, S. J. and Newbold, P. (1997). Testing the equality of prediction mean squared errors, *International Journal of Forecasting*, **13**, 281–291.

Hassani, H., Dionisio, A., and Ghodsi, M. (2010). The effect of noise reduction in measuring the linear and nonlinear dependency of financial markets, *Nonlinear Analysis: Real World Applications*, **11** (1), pp. 492–502.

Hassani H, and Zhigljavsky A. (2009a). Singular spectrum analysis: Methodology and application to economics data, *journal of system science and complexity*, 22(3), pp. 372–394.

Hassani H. Heravi S. and Zhigljavsky A. (2009b). Forecasting European industrial production with singular spectrum analysis, *International Journal of Forecasting*, 25(1), pp. 103–118.

Hassani, H. Soofi, A. and Zhigljavsky, A. (2009c). Predicting daily exchange rate with singular spectrum analysis. *Nonlinear Analysis: Real World Applications*, **11**(3), 2023–2034

Hassani, H. (2009d). Singular spectrum analysis based on the minimum variance estimator, *Nonlinear Analysis: Real World Applications*, **11**(3), pp. 2065–2077.

Hassani, H. (2007). Singular spectrum analysis: Methodology and comparison. *Journal of Data Science*, 5(2), pp. 239–257.

Heravi, S., Osborn, D. R. and Birchenhall, C. R. (2004). Linear versus neural network forecasts for european industrial production series. *International Journal of Forecasting, 20*, 435–446.

Hiemstra, C. and Jones, J. D. (1994). Testing for linear and nonlinear Granger causality in the stock price-volume relation. *Journal of Finance*, **49**(5), pp. 1639–1664.

Su, L., and White, H. (2008). A nonparametric hellinger metric test for conditional independence, *Econometric Theory*, 24(4), pp. 829–864

Patterson, K. Hassani, H. Heravi, S. and Zhigljavsky, A. (2010). Forecasting the final vintage of the index of industrial production, *Journal of Applied Statistics*, Forthcoming.

Patterson, K. D. (1992). Revisions to the components of the trade balance for the United Kingdom, *Oxford Bulletin of Economics and Statistics*, **54**, pp. 103–120.

Patterson, K. D. (1994). Consumers' expenditure on nondurables and services and housing equity withdrawal in the United Kingdom, *Manchester School of Economic and Social Studies*, **62**, pp. 251–274.

Patterson, K. D. (1995a). An integrated model of the data measurement and data generation processes with an application to consumers' expenditure, *Economic Journal*, **105**, pp. 54–76.

Patterson, K. D. (1995b). A state space approach to forecasting the final vinatege of revised data with an application to the index of industrial production, *Journal of Forecasting*, **14**, pp. 337–350.

Patterson, K. D. (1995c). Forecasting the final vintage of real personal disposable income: A state space approach, *International Journal of Forecating*, **11**, pp. 395–405.

Patterson, K. D. and Heravi, S. M. (1991a). Data revisions and the expenditure components of GDP, *Economic Journal*, **101**, pp. 887–901.

Patterson, K. D. and Heravi, S. M. (1991b). Are different vintages of data on the components of GDP cointegrated? *Economics Letters*, **35**, pp. 409–413.

Patterson, K. D. and Heravi, S. M. (1992). Efficient forecasts or measurement errors? Some evidence for revisions to the United Kingdom GDP growth rates, *Manchester School of Economic and Social Studies*, **60**, pp. 249–263.

Plaut, G. R. and Vautard, R. (1994). Spells of oscillations and weather regimes in the low-frequency dynamics of the Northern Hemisphere. *J. Atmos. Sci.*, 51, 210–236.

Soofi, A. and Cao, L. Y. (2002). Nonlinear forecasting of noisy financial data, in A. soofi and L. Y. cao (eds.), *Modeling and Forecasting Financial Data: Techniques of Nonlinear Dynamics*, Kluwer Academic Publishers, Boston.

Su, L. and White, H. (2008). Nonparametric Hellinger metric test for conditional independence, *Econometric Theory*, **24**, pp. 829–864.

Toda, H. Y. and Phillips, P. C. B. (1991). Vector autoregressions and causality: A theoretical overview and simulation study, *Econometric Reviews*, **13**, pp. 259–285.

Appendix A: Formal description of SSA

Stage 1: Decomposition

First step: Embedding

Embedding can be regarded as a mapping that transfers a one-dimensional time series $Y_T = (y_1, \ldots, y_T)$ into the multi-dimensional series $X_1, \ldots, X_K$ with vectors $X_i = (y_i, \ldots, y_{i+L-1})' \in \mathbf{R}^L$, where $K = T - L + 1$. Vectors X_i are called *L-lagged vectors* (or, simply, *lagged vectors*). The single parameter of the embedding is the *window length* L, an integer such that $2 \le L \le T$. The result of this step is the trajectory matrix $\mathbf{X} = [X_1, \ldots, X_K]$:

$$\mathbf{X} = (x_{ij})_{i,j=1}^{L,K} = \begin{pmatrix} y_1 & y_2 & y_3 & \cdots & y_K \\ y_2 & y_3 & y_4 & \cdots & y_{K+1} \\ \vdots & \vdots & \vdots & \ddots & \vdots \\ y_L & y_{L+1} & y_{L+2} & \cdots & y_T \end{pmatrix}. \tag{18.19}$$

Note that the trajectory matrix $\mathbf{X}$ is a Hankel matrix, which means that all the elements along the diagonal $i + j =$ const are equal. Embedding is a standard procedure in time series analysis. With the embedding performed, future analysis depends on the aim of the investigation.

Second step: Singular value decomposition (SVD)

The second step, the SVD step, makes the singular value decomposition of the trajectory matrix and represents it as a sum of rank-one bi-orthogonal elementary matrices. Denote by $\lambda_1, \ldots, \lambda_L$ the eigenvalues of $\mathbf{XX}'$ in decreasing order of magnitude ($\lambda_1 \ge \ldots \lambda_L \ge 0$) and by $U_1, \ldots, U_L$ the orthonormal system (that is, (U_i, U_j)=0 for $i \neq j$ (the orthogonality property) and $\| U_i \|$=1 (the unit norm property)) of the eigenvectors of the matrix $\mathbf{XX}'$ corresponding to these eigenvalues. (U_i, U_j) is the inner product of the vectors U_i and U_j and $\| U_i \|$ is the norm of the vector U_i. Set

$$R = \max(i, \text{ such that } \lambda_i > 0) = \text{rank}\,\mathbf{X}.$$

If we denote $V_i = \mathbf{X}' U_i / \sqrt{\lambda_i}$, then the SVD of the trajectory matrix can be written as:

$$\mathbf{X} = \mathbf{X}_1 + \cdots + \mathbf{X}_R, \tag{18.20}$$

where $\mathbf{X}_i = \sqrt{\lambda_i} U_i V_i'$ $(i = 1, \ldots, R)$. The matrices $\mathbf{X}_i$ have rank 1; therefore they are elementary matrices, U_i (in SSA literature they are called 'factor empirical orthogonal functions' or simply EOFs) and V_i (often called 'principal components') stand for the left and right eigenvectors of the trajectory matrix. The collection $(\sqrt{\lambda_i}, U_i, V_i)$ is called the ith eigentriple of the matrix $\mathbf{X}$, $\sqrt{\lambda_i}$ $(i = 1, \ldots, R)$ are the singular values of the matrix $\mathbf{X}$ and the set $\{\sqrt{\lambda_i}\}$ is called the spectrum of the matrix $\mathbf{X}$. If all the eigenvalues have multiplicity one, then the expansion (18.20) is uniquely defined.

SVD (18.20) is optimal in the sense that among all the matrices $\mathbf{X}^{(r)}$ of rank $r < R$, the matrix $\sum_{i=1}^{r} X_i$ provides the best approximation to the trajectory matrix $\mathbf{X}$, so that $\| \mathbf{X} - \mathbf{X}^{(r)} \|$ is minimum. Note that $\| \mathbf{X} \|^2 = \sum_{i=1}^{R} \lambda_i$ and $\| \mathbf{X}_i \|^2 = \lambda_i$ for $i = 1, \ldots, d$. Thus, we can consider the ratio $\lambda_i / \sum_{i=1}^{R} \lambda_i$ as the characteristic of the contribution of the matrix $\mathbf{X}_i$ to expansion (18.20). Consequently, $\sum_{i=1}^{r} \lambda_i / \sum_{i=1}^{R} \lambda_i$, the sum of the first r ratios, is the characteristic of the optimal approximation of the trajectory matrix by the matrices of rank r.

Stage 2: Reconstruction

First step: Grouping

The grouping step corresponds to splitting the elementary matrices $\mathbf{X}_i$ into several groups and summing the matrices within each group. Let $I = \{i_1, \ldots, i_p\}$ be a group of indices $i_1, \ldots, i_p$. Then the matrix $\mathbf{X}_I$ corresponding to the group I is defined as $\mathbf{X}_I = \mathbf{X}_{i_1} + \cdots + \mathbf{X}_{i_p}$. The spilt of the set of indices $J = 1, \ldots, R$ into the disjoint subsets $I_1, \ldots, I_m$ corresponds to the representation

$$\mathbf{X} = \mathbf{X}_{I_1} + \cdots + \mathbf{X}_{I_m}. \tag{18.21}$$

The procedure of choosing the sets $I_1, \ldots, I_m$ is called the eigentriple grouping. For given group I the contribution of the component $\mathbf{X}_I$ into the expansion (1) is measured by the share of the corresponding eigenvalues: $\sum_{i \in I} \lambda_i / \sum_{i=1}^{R} \lambda_i$.

Second step: Diagonal averaging

Diagonal averaging transfers each matrix I into a time series, which is an additive component of the initial series Y_T. If z_{ij} stands for an element of a matrix $\mathbf{Z}$, then the kth term of the resulting series is obtained by averaging z_{ij} over all i, j such that $i + j = k + 2$. This procedure is called *diagonal averaging*, or Hankelization of the matrix $\mathbf{Z}$. The result of the Hankelization of a matrix $\mathbf{Z}$ is the Hankel matrix $\mathcal{H}\mathbf{Z}$, which is the trajectory matrix corresponding to the series obtained as a result of the diagonal averaging.

The operator $\mathcal{H}$ acts on an arbitrary $L \times K$-matrix $\mathbf{Z} = (z_{ij})$ with $L \leq K$ in the following way: for $i + j = s$ and $N = L + K - 1$ the element $\widetilde{z}_{ij}$ of the matrix $\mathcal{H}\mathbf{Z}$ is

$$\begin{cases} \dfrac{1}{s-1} \displaystyle\sum_{l=1}^{s-1} z_{l,s-l} & 2 \leq s \leq L-1, \\ \dfrac{1}{L} \displaystyle\sum_{l=1}^{L} z_{l,s-l} & L \leq s \leq K+1, \\ \dfrac{1}{K+L-s+1} \displaystyle\sum_{l=s-K}^{L} z_{l,s-l} & K+2 \leq s \leq K+L. \end{cases}$$

Note that the Hankelization is an optimal procedure in the sense that the matrix $\mathcal{H}\mathbf{Z}$ is the nearest to $\mathbf{Z}$ (with respect to the matrix norm) among all Hankel matrices of the corresponding size (for more information see Golyandina *et al.* 2001, chap. 6, sec. 2). In its turn, the Hankel matrix $\mathcal{H}\mathbf{Z}$ uniquely defines the series by relating the value in the diagonals to the values in the series. By applying the Hankelization procedure to all matrix components of (18.21), we obtain another expansion:

$$\mathbf{X} = \widetilde{\mathbf{X}}_{I_1} + \ldots + \widetilde{\mathbf{X}}_{I_m} \tag{18.22}$$

where $\widetilde{\mathbf{X}}_{I_1} = \mathcal{H}\mathbf{X}$. This is equivalent to the decomposition of the initial series $Y_T = (y_1, \ldots, y_T)$ into a sum of m series:

$$y_t = \sum_{k=1}^{m} \widetilde{y}_t^{(k)} \tag{18.23}$$

where $\widetilde{Y}_T^{(k)} = (\widetilde{y}_1^{(k)}, \ldots, \widetilde{y}_T^{(k)})$ corresponds to the matrix $\mathbf{X}_{I_k}$.

Selection of parameters

Here we consider a version of SSA where we split the set if indicies $\{1, 2, \ldots, L\}$ into two groups only: $I = \{1, \ldots, r\}$ and $\bar{I} = \{r+1, \ldots, L\}$. We associate the group I with signal and the group $\bar{I}$ with noise. The SSA method requires then the selection of two parameters, the window length L and the number of elementary matrices r. There are specific rules for selecting these parameters; their choice depends on structure of the data and the analysis we want to perform. Detailed description of parameter selection procedures is given in Golyandina *et al.* (2001). Here we summarize a few general rules.

The window length L is the single parameter that should be selected at the decomposition stage. Selection of the proper window length depends on the problem in hand, and on preliminary information about the time series. For the series with a complex structure, too large window length L can produce an undesirable decomposition of the series components of interest, which may lead, in particular, to their mixing with other series component. Let us, for example, consider the problem of trend extraction in GCM. Since trend is a relatively smooth curve, its separability from noise requires small values of L. It should be noted that the values of L should not be smaller than the true eigenvalues r. The chosen L also should results good separability between the reconstructed series using $I = \{1, \ldots, r\}$ and $\bar{I} = \{r+1, \ldots, L\}$. In growth curve model that we are dealing with only trend extraction, usually the first or second eigenvalue is considered for reconstruction step.

The first elementary matrix $\mathbf{X}_1$ with the norm $\sqrt{\lambda_1}$ has the highest contribution to the norm of $\mathbf{X}$ in $\mathbf{X} = \mathbf{X}_1 + \cdots, \mathbf{X}_L$ and the last elementary matrix $\mathbf{X}_L$ with the norm $\sqrt{\lambda_L}$ has the lowest contribution to the norm of $\mathbf{X}$. The plot of the eigenvalues $\lambda_1, \cdots, \lambda_L$ gives an overall view concerning the values of the eigenvalues and is essential in deciding where to truncate the summation of $\mathbf{X} = \mathbf{X}_1 + \cdots, \mathbf{X}_L$ in order to build a good approximation of the original matrix. A slowly decreasing sequence of eigenvalues typically indicate the presence of noise in the series.

A group of r (with $1 \leq r < L$) eigenvectors determine an r-dimensional hyperplane in the L-dimensional space $\mathbb{R}^L$ of vectors X_j. The distance between vectors $X_j (j=1, \ldots, K)$ and this r-dimensional hyperplane can be rather small (it is controlled by the choice of the eigenvalues) meaning that the projection of **X** into this hyperplane is a good approximation of the original matrix **X**. If we choose the first r eigenvectors $U_1, \ldots, U_r$, then the squared L_2-distance between this projection and **X** is equal to $\sum_{j=r+1}^{L} \lambda_j$. According to the Basic SSA algorithm, the L-dimensional data is projected onto this r-dimensional subspace and the subsequent averaging over the diagonals allows us to obtain an approximation to the original series.

Forecasting algorithm

Let us formally describe the forecasting algorithm under consideration (for more information see Golyandina *et al.* 2001):

Algorithm input:

(a) Time series $Y_T = (y_1, \ldots, y_T)$.

(b) Window length L, $1 < L < T$.

(c) Linear space $\mathfrak{L}_r \subset \mathbf{R}^L$ of dimension $r < L$. It is assumed that $e_L \notin \mathfrak{L}_r$, where $e_L = (0, 0, \ldots, 1) \in \mathbf{R}^L$.

(d) Number M of points to forecast for.

Procedure:

(a) $\mathbf{X} = [X_1, \ldots, X_K]$ is the trajectory matrix of the time series Y_T.

(b) $U_1, \ldots, U_r$ is an orthonormal basis in $\mathfrak{L}_r$.

(c) $\widehat{\mathbf{X}} = [\widehat{X}_1 : \ldots : \widehat{X}_K] = \sum_{i=1}^{r} U_i U_i^T \mathbf{X}$. The vector $\widehat{X}_i$ is the orthogonal projection of X_i onto the space $\mathfrak{L}_r$.

(d) $\widetilde{\mathbf{X}} = \mathcal{H}\widehat{\mathbf{X}} = [\widetilde{X}_1 : \ldots : \widetilde{X}_K]$ is the result of the Hankellization of the matrix $\widehat{\mathbf{X}}$.

(e) For any vector $Y \in \mathbf{R}^L$ we denote by $Y_\Delta \in \mathbf{R}^{L-1}$ the vector consisting of the last $L-1$ components of the vector Y, while $Y^\nabla \in \mathbf{R}^{L-1}$ is the vector of the first $L-1$ components of the vector Y.

(f) We set $v^2 = \pi_1^2 + \ldots + \pi_r^2$, where π_i is the last component of the vector U_i $(i = 1, \ldots, r)$.

(g) Suppose that $e_L \notin \mathfrak{L}_r$. (In the other words, we assume that $\mathfrak{L}_r$ is not a vertical space.) Then $v^2 < 1$. It can be proved that the last component y_L of any vector $Y = (y_1, \ldots, y_L)^T \in \mathfrak{L}_r$ is a linear combination of the first $L-1$ components $(y_1, \ldots, y_{L-1})$:

$$y_L = a_1 y_{L-1} + \ldots + a_{L-1} y_1.$$

Vector $A = (a_1, \ldots, a_{L-1})$ can be expressed as

$$A = \frac{1}{1 - v^2} \sum_{i=1}^{r} \pi_i U_i^\nabla \tag{18.24}$$

and does not depend on the choice of a basis $U_1, \ldots, U_r$ in the linear space $\mathfrak{L}_r$. In the above notations, define the time series $Y_{T+M} = (y_1, \ldots, y_{T+M})$ by the formula

$$y_i = \begin{cases} \widetilde{y}_i & \text{for } i = 1, \ldots, T \\ \sum_{j=1}^{L-1} a_j y_{i-j} & \text{for } i = T+1, \ldots, T+M. \end{cases} \tag{18.25}$$

The numbers $y_{T+1}, \ldots, y_{T+M}$ from the M terms of the SSA recurrent forecast. Let us define the linear operator $\mathcal{R}^{(r)} : \mathfrak{L}_r \mapsto \mathbf{R}^L$ by the formula

$$\mathcal{R}^{(r)} Y = \begin{pmatrix} Y_\Delta \\ A^T Y_\Delta \end{pmatrix}, \quad Y \in \mathfrak{L}_r.$$

If we set

$$Z_i = \begin{cases} \widetilde{X}_i & \text{for } i = 1, \ldots, K \\ \mathcal{R}^{(r)} Z_{i-1} & \text{for } i = K+1, \ldots, K+M \end{cases} \tag{18.26}$$

the matrix $\mathbf{Z} = [Z_1, \ldots, Z_{K+M}]$ is the trajectory matrix of the series Y_{T+M}. Therefore, (18.26) can be regard as the vector form of (18.25).

The SSA recurrent forecasting algorithm can be modified in several ways. For example, we can base our forecast on the Toeplitz SSA or SSA with centering rather than on the basic SSA. Perhaps the most important modification is the so-called SSA vector forecasting algorithm developed in Golyandina *et al.* (2001).

So far we considered SSA recurrent forecasting algorithm. In the following we consider SSA vector forecasting algorithm. The SSA vector forecasting algorithm has the same inputs and conditions as the SSA recurrent forecasting algorithm. The notation in (a)–(g) is kept. Let us introduce some more notations. Consider the matrix

$$\Pi = V^\nabla (V^\nabla)^T + (1 - v^2) A A^T$$

where $V^\nabla = [U_1^\nabla, \ldots, U_r^\nabla]$. The matrix Π is the matrix of the linear operator that performs the orthogonal projection $\mathbf{R}^{L-1} \mapsto \mathfrak{L}_r^\nabla$, where $\mathfrak{L}_r^\nabla = \text{span}(U_1^\nabla, \ldots, U_r^\nabla)$. We define the linear operator $\mathcal{V}^{(v)} : \mathfrak{L}_r \mapsto \mathbf{R}^L$ by the formula

$$\mathcal{V}^{(v)} Y = \begin{pmatrix} \Pi Y_\Delta \\ A^T Y_\Delta \end{pmatrix}, \quad Y \in \mathfrak{L}_r$$

In the notation above we define the vectors Z_i as follow:

$$Z_i = \begin{cases} \widetilde{X}_i & \text{for } i = 1, \ldots, K \\ \mathcal{V}^{(v)} Z_{i-1} & \text{for } i = K+1, \ldots, K+M+L-1. \end{cases}$$

By constructing the matrix $\mathbf{Z} = [Z_1, \ldots, Z_{K+M+L-1}]$ and making its diagonal averaging we obtain a series $y_1, \ldots, y_{T+M+L-1}$. The numbers $y_{T+1}, \ldots, y_{T+M}$ form the M terms of the SSA vector forecast.

PART V
Natural sciences

·19·

Mechanism schemas and the relationship between biological theories

Tudor M. Baetu

Abstract

Current accounts of the relationship between classical genetics and molecular biology favour the 'explanatory extension' thesis, according to which molecular biology elucidates aspects of inheritance unexplained by classical genetics. I identify however an unresolved tension between the 'explanatory extension' account and examples of 'explanatory interference' (cases when the accommodation of data from molecular biology results in a more precise genotyping and more adequate classical explanations). This chapter provides a new way of analysing the relationship between classical genetics and molecular biology capable of resolving this tension. The proposed solution makes use of the properties of mechanism schemas and sketches, which can be completed by elucidating some or all of their remaining 'black boxes' and instantiated via the filling-in of phenomenon–specific details. This result has implications for the reductionism-antireductionism debate since it shows that molecular elucidations have a positive impact on classical explanations without entailing the reduction of classical genetics to molecular biology.

19.1 Introduction

It is widely accepted by scientists (Morange 1998; Muller 1951) and philosophers (Darden 1991; 2006; Schaffner 1969; Waters 2004) alike that the development of molecular biology was driven by an attempt to answer the questions: 'What is the physical nature of genes?' and 'How do genes determine phenotypes?' Such questions fall outside the immediate explanatory scope of classical genetics, which is mainly concerned with the transmission of inherited traits (Morgan 1935; Moss 2003; Waters 2004).

Given the seemingly distinct explanatory scopes of classical genetics and molecular biology, Philip Kitcher (Culp and Kitcher 1989; Kitcher 1982, p. 357; 1999, p. 199) proposed the 'explanatory extension' thesis, according to which molecular biology explains aspects of inheritance not explained by classical genetics. Kitcher (1989) embedded this thesis in the larger context of his own unificationist account of explanation, which relies on the notion that explanation requires a particular kind of deductive argument schema. However, it has been repeatedly pointed out that most theories and explanations in biology cannot be accounted for in terms of laws and logical derivations (Hull 1979; Rosenberg 1985; Sober 1993), but are best characterized as descriptions of productive mechanisms (Bechtel and Abrahamsen 2005; Machamer, Darden, and Craver 2000; Skipper, 1999; Wimsatt, 1976). Given this shortcoming, Lindley Darden articulated a mechanistic version of the 'explanatory extension' thesis, more readily applicable to explanations in biology. Darden argues that

> [t]he general knowledge in molecular biology is best characterized not in terms of laws or a theory but as a set of mechanism schemas [where a] mechanism schema is a truncated abstract description of a mechanism that can be easily instantiated by filling it with more specific descriptions of component entities and activities. (2006, pp. 111–112)

Building on this new approach, Darden (2006, p. 98) argues that classical genetics and molecular biology elucidate 'separate but serially connected mechanisms'. According to this account, Mendel provided a highly schematic outline of a series of events explaining inheritance phenomena. Then, classical genetics elucidated in more detail some elements contained in this scheme (e.g. the mechanism of allelic segregation, explained by meiosis, and recombination, explained by chromosomal crossing-over) while relegating other elements to 'black-boxes' (the ability of alleles to replicate and determine phenotypes), thus providing a first incomplete general schema of a series of mechanisms. Finally, molecular biology gradually filled in the remaining 'black-boxes' with more and more mechanistic details (most famously, the mechanisms of DNA replication and gene expression), until the present-day picture of genetics emerged. The 'serially connected mechanisms' account specifies the aspects of inheritance explained by classical genetics and those explained by molecular biology, and shows how the mechanisms postulated by classical and molecular explanations fit together without leaving gaps in the productive continuity from start (the genotype of the parents) to finish conditions (the expression of specific traits in the offspring).

While the 'serially connected mechanisms' account provides an adequate description of the relationship between classical genetics and molecular biology, it fails to explicitly address a major point of disagreement between reductionists and antireductionists: Does the elucidation of molecular details impose revisions of classical explanations? If the 'explanatory extension' thesis is true, then the intuitive answer is 'No.' Indeed, Hull (1974) and Kitcher (1982; 1989) suggest that the elucidation of molecular details must be neutral in respect to individual classical explanations: the molecular details contribute to a better understanding of aspects of inheritance not explained by classical genetics, but they do not provide a better explanation of the transmission patterns already explained by classical explanations. While tempting, this answer is incorrect. In this chapter, I discuss examples of 'explanatory interference', that is, cases when the accommodation of data from molecular biology results in a more precise genotyping and more adequate classical-style explanations of the transmission patterns associated with certain inherited conditions. Such examples constitute a problem, since they seem to contradict the 'explanatory extension' thesis and raise the possibility that a reductionist or eliminativist account might, after all, provide a better account of the relationship between classical genetics and molecular biology.

This chapter has two aims. First, I identify instances of 'explanatory interference' in contemporary research practice. This is an important achievement, since, despite vigorous debates in the past, it has never been conclusively shown that the elucidation of molecular details impacts specifically on the empirical adequacy of classical explanations dealing with issues of transmission. Second, I show that both the 'serially connected mechanisms' account and instances of 'explanatory interference' can be accommodated without contradiction. To this end, I propose a new way of analysing the relationship between classical genetics and molecular biology hinging on the properties of mechanism schemas, which can be completed, on the one hand, by elucidating some or all of the remaining 'black boxes' and, on the other, by instantiation via the filling-in of phenomenon-specific details. This result has implications for the reductionism–antireductionism debate since it shows that molecular elucidations have a positive impact on classical explanations, yet does not entail the reduction of classical genetics to molecular biology.[1]

The chapter is organized as follows: In section 19.2, I review presently available answers to the question 'Do molecular elucidations have a positive impact on classical explanations?' In section 19.3, I proceed to show that molecular elucidations have in fact a positive impact on classical-style explanations. In section 19.4, I provide a solution to the apparent incompatibility between the 'explanatory extension' thesis underlying the 'serially connected mechanisms' account and instances of 'explanatory interference' demonstrated in actual scientific practice. Finally, in section 19.5, I summarize my arguments and discuss implications for the reductionism-antireductionism debate in genetics.

[1] In response to initial attempts to model the relationship between classical genetics and molecular biology as a form of inter-theoretical reductionism, it has been convincingly argued that such a form of reductionism is not something biologists are actively interested in achieving (Darden 2006, 105–105; Schaffner 1974; 1993, p. 512; Waters 2008, p. 249).

19.2 Do molecular elucidations have a positive impact on classical explanations?

The issue of reductionism in genetics takes, in part, the form of a debate about whether the elucidation of molecular details requires a revision of previously accepted classical explanations. In Alex Rosenberg's words, antireductionism is the claim that

> nonmolecular biological explanations are adequate and need no macromolecular correction, completion, or grounding. (2007, p. 120)

All parties agree that molecular biology elucidates the lower-level mechanistic and structural details of the entities and activities hypothesized by classical explanations. Furthermore, both reductionists and antireductionists acknowledge the causal relevance of the molecular structures and mechanisms underlying cytological, developmental and other higher-level biological phenomena. For example, nobody is denying that a phenomenon such as recombination is explained at the cytological level by chromosomal crossing-over, itself explained at the molecular level by a mechanism involving Spo11-mediated double-strand DNA break followed by the formation of a Holliday junction (for a more detailed example and philosophical discussion, see levels of mechanisms, Craver 2007, chap. 5)). Rather, the disagreement is about the explanatory relevance of the molecular details to already successful nonmolecular explanations. Thus, the question is 'Does the elucidation of the molecular mechanisms underlying cytological entities and activities contribute to the ability of classical genetics to provide more adequate explanations of the transmission of inherited traits phenomena?'

Several attempts to assess the impact of molecular elucidations on classical explanations have been made during exchanges between reductionists and antireductionists, yet no definitive conclusion was ever reached. Kitcher (1984) claims that taking into account the molecular details muddles the crispy-clear explanations of classical genetics, while Rosenberg (1985, p. 101) sees an unbridgeable degree of complexity separating classical and molecular explanations. In both cases, the argument is that molecular analysis reveals a multiplicity of interacting and redundant gene products involved in the production of any single phenotype, while transmission genetics explains the same phenotype more simply and elegantly by assigning it a small number of alleles associated with a single locus. The argument is however rather vague, as no particular examples are discussed in thorough scientific detail. Equally problematic, the argument hinges on an alleged virtue of simplicity which, as I will show, is not reflected in the views and results of prominent geneticists such as T. H. Morgan or S. Benzer. Finally, neither Kitcher, nor Rosenberg discusses the possibility that taking into account molecular details sacrifices simplicity in favour of a much more valuable increase in empirical adequacy.

In his more recent work, Rosenberg claims that

> [m]olecular information about the location and structure of the genetic material [...] helps the Classical geneticist understand where Mendel's 'laws' go wrong, and what exceptions to these rules of thumb are to be expected. (1997, p. 447)

He concludes that

> molecular biology shows why Classical genetics is a useful instrument, even pedagogically indispensable, but is fundamentally flawed. (1997, p. 447)

It can be retorted though that Mendel's laws were corrected not so much by molecular biology, but within classical genetics itself after the discovery of linkage, recombination and complementation. Furthermore, claiming that classical genetics is 'fundamentally flawed' doesn't sit well with the generally accepted view that classical genetics offers satisfactory explanations of certain aspects of inheritance.

Kenneth Waters (1990, pp. 132–133) argues against Kitcher that knowledge of the molecular mechanisms underlying chromosomal crossing over must somehow contribute to our understanding of inheritance. This must be indeed the case, but the claim is too general to conclude something specifically restricted to the impact of molecular elucidations on classical explanations. More recently, Waters adopts a different approach to the reductionism–antireductionism debate. He argues that

> [t]he developments following Watson and Crick's discovery that mattered were not primarily theoretical. [...] What changed biology so dramatically was a retooling of the investigative strategies used in genetics. (2008, p. 239)

The example discussed (a RNAi knockout study) shows how molecular biology provides additional means of experimental investigation and control in the context of a classical-style (forward) genetic analysis. Waters is making a valid point, but this cannot be the whole story. As it stands, his account seems to entail that molecular biology is nothing but a set of experimental techniques and not a scientific field proper, endowed with its own theoretical and experimental resources. This is a counterintuitive conclusion that very few biologists would endorse. Furthermore, even if the contribution of molecular biology is primarily experimental, it is still not clear how the knowledge generated by classical and molecular techniques fits together. Do the two sources of information complement each other by providing knowledge about distinct aspects of the phenomena under study? Or do they make claims about the same aspects, and therefore there is a possibility that contradictions may arise? These questions remain unanswered.

19.3 Mendelian errors and molecular genotyping

My answer to the question 'Does the elucidation of molecular details have an impact on classical explanations?' is 'Yes.' The key element in understanding how molecular elucidations can have an impact on classical explanations rests on the notion of 'schema instantiation.'[2] For example, the 'Central Dogma' is an abstract mechanism schema highlighting the common elements of the mechanisms responsible for prokaryotic and eukaryotic gene expression in general.

However, if it is generally understood that genes determine phenotypes via a universal mechanism of gene expression involving transcription and translation, molecular explanations of individual phenotypes require the elucidation of many further details, such as the DNA sequences involved, the mechanisms regulating the expression of these sequences and the mechanisms by means of which the expressed gene products contribute to the phenotype under investigation. In other words, in order to provide a satisfactory explanation of the genetic underpinnings of any given trait, the 'Central Dogma' schema needs to be instantiated by elucidating and filling in phenomenon-specific details.

A similar comment applies to classical genetics. The Machamer–Darden (2000) characterization of mechanisms is compatible with the notion that classical genetics offers a general schema explaining the transmission of inherited traits by appealing to the segregation and recombination of alleles located at specific chromosomal loci (see Figure 19.1). However, this schema too needs to be instantiated before it can account for the peculiarities of any given inheritance phenomenon. To give a very striking example, as early as 1911, Morgan discovered the complementation of the white and pink eye mutants in *Drosophila* (Morgan 1911). Complementation refers to a situation whereby the crossing of two different kinds of homozygous recessive mutants yields a wild-type phenotype (Lewis 1951; Benzer 1955). Classical geneticists interpreted complementation as an indication that the two mutations affect two distinct 'genetic units', dubbed 'functional units.' If the mutations were in the same unit, then the offspring could not have received a copy of the wild-type unit since none of the parents had one to begin with. This immediately indicates that, in many cases, more than one functional unit is required for the expression of any given phenotype. Furthermore, mutations

[2] Kitcher (1989) proposed that general knowledge in science consists of 'schematic arguments' (sequences of 'schematic sentences' in which some non-logical expressions are replaced with dummy letters) that can be instantiated by means of a set of 'filling instructions' for each term of the schematic argument. In the case of mechanistic explanations, Darden (2006) argues that general knowledge is best described as a set of mechanisms schemas, often represented via diagrams, that can be instantiated by filling it with specific descriptions of component entities and activities.

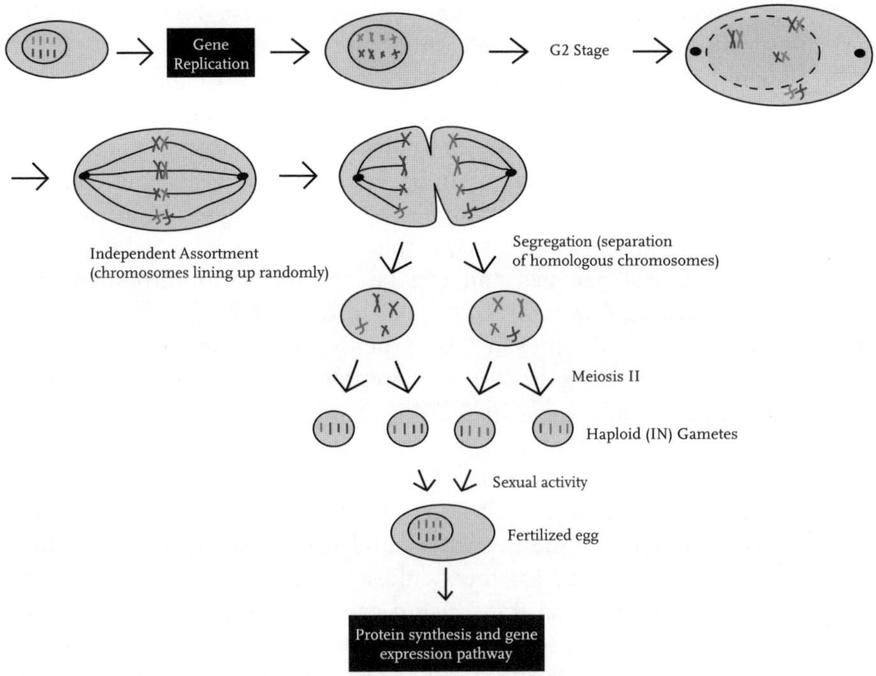

Fig. 19.1 Explanatory extension as serially connected mechanism schemas. (From Darden 2006, used with permission).

targeting apparently unrelated traits can complement, meaning that mutation in one gene can affect several traits/biological functions. Finally, since non-complementing mutations can map at distinct chromosomal loci it is possible to distinguish between mutations in the same functional unit that result in an identical phenotype, meaning that classical geneticists were able to distinguish mutations that cannot be directly differentiated by observing phenotypes before DNA sequencing techniques became available. This indicates that the general schema postulating segregation and recombination of alleles located at specific chromosomal loci provides only the rough guideline for a genetic explanation. A considerable amount of further research is required in order to work out genetic maps capable of accounting for every single instance

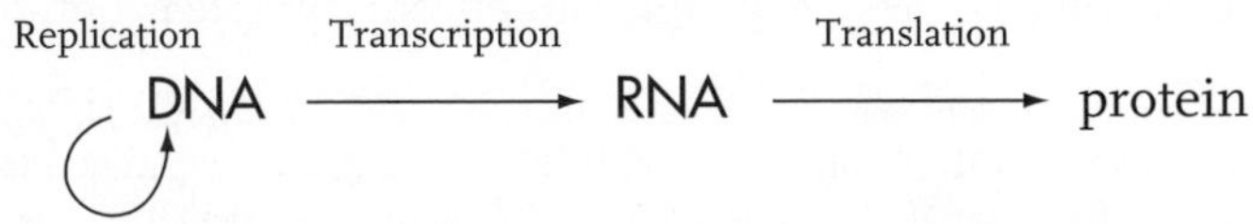

Fig. 19.2 The Central Dogma of molecular biology.

of linkage, recombination and complementation associated with the specific transmission patterns under investigation.[3]

Since schema instantiation is dependent on further information, it becomes interesting to investigate the nature (experimental data vs. theoretical assumptions) and origin (the theory's own internal vs. external theoretical and experimental resources) of this information. In the case of classical genetics, genes can be identified via classical techniques involving analyses of linkage and complementation (or *cis-trans*) assays; this strategy relies on the internal resources of classical genetics and can be referred to as intra-theoretical schema instantiation. Alternatively, genes can be identified as transcription units characterized by the presence of structural motifs (e.g. TATA box followed by an open reading frame) and homology with known gene product sequences (Altschul *et al.* 1990; Wain *et al.* 2002); this strategy relies on the external resources of molecular biology, and can be referred to as cross-theoretical schema instantiation. Cross-theoretical instantiations of the general explanatory schema of classical genetics are possible because loss of function mutations in the regulatory and coding sequences of distinct transcription units required for the synthesis of gene products involved in the same metabolic or signal transducing pathway complement each other. Hence, transcription units identified via molecular techniques behave like classical functional units. Furthermore, transcription units were shown to overlap extensively with functional units, as mapped via classical analysis (Benzer 1966; Mosig and Eiserling 2006).

Next, since the same general schema can be instantiated intra- and cross-theoretically, it becomes interesting to establish whether the two instantiations coincide, or whether they conflict with each other. If conflicts arise, it is important to find out how they are settled. In the case of classical genetics, it can be shown that taking into account theoretical assumptions and experimental data from molecular biology force distinct, and usually more complex, schema instantiations. Furthermore, such cross-theoretical instantiations are typically more adequate than straightforward intra-theoretical instantiations relying solely on the internal resources of classical genetics.

Let us imagine that a particular metabolic pathway involves three enzymes E1, E2 and E3. Given this piece of information from biochemistry, it becomes reasonable to hypothesize that at least three distinct genes must be expressed. Let us further assume that G_{E1}, G_{E2} and G_{E3} are then identified as alleles encoding functional products, while alternative alleles M_{E1}, M_{E2} and M_{E3}, naturally occurring or created in the lab, encode mutated, non-functional

[3] The discovery of complementation also shows that the simplicity advertised by Kitcher (1984) and Rosenberg (1985, p. 101) dissolves away in the kind of complexity typically associated with molecular analysis. By the same token, Rosenberg's (1997, p. 447) argument that classical genetics is false because nothing in the physical world corresponds to its level of simplicity is also considerably weakened.

products. In order for the normal/wild-type phenotype N_{Ph} to obtain, an organism must have at least one copy of the genes G_{E1}, G_{E2} and G_{E3} (in classical terms, the wild-type allele is dominant). A mutant phenotype M_{Ph} obtains when at least one gene is mutated on both chromosomes (the mutant allele is recessive). Consider now that G_{E1} is never mutated in the populations accessible to classical analysis and that G_{E2}/M_{E3} and M_{E2}/G_{E3} fail to complement because, it turns out, their respective products E_2 and E_3 form a functional hetero-dimer.[4] If a geneticist ignores these details from biochemistry and molecular biology, he or she ends up identifying a single gene required for normal metabolism and concludes that two alleles of the same gene are associated with metabolic function and dysfunction. This is a case when intra-theoretical schema-filling diverges from cross-theoretical schema-filling. One gene or three, the general explanatory scheme remains the same. However, in this hypothetical case, theoretical assumptions justified by a partial elucidation of the metabolic pathways underlying the phenotype under investigation (i.e. the one-enzyme one-gene assumption, Beadle and Tatum, 1941) prompts a distinct schema instantiation that does not coincide with the simpler one favoured by straightforward classical analysis.

Subtle, yet measurable differences in symptoms and responses to treatment always puzzled clinical geneticists. Simplified genotypes provide adequate explanations only by ignoring a host of minute variables, such as the severity of the symptoms, the onset of the disease, secondary complications, and differences in response to treatment. In contrast, a more minute analysis taking into account the molecular details is often able to account for the diversity of sub-phenotypes. A famous example is that of Huntington disease. Classical analysis notoriously failed to explain why 5% of the individuals inheriting the dominant allele for Huntington don't develop the disease while the remaining 95% are affected to various degrees (in classical terms, the allele is said to be partially penetrant and to display various degrees of expressivity). In contrast, the molecular analysis reveals that the affected gene (HTT) contains multiple repeats of the CAG sequence, coding for glutamine. The normal gene codes for less than 27 glutamine amino-acid repeats, while the mutated version codes for 36 or more (Kieburtz *et al.* 1994). The number of additional glutamines in the final gene product is related to the rate of neuronal decay, and thus to the severity of the symptoms (Chong *et al.* 1997). As it turns out, it is not the case that a single allele displays different degres of penetrance and expressivity; instead, a whole series of mutations is responsible for the variability of the observed symptoms (Figure 19.3). The molecular instantiation provides a more satisfactory explanation both from an empirical adequacy

[4] Two peptide chains, coded by distinct genes, combining to form a single functional protein; in such cases complementation experiments were shown to be inconclusive (Garen and Garen 1963).

point of view, as well as by avoiding giving metaphysical weight to the purely hypothetical properties of 'penetrance' and 'expressivity.' Note also that, for the time being, the mechanism underlying the disease, as well as the function of HTT, are not fully understood. The molecular analysis does not elucidate the 'black box' mechanism linking genotype and phenotype. Rather, it provides a more adequate 'classical-style' characterization of the genotype associated with the disease and its transmission patterns.

Subtle differences between the postulated genotypes also make an important difference when it comes to providing accurate predictions of offspring phenotypic frequencies required for medical applications such as genetic counseling. A very striking example where a more complex genotype resulting from cross-theoretical schema filling is more adequate than the simpler genotypes resulting from straightforward intra-theoretical schema filling is that of Marfan, Loeys-Dietz and Ehlers-Danlos syndromes. Since they are all autosomal dominant diseases characterized by similar symptoms, they were, and still are often confused as a unique genetic disease. It turns out Marfan syndrome is caused by mutations in the fibrillin-1 gene, coding for a glycoprotein found in the extracellular matrix. Although the exact mechanism of the disease is not known, the favoured explanation is that fibrillin-1 binds TGFβ (known to inhibit cell growth and induce apoptosis) keeping it inactive; dysfunctional or reduced levels of fibrilin-1 result in a TGFβ-induced inflammatory reaction leading to connective tissue degradation (Pereira *et al.* 1999). In contrast, Loeys-Dietz syndrome is caused by mutations in the TGFβ receptor genes resulting in enhanced TGFβ signaling (Loeys *et al.* 2005). Finally, the Ehlers-Danlos family of syndromes is due to mutations that affect the structure or production of collagen (reviewed by Beighton *et al.* 1998). It follows that TGFβ receptor inhibitors may help alleviate the symptoms of Marfan, but not those of Loeys–Dietz and Ehlers–Danlos syndromes. Targeting intracellular components of the TGFβ signaling pathway may provide a cure for the Loeys–Dietz and Marfan syndromes, but should have no impact on patients affected by Ehlers–Danlos syndromes. Finally, gene therapy targeting the collagen genes may provide a cure to the Ehlers–Danlos family of syndromes, but are expected to be ineffective in treating the Marfan and the Loeys–Dietz syndromes. An empirically adequate explanation must explain such minute clinical differences. In this respect, the simpler explanation postulated in light of the classical analysis (i.e. mutations in one gene are responsible for one disease) is less adequate than the more complex explanation taking into account partial knowledge from molecular biology.

In the above examples, a partial elucidation of the molecular details prompts a revision of the genotypes underlying the inherited condition. A revision of the genotype counts as an instance of 'explanatory interference' because it prompts a further revision of the predictions made by classical-style explanations (e.g. the patterns of transmission associated with that condition). Since

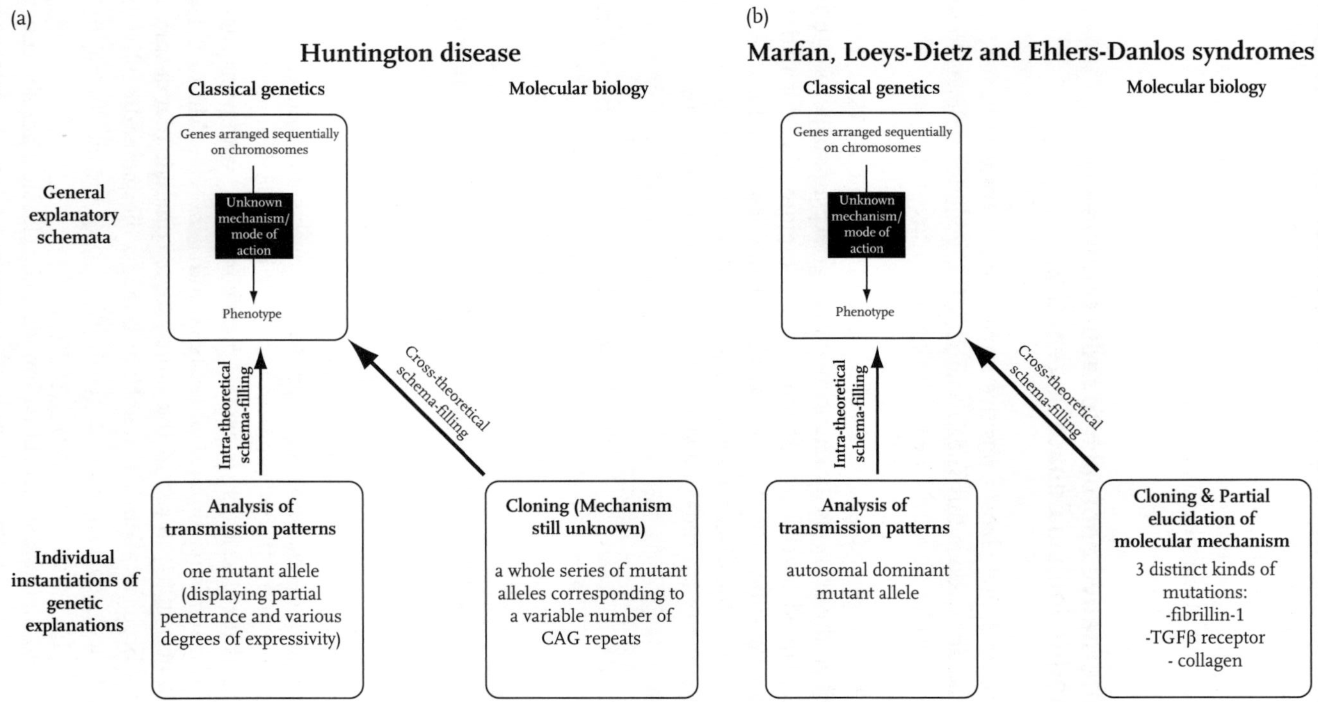

Fig. 19.3 Alternative, intra- vs. cross-theoretical schema instantiations in classical genetics. (The arrows indicate the import of specific details into the general schema of classical genetics.)

these revisions result in a better empirical fit (e.g. explain the spectrum of phenotypic differences associated with Huntington disease), more satisfactory explanations (e.g. dispense with the notions of 'penetrance' and 'expressivity') and an ability to explain potential anomalies (e.g. differences in response to treatment), I conclude that the elucidation of the molecular details has a positive impact on classical explanations.

19.4 An alternative approach to reductionism–antireductionism debate in genetics

If molecular elucidations have a positive impact on the empirical adequacy of classical explanation, does this mean that classical genetics reduces to or is replaced by molecular biology? The reductionism–antireductionism debate in genetics hinges, in part, on the issue of 'explanatory extension' vs. 'explanatory interference.' Some philosophers of biology (Hull 1974; Kitcher 1984) rejected reductionism on the grounds that molecular biology explains aspects of inheritance not explained by classical genetics and, therefore, the elucidation of molecular details shouldn't affect classical explanations. Paradoxically, this claim is falsified by the discussed examples, yet reductionism is by no means vindicated. Reductionism posits a stronger thesis, summarized by Waters as follows:

> Watson and Crick's discovery [...] led to a deeper and more fundamental, molecular-level theory. The new theory allegedly improves upon higher-level explanations of the classical theory by explaining its core theoretical principles in terms of molecular processes. (2008, p. 239)

Even if the examples discussed in this paper show that classical explanations need corrections and are enhanced by taking into account molecular elucidations, terms like 'meiotic segregation', 'recombination' or 'complementation' are not recast in molecular terms. Nor does there seem to be any motivation for such a recasting. Data from molecular biology are used to identify functional units, which are classical gene concepts (genes as 'difference makers', Griffiths and Stotz 2007; Rheinberger and Müler-Wille 2008; Waters 1994) and provide the basis for classical-style explanations concerned with transmission and not the mechanisms linking phenotype and genotype (Morgan 1935; Moss 2003; Waters 2004). It seems therefore safe to conclude that despite the improvements brought about by molecular biology, the general explanatory schema of classical genetics is not derived, inferred or reconstructed in any way from biochemistry or molecular biology.

The solution to this paradoxical situation whereby classical genetics doesn't seem to reduce to or be replaced by molecular biology, yet there is a clear sense in which molecular biology represents an improvement over classical genetics

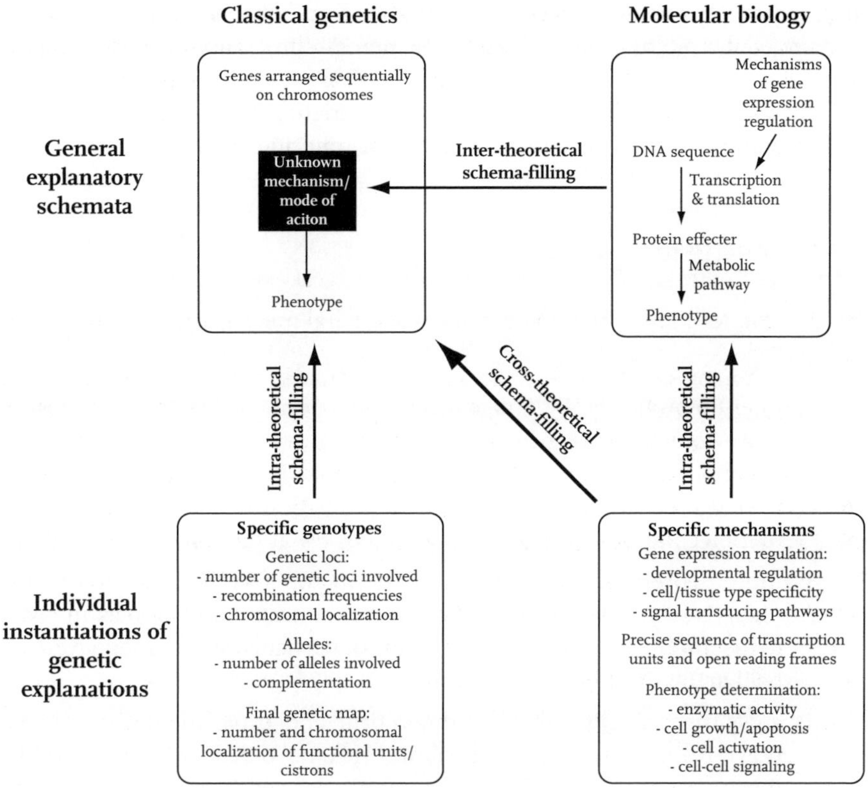

Fig. 19.4 Inter-, intra- and cross-theoretical schema-filling.

lies in the peculiarities of mechanism schemas. According to the 'serially connected mechanisms' account, the 'black-boxes' of the general schema of classical genetics are elucidated in order to generate the general explanatory schema of molecular biology. This is a form of inter-theoretical schema-filling accounting for the 'explanatory extension' aspect of the relationship between classical genetics and molecular biology. At the same time, the same general schema of classical genetics is also instantiated by filling in phenomenon-specific details in order to generate individual explanations of inheritance phenomena. This can be achieved intra-theoretically, by using the internal resources of classical genetics; or cross-theoretically, by taking into account data and assumptions from molecular biology (Figure 19.4).

As exemplified in the previous section, there can be a clash between intra- and cross-theoretical schema instantiation, typically resolved in favour of a cross-theoretical instantiation taking into account molecular findings. However, this clash has no bearing on inter-theoretical schema-filling and the 'explanatory extension' aspect of the relationship between classical genetics and molecular biology. Inter-theoretical schema-filling is about elucidating the 'black boxes' of the general schema of classical genetics, while intra-/cross-theoretical schema-filling is a matter of instantiating individual explanations by filling in the details specific to a particular inheritance phenomenon.

19.5 Conclusion

In this chapter I argue that the details of a general mechanism schema can be completed:

(i) inter-theoretically, by elucidating some or all of the 'black boxes' of a previous general explanatory schema in order to generate another general explanatory schema, as illustrated by Darden's 'serially connected mechanisms' account of the transition from classical genetics to molecular biology (horizontal bold arrow at the top of Figure 19.4);

(ii) intra-theoretically, by filling in phenomenon-specific details using the theory's own internal theoretical and experimental resources in order to instantiate individual explanations of particular phenomena (vertical bold arrows in Figure 19.4); and

(iii) cross-theoretically, whereby theoretical and experimental considerations from one theory contribute specific details required for the instantiation of individual explanations derived from another general explanatory schema without explicitly elucidating any of the 'black boxes' of the latter (diagonal bold arrow in Figure 19.4).

Distinguishing between the three types of schema-filling is crucial for a clear understanding of the complex relationship between classical genetics and molecular biology. As expounded by Darden (2006), and on occasions by Kitcher (1982; 1999), the 'explanatory extension' thesis is a claim about inter-theoretical schema-filling. In as much as molecular biology elucidates the 'black boxes' of the general schema of classical genetics, the relationship between the two sciences is neither reductive, nor eliminative, but rather a form of cumulative completion. To this day, genetics combines classical experimental and explanatory strategies, such as breeding and genetic linkage mapping of phenotypes, with cloning, sequencing and reverse genetic analysis later introduced by molecular biology (Falk 2003; Vance 1996; Waters 2008). Furthermore, the fact that experimental data and theoretical assumptions from molecular biology have a positive impact on classical-style explanations hints to a high level of integration of the two sciences in a unique field of research. In these respects, the transition from classical genetics to molecular biology is theoretically-cumulative, ruling out reductionism in favour of inter-field integration (Darden and Maull 1977; Darden 2006). Finally, since classical genetics and molecular biology provide explanations at different levels of organization (e.g. cytological vs. molecular), claims to a 'mosaic' of multilevel explanations in biological sciences (Craver 2007) are also vindicated.

In contrast, according to a non-integrative brand of antireductionism held sometimes by Kitcher (1984) and identified by Rosenberg (2007) as defining antireductionism in biology, molecular biology fails to contribute in a positive way to the ability of classical genetics to provide adequate explanations of inheritance phenomena. This thesis is a combination of the 'explanatory extension' thesis doubled by the claim that classical genetics generates its most successful explanations in virtue of intra-theoretical schema instantiation, while 'explanatory interference' from molecular biology (cross-theoretical instantiation) is impossible, irrelevant or damaging. If this is how antireductionism is construed, then antireductionism is false. I showed by means of examples that data from biochemistry and molecular biology needs to be accommodated by changes in genetic explanations resulting in a more precise genotyping. In turn, a more accurate knowledge of genotypes plays an important role in making more accurate predictions of phenotypic distributions, assessing the risk of disease and response to treatment, providing more accurate diagnosis and genetic counseling.

In order to resolve the apparent problem, I argued that instances of 'explanatory interference' do not conflict with the 'explanatory extension' aspect of the relationship between classical genetics and molecular biology as long as a distinction is made between two distinct degrees of abstraction. At the degree of abstraction associated with the general explanatory schemas of classical genetics and molecular biology, molecular biology extends classical genetics by elucidating some of its 'black boxes' (inter-theoretical schema-filling). In contrast, at the level of specific instantiations associated with individual explanations, data from molecular biology can and often is used to fill-in phenomenon-specific details (cross-theoretical schema-filling). As exemplified in the chapter, taking into account the molecular details results in more adequate explanations.

Although this complex issue transcends the more modest scope of this paper, I would like to conclude with a few words on the possible implications for the connection between causation and explanation. The 'explanatory extension' aspect of the relationship between classical genetics and molecular biology indicates that not all causally relevant factors are necessarily deemed explanatorily relevant. For most philosophers, this conclusion is hardly a surprise. What may come as a surprise is that the explanatory relevance of causally relevant factors seems to be context-sensitive. At the degree of abstraction associated with general explanatory schemas, the molecular details (e.g. the molecular mechanisms underlying chromosomal crossing-over) are acknowledged to be causally relevant for the production of inheritance phenomena, with all that this may entail in terms of experimental manipulability and practical applications, yet their elucidation did not prompt a rethinking of previously accepted classical explanations (e.g. recombination as an explanation of certain anomalies in offspring phenotypic frequencies). However, as shown in this paper, at the level of particular instantiations of classical-style explanations, some molecular details become highly relevant.

Acknowledgements

I would like to thank Lindley Darden, Carl Craver, Phyllis McKay Illari, Brendan Ritchie, discussants at the Kent Conference on Causality and Mechanisms, the DC History and Philosophy of Biology group, and two referees for helpful discussion and comments on earlier drafts. This work was supported by 'Fonds de la recherche sur la société et la culture', Québec, Canada [grant number 127231].

References

Altschul, S. F., W. Gish, W. Miller, E. W. Myers, and D. J. Lipman. (1990). Basic local alignment search tool. *Journal of Molecular Biology* 215: 403–410.

Beadle, G. W., and E. L. Tatum. (1941). Genetic control of biochemical reactions in neospora. *Proceedings of the National Academy of Science* 27: 499–506.

Bechtel, W., and A. Abrahamsen. (2005). Explanation: A mechanist alternative. *Studies in History and Philosophy of Biological and Biomedical Sciences* 36: 421–441.

Beighton, P., A. De Paepe, B. Steinmann, P. Tsipouras, and R. J. Wenstrup. (1998). Ehlers–Danlos syndromes: Revised nosology. *American Journal of Medical Genetics* 77: 31–7.

Benzer, S. (1955). Fine structure of a genetic region in bacteriophage. *Proceedings of the National Academy of Science* 41: 344–354.

Benzer, S. (1966). Adventures in the rII region. in J. Cairns, G.S. Stent and J.D. Watson (eds.), *Phage and the Origins of Molecular Biology*. Cold Spring Harbor: Cold Spring Harbor Laboratory Press.

Chong, S. S., E. Almqvist, H. Telenius, L. LaTray, K. Nichol, B. Bourdelat-Parks, Y. P. Goldberg, *et al.* (1997). Contribution of DNA sequence and CAG size to mutation frequencies of intermediate alleles for Huntington disease: Evidence from single sperm analyses. *Human Molecular Genetics* 6: 301–9.

Craver, C. (2007). *Explaining the Brain : Mechanisms and the Mosaic Unity of Neuroscience*. Oxford: Clarendon Press. http://www.loc.gov/catdir/toc/ecip077/2006103230.html.

Culp, S., and P. Kitcher. (1989). Theory structure and theory change in contemporary molecular biology. *British Journal for the Philosophy of Science* 40: 459–483.

Darden, L. (1991). *Theory Change in Science: Strategies from Mendelian Genetics*. New York: Oxford Univ. Press.

Darden, L. (2006). *Reasoning in Biological Discoveries: Essays on Mechanisms, Interfield Relations, and Anomaly Resolution*. Cambridge: Cambridge University Press.
Darden, L. and N. Maull. (1977). Interfield theories. *Philosophy of Science* 44: 43–64.
Falk, R. (2003). Linkage: From particulate to interactive genetics. *Journal of the History of Biology* 36: 87–117.
Garen, A., and S. Garen. (1963). Complementation *in vivo* between structural mutants of alkaline phosphatase from *E. coli*. *Journal of Molecular Biology* 7: 13–22.
Griffiths, P., and K. Stotz. (2007). Gene. In *The Cambridge Companion to the Philosophy of Biology*. Cambridge: Cambridge University Press.
Hull, D. (1974). *Philosophy of Biological Science*. Englewood Cliffs: Prentice Hall.
Hull, D. (1979). Reduction in genetics. *Philosophy of Science* 46: 316–320.
Kieburtz, K., M. MacDonald, C. Shih, A. Feigin, K. Steinberg, K. Bordwell, C. Zimmerman, J. Srinidhi, J. Sotack, and J. Gusella. (1994). Trinucleotide repeat length and progression of illness in Huntington's disease. *Journal of Medical Genetics* 31: 872–874.
Kitcher, P. (1982). Genes. *British Journal for the Philosophy of Science* 33: 337–359.
Kitcher, P. (1984). 1953 and all that: A tale of two sciences. *Philosophical Review* 93: 335–373.
Kitcher, P. (1989). Explanatory unification and the causal structure of the world. In *Minnesota Studies in the Philosophy of Science*, XIII: Minneapolis: Minnesota University Press.
Kitcher, P. (1993). *The Advancement of Science: Science without Legend, Objectivity without Illusions*. New York: Oxford University Press.
Kitcher, P. (1999). The hegemony of molecular biology. *Biology and Philosophy* 14: 195–210.
Lewis, E. B. (1951). Pseudoallelism and gene evolution. *Cold Spring Harbor Symposia on Quantitative Biology* 16: 159–174.
Loeys, B. L., J. Chen, E. R. Neptune, D. P. Judge, M. Podowski, T. Holm, J. Meyers, *et al.* (2005). A syndrome of altered cardiovascular, craniofacial, neurocognitive and skeletal development caused by mutations in TGFBR1 or TGFBR2. *Nature Genetics* 37: 275–281.
Machamer, P., L. Darden, and C. Craver. (2000). Thinking about mechanisms. *Philosophy of Science* 67: 1–25.
Morange, M. (1998). *A History of Molecular Biology*. Cambridge, MA: Harvard University Press.
Morgan, T. H. (1911). The origin of five mutations in eye color in *drosophila* and their modes of inheritance. *Science* 33: 534–537.
Morgan, T. H. (1935). The relation of genetics to physiology and medicine. *Les prix Nobel en 1933. Imprimerie Royale*: 1–16.
Mosig, G., and F. Eiserling. (2006). T4 and related phages: Structure and development. In *The Bacteriophages*. Oxford: Oxford University Press.
Moss, L. (2003). *What Genes Can't Do*. Cambridge: MIT Press.
Muller, H. J. (1951). The development of the gene theory. In *Genetics in the 20th Century: Essays on the Progress of Genetics During its First 50 Years*, pp. 77–99. New York: Macmillan.
Pereira, L., S. Y. Lee, B. Gayraud, K. Andrikopoulos, S. D. Shapiro, T. Bunton, N. J. Biery, H. C. Dietz, L. Y. Sakai, and F. Ramirez. (1999). Pathogenetic sequence for aneurysm revealed in mice underexpressing fibrillin-1. *Proceedings of the National Academy of Science* 96: 3819–3823.

Rheinberger, H.-J., and S. Müller-Wille. (2008). Gene concepts. In *A Companion to the Philosophy of Biology* 39, 3–21. Malden, MA: Blackwell.

Rosenberg, A. (1985). *The Structure of Biological Science*. Cambridge: Cambridge University Press.

Rosenberg, A. (1997). Reductionism redux: Computing the embryo. *Biology and Philosophy* 12: 445–470.

Rosenberg, A. (2007). Reductionism (and antireductionsim) in biology. In D. Hull and M. Ruse (eds.), *The Cambridge Companion to the Philosophy of Biology*, 349–368. Cambridge: Cambridge University Press.

Schaffner, K. F. (1969). The Watson–Crick model and reductionism. *British Journal of Philosophy of Science* 20: 325–348.

Schaffner, K. F. (1974). The peripherality of reductionism in the development of molecular genetics. *Journal of the History of Biology* 7: 111–139.

Schaffner, K. F. (1993). *Discovery and explanation in biology and medicine*. Chicago: University of Chicago Press.

Skipper, R. (1999). Selection and the extent of explanatory unification. *Philosophy of Science* 66: S196–S209.

Sober, E. (1993). *Philosophy of Biology*. Boulder: Westview Press.

Vance, R. E. (1996). Heroic antireductionism and genetics: A tale of one science. *Philosophy of Science* 63: S36–45.

Wain, H. M., E. A. Bruford, R. C. Lovering, M. J. Lush, M. W. Wright, and S. Povey. (2002). Guidelines for human gene nomenclature. *Genomics* 79: 464–470.

Waters, C. K. (1990). Why the anti-reductionist consensus won't survive: The case of classical Mendelian genetics. *Proceedings to the Biennial Meeting of the Philosophy of Science Association*: 125–139.

Waters, C. K. (1994). Genes made molecular. *Philosophy of Science* 61: 163–185.

Waters, C. K. (2004). What was classical genetics? *Studies in History and Philosophy of Science* 35: 783–809.

Waters, C. K. (2008). Beyond theoretical reduction and layer-cake antireduction: How DNA retooled genetics and transformed biological practice. In *The Oxford Handbook of Philosophy of Biology*, pp. 238–262. Oxford: Oxford University Press.

Wimsatt, W. C. (1976). Reductive explanation: A functional account. In *Boston Studies in the Philosophy of Science*, 30:671–710. Vol. 30. Dordrecht: Reidel.

·20·

Chances and causes in evolutionary biology: How many chances become one chance

Roberta L. Millstein

Abstract

As a number of biologists and philosophers have emphasized, 'chance' has multiple meanings in evolutionary biology. Seven have been identified. I will argue that there is a unified concept of chance underlying these seven, which I call the UCC (Unified Chance Concept). I will argue that each is characterized by which causes are considered, ignored, or prohibited. Thus, chance in evolutionary biology can only be understood through understanding the causes at work. The UCC aids in comparing the different concepts and allows us to characterize our concepts of chance in probabilistic terms, i.e. provides a way to translate between 'chance' and 'probability'.

20.1 Introduction

In everyday English language contexts, the term 'chance' has multiple meanings.[1] For example:

> I'd give that horse a 50–50 chance (degree of belief) of winning.
> I hope I get the chance (i.e. opportunity) to see you.
> I found this great restaurant just by chance (i.e. accident).
> I'm sorry, but I just can't take the chance (i.e. risk).

As a number of biologists and philosophers have emphasized, 'chance' also has multiple meanings in *evolutionary biology* (see, e.g. Monod 1971; Beatty 1984; Eble 1999; Millstein 2000a, 2006; Gayon 2005; Lenormand *et al.* 2009; Merlin 2009). Elsewhere (Millstein 2000a, 2006), drawing on Beatty (1984) and Eble (1999), I argue that there are at least six conceptions

[1] It also has multiple synonyms, or almost synonyms, 'stochasticity' and 'randomness' in particular. However, for the most part I will restrict myself to the broader term 'chance', since both stochasticity and randomness often have particular mathematical connotations that not all concepts of chance have.

of chance that are potentially relevant to evolutionary theory. To that list, I add a seventh, chance as contingency (Gould 1989; Beatty 1995, 2006a), as follows:

1. indeterministic chance ('pure' chance)
2. chance as ignorance of the real underlying causes
3. chance as not designed
4. chance as sampling (both discriminate and indiscriminate)
5. chance as coincidence
6. evolutionary chance (independent of the generally adaptive direction of natural selection)
7. chance as contingency.

This list may still not be exhaustive; however, I will limit my comments here to these seven. As I will show, each of these seven concepts of chance has a distinct meaning, and each plays a role[2] within evolutionary biology (although some play a greater role than others).[3]

The question of the nature of chance, and related questions of determinism and indeterminism, are longstanding philosophical problems that have occupied philosophers for centuries. More recently, philosophers of science in general and philosophers of physics in particular have looked to quantum mechanics to settle issues concerning the nature of chance and to settle questions concerning the fundamentally probabilistic nature of the universe. However, evolutionary theory itself has a decidedly probabilistic character (indeed, probabilities are ubiquitous in evolutionary theory), one which in some sense does not seem to rest on any new discoveries in quantum mechanics; evolutionary biology was given a probabilistic formulation prior to and independently of the development of quantum mechanics. And as will be discussed below, chance played a role in Darwin's evolutionary thinking prior to evolutionary theory's twentieth-century probabilistic formulations. Thus, the study of the various concepts of chance in evolutionary biology may shed light on the study of chance in quantum mechanics and in other areas of science and philosophy.

The fact that there are many concepts of chance within evolutionary biology raises the following questions: Is there something that all of the concepts

[2] Possible roles include explanatory, instrumental, representational, and/or justificatory (see Millstein 2006 for discussion).

[3] One possible exception here is indeterministic chance. I myself don't think that it plays a direct role in evolutionary theory, in part for philosophical reasons I have discussed elsewhere (Millstein 2000b) and in part because it is simply my impression that it is not what biologists usually *mean* when they invoke chance and probability. However, as we shall see below, others do think that indeterminism plays a role in evolution. Moreover, it at least plays an indirect role in the sense that it is often explicitly rejected in order to begin to clarify which sense of chance *is* being used in a particular context.

have in common that can serve as a unified concept of chance? Or is this a heterogeneous collection? If there is a unified conception of chance, what is it? What makes all the chance concepts 'chance'? And what can a unified concept of chance be used for?

I will argue that there *is* a unified concept of chance, which I call the UCC (Unified Chance Concept). However, it will turn out to be quite general, almost trivial; indeed, like many unifications, it is achieved only through a loss of content.[4] Because of this loss of content, it cannot replace the other concepts of chance. In other words, though I seek one concept of chance, the end result is not monism, but pluralism.

Nonetheless, I will argue, the UCC is not without utility: It tells us that the seven different concepts do have something in common, and identifies that common core, perhaps aiding in identifying future concepts both within and outside of evolutionary biology. It can be used to aid in comparisons among the different concepts. And perhaps most importantly, it will allow us to characterize our concepts of chance in probabilistic terms (i.e. provide a way to translate between 'chance' and 'probability').

20.2 Characterization of the Unified Chance Concept (UCC)

We will see each of the seven concepts of chance has a somewhat different meaning (although they can be hard to differentiate because a particular biological phenomenon might manifest more than one concept simultaneously). However, one striking commonality between them, as will become clearer below, is that there are all described in the negative, in contrast to various causes. What is interesting about this is that *indeterminism* is – though a bit misleadingly, I think – sometimes characterized as meaning 'uncaused'.

To see why 'uncaused' is a misleading definition of indeterminism, consider, for example, radioactive decay under the assumption that it is an indeterministic phenomenon.[5] When one compares the half-lives of two elements (e.g. carbon-14 and uranium-238), it is their different structures that give rise to (cause) their different half-lives. Even a particular decay event is caused by the structure of the particular atom. It may be, however, that two identical atoms in identical environments may not (indeed, are likely not to) decay at the same time. This is the intuition behind the idea that indeterministic events are 'uncaused', but again, that overlooks the causal role played by the structure of the atom.

[4] My favourite example: All scientific theories could be unified under the assertion that 'things happen'. (Slightly modified from a version suggested to me by Frédéric Bouchard.) This would be true, but so general as to be completely uninformative.

[5] See Earman (1986) for a far more detailed and sophisticated discussion for why 'uncaused' is an inaccurate description of the sort of indeterminism suggested by quantum mechanics.

Thus, even as indeterministic events are not fully caused, they are still (to an extent) caused. Similarly, it will turn out that 'chance' is defined both by what causes are left out and what causes are included (this will be explained further below).

Another commonality (again, as will become clearer below), is that both indeterminism and the various concepts of chance imply more than one possible outcome. These commonalities suggest that an analogy between indeterminism, as an empirical claim about the world, might prove fruitful for understanding the various concepts of chance. That is, examining the commonalities between each concept of chance and indeterminism gives rise to a set of characteristics that are common to all of the concepts of chance and also helps illuminate why they are considered to be concepts of chance (i.e. because they are similar to indeterminism). I won't walk the reader through that particular exercise, but rather, describe the analogy and the UCC and then in subsequent sections show how each concept of chance fits.

But we need a new definition of indeterminism. Here is a slightly better[6] one: Given the *complete state* of the world at one point in time, the state of the world at every future point in time is not uniquely determined; for a given point of time in the future, more than one state is possible.[7] Now suppose that the UCC were analogous – but not identical – to the definition of indeterminism. In particular, suppose we consider *not the complete state* of the world, but rather, some *subset* of it. This yields, I will argue, the UCC:

UCC: Given a specified subset of causes, more than one future state is possible.

This might seem trivial. But the key will be to identify the subset of causes for each type of chance. Identifying the subset of causes involves identifying, for each concept of chance, which causes are taken into account (what I will call the *considered* causes), which causes are operating but *ignored*,[8] and which causes are *prohibited* from operating altogether if the particular concept of chance is to be manifested (if any). The particular chance concepts differ in the types of causes that are considered, ignored, and prohibited; they also differ in the relevant types of *possible outcomes*. Figure 20.1 will serve as a basic template for the UCC, to be filled in with specifics for each concept of chance.

[6] Most importantly, the definition is an ontological one (and thus is a claim about the world), rather than an epistemological one (which would be a claim about our, or a Laplacean demon's, ability to make predictions about the future).

[7] That is, *physically* possible. Here, I leave the characterization of physical possibility open; for example, one may do so in terms of laws if one is convinced that laws will clarify the notion. I am not so convinced.

[8] As should become clear below, the fact that we *ignore* certain causes does not necessarily imply that we are *ignorant* of them.

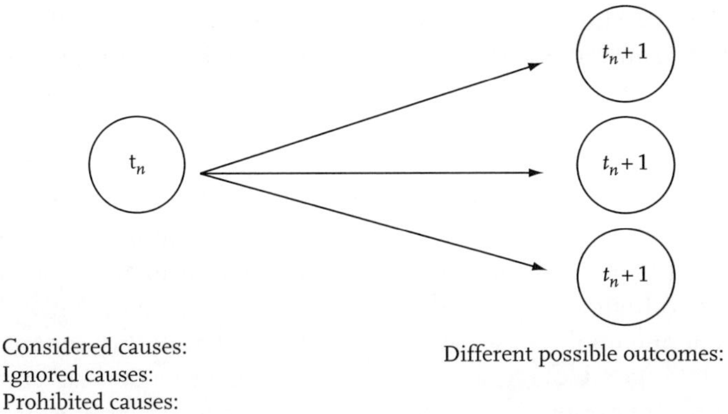

Fig. 20.1 The unified conception of chance.

20.3 The seven chance concepts

Now I need to show that the UCC genuinely unifies the seven concepts of chance. I will explain each concept, give examples from evolutionary biology, and then show how the concept fits the UCC.

20.3.1 Indeterministic chance ('pure' chance)

Above, I discussed indeterminism as a thesis about the world in general. Here I focus more locally, on the indeterminism of particular processes or types of processes. Indeterminism is said to be a true description of microlevel processes by, e.g. those who argue for the Copenhagen interpretation of quantum mechanics. A typical example of a such a process is radioactive decay. But what about macrolevel processes, and evolutionary processes in particular – are they indeterministic? For example, if cloned plants grown under (purportedly) identical conditions nonetheless differ considerably in height, weight, etc., is this an example of an indeterministic macrolevel process? Brandon and Carson (1996) argue, on the basis of examples such as these, that a scientific realist ought to conclude that the evolutionary process is indeterministic.[9]

Of course, if it were the case that the height of a cloned plant were due to indeterministic chance, it would not follow that the plant could be any height at all. A California poppy, which typically grows to a height of about 5–60 centimeters, would not grow to a height of 1 meter. Rather, the claim

[9] As will become clearer below, I do not find these arguments persuasive; my reason for including this example is to incorporate the full range of the types of claims that have been made about chance in evolution. Indeed, I think it is rare – if ever – that indeterministic chance is invoked in evolutionary contexts. However, it is important to leave open the conceptual possibility, as well as to emphasize that the other concepts of chance do *not* assume indeterministic chance.

is that given certain physical characteristics of the plant and the conditions in which it grows, there is a range of possible heights that it can achieve. If the plant's height really is due to indeterministic chance, then identical conditions can yield more than one possible height, with nothing further to be said (there are no 'hidden variables' to distinguish one case from another). Thus, indeterministic chance, unsurprisingly, fits the UCC quite cleanly: all the causes at a given point in time are considered, yielding a range of possible future outcomes with respect to the phenomenon of interest. No causes are ignored or prohibited.

However, one might reasonably ask just how relevant this notion of chance is for evolutionary biology, a point that Richardson (2006) has emphasized. Indeed, a recent exploration by three biologists of the meanings of chance in evolution states very clearly that describing evolution as a stochastic process 'has nothing to do with the claim that the natural world is, *in fine*, deterministic or not' (Lenormand *et al.* 2009, 158). Moreover, Brandon and Carson (1996) arguments for the indeterminism of evolution are not by any means universally accepted. Graves, Horan, and Rosenberg (1999) argue that at the macrolevel of evolutionary processes, there is '*asymptotic determinism*'. And as I argue (Millstein 2000b), given our current state of knowledge, even a scientific realist ought to be an agnostic on debate between the indeterminists and the 'asymptotic determinists'; most discussions end up trading one philosophical intuition for another rather than engaging in the painstaking task of tracing uncontroversially indeterministic microlevel phenomena to widespread macrolevel evolutionary processes (see Millstein 2003a for further discussion).

Even if evolutionary processes are to some extent indeterministic (as is likely to be the case if there is indeterminism at the microlevel – even the determinists concede that microlevel phenomena could occasionally 'percolate up' to the macrolevel), it does not seem plausible that observed statistical outcomes of evolutionary processes (such as variations in cloned plants given purportedly identical treatment) could be *fully* explained by indeterminism (Weber 2001). It is unlikely that in either our models or in any particular case we know all of the relevant causes; surely *some* of the observed statistical outcomes are due to these unknown causes. Thus, for the remainder of this paper, I will remain neutral on the indeterminism question, i.e. the rest of the chance concepts assume neither determinism nor indeterminism. Some might think that the indeterministic chance concept is the only 'real' chance concept; the subsequent discussion will show that the other meanings of chance in evolutionary biology are equally 'real' and equally useful, just different in meaning. That is, each chance concept describes a particular way that the world could be, and when a concept of chance is ascribed to a particular phenomenon, it is an empirical matter whether or not the phenomenon in fact manifests that description.

20.3.2 Chance as ignorance of the real underlying causes

Sometimes, we say that a future event is a matter of chance because we are unaware of some of the causes. To use an everyday example, we think of the flip of a fair coin as having a 50% chance of turning up heads. And yet (again, barring the occasional percolation of microlevel indeterminism), we generally think that is because there are numerous unknown causes, such as the way the coin is flipped or wind resistance, that are responsible for the outcome on each toss. In other words, we might say that the our ascription of the coin's 50% chance of turning up heads is really just a reflection of our ignorance of these other causes.

In the evolutionary realm, Darwin invoked chance as ignorance when he said that new variations were due to chance (though this was not the only sense of chance that Darwin ascribed to new variations):

> I have hitherto sometimes spoken as if the variations—so common and multiform in organic beings under domestication, and in a lesser degree in those in a state of nature—had been due to chance. This, of course, is a wholly incorrect expression, but it serves to acknowledge plainly our ignorance of the cause of each particular variation (Darwin 1859, p. 131).

And the use of this sense of chance, chance as ignorance, persists among biologists today, as Lande *et al.* make clear:

> Fluctuations in population size often appear to be stochastic, or random in time, reflecting our ignorance about the detailed causes of individual mortality, reproduction, and dispersal (Lande *et al.* 2003, p. 1).

Of course, we are not completely ignorant about the causal factors influencing fluctuations in population size. Known causal factors include the current population size, the typical life span of organisms of the species in question, and density dependent population regulation (Lande *et al.* 2003). These causal factors are the considered causes, and they determine which outcomes (in this cause, which range of fluctuations in population size) are possible. The causes we are ignorant of are ignored. No type of cause is prohibited from operating altogether.

Here one might worry that this is just the Laplacean notion of chance, and so only a deterministic notion. However, this is not the case. Even if evolution is indeterministic, we still might be ignorant of some of the underlying causes, and so, this concept of chance might still be relevant. In such a case, the range of possible outcomes is mediated both by the unknown causal factors and the inherent indeterminism.

A related worry is that determinism, combined with 'chance as ignorance', is sometimes seen as exhausting all the possible meanings of chance. In other words, some authors have written as though 'chance as ignorance' is the only concept of chance that makes sense under determinism (e.g. Rosenberg 1994;

but see Bouchard and Rosenberg 2004 for an alternate view). As Henri Poincaré noted, such a position is wrong-headed.[10] The fact that the 'laws of chance' can correctly predict phenomena such as the motions of molecules of a gas shows that ignorance does not exhaust the meaning of chance and that 'the information given us by the calculus of probabilities will not cease to be true upon the day when these phenomena shall be better known' (Poincaré 1921: p. 396). Indeed, we will see that each of the following concepts of chance is sensible in deterministic (as well as indeterministic) contexts.

20.3.3 Chance as not designed

Events that appear planned or designed, but are not in fact so, are often attributed to chance. For example, one might say that the shape of a running horse appearing in the clouds was due to chance in this sense.[11] For Darwin, new variations were chance in this sense:

> ...if we do not admit that the variations of the primeval dog were intentionally guided in order that the greyhound, for instance, that perfect image of symmetry and vigour, might be formed,—no shadow of reason can be assigned for the belief that variations, alike in nature and the result of the same general laws...were intentionally and specially guided"(Darwin 1868, pp. 431–432).

Here we see that the same phenomenon can be chance in more than one sense: in the previous section, I noted that Darwin also held that new variations were due to chance in the sense that he was ignorant of the true causes.[12] This form of chance is often not explicit in contemporary practice, but it is made explicit in responding to proponents of creationism or so-called 'Intelligent Design'. Neither the new variations nor selection itself (nor any evolutionary process, for that matter) are thought to be designed; thus, they are all due to chance in this sense, though this is not to imply the other senses of chance are relevant. One of the sources of confusion in the creationist debates is the conflation between 'chance as not designed' and the other senses of chance, but clearly, the former sense does not imply the latter, as Richard Dawkins notes:

> The argument from improbability states that complex things could not have come about by chance. But many people *define* 'come about by chance' as a synonym for 'come about in the absence of deliberate design'. Not surprisingly, therefore, they think

[10] See also Weber (2001) and Millstein (2003b).

[11] The famous 1965 'tomahawk toss' on *The Tonight Show Starring Johnny Carson* is a classic example of an unplanned event that looks like it could have been planned. Another example is the rock formation in New Hampshire that was known as 'The Old Man of the Mountain'.

[12] Beatty (1984) has emphasized the fact that Darwin used a multiplicity of chance concepts in characterizing the origin of new variations.

improbability is evidence of design. Darwinian natural selection shows how wrong this is with respect to biological improbability (Dawkins 2006, p. 139; emphasis in original).

In order for an undesigned process to be operating, there must be a complete absence of any 'intentionally guided' causes; thus, these causes are *prohibited* from this concept of chance. Any other cause may be considered; none need be ignored (though there may be other reasons, such as pragmatic reasons, for ignoring certain causes). Undesigned processes may give rise to outcomes that appear designed (what Dawkins calls 'designoids'), but they may not.

If no causes are ignored, under determinism a *token* non-intentional cause will of course uniquely determine one future outcome. In such cases, chance as undesigned can be construed as being true of a *type* of non-intentional cause whose tokens can give rise to different possible outcomes, some of which appear designed and some of which do not.

20.3.4 Chance as sampling (both discriminate and indiscriminate)

Discriminate sampling processes are processes in which physical differences among entities are causally *relevant* to differences in which entities are 'picked'. It can be thought of as 'sloppy' picking; the physical characteristics of some of the entities are the reason that they get picked, but they will not necessarily get picked; other entities that lack the characteristic in question might get picked instead. *Natural selection* is a chance evolutionary process in this sense. That is to say, natural selection is a process in which heritable physical differences among entities (e.g. organisms) are causally relevant to differences in reproductive success.

Indiscriminate sampling processes, on the other hand, are processes in which physical differences among entities are causally *irrelevant* to differences in which entities are 'picked'. This is the sort of picking that would occur if one were picking while blindfolded and if the physical differences in question were color differences only. *Random drift* is a chance evolutionary process in this sense. That is to say, as I have argued elsewhere (Millstein 2002, 2005), random drift is a process in which heritable differences among (for example) gametes[13] are causally irrelevant to which gametes are successfully joined (i.e. which gametes participate in successful fertilizations that yield zygotes). Biologists have also developed macroevolutionary models in which physical differences between taxa are causally irrelevant to difference in rates of branching and extinction within the taxa (see Millstein 2000a for a discussion).

[13] Note that both discriminate and indiscriminate sampling processes can occur at any level of the biological hierarchy, e.g. DNA bases, gametes, organisms, groups, etc. Also, they can occur simultaneously, as in the nearly neutral theory of molecular evolution (see Dietrich and Millstein 2008 for a discussion).

A sampling process takes into account the proportion of types within the population that is being sampled, the size of the sample, and the picking mechanism(s); these are the considered causes. Given those causes (and possibly indeterminism), there are different samples that can be produced; the samples differ in the proportion of types. A discriminate sampling process will also consider the physical characteristics that give rise to relative capacities to be 'picked'; this will further constrain the possible outcomes, or at least their expected frequency of appearance. 'Petty influences', such as the locations of entities within the population, are ignored; no causes are prohibited from operating. Sampling processes give rise to different possible proportions of types subsequent to picking (in evolutionary cases, this is the next generation).

20.3.5 Chance as coincidence, i.e. 'accident'

This concept of chance is associated with Aristotle; it implies the confluence of independent causal chains. For example, there may be a causal chain that leads to a white Toyota Prius being in the intersection of 3rd St. and B St. at 1:02 PM – and an entirely different and independent causal chain that led a green Ford Expedition to be in the same intersection at the same point in time. The collision of the cars was due to chance; it was a coincidence that they were in the intersection at the same time.

Note that under determinism, no two causal chains are truly independent, i.e. it is likely that there is a common cause if one looks back far enough in time (to the Big Bang, if necessary). Cournot (1843) provides a useful way of handling chance as coincidence under determinism. He describes two pairs of brothers – one pair serves in the same army, one pair serves in different armies, and yet in both cases the brothers perish on the same day. The brothers' deaths are both independent to a degree, yet the latter is more independent than the former, and thus chance to a greater degree.

In evolutionary biology, the question has arisen as to whether extinction is a chance process in this sense; for example, David Raup writes:

> The main question, to be visited again and again, is whether the billions of species that died in the geologic past died because they were less fit (bad genes) or merely because they were *in the wrong place at the wrong time* (bad luck) (Raup 1992, p. xi; emphasis added).

So, for example, consider a causal chain that ends with an asteroid impact on the Earth and the causal chain of the persistence of a particular species. If these two causal chains intersect without having had a common cause, then their confluence was a coincidence. Genetic draft (Gillespie 2000ab; Skipper 2006) also exhibits this sense of chance. Genetic draft[14] is a process of linked selection where it is a matter of chance which of two neutral alleles

[14] The phenomenon is called 'genetic draft' because some of its predicted outcomes are similar to those of genetic drift and because it involves 'hitchhiking', i.e. linked selection.

(in a two-locus model) happens to be linked to a site that undergoes an advantageous mutation, and where the timing of these mutations, followed by a rapid selective 'sweep' to fixation, is random. In models of genetic draft, the linking of a particular neutral allele to a locus where an advantageous mutation occurs represent two independent causal chains that intersect in a point in space-time.

Thus, chance as coincidence takes into account the two (or more) causal chains, while prohibiting those causal chains that have a (recent, under determinism) common cause. By ignoring the timing and/or location of the causal chains, as possible outcomes the chains may or may not intersect. Returning to the example I gave at the outset of this section, this captures the sense that had things been just a little bit different (had I left a little earlier, or gone a different route), the accident would not have occurred.

20.3.6 Evolutionary chance

Evolutionary chance (the term is due to Eble 1999) is exhibited when phenomena are independent of the generally adaptive direction of natural selection. This sense of chance has its origins in Darwin's thinking; it is one of the several senses in which he held new variations to be chance (we have seen two others so far, chance as ignorance and chance as not designed) but it is perhaps the sense that has been most influential and persistent in evolutionary biology. That is to say, Darwin (in his non-Lamarckian moments) believed that new variations were not directed, but rather that they were due to chance in the sense that they did not arise because they would be beneficial for the organism. Today, biologists generally believe that mutations, as a source of new variations in a population, are chance in this sense – mutations may be adaptive, maladaptive, or neutral – although there is some debate over whether all mutations are chance mutations.[15] Recombination (the chromosomal crossover between chromosome pairs that occurs during meiosis, giving rise to new gene combinations) is another source of new variation in populations and is similarly conceived of in terms of evolutionary chance.

Random drift, which as we saw earlier is a form of chance as indiscriminate sampling, also exhibits evolutionary chance. Although drift may sometimes proceed in an adaptive direction, it is no more likely to do so than it is to

[15] See Millstein (1997) and Merlin (2009) for a discussion. In order to account for the nuances of this complicated debate, the concept of chance mutation needs further refinement. Millstein (1997) argues that a mutation is directed if and only if it is specifically caused by environmental stress in an exclusively adaptive manner. Otherwise (if the mutation is non-specific, or specific but not exclusively adaptive, or not caused by environmental stress) it is a chance mutation. Merlin (2009) modifies this account of chance mutation. However, these refinements, while crucial for a full understanding of the debates over directed mutation, are not essential for the discussion here.

proceed in a maladaptive or neutral direction, in contrast to natural selection, which is weighted in an adaptive direction (and thus natural selection does *not* exhibit evolutionary chance, though it *is* chance in the sense of discriminate sampling, as we saw above). Similarly, the stochastic models of macroevolution that were mentioned earlier manifest evolutionary chance.

An anonymous referee has raised the concern that if drift exhibits more than one type of chance, the concepts of chance as indiscriminate sampling and evolutionary chance are not actually different from one another. However, such a concern is misplaced. Clearly, indiscriminate sampling is manifested in many non-evolutionary contexts, such as the sampling of coloured balls from an urn (a model that is used to aid in understanding drift, but which is not itself random drift; to name one obvious difference, there is no reproduction involved in sampling balls from an urn as there is with random drift). Evolutionary chance, on the other hand, can be manifested in phenomena where indiscriminate sampling is not manifested, such as chance mutation, where it is not the case that there is some standing variation from which some variants are 'picked' and some are not. Rather, mutations are 'mistakes' made during the DNA replication process with the consequence that the new nucleotide sequences differ from the previous nucleotide sequences. The fact that there are phenomena that exhibit indiscriminate sampling, but not evolutionary chance, and phenomena that exhibit evolutionary chance, but not indiscriminate sampling, shows that the two concepts are different – as does the fact that they differ in their considered causes, ignored causes, and prohibited causes, as previously described. (See Millstein 2000 for arguments concerning the distinctiveness of some of the other concepts of chance.)

Thus, evolutionary chance is primarily characterized by the causes that it prohibits entirely, namely, causes that proceed primarily in an adaptive direction. All other causes are taken into account, except for those that might be ignored for other (e.g. pragmatic) reasons. As a consequence, outcomes may be in an adaptive direction, but they may also be in maladaptive or neutral directions.

If no causes are ignored, under determinism a *token* not-adaptively-biased cause will of course uniquely determine one future outcome. In such cases, evolutionary chance can be construed as being true of a *type* of not-adaptively-biased cause whose tokens can give rise to adaptive, maladaptive, or neutral outcomes.

20.3.7 Chance as contingency

Stephen Jay Gould's *Wonderful Life: The Burgess Shale and the Nature of History* makes the case for the role of contingency in the evolution of life on his planet, using the movie *It's a Wonderful Life* as a metaphor.[16] Gould explains that when Clarence (a guardian angel) shows George what life in the town of Bedford Falls would have been like without him, the movie gave 'the finest illustration that I [Gould] have ever encountered for the basic principle of contingency – a replay of the tape yielding an entirely different but equally sensible outcome; *small and apparently insignificant changes*, George's absence among others, lead to cascades of accumulating difference' (Gould 1989, p. 287; emphasis added). Similarly, Gould says, '[a]lter any event, *ever so slightly and without apparent importance at the time*, and evolution cascades into a radically different channel' (Gould 1989, p. 51; emphasis added). In other words, if we replay the tape of life with small (and seemingly 'insignificant') changes at the outset, a radically different outcome will result: this is an example of what Gould calls (and what I will call) contingency.[17] Contingency thus involves sensitivity to initial conditions.

Other examples of chance as contingency occur in evolutionary biology. For example, Beatty describes how differences in the order of mutations in similar ancestral orchid populations can account for the vast diversity of orchids species (Beatty 2006b). Genetic draft, mentioned above, may be analogous here, if different timing of mutations to advantageous alleles that are linked to neutral alleles would lead to very different outcomes (Gillespie 2000ab; Skipper 2006). These examples show that the initial conditions to which the relevant processes are sensitive can be small changes in *timing* as well as small qualitative changes, e.g., 'If *Pikaia* does not survive in the replay, we are wiped out of future history – all of us, from shark to robin to orangutan' (Gould 1989, 323).

If a causal process is sensitive to initial conditions (the considered causes), then small differences in initial conditions (the ignored causes) will yield very different possible outcomes.[18] These processes are contingent. Causal

[16] Beatty (1995, 2006a) analyses Gould's concept of 'contingency' in detail; my presentation of it differs slightly from his. In particular, Beatty describes two meanings of contingency: unpredictability and causal dependence. Beatty describes unpredictability as 'different, unpredictable outcomes from the same or indistinguishable prior states' (Beatty 2006a, p. 339). If the prior states are truly identical, then contingency is the same concept as what I have called 'indeterministic chance' above. It seems to me that Gould denies this meaning when he differentiates contingency from the 'truly random' (Gould 1989, p. 284). Causal dependence, on the other hand, is described as 'the particular outcome depends strongly on which particular states preceded it' (Beatty 2006a, p. 339). That is closer to the view that I will describe.

[17] Sixty-eight years earlier, Poincaré describes chance in the same way, where 'slight differences in the initial conditions produce great differences in the final phenomena' (1921, p. 397). One of his examples is a perfectly symmetrical cone on a perfectly vertical axis, with no forces acting on it. I do not know whether Gould read Poincaré; no doubt this idea has been articulated many times.

[18] Chaotic processes are examples of contingent processes; however, being sensitive to initial conditions is only one characteristic of chaotic processes. That is, all chaotic processes are contingent, but not all contingent processes are chaotic.

processes that are not sensitive to initial conditions (i.e. that will tend to yield the same outcome even if the initial conditions differ slightly), are not contingent (so, these processes are 'prohibited').

An anonymous referee has raised the concern 'that the notion of chance as sensitivity to initial conditions does not directly refer to the unpredictability of the final result due to our ignorance of little differences at the level of initial conditions. Rather, it refers to the disproportion between little changes at the level of causes (initial conditions) and big changes at the level of effects (result). Chance is this disproportion, which makes the prediction very difficult and even impossible in the long term. If it is so, there are no ignored causes in this case (but there can be always ignored causes for other reasons, like in the case of other concepts of chance).' Here I would reply that what makes a phenomenon contingent is not our *ignorance* about the effects of small changes in initial conditions; the phenomenon would still be sensitive to initial conditions even if we knew those differences in initial conditions and their effects. Rather, I draw attention to the wide range of possible outcomes from similar initial conditions, a characteristic that chance as contingency shares with the other concepts of chance.

See Table 20.1 for a summary of the seven concepts of chance.

20.4 Connecting the UCC to probability

I have shown how each of the seven concepts of chance can be characterized in terms of the UCC. Note, however, that the seven concepts of chance are not to be 'eliminated' or made unnecessary simply because a common conception has been found. Each of them is more precise than the general definition; each 'gets at' a different aspect of chance. To put the point another way – each of the concepts of chance entails the general concept, but not vice versa. That is not to say that a particular phenomenon can't manifest more than one concept of chance. For example, as we have seen, random drift exhibits more than one concept of chance (chance as sampling and evolutionary chance), as do new variations (chance as ignorance, at least for Darwin; chance as not designed; and evolutionary chance). Which concepts of chance a given phenomenon manifests (if any) is an empirical question (though there may of course be a conceptual/theoretical component); again, note in particular that indeterministic chance may or may not be appropriate for a given phenomenon, and that it may be manifested in conjunction with one or more of the other concepts.

One benefit of characterizing the seven concepts of chance in terms of the UCC is that it provides a straightforward way for us to translate these more colloquial uses of chance into probabilistic terms. The translation between

Table 20.1 Summary of the seven concepts of chance, using the parameters of the UCC as illustrated in Figure 20.1

Concept of chance	Considered causes	Ignored causes	Prohibited causes	Different possible outcomes
Indeterministic chance	All (more precisely, the complete state of the world at t_n).	None.	None.	Any type of outcome in principle, though possibilities will be restricted by the included causes.
Chance as ignorance	All the causes that we know (or whose effects can calculate easily).	All other causes.	None.	Any type of outcome, though possibilities will be restricted by the included causes.
Chance as not designed	Any type of causes other than 'intentionally guided' causes.	None by definition, though some may be ignored for other reasons.	Intentionally guided causes.	Events that appear designed or events that do not appear designed.
Chance as sampling	Proportion of types within the population, size of sample, 'picking' mechanism. If discriminate, also includes relative capacities to be picked.	'Petty influences', such as the locations of entities.	None.	Different proportions of types in the population subsequent to the 'picking'.
Chance as coincidence	Two or more causal chains.	The timing and/or location of the causal chains.	A common cause for the causal chains.	The causal chains intersect or the causal chains do not intersect.
Evolutionary chance	Any cause that does not proceed primarily in an adaptive direction.	None by definition though some may be ignored for other reasons.	Any cause that proceeds primarily in an adaptive direction.	Outcomes may be adaptive, maladaptive, or neutral.
Chance as contingency	Causal processes that are sensitive to initial conditions.	Small, seemingly 'insignificant' causes.	Causal processes that are not sensitive to initial conditions.	Very different outcomes, depending on which small causes are ignored.

UCC and a (conditional) probability[19] is as follows. The UCC corresponds to the *probability of a particular outcome given the specified subset of causes.* Or, more formally, a given instantiation of the UCC can be translated to *Pr(outcome/subset of causes)*, where 'subset of causes' are the considered causes and 'outcome' is one of the possible outcomes, for that instantiation of the UCC.[20]

Here is an example of how this works. As described above, for chance as indiscriminate sampling, the considered causes are the proportion of types within the population, the size of sample, and the 'picking' mechanism. For chance as sampling petty influences are ignored, and so are not included in the subset of causes; no causes are prohibited, however. For random drift in particular, the UCC might be instantiated with the current proportions of phenotypes with respect to a given heritable trait, a given size of the population, and the particular environmental factor(s) that interact with that trait. This subset of causes yields different possible outcomes, where the different outcomes are different proportions of phenotypes in the subsequent generation.

So, for this sort of random drift, chance as indiscriminate sampling can be characterized as:

Pr (a particular proportion of phenotypes in the subsequent generation / current proportions of phenotypes with respect to a given heritable trait & size of population & environmental factor(s) interacting with those traits)

or, in a particular case, where the population consists of yellow, brown, and pink snails:

Pr (0.6 yellows, 0.3 pinks, and 0.1 browns in the subsequent generation /
0.5 yellows, 0.4 pinks, and 0.1 browns (heritable colors) &
200 snails & drought to which all snails are equally susceptible).

Once the translation is effected, it may now be possible to characterize our fairly colloquial uses of chance in quantitative terms for a particular case. For example, for random drift and chance as indiscriminate sampling, we could calculate the probability of a particular change in the population through

[19] Hájek (2007) argues that all genuinely informative theories of probability suffer from the reference class problem, a problem that can (in its metaphysical form) be dissolved by recognizing that conditional probabilities are the proper primitive of probability theory. I have much sympathy with his arguments, although nothing turns on them here.

[20] Here I make no claim as to whether the product of such a translation would satisfy all of Kolmogorov's axioms; such a demonstration would take us astray from the main points of this paper. Here I will simply note that some of the main candidates for interpretations of probability, such as the propensity interpretation, do not satisfy all of the Kolmogorov axioms, either (see, e.g. Hájek 2009 for a discussion).

the usual means of calculating such probabilities (laboratory experiments, observations of similar populations, etc.).

The other concepts of chance can be translated into probabilities in the same way: we can describe the probability of a particular outcome given the subset of causes for the concept of chance at hand (including the considered causes and excluding the ignored and prohibited causes), and then determine the values of those probabilities in the usual way that such probabilities are estimated. Although it might seem strange to quantify what seem like colloquial concepts of chance, this is not completely new for evolutionary biology. For example, Lenormand, Roze, and Rousset (2009) describe mathematical models of evolutionary chance. In principle, we could use the quantitative probability measures to compare different colloquial senses of chance. For example, we could compare an instance of chance as sampling (say, an outcome with a probability of 0.3) to an instance of chance as contingency (say, an outcome with a probability of 0.6). It remains to be seen, however, whether such comparisons would prove useful.

Also, if we had reason to think that a particular interpretation of probability was the appropriate interpretation for a particular case, we could link it to our chance concepts via the UCC. For example, we have good reason to think that our random drift probabilities are propensities, either deterministic or indeterminisitic (Millstein 2003b). Propensities are grounded in, and arise from, the physical characteristics of a system. The physical characteristics of the system, however, are just what I have been calling the considered causes, and it is their dispositions that are the source of the probability measures. In such cases, we would have good reason to think that this form of chance as indiscriminate sampling was a (deterministic or indeterministic) propensity.

However, the concepts of chance themselves can in principle be understood through any defensible interpretation of probability (with the exception of indeterministic chance, which is most naturally understood in terms of indeterministic propensities). Consider, for example, chance as ignorance. This might seem like an epistemic probability (that is, a probability that is concerned with the knowledge or beliefs of human beings), and it could indeed be interpreted that way; however, it need not be. Consider again the case of the coin flip; we need not be truly ignorant of the causes apart from the flipping mechanism and the coin itself (such as wind resistance). More to the point, we can understand the 50% probability we ascribe as arising from the physical characteristics of that type of setup (the considered causes) or the possible outcomes of that type of setup, making one of the objective probability interpretations (propensity or frequentist, respectively) a possible interpretation. On the other hand, by construing the relevant considered causes and observed outcomes as Bayesian evidence, one can likewise translate any of the concepts of chance into subjective probabilities.

20.5 Conclusion

There are at least seven particular, colloquial uses of chance in evolutionary biology. With the exception of indeterministic chance, each is meaningful regardless of whether the evolutionary process is deterministic or indeterministic. Each of the seven can be translated into the Unified Chance Concept (UCC) by specifying the types of causes that are taken into account (i.e. considered), the types of causes that are ignored or prohibited, and the possible types of outcomes. Again, however, let me emphasize that the existence of a more general concept does not entail that the more specific meanings are eliminated; the plurality of concepts is useful in illuminating different aspects of chance phenomena.

The UCC has the following benefits: The UCC reveals what is in common to the different chance concepts (what makes 'chance' chance). The UCC may aid in finding other concepts of chance, perhaps even outside of evolutionary biology. And most importantly, the UCC connects colloquial concepts of chance to probability, permitting them to be quantified, compared, and understood in terms of more formal interpretations of probability.

Acknowledgements

I would like to thank Ken Waters for continuing to insist that I address this question and for useful discussion over the years; the Griesemer/Millstein Lab at UC Davis for vetting the basic ideas behind the paper; the audience at the l'Institut d'Histoire et de Philosophie des Sciences et des Techniques (IHPST) for helpful comments on an earlier presentation; Rob Skipper, Bob Richardson, and their philosophy of biology class and attendees at the University of California Philosophy of Science Retreat for comments on an earlier draft; and Francesca Merlin for many fruitful conversations about chance and about this paper in particular. Three anonymous reviewers also provided helpful feedback.

References

Beatty, John. (1984). Chance and natural selection. *Philosophy of Science* 51: 183–211.

Beatty, John. (1995). The evolutionary contingency thesis. In *Concepts, Theories, and Rationality in the Biological Sciences: The Second Pittsburgh-Konstanz Colloquium in the Philosophy of Science, University of Pittsburgh, October 1–4, 1993*, edited by G. Wolters and J. Lennox. Pittsburgh, PA: University of Pittsburgh Press.

Beatty, John. (2006a). Replaying life's tape. *Journal of Philosophy* 103: 336–362.

Beatty, John. (2006b). Chance variation: Darwin on orchids. *Philosophy of Science* 73: 629–641.

Bouchard, Frédéric, and Alex Rosenberg. (2004). Fitness, probability, and the principles of natural selection. *The British Journal for the Philosophy of Science* 55: 693–712.

Brandon, Robert, and Scott Carson. (1996). The indeterministic character of evolutionary theory: No 'no hidden variables' proof but no room for determinism either. *Philosophy of Science* 63 (3): 315–337.

Cournot, Antoine Augustin. (1843). *Exposition de la Théorie des Chances et des Probabilités*. Paris: Hachette.

Darwin, Charles. (1859). *On the Origin of Species by Means of Natural Selection, or the Preservation of Favoured Races in the Struggle for Life*. 1st edn. London: John Murray.

Darwin, Charles. (1868). *The Variation of Animals and Plants Under Domestication*. 1st edn. London: John Murray.

Dawkins, Richard. (2006). *The God Delusion*. Boston: Houghton Mifflin Harcourt.

Dietrich, Michael R. and Millstein, Roberta L. (2008). The role of causal processes in the neutral and nearly neutral theories. *Philosophy of Science* 75: 548–559.

Earman, John. (1986). *A Primer on Determinism*. Dordrecht: D. Reidel Publishing.

Eble, Gunther J. (1999). On the dual nature of chance in evolutionary biology and paleobiology. *Paleobiology* 25: 75–87.

Gayon, Jean. (2005). Chance, explanation, and causation in evolutionary theory. *History and Philosophy of the Life Sciences* 27: 395–405.

Gillespie, John. (2000a). Genetic drift in an infinite population. *Genetics* 155: 909–919.

Gillespie, John. (2000b). The neutral theory in an infinite population. *Gene* 261: 11–18.

Gould, Stephen Jay. (1989). *Wonderful Life: The Burgess Shale and the Nature of History*. New York: W. W. Norton & Company.

Graves, Leslie, Barbara L. Horan, and Alex Rosenberg. (1999). Is indeterminism the source of the statistical character of evolutionary theory? *Philosophy of Science* 66: 140–157.

Hájek, Alan. (2007). The reference class problem is your problem, too. *Synthese* 156: 563–585.

Hájek, Alan. (2009). Interpretations of probability. In *The Stanford Encyclopedia of Philosophy (Spring 2009 Edition)*, edited by E. N. Zalta, URL = <http://plato.stanford.edu/ archives/spr2009/entries/probability-interpret/>.

Lande, Russell, Steinar Engen, and Bernt-Erik Saether. (2003). *Stochastic Population Dynamics in Ecology and Conservation: An Introduction*. Oxford: Oxford University Press.

Lenormand, Thomas, Denis Roze, and François Rousset. (2009). Stochasticity in evolution *Trends in Ecology and Evolution* 24 (3): 157–165.

Merlin, Francesca. (2009). *Le hasard et les sources de la variation biologique: analyse critique d'une notion multiple*. Ph.D. Thesis, Institut d'Histoire et de Philosophie des Sciences et des Techniques, University of Paris 1, Paris.

Millstein, Roberta L. (1997). *The Chances of Evolution: An Analysis of the Roles of Chance in Microevolution and Macroevolution*. Ph.D thesis, Department of Philosophy, University of Minnesota, Minneapolis, MN.

Millstein, Roberta L. (2000a). Chance and macroevolution. *Philosophy of Science* 67 (4): 603–624.

Millstein, Roberta L. (2000b). Is the Evolutionary Process Deterministic or Indeterministic? An Argument for Agnosticism. Presented at the Biennial Meeting of the Philosophy of Science Association, Vancouver, Canada, November 2000, http://philsci-archive. pitt.edu/archive/00004544/.

Millstein, Roberta L. (2002). Are random drift and natural selection conceptually distinct? *Biology and Philosophy* 17 (1): 33–53.

Millstein, Roberta L. (2003a). How not to argue for the indeterminism of evolution: A look at two recent attempts to settle the issue. In *Determinism in Physics and Biology*, edited by A. Hüttemann. Paderborn: Mentis.

Millstein, Roberta L. (2003b). Interpretations of probability in evolutionary theory. *Philosophy of Science* 70 (5): 1317–1328.

Millstein, Roberta L. (2005). Selection vs. drift: A response to Brandon's reply. *Biology and Philosophy* 20 (1): 171–175.

Millstein, Roberta L. (2006). Discussion of 'Four case studies on chance in evolution': Philosophical themes and questions. *Philosophy of Science* 73: 678–687.

Monod, Jacques. (1971). *Chance and Necessity: An Essay on the Natural Philosophy of Modern Biology*. New York: Knopf.

Poincaré, Henri. (1921). *The Foundations of Science: Science and Hypothesis, The Value of Science, Science and Method*. Translated by G. B. Halsted. New York: The Science Press.

Raup, David M. (1992). *Extinction: Bad Genes or Bad Luck?* New York: W.W. Norton & Co.

Richardson, Robert C. (2006). Chance and the patterns of drift: A natural experiment. *Philosophy of Science* 73: 642–654.

Rosenberg, Alex. (1994). *Instrumental Biology or the Disunity of Science*. Chicago: University of Chicago Press.

Skipper, Robert A. (2006). Stochastic evolutionary dynamics: Drift versus draft. *Philosophy of Science* 73: 655–665.

Weber, Marcel. (2001). Determinism, realism, and probability in evolutionary theory. *Philosophy of Science* 68 (Proceedings): S213–S224.

·21·

Drift and the causes of evolution

Sahotra Sarkar

Abstract

This chapter defends a stochastic dynamical interpretation of evolution under which drift does not emerge as an evolutionary cause, unlike mutation and selection. Rather, whether drift occurs in a population model depends on its constitutive assumptions, namely, whether the population size is finite. The same amount of selection makes quantitatively and qualitatively different predictions in finite and infinite population models: this is all that there is to drift. This argument is illustrated through the explicit solution of a haploid model in which differences in vital parameters lead to drift in the presence of selection. In the absence of these differences, the model reduces to a neutral model. In the infinite population limit, the standard results without drift also obtain. The stochastic dynamical interpretation is contrasted with the views that evolution is a theory of forces and the statistical interpretation of evolution.

21.1 Introduction

Consider an infinite population obeying Mendel's rules of inheritance. We will track evolution at a single locus at which there may be several alleles.[1] We will assume that the population is of the simplest type satisfying all the criteria listed in Table 21.1. This simple model, often called the 'standard selection' model, is used for didactic simplicity in elementary textbooks of population genetics – in fact much of the technical discussion of this paper will use an even simpler model in which we will ignore diploidy (paired chromosomes) and assume that the population is haploid (chromosomes occur singly) as, for instance, in bacterial species.

[1] An *allele* is a version of a gene. So, in a diploid species, which form *Mendelian* systems, an individual has two alleles at each locus not on a sex chromosome. This is because these chromosomes occur in *homologous* pairs. Depending on the taxon, in sexual species, one of the sexes has a homologous pair of sex chromosomes (females are *XX* in *Homo sapiens*) while the other does not (males are *XY* in *Homo sapiens*). A *locus* corresponds to a 'position' on a chromosome that can be occupied by an allele. The two alleles at a locus specify the genotype at that locus. Subtleties about the abstract organization of the genome and its physical realization on chromosomes are being ignored here–for a discussion, see Sarkar (1998, Chapter 5). Stephens (2008) provides a good introduction to population genetics for philosophers.

Table 21.1 Assumptions of the standard selection model. For further discussion of these assummptions, see, e.g. Nagylaki (1992).

	Condition	Further explanation
1	Infinite population	
2	Full diploidy	Diploid organism; the locus being considered is not on a sex chromosome (which would have no locus homologous to it).
3	Two sexes	
4	Dioecy	Each individual will be of only one sex.
5	Initial genotypic frequencies same for both sexes	
6	Random mating	Only necessary for the locus under consideration.
7	Co-equal segregation of alleles into gametes	
8	Independent assortment of alleles	At this locus; together with Condition (7) this is Mendel's first rule of inheritance.
9	Discrete generations	
10	Non-overlapping generations	
11	Selection operating at one locus	
12	Frequency-independent genotypic fitnesses	
13	Time-independent genotypic fitnesses	
14	No mutation	
15	No migration	

Two results make the standard selection model particularly salient in philosophical discussions of evolutionary theory. First, assume that all fitnesses are the same.[2] Then the frequency of every allele remains constant through time. Moreover, after the first generation, the frequency of each genotype also does not change and is equal to the product of the frequencies of the two alleles that comprise that genotype. Let p_i and p_j be the frequencies of the i–th and j–th alleles, respectively. Then, this constant genotypic frequency for the $\{i, j\}$–th genoptype, P_{ij}, is equal to $p_i p_j$. This is known as the Hardy–Weinberg rule after its independent discoverers, the British mathematician,

[2] Throughout this chapter, selection will be viewed as synonymous with dynamical change under differential fitness of types in a population. Thus, if selection is a cause of evolution, so are fitness differences. This formulation ignores many subtleties but these do not surface in the models discussed in the paper either because they have infinite populations with fitnesses depending on a single factor, or they explicitly model reproduction stochastically (that is, with an element of chance) while incorporating differential reproduction and survival. (See Sober (1984), for arguments designed to show that fitness is causally inert.) However, what is being assumed about fitness is not compatible with the propensity interpretation of fitness (Brandon 1978; Mills and Beatty 1979). For a philosophical discussion of the relation of selection to fitness, see Sober (1984) though this is now somewhat dated.

G. C. Hardy, and German physician, Wilhelm Weinberg.[3] The deification of this rule will be criticized later in this chapter (see Section 21.4) but that is because there is much more to evolution than the standard selection model. In the context of this model the constancy of genotypic frequencies is, indeed, quite remarkable.

Second, when not all fitnesses are equal, the rate of change of the mean fitness of the population (from one generation to the next) is proportional to the variance in fitness in the original generation. R. A. Fisher (1930), one of the founders of theoretical population genetics, regarded this result (which he was the first to derive) as the 'fundamental theorem of natural selection'. How important this result really is for biology remains controversial. However, there should be little doubt that it is philosophically interesting because it shows that the power of selection depends on how much variation there is in the population, and this conclusion is not challenged by any of the controversies surrounding the so-called fundamental theorem. In any case, issues regarding the 'proper' interpretation of Fisher's result (and its many extensions) are not germane to the concerns of this chapter.

Returning to the Hardy–Weinberg rule and the standard selection model, it is instructive to see what happens when the assumptions of Table 21.1 are relaxed. Perhaps what is most interesting is that many types of non-random mating may change genotypic frequencies from one generation to the next but will not change allele frequencies. Non-random mating includes assortative mating ('like' mates preferentially with 'like'), its opposite (disassortative mating), and inbreeding.

What changes allele frequencies? The usual answer – and we are no longer limiting ourselves to the standard selection model – lists five mechanisms: drift (random changes during reproduction), meiotic drive, migration, mutation, and selection. Meiotic drive is irrelevant to this analysis: it is the result of non-Mendelian mechanisms, for instance, preferential segregation of one of the alleles into the gametes. The discussions of this paper are restricted to Mendelian and haploid populations. Migration will also be ignored; instead, it will be assumed that populations are closed, as in the standard selection model. All the conceptual problems that concern this chapter remain unmitigated in spite of this simplifying assumption. This leaves drift, mutation, and selection to be discussed. The relative role of each in evolution has been one of the major debates within recent evolutionary theory (Kimura, 1983; Gillespie, 1994).

This chapter is about the question whether drift is a 'cause' of evolution. Section 21.2 is about how 'cause' will be construed. However, what constitutes evolution also remains a live question. First, there is the controversial relation

[3] The Hardy–Weinberg rule dates to 1908; for an account of this history, see Provine (1971) and Sarkar (2007).

between macroevolution (evolution at taxonomic levels higher than that of species) and microevolution (within species) (Gould, 2002; Plutynski, 2008). It will be assumed here that macroevolution can be reduced (*sensu* Sarkar (1998)) to microevolution, that is, the facts of macroevolution can be explained by the operation of microevolutionary processes over long times. Those who are uncomfortable with this assumption should simply regard this chapter as being about whether drift is a cause of microevolution. After all, if drift is a cause of microevolution, *ipso facto*, it is a cause of evolution because the former is part of evolution, whether or not macroevolution can be reduced to microevolution.

Second, within the microevolutionary context, Dobzhansky (1937) popularized the definition that evolution consists of changes of allele frequencies and this definition has often been endorsed by biologists.[4] Note that this is a very restrictive definition in the sense that phenotypic changes in populations resulting, for instance, from genotypic, but not allelic, changes do not constitute evolution. Dobzhansky's restrictive definition will not be endorsed here. Rather we will explore the consequences for the definition of evolution to require either a change of allele frequencies or, less restricively, a change of genotype frequencies.[5]

As mentioned earlier, Section 21.2 of this chapter will elaborate the causal framework adopted for this analysis. It will be assumed that evolving populations are dynamical systems that are described by a minimal set of assumptions. This set will be partitioned into constitutive assumptions which establish the identity of a system, and facultative assumptions which can vary without challenging this identity. The latter embody the relevant *causes* which operate against the background *conditions* provided by the constitutive assumptions. In order to explore the relation between drift and selection, Section 21.3 will construct and solve a haploid model with a finite population. Stochastic reproduction will be modelled explicitly. In this model, death rates of different types will give rise to fitness differences and, thus, to selection. Fitnesses are summary parameters incorporating birth and death rates; all uncertainties arise from the finite population size, that is, because finite samples are being drawn from a finite population.[6] Thus, drift is a result of population size. If we take as constitutive the assumption whether a

[4] It has on occasion also been endorsed by philosophers, most notably, by Sober (1984) in his influential analysis of evolution. Magnus (1998) provides an interesting counter-example.

[5] Even this is a restrictive definition. As Kimura (1983, p. xiv, emphasis in the original) has correctly pointed our, evolution includes '*all* changes, large and small, visible and invisible, adaptive and nonadaptive'. However, the restrictive definition used here suffices to capture what is controversial within microevolutionary theory.

[6] Thus, as noted earlier, this framework is incompatible with a propensity interpretation of fitness – a discussion of this issue will be left for another occasion; it is beyond the scope of this chapter.

population is finite or infinite, drift is not a cause of evolution but results from the conditions in which a population is evolving; selection remains a cause. Finally, Section 21.4 turns to how this analysis coheres with the most common philosophical interpretations of evolution. It is almost entirely at odds with the so-called 'statistical' interpretation which has recently gained some prominence. It also does not agree with details of most versions of the dynamical interpretation. But it *is* a dynamical interpretaion, perhaps best called a *stochastic dynamical interpretation* of evolution.

21.2 Causality and evolutionary dynamics

Potentially evolving populations are 'dynamical systems' in the minimal sense that they are modelled by a set of equations that specify (deterministically or probabilistically) the future state of a system, given its present state (Hirsch and Smale 1974).[7] Defined in this way, population models satisfy Lewontin's (1974) requirement of dynamic sufficiency, and we will assume that they do so by embodying within them a minimal set of assumptions for that purpose.[8] Minimality will be construed as meaning that the assumptions are jointly (dynamically) sufficient and each one is necessary. We will make a distinction between 'constitutive' and 'facultative' assumptions. The former are privileged in the sense that they cannot be changed without changing the identity of the sytem (i.e. changing what is being modeled): this is the sense in which they are constitutive. The latter may change without changing the identity of the system.

Three subtleties about constitutive assumptions must be noted. First, what a system is (obviously, as represented in a model – no position will be taken here on realism or related issues) depends on the context of inquiry. This means that what should be taken as the constitutive assumptions of a model is context-dependent. Consider mating patterns. Suppose what interests us is what happens to a population as mating becomes more assortative progressively.[9] In that case the assumption about mating behaviour will be a facultative assumption rather than a constitutive one: we would be studying the same

[7] Given the focus of this chapter, the discussion will largely be limited to microevolutionary models. However, the causal framework that is being introduced here is intended to be generally applicable to dynamical systems. For expository simplicity we ignore the fact that models, especially computer models for simulation, may sometimes not be explicitly presented using equations. It does not make any difference to the conclusions reached here. Finally, objections to the dynamical interpretation of evolution will be taken up in Section 21.4.

[8] In this framework, theories are construed as more general models, that is, as models intended to apply to larger domains. Theories are thus of the same logical type as models – however, the results of this analysis do not depend on this assumption.

[9] See, for example, Williams and Sarkar (1994) who show that assortative mating allows traversals of valleys in adaptive surfaces in a two-locus model. Similarly, we may want to explore

population as mating behaviour changes. In contrast, in the standard selection model, the assumption of random mating is constitutive, for instance, when we are mostly interested in how rapidly a system may respond to different types of selection pressure.

Second, in many situations, the constitutive assumptions alone will not be dynamically sufficient. Return to the case of mating behaviour. For dynamic sufficiency in models of Mendelian populations, we must make some assumption about mating, whether it be random, assortative, disassortative, inbreeding, obligate outcrossing, etc. But, as we noted earlier, in many situations we may want to let mating behaviour vary in what we regard as the same system – indeed we may even be interested in the evolution of mating behavior. For instance, Otto *et al.* (2008) explore the evolution of assortative mating in a two-locus model in which this mating pattern is indirectly selected for whenever heterozygotes are less fit than homozygotes at one of the two loci.[10] They show that, if assortative mating increases in frequency, it can lead to reproductive isolation of different groups in a sympatric population (one that is geographically localized 'at the same place') potentially leading to speciation without geographical separation. Thus, the mating assumption is not constitutive, and the constitutive assumptions are not dynamically sufficient.

Third, when the constitutive assumptions are dynamically sufficient, because of the presumed minimality of the set of assumptions, there is no further facultative assumption made in the model. This would mean that the system cannot be conceptualized in any other way. Suffice it here to note this logical possibility; we will return below to the question whether this case is of any biological relevance: briefly, if we take the finiteness of a population to be a constitutive assumption, we will have a dynamically sufficient set of constitutive assumptions if we want to explore neutral evolution (that is, evolution in the absence of fitness differences).

Is there a canonical list of constitutive assumptions for evolutionary models? There cannot be. We – all entities of the biosphere – have evolved from some rather nondescript molecules of the distant past. Every aspect of living systems is a product of evolution and the history of its emergence presents questions to be potentially explored. What is constitutive about a living system depends on the stage of evolutionary history and the questions we choose to ask. We owe this insight primarily to Fisher (1928) who realized that even the properties that define Mendelian systems may evolve and are potentially subject to modification through natural selection.[11]

how – and to what extent – assortative mating may encourage speciation – see, for instance Otto *et al.* (2008) which is discussed later in the text.

[10] A *heterozygote* has different alleles at a locus; a *homozygote* does not.

[11] Fisher (1928) was interested in the evolution of dominance in Mendelian systems when the heterozygote at a locus has the same phenotype as one of the two homozygotes. It is clear

We will restrict ourselves to Mendelian and haploid systems. The restriction to Mendelian systems is standard in much of evolutionary theory though this may reflect little more than the fact that we are Mendelian organisms and most of our biological intuitions are developed in the context of large animals which share the same property.[12] We will leave further discussion of this issue for another occasion though a consideration of polyploidy (and around half of flowering plant species are polyploid) may well lead to a different assessment of the causes of evolution.[13] Haploid systems are also included here because the mathematics of drift (see Section 21.3) is much simpler for haploid systems than for Mendelian systems and the conceptual problems that concern this chapter remain unmitigated in the simpler context. Moreover, because Mendelian organisms go through a haploid phase (as gametes) in their life-cycle, considerable insight into the evolution of Mendelian systems can still be obtained from haploid models (Kimura, 1983).

Evolutionary models are typically constructed to explain dynamical patterns of systems, more often than not differences in dynamical patterns of the same system in different circumstances. Here, the 'causes' of those patterns will be construed as the factors that are embodied in the facultative assumptions of a model. The constitutive assumptions provide the 'background conditions' against which these causes operate.[14] On this account, causes make differences but what counts as causes depends on the context in two ways: what we are trying to explain, and in which system at what stage of its history.[15] The former correspond to what Menzies (2004), following Gorovitz (1965), calls the context of inquiry, the latter to what he calls the context of occurrence. Both, but especially the latter, determine what the constitutive assumptions are.

But, somewhat counterintuitively, on this account, systems may undergo changes of state variables without any causal story to be told. A system may change just because of what it is. The standard way in which this may happen is when a system consists of a finite number of constituent parts and there is some sampling of the parts during dynamical evolution.[16] We may want to

now that Fisher's selectionist model for the origin of dominance is incorrect (Orr, 1991) but that does not detract from the insight that constitutive features of Mendelian systems are subject to evolutionary explanation.

[12] A similar point is made by Kimura (1983, pp. 5–6).

[13] A *polyploid* has more than two homologous chromosomes; for instance, a *tetraploid* has four.

[14] This distinction between causes and background conditions goes back at least to Mackie (1974) but, see, especially, Hart and Honoré (1985) for the explication most relevant to the discussion here.

[15] While this claim is intended to be read counterfactually, it differs from counterfactual accounts that go back to Lewis (1973) influential treatment because of the explicit context-dependence – see Menzies (2004) for more detail on this issue.

[16] Whether this is the only way in which such a situation – change without causes – may happen is an issue that will be set aside for another occasion. The quantum domain provides potential examples of such changes without sampling being involved.

say that randomness is a cause of change, or that what a system is includes causes of its change. The former possibility has generated much philosophical work (Forber and Reisman, 2007) but it has largely remained imprecise: 'randomness' does not specify any single thing. We need to specify what distribution random variables must be drawn from, that is, the sampling process: if there are alternatives we will have to make a facultative assumption. The latter possibility amounts to maintaining that even when we have explained a process of change, and have incorporated the results into our definition of a system, we are still searching for causes; rather, in the causal framework of this paper, we should be construing these assumptions as the conditions against which causal analysis takes place.

In the case in which there is no facultative assumption in a model, its constitutive assumptions are dynamically sufficient. For the reasons mentioned in the last paragraph, the account of causality used here has the implication that there are no causes to investigate. But, once again, does that make sense? Digressing, briefly, into physics, consider a Newtonian system consisting of particles with no mass, no charge, and so on, so that no force is acting on it or any of its constituent parts.[17] Assume that what we are interested in is the dynamical evolution of the system. The constitutive assumptions are sufficient to determine the system's dynamics: each particle continues in a state of rest or of uniform rectilinear motion indefinitely. It is clear that there are no causes to investigate. Now, consider, instead, a system in which the particles just have some mass. What are the constitutive assumptions about the system? In a classical context, presumably the number of particles and their masses would fall under this category. Almost certainly (depending on the context of inquiry), the distances between particles and their velocities would not, and the constitutive assumptions would no longer be dynamically sufficient. We are allowed to investigate causes. At least Newtonian physics presents no problem.[18]

In microevolutionary models, presumably (once again depending on the context of inquiry), a system would not lose its identity just because it is placed in a different environment. This means that fitness values are not constitutive

[17] The particles must have no mass because, otherwise (in Newtonian physics) gravitational forces would act between them. However, once we specify that a particle has no mass, it is far from clear that we have a coherent concept of a particle in Newtonian physics – at the very least, Newton required particles to have mass. What adds to the potential incoherence is that the moment we attribute any physical property to a particle, it seems amenable to being subject to forces. But if a particle has no physical properties, it is unclear how it is to be individuated. So, this example is contrived, perhaps illegitimate, even as an idealization, and philosophical conclusions drawn from it remain suspect.

[18] This discussion requires little modification when extended to classical fields or to special relativity (masses change with velocity – so, we would have to specify rest masses as being constitutive); general relativity and the quantum domain present problems which will be left for a different occasion. Especially in the latter case, questions of identity are much more complicated.

assumptions even if they remain constant in a particular environment over generations, as in the standard selection model. Consequently, in all contexts in which we are interested in selection, we will not face the problem that constitutive assumptions are dynamically sufficient and we are thus left with no cause to investigate. Neutral evolution is different. In diploids, if the context allows some non-random mating assumption to be constitutive, genotypic frequencies may change from the constitutive assumptions alone. Much of this chapter will be about how allele frequencies change in finite populations even without selection. If the finiteness of a population is constitutive, we have change without causes. Evolution is rather different from the Newtonian world.

It is time to turn to the causes of evolution. Given the characterization of evolution in Section 21.1, there are two cases to consider, corresponding to whether we define evolution as solely a change in allele frequencies, or we also allow changes of genotypic frequencies in the definition:

- If evolution is viewed as requiring a change in allele frequencies, what was said earlier about fitness values implies that selection is a cause of evolution. This is not controversial; in fact, any analysis of the causes of evolution that denies such a role to selection would be absurd.[19] What about mutation? Because mutations arise stochastically (that is, random processes are involved), it would be odd to suggest that which mutations are supposed to occur should be part of the definition of a system. Mutation trivially changes allele frequencies and are, thus, a cause of evolution.[20] The dialectic between mutation and selection has been extensively explored, starting with the pioneering work of another of the founders of theoretical population genetics, J. B. S. Haldane (1927).
- Allele frequency changes will lead to genotypic frequency changes except in highly contrived examples. Consequently, if genoypic frequency changes are also presumed to constitute evolution, selection and mutation remain causes of evolution. What is more interesting is mating behaviour. As noted earlier, a large array of mating patterns (assortative and disassortative mating, inbreeding, etc.) can lead to genotypic frequency changes without allele frequency changes. Should assumptions about mating patterns be regarded as constitutive or facultative? As usual,

[19] But this is what the statisticalist interpretation (Walsh, Lewens and Ariew, 2002; Walsh 2007) asserts; we will turn to it in Section 21.4.

[20] For want of space, some subtleties are being glossed over in this assessment. Briefly, some organisms are more prone to mutation than others (e.g. laboratory strains specially created with this property in mind), and this property should then be part of the system's identity. Moreover, we can often predict which class of mutations is more likely than others. The point is that we cannot predict which mutation *will* occur.

the answer will depend on context. For instance, as was noted earlier in the case of assortative mating, if the evolution of a mating pattern is itself the focus of inquiry, the assumptions are bound to be facultative. There are also systems in which some mating pattern is obligatory: for instance, plants with self-incompatibility alleles cannot self-fertilize and two plants may not be able to cross-fertilize if they share as few as one self-incompatibility allele (Levin, Kelley and Sarkar, 2009). In almost all such contexts, highly constrained mating pattern assumptions will be constitutive, part of the definition of the system. However, allele frequencies may also change, for instance when the same locus admits both self-incompatibility and 'ordinary' alleles.[21]

We have not yet discussed drift. The crucial issue will be whether the initial size of a population – at the very least, whether it is finite or infinite – is a constitutive assumption. There is no trivially obvious answer. If initial allelic or genotypic frequencies are different in the different populations, we arguably have different systems. But the interesting situation, as we shall see in Section 21.3, is the one in which such frequencies remain the same even as the population size changes (including an increase to the idealized infinite limit). Should the initial size of a population be viewed in analogy with the initial conditions from which a physical dynamical system evolves? If so, then the same system may have different initial population sizes. However, for evolutionary systems the analogy seems misleading for two reasons:

1. Evolutionary changes, and many other biological processes, are much more history-dependent than typical physical systems.[22] Some processes cannot even be modelled accurately as Markov processes in which the state of a system at one time stage only depends on its state at the previous time stage though, because of mathematical complexities, non-Markov processes remain poorly explored (Iizuka and Matsuda, 1982). Consequently, the initial size of a population may be definitive to what it will be, part of what establishes its identity.
2. Even without taking drift into account, we know that small populations are very different from large ones in the context of both evolution and ecology. For instance, a moderate amount of inbreeding may have qualitatively different effects in the two situations (Crow and Kimura, 1970). Sometimes entirely different methods have to be used to model processes in the two situations. For instance, in large populations mutation may be sometimes modeled deterministically (Crow and Kimura,

[21] It is also not completely clear that we should not view such phenomena as selection – we will return to this topic in another occasion.

[22] There are exceptions – see, for instance, Rabotnov (1980) for a discussion of Boltzmann–Volterra hereditary mechanics. These apply to elastic materials which are also history-dependent in their physical properties.

1970); in contrast, models of mutation in small populations require a stochastic treatment of the mutation process (Stewart, Gordon and Levin 1990; Ma, Sandri and Sarkar 1992; Sarkar, Ma and Sandri 1992). In ecology, the techniques of population viability analysis are even more different in the two situations, with large populations typically modeled as if they are infinite in size (Caughley, 1994; Sarkar, 2005).

In what follows, assuming that the context of inquiry specifies that we are interested in the eventual composition of a population, we will treat whether the initial size of a population is finite or infinite as a constitutive assumption. This is an intermediate position between treating the exact initial size as constitutive and not regarding size as constitutive at all.

It is time to model evolution in a finite population explicitly and see what happens. Though the philosophical literature on drift is large (and defies easy summary), there has been surprisingly little precise analysis of the evolutionary dynamics of a fully specified model – for instance, even as a *Gedankenexperiment*.[23] We will consider a very simple model. Nevertheless, the mathematics is a little complex. Moreover, only a few parts of the mathematical analysis are new. Those not interested in technical detail should skip to Section 21.3.3. As noted earlier, for mathematical simplicity, we will use a haploid model.

21.3 The haploid model

Besides requiring that the population is haploid and finite, this model differs somewhat subtly from the standard selection model in how selection is imposed. An informal description here will be followed by an exact specification in Section 21.3.1. We will assume the population size is fixed, for instance, due to resource constraints.[24] Fitness differences will be modeled as viability differences: at each temporal stage (which may loosely be interpreted as a generation), individuals of different types have different probabilities of dying. All individuals have the same probability of reproducing. These probabilities all remain constant over time. In this sense fitnesses are constant. However, in disanalogy to the standard selection model, because the population size

[23] What is perhaps even more unfortunate is that the few precise discussions that there have been – see, for instance, Walsh (2007) – typically make use of coin tossing and similar examples which have no clear biological interpretation. Brandon (2005) is an exception.

[24] In ecology this corresponds to the situation in which the population has reached the carrying capacity of the environment. Abrams (2007) partially analyses a related slightly more complex model. His model leads to the same conclusions if the mathematical analysis is completed. For a set of similar models, used very effectively to describe evolutionary dynamics, see Nowak (2006, Chapter 6).

is fixed, relative fitnesses of the types will not remain constant over time.[25] In order to compare predictions of this model to one in which the population is infinite, we will be interested in computing the probability that one type will eventually prevail over the others. We now turn to this model in detail.

21.3.1 Model specification

Let the fixed size of the population be N. As in the discussion of the standard selection model, we will assume non-overlapping time stages indexed by time, t_i, $i \in \mathbb{Z}^+$.[26] Evolution will be tracked at one locus and, for simplicity, it will be supposed that there are only two alleles, A and a; because this is a haploid model, alleles and genotypes will not be distinguished.[27] At t_1 it is assumed that there are k individuals of type A and, therefore, $N - k$ individuals of type a. The two types have different fitnesses because of different death rates (that are constant over time): at any time stage, let these death rates of A and a be μ_A and μ_a, respectively. The model assumes no difference in reproductive capabilities for A and a, that reproduction occurs at the beginning of each time stage (that is, before potential death), and that each individual reproduces at most once. Thus, if $\mu_A > \mu_a$, a has higher fitness than A and vice versa.

Because the population is finite, one of the two types will prevail over time (that is, it will be the only one left after an infinite number of time steps). The parameter that is of most interest is $p(a_\infty)$, the probability that this type is a. At time stage, t_n, let there be j individuals of type A. Then there are $N - j$ individuals of type a. Then, at the stage, t_n, the probability of an A death is $\frac{\mu_A j}{\mu_A j + \mu_a (N-j)}$, while for a, it is $\frac{\mu_a (N-j)}{\mu_A j + \mu_a (N-j)}$. The probability of replacement by a new A individual is $\frac{j}{N}$ and that by an a individual is $\frac{N-j}{N}$. It is trivial to check that the total number of individuals will remain at N for stage t_{n+1} (because the probabilities of replacement by either an A or an a individual add up to 1).

In this model the state of a population is described by the number of A individuals (or, alternatively, the number of a individuals because the sum of the two numbers must always add up to N). Given the arguments of the last paragraph, state changes of the population are described by a Markov chain with transition probabilities:[28]

[25] The reasons for specifying the selection process in such – perhaps excruciating – detail will become obvious in Section 21.4 when we turn to the issue of 'statisticalism' in the interpretation of modern evolutionary theory.

[26] This formulation of the model (Equation 21.1) and its solution (Equation 21.2), are due to Karlin and Taylor (1974, 114).

[27] More alleles only make the mathematical analysis more algebraically complicated without providing new insight.

[28] The transitions are from one state (parameterized by j) to another at any time stage.

$$
\begin{aligned}
P_{j,j-1} &= \frac{\mu_A j}{\mu_A j + \mu_a(N-j)} \frac{N-j}{N};\\
P_{j,j+1} &= \frac{\mu_a(N-j)}{\mu_A j + \mu_a(N-j)} \frac{j}{N};\\
P_{j,j} &= 1 - P_{j,j-1} + P_{j,j+1};\\
P_{i,j} &= 0 \quad \text{for} \quad |i-j| > 1,
\end{aligned} \tag{21.1}
$$

where $P_{i,j}$ is the probability of transition from a state in which A individuals number i to one in which they number j. (Since, at any stage, the number of A individuals may change at most by 1, there are only three non-zero transition probabilities.)

21.3.2 Analysis

The solution of the Markov chain model described by Equation (21.1) to obtain $p(a_\infty)$ is straightforward.[29] It depends critically on the initial number, k, of A individuals at t_1:

$$
p(a_\infty) = \begin{cases} \dfrac{(\frac{\mu_A}{\mu_a})^N - (\frac{\mu_A}{\mu_a})^k}{(\frac{\mu_A}{\mu_a})^N - 1} & \text{if } \mu_A \neq \mu_a \\ 1 - \dfrac{k}{N} & \text{if } \mu_A = \mu_a. \end{cases} \tag{21.2}
$$

The exact form of the result when there is selection, $\mu_A \neq \mu_a$, is sensitive to the details of the model. If we had modeled selection differently, a different expression would be found. (The implications of this aspect of Equation (21.2) will be noted in Section (21.3.3).) The result for the case when there is no selection, $\mu_A = \mu_a$, is more robust and remains the same for a large class of similar models.

To connect Equation (21.2) to the haploid version of the standard selection model of Section 21.1 (that is, to an infinite population haploid model), consider the limit $N \to \infty$ with $\frac{k}{N}$ held constant. First, let $\alpha = \frac{\mu_A}{\mu_a}$ and $f_0 = \frac{k}{n}$. Then Equation (21.2) can be rewritten as:

$$
p(a_\infty) = \begin{cases} \dfrac{\alpha^N - \alpha^k}{\alpha^N - 1} & \text{if } \mu_A \neq \mu_a \\ 1 - f_0 & \text{if } \mu_A = \mu_a. \end{cases} \tag{21.3}
$$

Now, consider the case of $\mu_1 = \mu_2$ first, that is, when there is no selection. Then $p(a_\infty)$ is the initial frequency of a which is not surprising. The case with selection, $\mu_1 \neq \mu_2$, is a little more interesting. Now,

[29] For details of the calculation, see Karlin and Taylor (1974, p. 114).

$$\frac{\alpha^N - \alpha^k}{\alpha^N - 1} = \frac{\alpha^N}{\alpha^N - 1} - \frac{\alpha^k}{\alpha^N - 1} \tag{21.4}$$

$$= \frac{\alpha^N}{\alpha^N - 1} - (\alpha^{f_0 - 1})^N \frac{\alpha^N}{\alpha^N - 1}. \tag{21.5}$$

If $\alpha > 1$, $\lim_{N\to\infty} \frac{\alpha^N}{\alpha^N - 1} = 1$ and, because $0 \leq f_0 \leq 1$, $\lim_{N\to\infty} (\alpha^{f_0-1})^N = 0$. Consequently, $\lim_{N\to\infty} p(a_\infty) = 1$. Similarly, if $\alpha < 1$, $\lim_{N\to\infty} p(a_\infty) = 0$. This means that, if a has higher fitness than A (which corresponds to $\alpha > 1$), then a eliminates A from the population if the initial population size was infinite (and vice versa).

21.3.3 Interpretation and implications

Though neither model specification nor analysis mentioned drift, we have modeled drift fully (and exactly, that is, without recourse to approximations to solve our model). In particular, we obtained the well-known result – usually obtained using a diffusion approximation (Kimura, 1983) – that, in the absence of selection, the probability that a type gets fixed in a population (that is, it is the only one that remains after an infinite number of time steps) is equal to its initial frequency (Equation 21.2). As noted earlier, the expression for this probability when there is selection is quite sensitive to the details of the model of selection whereas the result in the absence of selection is robust. For instance, Moran (1958, 1962) analylsed a haploid model with fixed finite population. At each time step, two individuals are selected, one for death and the other for reproduction. The one that dies is replaced by the one that is reproduced. Selection can be imposed through the way these individuals are chosen. In the absence of selection we get the same result as Equation (21.2); when selection is present we get something very different.[30]

From Equation (21.3), in the infinite population limit, we encounter another familiar result: the type that has the higher fitness has a probability of 1 of being fixed, irrespective of its initial frequency. Thus, with one notable exception discussed at the end of this paragraph, our stochastic haploid model is seamlessly transformed into a haploid version of the standard selection model. In fact, what is distinctive about this model is the probability of a type reaching fixation in a finite model with selection which depends both on the initial frequencies and the fitnesses (Equation 21.2) – and this is the parameter that is sensitive to the details of the model. The exception mentioned earlier in this paragraph is the infinite population limit after an infinite number of time steps: the probability that a type gets fixed in a population remains equal to its initial frequency rather than being 0 as one would get from a deterministic model which predicts that no type changes in frequency at all.

[30] For a penetrating dicussion of this model – and many others – see Nowak (2006).

To see if drift is a cause of evolution in this model, we must examine the facultative assumptions. There was only one, that which specified the fitnesses of the two types. There was not even a constitutive assumption – a background condition – that explicitly mentioned drift. There was only the condition that the population had a finite size, and that was all that was necessary. The mythology of drift as a *cause* of evolution needs some deflation.

In contrast to treating drift as a cause, the story that should be told is the following: the finite size of a population is part of the conditions under which evolutionary causes – selection and mutation – operate.[31] What these causes can achieve under such conditions is different from what they could have achieved were the population infinite. That is all there is to drift.[32] Another way of putting it is that, because reproduction involves sampling from the population in evolutionary models, stochastic models needed to capture the dynamics of populations make different predictions than deterministic models needed for infinite models. This happens irrespective of whether all fitnesses are the same (no selection) or not (selection). Drift and selection commingle necessarily.[33] Moreover, the quantitative difference between models allowing drift (e.g. Equation 21.2) and infinite population models with the same facultative assumptions provide a natural quantitative measure of drift.

The reason (background condition) why type frequencies change in the absence of selection in stochastic models is that there is sampling during reproduction, exactly the same assumption that, in the standard selection model, generates the Hardy–Weinberg ratios. *Drift occurs for the same reason that Hardy–Weinberg ratios obtain in the standard selection model.* Finally, our haploid model set up reproduction and state transitions of the population in a standard way – the conclusions reached here about the nature of drift are not a result of some idiosyncratic feature of the model.

Before concluding this chapter with some defence of the framework for evolutionary dynamics it uses, an interesting technical point warrants brief mention. Holding $\frac{k}{N} = f_0$ constant as $N \to \infty$ means that the frequency

[31] Recall that in Section 21.1 we decided to exclude migration and all non-Mendelian mechanisms from the discussions of this paper. For a different route to the same conclusion, see Rice (2004) who points out that the expectation of the change of type frequencies due to drift is 0. Consequently, for Rice, drift cannot be a cause of evolution. The intended contrast is with the effects of mutation and selection which are non-zero. This argument coheres well with those made in this paper.

[32] For what appears to be an almost diametrically opposite interpretation of drift, see Brandon (2005) who seems to identify drift with neutrality.

[33] Beatty (1984) seems to have been the first philosopher to have noted this point. Sober (1984, 38n) also noted that drift always occurs in 'real' populations because they are necessarily finite. However, this obvioulsy does not make drift and selection identical, a point emphasized by Pfeifer (2005) who continues to view drift as a cause with no explicit consideration of what constitutes causes in dynamical systems.

of A (and, therefore, of a) remains constant as the size of the population changes.[34] This limiting process is a direct analogue of the thermodynamic limit in statistical physics in which the number of particles in the system is taken to infinity while the density remains constant (Thompson, 1972). This is the analogue of using Equation (21.4) to see if the infinite population limit gives the same results as deterministic models, with frequencies rather than densities kept invariant during the limiting process. The question of interest is whether the expressions for physical parameters at the level of statistical mechanics converge to their thermodynamic counterparts at the limit. This suggests that we may usefully think of deterministic (infinite) population models as 'macroscopic' analogues of 'microscopic' stochastic (finite) population models, at least for the purpose of formal analysis. But, recall that there was some discrepancy between the infinite population limit and predictions of deterministic models in the case of no selection. Analogous problems abound in *classical* statistical mechanics: the existence of the thermodynamic limit has been proved for precious few systems.[35]

21.4 Discussion

Walsh (2007) has recently distinguished 'dynamical' and 'statistical' interpretations of modern evolutionary theory (the so-called 'synthesis' – but see Sarkar (2004) for skepticism on this point – of the early 1930s). According to him, dynamical interpretations invoke causes such as selection and drift to explain patterns of evolutionary change. This is a somewhat eccentric construal of 'dynamic', a point that will be discussed below since we have taken it to be trivially unproblematic that evolutionary models are models of dynamical systems (at the beginning of Section 21.2). In contrast, two assumptions are supposed to define the statistical interpretation: evolutionary explanations invoke only statistical properties of populations, and selection and drift are not causes but 'mere statistical effects' (which are not defined but illustrated by an example that has no bearing on evolutionary models).

Assuming that statistical properties include stochastic ones (those defined using a random variable), the first assumption is not controversial:[36] most dynamical interpretations – including the influential one of Sober (1984) – make the same assumption. According to Walsh (2007, p. 292), drift and

[34] Note that this does not contradict the fact that these frequencies change over time. The limit is being taken at a given time.

[35] See Thompson (1972); there has been little progress since in classical statistical mechanics. For once, the situation is a little better in the quantum domain, at least in quantum electrodynamics (QED).

[36] However, the consensus is not unanimous – see, for instance, Bouchard and Rosenberg (2004).

selection are not causes because they are not 'description-independent'. Now, description-independence is not the same as context-independence: rather, the idea is that, even after the context is fully specified, the quantitative consequences of drift and selection are underspecified. This is tantamount to claiming that, in a model such as the one described by Equation (21.1), the sampling distribution for each type is not unique.

But this claim is manifestly false for any fully specified model no matter whether the population is infinite (in which case we have a deterministic model) or finite (in which case we have a stochastic model). Strangely, Walsh (2007) provides no biological example.[37] The model described by Equation (21.1) included a full description of the reproductive properties of the types to make the point that, in a fully specified model, these properties, which Walsh (2007) accepts as causal, uniquely specify fitness differences which provide the basis for selection. If reproduction were to occur any differently than in the model of Equation (21.1), except in accidentally degenerate cases, the formulae in Equation (21.2) would be different. As was noted earlier, Moran's (1958, 1962) model provides an example.

Let us turn to the dynamical interpretation. The sense in which this chapter has interpreted evolving populations as dynamical systems was indicated at the beginning of Section 21.2 and goes back to the pioneering discussion of Lewontin (1974). This minimal dynamical interpretation only assumes that the causes of evolution (the facultative assumptions) and the conditions under which it occurs (the constitutive assumptions) are jointly dynamically sufficient to specify the future trajectory of the system (at least probabilistically). Once we embed our account of causes in the minimal interpretation, selection is a cause of evolution because it arises from fitness differences between types

[37] Instead he offers the following coin-tossing case the relevance of which to evolution is at best obscure. The experiment has a paired design involving two coins and five paired treatments, with each treatment being a sequence of 10 tosses by a different experimenter. Each coin is simultaneously tossed 10 times in parallel by two different experimenters, and then a new pair of experimenters takes over, with the process repeated five times. According to Walsh, there are three possible ways to describe the experiment: a single series of 100 tosses, two series of 50 tosses, and 10 series of 10 tosses. According to him choosing between them is entirely arbitrary. In a sense he is correct but only because none of these interpretations is appropriate. The third comes closest but ignores the fact that there are two coins. The veridical interpretation – the one that is isomorphic in structure to the experiment – is that of 5 paired treatments of two systems. What techniques should be used to analyse the data (including the type of simulations which Walsh uses, even though analytic computation of the sampling distributions is straightforward) depends on the question being asked which was not specified in the first place. If it is to test whether a single coin is fair, this is an inappropriate experimental design: why use two coins? If it is to test whether there is a difference between the coins, the array of tests developed by Fisher (1935) remain the most appropriate among relatively simple techniques. We are simply not told. The 'statisticalistas' – this is Walsh's (2007, p. 281*n*) term – seem to have some rather peculiar views about statistics. But the most important objection to this example is that it has no biological relevance.

in a population; the quantitative effects of selection are unequivocally assigned in fully specified evolutionary models. Like the statistical interpretation, our account denies that drift is a cause of evolution; however, the quantitative effects of finite population size – that is, drift – are also unequivocally assigned in fully specified evolutionary models. Most adherents of the dynamical interpretation would probably accept the minimal interpretation, whether or not they accept the causal framework adopted here.

However, especially in philosophical circles, the dynamical interpretation is typically construed as endorsing one or more of three additional claims of increasing strength: (1) that populations persist unchanged (in some specified sense) in the absence of causal factors responsible for evolution; (2) the list of such factors that are all on par with each other include selection and drift; and (3) evolutionary theory is a theory of forces. We will consider each of them in turn:

1. The most plausible version of this claim is that due to Kimura (1983): allele frequencies do not change. The trouble with this is that it is only true if populations are infinite. Kimura's (obvious) response is to allow drift to be a factor of evolution which, roughly, would be a cause of evolution in this construal. The question boils down to whether being finite or infinite is a constitutive assumption about a population, setting conditions against which causal factors operate. Kimura's claim is also at odds with his assertion, quoted earlier, that evolution includes '*all* changes, large and small, visible and invisible, adaptive and nonadaptive (1983, p. xiv)', unless we interpret this claim to be implicitly restricted to allelic changes. The analysis of this paper denies Kimura's claim: finite populations may evolve over time even in the absence of causal factors operating on them, simply because of one of their constitutive features: sampling during reproduction. This analysis thus also denies the much stronger claim of Sober (1984) for whom populations persist in Hardy–Weinberg equilibrium (that is, with the Hardy–Weinberg genotypic ratios) in the absence of causes of evolution operating on them.[38] Here, Sober is following a venerable textbook going back to the third edition of Dobzhansky's (1951) highly influential *Genetics and the Origin of Species.*[39] Like Sober (1984), textbooks often claim that evolution consists of departures from Hardy–Weinberg evolution while, sometimes, inconsistently also defining evolution as consisting of changes in allele frequencies. Kimura (1983, pp. 5–6) has correctly criticized the deification of the Hardy–Weinberg rule on the ground of its lack of generality. Against Sober's position, we may also add that it assumes *ex cathedra*

[38] Sober's explicit claim is obviously only applicable to diploid populations but analogues exist for other types of ploidy.

[39] Dobzhansky (1951) was the first to refer to a Hardy–Weiberg 'rule'.

that random mating is constitutive (in our terms, not his) of what an evolving population is, and it is the only type of mating that can play this role.

2. Once again, Dobzhansky (1937) seems to be responsible for introducing 'parity' between drift and selection (with 'parity' being construed as being the same type of factor). Earlier statements were much more careful. In the paper that introduced the term 'drift' to evolutionary biology, Sewall Wright (1931) (who, along with Fisher and Haldane, was the third founder of theoretical population genetics) wrote of genes and gene frequencies drifting, not of drift as a factor of evolution. The next year, in the first relatively complete statement of the shifting balance theory of evolution, Wright (1932, 356) wrote that 'evolution depends on a certain balance of its factors' but these factors did not include drift; rather, they included mutation, selection, inbreeding or outbreeding, and population structure. Wright and his followers – and his critics – should have stuck to these early formulations. Both Fisher (1922) and Haldane (1927) also modelled finite populations during this period but neither of them thought of drift (using this term or some equivalent one) as a 'cause' of evolution.[40] In the analysis presented here, drift results from the (background) conditions against which a population evolves, whereas selection is a cause of evolution. Thus, not only are drift and selection different, they are different kinds of things – there is no question of parity. Moreover, as in the model discussed above (Equation 21.1), drift and selection act together and a quantitative measure of the consequences of drift can be the difference between what would happen in an infinite population and what happens in the (finite) population being modelled. There is no clear sense in which drift and selection should be regarded as opposing factors, let alone 'forces' – which brings us to the third point.
3. The view that evolutionary dynamics should be viewed as being analogous to Newtonian mechanics and that the factors of evolution should be viewed as 'forces' goes back to Sober (1984). It has been highly influential among philosophers but has also been extensively criticized, especially by the 'statisticalists' (Walsh, 2007). It is the presumed Newtonian analogy that led Sober to venerate the Hardy–Weinberg ratios which, in his view, serve the analogue of the Newtonian state of inertia, not disturbed unless some external force acts on it. We have already noted the problems with giving such a privileged role to random mating. For Sober, the operative forces in evolutionary contexts include selection, mutation, and drift. In the case of drift, a change in population size – and

[40] There is thus some additional historical support for treating the finiteness (or not) of a population as a constitutive assumption.

not even what led to such a change – is supposed to be analogous to a Newtonian force: this is counterintuitive. Ultimately, however, there are two reasons for rejecting this view of evolutionary theory: (a) thinking of evolution as a departure from Hardy–Weinberg equilibrium is a myopic view of the subject (for the reasons mentioned above); and (b) the Newtonian analogy does not work. With respect to (a) Sober and his followers do have the option to resort to Kimura's (1983, 6) alternative – define the 'no–force state' as one of unchanging allele frequencies. The difficulties of this position were noted in (1) above. With respect to (b) no credible response is forthcoming. We do not build evolutionary models by beginning an analog of the force equation expressing Newton's second law of motion ($\vec{F} = m\vec{a}$, where $\vec{F}$ is the force, m is the mass, and $\vec{a}$ is the acceleration) and substituting for the force term. Rather, we use the strategy of Section 21.3 and follow a model construction protocol that integrates constitutive and facultative assumptions which are far more interlinked than Newtonian dynamical laws and their initial conditions.[41]

For future reference, the interpretation of modern evolutionary theory advocated in this chapter will be called the *stochastic dynamical interpretation*. It includes: (i) the minimal dynamical interpretation; (ii) a distinction between constitutive and facultative assumptions; (iii) an identification of the former with the conditions under which a population evolves and uses the latter to identify the causes operating in those conditions; and, perhaps more controversially, (iv) that whether or not a population is finite is a constitutive assumption; and (v) fitnesses used in models are summary parameters incorporating causes of differential survival and reproduction. This interpretation is stochastic simply because the dynamics of finite populations must be modeled stochastically, with infinite populations treated as (mathematically) degenerate cases of finite populations. This last insight goes back to the pioneering work of Bartlett (1955) and Moran (1962).

Finally, denying drift to be a cause of evolution is not intended to suggest that it is not important for evolutionary change. Rather we must interpret evolution differently than received philosophical analyses, for instance, those of Sober (1984) and Brandon (1990). The future dynamics of a system depends on both causes and conditions, with the latter defining what the system is in the context of any inquiry. The same causes may result in radically different consequences under different conditions. Consider evolutionary history. There have been two major debates since modern evolutionary theory was formulated: that between the followers of Fisher and Wright (Provine, 1985), and

[41] Abner Shimony (personal communication) has often emphasized this point. For a recent defense of the Newtonian analogy, see Stephens (2004a).

the debate between the selectionists and neutralists (with respect to fitness differences) (Lewontin, 1974).

In the first debate, for Fisher, evolutionary change occurred most easily when selection acts on small fitness differences in large populations with random mating ('panmictic' populations). For Wright, evolution occured most easily in small sub-divided populations. To the extent that these two proposals can be investigated analytically, what the various causes – that is, mutation and selection – can accomplish was resolved quite quickly with little disagreement between Fisher and Wright. Since then, because of the mathematical complexities involved, the dispute has largely been explored by simulation with no final resolution immediately forthcoming (Coyne, Barton and Turelli, 1997; Peck, Ellner and Gould, 1998; Peck, Ellner and Gould, 1998; Wade and Goodnight, 1998; Coyne, Barton and Turelli, 2000). But the continuing dispute is one about the conditions in which evolution occurs, large or small sub-divided populations, with obvious implications for the reconstruction of the evolutionary history of life on Earth.

In the debate over the neutral theory of evolution (which holds that most mutations are either neutral with respect to fitness or slightly deleterious; the latter is often called the 'nearly neutral' theory), once again what is at stake are both causal differences – the mutation rates and selective differences – and the conditions in which these causes operate: the size of the population. Most biologists would probably view this debate as having swung in alternating directions over the last four decades, with the selectionists currently in ascendancy (Dietrich, 2008). But the point is that what selection can do also depends on the size of the population as was emphasized in Section 21.3.3. Conditions – the constitutive assumptions of our models – matter. Perhaps the most important moral to draw is that evolution is a historical process; the sooner we move beyond the simplistic Newtonian analogy, the better. Drift may not be a cause of evolution, but it may well have been one of the most important determinants of the remarkable diversity of life we see around us.

21.5 Acknowledgments

For discussions, sometimes over many years, thanks are due (in alphabetical order) to Robert Brandon, James F. Crow, Mark Kirkpatrick, Manfred Laubichler, Richard Lewontin, the late John Maynard Smith, Samir Okasha, Jessica Pfeifer, Anya Plutynski, Mike Singer, Mike Wade, and Bill Wimsatt. This work was first presented at the International Summer School on Probability and the Special Sciences, Universität Konstanz, Summer 2003. Thanks are due to members of that audience, in particular, Stephan Hartmann and Pat Suppes. Finally, thanks are also due to David Frank, Samir Okasha, and Stavana Strutz for comments on an earlier version of this paper. This work was supported by the United States NSF Grant No. SES-0645884, 2007–2009.

References

Abrams, M. (2007). How do natural selection and random drift interact? *Philosophy of Science*, **74**, 666–679.

Bartlett, M. S. (1955). *An Introduction to Stochastic Processes*. Cambridge University Press, Cambridge, UK.

Beatty, J. (1984). Chance and natural selection. *Philosophy of Science*, **51**, 183–211.

Bouchard, F. and Rosenberg, A. (2004). Fitness, probability and the principles of natural selection. *British Journal for the Philosophy of Science*, **55**, 693–712.

Brandon, R. (1978). Adaptation and evolutionary theory. *Studies in the History and Philosophy of Science*, **9**, 181–206.

Brandon, R. (1990). *Adaptation and Environment*. Princeton University Press, Princeton, NJ.

Brandon, R. (2005). The difference between selection and drift. *Biology and Philosophy*, **20**, 153–170.

Caughley, G. (1994). Directions in conservation biology. *Journal of Animal Ecology*, **63**, 215–244.

Coyne, J. A., Barton, N. H., and Turelli, M. (1997). Perspective: A critique of Sewall Wright's shifting balance theory of evolution. *Evolution*, **51**, 643–671.

Coyne, J. A., Barton, N. H., and Turelli, M. (2000). Is Wright's shifting balance balance process important in evolution. *Evolution*, **54**, 306–317.

Crow, J. F. and Kimura, M. (1970). *An Introduction to Population Genetics Theory*. Burgess, Minneapolis.

Dietrich, M. (2008). Molecular evolution. In *A Companion to the Philosophy of Biology* (ed. S. Sarkar and A. Plutynski), pp. 157–168. Blackwell Publishing, Malden, MA.

Dobzhansky, T. (1937). *Genetics and the Origin of Species*. Columbia University Press, New York.

Dobzhansky, T. (1951). *Genetics and the Origin of Species* (3rd edn). Columbia University Press, New York.

Fisher, R. A. (1922). On the dominance ratio. *Proceedings of the Royal Society of Edinburgh*, **42**, 321–341.

Fisher, R. A. (1928). The possible modification of the response of the wild type to recurrent mutations. *American Naturalist*, **62**, 115–126.

Fisher, R. A. (1930). *The Genetical Theory of Natural Selection*. Clarendon Press, Oxford, UK.

Fisher, R. A. (1935). *The Design of Experiments*. Oliver & Boyd, Edinburgh.

Forber, P. and Reisman, K. (2007). Can there be stochastic evolutionary causes? *Philosophy of Science*, **74**, 616–627.

Gillespie, J. H. (1994). *The Causes of Molecular Evolution*. Oxford University Press, Oxford, UK.

Gorovitz, S. (1965). Causal judgements and causal explanations. *Journal of Philosophy*, **62**, 695–711.

Gould, S. J. (2002). *The Structure of Evolutionary Theory*. Harvard University Press, Cambridge, MA.

Haldane, J. B. S. (1927). A mathematical theory of natural and artificial selection. Part V. Selection and mutation. *Proceedings of the Cambridge Philosophical Society*, **23**, 838–844.

Hart, H. L. A. and Honoré, A. M. (1985). *Causation and the Law*. Clarendon Press, Oxford, UK.

Hirsch, M. W. and Smale, S. (1974). *Differential Equations, Dynamical Systems, and Linear Algebra*. Academic Press, New York.

Iizuka, M. and Matsuda, H. (1982). Weak convergence of discrete time non-Markovian processes related to selection models in population genetics. *Journal of Mathematical Biology*, **15**, 107–127.

Karlin, S. and Taylor, H. M. (1974). *A First Course in Stochastic Processes* (2nd edn). Academic Press, San Diego.

Kimura, M. (1983). *The Neutral Theory of Molecular Evolution*. Cambridge University Press, Cambridge, UK.

Levin, D. A., Kelley, C. D., and Sarkar, S. (2009). Enhancement of Allee effects in plants due to self-incompatibility alleles. *Journal of Ecology*, **97**, 518–527.

Lewis, D. (1973). Causation. *Journal of Philosophy*, **67**, 556–567.

Lewontin, R. C. (1974). *The Genetic Basis of Evolutionary Change*. Columbia University Press, New York, NY.

Ma, W. T., Sandri, G. V., and Sarkar, S. (1992). Analysis of the Luria–Delbrück distribution using discrete convolution powers. *Journal of Applied Probability*, **29**, 255–267.

Mackie, J. L. (1974). *The Cement of the Universe*. Clarendon Press, Oxford, UK.

Magnus, D. (1998). Evolution without change of gene frequencies. *Biology and Philosophy*, **13**, 255–261.

Menzies, P. (2004). Difference-making in context. In *Causation and Counterfactuals* (ed. J. Collins, N. Hall, and L. A. Paul), pp. 139–180. MIT Press, Cambridge, MA.

Mills, S. and Beatty, J. (1979). The propensity interpretation of fitness. *Philosophy of Science*, **46**, 263–286.

Moran, P. A. P. (1958). Random processes in genetics. *Proceedings of the Cambridge Philosophical Society*, **54**, 60–71.

Moran, P. A. P. (1962). *The Statistical Processes of Evolutionary Theory*. Clarendon Press, Oxford, UK.

Nagylaki, T. (1992). *An Introduction to Theoretical Population Genetics*. Springer, Berlin.

Nowak, M. A. (2006). *Evolutionary Dynamics: Exploring the Equations of Life*. Harvard University Press, Cambridge, MA.

Orr, H. A. (1991). A test of Fisher's theory of dominance. *Proceedings of the National Academy of Sciences (USA)*, **88**(24), 11413–11415.

Otto, S. P., Servedio, M. R., and Nuismer, S. L. (2008). Frequency-dependent selection and the evolution of assortative mating. *Genetics*, **179**, 2091–2112.

Peck, S. L., Ellner, S. P., and Gould, F. (1998). A spatially explicit stochastic model demonstrates the feasibility of Wright's shifting-balance theory. *Evolution*, **52**, 1834–1839.

Pfeifer, J. (2005). Why selection and drift might be distinct. *Philosophy of Science*, **72**, 1135–1145.

Plutynski, A. (2008). Speciation and macroevolution. In *A Companion to the Philosophy of Biology* (ed. S. Sarkar and A. Plutynski), pp. 169–185. Blackwell Publishing, Malden, MA.

Provine, W. B. (1971). *The Origins of Theoretical Population Genetics*. University of Chicago Press, Chicago.

Provine, W. B. (1985). The R.A. Fisher—Sewall Wright controversy and its influence upon modern evolutionary biology. *Oxford Surveys in Evolutionary Biology*, **2**, 197–219.

Rabotnov, Y. N. (1980). *Elements of Hereditary Solid Mechanics*. Mir Publishers, Moscow.

Rice, S. H. (2004). *Evolutionary Theory: Mathematical and Conceptual Foundations*. Sinauer Associates, Sunderland, MA.

Sarkar, S. (1998). *Genetics and Reductionism*. Cambridge University Press, Cambridge, UK.

Sarkar, S. (2004). Evolutionary theory in the 1920s: The nature of the synthesis. *Philosophy of Science*, **71**, 1215–1226.

Sarkar, S. (2005). *Biodiversity and Environmental Philosophy: An Introduction to the Issues*. Cambridge University Press, Cambridge, UK.

Sarkar, S. (2007). Haldane and the emergence of modern evolutionary theory. In *Handbook of the Philosophy of Biology* (ed. M. Matthen and C. Stephens), pp. 49–86. Elsevier, New York.

Sarkar, S., Ma, W. T., and Sandri, G. V. (1992). On fluctuation analysis: A new, simple and efficient method for computing the expected number of mutants. *Genetica*, **85**, 173–179.

Sober, E. (1984). *The Nature of Selection: Evolutionary Theory in Philosophical Focus*. MIT Press, Cambridge, MA.

Stephens, C. (2004). Selection, drift, and the 'forces' of evolution. *Philosophy of Science*, **71**, 550–570.

Stephens, C. (2008). Molecular evolution. In *A Companion to the Philosophy of Biology* (ed. S. Sarkar and A. Plutynski), pp. 119–137. Blackwell Publishing, Malden, MA.

Stewart, F. M., Gordon, D. M., and Levin, B. R. (1990). Fluctuation analysis: The probability distribution of the number of mutants under different conditions. *Genetics*, **124**, 175–185.

Thompson, C. J. (1972). *Mathematical Statistical Mechanics*. Princeton University Press, Princeton, NJ.

Wade, M. J. and Goodnight, C. J. (1998). Perspective: The theories of Fisher and Wright in the context of metapopulations: When nature does many small experiments. *Evolution*, **52**, 1537–1548.

Walsh, D. M. (2007). The pomp of superfluous causes: The interpretation of evolutionary theory. *Philosophy of Science*, **74**, 281–303.

Walsh, D. M., Lewens, T, and Ariew, A. (2002). The trials of life: Natural selection and random drift. *Philosophy of Science*, **69**, 452–473.

Williams, S. M. and Sarkar, S. (1994). Assortative mating and the adaptive landscape. *Evolution*, **48**, 868–875.

Wright, S. (1931). Statistical theory of evolution. *Journal of the American Statistical Association*, **22**, 201–208.

Wright, S. (1932). The roles of mutation, inbreeding, crossbreeding and selection in evolution. In *Proceedings of the Sixth International Congress of Genetics* (ed. D. F. Jones), Volume 1, pp. 356–366. Brooklyn Botanic Garden, Monisha, WI.

·22·

In defense of a causal requirement on explanation

Garrett Pendergraft

Abstract

Causalists about explanation claim that to explain an event is to provide information about the causal history of that event. Some causalists also endorse a proportionality claim, namely that one explanation is better than another insofar as it provides a greater amount of causal information. In this chapter I consider various challenges to these causalist claims. There is a common and influential formulation of the causalist requirement – the 'Causal Process Requirement' – that does appear vulnerable to these anti-causalist challenges, but I argue that they do not give us reason to reject causalism entirely. Instead, these challenges lead us to articulate the causalist requirement in an alternative way. This alternative articulation incorporates some of the important anti-causalist insights without abandoning the explanatory necessity of causal information. For example, proponents of the 'equilibrium challenge' argue that the best available explanations of the behaviour of certain dynamical systems do not appear to provide any causal information. I respond that, contrary to appearances, these equilibrium explanations are fundamentally causal, and I provide a formulation of the causalist thesis that is immune to the equilibrium challenge. I then show how this formulation is also immune to the 'epistemic challenge' – thus vindicating (a properly formulated version of) the causalist thesis.

22.1 Introduction

Let us identify a *causalist* about scientific explanation as one who is committed, in some way or other, to the following thesis from David Lewis: 'to explain an event is to provide some information about its causal history' (Lewis, 1986 p. 217). The tricky part, of course, is specifying what sort of causal information we are talking about. One important explication of 'causal information' comes from Wesley Salmon, a causalist who developed the original and influential *causal-mechanical* model of explanation.[1] On this model, to explain an event

[1] Salmon's view is laid out in, among other places, his (1993b). For a book-length treatment of his view, see his (1984).

is to provide some subset of the causal processes (and interactions between causal processes) that brought about that event. The causal-mechanical model falls under a general conception of explanation that Salmon calls the 'ontic' conception – according to which explaining an event involves locating that event (the explanandum event) within certain nomologically necessitated regularities, or patterns, in the world.[2] Moreover, he views these regularities as causal regularities governing interactions between causal processes.[3]

Salmon's approach to causal information (as information about causal processes) gives us a slightly better handle on what exactly the causalist is committed to. As a first approximation, then, in deference to Lewis and with help from Salmon, we can say that a causalist is committed to something like the following causal requirement:

> *The Causal Process Requirement*: An explanation of an event must specify some of the causal processes that constitute that event's causal history.

The Causal Process Requirement (CPR) is a plausible way of spelling out the general causal requirement on explanation to which all causalists are committed.[4] It also gives us with an initial gloss on 'providing causal information': to provide causal information about an event is to list some of the causes of that event.

If CPR is apt, then it seems reasonable to suppose that, given two competing explanations of the same event, the one that provides *more* causal information will be better than the other. Hence, the proponent of CPR will often be committed to this additional thesis, which I will call *Proportionality:*

> *Proportionality*: Explanatory power increases in direct proportion to the *amount* of causal information provided.[5]

(As with the notion of 'causal information' itself, it is difficult to specify exactly what we mean by the 'amount' of causal information. I will say more about this problem below.) If providing causal information involves providing the

[2] Salmon (1993, pp. 79–81) characterizes the other conceptions of explanation as either *epistemic* (i.e. as arguments in which the explanandum statement follows deductively from the statements in the explanans) or *modal* (i.e. as exhibiting the physical necessity of the explanandum fact, given the facts in the explanans). There is reason to question this taxonomy of the different conceptions of explanation, but for now the important part is Salmon's focus on causal processes.

[3] For more on the important notions of 'causal process' and 'causal interaction,' see Salmon's (1993a).

[4] Note that CPR is relatively weak, at least insofar as it leaves open the question of which (and how many) causal processes must be cited. Nevertheless, as we will see, there are certain explanations that do not appear to satisfy even this weak version of CPR.

[5] This dictum is perhaps most evident in Railton's deductive-nomological-probabilistic (DNP) account, which adverts to the 'ideal explanatory text' – the full-fledged explanation of an event, including *all* of the relevant causal detail. For more on the DNP model and the ideal text, see Railton's (1978).

causes themselves, then we can see that Proportionality presupposes CPR.[6] (And in general we can say that the proponent of Proportionality is also committed to some variation or instantiation of the causalist requirement – whether it be CPR or some alternative formulation.) With these two theses in hand, we can define an *anti-causalist* as one who rejects CPR, and thus by extension Proportionality as well. Most commonly, anti-causalists base this rejection on putative counterexamples in which the best explanation of some phenomenon appears to be completely non-causal.

My claim is this: the anti-causalists are being too hasty if they let their rejection of CPR lead them to eschew causalism in general – because the causalist requirement can be reformulated in a way that renders it immune to the anti-causalist challenge. Proportionality, on the other hand, turns out to be unsalvageable. The examples cited by anti-causalists do indeed show that more is not always better, when it comes to causal information; but to extend this conclusion to a complete rejection of causal requirements more generally is not warranted. Instead, I argue, these examples lead us to articulate the causalist requirement in an alternative way. This alternative articulation incorporates some of the important anti-causalist insights without giving up on the explanatory necessity of causal information.

As I defend my claim, I will consider what I take to be two of the strongest challenges to the causal requirement. The first challenge, which I will characterize as an 'equilibrium challenge', comes from Elliott Sober. He argues that the best available explanations for the behaviour of certain dynamical systems do not appear to provide any causal information, thus refuting CPR (Sober, 1983). In response to the equilibrium challenge, I argue that, despite appearances, these equilibrium explanations are fundamentally causal. Thus, even if equilibrium explanations do not satisfy CPR, there will be an alternative formulation of the causalist thesis that does apply to those explanations. I will propose just such a formulation. I will then take the conceptual apparatus developed in response to the equilibrium challenge and apply it to instances of the second challenge, which I will characterize as an 'epistemic challenge'. Proponents of this challenge point out that understanding can actually be obscured when we focus on providing causal information. The insights gleaned from the equilibrium challenge provide a way of responding to the epistemic challenge as well – once again vindicating (a revised version of) the causalist thesis.

My project here is, in short, a focused attempt to trace and develop the dialectic surrounding causal requirements on scientific explanation. The general causalist requirement comes from Lewis, and is fleshed out in a particular

[6] Although, as we will see below (cf. note 23), we can also leave open the question of whether all explanation is causal explanation and ask whether Proportionality is true if we restrict our discussion to causal explanations only.

way by Salmon. Sober presents a criticism of this fleshed out requirement, which inspires various revisions and reformulations. Along the way, I will be strengthening my defense (and revision) of the causalist requirement by pointing out how other, more recent authors have arrived at similar conclusions via different routes. I will consider, for example, James Woodward's (2003) influential manipulationist account of explanation. One of the lessons learned from the equilibrium challenge is that an important part of the explanation of a dynamical system is a description of the system that provides some set of constraints on its behavior. These constraints will indicate what sort of factors need to be present (or absent) for the system in question to reach equilibrium. Thus, when the system satisfies these constraints, it can be characterized by a certain sort of *invariance* – and for Woodward (2003, p. 183), 'invariance is the key to explanatoriness.' I will also consider Michael Strevens's (2008) 'kairetic' account of explanation, according to which one of the key explanatory virtues is *depth*. The equilibrium challenge forces us to reject Proportionality, and Strevens's notion of explanatory depth provides, among other insights, an elegant way of characterizing the motivation for that rejection.

22.2 Equilibrium challenges to causalism about explanation

Perhaps the most difficult examples for causalists to deal with are those involving *equilibrium explanations* – which explain an observed equilibrium state of a dynamical system by providing a range of possible initial states and possible causal trajectories. Given the possible causal trajectories of the system, each of the possible initial states would have led to the observed equilibrium.[7] Even if providing an explanation *usually* consists in providing causal information (i.e. consists in listing some of the causes of the explanandum event), equilibrium explanations appear to be an exception. They explain the equilibrium state of a dynamical system by providing a disjunction of possible causal trajectories – and, as Sober points out, 'disjunctions of causal scenarios will sometimes fail to say what the cause is' (Sober, 1983, p. 205). Thus, it would seem that we have a counterexample to CPR, in which the explanatory work is done *sans* causal information.

This, at least, is the argument as advanced by Sober.[8] And although it may succeed against CPR as stated above, it does not succeed against an alternative formulation of the causalist commitment – or so I claim. In other words, the

[7] For a representation of this viewpoint regarding equilibrium explanations, see Batterman (1992).

[8] Another instance of the anti-causalist strategy can be found in Berger (1998). In her example, the unpredictable growth patterns of Dungeness crabs are explained by the linear distribution of their eggs.

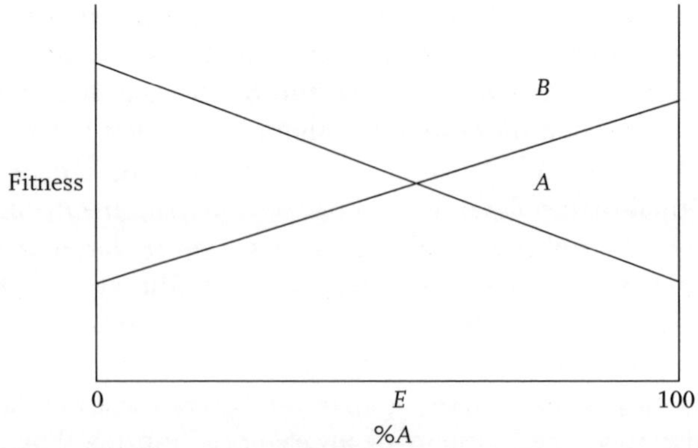

Fig. 22.1

argument is too hasty if meant to apply to causal requirements in general. In order to see why, we need to have a closer look at the example Sober utilizes.

Sober's example of an equilibrium explanation involves a fitness function from population genetics. Given some population with two traits (A and B), the fitness function for each trait specifies the expected number of offspring for that trait, according to its frequency in the population. The system can be modeled as in Figure 22.1.

Among other things, this diagram represents the selection forces at work in this population: each of the two traits is favoured by selection when it is in the minority.[9] Several points on this diagram are especially salient. Note first that E is an equilibrium value, at which the proportion of type A to type B does not change. Furthermore, E is a *stable* equilibrium, because any deviation from E will trigger selection forces that will push the system back toward E. (An unstable equilibrium would be the opposite – e.g. if the fitness functions of A and B were reversed. In that case, given a deviation from E, the selection forces would move the system away from E.) The other salient states of the system are the *absorbing* states, as represented by the four endpoints (or absorbing points) of the two fitness functions. If these absorbing states are reached, the forces represented in the model will not be able to move the system to a different state. In the system modeled above, then, absorbing points are points of no return.

In addition to the distinction between stable and unstable equilibria, we can also draw a distinction between *global* and *local* equilibria. This distinction is understood in terms of the range of initial conditions that will lead to the equilibrium state (Sober, 1983 p. 204). A global equilibrium is such that the system

[9] The example, including the diagram, is taken directly from Sober (1983, pp. 207–8). Referring to natural selection as a 'force' is somewhat contentious, but for simplicity of exposition I will nevertheless continue to do so. (But see Walsh et al. (2002)).

will end up in that state no matter which initial conditions obtain, whereas a local equilibrium is such that a range of initial conditions must be specified, outside of which the system might not reach that particular equilibrium. Thus we can think of an explanation of a dynamical system in terms of the *degree to which initial conditions must be specified:* no specification is required in the case of global equilibria, whereas some specification is required in the case of local equilibria. And of course local equilibria themselves come in degrees – some equilibria are more local than others. So the 'locality' of an equilibrium is directly proportional to the amount of information that needs to be specified, with respect to the system's initial conditions, in order to explain that system's equilibrium state (Sober, 1983 pp. 208–209).

We can now return to the received anti-causalist wisdom, as nicely encapsulated in the following passage from Sober (1983, p. 209).

> When we are at one end of the continuum – when the equilibrium is a global one – an event can be explained in the face of considerable ignorance of the actual forces and initial conditions that in fact caused the system to be in its equilibrium state. In this circumstance, we are, in one natural sense, ignorant of the event's cause, but explanation is possible nonetheless.

This received wisdom maintains a sharp dichotomy between causal explanation and equilibrium explanation, on the basis of which CPR, among other causal requirements, is rejected. As we will see, equilibrium explanations apparently do show that CPR is inadequate. However, rather than leading us to reject causal requirements altogether, I will attempt to show how Sober's treatment actually provides the resources for a *reformulation* of the causalist thesis – one which renders it immune to challenges from equilibrium explanations (and to related challenges). The crucial element of this reformulation project, as we will see below, is the claim that equilibrium explanations can actually be characterized in terms of two distinct continua: in addition to the continuum of locality (i.e. the continuum of information about the initial conditions) that Sober identifies, there is also a continuum of causal information.

22.3 Causal information

The first step in this reformulation project is to acknowledge that, if causal information is construed in terms of the processes that constitute an event's actual causal history, then it does appear to be true that equilibrium explanations do not provide any causal information, and thus constitute a class of counterexamples to CPR.[10] In the face of this problem, I propose that the

[10] There is room, perhaps, for the causalist to resist the conclusion that equilibrium explanations do not provide causal process information. For example, if one subscribes to something like Railton's DNP model of explanation, then one might think that a disjunction of possible

causalist can modify his requirement so that it will no longer be vulnerable to such counterexamples, while at the same time retaining the spirit of causalism. I will begin fleshing out this proposal by examining the notion of the *amount of causal information* that an explanation provides, in light of what we have learned about equilibrium explanations. This examination will point to a replacement for CPR that is immune to the equilibrium challenge.

As noted above, it is difficult to know what exactly it means to say that one explanation provides a greater amount of causal information than another. One clear case is when the explanans of one explanation entails the explanans of a different explanation; in this case the logically stronger explanans provides a greater amount of information than the logically weaker explanans. But when there is no entailment relation, it is less clear how to compare two different explanans.[11] We could begin the comparison, however, by drawing upon the resources of information theory.[12] We could say that an explanans provides more information to the extent that it rules out possibilities. If we construe 'causal history' as it is construed in CPR – i.e. as incorporating information about the actual causal processes leading up to and bringing about the explanandum event – then this information-theoretic suggestion amounts to the following claim: the narrower the range of possible causal histories (of the explanandum event) allowed by the explanans, the more information it will provide. But this suggestion will not do, because explanations are always given *in a context*. This point is strongly emphasized in Wright (n.d.), and can be better understood if we consider one of his examples: an explanation of one's house burning down.[13] He imagines that a contributing cause to this fire was a candle falling down on a curtain, and we might think it straightforward to say that an explanation citing the candle falling on the curtain would provide more information than, say, an explanation merely citing the curtain's catching fire. After all, the explanation citing the candle provides more restrictions on the causal history leading up to the event (for example, it rules out the curtain's catching fire as a result of an electrical short). However, as Wright points out, the context of inquiry might affect which of the two competing explanations is more informative. For example, if the curtain has been treated with fire retardant, and we are not surprised that the candle fell on the curtain (perhaps this has happened before; perhaps it is *because* this has happened before that

causal trajectories could be incorporated into the ideal explanatory text, and thus count as causal information in some minimal sense. For an implementation of a similar strategy, see note 27.

[11] Thanks to an anonymous reviewer for helping me with this point.

[12] Salmon very briefly considers an information-theoretic approach to unification theories of explanation in Chapter 4 (p. 131) of his (2006). Railton briefly discusses the usefulness (as well as some of the shortcomings) of information theory in his (1981).

[13] The context-sensitivity of explanation is also emphasized in Scriven (1962). (See especially Section 3, pp. 52–3.)

the curtain was treated with fire retardant), then even a detailed and colourful causal history of the candle's fall on the curtain might not be as informational as the simple fact that the retardant was abraded by a recent cleaning of the curtains. In short, what counts as a greater amount of causal information is going to vary from explanatory context to explanatory context. As a result, there may not be any simple, absolute, one-dimensional scale on which we can map the various 'amounts' of causal information that an explanation might provide. If there were a general scale that we could use to measure differing amounts of causal information, it would likely involve several dimensions, and thus be quite complex.

Nevertheless, I think our consideration of equilibrium explanations does point us toward a helpful way of providing some restrictions on what we mean by a greater (or lesser) amount of causal information. First, however, let us try to quantify the information about initial conditions that is provided by an equilibrium explanation of a dynamical system. Recall that, because the equilibrium point E in the diagram above is a stable, global equilibrium, no information about the initial conditions need be provided in the equilibrium explanation (apart from the stipulation that the population cannot begin in one of the absorbing states).[14] But notice that if E were instead a *local* equilibrium, then the equilibrium explanation would have to provide a range of possible initial conditions, within which selection forces would move the system toward E. Thus, returning to information theory, we can say that in the context of the dynamical system being modelled, an explanation of a local equilibrium will provide a greater amount of information about initial conditions – will rule out more possibilities – than will an explanation of a global equilibrium. This point extends to a comparison of two local equilibria, about which we can say that the more local the equilibrium (i.e. the more detail required when specifying the initial conditions), the more information provided by the explanation. Thus, we can flesh out Sober's notion of the continuum of locality as follows. At the extreme local end of this continuum, the maximal amount of information is required: the range of initial conditions specified will consist simply of the *actual* initial conditions. At the other

[14] The parenthetical qualification, particularly in light of what I say below, might lead one to ask whether we have properly identified the endpoint of the continuum of information about the initial conditions. (Recall that a global equilibrium is defined as one in which there is only one restriction on initial conditions, namely that the population not begin in one of the system's absorbing states.) It could be argued, for example, that there might be some dynamical systems that will reach equilibrium no matter what state they start in (which would preclude those systems from having any absorbing states). If that is correct, then the specification of a true global equilibrium need not refer to absorbing states. But such a claim would not affect my own argument. In principle I can accept either definition of 'global equilibrium,' since I am only concerned to argue that equilibrium explanations cannot appeal to an equilibrium that falls on the extreme minimal endpoint of the *causal information* continuum. (Thanks to an anonymous referee for helping me clarify this point.)

endpoint of this continuum (i.e. the point represented by global equilibria), the minimal amount of information is required: the only restriction on the initial conditions of the system is that the population cannot start in one of the absorbing states. So the equilibrium point of any given dynamical system can be mapped on this continuum according to the amount of information required when specifying the initial conditions.

Within the context of equilibrium explanations, then, we have a straightforward and illuminating way of cashing out the notion of a greater or lesser amount of information. A global equilibrium can be specified using the minimal amount of information regarding the initial conditions, whereas a maximally local equilibrium will have to be specified using the maximal amount of information: the actual initial conditions. The amount of information that an equilibrium explanation provides can thus be measured in terms of the locality of the relevant equilibrium point.

So far we are in agreement with Sober. But recall that he uses this continuum to represent two kinds of information: information about initial conditions as well as information about causal forces: 'When we are at one end of the continuum – when the equilibrium is a global one – an event can be explained in the face of considerable ignorance of the *actual forces and initial conditions* that in fact caused the system to be in its equilibrium state' (Sober, 1983 p. 209 [emphasis mine]). My claim here is that there are actually two separate continua: one corresponding to information about initial conditions, and one corresponding to information about causal forces. Moreover, the equilibrium points of different dynamical systems (and thus the equilibrium explanations of those systems) can vary independently with respect to these two continua.

If this is right, then the next natural step is an attempt to use the results of the above discussion to make more sense of the notion of a greater or lesser amount of *causal* information. Here again we can learn from equilibrium explanations, which, in addition to providing information about initial conditions, also provide a disjunction of possible causal trajectories. In other words, equilibrium explanations refer, not just to the relevant initial conditions, but also to the relevant *causal forces* – forces that, up to this point, I have been largely ignoring (in order to focus on local and global equilibria, which are distinguished according to initial conditions). But now we can (and should) extend our treatment of the initial conditions to cover the causal trajectories as well. For just as we can locate various equilibrium explanations on a continuum from minimal to maximal amounts of information about initial conditions, we can also locate these explanations on a continuum from minimal to maximal amounts of information about the causal forces leading to equilibrium. Explanations toward the minimal end of this continuum will provide very little information about the causal trajectories; i.e. they will only rule out a relatively small number of the possible trajectories that might have led to the observed equilibrium. Explanations toward the maximal end of this continuum, on the other hand, will provide a great deal of information; i.e. they will provide significant constraints on the ways in which the system reaches equilibrium. At the extreme maximal end of this continuum, all possible causal trajectories will be ruled out save one: the *actual* causal trajectory taken through the phase space of the system in which the equilibrium was observed. Moreover, in this (admittedly unlikely) scenario, the equilibrium explanation will after all satisfy the Causal Process Requirement. Most equilibrium explanations will in fact violate CPR, but nothing about the structure of such explanations dictates that they *must* violate CPR.[15]

Let us now take stock of our progress. We have seen, first of all, that the equilibrium explanation that Sober provides does not satisfy CPR; it offers a disjunction of possible causal histories, but it does not cite any of the causal processes that constitute the actual causal history. We have also seen that in general we can describe equilibrium explanations as offering more or less information about (1) the initial conditions of the dynamical system and (2) the possible combinations of causal forces leading to equilibrium. Finally, we have seen that a hypothetical equilibrium explanation at the maximal information end of the causal forces continuum will specify the exact causal forces that brought about the equilibrium – i.e. the actual causal processes that constituted the history of the equilibrium event. This hypothetical equilibrium explanation *would* satisfy CPR.

I suggest that the most reasonable move, in light of these considerations, is not a rejection of causalism but a replacement of CPR with a closely related variant of the causalist requirement. We cannot require a specification of the actual causal trajectory (i.e. we cannot require the maximum amount of information on the causal forces continuum), because equilibrium explanations typically explain without providing that level of detail. So why not discard CPR (and related causal requirements) entirely? Because even equilibrium explanations must provide *some sort* of causal information. Since any given equilibrium explanation can be located on a continuum that represents the amount of causal information provided, the only way in which an equilibrium explanation could do its explanatory work *sans* causal information is if it fell on the extreme minimal endpoint of the causal forces continuum: if it provided absolutely no information about the causal forces at work in the system. Such an explanation would tell us only that there is a certain range of initial conditions needed to reach equilibrium, and when the system begins within those initial conditions, it somehow reaches equilibrium.

[15] Note also that since we are restricting our consideration of possible causal trajectories to a *particular system*, we are respecting Wright's (n.d.) claims about the contextual nature of explanation and avoiding the difficulties that would plague a more general information-theoretic attempt to specify the amount of causal information provided by various competing explanations.

This 'explanation', however, is next to useless, as it tells us only that we are dealing with a dynamical system – a fact which presumably we already knew.

Perhaps, though, we presume too much. Perhaps an equilibrium explanation would be useful despite providing no information about the causal forces at work in the relevant system. (Imagine that the relevant system, as above, is a relatively simple one consisting of a population with two traits.) Such an explanation would identify some distribution of the population as the equilibrium state, and then provide a range of initial conditions within which the population would reach that state. From these two pieces of information (the equilibrium state and the initial conditions) would follow a third: that the population we are trying to explain is (or at least can be represented as) a dynamical system. This might seem to be enough to make the explanation a useful one. But if we consider an analogy, then I think it will become clear that the proposed (non-causal) equilibrium explanation is no good.

Consider an explanation of the state of a gas in terms of the behaviour of its component molecules. While it does not seem explanatorily useful (or even feasible) to attempt to describe the causal history of any (much less all) of the component molecules, it also does not seem useful to explain the state of the gas by merely pointing out that it is composed of molecules, which collide with each other according to certain causal laws (e.g. the laws of Newtonian mechanics), and that those collisions somehow produce the relevant state. Intuitively, this latter explanation seems unsatisfactory. (As Woodward says, this would appear to be a 'trivial, non-serious explanation of the behavior of the gas'.) If this intuition is correct, then the same can be said of the equilibrium explanation that we are considering. For it seems that this trivial and non-serious explanation of the state of the gas is structurally similar to the equilibrium explanation in question. The proposed equilibrium explanation essentially says that, given certain initial conditions, features of the system somehow produce the equilibrium state. If the imagined explanation of the state of the gas is trivial and non-serious, then surely the proposed non-causal equilibrium explanation is trivial and non-serious as well.

Can we put any flesh on the bones of this triviality intuition? I think we can, if we appeal to one of Strevens's (2008) criticisms of an explanation similar to the wholly non-causal equilibrium explanation we have been considering. He claims, in short, that explanations that include 'black boxes' are unacceptably 'shallow' (Strevens, 2008 pp. 130–132). (For Strevens, a black box is essentially a functional definition, 'which explains *c*'s causing *e* by citing only the fact that *c* has the property of being *e*-producing' (Strevens, 2008 p. 131)). And it seems that the proposed non-causal equilibrium explanation is a black box explanation of the equilibrium state. It makes no claims about causes, but it does explain a system *S*'s ending up in its equilibrium state *E* by citing a function. This function tells us only that the system has the property of being *E*-producing (given certain initial conditions). We might conclude that insofar as a completely non-causal equilibrium explanation relies on a black box to do its explanatory work, it should be dismissed as shallow.

Thus, once we see that equilibrium explanations can be classified according to where they fall on a continuum representing the amount of causal information provided, we can also see that equilibrium explanations falling on the minimal endpoint of the continuum will not really serve as explanations at all. Such explanations will be trivial and non-serious, not to mention shallow (among perhaps other explanatory vices).

In addition to these negative reasons for construing even equilibrium explanations as causal, there are – returning to Strevens – also positive reasons for making the same claim. Consider a different equilibrium explanation, of the fact that a ball released on the lip of a basin will end up resting at the lowest point of the basin no matter where on the lip the ball is released. Strevens (2008, p. 268) points out that

> while a casual inspection of the equilibrium model might give the impression that it says nothing about the particular causal process leading to the explanandum event, in fact the model is exclusively concerned to describe this very token process, but at an extremely abstract level, so abstract that the description is satisfied by every process by which the ball might have reached the bottom of the basin.

In other words, as we move toward the minimal end of the causal information continuum, we are moving in the direction of increasingly abstract description. Nevertheless, what is being described is still a causal process (or causal force, as the case may be).

In short: we have both negative and positive reasons in support of the claim that even equilibrium explanations are in some sense causal explanations.

22.4 From causal processes to causal factors

Since the Causal Process Requirement (CPR) is inadequate, and since we cannot entirely eschew causal information, we should reformulate the causalist requirement while insisting that an explanation provide some sort of causal information. How should we specify the sort of causal information required in a way that is general enough not to exclude equilibrium explanations? My suggestion is that we avoid reference to the actual causal history leading up to the explanandum event, and instead require simply that an explanation give us information about the *causal factors* that influence, in one way or another, whether or not the explanandum event occurs. (I say more about what counts as a 'causal factor' below.) Thus, I propose the following alternative causalist requirement:

> *The Causal Factors Requirement*: An explanation of an event must provide information about the causal factors that influenced whether or not that event occurred.

Causal-mechanical explanations that provide actual causal history will easily meet this requirement; but what about equilibrium explanations? Well, since the function describing the behaviour of the dynamical system being explained is going to represent the interplay of causal forces, it will provide some constraints, however minimal, on the ways in which the objects in the system interact with each other. Thus, it strikes me as perfectly legitimate to describe this function as providing information about the causal factors that influence whether or not the explanandum event occurs.

In fact, thinking in terms of constraints on the behaviour of a system is a useful way of fleshing out the admittedly vague notion of 'causal factors'. Thus I also propose that we think of providing 'causal information' – i.e. providing 'information about causal factors' – as providing some sort of *causal law* governing the behavior of the system within which the explanandum event occurs. Within the context of the relevant system, an explanatory causal law, at least in the sense I intend, should indicate which combinations of causal interactions will result in the occurrence of the explanandum event, and likewise which combinations of causal interactions will fail to result in the occurrence of the event (within a certain range of initial conditions). The force of this causal law will simply be that if it is violated, then, *ceteris paribus*, the explanandum event will not occur.[16]

In Sober's equilibrium example, the fitness function, governing selection forces, is serving as the sort of causal law I am suggesting. What this function is telling us is that certain systemic changes to the selection forces (including the introduction of different forces, or the deletion of selection forces entirely) will have the result that equilibrium is not reached. When it comes to more standard causal explanations – as, for example, in Wright's explanation of the house burning down in terms of the candle falling on the curtain – the causal law will be much more specific (in keeping with that explanation's placement toward the maximal end of the causal information continuum). In that case,

[16] The *ceteris paribus* clause is designed to address, among other things, the concern that this counterfactual construal of causal laws is too strong – because there could be backup processes, operating according to different laws, that would bring about the explanandum event even if the explanatorily relevant causal factors were subtracted. In the context of Sober's equilibrium explanation, for example, mutation could serve as a backup process: a mutation could occur which would move the population toward equilibrium even if the selection forces were altered significantly. Thus, one of the 'other conditions remaining the same' is the absence of mutation (cf. Sober (1983, p. 207), where he explicitly rules out mutation as part of the model of the system). Thanks to an anonymous referee for bringing this concern to my attention.

the information about causal factors is telling us that, holding certain initial conditions fixed, the house will not burn (*ceteris paribus*) unless the candle falls on the curtain.

My suggestion here resembles something that Woodward says in his important and immediately influential (2003).[17] He rejects any sort of nomothetic model of explanation according to which explanation involves subsumption under laws, but he does want to appeal to explanatory generalizations – where a generalization counts as explanatory if it is invariant in the right way. (For Woodward, recall, invariance is the key to explanatoriness.) Although Woodward's treatment of invariance is certainly worth considering in more detail, all I will say here is that for him a generalization has the right kind of invariance if it represents a pattern of counterfactual dependence – which is to say that it is stable under some specifiable range of interventions (2003, p. 17).[18] The 'causal laws' that, I claim, are required for explanation share this invariance that Woodward finds crucial to explanatoriness, but they also include a contextual parameter that indicates which features of the relevant system must be held fixed (e.g. the initial conditions) in order for the necessary invariance to obtain.

The schema I am proposing, then, is one in which an explanation will provide more or less information about the initial conditions of the relevant system or context, and more or less – but some – information about the causal trajectory of that system. At this point it is worth noting that the causal requirement I favour suggests a model of explanation that closely resembles the one that Hempel and Oppenheim (1948) proposed in their groundbreaking work on scientific explanation. Recall, briefly, that their deductive-nomological (DN) model of explanation required a set of explanans statements, which included a statement of the antecedent conditions and a statement corresponding to each applicable general law. An explanation consisted in providing these explanans statements such that the explanandum – a sentence describing the phenomenon to be explained – could be logically derived from the explanans. Although I do not wish to attempt a rehabilitation of the DN model (the counterexamples are numerous and well-established), I do find it interesting that consideration of equilibrium explanations, as providing a challenge to models of explanation that endorse CPR, leads us to reformulate the causalist requirement in a way that evokes the DN model. The important difference, of course, is that my proposal replaces the DN model's inadequate notion of a 'general law' with a notion that requires causal information (and more closely resembles a generalization with the right sort of invariance, in Woodward's

[17] Woodward's theory is a causal theory, but he is not interested in arguing that all explanation is causal in nature; instead, he explicitly restricts his discussion to causal explanations. Nevertheless (2003, p. 6), he thinks that Sober's explanation would count as causal by his criteria.

[18] *Ibid.*, p. 17. See also Chapter 6 for Woodward's extended discussion of invariance.

sense). But whereas a view that endorses CPR requires that the causal information come in the form of information about the actual causal history of the explanandum event, my proposal (CFR) does not require this. Instead, my proposal allows the required causal information to be represented in a wide variety of ways, depending upon the nature of the explanandum. The causal information might (and often will) come in the form of a causal history, in which case the causal law will be very specific. But the relevant causal law might also prescind from individual causal processes in order to make a more abstract statement about the system in question: it might come in the form of a fitness function, or a probabilistic description of molecular movement, or some other specification that represents a move toward the minimal end of the causal information continuum. (This flexibility of representation is in part what I am trying to suggest by using the 'causal factors' locution.) To summarize, we might say that in moving from CPR to CFR, we have moved from providing the causes themselves to providing *information about* those causes. Along the way, we have learned a few lessons from Sober's treatment of equilibrium explanations. The first lesson is one we should have learned from the DN model: initial conditions are important. The second lesson that we learn from equilibrium explanations is that CPR is inadequate. Together, these two lessons suggest CFR – which replaces CPR while preserving the importance of both initial conditions and causal information.

The revised causal requirement I am suggesting also has affinities with certain aspects of Strevens's kairetic account of explanation. One of the building blocks of his account is a 'causal model', which has the form of a DN explanation but also 'purports to represent a chain of causal influence running from the states of affairs identified by the premises to the event identified by the conclusion' (Strevens, 2008 p. 72).[19] Thus, Strevens's approach, like mine, is crucially different from the DN approach in that the relevant entailment represents a causal, rather than a logical relation (Strevens, 2008 pp. 92–93). But we take different routes to this destination. Strevens begins with some of the famous counterexamples to the DN model, and points out that, in these cases, there is an *explanatory* asymmetry where there is no *logical* asymmetry.[20] The basis for this asymmetry is the causal relation, which suggests causal entailment rather than logical entailment. I, on the other hand, have built in causal relations from the beginning, and attempted to show that these

[19] His 2008 discussion of the notion of a causal model begins on p. 71.

[20] Strevens discusses this in his (2008, section 1.4). The two examples he considers are the famous flagpole and barometer examples. From the length of a flagpole's shadow, together with the position of the sun (along with some laws about how light behaves), we can logically deduce the height of the flagpole; but the length of its shadow (etc.) does not explain the height of the flagpole. Similarly, we can deduce from a certain barometer reading that a storm is approaching; but the barometer reading does not explain the occurrence of the storm. In both of these cases, what is wrong with the logical derivation is that it runs counter to the direction of causation.

causal relations are an essential part of even equilibrium explanations. In so doing, I have revised the causalist requirement in a way that evokes the DN model.

We are now in a position to see whether the revised causal requirement CFR can hold up in the face of further challenges. But before we move beyond the equilibrium challenge, I would like to draw out a third lesson from equilibrium explanations.

22.5 Explanatory depth and the rejection of the Proportionality thesis

The third thing we learn from equilibrium explanations is that we should jettison the Proportionality thesis. In order to see why, let us return once again to the dual continua that characterize an equilibrium explanation: information about initial conditions and information about causal forces. In both cases, less is more; in both cases, reducing the amount of information increases the power of the explanation. If the equilibrium is global, then minimal specification of initial conditions is necessary; in this case, we have an explanation that applies (almost) no matter what the system's initial state is. To provide more information about the initial conditions is to offer an explanation that does not apply as broadly, and hence is not as powerful. And the same holds for the causal information continuum: at least some causal information is required, but the more general the causal law, the more possible trajectories (and hence states of the system) will be covered by the explanation. A more specific causal law, which provides more information about (and hence more restrictions on) the possible causal pathways leading to the explanandum event, will not apply as broadly and thus will not be as powerful.

There are various ways to understand this claim that, *contra* the Proportionality thesis, a more general causal law is preferable. The first is in terms of the unification approach to explanation. According to unification approaches, an explanation is better insofar as it has greater unifying power; and one of the key elements of unifying power is generality.[21] Despite the fruitfulness of thinking about explanation as unification, I will focus on a different way of understanding the claim: in terms of *explanatory depth*.

Another way to put the point, then, is to say that the virtue of explanatory depth sometimes conflicts with the Proportionality requirement. And in cases of conflict, we should choose greater explanatory depth over greater amounts of causal information. And just what is explanatory depth? Strevens (2008, p. 137) provides an elegant characterization:

[21] For more on unification approaches to explanation, see Friedman (1974). See Kitcher (1981, 1989).

Explanations having depth . . . strip away vast quantities of apparently relevant, large-scale causal detail, showing thereby that the phenomenon to be explained depends on only a kind of 'deep causal structure' of the system in question, a structure that is deep now not because it is so physical (though it is that) but because it is so abstract. The salient but irrelevant causal details are the shallows, then, and the more abstract – that is, more general – properties of the system are its depths, fleshed out by the details but inconsequentially so.[22]

As Strevens (2008, p. 137) points out, equilibrium explanations in particular represent the ideal of explanatory depth,

combining as they do two monumental abstractions: first, the abstraction of a high-level dynamics from the physical underpinnings, . . . and second, in the equilibrium stage, the abstraction of a certain even higher-level property of the dynamics – the universality of a particular end point – from the high-level dynamics obtained in the first step.

Thus, consideration of doubly and elegantly abstract equilibrium explanations, in light of the explanatory virtue of depth, leads us to reject Proportionality.

Of course, one need not go quite so far in rejecting Proportionality. One could pursue a more ecumenical or pluralistic approach to the question of whether higher-level explanations are preferable to lower-level explanations (which provide more causal information at the expense of explanatory depth). Frank Jackson and Philip Pettit (1992), for example, argue that the choice between the two levels of explanation is pragmatic: whether one prefers the higher- or lower-level explanation depends on one's perspective or purpose: 'Explanations of different levels provide complementary bodies of information on one and the same topic; we do not throw any explanation away just because we have access to another' (Jackson and Pettit, p. 16).[23] Sober himself (1999, p. 560) also advocates a similar sort of pluralism, when he points out that

higher-level sciences 'abstract away' from the physical details that make for differences among the micro-realizations that a given higher-level property possesses. However, this does not make higher-level explanations 'better' in any absolute sense. . . . The reductionist claim that lower-level explanations are always better and the antireductionist claim that they are always worse are both mistaken.[24]

[22] Strevens (2008, p. 137). Strevens also advocates (pp. 147–148) small tradeoffs in the accuracy of a particular explanatory model in exchange for greater generality – i.e. greater explanatory depth.

[23] Jackson and Pettit (1992) endorse causalism about explanation (cf. p. 13), but they argue against 'explanatory or methodological fundamentalism' (p. 7), which always recommends the lower-level explanation and thus considers a micro-physical explanation to be objectively superior. Insofar as lower-level explanations provide greater amounts of casual detail, explanatory fundamentalists will be very much in sympathy with proponents of proportionality.

[24] Insofar as I have construed Proportionality as presupposing CPR (or some alternate formulation of the causalist requirement), then it is obvious that Sober rejects Proportionality –

I join these authors in rejecting Proportionality. I also lean toward Strevens's view, according to which depth (or generality, or a higher level of abstraction) is a greater explanatory virtue than causal detail – but that tendency is not essential to my defense of causalism. The causalist can opt for a pluralist approach instead.

22.6 Applying the causal factors requirement

If what I have said in response to the challenge from equilibrium explanations is right, then we should expect a parallel response (i.e. a response that appeals to CFR) to succeed in the face of other sorts of challenges to CPR. Thus, as a way of vetting my proposal, I will briefly consider two examples of another type of challenge.

The first example I will consider involves another dynamical system, but the challenge is more of an *epistemic* challenge, rather than an equilibrium challenge. The basic point of this challenge is that additional causal detail can actually *obscure* understanding, and hence an explanation[25] – especially when one wants to use the explanation in question as a part of practical reasoning. Alan Garfinkel (1991, p. 53) uses an example from population ecology to make the point clear:

> Suppose we have an ecological system composed of foxes and rabbits. There are periodic fluctuations in the population levels of the two species, and the explanation turns out to be that the foxes eat the rabbits to such a point that there are too few rabbits left to sustain the fox population, so the foxes begin dying off. After a while, this takes the pressure off the rabbits, who then begin to multiply until there is plenty of food for the foxes, who begin to multiply, killing more rabbits, and so forth.[26]

As Garfinkel points out, a rabbit that is trying to avoid a predator will not be interested in an explanation of another rabbit's death that includes the specific details of the causal history of that rabbit's death (eaten by which fox, at what time, etc.). In fact, the particular fox and the particular time do not make a difference; had it not been that fox, chances are good (given a large enough

since he rejects CPR on the basis of equilibrium explanations. But in this article he is restricting his discussion to causal explanations, and even in that context he rejects (or would reject) Proportionality.

[25] I am not here advocating an identification between explanation and understanding, merely pointing out that one way to weaken an explanation is to reduce the amount of understanding it generates.

[26] Garfinkel is arguing specifically against reductionism (pp. 53–57), but his reasons for eschewing reductionism also militate against Proportionality. Hilary Putnam also argues against reductionism, particularly in his (1975). Putnam says (p. 296) that a micro-story about why a square peg won't fit in a round hole is either a terrible explanation or no explanation at all.

fox population) that it would have been another.[27] A more useful explanation would instead point to the large fox population as a whole. Additional causal detail is irrelevant and therefore unhelpful.

Garfinkel's example is clearly telling against Proportionality. An explanation of a rabbit's death that cites the large fox population provides less causal information than an explanation that cites the specific causal trajectory leading up to the event. And yet this explanation, given that it applies to a much broader range of rabbit deaths (and would be more useful for rabbits trying to avoid getting eaten [or humans trying to preserve the rabbit population]), is at the same time more powerful than an alternative explanation that provides a greater amount of causal information. It is, in short, the deeper of the two alternatives. Moreover, as also evidenced in this example, less causal information can be better if it allows us to explain not only what happened, but what *could* have happened, had certain things been otherwise. The higher-order explanation of the rabbit's death (in terms of the large fox population) makes it plain that had the first fox not eaten him, the second (or third, etc.) likely would have. Similarly, as we saw above, Sober's equilibrium explanation is superior to a maximal causal detail (i.e. actual causal history) explanation: it explains not just why the population took the actual trajectory it did, but why the population would have still ended up at the same place, even had it taken a different trajectory. If Proportionality were correct, then the explanation at the maximal causal detail end would be the better explanation; but, as we have seen, it is clearly not. Sometimes, then, *pace* proponents of Proportionality, less causal detail makes for a *better* causal explanation – in particular, when less causal detail allows us to pick out modal features of the system being explained.

It seems, then, that we have another good reason to give up on Proportionality. Are these considerations from Garfinkel similarly telling against CPR? And what about CFR? Hopefully it is clear by now that even if the answer to the first question is 'Yes', the answer to the second is 'No'. As we have already noted, the best explanation of the rabbit's death (i.e. the one that is most useful for those concerned with keeping rabbits safe from predators) is one that abstracts away from causal history, and instead points to a structural feature of the system: the large fox population. Hence, the best explanation of this system violates CPR, and thus endangers causalism – unless there is an alternative to CPR that can accommodate examples such as Garfinkel's. Fortunately for the causalists, there is such an alternative, and it is represented by CFR – which, recall, says that explanation requires only some information about the causal factors influencing the occurrence of the explanandum event

[27] In certain 'difference-making' respects, Garfinkel's approach anticipates Strevens's kairetic account of explanation, especially as laid out in his (2004) – material from which was incorporated into *Depth*.

(about the 'causal laws' governing the behaviour of the system). And this is precisely what Garfinkel's explanation is providing. His explanation is telling us that without a certain distribution of rabbit and fox populations, the rabbit's death would not have occurred (or, more precisely, it says that a certain *number* of rabbit deaths would not have occurred.) The behaviours being modelled – predation and reproduction chief among them – most certainly involve causal processes, which means that information about the various relationships between those behaviours reasonably counts as information about the relevant causal factors. It seems, then, that this explanation of the rabbit's death satisfies CFR for much the same reason that Sober's equilibrium explanation satisfies CFR.

Another example of an epistemic challenge comes from an earlier article by Woodward, who is addressing Salmon's causal-mechanical model of explanation (as utilized above when formulating CPR). His criticism (1989, pp. 362–363) thus pertains to both CPR and Proportionality:

> More also needs to be said about how Salmon's model applies to complex physical systems which involve large numbers of interactions among many distinct fundamental causal processes. In such cases it is often hopeless to try to understand the behavior of the whole system by tracing each individual process. Instead one needs to find a way of representing what the system does on the whole or on average, which abstracts from such specific causal detail.

Woodward goes on to apply this point to an explanation of the behaviour of a gas that appeals to the ideal gas laws. The reason why the ideal gas laws are useful for explaining the behaviour of a gas is precisely because they omit (abstract from) the individual causal processes that constitute the gas's behaviour. With respect to a particular state of the gas, we could say that the ideal gas laws abstract from the individual causal processes that constitute the causal history of that state.

Notice first of all that Woodward makes another strong case against Proportionality. The explanation of the behaviour of a gas is yet another example of many in which (1) there are two competing explanations, (2) which differ markedly in terms of the amount of causal information they provide, and yet (3) the explanation with less causal information is clearly the superior explanation (i.e. the explanation with greater power). In other words, there will be many cases in which, as Woodward says, it would be 'hopeless to try to understand the behavior of the whole system by tracing each individual process.' (Woodward, 1989 p. 363) And since tracing individual processes is precisely what CPR requires, it seems that Woodward's example is also telling against CPR.

As with Garfinkel's case, however, a retreat to C*F*R remains available. If we consider the behaviour of the gas in question as the behaviour of a system, then an explanation of any particular state must posit some restriction on the initial conditions, and moreover will posit the ideal gas laws as accurately governing the behaviour of this system. Considered in light of the continuum of causal information, it does appear that this explanation provides less causal information than does, for example, Sober's equilibrium explanation. Nonetheless, the interactions represented by the gas laws are causal interactions, and thus even this explanation is not completely devoid of causal information. CFR once again appears vindicated.

There might be a concern here that the move I am suggesting is just as trivial and non-serious as the move we saw Woodward criticizing above. According to Woodward, recall, a putative explanation of the behaviour of a gas that posits molecules that collide with each other according to the laws of Newtonian mechanics – and says only that these collisions somehow produce the behavior in question – is a trivial and non-serious causalist explanation. (We also saw that an equilibrium explanation completely devoid of causal information is akin to this trivial and non-serious explanation of the behaviour of a gas.) Since Woodward directs this criticism against a (hypothetical) causalist view that appeals to an objectionable form of abstraction, it might seem as though my own proposal is vulnerable to this criticism as well. But it is not. To see why, first recall the equilibrium explanations considered above. The causal approach I am suggesting does not *replace* those equilibrium explanations with corresponding causal explanations; rather, it simply points out that equilibrium explanations (or at least the ones in the examples given) are already causal to begin with. So the point I am making here is not analogous to any attempt to provide an abstract (but unhelpful) version of a 'causal processes' explanation. Instead, I am suggesting a slightly, but crucially, different treatment of Woodward's example. This treatment points out that the best explanation of the behaviour of a gas, whatever that turns out to be, is already causal to begin with. This is because the equation(s) used in the best explanation, much like the fitness function in Sober's equilibrium explanation, represent causal processes. The causal information that satisfies CFR can be abstracted away from, or selectively highlighted, or what have you; but such information remains, even if under the surface, a crucial part of any explanation. The causal nature of explanation is inescapable.

22.7 Conclusion

I will conclude by briefly returning to Sober's equilibrium explanation. It is true that his example, and related examples, do cause trouble for certain formulations of the causal requirement on explanation – in particular, the Causal Process Requirement. However, as I have tried to show, Sober's distinction between local and global equilibria provides an explanatory framework that we can fruitfully extend by taking into account the amount of causal detail as regards not just initial conditions, but also causal trajectories. And if we do take into account causal trajectories, then once we are told that the graph is about selection, there is a sense in which *we have already got a causal explanation*. Moreover, this point complements certain themes in Woodward and Strevens – two different authors who have converged upon a similar conclusion via different routes.

Therefore Sober's equilibrium explanation, contrary to the received wisdom, is after all a causal explanation (in virtue of its reference to selection forces). Whether he originally intended it or not, Sober appears to have pointed the way toward a causal gloss on equilibrium explanation.[28] Moreover, this treatment of equilibrium explanations can be extended to other sorts of explanations in a way that supports a revised causal requirement – the Causal Factors Requirement.

Acknowledgements

I would like to thank Zachary Ernst, Eric Schwitzgebel, and two anonymous referees for Oxford University Press for helpful comments on earlier drafts. I would also like to thank André Ariew (who nurtured this paper in its infancy) and Erich Reck (who stepped in and helped out during the difficult teenage years) for many detailed and helpful comments on numerous drafts – as well as extended discussion that has greatly increased my understanding of these issues.

References

Batterman, R.W. (1992). Explanatory instability. *Noûs* 26: 325–348.

Berger, R. (1998). Understanding science: why causes are not enough. *Philosophy of Science* 65: 306–332.

Ernst, Z. (2002). *Evolutionary Game Theory and the Origins of Fairness* (PhD dissertation: University of Wisconsin-Madison).

Friedman, M. (1974). Explanation and scientific understanding. *Journal of Philosophy* 71: 5–19.

Garfinkel, A. (1991). *Forms of Explanation: Rethinking the Questions in Social Theory*. New Haven, CT: Yale University Press.

[28] Sober has acknowledged as much in personal correspondence. Zachary Ernst, in his defense of Railton's DNP model (2002) provides another causal gloss on this type of equilibrium explanation: he construes it as providing information about the 'ideal explanatory text' (which is a crucial element of the DNP model; see note 5), and thus causal even if indirectly so.

Hempel, C.G. and Oppenheim, P. (1948). Studies in the logic of explanation. *Philosophy of Science* 15: 135–175.

Jackson, F. and Pettit, P. (1992). In defense of explanatory ecumenism. *Economics and Philosophy* 8: 1–21.

Kitcher, P. (1981). Explanatory unification. *Philosophy of Science* 48: 507–531.

Kitcher, P. (1989). Explanatory unification and the causal structure of the world. In P. Kitcher and W. C. Salmon (eds.), *Minnesota Studies in the Philosophy of Science, Volume XIII: Scientific Explanation*. Minneapolis: University of Minnesota Press, pp. 410–505.

Lewis, D. (1986). Causal explanation. In D. Lewis, *Philosophical Papers: Volume II*, pp. 214–240. New York: Oxford University Press.

Putnam, H. (1975). Philosophy and our mental life. In H. Putnam, *Mind, Language, and Reality*. Cambridge: Cambridge University Press, pp. 291–303.

Railton, P. (1978). A deductive-nomological model of probabilistic explanation. *Philosophy of Science* 45: 206–226.

Railton, P. (1981). Probability, explanation, and information. *Synthese* 48: 233–256.

Salmon, W.C. (1984). *Scientific Explanation and the Causal Structure of the World*. Princeton, NJ: Princeton University Press.

Salmon, W.C. (1993a). Causation: Production and propagation. In E. Sosa and M. Tooley (eds.), *Causation*, pp. 154–171.

Salmon, W.C. (1993b). Scientific explanation and the causal structure of the world. In D-H. Ruben (ed.), *Explanation*, pp. 78–112. New York: Oxford University Press.

Salmon, W.C. (2006). *Four Decades of Scientific Explanation*. Pittsburgh: University of Pittsburgh Press.

Scriven, M. (1962). Explanation, predictions, and laws. In H. Feigl and G. Maxwell (eds.), *Minnesota Studies in the Philosophy of Science, Volume III: Scientific Explanation, Space, and Time*. Minneapolis: University of Minnesota Press, pp. 51–74.

Sober, E. (1983). Equilibrium explanations. *Philosophical Studies* 43: 201–210.

Sober, E. (1999). The multiple realizability argument against reductionism. *Philosophy of Science* 66: 542–564.

Strevens, M. (2004). The causal and unification approaches to explanation unified – causally. *Noûs* 38: 154–176.

Strevens, M. (2008). *Depth: An Account of Scientific Explanation*. Cambridge, MA: Harvard University Press.

Walsh, D.M., Lewens, T., and Ariew, A. (2002). The trials of life: natural selection and random drift. *Philosophy of Science* 69: 429–446.

Woodward, J. (1989). The causal mechanical model of explanation. In P. Kitcher and W.C. Salmon (eds.), *Minnesota Studies in the Philosophy of Science, Volume XIII: Scientific Explanation*. Minneapolis: University of Minnesota Press, 357–383.

Woodward, J. (2003). Making Things Happen: *A Theory of Causal Explanation*. New York: Oxford University Press.

Wright, L. (n.d.). Causal explanations. Unpublished typescript.

·23·

Epistemological issues raised by research on climate change

Paolo Vineis, Aneire Khan, and Flavio D'Abramo

Abstract

Climate change has become a reality, and much research on its causes and consequences is currently conducted. To our knowledge, very little attention has been paid to epistemological issues raised by climate change research. Randomized experiments cannot of course be done, so that climate change research needs to be observational, usually spanning over many decades or centuries. The amount and quality of information is often limited, at least as far as extrapolation to the remote past or future is concerned. In general causality assessment poses special problems, both in attributing meteorological events like tornados to man-made climate change, and in attributing health effects to meteorological changes. We exemplify some of the major epistemological challenges in this chapter. We stress that climate change leads to extreme consequences the application of the Precautionary Principle: the consequences of certain forecasts would be so devastating (e.g. the melting of permafrost, that would free enormous quantities of CO_2) that we have to act to prevent them, though their likelihood is extremely low. The usual balancing of the seriousness of the consequences vs. their likelihood of occurrence becomes very challenging.

23.1 The relevance of epistemological issues to different areas of climate change

The Intergovernmental Panel on Climate Changes considers climate change from three angles: the physical science of climate change; the impact of climate change on human populations, including health; and mitigating strategies. Epistemological issues refer to each of these chapters, but we mainly refer to the health field, and make a comparison with epistemological approaches in biomedical sciences.

23.2 New sources of uncertainty in observational climate change research

The case of climate change exemplifies the challenges posed by recent environmental and social changes. Discussing 'uncertainty' is insufficient in climate change research, not only because the term has become too vague, but also for more specific reasons. We will make a comparison with biomedical sciences. Uncertainties are very common in medicine, and they are related to the small size of samples, or their biased nature (opportunistic sampling in selected subpopulations). But with climate change uncertainty rises to a different scale, another order of magnitude. First, in biomedicine we can do limited randomized experiments, mainly with preventive or curative purposes. Should we stick to the Galilean experimental paradigm, we would probably stop making causal inferences on climate change. We do not have two worlds to randomize to climate change; the most we can do is to set up micro-experiments, created artificially in the laboratory, where microenvironments are subject to experimental change. But the results of such experiments could not be easily extrapolated to the real world.

Second, even if we stick to an *observational paradigm*, still we face unprecedented challenges when dealing with climate change. One problem is the time scale: we are used (particularly in the biomedical sciences) to making observations on a very limited time scale, usually a few years, while climate change spans a much longer period (particularly if we want to make long-term predictions, for example on the health implications for humankind, or for species diversity). Another reason is that, in additional to being observational – like most biomedical research – the science of climate change cannot easily rely upon repetition of the observations in similar settings (for example, smokers have an increased risk of lung cancer in all continents and in different time periods). It is difficult to appreciate regularities in climate, particularly at the local level.

23.3 Causality assessment in observational climate change research on health: An example

Causality assessment has always been difficult for observational disciplines, including epidemiology (Vineis 2009). The main objection raised by supporters of experimental science was that observation alone does not allow the researcher to rule out confounders, i.e. unobserved or observed variables that underlie the association and can explain it away. Consider the following example related to climate change: salt-water intrusion in Bangladesh. For decades, salinity levels in surface and ground water in coastal Bangladesh have been rising at unprecedented rates (Mirza *et al.* 2004, Tanner *et al.* 2007), and currently higher sea levels are likely to further increase salinization (Mondal *et al.* 2001, Salim *et al.* 2007). Salt water from the Bay of Bengal is reported to have penetrated over 100 km along tributary channels currently affecting 20 million people and 830,000 ha of arable land by varying degree of salinity in Bangladesh. This has raised serious public health concerns as salt-related diseases have been reported in those areas, in particular hypertension, eclampsia in pregnancy (Khan *et al.* 2008), and cholera outbreaks as a consequence of changes in water quality and temperature that facilitate the proliferation of *Vibrio cholerae*. One would thus conclude that at least three kinds of diseases may be the direct consequence of climate change (without considering heat waves and flood-related deaths): hypertension, eclampsia and outbreaks of cholera. (In fact, based on my recent experience, even tiger assaults can be attributed to climate change! I recently visited an area that was heavily hit by the Aila cyclone in South Bangladesh. At the Chalna hospital there were two fishermen attacked by tigers. In fact, the cyclone had killed all the prey of tigers – mainly deer – on one island, so that the hungry tigers attacked fishermen on their boats, something that did not happen before.)

However, other factors also contribute to the effects of climate change and can confound them. In particular, the shrimp farming business, which requires high levels of salt in pond water for cultivation, has risen in the same region and has become a major export industry (Mondal *et al.* 2001), further worsening the ecological situation. In addition, farming of the freshwater prawn (*Macrobrachium rosenbergii*) has spread in the region, and the latter group of farmers have started using banned antibiotics like nitrofurans (Ahmed *et al.* 2008), due to temperature-related bacterial proliferation. The diseased prawns that are rejected from international markets (sometimes found to be contaminated with bacteria like *Vibrio cholerae*) are distributed and consumed by the local communities, resulting in outbreaks of cholera, diarrhoea, dysentery and skin diseases on the one hand (Ahmed *et al.* 2008, USAID 2006), and antibiotic resistance on the other hand. On top of all that, and to complicate the picture even further, India has built a dam (Farraka) that diverts towards Calcutta the fresh waters of the big rivers flowing from Himalaya to Bangladesh. This also interacts with the impact of climate change.

The causal pathway between the rise in sea levels and salinity-related health effects is therefore confounded by this complex scenario, since cholera outbreaks can either be due to climate change (via modifications in water salinity, pH and temperature) or to the consumption of infected shrimps, and can even be exacerbated by exposure to antibiotics. The association becomes even more complex when we address other variables or 'effect modifiers' that change the salinity and health relationship.

To be clearer, salt water shrimp farming can be a confounder of the association between climate change and cholera outbreaks, because of the spatial and temporal association between farming and climate change in certain areas, but it can also be an effect modifier, since it adds upon the conditions created by climate change. The case of Bangladesh exemplifies the fact that sticking to causality models that address single variables is becoming more and more difficult in contexts in which experiments cannot be conducted.

23.4 Problems in inference: Direct and indirect health effect

On epistemological grounds, the work done by the Intergovernmental Panel for Climate Change (IPCC) is remarkable, because it introduces the management of uncertainty into science. IPCC has introduced a 'graduation of evidence' similar to the one used in the *Monographs on the Carcinogenicity of Chemicals* of the International Agency for Research on Cancer (www.iarc.fr) to assess the carcinogenicity of chemicals. The basic idea is to use a 'weight-of-evidence' language that summarizes the process of literature search and evaluation done by Working Groups. As said, this work is much more standardized and well-defined in the methodological Preamble of the IARC Monographs, but the basic idea is the same. Thus, the reader knows exactly what a 'Group 3' carcinogen is, i.e. not a chemical that is free of carcinogenic activity, but a chemical that has been studied inadequately or with contradictory findings. Similarly, the reader of the IPCC reports knows what the graduation of the evidence means and approximately what the level of evidence is for a given problem, thus avoiding gross misinterpretations. We believe this is an important epistemological principle, which allows a transparent communication between scientists, decision-makers and the community at large (see also Sheila Jasanoff 2005, on related issues).

If we accept, as IPCC does with a 'high level of confidence', that C_{O2} is increasing, and this is due to human activity, what will be the time scale of effects on human health and well-being? Possible events range from rapid and catastrophic to very mild: for example, IPCC does not rule out (though it is very unlikely) a cascade of events leading for example to Bangladesh being swept away because of rapid Himalaya glacier melting; or an increase of the sea level of several meters in a few decades; or even an inversion of the Gulf Stream.

Strata of complexity and uncertainty further increase when we come to human health. Direct consequences of certain climate events are simple and can be easily perceived, such as the deaths related to heat waves in Europe in 2003, or catastrophic floods, or the Katrina storm. The death toll in these cases is clear, it does not require any sophisticated epidemiological technique. But still: were these events all due to climate change? Was the flood in Bangladesh in 1974 the first attributable to climate change, or the last not due to it? And what about the one in 1998? What about tigers in Bangladesh?

Uncertain inferences on the causal nature of events also concern attributing indirect health effects, such as infectious disease outbreaks, changes in food quality and availability, water salinization and the ensuing epidemic of hypertension. Even wars and conflicts (like in Darfur), mass migrations and effects on mental health have been attributed to climate change. In a prospective cross-sectional survey conducted among children aged between 2 and 9 in Bangladesh, Durkin *et al.* (1993) found post-flood changes in behaviour and bedwetting. Children were reported to have 'very aggressive behaviour' after floods, with a significant increase compared to the pre-flood situation ($p < 0.0001$). While 16% of children wet their beds before the flood, the proportion became 40% post flood ($p < 0.0001$) . A qualitative study explored the experiences of female adolescents during the 1998 floods in Bangladesh, focusing on the implications of socio-cultural norms related to notions of honour, shame, purity and pollution. A number of the girls were vulnerable to sexual and mental harassment through exposure to unfamiliar environment of flood shelters and relief camps. Their difficulty in trying to follow social norms had far-reaching implications on their health, identity, family and community relations.

The spectrum of health consequences related to trauma that could occur in a demoralized population following climate-induced displacement need to be better investigated. Common mental health disorders include anxiety, depression, post-traumatic stress disorder, irritability, sleeplessness and suicide. There is a huge psychological burden associated with losing a child, a sibling or a family member during or after natural disasters. During the recent cyclone *Sidr* in 2007, even with warnings, those 'washed away' by the tidal bore were mostly children. For example, among 200 children in *Majher Char*, a remote island in the southern district of Bangladesh, only 12 survived the devastating cyclone, leaving the population deeply traumatized. Moreover, conflict situations that may arise among farmers in times of climate-induced natural disasters like droughts and floods need to be addressed.

Is all of this attributable to climate change? Where are the borders between the burden of events that would occur anyhow, particularly in low-income countries, even in the absence of climate change, and those attributable to the latter? It should be noticed that the effects we have described are mainly occurring or foreseen in low-income countries, where disentangling new threats from the old ones is not straightforward. However, perhaps there is no need to dichotomize the issue, i.e. asking 'is this due to climate change or not'. Climatology is in fact much more developed than biomedicine in considering reality as a continuum and in studying fluctuations rather than dichotomies. Thus, floods in Bangladesh, for example, could be depicted as a fluctuating (e.g. sinusoidal) curve. If climate change increases the probability of floods, this will be observed as a gradually increasing fluctuating curve. There is no need of dichotomies.

23.5 Proximate vs distant causes

Part of the problem with attribution involves a well known epistemological issue, that of proximate vs distant causes. The most convincing causal factors belong to proximate causes: in medicine Mycobacterium is 'the' cause of tuberculosis, though it is well known that it does not explain why only some people develop the disease, and the geographic and temporal variation of the disease occurrence. Causal webs in medicine require a more sophisticated approach than looking for necessary and proximate causes, and this holds true for climate change as well.

Most causal inferential procedures in biomedicine have dealt with one or another approach aiming at an evaluation of the strength of evidence, partly relying on statistical tools. However, when we come to complex causal pathways like the ones described in Bangladesh, a 'strength-of-evidence' reasoning plus statistical inference are insufficient, because they do not incorporate the direction and strength of the different vectors that are operating, such as the Farraka dam, shrimp farms, increasing sea levels, etc. The main objective of causal inference is in fact to connect the language of statistical association with the language of causality. To this end graphical models for causal inference have been developed, that have the goal of elucidating the web of relationships between different potential risk factors, other circumstances (such as vulnerability) and disease. Directed acyclic graphs (DAGs) are graphical representations of relationships among variables, and they can be used to disentangle causal relationships from simple statistical associations. However, causal assumptions must be applied to distinguish the causal edges from the merely associational ones, and this external information needs to be explicitly stated as it is not in itself contained in a DAG. In particular DAGs derived from observational data without interventions are just representations of *conditional independences* and are thus purely probabilistic until additional causal information can be brought to bear (see e.g. Dawid 2002, and Geneletti 2005).

23.6 Experiments on climate: From cloud seeding to the control of climate change

It is in fact not totally true that experiments are not done with climate, but their purpose is not scientific. An influential but yet unrecognized cause of climate change is represented by cloud seeding and geo-engineering, a series of techniques used to control the weather. These techniques, consisting in dispersing chemicals into the atmosphere, have been developed since the first years of the last century when rain making and rain enhancement was funded by farmers, to irrigate their crops, and sometimes also by municipalities to fill their reservoirs with rain. Facing the dry weather, in 1951 New York City asked the intervention of Wallace Howell, a 'rain maker'. The effect was a dreadful flood. There were 169 damage claims totaling over $2 million of damages. Catskill communities and citizens obtained a permanent injunction against New York City, which ceased further cloud seeding activities (Fleming 2006).

Scared by lawsuits, General Electric Research Laboratory decided to transfer its research on cloud seeding to the military. These techniques were then utilized and developed also as weapons. Between 1967 and 1972 weather modification took place in a huge area between Vietnam, Laos and Cambodia. After some accusations, in 1973 the American Senate adopted a resolution 'to prohibit and prevent, at any place, any environmental or geophysical activity as a weapon of war' (Fleming 2006, p. 14).

At least since 1965, with president Johnson, the possibility of modifying climate and to restore 'the quality of our environment' through geo-engineering, by dispersing buoyant reflective particles on the sea surface, was taken into account as a realistic opportunity (Fleming 2006).

In 2003 a Pentagon report concluded with the recommendation that the government 'explore geo-engineering options that control the climate' (Fleming 2006, p. 21). In the symposium '*Macro-engineering Options for Climate Change Management and Mitigation*' held in Cambridge in January 2004, the Tyndall Centre for Climate Research and the MIT identified, debated and evaluated macro-engineering options for the management of climate to put this plan into operation. Among the techniques used, one was the *albedo modification* on planetary scale, for example, by launching mirrors of reflective particles into orbit, adding aereosols to the stratosphere, enhancing cloud reflectivity, and modifying land surface (http://www.tyndall.ac.uk).

Proposed as an option for CO_2 reduction, these techniques are very similar to those considered in restoration ecology (Yearley 2007), and raise a number of concerns. For instance, they create problems for the modeling of climate change as they modify empirical factors taken into account within the GCM (General Circulation Model) used by IPCC scientists. The military management of geoengineering techniques means that this kind of technology is not shared within the scientific communities, is not open to criticism and to the judgment of social institutions, in a word it is black-boxed. Moreover, if these techniques are already widespread, every single model of forecast, based on the evaluation of classical factors cannot be considered reliable.

23.7 Climate change: Changes in the image of science

Climate change is also changing the image of science and of humankind itself. First, while we cannot stick to a strict Galilean paradigm, still we have to make sound inferences on climate change for very practical reasons. Inferences on climate change are based on a counterfactual logic, i.e. they consider 'how things would have gone had they not gone just like that'. This kind of inference is often based on thought experiments, not on real, well planned experiments, which has a broad historical and methodological impact, i.e. we have to give up with the deeply rooted Galilean ideal of controlled experiments.

Another impact has to do with our own human identity: there is no longer on the Earth any place that has not been touched by civilization and technology, including climate. Finally, climate change leads to extreme consequences the application of the Precautionary Principle: the consequences of certain forecasts would be so devastating (e.g. the melting of permafrost, that would free enormous quantities of CO_2) that we have to act to prevent them, though their likelihood is extremely low. The usual balancing of the seriousness of the consequences vs. their likelihood of occurrence becomes almost impossible.

Even the rules of confirmation and falsification within the scientific community are changing. The strong emphasis on the consensus obtained around the climate change forecasts (in particular within IPCC) has in practice exhausted the number of available experts and therefore of potentially dissenting voices or critical opinions. Should IPCC's work be submitted to 'peer review', according to the best tradition of science, it would not be easy to find prominent scientists who were not involved in the exercise. While IPCC is an extremely good example of broad consensus, it also risks undermining the basic principle of open and critical scientific practice.

Acknowledgements

This research has been made possible by a contribution of the Grantham Institute for Climate Change to Aneire Khan. We are grateful to Sara Geneletti and Mike Joffe for thoughtful comments.

References

Ahmed, N., Demaine, H. and Muir, J. (2008). Freshwater prawn farming in Bangladesh: History, present status and future prospects. *Aquaculture Research* **39**, 806–819.

Dawid, A.P. (2002). Influence diagrams for causal modelling and inference. *International Statistical Review* **70**, 161–189.

Durkin, M.S., Khan, N., Davidson, L.L., Zaman, S.S., and Stein, Z.A. (1993). The effects of a natural disaster on child behavior: Evidence for post-traumatic stress. *American Journal Public Health* **83**(11), 1549–53.

Few, R., Ahern, M., Matthies, F., Kovats, S. (2004). Floods, health and climate change: A strategic review. Tyndell centre for climate change Research 63: working paper.

Fleming, J.R. (2006). The pathological history of weather and climate modification: Three cycles of promise and hope. *Historical Studies in the Physical and Biological Sciences,* **37**, 3–25.

Geneletti, S.G. (2005). Aspects of causal inference without counterfactuals. PhD thesis. University of London.

IPCC (2007). Intergovernmental Panel for Climate Change. Climate Change. http://www.ipcc.ch.

Khan, A., Mojumder, S.K., Kovats, S. and Vineis, P. (2008). Saline contamination of drinking water in Bangladesh. *The Lancet* **371**(9610), 385.

Jasanoff, S. (2005). *Designs on Nature*. Princeton University Press.

Mirza, M. (2004). *The Ganges water diversion: Environmental effects and implications – An introduction*. Springer Netherlands, Dordrecht, Netherlands.

Mondal, M., Bhuiyan, S. and Franco, D. (2001). Soil salinity reduction and prediction of salt dynamics in the coastal ricelands of Bangladesh. *Agricultural Water Management* **47**, 9–2.

Nazrul, I. (2007). *A Tale of Two Traumatised Isles*. New Age, Nov. 20.

Rashid, S.F., Michaud, S. (2000). Female adolescents and their sexuality: Notions of honour, shame, purity and pollution during the floods. *Disasters* **24**(1), 54–70.

Salim, M., Maruf, B.U., Chowdhury, A.I., Babul, A.R. (2007). Increasing salinity threatens productivity of Bangladesh: COAST Trust. COAST positon papers 3. COAST Trust, Bangladesh. http://www.coastbd.org.

Tanner, T., Hassan, A., Islam, K., Conway, D., Mechler, R., Ahmed, A. and Alam, M. (2007). ORCHID: Piloting climate risk screening in DFID Bangladesh. Detailed research report April 2007. Institute of Development Studies (IDS), Dhaka.

USAID (2006). A pro-poor analysis of the shrimp sector in Bangladesh. The United States Agency for International Development (USAID). Development and Training Services. USAID Bangladesh, Arlington, Virginia, USA.

Vineis, P. (2009). Causal models in carcinogenesis. B. Fantini (Ed.) Medicina e Storia (2010, Special issue on Casual Models; in press).

Yearley, S. (2007). 'Nature and the Environment in Science and Technology Studies', in E. J. Hackett et al. (eds.), *The Handbook of Science and Technology Studies*, Third Edition. The MIT press.

·24·

Explicating the notion of 'causation': The role of extensive quantities

Giovanni Boniolo, Rossella Faraldo, and Antonio Saggion

Abstract

We propose an empirical explication of the notion of 'causation', which we call a *generalized explication of 'causation'* (GEC), based on the numerical balance between instantiations of extensive quantities. In this way, it will be shown that both the conserved and the non-conserved quantities have a role. It follows that the Salmon–Dowe approach should be considered valid only in particular cases.

24.1 Introduction

Nowadays causation is widely discussed. Philosophers of mind, philosophers of action, philosophers of science, epistemologists and metaphysicians all deal with the subject. We wish to enter this field of investigation by trying to improve understanding of what we mean when we speak of causation and causal interactions. In particular, we will offer an explication *à la* Carnap.[1] Note that Carnap (1950) discussed two kinds of explication. Both of them are obtained by turning a somewhat imprecise and ambiguous term (the *explicandum*) into a precise and unambiguous term (the *explicatum*). In the first case, the explication is reached by inserting the *explicatum* into a well-constructed system of logical-mathematical concepts, i.e. by adopting logical-mathematical language. Whereas in the second case, the explication is obtained by inserting the *explicatum* into a system of empirical concepts, i.e. by adopting an empirical language pertaining to empirical sciences such as physics or biology. It is this second kind of Carnapian explication that we will apply to the notion of 'causation'. Furthermore, it was this way of explicating that was implicitly or explicitly used, for example, by Russell (1912–1913, 1948), who introduced the notion of 'causal line'; by Aronson (1971) and Fair (1979), who proposed the so-called 'transference theory of causation'; by Salmon (1984, 1997), who

[1] We are well aware of the objections of Bontly (2006) to the explication process, but this is not the right place to discuss them as deeply as they would merit.

suggested the 'mark transmission theory of causation'; and by Dowe (1992, 2000), who argued for a 'conserved quantity theory of causation', creatively putting together the best part of the transference theory and the best part of the mark theory. Our proposal for a *generalized explication of 'causation'* (GEC), as we wish to call it, is an attempt to continue this tradition focused on the empirical explication of concepts.

Note that this way of clarifying philosophical concepts has a long history that we cannot go into here but that started with Aristotle, passed through Kant and Husserl and arrived, for example, at Carnap (recall his explication of the notion of 'confirmation'), Tarski (recall his explication of the notion of 'truth'), and Hempel (recall his explication of the notion of 'explanation').

At this point, two aspects should be considered. *Firstly*, even if, in the case of empirical explication, we use a physical language, this, of course, does not imply that the resulting empirical explication works only in the physical field. For example, in what follows we will adopt a physical language to explicate 'causation', but our explication will work also for biological causation and not only for physical causation. *The formal language adopted does not fix the application field, but the way of 'speaking' about that field.* This means that what we will present can be applied to any field, satisfying only the one condition, as we will show: that there must be extensive quantities.

Secondly, the empirical explication of a notion has no direct implications for the reality referred to in that notion. That is, *the philosophical aim of an empirical explication does not lie in telling us what the ontological structure of the world is*, or in particular what the causal structure of the world is (if any). Rather, as has already been said, its philosophical aim concerns the clarification of the notion (in particular of 'causation') by using a scientific language. This means that, hereinafter, the reader will not find any ontological commitment to the causal structure of the world. And it cannot be otherwise, considering the philosophical level at which the explication is placed. In other words, the 'real and intrinsic' causal structure of the world cannot be our concern. We limit ourselves to showing how the physical language enables us to explicate the concept of 'causation'. We deal with an empirical conceptual analysis and not with the epistemological status of the physical representations we use.[2]

[2] That is, we do not discuss if how the physical theories represent the world tells us something on how the real world is, or should be. Note, however, that those who took this step should solve the many problems it brings up. In particular those regarding (1) the *historical awareness* that in different times different scientific theories have told us different things about 'the structure of the world'; (2) the *epistemological awareness* that our knowledge of the external world is theory-laden; (3) the *methodological awareness* that our theories are not true but, at best, only well confirmed (if you prefer a Carnapian-like approach) or well corroborated (if you prefer a Popperian-like approach); (4) the *rhetorical awareness* that an a posteriori argument based on scientific concepts and results cannot be strong enough to unquestionably support a metaphysical claim, as the whole history of western philosophy has taught us.

We want to illustrate how a precise and unambiguous discussion of causation developed through a physical language can clarify imprecise and ambiguous discussions developed through ordinary language. All of this, of course, does not at all imply any commitment to a reconstructionist approach to ordinary language.

Summing up, we do not want to propose a theory of causation, especially at a metaphysical level, but an empirical explication of 'causation'; something totally different.

Let us come to a methodological issue. Our proposal will be proffered via a step-by-step preparation of the required explicative framework, in order to arrive at the concept-*explicatum* in a self-contained way. This means, first the system of empirical concepts by means of which we explicate is prepared, and then the concept-*explicatum* is proposed. This is strictly coherent, as is well-known, with the four-step process designed by Carnap (1950, ch. I). The process must comprehend: (1) the identification, as we have done in this introduction, of the ambiguous and imprecise concept-*explicandum*; (2) the setting up of a system of empirical concepts (Section 24.2); (3) the explication in terms of the empirical concepts just discussed (Section 24.3); and (4) the conclusion showing that the unambiguous and precise concept-*explicatum* is fruitful and similar enough to the ambiguous and imprecise concept-*explicandum* (Section 24.4).

Before beginning, it is worth adding two further remarks, which are essential in order to better understand what we are going to propose. The first concerns the kind of empirical language we will adopt. It is the typical language used to describe the continuous variations of quantities, especially adopted in fluid dynamics but which can be found in all those physical domains where we have to speak in terms of continuity equations.

The second remark regards the conceptual framework in which we will move. Our conceptual background, even if we cannot discuss it here, is Salmon and Dowe's conserved quantity theory of causation. As is well known, it is based on the idea that we have a causal interaction whenever we have an intersection of the world-lines of two bodies with an exchange of conserved quantities. Following this thread, a causal process should be described by a world-line possessing a conserved quantity. This theory has been discussed a lot from a philosophical point of view, but few papers have paid attention to the scientific details. Among these few, Lupher (2009) heavily criticizes it mainly because not all conserved 'quantities' could be properly used and because Salmon and Dowe's account would fail in explicating causal interactions in stationary situations. If Lupher's paper may be considered as a *pars destruens*, what we are going to propose might be thought of as a *pars construens*. For, in our empirical explication we will show that we can understand causation independently from the conserved quantities but in terms of the so-called *extensive quantities* (and we will define what they are) instead, some of which could also be conserved quantities. By the way, it is precisely this more comprehensive account of causation than the one offered by the conserved quantity approach that has induced us to speak of a *generalized explication of 'causation'* (GEC).

The chapter has two appendixes. The first concerns the discussion of some aspects involving time. The second regards the analysis of the case of the stationary states mentioned by both Dowe (2000) and Lupher (2009).

24.2 The explicative framework

Def$_1$*: System and state of the system*
Let us assume a theoretical context T and a *system* P. We can know, inside a given theoretical context T, which system we are speaking of by indicating the set of quantities $\{P_1, \ldots, P_r\}$ (e.g. velocity, mass, charge, etc.) defining it in T. Let $P^{\mathrm{T}} = \{P_1, \ldots, P_r\}$ be the system at stake. Since P^{T} is an empirical system, knowing, via measurements, the instantiations of $\{P_1, \ldots, P_r\}$ becomes necessary. This means knowing the *state of the system* in a given space-time point whose coordinates are $x^i = (x, y, z, t)$, where (x, y, z) are the spatial ones and t the temporal one. Therefore, $P^T_{x^i} = \{p_1, \ldots, p_n\}$ is the state of the system P^{T} in x^i, where $\{p_1, \ldots, p_n\}$ are the instantiations of the quantities $\{P_1, \ldots, P_r\}$.[3]

Remark 1
Clearly the number r indexing the $\{P_1, \ldots, P_r\}$ can be rather large if we want to specify all the quantities we are interested in, but, in general, there are relations among some of them. For instance, in a black body the peak frequency of the emitted radiation, the energy density, and the temperature are mutually dependent. Let us agree to select one set of *independent* quantities that are necessary and sufficient to define the system in T. In general there are various different possible sets and the specific choice will depend on the nature of the problem the observer is dealing with.

As mentioned, *'system' and 'state of the system' are theoretical context dependent notions*. By this we mean simply that different theories deal with these notions differently, providing, therefore, different representations. The particular choice of the theoretical context may be naturally suggested by what we want to observe, or by the constraints on the system. In other situations, it may be determined by more fundamental conditions, as in the case of physics in which we have to shift from a classical to a relativistic, or to a quantum-mechanical, description. A paradigmatic example of context dependency of the notion of 'state of the system' concerns neutrons and protons. They have to be considered as two different systems in a theoretical context emphasizing

[3] Note that $n \leq r$ since it is not said that all the quantities can be instantiated.

the electromagnetic interactions, but they have to be regarded as two quantum states of the same system (the nucleon) if we move to a theoretical context emphasizing the strong interactions.[4]

Def₂: Extensive quantities

Our approach to causation is based on the notion of extensive quantity. There are various definitions of this later. Some authors define it as a quantity which scales with mass; others make reference to additivity: additive quantities are defined as those quantities whose value for the entire system is the sum of the corresponding values for any partition into 'subsystems' (see Guggenheim 1950; Tisza 1966; Callen 1985). Let us adopt the most general definition, since it is the only one which is suitable for generalizations on arbitrary space-time metrics:

Given a system P^{T} and given the quantities defining it, by *extensive quantity* we mean any quantity E whose value is given by the volume integral of a function $\varepsilon(x^i)$, called *density*, of E in the point x^i.

[4] The notions of 'system' and 'state of the system', and the fact they are theoretically context dependent are extremely useful for our aims. In particular the usefulness concerns the problems regarding the so-called 'timewise gerrymanders' and 'spacewise gerrymanders'. The former would be 'putative objects defined over a time interval where the definition changes over time', while the latter would be 'putative objects consisting of many independent objects' (Dowe 2000, p. 99). We could avoid the difficulties concerning them simply by ruling them out of our analyses. But it is not that easy. According to the definition above, a particle beam is a spacewise gerrymander, and, of course, we cannot eliminate it from our discussions. Similarly we cannot eliminate another spacewise gerrymander, that is, a population of living beings such as a flock of birds or a school of fish. Nevertheless we must eliminate a spacewise gerrymander such as 'my pen + my jacket + the light bulb in front of me + the hostess in a restaurant 2000 km away from me'. But we cannot eliminate 'my pen in my pocket + my jacket + the light bulb in my hand + the hostess in the hall of the restaurant I am just entering' if I must physically discuss the impact between me and the hostess. I must take into account all the masses, all the velocities, etc. How can I demarcate the 'good' spacewise gerrymanders from the 'not-good' ones? The same goes for the timewise gerrymanders. Usually, the timewise gerrymanders should be ruled out. But why? Simply because we do not like them? Should all of them be banned? Let us think about the whole phylogenetic tree starting more or less $3.5 \cdot 10^9$ years ago and arriving at our age. Are not the definitions of the different living species valid for a certain time period (of course here at issue is not the notion of species *qua* category but *qua* taxon)? Do they not change over time? How can we demarcate a 'good' timewise gerrymander from a 'not-good' one? The solution is pragmatically contained in the proposed notion of 'system' and 'state of systems'. Both are theoretically contextualized, and note that the system is something which is structured in a certain way, that is, it is characterized by its elements and by its relations among the elements. Yet which are the relations that can be considered 'good' relations so that the system can be considered a 'good' system, and the timewise or spacewise gerrymander a 'good' timewise or spacewise gerrymander? The answer is: the theoretical context. That is, if we have a 'good' theory about that system (be it or not a timewise or spacewise gerrymander), then that system is a 'good' system. We are perfectly aware that the way out we are proposing could suffer of the same objections that could be raised against Goodman's solution to the green-blue theory *versus* grue-bleen theories on emeralds. But this is not the right place to embark upon this question ourselves.

Remark 2

Note that in the *definiens* above, volume plays the main role. For volume is, by definition, extensive and homogeneous, thanks to the homogeneity of space-time. Keeping this definition in mind, the 'additivity property' often used to define extensive quantities acquires a rigorous and general meaning: if we considered the system P^{T} as if it were subdivided into n sub-systems, in the sense of volume (n sub-volumes), the value of E would be given by the 'sum' of the n values of E within the n sub-systems (by the additivity of integrals).

The mass of a system is a paradigmatic example of extensive quantity, since it can be thought of as the volume integral of the mass density. Only in this sense can we affirm that the mass of the whole is the 'sum' of the parts; this would no longer be true, in general, if we 'put together' the parts. Other extensive quantities are the total energy, the momentum and the angular momentum, the electric charge, the number of moles, entropy, etc.[5]

Def$_3$: Intensive quantities

Given a system P^{T} and given the quantities defining it, by *intensive quantity* we mean any quantity I whose value is defined in every point of the system, but not for the system as a whole.

Remark 3

An instance of intensive quantity is given by the velocity, which is defined at a point. For, strictly speaking, we cannot talk of the velocity of an extended body unless it is rigid and in translational motion. Likewise for the temperature: it is defined in every point. Only in the case of the isothermal body do we use the convention of speaking of 'the temperature of the body'. The pressure, the electric potential, the field intensities, the chemical potential, and, of course, all the densities of the various extensive quantities are examples of intensive quantities.

Each intensive quantity is defined as the partial derivative of an extensive quantity with respect to another extensive quantity (of course keeping some others constant; the latter depends on the choice of the variables we consider as independent). That is, given an intensive quantity I, two extensive quantities E and E' exist so that

$$I = \frac{\partial E}{\partial E'}.$$

For example, temperature τ is defined as

$$\tau = \frac{\partial U}{\partial S},$$

[5] For the discussion of their 'extensiveness', see Guggenheim (1950); Callen (1985); Tisza (1966); Herrmann (2000).

where U is the energy and S is the entropy (U and S are two extensive quantities and the derivative must be carried out at constant volume).

This is a crucial point for our proposal and we must spend some time on it. It appears that the E-quantities play a fundamental role in the definition of a system while the I-quantities can be considered as sort of 'second rank' parameters. Actually the latter tell us how rapidly an extensive parameter varies when another is varied in a given way. Nevertheless, an extremely relevant consequence of the above regards the definition of the state of the system. Now it can be fixed by resorting only to extensive quantities. For a quantity can be either extensive or intensive and in the latter case it can be defined as a function of the former. That is, $P_{x^i}^T = \{E_1, \ldots, E_n\}$ (where the $E_1, \ldots, E_n$ are mutually independent).

Needless to say, what we are affirming, and what we will affirm, in terms of extensive and intensive quantities can be applied also to non-physical domains, such as, for example, chemistry and biology. It is sufficient to individuate, for that particular chemical or biological system, the suitable sets of extensive and intensive quantities.

Th_1: Extensive quantities and the balance equation
Given a theoretical context T and a system P^{T}, any variation of an extensive quantity E of P^{T} can be described by a *balance (or continuity) equation*, whose integral form is

$$\frac{dE}{dt} = C_E + \Sigma_E,$$

where C_E is the ingoing *flux* of the extensive quantity E, and Σ_E is its *production rate within the volume occupied by the system.*

Remark 4
The proof concerning the above balance equation comes as an application of a general theorem called *Transport Theorem* or *Reynolds' Theorem,* which is well known in the mathematical theory of continuous media (Gurtin and Morton 1981). The flux term of the balance equation, C_E, is an integral over the surface K surrounding the physical system P^{T}, where the surface K is oriented inward. Instead the production rate term, Σ_E, is an integral over the volume V bounded by the surface K. Both terms contribute to representing the observed change in the extensive quantity E but in a substantially different way. The production rate term, Σ_E, describes the contribution due to processes occurring in the system. It depends *only* on the state of the system and *does not depend on the state of the external environment* (other system). The C_E term describes a 'flux' of the quantity E across the surface and it describes the 'exchange' of the quantity E with the external environment in which P^{T} is embedded.

Note that, actually, 'flux', 'exchange' and 'transmission' must be interpreted metaphorically: there is no quantity which flows, nor which is exchanged or transmitted! Hereinafter this extremely important question will be better clarified.

Def$_4$*: Conserved quantity*
Given a theoretical context T and a physical system P^{T}, an extensive quantity E of P^{T} is a *conserved quantity* if and only if

$$\Sigma_E = 0.$$ [6]

Remark 5
One of the most paradigmatic examples of a conserved quantity is given by energy. In our account, the principle of energy conservation states that the total energy of any system P^{T} can vary only by 'exchanges' with the environment in which P^{T} is embedded. Hence any change of energy in P^{T} must be exactly and numerically balanced by the opposite change in its environment. The same applies to momentum, angular momentum, electric charge, etc.

Among the non-conserved quantities we may recall the total number of moles n and the numbers of moles of the various components n_i. In this case, if the measured variation of n_i is not numerically balanced with the C_n term (in particular, this happens when the system cannot exchange matter), then something must be occurring within the system and accounting for the difference.

Entropy (S) is a non-conserved quantity that is particularly relevant. In this case the Σ_S term of the balance equation contains the core of the second principle of thermodynamics. For

(a) $\Sigma_S = 0$ for changes in a system in *internal equilibrium*. Notice that this works only in an ideal and asymptotic case. In real processes, equilibrium and change are incompatible concepts, at least on a macroscopic scale;

(b) $\Sigma_S > 0$ for any real case;

(c) Σ_S can be written as the sum of various terms, that is,

[6] Note that the definition above could be rethought, in a different formal approach, as a theorem. That is, given a theoretical context T and a physical system P^{T}, an extensive quantity E of P^{T} is a conserved quantity if and only if $\Sigma_E = 0$. In this case the proof goes along these lines: (1) Given a theoretical context T, in it the system P^{T} is characterized by a particular symmetry group. (2) Thanks to Nöther's theorem, we find the correlated conserved quantities E_i. (3) We prove that for any conserved quantity E, $\Sigma_E = 0$ holds. Moving along the opposite way the sufficient clause could be proven. Note, again, that this is not an 'intuitive demonstration', but a way of sparing the reader from complex calculi involving Nöther's theorem.

$$\Sigma_S = \sum_{i=1}^{n} \frac{d\,E_i}{dt} \chi_i.$$

Though with different notations this is a fundamental relation in non-equilibrium thermodynamics (see, for instance, Callen 1985, p. 30; Prigogine 1955, p. 40). Note that each term on the right-hand side is written as the product of the time derivative of an extensive quantity E_i times a suitable intensive quantity χ_i, called *the entropy-conjugate quantity* of the extensive quantity E_i. This equation is extremely general, since it connects entropy and each possible extensive quantity E (to be more precise, their time variations). For example, if the extensive quantity is the energy U, one of the terms can be given by the scalar product $\vec{j} \cdot \mathrm{grad}\left(1/_\tau\right)$, where $\vec{j}$ is the flux of energy (supposed, in this case, as a heat flux), and τ is the absolute temperature.[7]

The three points above are of fundamental importance. Point (a) characterizes *equilibrium states*, that is, those ideal states in which every small change is, by definition, reversible. Actually reversible changes do not exist in the real processes, but the former can approximate the latter with the highest accuracy in a very large number of cases. Point (b) separates allowed directions from forbidden directions in all real processes. It states a sort of thermodynamical arrow of time. Point (c), as just observed, explains the profound meaning of the distinction between extensive quantities and intensive quantity.

Remark 6

The 'exchange' of an extensive quantity is completely symmetrical between the two systems, owing to the balance condition. The direction of the 'exchange' is merely conventional since any situation can be depicted as an 'exchange' of the opposite quantity in the opposite direction. To illustrate this symmetric scenario, one could note the analogy with the content of the third law of Newtonian physics for a mechanical context, in which interactions consist in the 'exchange of quantity of movement' (momentum). As we will show, our explication of the 'causal interaction' will be given in a more general sense, and *the explication will not imply a time succession,* as can also be intuitively inferred from what has just been said on the symmetry property of the 'exchanges'.

[7] It is worth recalling that in scientific formalism the time derivative of extensive quantities ($d\,E_i/dt$) is named 'generalized flux' while the intensive entropy-conjugated quantity χ_i is named 'generalized force' or 'generalized affinity' in order to underline its role of 'cause' (like the Newtonian force) in governing the evolution of the system.

At this point, in order to avoid misunderstandings, it is worth recalling some useful definitions, in particular those of 'open', 'closed' and 'isolated' systems,[8] and those of 'stationary' and 'rest state'.

Given a theoretical context T and a physical system P^{T}, let $P^{T}_{x^i}$ and $P'^{\mathrm{T}}_{x'^i}$ be the states of the system at t and t', respectively. In $\Delta t = |t - t'|$,

(1) P^{T} is an *open system* relatively to an extensive quantity E if that extensive quantity can be 'exchanged' with the environment; in this case we may have $C_E \neq 0$.

(2) P^{T} is a *closed system* relatively to an extensive quantity E, if C_E is bound to be always equal to zero; that is, if there is a constraint such that this extensive quantity cannot be 'exchanged' with the environment;

(3) P^{T} is an *isolated system* if C_E is bound to be zero for any extensive quantity E; that is, P^{T} is isolated if it is closed with respect to all the extensive quantities defining its state.

And,

(1) the system P^{T} is said to be in a *stationary state* in the time interval Δt if the initial state $P^{T}_{x^i}$ (defined in x^i) does not vary, that is, we have

$$\frac{dE}{dt} = 0$$

for every extensive quantity E in Δt. Note that, as a consequence of what is stated above, in a stationary state even $\frac{dI}{dt} = 0$ for any intensive quantity I;

(2) the system P^{T} is said to be in a *rest state* in the time interval Δt if for every point of P^{T} there is an intensive quantity $\vec{v}$, called *velocity*, so that $\vec{v} = 0$. The notion of 'rest state' is observer-dependent and therefore, from this point of view, it can be naturally considered as a synonym of 'inertial state' if the observer is an inertial observer.

24.3 Causation: The explication

So far, step by step, we have pointed out all we need to arrive at the conclusion of the explication, that is, to arrive at the concept-*explicatum*. Up to now we

[8] We are perfectly aware that there is not a unanimous consensus on the definitions of these notions. Our proposal is grounded on how they are commonly used and defined in thermodynamics. Other definitions, however, could be proposed. What is important is that they are coherent amongst themselves and not erroneously interpreted. For, an erroneous or non-coherent interpretation of them can lead to an erroneous explication of causation, as happens in Choi (2003). For a critique, see Sant'Anna (2005).

should have accepted the idea that all causal interactions, connected with changes of states, are characterized by the 'exchange' or the 'transference' of some physical extensive quantity since we defined the state in terms of those quantities. Nevertheless, differently from Salmon and Dowe, we claim that we do not rely on conserved quantities but on extensive quantities, since by the former we cannot explicate all the uses of the term 'causation'.

But now the time has come for the last step in the explicative process and we must give the concept-*explicatum*. Actually, we shall give it in two different forms. In the first, that will offer an 'integral *explicatum*', we will consider the systems at stake as a whole; in the second, that will give a 'local *explicatum*', we will take into consideration only small (infinitesimal) parts of the systems.

24.3.1 Causation: The integral *explicatum*

Let us fix a theoretical context T and let us have two systems P^{T} and P'^{T} in the states.

$P^{T}_{x^i}$ and $P'^{T}_{x'^i}$, respectively, determined by the respective instantiations of the extensive quantities $E_1, \ldots, E_n$. We are allowed to speak of *integral causal interaction* between P^{T} and P'^{T}, in the time interval $\Delta t = |t - t'|$, *if and only if*:

(1) We observe, relatively to an extensive quantity E_i, that $C_{E^i} = -C'_{E^i}$ and $C_{Ei} \neq 0$;

(2) P^{T} and P'^{T} form a system G that is closed with respect to the quantity E_i.

We call this 'integral *explicatum*', since we want to stress that it is written in integral form and by this we mean that the 'exchange' of the extensive quantity E_i between the two systems is given by the total amount of E 'flowing' through the whole surface separating the two systems.

Let us dwell on the relations between the two requirements above. Suppose that, as a result of our observations, we measure the two fluxes C_{Ei} and C'_{Ei} and we see that (1) is fulfilled. Is this sufficient to claim that P^{T} and P'^{T} are in causal interaction? The answer is negative: (2) also has to be satisfied, for we have to be sure that the 'exchange' of E takes place *between* P^{T} and P'^{T} only.

Suppose now that (2) is fulfilled. We expect that (1) is satisfied as a natural consequence if we have that $C_{E^i} \neq 0$. Indeed it states simply that the amount 'transferred' from P^{T} to P'^{T} is equal and opposite in value to the amount 'transferred' from P'^{T} to P^{T} and this is well known in the case of conserved quantities. For instance if two bodies P^{T} and P'^{T} form an isolated system G with respect to energy, the non-zero variation of the energy in one of the two is compensated by the opposite variation in the other, and (1) is automatically satisfied. The same does not necessarily happen for non-conserved quantities too, but in this case a deeper knowledge of the occurring processes will be necessary, i.e. we should know the values of $\sum_{Ei}$ and $\sum'_{Ei}$, which may now be non-zero. Hence (1) is fulfilled provided that the adopted theoretical context is *safe*, which means: first, the closure of the total system G with respect to E_i is assured and, second, our knowledge of the two source terms is also secure. If (1) is not verified then the entire description is to be re-examined and nothing can be said about causal interactions.

One further comment devoted to the 'closure condition' explicitly required in point (2) of the above integral *explicatum* must be made. *Prima facie* the condition seems to be more restrictive than what is required by Salmon and Dowe's account where nothing is said concerning the closure condition. Actually, Salmon and Dowe's account does not offer an operational expression telling us how the exchange of some conserved quantity between two systems can be recognized. On the other hand, we want to face this aspect (in a more generalized framework) and this means entering into details, including having to deal with the 'closure condition'.

Let us consider one example with conserved quantities and one with non-conserved quantities. With regards the first case, let us suppose we have two systems P^{T} and P'^{T} and we observe variations of their respective momenta $\vec{m}$ and $\vec{m}'$. Let us call those variations $\Delta\,\vec{m}$ and $\Delta\,\vec{m}'$ respectively. If we think that momentum is a *conserved quantity* and *the system* $G = P^{T} + P'^{T}$ *is closed respect to momentum,* then we are allowed to speak in terms of causal interaction between P^{T} and P'^{T}. For we have:

(a) the momentum of the entire system $\vec{m}_G$ is, by definition, the sum of $\vec{m} + \vec{m}'$;

(b) the variation of $\vec{m}_G$ is zero by the above closure condition; hence $\Delta(\vec{m} + \vec{m}') = 0$;

(c) we have $\Delta\,\vec{m} = C_m$ and $\Delta\,\vec{m}' = C'_{m'}$, because we assume that the momentum is a conserved quantity ($\Sigma_m = 0$);

(d) hence, we have $C_m = -C'_m$ which is precisely condition (1) which *proves* that some extensive quantity (momentum in this case) has been 'transferred' from one system to the other .

To further exemplify what has just been said, let us consider two colliding billiard balls within the theoretical context of classical mechanics. *Prima facie,* the system can be considered isolated with respect to momentum, but we could wonder how our account works if we release the closure condition. This is exactly what happens in real situations if we consider that something else is interfering with the two balls' interaction; for instance, the friction with the billiard table, the presence of the air, etc. In this case, the balance equation between the two balls is no longer satisfied and we have to consider a more complex situation in which the 'exchange' of the extensive quantity is among three, or more systems. Some effect on the new partners can be estimated by observing their modifications and, hopefully, we can argue that some momentum has been 'exchanged' between the two balls; in any case we must reach a quantitative estimation no matter how large the error will be.

One further point concerning the closure condition is worth discussing. It appears that closure condition is necessary to our causation account but, at the same time, the closure condition is equivalent to an 'absence of causal interaction with the external world' and it seems that we are in a circular situation. Actually, in every theoretical context we use there *have to be* principles stating what the *non-interaction situation* is. For instance, Newton's first law defines the non-interaction situation in mechanical systems, and *with respect to that* we are entitled to acknowledge interactions. Only successively are we able to formulate the concept of 'isolated system'. The set of all known 'things', which is frequently called *the universe*, is isolated *by definition*; then a system is isolated from 'the rest of the universe' (external world) if, *according to the theoretical context in use*, no variation from the 'inertial' state is observed outside, whatever happens within the system. Finally, we are now able to speak of *conserved quantities*, their existence and their fundamental role with regards the theoretical context at issue.

As a final remark on the supposed restrictive nature of the closure condition, let us anticipate that this condition will be completely abandoned below, when we shall formulate our account in the local form. This latter statement will represent the complete generalisation of the integral account we are discussing here in the sense that we shall seek 'fluxes' of extensive quantities point by point instead of global 'fluxes'. So we may have local interactions even if we have zero global interactions, and we do not need any closure requirement. However concepts like 'closure' or 'isolation' are still included in the definition of conserved and non-conserved quantities.

In the case of non-conserved quantities E, the determination of the internal production rate Σ_E also has to be considered. Suppose we are given a container (for instance, a room), which is exchanging molecular oxygen with another container (another room) in such a way that the two containers form a system which is closed with respect to O_2. Let us suppose that the extensive quantity we are observing is the amount of molecular oxygen O_2 inside it. Actually, for our convenience, we will deal with the mole number n_{O_2} and measure its variation, say Δn_{O_2}, in a definite time interval. Suppose, *further*, that we know that the following chemical reaction is at work inside the system

$$2C + 2Ca + 3O_2 \rightarrow 2CaCO_3.$$

It is important, for what follows, that we know that a given process (in this case 'that' chemical reaction) is at work within the system. For instance, in the above example, we could have a monitoring of the temporal growth (positive or negative) of both calcium and calcite. Then we are able to measure the production rate of oxygen *within the system* by measuring the amount of calcium, Δn_{Ca}, destroyed in the same time interval (negative production). Now we can evaluate the flux term, i.e. the C_{no_2}, and have a complete understanding of the balance equation *with respect to the variations in molecular oxygen*. At this point, according to our proposal, we are able to speak in terms of causal interaction since the two requirements indicated above regarding the concept-*explicatum* 'causation' are satisfied. This means we are allowed to claim that there is a causal interaction *even if there are no conserved quantities at stake*.

The above also enables us to understand precisely what we mean by 'exchange' and by the correlated notions. As we have already seen, we are allowed to speak of 'causal interaction' if we have two systems P^{T} and P'^{T} in the states $P^{T}_{x^i}$ and $P'^{T}_{x'^i}$ respectively and we note (1) a variation in the numerical values of the instantiations of E in P^{T} and P'^{T}, where E is the extensive quantity relative to which P^{T} and P'^{T} form a closed system; (2) that $C_E = -C'_E$. At this point, any time we affirm that there has been an 'exchange' of the quantity E, or a 'flux' of E, between the two systems, *we actually mean that we have measured a numerical balance between the values of E in the two systems.* Therefore, if we were to be extremely punctilious and reconstructivist, the exchange-talk, or the flux-talk could be replaced completely with the more correct talk in terms of numerical balance and balance equation. Nevertheless, there is no point in doing this if the real meaning of the term 'exchange' and its metaphorical use have been grasped.

It should be noted that the approach developed so far uses concepts and terminology from fluid dynamics. This is quite common in many domains of physics (electrodynamics and relativity theory are two examples), and in many mathematical models of chemical and biological situations. It emphasizes the nature of extensive quantities of a system (see Herrmann 1986). There is no need to be reminded that the fluid-dynamical approach is characterized by a flux-talk and an exchange-talk, according to which, for example, we may say that bodies 'contain a certain amount of' electric charge, which 'flows', is 'accumulated', 'concentrated', 'diluted', 'distributed', 'lost', 'collected', 'transferred', etc. But, as we have just argued, this could be replaced by a different and more correct balance equation-talk. Nevertheless, this should not be a problem if we really know what we are referring to when we metaphorically use the former.

24.3.2 Causation: The local *explicatum*

Let us move on to the inquiry into how the GEC deals with complex situations which also involve more than two systems. In this way, we will provide a deeper insight into the explication of the notion of 'causation'. In particular, we could offer what we call a 'local *explicatum* of causation', since it allows us to speak about causation considering only (differential) parts of the systems at stake, and not necessarily the systems as a whole, as occurred in the integral *explicandum*.

Let us begin by reconsidering the example given above of the two communicating rooms, let them be A and B. Let us suppose that there is a third room C communicating with A but not with B. Let us call $C^{A}{}_{O2}$, $C^{B}{}_{O_2}$ and $C^{C}{}_{O_2}$ the integral fluxes of O_2 entering rooms A, B and C respectively. We have to measure the variation of n_{O_2} and monitor the source terms in A, B and C. Then, by using the balance equation, we can instantiate the three fluxes. At this point, a system of simple equations must be solved and the flux terms can be obtained. If requirements (1) and (2) above are satisfied we can speak of causal interactions.

However, if the requirements are not satisfied, we could go deeper and analyse what happens. Let us begin.

In the balance equation introduced in Section 24.2, the flux term C_E is, by definition, an integral quantity or, in other words, it takes into account the amount of extensive quantity E 'transferred', per time unit, as a whole to the system under consideration. In the example of the three rooms it was natural to separate the flux through door A-B from the flux through door A-C and to solve the consequent system of equations. In this case C_E^{A} has a structure: it is the sum of two contributions, each describing the flux of the quantity E through two different regions of the boundary ($C^{\mathrm{A}}{}_{O_2} = -C^{\mathrm{B}}{}_{O_2} - C^{\mathrm{C}}{}_{O_2}$). We should note that this separation of the different contributions to the integral flux term can be thought about at a more detailed and deeper level by considering all the infinitesimal contributions. In this case we have

$$C_E = \int_K \vec{\Gamma}_E \cdot \vec{n}\; d\kappa,$$

where $\vec{\Gamma}_E$ is the flux density (a vector); K denotes the surface surrounding the volume occupied by one system; $d\kappa$ the area of the surface element; and $\vec{n}$ the unit vector normal to the element $d\kappa$ and directed inward towards the volume.

Suppose that we know the flux density $\vec{\Gamma}_E$ at every point of the surface. In this case, we can describe the 'exchange' of a given extensive quantity point by point through the entire surface, limiting the volume. It follows, and by taking into account all that has been said until now, that we are able to speak in terms of 'local causation', and thus to study causal interactions between small (as small as we wish) subsystems.

Let us take a step forward. To deal with flux densities allows us to speak of local (differential) causal interactions as a function of the different points of the boundary. Given a certain area of the surface, there we can have some definite flux of the extensive quantity through it; but that local flux could be the superposition of various fluxes flowing through the same area. In this case, how can we distinguish between two sub-fluxes flowing through the same surface?

We can do it *only if we know* (and this depends on our T) that the different sub-fluxes are 'exchanging' the quantity E between *different parts* of our system P^{T} and *different parts* of the other system P'^{T}, the candidate to be in causal interaction with P^{T}. *We could say, more simply, that the sub-fluxes are "connecting"* P^{T} *with different subsystems.*

Evidently this discussion implies that the ability to separate the total flux C_E into different fluxes and each flux into different sub-fluxes is entirely dependent on our theoretical context T. However, now we can propose the local *explicatum.*

Let us fix a theoretical context T and let us have two systems P^{T} and P'^{T} in the states $P^{\mathrm{T}}_{x^i}$ and $P'^{T}_{x'^i}$ respectively, determined by the respective instantiations of the extensive quantities $E_1, \ldots, E_n$. Let $d\kappa$ be the area of an infinitesimal element of the surface separating P^{T} and P'^{T}; let $\vec{n}$ *and* $\vec{n}'$ be the two corresponding normal unit vectors both perpendicular to the surface in that point and directed inward towards P^{T} and P'^{T} respectively; and let $\vec{\Gamma}_E$ be the flux density in that point x^i. By defining the (infinitesimal) flux entering P^{T} and P'^{T} we have that $\vec{\Gamma}_E \cdot \vec{n} = -\vec{\Gamma}_E \cdot \vec{n}'$. Thus, we are able to speak of *local causal interaction* between P^{T} and P'^{T}, *if and only if*:

$$\vec{\Gamma}_E \cdot \vec{n} \neq 0.$$

This implies $dC_E = -dC'_E$, where dC_E and dC'_E are the infinitesimal fluxes entering P^{T} and P'^{T} respectively, through the element $d\kappa$ of the surface K at point x^i. Needless to say, now we no longer need to introduce the condition according to which P^{T} and P'^{T} form an isolated system G together.

24.4 Discussion

We wish to emphasize that, according to our GEC, we need not restrict ourselves to conserved quantities in determining what we mean when we talk about two causally interacting systems. The non-conserved quantities are also introduced. And this generalization is really one of the innovative aspects of our approach. It permits us to cope with all those situations, in particular those concerning complex systems, in which, if we want to speak about causation unambiguously and precisely, we are forced to introduce the non-conserved quantities, since by using solely the conserved quantities we would not be able to account for it.

Generally speaking, according to the GEC, if there is a variation of an extensive quantity E, there are two possibilities:

(1) E is a conserved quantity, i.e. we must have $\Sigma_E = 0$, and the possibility of speaking about causation is rooted only in the determination of the variation of said E; then the consideration of the balance equation containing only the C_E term becomes almost trivial;

(2) E is a non-conserved quantity, i.e. we may have $\Sigma_E \neq 0$, and the possibility of speaking about causation is rooted in the determination of the variation of said E and in the determination of the 'source' term $\sum_E$. The latter, in general, requires the determination of the variation of some other extensive quantities of the system, and then, by considering the balance equation containing both the dE/dt term and the Σ_E term, we are able to determine C_E, which is the fundamental quantity in order to acknowledge the causal interaction.

To better clarify our GEC let us return to the example above where the extensive quantity was the amount of molecular oxygen O_2 which we described, as usual, with the mole number n_{O_2}. In particular it allows us to illustrate the reason why the generalization to non-conserved quantities is conceptually (and not only practically) necessary. In the process under consideration, the extensive quantity, the 'mole number', is obviously non-conserved, while the total number of atoms is conserved. Then a question could be raised; whether a non-conserved quantity could be, at a different level of analysis, translated into conserved quantities. If this were the case, the generalisation regarding the non-conserved quantities would be useful in practice (since it would solve some technical difficulties) but not conceptually necessary (since the non-conserved description could be reduced to the conserved description). Actually this is not the case, and for two reasons:

(1) The statement affirming that the atom number is a conserved quantity can be assumed to be true (with a good degree of approximation) only in a relatively limited number of cases. It ceases working if, for instance, the energy involved in the process is high enough. So it works for our ordinary life, but not in other less ordinary physical situations. Hence we see that the property of being conserved is not an 'absolute' property but depends on the theoretical context inside which the problem is considered, and on the degree of depth of analysis permitted by that theoretical context.

(2) Let us suppose we are in a case in which we accept the conservation of the atom number. It is commonly said that a molecule of oxygen is 'composed of' two atoms of oxygen. In general, expressions of the type: object A is 'composed' of object B 'plus' object C are widely used in chemistry as well as in physics and in other sciences. They are, however, very gross and approximate, and are to be accepted only within a definite theoretical context. One molecule of oxygen is different from a set of two

atoms of oxygen: indeed we need a certain density of O_2 to live and we would not be satisfied with an equivalent quantity of oxygen atoms in different states of aggregation. Generally speaking, a complex system is not equivalent to the 'sum'[9] of its constituent parts. As far as the causal interaction is concerned, we may easily conceive a situation in which molecular oxygen is transferred in one direction while atomic oxygen is replaced at the right rate backwards. In such a case, no interaction would be present if we rely on one extensive quantity only (the so-called 'conserved' one in this example, but this is irrelevant). Hence we see that the presence of a causal interaction is unveiled only if we explore all possible extensive state parameters. Notice, however, that we may speak of causal interaction only in relation to a definite extensive quantity.

The two points above allow us to set out an important feature of our account properly: they show that *the possibility of speaking in terms of causal interaction between two systems depends on the choice of both the extensive quantity and the theoretical context*. Therefore another conceptual improvement offered by our explication begins becoming clearer: we understand that the ordinary way of speaking in terms of *tout court* causally interacting systems is not only ambiguous and imprecise, but also misleading. *Actually we must speak of systems causally interacting with respect to the extensive quantity we have decided to observe inside that given theoretical context*. Moreover, our explication shows that even the property of being conserved or non-conserved depends on the theoretical context adopted.[10]

In conclusion, the generalization of the explication of 'causation' in terms both of conserved quantities and non-conserved quantities is to be considered a real improvement not only for practical reasons but, most importantly, for conceptual reasons. Therefore, to also satisfy the four steps of the explicative process, we may claim that the explication proposed is fruitful, since it permits us a deeper comprehension of what we should really mean by 'causation'.

Moreover, not only is the concept-*explicatum* 'causation' fruitful but it is also similar enough to the starting concept-*explicandum* 'causation', as to be rather intuitive.

24.5 Conclusion

In the above, we have given a rigorous and unambiguous empirical explication of the notion of 'causation'. We have arrived at this result by offering a strongly

[9] Note that the notion of 'sum' is totally undefined; usually we mean simply that if we break the whole we see some pieces.

[10] On theoretical context dependency, see Boniolo, Faraldo and Saggion (2009).

consistent analysis and by resorting to an approach based on extensive quantities and on the balance equation. In this way we have proposed what we have called a *generalized explication of 'causation'* (GEC), both in the integral and less complex form and in the local form in which both conserved and non-conserved quantities have the same role.

Note that the possibility of dealing with non-conserved quantities is of fundamental importance. We can easily consider several examples in which we naturally have to deal with non-conserved quantities, as occurs in chemical processes or in population dynamics, but there is at least one more fundamental reason which makes this generalization necessary. For the GEC, which is a *theoretical context sensitive approach*, enables us to assert that two systems, discussed inside a given theoretical context, are causally connected if and only if a particular balanced change of their respective extensive quantities is observed.

To conclude, we believe that the GEC provides a well-grounded platform allowing us to understand precisely and unambiguously what we are affirming whenever we speak of causation in many different contexts, provided that there is the possibility of characterizing the systems under discussion in terms of extensive quantities. In these cases, our GEC can be applied (and it can be applied without any reductionist or reconstructionist purpose).

Acknowledgements

We would like to express our sincere thanks to the two unknown referees whose precious suggestions have allowed us to improve the paper. We would acknowledge also the many friends and colleagues who have commented on the content of the paper during seminars, talks and discussions. In particular, our gratitude goes to M.J. Bisset for her invaluable help in improving the style of the paper.

References

Aronson, J. (1971). On the grammar of cause, *Synthese*, 22, 414–430.

Ben-Yami, H. (2006). Causality and temporal order in special relativity, *The British Journal for the Philosophy of Science*, 57, 459–479.

Boniolo, G., Faraldo, R. & Saggion, A. (2009). On spatial and temporal *ex mensura* boundaries, *Foundations of Science*, 14, 181–193.

Bontly, T.D. (2006). What is an empirical analysis of causation? *Synthese*, 151, 177–200.

Callen, H.B. (1985). *Thermodynamics and an Introduction to Thermostatistics* (Hoboken, (NJ): John Wiley & Sons).

Carnap, R. (1950). *Logical Foundations of Probability* (Chicago: Chicago University Press).
Choi, S. (2003). The conserved quantity theory of causation and closed systems, *Philosophy of Science*, 70, 510–530.
Dowe, P. (1992). Wesley Salmon's process theory of causality and the conserved quantity theory, *Philosophy of Science*, 59, 195–216.
Dowe, P. (2000). *Physical Causation* (Cambridge: Cambridge University Press).
Fair, D. (1979). Causation and the flow of energy, *Erkenntnis*, 14, 219–250.
Fuchs, H.U. (2002). *Modeling of Uniform Dynamical Systems* (Zurich: Orell Füssli).
Garbois, M. & Hermann, F. (2000). Momentum flow diagrams for just-rigid static structures, *European Journal of Physics*, 21, 591–601.
Guggenheim, E.A. (1950). *Thermodynamics* (Amsterdam: North Holland).
Gurtin, M.E. (1981). *An Introduction to Continuum Mechanics*. (New York: Academic Press).
Heiduck, G., Herrmann, F. & Schmid, G.B. (1987). Momentum flow in the gravitational field, *European Journal of Physics*, 8, 41–43.
Herrmann, F. & Schmid, G.B. (1985). Momentum flow in the electromagnetic field, *American Journal of Physics*, 53, 415–420.
Herrmann, F. (1986). Is an energy current energy in motion?, *European Journal of Physics*, 7, 198–204.
Herrmann, F. (2000). The Karlsruhe Physics Course, *European Journal of Physics*, 21, 49–58.
Prigogine, I. (1955). *Thermodynamics of Irreversible Processes* (Hoboken (NJ): John Wiley).
Russell, B. (1912–1913). On the notion of cause, *Proceedings of the Aristotelian Society*, 13, 1–26.
Russell, B. (1948). *Human Knowledge* (New York: Simon and Schuster).
Salmon W. (1997). Causality and explanation: A reply to two critiques, *Philosophy of Science*, 64, 461–477.
Salmon, W. (1984). *Scientific Explanation and the Casual Structure of the World* (Princeton: Princeton University Press).
Sant'Anna, A.S. (2005). The definability of physical concepts, *Boletim da Sociedade Paranaense de Matematica*, 23, 163–175.
Shavit, A. & Gutfinger, C. (1995). *Thermodynamics* (Englewood Cliffs (NJ): Prentice Hall).
Super, T. (2009). A physical critique of physical causation, *Synthese*, 167, 67–80.
Tisza, L. (1966). *Generalized Thermodynamics* (Cambridge (MA): The M.I.T. Press).

Appendices

A1: On 'action at a distance' and time

Some could object that, according to the GEC, if a molecule on Alpha Centauri and a molecule on Earth each happen to lose and gain, respectively, the same magnitude of momentum, then they are in causal interaction. *Prima facie*, this might appear strange since some sort of spatial contiguity condition (and also some sort of 'time contiguity' or 'time correlation') should be called for. Actually, this is exactly what we defend, provided that all the mentioned requirements are fulfilled. In particular we require that: (1) the two molecules form a system which is closed with respect to the extensive quantity E under consideration (in this case the momentum); (2) we know that two molecules undergo those changes of momentum.

Notice, on the other hand, that the spatial contiguity remains a part of the very definition of the two (or more) systems under consideration. In our case, if we focus on the two molecules mentioned above, there certainly are some (infinite in number) surfaces between here and Alpha Centauri that we may select as separating them. Any of them would be suitable. However it is not necessary to go so far apart. The same objection could be raised for two molecules some μm apart. This seems to be a severe limitation because it substantially considers actions at a distance. But it is not so, and the objection could be rebutted if we refer to interaction as mediated by some field, so that the case can be treated as a chain of more than one 'causal interaction': (System A, e.g. a molecule on Alpha Centauri)-(field)-(...)-(field)-(System B, e.g. a molecule on Earth).

Along the same line of thought, we would like to comment very briefly on the connection between causation and time ordering. As is well known, the problem has been widely discussed (see Ben-Yami 2006) and, of course, in solving it we cannot disregard the constraints posed by the postulates of relativity. As previously mentioned, in the GEC there is no specific role for time. In particular no relation is required between the times of observation of the two events in the two systems P^{T} and P'^{T}.

One possible and frequently adopted perspective could be a Newtonian one, in which the two events occur simultaneously according to an 'action at a distance' assumption. This assumption is equivalent to the requirement that interactions propagate at infinite speed and therefore causation can refer only to simultaneous events. Instead, according to a relativistic approach, interactions propagate at finite speeds and, accordingly, events have to occur at different times if they are to be causally connected. In this case the time relations are constrained by the requirements imposed by relativity. Nevertheless, in both perspectives the balance between the changes of extensive quantities has to be satisfied. And the GEC focuses on this aspect, decoupling, in such a way, causal connection from time ordering. Please note, but unfortunately we cannot spend any longer on this point, that this view, according to which 'the notion of causation is a-temporal', has had a long history exemplified, for instance, by the positions of J.F. Herbart, in the philosophical field, and B. Riemann, in the scientific field.

A2 : On causation in equilibrium and non-equilibrium stationary states

Let us think about systems in stationary states (remembering the definition of 'stationary state' in Remark 6), in particular the case in which they are in mechanical equilibrium. Should we claim that they are not, *by definition*, in a causal interaction? This could be one possible outcome based on the fact that systems in stationary states do not undergo any change and one would possibly want to associate causal interactions with some modifications. *Prima facie*, this would not seem so odd. Indeed we are used, *from a commonsense point of view*, to associating a cause to some 'visible' change, thought of as effect. This, however, would lead to the consequence that in order to keep a bow drawn no causal interaction is needed, and this type of account would appear unsatisfactory for many reasons. Let us discuss why and how our GEC can cope with these situations.

It should be remembered that by 'mechanical systems' we mean those systems whose dynamics are entirely determined by the action of Newtonian forces. Even though they can be of various origins (contact forces, gravitational forces, electromagnetic forces, etc.) they have their Newtonian definition in common: the application of a force to a body is equivalent to the 'transfer' to that body, at the point where the force is applied, of momentum at the rate given by Newton's second law. Therefore, a continuous distribution of forces, i.e. a field of forces, is equivalent to a field of momentum currents 'flowing' with certain rates and lines of flow in complete analogy with the representations of fluid dynamics.

By current methods in potential theory we can describe the field lines for gravitational, electrostatic, magnetostatic, etc. fields. Each field line corresponds to a line of force, and each line of force to a flux of momentum of the right density. The description of fields of forces as distributions of momentum currents has been already introduced by Herrmann and Schmid (1985) for electrostatic and magnetostatic fields, and by Heiduck, Herrmann and Schmid (1987) for weak gravitational fields.

At this point we can consider a couple of examples. We would like to start with the one proposed by Dowe (2000, p. 177). Let us consider the case of two wooden planks leaning on each other on a vertical plane and, therefore, staying in a position of equilibrium. Is it true, as Dowe claims, that each plank causally interacts with the other? Let us apply our account on causation in relation with momentum, selected as the extensive quantity under consideration. Our observation shows that for each plank its momentum is constant and, in this case, equal to zero. However, each plank may be exerted by two contact forces in two different parts: one at the surface shared with the soil, and the other at the surface shared with the other plank. Besides these two forces, each plank is acted upon by the gravitational field of the Earth and this is a 'volume force'. Therefore, each plank is traversed by a complicated system of momentum currents. To our aims, a simplified description suffices: one momentum current is the almost parallel and vertical bundle of current-lines describing the volume force (gravity) acting between the plank and the Earth. The other two momentum flows take place in the two contact regions between a plank (plank 1) and the soil and the other plank (plank 2). The vector sum of these three momentum-currents must be equal to zero, as required by our observation and because we assume that momentum is a conserved quantity (see balance equation), but each momentum current may be non-zero. Now in order to recognize the existence of causal interaction between the two planks we *need* to know enough about the existence *and* the structure of the other two momentum currents; the one generated by gravity is sufficiently known but we need *further* information concerning the plank–soil momentum flow. By playing with these two parameters and by using the balance equation we can get almost any value for the momentum flux across the plank–plank surface. For instance, we may imagine being in absence of gravity in a space shuttle, or in a situation in which the two planks are fixed to the soil in a particular way. Then we may prove that the momentum flux across the plank–plank surface is zero and hence they *are not* in mutual causal interaction. Otherwise it can be easily proven that in the contact region between the two planks, the momentum flux from one plank to the other has to be *non-zero* and it satisfies the Newtonian required symmetry condition (i.e. the flux of momentum from plank 1 to plank 2 is equal in magnitude but opposite in sign to the previous one). In general it is worth recalling that our ability to recognize causal interactions between systems relies on our theoretical context (theories and information).

Thanks to our GEC, in particular to the local *explicatum* of causation, we can claim that the two planks are in mutual causal interaction, but it would not be correct to affirm: 'The equilibrium of the plank 1 is caused by plank 2 (and vice versa)'. Actually, the equilibrium of plank 1 is ensured by the combined contribution of plank 2 *and* other systems like, for instance, the ground and the Earth (Garbois and Hermann 2000). So in Dowe's example, each plank is in causal interaction with various systems and this is easily visible because the three momentum fluxes are spatially separated.

Likewise in the example of the bow we would be unsatisfied with the 'non-causal interaction hypothesis'. We have *previously learned* that some muscular sensations are equivalent to the onset of some momentum fluxes (i.e. forces); for instance they can balance a weight, etc.

Indeed we could draw a bow by applying a weight and, as in the case of the two planks, the interaction with the Earth gravity creates a flux of momentum.

In other words in both examples *we (should) know* that there is some external agent taking part in the game (the gravitational field, for instance). Seeing the two planks in contact does not mean, by itself, that they are in causal interaction. We can say whether they are or not only because we know enough about the Earth's gravity, or about our muscular sensations and we are able to apply our balance condition. In other words, our theoretical context is 'too highly developed' and the 'non-causal hypothesis' gives rise to a conceptual emergency. It is possible that there might be some other causal connections that we are not able to recognise only because our theoretical context is insufficient.

So that's it for mechanical equilibrium but, for the sake of thoroughness, our discussion about causation in stationary states must also concern stationary non-equilibrium configurations. It is well known that this possibility does not occur in mechanical systems (fluctuations cannot be reduced to zero) but it is quite common in thermodynamical (macroscopical) systems. The example of an electric heater powered by a constant electromotive force and observed after a long enough time for it to have reached a stationary state well explains the situation. In this example the resistor (our system) has a constant volume, temperature, energy, mass, entropy, and so on. Therefore it can be said to be in a stationary state.

Does this mean that such a system is not causally interacting with other systems? Not at all. If we examine the fluxes of various extensive quantities entering and leaving the heater we can easily estimate the following: energy 'enters' at a certain rate from the electric plant, and it 'goes out' at the same rate into the room. In this case we are in an analogous situation to the two planks in Dowe's example, but if we look for entropy we have a net outflow of it towards the room while its content in the heater is constant (stationary condition). This is due to the fact that entropy is a non-conserved quantity:

it is continuously produced within the system and therefore it can only be maintained constant within the system (stationary state) thanks to the possibility of it flowing towards the room at the right rate. Hence the stationary state can be maintained under the condition that our system interacts causally with the external world: in this case the extensive quantities accounting for the existence of a causal interaction between the resistor and the room are energy and entropy. Notice that the latter can be evidenced simply by the global explicatum (lower complexity account) since the global flux of entropy, C_S, is different from zero while the causal interaction with respect to the former can be recognized only through the local *explicatum* since the total ingoing flux of energy is zero ($C_E = 0$) *but* it is the result of two equal and opposite fluxes which *can be separated* in a higher complexity theoretical context.

This is a typical non-equilibrium stationary state.

We can conclude this appendix regarding equilibrium and, in general, stationary states with the following: our account on causation, based on an accurate evaluation of the balance equation for the exchanges of extensive quantities among systems, can also successfully be employed to give a satisfactory representation of causal interactions in the case of mechanical equilibrium and of non-equilibrium stationary states. Causal interactions are once again recognized by the existence of some fluxes (or more precisely flux densities) of some extensive quantity in some portion of the surface separating two parts.

·25·

Causal completeness of probability theories – Results and open problems

Miklós Rédei and Balázs Gyenis

Abstract

A classical (Kolmogorovian) probability measure space is defined to be causally closed with respect to a causal independence relation between pairs of random events if the probability space contains a Reichenbachian common cause of every correlation between causally independent random events. A number of propositions are presented that characterize causal closedness. Generalizing the notion of Reichenbachian common cause in terms of non-classical probability spaces, where the Boolean algebra of random events is replaced by a non-distributive orthocomplemented lattice, the notion of causal closedness is defined for non-classical probability spaces and propositions are presented that state causal closedness of certain non-classical probability spaces as well. Based on the generalization of the notion of common cause to a common cause system containing N random events, causal N-closedness is defined with respect to a common cause system both in classical and non-classical probability spaces, and the problem of causal N-closedness is formulated. Characterizing causal N-closedness remains a largely open problem.

25.1 The Common Cause Principle and causal completeness informally

The aim of this chapter is to investigate the problem of causal completeness (closedness) of classical and non-classical (quantum) probability spaces. Causal closedness of a probabilistic theory means that the theory is causally rich enough to be able to explain causally all the correlations it predicts. It is natural to ask whether probabilistic theories are causally closed if one assumes that Reichenbach's Common Cause Principle holds: this principle states that if two events A and B are probabilistically correlated then either the correlation is due to a causal interaction between A and B, or, if A and B are causally independent, $R_{\text{ind}}(A, B)$, then there exists a third event C, a so-called common cause, that explains the correlation by being related to A and B probabilistically in a specific way (see Definition 25.1). Causal closedness of a probabilistic theory is intended to express that the theory complies with the Common Cause Principle; accordingly, and more precisely, a probability theory is defined to be causally closed with respect to a causal independence relation R_{ind} defined between pairs of random events if for any correlation between elements A and B such that $R_{ind}(A, B)$ holds, there exists a common cause C in that theory of the correlation between A and B (Definition 25.3). The problem is then: under what conditions on the probability space and on R_{ind} is the probability space causally closed?

We will see that causal closedness is non-trivial and not impossible in classical probability spaces, not even if the probability space contains a finite number of random events only – but causal completeness is not typical either. There does not seem to be any regular and easily characterizable behaviour of probability spaces from the perspective of causal closedness; one has to check in each and every case by brute force whether the causal closedness holds.

The notion of causal closedness of classical probability spaces was introduced by Gyenis and Rédei (2004) but the notion of common cause can be naturally defined in non-classical (quantum) probability spaces as well; hence the notion of causal closedness also makes perfect sense for such probability theories. Little is known about causal closedness of such general probability spaces. The known results are summarized in Section 25.4.

The status of the Common Cause Principle and the notion of the common cause has been widely discussed in the philosophical literature. A significant part of this discussion focuses on evaluating the validity of the Principle by means of analysing informally constructed putative counterexamples; for a review the Reader is referred to Arntzenius (2005). It quickly becomes apparent, however, that the results of these analyses are highly sensitive to the way one identifies the space of relevant events, the probability structure, the definition of correlation between events, and the notion of the common cause. In addition different authors operate with different implicit assumptions about the importance of temporal precedence of causes, issues of locality, the importance of a distinction between microscopic and macroscopic events, and so on. It is difficult to assess, then, the validity of the Principle without being more cautious about the notions and requirements with which we operate with. Formal clarity becomes especially important when one discusses the case of Algebraic Quantum Field Theory, where relying on an informal characterization may easily lead us astray.

With this cautionary note we now turn to Reichenbach's original definition of the common cause.

25.2 The notion of common cause and some terminology

In what follows $(X, \mathcal{S}, p)$ denotes a classical (Kolmogorovian) probability space with Boolean algebra $\mathcal{S}$ of subsets of a set X (with respect to the set theoretic operations $\cap$, $\cup$ and $A^{\perp} = X \setminus A$ as Boolean algebra operations) and with the probability measure p on $\mathcal{S}$. To simplify notation, occasionally we write $(\mathcal{S}, p)$ instead of $(X, \mathcal{S}, p)$, when the precise nature of X is not important. For instance, when $\mathcal{S}$ has a final number of elements, then $\mathcal{S}$ is the set of all subsets of a finite set having $n < \infty$ elements, in which case we write $(\mathcal{S}_n, p)$.

Given $(\mathcal{S}, X, p)$, the quantity $\mathrm{Corr}_p(A, B)$ defined by

$$\mathrm{Corr}_p(A, B) \equiv p(A \cap B) - p(A)p(B) \tag{25.1}$$

is called the *correlation* of A, B in p. Events A and B are said to be positively correlated if $Corr_p(A, B) > 0$. A correlation $\mathrm{Corr}_p(A, B) \neq 0$ is called *non-degenerate* (and (A, B) a non-degenerate correlated pair) if $A \neq B$. A correlation is called *maximal* if

$$p(A|B) = p(B|A) = 1. \tag{25.2}$$

The next definition specifies the notion of common cause. Since the definition was first given by Reichenbach (1956), this type of common cause is called 'Reichenbachian'; however, since in this chapter only Reichenbachian common causes feature, the qualifier 'Reichenbachian' will be omitted.

Definition 25.1. C is a *common cause* of the correlation (25.1) if the following (independent) conditions hold:

$$p(A \cap B|C) = p(A|C)p(B|C) \tag{25.3}$$

$$p(A \cap B|C^{\perp}) = p(A|C^{\perp})p(B|C^{\perp}) \tag{25.4}$$

$$p(A|C) > p(A|C^{\perp}) \tag{25.5}$$

$$p(B|C) > p(B|C^{\perp}) \tag{25.6}$$

where

$$p(X|Y) = \frac{p(X \cap Y)}{p(Y)}$$

denotes the conditional probability of X on condition Y, $C^{\perp}$ denotes the complement of C and it is assumed that none of the probabilities $p(X)$ $(X = A, B, C, C^{\perp})$ is equal to zero.

Since the notion of common cause is a measure theoretic one, measure zero sets have to be dealt with. The next definition takes care of this and summarizes some terminology used later.

Definition 25.2.

1. Let

$$\Delta(X, Y) \equiv (X \setminus Y) \cup (Y \setminus X)$$

be the symmetric difference of sets X, Y. A common cause C of the correlation between A, B is called *proper* if

$$p\,(\Delta(B, C)) \neq 0 \neq p\,(\Delta(A, C)) \tag{25.7}$$

That is to say, a common cause C of the correlation $\mathrm{Corr}_p(A, B) > 0$ is proper if the common cause differs from the correlated events by more than a measure zero event. Otherwise C is called *improper.*

2. It can happen that, in addition to being a probabilistic common cause, the common cause event C logically implies both A and B, i.e. $C \subseteq A \cap B$. If this is the case then we call C a *strong* common cause. If C is a common cause such that $C \not\subseteq A$ and $C \not\subseteq B$ then C is called a *genuinely probabilistic* common cause.
3. A common cause C will be called *deterministic* if

$$p(A|C) = 1 = p(B|C) \tag{25.8}$$

$$p(A|C^{\perp}) = 0 = p(B|C^{\perp}). \tag{25.9}$$

25.3 Causal closedness of classical probability theories

Given the notion of common cause one can define the concept of common cause closedness in a very natural manner:

Definition 25.3. Let $(X, \mathcal{S}, p)$ be a probability space and R_{ind} be a two-place causal independence relation between elements of $\mathcal{S}$. The probability space $(X, \mathcal{S}, p)$ is called *common cause closed* with respect to R_{ind}, if for every correlation $\mathrm{Corr}_p(A, B) > 0$ with $A \in \mathcal{S}$ and $B \in \mathcal{S}$ such that $R_{\text{ind}}(A, B)$ holds, there exists a common cause C in $\mathcal{S}$. If there are no elements A, B in $\mathcal{S}$ that are positively correlated, then $(X, \mathcal{S}, p)$ is called *trivially common cause closed.*

Proposition 1 (Gyenis and Rédei 2004) *Let $(\mathcal{S}_n, p)$ be a finite probability space. If R_{ind} contains all the pairs of events A, B in $\mathcal{S}_n$ that are correlated in p, then $(\mathcal{S}_n, p)$ is not non-trivially common cause closed with respect to R_{ind}.*

Proposition 1 shows that a probability space containing a finite number of random events contains more correlations than it can account for exclusively in terms of common causes. But this is not surprising because common cause closedness with respect to a causal independence relation that leaves no room for causal dependence is unreasonably strong. One can of course make a probability space $(X, \mathcal{S}, p)$ causally closed by stipulating that $R_{\text{ind}}(A, B)$ does *not* hold (i.e. that A and B *are* causally related) whenever A and B are correlated but there exists no common cause $C \in \mathcal{S}$ of the correlation. But this is unacceptable in general since this move makes the notion of causal closedness trivial and the causal dependence so defined (and the causal independence relation R_{ind} so defined) may turn out not to have reasonable features. One needs a disciplined, independent definition of the causal independence relation.

In general, the causal independence relation R_{ind} depends on the characteristics of the probabilistic theory predicting the correlations and little can be said in advance about its structure. However, on the basis of general considerations, some reasonable conditions can be imposed on R_{ind}. Intuitively, causal independence of A and B should imply that from the presence or absence of A one should not be able to logically infer either the occurrence or non-occurrence of B. Conversely, when from the presence or absence of A one is able to logically infer either the occurrence of non-occurrence of B, the two events are suspect to be in direct causal relation and hence their correlation doesn't need to be explained by a common cause. Hence there is a strong connection between the notions of causal independence and logical independence. Taking then, as it is common, the partial ordering $\subseteq$ in the Boolean algebra $\mathcal{S}$ as the implication relation between events (equivalently: between propositions that the corresponding events occur), this requirement about R_{ind} can be expressed by the demand that $R_{\text{ind}}(A, B)$ should imply all of the following relations

$$A \not\subseteq B, \quad A^{\perp} \not\subseteq B\,, \; A \not\subseteq B^{\perp}, \quad A^{\perp} \not\subseteq B^{\perp}$$
$$B \not\subseteq A, \quad B^{\perp} \not\subseteq A\,, \; B \not\subseteq A^{\perp}, \quad B^{\perp} \not\subseteq A^{\perp}.$$

This requirement can be expressed compactly by saying that $R_{\text{ind}}(A, B)$ implies that A and B are logically independent; equivalently, that

$$\{\emptyset, A, A^{\perp}, X\} \text{ and } \{\emptyset, B, B^{\perp}, X\}$$

are logically independent Boolean subalgebras of $\mathcal{S}$ in the sense of the following Definition 25.4:

Definition 25.4. Two Boolean subalgebras $\mathcal{L}_1$, $\mathcal{L}_2$ of the Boolean algebra $\mathcal{S}$ are called *logically independent* if

$$A \cap B \neq \emptyset \qquad \text{whenever } \emptyset, X \neq A \in \mathcal{L}_1 \text{ and } \emptyset, X \neq B \in \mathcal{L}_2. \tag{25.10}$$

The pair $(\mathcal{L}_1', \mathcal{L}_2')$ of Boolean subalgebras of Boolean algebra $\mathcal{S}$ is called a *maximal* logically independent pair, if logical independence of Boolean subalgebras $\mathcal{L}_1$ and $\mathcal{L}_2$ containing respectively $\mathcal{L}_1'$ and $\mathcal{L}_2'$ as Boolean subalgebras implies $\mathcal{L}_1' = \mathcal{L}_1$ and $\mathcal{L}_2' = \mathcal{L}_2$.

For later purposes we also need the following notions:

Definition 25.5. The pair (A, B) is called *logically independent modulo zero probability* if there exist A', B' such that

$$p(A') = p(B') = 0 \tag{25.11}$$

and $(A \setminus A')$ and $(B \setminus B')$ are logically independent.

This motivates the following definition, which formulates a natural notion of causal closedness.

Definition 25.6. $(X, \mathcal{S}, p)$ is called common cause closed with respect to the pair $(\mathcal{L}_1, \mathcal{L}_2)$ of logically independent Boolean subalgebras of $\mathcal{S}$, if for every $A \in \mathcal{L}_1$ and $B \in \mathcal{L}_2$ that are correlated in p, there exists a common cause C in $\mathcal{S}$ of the correlation between A and B.

Proposition 2 (Gyenis and Rédei 2004) *Let $(\mathcal{S}_5, p_u)$ be the probability space with the Boolean algebra $\mathcal{S}_5$ generated by five atoms and with p_u being the probability measure defined by the uniform distribution on atoms of $\mathcal{S}_5$. Then $(\mathcal{S}_5, p_u)$ is common cause closed with respect to every pair of logically independent Boolean subalgebras $(\mathcal{L}_1, \mathcal{L}_2)$ of $\mathcal{S}_5$.*

The next proposition shows that the behaviour of the probability space $(\mathcal{S}_5, p_u)$ described in Proposition 2 is exceptional.

Proposition 3 (Gyenis and Rédei 2004) *If the probability space $(\mathcal{S}_n, p)$ is not $(\mathcal{S}_5, p_u)$, then it is not non-trivially common cause closed with respect to every pair of logically independent Boolean subalgebras.*

But causally not closed probability spaces can be extended in such a manner that the extension contains common causes of a finite number of correlations in a given pair of logically independent sublattices:

Proposition 4 (Gyenis & Rédei 2004; Hofer-Szabó *et al.* 1999) *If $(X, \mathcal{S}, p)$ with finite $\mathcal{S}$ is not common cause closed with respect to a logically independent pair $(\mathcal{L}_1, \mathcal{L}_2)$, then it can be extended into a $(X', \mathcal{S}', p')$, with $\mathcal{S}'$ being also finite, in such a manner that $(X', \mathcal{S}', p')$ is common cause closed with respect to the logically independent pair $(h(\mathcal{L}_1), h(\mathcal{L}_2))$, where $h(\mathcal{L}_i)$ is the homomorphic image in $\mathcal{S}'$ of $\mathcal{L}_i$ $(i = 1, 2)$.*

By an extension is meant here that there exists a Boolean algebra embedding h of $\mathcal{S}$ into $\mathcal{S}'$ that preserves the probability in the sense that $p(X) = p'(h(X))$ for all $X \in \mathcal{S}$. Note that the images of $\mathcal{L}_1, \mathcal{L}_2$ under h will not necessarily be *maximally* logically independent, not even if $\mathcal{L}_1, \mathcal{L}_2$ is a maximally logically independent pair; so we have the following Problem, which is open:

Problem 1. *Does Proposition 4 remain true if 'logically independent' means* maximal *logically independent?*

The following problem is also open:

Problem 2. *Let $(\mathcal{S}, p)$ be a probability space with $\mathcal{S}$ having an infinite number of elements and assume that $(\mathcal{L}_1, \mathcal{L}_2)$ is a logically independent pair of Boolean subalgebras of $\mathcal{S}$ such that there exist an infinite number of pairs (A_i, B_i) of* *events $A_i \in \mathcal{L}_1$ and $B_i \in \mathcal{L}_2$ that are correlated in p. Does there exist and extension $(\mathcal{S}', p')$ of $(\mathcal{S}, p)$ such that $(\mathcal{S}', p')$ is common cause closed with respect to $(h(\mathcal{L}_1), h(\mathcal{L}_2))$? Does there exist such an extension so that $(h(\mathcal{L}_1), h(\mathcal{L}_2))$ is a* maximal *pair of logically independent sub-Boolean algebras in $\mathcal{S}'$?*

Probability spaces with infinite Boolean algebras can however be causally closed as the next proposition shows. Before stating the proposition recall that a probability space $(X, \mathcal{S}, p)$ is called *atomless* if for any $A \in \mathcal{S}$, $p(A) \neq 0$ there exists $B \subseteq A$, $B \in \mathcal{S}$ such that $0 < p(B) < p(A)$.

Proposition 5 (Gyenis and Rédei 2004) *If $(X, \mathcal{S}, p)$ is an atomless probability space, then it contains uncountably many proper common causes of every non-degenerate correlation in it. Moreover if A and B are correlated, logically independent modulo measure zero events, then $\mathcal{S}$ contains both uncountably many strong and uncountably many genuinely probabilistic common causes of the correlation between A and B.*

The notion of common cause can be naturally generalized to cover the case when the correlation is not explainable by a single common cause but by *system* of common cause like events. One such generalization was given by Hofer-Szabó and Rédei (2004, 2006):

Definition 25.7. Let $(X, \mathcal{S}, p)$ be a probability space and A, B be two events in $\mathcal{S}$. The partition $\{C_i\}_{i \in I}$ of $\mathcal{S}$ is said to be a *Reichenbachian common cause system* (RCCS for short) for the pair (A, B) if the following two conditions are satisfied

$$p(A \cap B|C_i) = p(A|C_i)p(B|C_i) \qquad \text{for all } i \in I \qquad (25.12)$$

$$[p(A|C_i) - p(A|C_j)][p(B|C_i) - p(B|C_j)] > 0 \ (i \neq j). \qquad (25.13)$$

The cardinality of the index set I (i.e. the number of events in the partition) is called the *size* of the RCCS. Since $C, C^{\perp}$ with a Reichenbachian common cause C is a RCCS of size 2, we call a RCCS *proper* if its size is greater than 2.

The motivation behind the definition of RCCS is that the correlation between A and B may not be explainable by displaying a single common cause but may be the cumulative result of a number of different 'partial common causes', none of which can in and by itself yield a complete common-cause-type explanation of the correlation, but all of which, taken together, can account for the entire correlation. Explaining a correlation by such a system of 'partial common causes' means that one can partition the statistical ensemble into more than two subensembles in such a manner that (i) the correlation disappears in each of the subensembles, (ii) any pair of such subensembles behaves like the two subensembles determined by a common cause and its negation and (iii) the totality of 'partial common causes' explains the correlation in the sense of entailing it.

It was shown by Hofer-Szabó and Rédei (2004, 2006) that Reichenbachian Common Cause systems of arbitrary finite size exist for *any non-maximal* correlation in the sense that for any such correlation in any probability space there exists an extension of that probability space that contains a Reichenbachian Common Cause system of the prescribed size.

In view of this, a natural refinement of the definition of causal closedness of $(X, \mathcal{S}, p)$ is obtained if one replaces the notion of common cause with the concept of common cause system:

Definition 25.8. $(X, \mathcal{S}, p)$ is called *causally N-closed* with respect to a causal independence relation R_{ind} if for any correlation $Corr_p(A, B) > 0$ such that $R_{\text{ind}}(A, B)$ holds, there exists in $(X, \mathcal{S}, p)$ a Reichenbachian common cause system of size N for the correlation.

There are a number of questions one can ask in connection with causal N-closedness:

Problem 3. On what condition on $(X, \mathcal{S}, p)$ and R_{ind} is $(X, \mathcal{S}, p)$ causally N-closed for a fixed N?

We have seen that probability spaces may or may not be causally 2-closed – causal 2-closedness depends sensitively on how R_{ind} is defined. This leads to the following open problems:

Problem 4. Can a probability space which is not causally 2-closed be causally N-closed for some fixed $N > 2$ (with respect to some non-trivial causal independence relation R_{ind})?

Problem 5. Can a probability space be causally N-closed for *every* N (with respect to some non-trivial causal independence relation R_{ind})?

We conjecture that atomless probability spaces are causally N-closed for every N with respect to every pair of logically independent Boolean subalgebras.

25.4 Causal closedness of non-classical probability spaces

The notion of common cause can be defined in non-classical (quantum) probability spaces $(\mathcal{L}, \phi)$, where an orthomodular lattice $\mathcal{L}$ takes the place of the Boolean algebra and ϕ is an additive (or σ additive) map from $\mathcal{L}$ into [0,1] (generalized probability measure), replacing a classical probability measure. Special examples of such spaces are the quantum probability spaces $(\mathcal{P}(\mathcal{N}), \phi)$ where $\mathcal{P}(\mathcal{N})$ is the projection lattice of a von Neumann algebra $\mathcal{N}$ and ϕ is a (normal) state on $\mathcal{N}$ (see Kadison & Ringrose 1986 and Takesaki 1979 for the theory of von Neumann algebras). An even more specific example of the latter is the probability space when $\mathcal{P}(\mathcal{N})$ is the von Neumann lattice of *all* projections on a Hilbert space $\mathcal{H}$ (in this case we write $(\mathcal{H}, \mathcal{P}(\mathcal{H}), \phi)$; this latter non-classical probability space describes standard, non-relativistic quantum systems.

Two elements $A, B \in \mathcal{L}$ are called *compatible* if they belong to the same Boolean subalgebra of $\mathcal{L}$. This condition is equivalent to

$$A = (A \wedge B) \vee (A \wedge B^{\perp}).$$

If A, B are compatible and

$$\mathrm{Corr}_{\phi}(A, B) \equiv \phi(A \wedge B) - \phi(A)\phi(B) > 0 \tag{25.14}$$

then A and B are called (positively) correlated with respect to the state ϕ.

Definition 25.9. If A and B are positively correlated, then $C \in \mathcal{L}$ is called a *common cause* of the correlation (25.14) if C is compatible with both A and B and the following conditions hold.

$$\phi(A \wedge B|C) = \phi(A|C)\phi(B|C) \tag{25.15}$$

$$\phi(A \wedge B|C^{\perp}) = \phi(A|C^{\perp})\phi(B|C^{\perp}) \tag{25.16}$$

$$\phi(A|C) > \phi(A|C^{\perp}) \tag{25.17}$$

$$\phi(B|C) > \phi(B|C^{\perp}) \tag{25.18}$$

where

$$\phi(X|Y) = \frac{\phi(X \wedge Y)}{\phi(Y)}$$

denotes the conditional probability of X on condition Y and it is assumed that none of the probabilities $\phi(X)$, $(X = A, B, C, C^{\perp})$ is equal to zero.

Extension of $(\mathcal{L}, \phi)$, logical independence of events in $\mathcal{L}$ and causal independence relation R_{ind} on $\mathcal{L}$ can all be defined in complete analogy with the classical definitions, which makes it possible to define causal completeness as well in complete analogy with the classical case:

Definition 25.10. Let $(\mathcal{L}, \phi)$ be a non-classical probability space and R_{ind} be a two-place causal independence relation between elements of $\mathcal{L}$. The probability space $(\mathcal{L}, \phi)$ is called *common cause closed* with respect to R_{ind}, if for every correlation $\mathrm{Corr}_{\phi}(A, B) > 0$ with $A \in \mathcal{L}$ and $B \in \mathcal{L}$ such that $R_{\mathrm{ind}}(A, B)$ holds, there exists a common cause C in $\mathcal{L}$ (in the sense of Definition 25.9). If there are no compatible elements A, B in $\mathcal{L}$ that are positively correlated, then $(\mathcal{L}, p)$ is called *trivially common cause closed.*

Problem 6. On what conditions on $(\mathcal{L}, \phi)$ and R_{ind} is the probability space $(\mathcal{L}, \phi)$ common cause closed with respect to R_{ind}?

This problem is largely open in this generality. The only general result known is

Proposition 6 (Kitajima 2007) *If $\mathcal{L}$ is an atomless, complete, orthomodular lattice and ϕ is a faithful state then $(\mathcal{L}, \phi)$ is causally closed with respect to every pair of logically independent sublattices.*

The above result is the quantum counterpart of Proposition 5; and the key fact that it rests on is that if $\mathcal{L}$ is an atomless lattice and ϕ is a faithful state on $\mathcal{L}$ then $(\mathcal{L}, \phi)$ is atomless as a measure space in the sense that for any $0 \neq A \in \mathcal{L}$, and for any real number $0 \neq r < p(A)$ there exists $B \leq A$, $B \in \mathcal{L}$ such that $p(B) = r$. This latter fact was proved by Rédei & Summers (2002) (see also Rédei & Summers 2007) for the specific quantum probability space $(\mathcal{P}(\mathcal{N}), \phi)$ where $\mathcal{N}$ is a type III von Neumann algebra and ϕ is a faithful normal state on $\mathcal{N}$, and Kitajima (2007) showed that the proof of can be carried over from the von Neumann algebra framework to more general non-classical probability spaces. It is known that the projection lattices of type II von Neumann algebras are also atomless; so one has as a specific case of Proposition 6 the following

Proposition 7. *Let $(\mathcal{P}(\mathcal{N}), \phi)$ be a quantum probability space with $\mathcal{N}$ as a type III or type II von Neumann algebra and ϕ as a faithful normal state on $\mathcal{N}$. Then $(\mathcal{P}(\mathcal{N}), \phi)$ is causally closed with respect to every pair of logically independent sublattices.*

Note that the lattice $\mathcal{P}(\mathcal{H})$ of all projections on a Hilbert space $\mathcal{H}$ is not atomless (it is atomic) irrespective of the dimensionality of the Hilbert space $\mathcal{H}$ (Rédei, 1998); moreover, the quantum probability spaces $(\mathcal{H}, \mathcal{P}(\mathcal{H}), \phi)$ are *not* atomless in a measure theoretic sense; consequently, Propositions 6 and 7 do not say anything about the causal closedness of the quantum probability space $(\mathcal{H}, \mathcal{P}(\mathcal{H}), \phi)$ and it is not known under what conditions such quantum probability spaces are causally closed (with respect to some R_{ind}).

Just like in the classical case, the notion of a (Reichenbachian) common cause system also can be formulated in a non-classical probability space, and one can define naturally a more general notion of causal N-closedness of a non-classical probability space: The set $\{C_i, \ i \in J\}$ of elements (J being an index set, $C_i \in \mathcal{L}$) is called a partition in $\mathcal{L}$ if $\vee_i C_i = I$ and C_i and C_j are orthogonal whenever $i \neq j$; i.e. $C_i \leq C_j^{\perp}$ for $i \neq j$.

Definition 25.11. A partition $\{C_i, \ i \in J\}$ is a (Reichenbachian) common cause system for the correlation (25.14) between compatible elements A and B if C_i is compatible with both A and B for every $i \in J$ and the following conditions (analogous to (25.12)–(25.13)) hold

$$\phi(A \wedge B|C_i) = \phi(A|C_i)\phi(B|C_i) \qquad \text{for all } i \in J \tag{25.19}$$

$$[\phi(A|C_i) - \phi(A|C_j)][\phi(B|C_i) - \phi(B|C_j)] > 0 \ \ (i \neq j) \tag{25.20}$$

The cardinality of the index set J is called the size of the common cause system.

Definition 25.12. The probability space $(\mathcal{L}, \phi)$ is called causally N-closed (with respect to some causal independence relation R_{ind}) if for any correlation between elements that stand in the causal independence relation there exists in $(\mathcal{L}, \phi)$ a Reichenbachian common cause system of size N.

There are a number of open problems in connection with Reichenbachian common cause systems in non-classical probability spaces and causal N-closedness of such probability theories:

Problem 7. *Given a correlation in a general probability space $(\mathcal{L}, \phi)$ that does not have a common cause system of a given size $N > 2$ of the correlation, does there exist an extension $(\mathcal{L}', \phi')$ of $(\mathcal{L}, \phi)$ such that there exists a Reichenbachian common cause system of size N of the correlation in the extension $(\mathcal{L}', \phi')$?*

We conjecture a positive answer to the above question.

Problem 8.

1. *Do there exist non-classical probability spaces that are causally N-closed for some fixed N (with respect to some nontrivial R_{ind})?*
2. *Do there exist non-classical probability spaces that are causally N-closed for* every *N (with respect to some decent R_{ind})?*
3. *Do there exist non-classical probability spaces that are causally closed (with respect to some non-trivial causal independence relation R_{ind}) in such a way that every correlation in the space has a common cause system of countably infinite size?*

These questions have not been investigated.

25.5 Closing comments on causal closedness

Further generalization can be achieved by treating the specific form the correlation measure Corr takes as a variable of the notion of the common cause. By allowing Corr to measure correlation between pairs of ordered partitions it becomes possible to handle the case of common cause-type explanations of correlating *variables*, not just that of events. This allows a more detailed analysis of causal closedness and of falsification attempts against the Common Cause Principle. It can be shown that Reichenbachian common cause systems are special cases of the resulting notion of a *generalized Reichenbachian common cause*. By imposing mild conditions on Corr extension theorems analogous to Proposition 4 can be proven. However the question of common cause closedness of general probability spaces with respect to generalized Reichenbachian common causes is still open. For further details the Reader is referred to Gyenis & Redei (2008).

One can strengthen Reichenbach's notion of common cause by requiring the common cause to satisfy some additional conditions. The additional conditions can be motivated by physics: after all, the probability measure spaces in terms of which the concept of common cause is formulated are not just abstract mathematical entities in physics but physically interpreted structures. Being organic parts of specific physical theories, these measure spaces offer means to express a possibly large variety of physical facts and principles. Two of such important principles are *locality* and *causality*. Both locality and causality are rich and many-layered concepts and there is no unique way of expressing them in terms of probability measure spaces. But it can happen that a physical theory entails both some additional conditions as necessary for the common cause C of a correlation between A and B to be 'local' and a causal dependence relation between random events. In such a situation the problem of causal closedness should be reformulated by taking into account these further restrictions.

This happens in local, algebraic, relativistic quantum field theory (AQFT). The theory predicts correlations between localized, causally independent (specelike separated) observables (Summers & Werner 1985, 1987a, 1987b, 1987c, 1988; Summers 1990a, 1990b), and the common causes of these correlations have to be localized (Rédei 1997). It turns out that there is no unique way of defining locality of the common cause and, consequently, causal closedness of AQFT can also be specified in different ways (Rédei & Summers 2007): strong, weak and right ('desirable') localizability of common causes lead to the concepts of 'strong', 'weak' and 'desirable' causal closedness of AQFT. To decide which of these causal closedness hold for AQFT is a difficult matter. While it is easy to see that strong causal closedness is violated in AQFT (Rédei and Summers 2007), and it could be shown that AQFT is weakly causally closed (Rédei 2002) (see also the review Rédei & Summers 2002), it remains an open problem whether AQFT is causally closed in the most desirable sense.

Acknowledgement

This work was supported in part by the Hungarian Scientific Research Found (OTKA), contract number: K68043. The authors would like to thank two anonymous referees for their valuable suggestions.

References

F. Arntzenius. Reichenbach's common cause principle. 2005. Available at http://plato.stanford.edu/entries/physics-Rpcc.

B. Gyenis and M. Rédei. When can statistical theories be causally closed? *Foundations of Physics*, 34: 1285–1303, 2004.

B. Gyenis and M. Redei. Causal completeness of general probability theories. In M. Suarez, editor, *Probabilities, Causes and Propensities in Physics*, Synthese Library, vol. 347 pages 157–171 Springer, 2008. forthcoming.

G. Hofer-Szabó and M. Rédei. Reichenbachian common cause systems. *International Journal of Theoretical Physics*, 43: 1819–1826, 2004.

G. Hofer-Szabó and M. Rédei. Reichenbachian common cause systems of arbitrary finite size exist. *Foundations of Physics Letters*, 35: 745–746, 2006.

G. Hofer-Szabó, M. Rédei, and L.E. Szabó. On Reichenbach's Common Cause Principle and Reichenbach's notion of common cause. *The British Journal for the Philosophy of Science*, 50: 377–398, 1999.

R.V. Kadison and J.R. Ringrose. *Fundamentals of the Theory of Operator Algebras*, volumes I. and II. Academic Press, Orlando, 1986.

Yuichiro Kitajima. Reichenbach's Common Cause in an atomless and complete orthomodular lattice. *International Journal of Theoretical Physics*, 46: 511–519, 2008.

M. Rédei. Reichenbach's Common Cause Principle and quantum field theory. *Foundations of Physics*, 27: 1309–1321, 1997.

M. Rédei. *Quantum Logic in Algebraic Approach*, volume 91 of *Fundamental Theories of Physics*. Kluwer Academic Publisher, 1998.

M. Rédei. Reichenbach's common cause principle and quantum correlations. In T. Placek and J. Butterfield, editors, *Modality, Probability and Bell's Theorems*, volume 64 of *NATO Science Series, II.*, pages 259–270. Kluwer Academic Publishers, Dordrecht, Boston, London, 2002.

M. Rédei and S.J. Summers. Local primitive causality and the common cause principle in quantum field theory. *Foundations of Physics*, 32: 335–355, 2002.

M. Rédei and S.J. Summers. Remarks on causality in relativistic quantum field theory. *International Journal of Theoretical Physics*, 46: 2053–2062, 2007.

H. Reichenbach. *The Direction of Time*. University of California Press, Los Angeles, 1956.

S.J. Summers. Bell's inequalities and quantum field theory. In *Quantum Probability and Applications V.*, volume 1441 of *Lecture Notes in Mathematics*, pages 393–413. Springer, 1990a.

S.J. Summers. On the independence of local algebras in quantum field theory. *Reviews in Mathematical Physics*, 2: 201–247, 1990b.

S.J. Summers and R. Werner. The vacuum violates Bell's inequalities. *Physics Letters A*, 110: 257–279, 1985.

S.J. Summers and R. Werner. Bell's inequalities and quantum field theory, I. General setting. *Journal of Mathematical Physics*, 28: 2440–2447, 1987a.

S.J. Summers and R. Werner. Bell's inequalities and quantum field theory, II. Bell's inequalities are maximally violated in the vacuum. *Journal of Mathematical Physics*, 28: 2448–2456, 1987b.

S.J. Summers and R. Werner. Maximal violation of Bell's inequalities is generic in quantum field theory. *Communications in Mathematical Physics,* 110: 247–259, 1987c.

S.J. Summers and R. Werner. Maximal violation of Bell's inequalities for algebras of observables in tangent spacetime regions. *Annales de l'Institut Henri Poincaré – Physique théorique,* 49: 215–243, 1988.

M. Takesaki. *Theory of Operator Algebras,* volume I. Springer Verlag, New York, 1979.

PART VI

Computer science, probability, and statistics

·26·

Causality Workbench

Isabelle Guyon, Constantin Aliferis, Gregory Cooper, André Elisseeff, Jean-Philippe Pellet, Peter Spirtes, and Alexander Statnikov

Abstract

The Causality Workbench project provides an environment to test causal discovery algorithms. Via a web portal (http://clopinet.com/causality), we provide a number of resources, including a repository of datasets, models, and software packages, and a virtual laboratory allowing users to benchmark causal discovery algorithms by performing virtual experiments to study artificial causal systems. We regularly organize competitions.

26.1 Introduction

Uncovering cause–effect relationships is central in many aspects of our everyday life: what affects our health, the economy, climate changes, world conflicts, and which actions have beneficial effects? Causal discovery is a problem of fundamental and practical interest in many areas of science and technology, including biology, medicine and pharmacology, epidemiology, climatology, economy, sociology, psychology, law enforcement, neurosciences, manufacturing, computer security, and marketing. The need for assisting policy making while reducing the cost of experimentation and the availability of massive amounts of 'observational' data prompted the proliferation of proposed causal discovery techniques, mostly evaluated on toy problems. Each scientific discipline has its favourite approach, e.g. Bayesian networks in biology (Friedman *et al.* 2000) and structural equation modelling in social sciences (Kaplan 2000), not necessarily reflecting better match of techniques to domains, but rather historical tradition. Hence, standard benchmarks are needed to foster scientific progress, but the design of good benchmarks, not biased in favour of a particular model or approach, is not trivial. Difficulties include finding large representative datasets with known ground truth and devising methods for evaluating the validity of causal relationships in datasets where the causal structure is unknown. This chapter describes our contribution to that endeavour.

One important goal of causal modelling is to predict the consequences of given *actions*, also called *interventions, manipulations*, or *experiments*. This is fundamentally different from the classical machine learning, statistics, or data mining setting, which focuses on making predictions from observations. Observations imply no manipulation on the system under study whereas actions introduce a disruption in the natural functioning of the system. In the medical domain, this is the distinction made between 'diagnosis' (prediction from observations) and 'treatment' (intervention). For instance, smoking and coughing are both predictive of respiratory disease and helpful for diagnosis purpose. However, acting on the cause (smoking) can change your health status, but not acting on the symptom or consequence (coughing). Thus it is extremely important to distinguish between causes and symptoms to predict the consequences of actions like predicting the effect of forbidding smoking in public places. Part of our effort is dedicated to bring such problems to the awareness of researchers in machine learning who have mostly focused recently on i.i.d. data (identically and independently distributed data), neglecting (with some notable exceptions, such as, Quiõnero-Candela *et al.* 2009) problems of distribution shifts between training and test set and the study of the mechanisms underlying the generation of the data. To that end, we devised datasets and tasks, which closely resemble machine learning problems and extend the problem of feature or variable selection to that of finding variables influencing a target variable.

Despite recent progresses in causal discovery algorithms, it is fair to say that causal models in science and engineering are still not widespread. This points to the need of illustrating the power of such methods on a variety of applications and addressing problems of efficiency. Part of our benchmarking effort is dedicated to collecting problems from diverse application domains. To address problems of efficiency, we are making available datasets with a large number of variables. We are planning future events in which costs will be associated with acquiring observational and experimental training data.

We have successfully channelled the effort of dozens of researchers to solve new problems of scientific and practical interest and identified effective methods. However, competition without collaboration is sterile. Recently, we have started introducing new dimensions to our effort of research coordination: stimulating creativity, collaborations, and data exchange. We are organizing regular teleconference seminars. We have created a data repository for the Causality Workbench already populated by 15 datasets. All the resources, which are the product of our effort, are freely available on the Internet at http://clopinet.com/causality.

26.2 What are 'causal problems'?

Causal discovery is a multi-faceted problem. The definition of causality itself has eluded philosophers of science for centuries, even though the notion of causality is at the core of the scientific endeavour and also a universally accepted and intuitive notion of everyday life. But, the lack of broadly acceptable definitions of causality has not prevented the development of successful and mature mathematical and algorithmic frameworks for inducing causal relationships.

The type of causal relationships under consideration have often been modeled as Bayesian causal networks or structural equation models (SEM) (Pearl 2000; Spirtes *et al.* 2000; Neapolitan 2003). In the graphical representation of such models, an arrow between two variables $A \rightarrow B$ indicates the direction of a causal relationship: A causes B. A node in of the graph, labelled with a particular variable X, represents a mechanism to evaluate the value of X given the parent node variable values. For Bayesian networks, such evaluation is carried out by a conditional probability distribution $P(X|\,Parents(X))$ while for structural equation models it is carried out by a function of the parent variables, plus some noise. Learning a causal graph can be thought of as a model selection problem: Alternative graph architectures are considered and a selection is performed, either by ranking the architectures with a global score (e.g. a marginal likelihood, or a penalty-based cost function), or by retaining only graphs, which fulfil a number of constraints such as dependencies or independencies between subsets of variables. Bayesian networks and SEMs provide a convenient language to talk about the type of problem we are interested in, but we made an effort to design tasks, which do not preclude of any particular model. Our objective is not to reduce causality to a simple or convenient definition or to a family of models, but rather to define tasks with clear objectives and give ourselves means of assessing how well these objectives are reached.

In designing our first benchmark tasks we have focused on some specific aspects of causal discovery:

Causality between random variables. We have so far addressed mostly causal relationships between random variables, as opposed to causal relationships between events, or objects.

Multivariate problems. Many early efforts in causal studies have concentrated on the study of cause–effect relationships between two or just a few variables. The availability of large observational datasets with thousands of recorded variables (in genomic studies with microarray data, in pharmacology with high throughput screening, in marketing with logs of internet customers, etc.) has drawn our attention to multivariate problems in which an array of eventually weak causes might influence an outcome of interest, called 'target'.

Time dependency. Our everyday-life concept of causality is very much linked to time dependencies (causes precede their effects). However, many machine learning problem are concerned with stationary systems or 'cross-sectional studies', which are studies where many samples are drawn at a given point in time. Thus, sometimes the reference to time is replaced by the notion of 'causal ordering'. Causal ordering can be understood as fixing a particular time scale and considering only causes happening at time t and effects happening at time $t + \delta t$, where δt can be made as small as we want. In practice, this means that the samples in our various training and test sets are drawn independently, according to a given distribution, which changes only between training and test set versions.[1] We are offering tasks with or without time dependencies.

Learning from observational or experimental data. We call *observational data*, data collected from a system let to evolve according to its own dynamics. In contract, *experimental data* is obtained as a result of interventions on the system of interest by an external agent who disrupts the system by imposing values to certain variables. Generally, experimenting is the only way to ascertain causal relationships. However, in many domains, experimenting is difficult and costly compared to collecting observational data. Hence, we have investigated settings in which only observational data are available for training. The tasks we collected also include settings in which both observational and experimental data are available.

We have so far mostly addressed two tasks of interest:

Predicting the consequences of manipulations. In one challenge we organized, our data included training samples drawn from a 'natural' pre-manipulation distribution and test data drawn from various post-manipulation distributions (in which the values of a subset of variables has been set to given values by an external agent, bypassing the natural functioning of the system). The objective was to predict withheld values of a target variable.

Discovering causal structures. Causal graphs (e.g. Bayesian networks or structural equation models) are powerful to represent mechanisms at a level sufficient to reason and plan for future actions. A common exercise is to investigate whether the structure of such models can be reconstructed from artificial data generated by the models, in an effort to reassure ourselves that structures generated from real data may be meaningful.

The first task has a clear objective and it does not preclude of any particular modelling technique. In particular, it is not required to produce a causal

[1] When manipulations are performed, we must specify whether we sample from the distribution before or after the effects of the manipulation have propagated. Here we assume that we sample after the effects have propagated.

graph. Operational definitions of causality (Glymour and Cooper 1999) use the notion of manipulation to evidence cause–effect relationships. Hence, predicting the consequences of manipulations is a 'causal question' that can serve to evaluate causal models against non-causal models. The second task is more explicitly 'causal' but it lacks of a generic mathematical statement of the objective, free of modelling assumptions. Evaluating causal structures poses major challenges (see Section 26.5), and such evaluations may be biased in favour of the class of models that generated the data.

There are many other causal questions, which we will progressively address:

- Determine what **manipulations** are needed to reach a desired system state with maximum probability (e.g. select variables and propose values to achieve a certain value of a response/target variable).
- Propose system queries to acquire training data, i.e. **design experiments**, with perhaps an associated cost per variable and per sample and perhaps with constraints on variables, which cannot be controllable.
- Determine a **local causal region** around a target variable (causal adjacency).
- Determine the **source cause(s)** for a target variable.
- Predict the existence of unmeasured variables (not part of the set of variables provided in the data), which are potential **confounders** (are common causes of an observed variable and the target).
- Predict which variables called 'relevant' by feature selection algorithms are potentially causally irrelevant because their correlation to the target is due to an **experimental artifact** (e.g. sampling bias or systematic error).
- Determine causal direction in time series data in which one variable is causing the other.
- **Estimate counterfactuals**. What would have occurred had a given variable taken a value that is different then the one it actually took? We cannot rewind time and find out. But, our evaluation platform using artificial systems will allow us to evaluate algorithms who can deal with counterfactuals.

26.3 Lessons from past challenges

Before entering into the technical details of challenge design, we wish to briefly outline some important results and lessons learned from past challenges:

- **Causation and prediction: a negative result.** In our first challenge we wanted to demonstrate that identifying causal relationships helped building better models to predict the consequences of interventions. Indeed, the knowledge of the *true* causal relationships allowed the organizers to build baseline predictive models performing considerably better than regular non-causal machine learning models. In contrast, the challenge participants had to face the problem of uncovering the causal relationships unknown to them, from observational data only, and built predictive models with such data. Although on average over all participants and all datasets, *better knowledge of the causal structure* correlated with *better prediction performance*, the variance of the results was so large that one could not claim that trying to unravel the causal structure provided an advantage over building a predictive model ignoring causal relationships.

- **Cause–effect pairs: a positive result.** In our second challenge, one task proposed by a participant drew a lot of attention: the cause–effect pair task. The problem was to try to determine in pairs of variables (of known causal relationships), which one was the cause of the other. This problem is hard for a lot of algorithms, which rely on the result of conditional independence tests of three or more variables. Yet the winners of the challenge succeeded in unraveling 8/8 correct causal directions.

From a challenge design perspective, we discovered that there is a lot of value in offering to the participants the possibility of contributing problems, which can evidence the power of causal discovery algorithms (Guyon *et al.* 2008b). From an algorithmic perspective, we stubbled on the multivariate problem. Moving from a multivariate variable selection task in an i.i.d. setting (Guyon *et al.* 2006) to a similar task in a non-i.i.d. setting (Guyon *et al.* 2008a) magnified the problem of overfitting, familiar to machine learning scientists: in the i.i.d. setting, multivariate algorithms struggled to outperform univariate algorithms (selecting variables for their individual predictive power); in the *causation and prediction challenge*, 'causal' variable selection algorithms struggled to outperform non-causal algorithms. From a methodology perspective, we realized that learning causal relationships reliably from observational data only may not be realistic. Experiments are needed to firm up hypotheses made by analysing observational data. This is particularly critical in a multivariate setting where errors cumulate and propagate. Finally, from a practical point of view, we learnt that causal discovery algorithms might serve better the Industry by ranking factors that might significantly influence a given outcome than by unraveling in details the causal structure of a web of interrelated variables.

These findings had a direct impact on the design of new benchmarks described in the remainder of this chapter.

26.4 Important problems and good benchmarks

Uncovering cause–effect relationships is of great practical importance in many domains, however, many real world problems are not suitable benchmarks. A good benchmark problem should be a problem featuring either:

1. A **data generative system** available to generate data as needed (for example an electrical circuit).
2. A **low-cost experimental setup** to identify the system (for example a temperature regulation problem).
3. **Data available from a large study**, including many examples drawn from the 'natural distribution' and from relevant 'post-manipulation distributions' (for example a large medical trial).

In addition, the task should be such that guessing causal relationships at random or ignoring them altogether yields performances significantly worse than accurately uncovering them. Hence a good benchmark requires uncovering many causal relationships. Problems of practical importance not suitable for benchmarking include:

- **Non-reproducible single events**, e.g. what caused a candidate to lose an election, what caused an economic or a social crisis, what caused a person to commit suicide?
- **Unrealistic experimental setups**. For instance, we cannot manipulate sunspots to verify their alleged causal influence on the Earth's climate. For other problems experimentation is too limited or scarce. This is the case of economic policies like tax cuts or subsidies.
- **Unobservable or uncontrollable systems**. Problems with too few directly observable and actionable variables to establish causation with confidence, for example, non-invasive neuroscience.
- **Problems with too few variables**. If only a few causal relationships must be uncovered, it is easy to win by chance (e.g. determining whether smoking causes lung cancer is important, but you have 50% chance of being right); so, unless many instances of the same or similar problems are available, problems with few variables are unfit for benchmarking.

We are making an effort to collect in the workbench repository a variety of tasks using real data sets (see Tables 26.1 and 26.2). But, when practical problems of interest do not fulfil all the requirements of good benchmarks, because of one of the reasons above mentioned, we complement these resources with designed semi-artificial datasets. We have adopted two strategies:

Table 26.1 Atemporal datasets. 'TP' is the data type, 'NP' the number of participants who returned results and 'V' the number of views as of December 2008. The semi-artificial datasets are generally 're-simulated' data, i.e. data obtained from simulators of real tasks, usually trained with real data. Two datasets of LOCANET are made of real data augmented with artificial 'probe' variables (SIDO and CINA). N is the number of variables and P is the number of examples (in training data; some datasets have test data too)

Name (TP; NP; V)	Size	Description	Objective
CEP (Real; 5; 218)	$P = 8$ pairs of var. $N = 2$ variables.	**Cause Effect Pairs**. Pairs of real variables with known causal relationships.	Find the causal direction.
CYTO (Real; 2; 394)	$P \simeq 800$ samples per experimental condition* 9 conditions. $N = 11$ proteins.	**Causal Protein-Signalling Networks in human T cells.** Protein activity monitored by flow cytometry. 'Heavy-handed' manipulations are performed using chemical activators or inhibitors.	Learn the architecture of the protein signalling network.
LOCANET (Semi-artificial; 10; 558)	REGED & MARTI: $P = 500$ patients; $N = 999$ genes + target (disease). CINA: $P = 16,033$ persons; $N = 132$ attributes + target (earnings). SIDO: $P = 12,678$ drugs; $N = 4932$ descriptors + target (activity).	**LOcal CAusal NETwork.** Four datasets: REGED and MARTI (genomics), CINA (marketing), and SIDO (drug discovery). The datasets also include large test sets that were used in the 'causation and prediction challenge' (Guyon *et al.* 2008a).	Find the local causal structure around a target variable.
SECOM (Real; NA; 59)	$P = 1567$ wafers. $N = 591$ QC measurements + 1 binary target (pass/fail) and 1 date of processing	**Semiconductor manufacturing.** Production entities (wafers) are associated with quality control (QC) measurements on a fabrication line. The labels represent a pass/fail yield in line testing (classification problem).	Predict pass/fail and identify predictive features.
TIED (Artificial; 1; 330)	$P = 750$ training ex. $N = 1000$ variables (including target).	**Target Information Equivalent Dataset.** A Bayesian network with 72 equivalent Markov boundaries of the target variable.	Find all Markov boundaries.

Table 26.2 Time dependent datasets. 'TP' is the data type, 'NP' the number of participants who returned results and 'V' the number of views as of December 2008. The semi-artificial datasets are obtained from simulators of real tasks. N is the number of variables, T is the number of time samples (not necessarily evenly spaced) and R the number of simulations with different initial states or conditions

Name (TP; NP; V)	Size	Description	Objective
MIDS (Artificial; NA; 65)	$T = 12$ sampled values in time (unevenly spaced); $R = 10000$ simulations. $N = 9$ variables.	**Mixed Dynamic Systems.** Simulated time-series based on linear Gaussian models with no latent common causes, but with multiple dynamic processes.	Use the training data to build a model able to predict the effects of manipulations on the system in test data.
NOISE (Real + artificial; NA; 43)	**Artificial:** $T = 6000$ time points; $R = 1000$ simul.; $N = 2$ var. **Real:** $R = 10$ subjects. $T \simeq 200{,}000$ points sampled at 256Hz. $N = 19$ channels.	**Real and simulated EEG data.** Learning causal relationships using time series when noise is corrupting data causing the classical Granger causality method to fail.	**Artificial task:** find the causal dir. in pairs of var. **Real task:** Find which brain region influences which other one.
PROMO (Semi-artificial; 3; 570)	$T = 365*3$ days; $R = 1$ simulation; $N = 1000$ promotions + 100 products.	**Simulated marketing task.** Daily values of 1000 promotions and 100 product sales for three years incorporating seasonal effects.	Predict a 1000x100 Boolean matrix of causal influences of promotions on product sales.
SEFTI (Semi-artificial; NA; 35)	$R = 4000$ manufacturing lots; $T = 300$ async. operations (pair of values {one of N=25 tool IDs, date of proc.}) + cont. target (circuit perf. for each lot).	**Semiconductor manufacturing.** Each wafer undergoes 300 steps each involving one of 25 tools. A regression problem for quality control of end-of-line circuit performance.	Find the tools that are guilty of performance degradation and eventual interactions and influence of time.
SIGNET (Semi-artif.; 2; 415)	$T = 21$ asynchronous state updates; $R = 300$ pseudodynamic simulations; $N = 43$ rules.	**Abscisic Acid Signalling Network.** Model inspired by a true biological signalling network.	Determine the set of 43 Boolean rules that describe the network.

- **Re-simulated data:** We train a causal model (e.g. a causal Bayesian network) with real data. The model is then used to generate artificial training and test data for the challenge. Truth values of causal relationships are known for the data generating model and used for scoring causal discovery results.[2]
- **Real data with probes:** We use a dataset of real samples. Some of the variables may be causally related to the target and some may be predictive but non-causal. The nature of the causal relationships of the variables to the target is unknown (although domain knowledge may allow us to validate discoveries to some extent). We add a number of distractor variables called 'probes', which are generated by an artificial stochastic process, including explicit functions of some of the real variables, other artificial variables, and/or the target. All probes are non-causes of the target, some are completely unrelated to the target. The identity of the probes in concealed.

Both strategies have advantages and disadvantages. The first approach allows us to generate as much data as desired and to simulate experiments. In addition, the structure of the data generative model is known. Hence a wide range of causal questions can be posed and evaluated. The disadvantages are that the model used to generate data may not be realistic and that the benchmark may be biased in favor of the family of models to which the data generative model belongs. The second approach has the advantage of including real data, with all the potential complexity of distributions generated by a real process. The disadvantages are that only the probes can be manipulated (not the real variables), that truth values of causal relationships are known only for the probes, and that we are limited to the available sample size. Furthermore, even though the distribution of the probes may be designed to mimic that of the real variables, the data generative process of the probes is artificial and not necessarily realistic.

26.5 Methods of evaluation

The best established way of assessing causal theories is to carry out **randomized controlled experiments** to test hypothetical causal relationships. In the 1930s, Fisher laid the mathematical foundations for experimental design (Fischer 1953). The central idea is the systematic use of randomization to avoid confounding, that is to avoid confusing mere correlation due to the

[2] Are the models realistic? Does it matter? Artificial problems might be arbitrarily too hard, too easy, or very different from real problems. We propose realistic problems and use the literature and domain knowledge to verify the model structure and parameters. But artificial data will never be as good as real data.

existence of a common cause with real causation. For example, in the medical domain, a causal relationships $C \to E$ between a treatment (a cause) C and an effect E may be tested in a **Randomized Controlled Trial** (RCT). Variable C may be the choice of one of two available treatments for a patient with lung cancer and E may represent five-year survival. If we randomly assign a large number of patients to the two treatments by flipping a fair coin and observe that the probability distribution for five-year survival differs between the two treatment groups, it may be concluded that the choice of treatment causally determines survival in patients with lung cancer. The double blind placebo-controlled Randomized Controlled Trial, where allocations are randomized and neither patient nor doctor knows which treatment has been assigned, is now standard in clinical trials.

Systems to stratify evidence by quality have been developed, such as those by the US Preventive Services Task Force and by the Oxford Centre for Evidence-based Medicine. Both rank evidence about the effectiveness of treatments or screening in a hierarchical way: (1) **Experimental evidence**: Evidence obtained from at least one properly designed Randomized Controlled Trial. (2) **Statistical analysis of observational data**: Evidence obtained from well designed retrospective studies (observational data) from more than one research group. (3) **Expert opinions**: Opinions of respected authorities, based on clinical experience, descriptive studies, or reports of expert committees.

An ideal benchmark of causal discovery methods (uncovering causal relationships from observational data) would compare predictions obtained by applying algorithms to large observational databases (level 2 evidence) with the outcome of well designed experimental studies (level 1 evidence). Because of the rarity of large observational data sets paired with appropriate randomized experiments, to our knowledge no such comparisons have been made. The Causality Workbench project is exploring alternative methods of evaluation:

- Comparing causal relationships inferred from observational data (level 2 evidence) to 'ground truth' established from human expertise (level 3 evidence).
- Using non-parametric statistical methods involving artificial variables (probes) added to real variables.
- Using realistic simulated systems to generate observational data and perform virtual experiments.

Such methods of evaluation must come with metrics to quantitatively assess causal models. The metrics evidently depend on the tasks. For instance, in prediction tasks, we have used the *prediction error* on manipulated test data of a given target variable. In tasks involving structure discovery, we have used the *precision* and *recall* of the set causes of a given target variable.[3] In artificial data any variable may be used for evaluation while in real data with probes, only the probes may be manipulated and/or used to evaluate the causal structure.

In the following section, we give several examples.

26.6 Tasks proposed

We have made available on our website a repository of causal problems. We summarize in Tables 26.1 and 26.2 the tasks proposed. These tasks span a wide range of domains (bioinformatics, medicine, pharmacology, marketing, manufacturing) and difficulties:

- Several assumptions commonly made in causal discovery are violated, including 'causal sufficiency',[4] 'faithfulness',[5] 'linearity', and 'Gaussianity'.
- Relatively small training sets are provided, making it difficult to infer conditional independencies and learning distributions.
- Large numbers of variables are provided, a particular hurdle for some causal discovery algorithms that do not scale up.
- Some problems are time independent, others involve time series.

To evaluate prediction under manipulation, some tasks use metrics such as the Area Under the ROC Curve (AUC),[6] commonly used to assess classification problems (e.g. the LOCANET task), others used the mean square error, commonly used to assess regression problems (e.g. the MIDS task).

To evaluate structure discovery, some tasks use 'ground truth' established from human expertise.[7] For instance the Cause Effect Pairs (CEP) dataset includes the pairs $Altitude \rightarrow Temperature$, $Longitude \rightarrow Precipitation$ in German cities, and $Age \rightarrow Length$ for the snail Abalone. In biology, regulatory pathways obtained by curating thousands of peer reviewed papers

[3] Precision is the ratio of correctly retrieved causes over the total number of variables 'called' causes of the target, i.e. TruePositive/(TruePositive+FalsePositive). Recall is the ratio of correctly retrieved causes over the total number of real causes of the target, i.e. TruePositive/(TruePositive+FalseNegative).

[4] 'Causal sufficiency' roughly means that there are no unobserved common causes of the observed variables.

[5] 'Faithfulness' roughly means that every conditional independence relation that holds in the population is entailed to hold for all values of the free parameters.

[6] The AUC is the Area Under the ROC Curve plotting sensitivity *vs.* (1− specificity) when the threshold is varied on the classifier discriminant value. We call 'sensitivity' the error rate of the positive class and 'specificity' the error rate of the negative class.

[7] In the pattern recognition jargon, 'ground truth' refers to verified information obtained by scouting the terrain *on the ground* as opposed to information collected from far away observations, like satellite images.

constitute reference human knowledge for discovery studies performed with genomic and proteomic observational data (Kanehisa *et al.* 2006). CYTO is an example of dataset using this type of 'ground truth'. It is to be noted however that due to many inconsistencies in the biological literature there is a lot of uncertainty in the reference regulatory pathways. Using artificially generated data is another way of having access to an established ground truth (i.e. the structure of the data generative model). For example, the dataset TIED is purely artificial and was designed to illustrate a particular technical difficulty. The datasets REGED and MARTI were build from a simulator of a gene regulatory network influencing lung cancer, trained with real data. The dataset SIGNET was simulated from a set of Boolean rules representing knowledge of a plant regulatory pathway gathered from several published papers. Yet another way of assessing the validity of a proposed set of causes of a target variable is to compute the fraction of probes (all non-causes of the target) in that subset. Large fractions of probes cast doubt to the validity of the proposed causes. The probability of getting a number of probes smaller than a certain threshold can serve as a basis for a statistical test. Such methods were used for the datasets SIDO and CINA.

26.7 Events organized

We have organized two events so far: the 'causation and prediction challenge', with a workshop at WCCI 2008 (Guyon *et al.* 2008a) and the 'causality pot-luck challenge', with a workshop at NIPS 2008 (Guyon *et al.* 2008b).

Causation and prediction challenge
This first event achieved a number of goals: familiarizing many new researchers and practitioners with causal discovery problems and existing tools to address them; pointing out the limitations of current methods on some particular difficulties; fostering the development of new algorithms.

The setting of the challenge purposely resembled a classical machine learning competition (with a training set and a test set, with omitted labels) to encourage the participation of data mining and machine learning researchers. We adopted such a *predictive modelling* perspective in an effort of distancing ourselves from the interpretation of causal models as data generative models. The goal of this first challenge was not to reverse engineer a data generative process, it was to make accurate predictions of a target variable. To sharpen this distinction, we made available only a limited amount of training data, such that the learner may not necessarily be able to reliably determine all conditional dependencies and independencies. Hence, modelling strategies making radical simplifying assumptions might do better than strategies trying to be faithful to the data generative process, because of the well-known fit *vs.* robustness (or bias *vs.* variance) tradeoff. The participants were asked to return *prediction scores* or **discriminant values** v for the target variable on test examples, and a **list of features** used for computing the prediction scores, sorted in order of decreasing predictive power, or unsorted. The classification decision is made by setting a threshold θ on the discriminant value v: predict the positive class if $v > \theta$ and the negative class otherwise. The participants were ranked according to the area under the ROC curve (AUC) computed for test examples (called Tscore). If results were provided for nested subsets of features, the best Tscore was retained. We also computed other statistics, which were not used to rank participants, but used in the analysis of the results. Those included the number of features used by the participants, and a statistic assessing the quality of causal discovery in the feature set selected (called Fscore).

We proposed four datasets from various domains (genomics, pharmacology, and marketing), two of which came from re-simulated data (REGED and MARTI), and two were real data with 'probes' (CINA and SIDO). Each task had three test sets, with increasing levels of difficulty. The first one was identically distributed as the training set. The two other test sets simulated manipulations by external agents, and thus were not distributed like the training set. In this way we illustrated the relationships between causation and prediction under manipulations and investigated whether causal models using 'causally relevant' features would perform better than regular statistical models on manipulated test sets.

Several algorithms demonstrated the effectiveness of discovering causal relationships, as indicated by a large Fscore. On average over all datasets and tasks, the Fscore correlated significantly with the Tscore, confirming the link between causation and prediction. As anticipated, non-causal feature selection methods did well on the first type of datasets (training and test data identically distributed). For the other two types of datasets (test data manipulated) designed to make non-causal methods fail, some causal method obtained good results, but many failed. Non-causal methods performed better than anticipated. The results indicate that informative causal prediction from observational data is possible, but it remains challenging. See Guyon *et al.* (2008a) for a detailed analysis.

Causality challenge pot-luck

The first challenge performed a quantitative evaluation of algorithms on four problems from various application domains. While quantitative evaluations are needed, there are so many constraints placed on their design that they rule out a lot of interesting problems. The second challenge aimed at enlarging the scope of causal discovery algorithm evaluation by encouraging creativity and diversity. We invited members of the community to submit their own problems and/or solve problems proposed by others. While we expected that the evaluation may not be as quantitative because of variance in data quality and number of entries on each problem, we achieved other important goals: the identification of new problems of interest, the formulation of new and interesting causal questions, and the development of new methods of evaluation.

We bootstrapped the competition by proposing five tasks (CYTO, LOCANET, PROMO, SIGNET and TIED) and the participants then added their own contributions (see Tables 26.1 and 26.2). The participant-contributed task Cause Effect Pairs (CEP) was particularly popular. Innovative solutions were proposed to:

- reverse engineering Boolean networks (the SIGNET task),
- finding local causal relationships around a target variable (the LOCANET task),
- finding all possible Markov blankets, when there is a large number of possible solutions (the TIED task),
- learning a causal network from 'heavy handed' manipulations affecting several variables simultaneously (the CYTO task),
- learning causal relationships among pairs of variables isolated from their context – therefore making impossible the use of conditional dependencies to unravel causal direction (the Cause Effect Pairs task),
- quantifying the causal effect of promotions on sales (the PROMO task).

See Guyon *et al.* (2008b) and references therein for more details. Some datasets, which were contributed too late into the challenge, will be used as part of future evaluations.

26.8 Planned events

One of the objectives of the Causality Workbench is to provide an interactive platform, which will allow researchers to conduct virtual experiments: a virtual laboratory. We are presently implementing such a platform. Several data generative models will be interfaced to the platform, including some of those proposed by participants of the pot-luck challenge. Using this platform, we are planning two competitions briefly described in this section.

Experimental Design in Causal Discovery (ExpDeCo)
Methods for learning cause–effect links without experimentation (learning from observational data) are attractive because observational data is often available in abundance and experimentation may be costly, unethical, impractical, or even plain impossible (London and Kadane 2002). Still, many causal relationships cannot be ascertained without the recourse to experimentation

and the use of a mix of observational and experimental data might be more cost effective. Since standard experimental design has not concentrated on discovering thousands of causal relations, as in a genetic regulatory network, there is a need for developing solutions suitable to such large dimensional problems. The challenge ExpDeCo (Experimental Design in Causal Discovery) will benchmark methods of experimental design (Rubin 1974; Shadish 2001; Quinn 2002; Montgomery 2004; Meganck *et al.* 2006; Rubin 2007; Eberhardt 2009) and query/active learning (Spirtes *et al.* 2000; Tong and Koller 2001; Murphy 2001; Eberhardt 2006), in application to causal modelling. The goal is to identify effective methods to unravel causal models, requiring a minimum of experimentation. There are several ways in which the models of the participants could be evaluated, including verifications of the correctness of the causal structure of their model, and verification of the predictive power of the model. We favour this last type of metrics because they do not preclude of any particular causal model and do not even require defining causality.

A budget of virtual cash will be allocated to participants to buy the right to observe or manipulate certain variables, manipulations being more expensive than observations. The participants will place queries for data records, setting certain variables to given values and requesting to observe certain variables. They will have to plan for an optimal way of spending their budget to make optimal predictions on test data.

Causal Models for System Identification and Control (CoMSICo)

The second planned challenge called CoMSICo for 'Causal Models for System Identification and Control' is more ambitious in nature because we will perform a continuous evaluation of causal models rather than separating training and test phase. In contrast with ExpDeCo in which the organizers will provide test data with prescribed manipulations to test the ability of the participants to make predictions of the consequences of actions, in CoMSICo, the participants will be in charge of making their own plan of action (policy) to optimize an overall objective (e.g. improve the life expectancy of a population, maximize profit, etc.) and they will be judged directly with this objective, on an on-going basis, with no distinction between 'training' and 'test' data. The participants will be given an initial amount of virtual cash, and, as previously, both actions and observations will have a price. New in CoMSICo, virtual cash rewards will be given for achieving good intermediate performance, which the participants will be allowed to re-invest to conduct additional experiments and improve their plan of action (policy). The winner will be the participant ending up with the largest amount of virtual cash. Both time independent tasks (e.g. assuming a static system, a stationary system or a population averaged within a short time period) and time dependent tasks (assuming a dynamic system) will be considered. As indicated by the name of the challenge, there are obvious ties between the problems we are interested in and control problems. We very much hope that this will trigger cross-fertilization between the control community and the causal discovery community. Unlike in classical control problems (like temperature regulation, navigation, robot arm control), the system identification part of the problem includes a causal variable selection component, namely finding those variable, which will significantly affect the objective in a desirable manner, among a possibly very large number of candidates. Another difference with classical control problems is that we also consider time independent tasks.

26.9 Conclusion

Standard benchmarks foster scientific and technical progress, but the design of good benchmarks for causal problems is not trivial. Our program of data exchange and benchmark challenges the research community with a wide variety of problems from many domains and focuses on realistic settings. Causal discovery is a problem of fundamental and practical interest in many areas of science and technology and there is a need for assisting policy making in all these areas while reducing the costs of data collection and experimentation. Hence, the identification of efficient techniques to solve causal problems will have a widespread impact. By choosing applications from a variety of domains and making connections between disciplines as varied as machine learning, causal discovery, experimental design, decision making, optimization, system identification, and control, we anticipate that there will be a lot of cross-fertilization between different domains. Our activities, such as teleconference seminars, data and tool exchange, competitions and post-competition collaborative experiments, will cement collaborations between researchers and ensure a rapid and broad dissemination of the results. Our project has also several educational components, the playful nature our competition program is attractive to students and encourages them to work on difficult high-impact problems; teachers will be able to use our interactive platform for practical work on causal discovery.

Acknowledgements

This project is an activity of the Causality Workbench supported by the Pascal network of excellence funded by the European Commission and by the US National Science Foundation under Grant N0. ECCS-0725746. Any opinions, findings, and conclusions or recommendations expressed in this material are those of the authors and do not necessarily reflect the views of the National Science Foundation.

References

F. Eberhardt. Active structure search using interventions. In *Workshop on Large Graphical Models and Random Matrices*, North Carolina, November 2006. Statistical and Applied Mathematical Sciences Institute.

F. Eberhardt. Almost optimal intervention sets for causal discovery. In *24th Conference in Uncertainty in Artificial Intelligence (UAI)*, 2009.

R. A. Fischer. *The Design of Experiments*. Oliver & Boyd, Edinburgh, 1953.

N. Friedman, M. Linial, I. Nachman, and D. Pe'er. Using Bayesian networks to analyze expression data. In *RECOMB*, pages 127–135, 2000.

M. J. Keough and G. P. Quinn. *Experimental Design and Data Analysis for Biologists*. Cambridge University Press, 2002.

C. Glymour and G. F. Cooper, editors. *Computation, Causation, and Discovery*. AAAI Press/The MIT Press, Menlo Park, California, Cambridge, Massachusetts, and London, 1999.

I. Guyon, C. Aliferis, G. Cooper, A. Elisseeff, J.-P. Pellet, P. Spirtes, and A. Statnikov. Design and analysis of the causation and prediction challenge. In *JMLR W&CP*, volume 3, pages 1–33, WCCI2008 workshop on causality, Hong Kong, June 3–4, 2008a.

I. Guyon, S. Gunn, M. Nikravesh, and L. Zadeh, editors. *Feature Extraction, Foundations and Applications*. Studies in Fuzziness and Soft Computing. With data, results and sample code for the NIPS 2003 feature selection challenge. Physica-Verlag, Springer, 2006.

I. Guyon, D. Janzing, and B. Schölkopf. Causality: objectives and assessment. In *JMLR W&CP*, volume NIPS 2008 causality workshop, to appear, Whistler, Canada, December 12, 2008b.

J. Quiõnero-Candela *et al.*, editors. *Dataset Shift in Machine Learning*. MIT Press, 2009.

M. Kanehisa, S. Goto, M. Hattori, K. F. Aoki-Kinoshita, M. Itoh, S. Kawashima, T. Katayama, M. Araki, and M. Hirakawa. From genomics to chemical genomics: new developments in kegg. *Nucleic Acids Research*, 34: D354–357, 2006.

D. Kaplan. *Structural equation modeling: Foundations and extensions*, volume 10 of *Advanced Quantitative Techniques in the Social Sciences series*. Sage, 2000.

A. J. London and J. B. Kadane. Placebos that harm: sham surgery controls in clinical trials. *Statistical Methods in Medical Research*, 11: 413–427, 2002.

S. Meganck, P. Leray, and B. Manderick. Learning causal Bayesian networks from observations and experiments: A decision theoretic approach. In V. Torra *et al.* editor, *MDAI 2006*, volume 3885, pages 58–69. LNAI, 2006.

D. C. Montgomery. *Design and Analysis of Experiments*, 6th edition. John Wiley, 2004.

K. P. Murphy. Active learning of causal Bayes net structure. Technical report, UC Berkeley, http://http.cs.berkeley.edu/ murphyk/Papers/alearn.ps.gz, 2001.

R. E. Neapolitan. *Learning Bayesian Networks*. Prentice Hall series in Artificial Intelligence. Prentice Hall, 2003.

J. Pearl. *Causality: Models, Reasoning, and Inference*. Cambridge University Press, 2000.

D. Rubin. Estimating causal effects of treatments in randomized and nonrandomized studies. *Journal of Educational Psychology*, 66(5): 688–701, 1974.

D. Rubin. *Matched Sampling for Causal Effects*. Cambridge University Press, 2007.

P. Spirtes, C. Glymour, and R. Scheines. *Causation, Prediction, and Search*. The MIT Press, Cambridge, Massachusetts and London, 2000.

S. Tong and D. Koller. Active learning for structure in Bayesian networks. In *International Joint Conference On Artificial Intelligence*, 2001.

D. T. Campbell, W. R. Shadish, and T. D. Cook. *Experimental and Quasi-Experimental Designs for Generalized Causal Inference*. Houghton Mifflin, 2001.

·27·

When are graphical causal models not good models?

Jan Lemeire, Kris Steenhaut, and Abdellah Touhafi

Abstract

The principle of Kolmogorov minimal sufficient statistic (KMSS) states that the meaningful information of data is given by the regularities in the data. The KMSS is the minimal model that describes the regularities. The meaningful information given by a Bayesian network is the directed acyclic graph (DAG) which describes a decomposition of the joint probability distribution into conditional probability distributions (CPDs). If the description given by the Bayesian network is incompressible, the DAG is the KMSS and is faithful. We prove that if a faithful Bayesian network exists, it is the minimal Bayesian network. Moreover, if a Bayesian network gives the KMSS, modularity of the CPDs is the most plausible hypothesis, from which the causal interpretation follows. On the other hand, if the minimal Bayesian network is compressible and is thus not the KMSS, the above implications cannot be guaranteed. When the non-minimality of the description is due to the compressibility of an individual CPD, the true causal model is an element of the set of minimal Bayesian networks and modularity is still plausible. Faithfulness cannot be guaranteed though. When the concatenation of the descriptions of the CPDs is compressible, the true causal model is not necessarily an element of the set of minimal Bayesian networks. Also modularity may become implausible. This suggests that either there is a kind of meta-mechanism governing some of the mechanisms or a wrong model class is considered.

27.1 Introduction

Inductive inference comes to modelling the patterns in the data. Patterns or regularities in observations are – most likely – not coincidences, but give us valuable information about the system under study. A regularity is identified by its ability to compress the data, i.e. to describe the data using fewer symbols than the number of symbols needed to describe the data literally. Compressiveness is objectively defined by the Kolmogorov complexity. The concept is, however, not directly applicable since there does not exist an algorithm that computes the shortest program for a string. Kolmogorov complexity is therefore mainly used for giving preference within a given set of models.

This has given rise to different methods for inductive inference, such as minimum message length and minimum description length. These methods are used for selecting the best model from a given set of models, the model class. The choice of model class, however, determines the regularities under consideration.

For analysing the validity of causal inference, we do not want to stick to an a priori chosen set of regularities, but search for *all* relevant regularities. This idea is captured by the concept of Kolmogorov minimal sufficient statistic (KMSS). The KMSS is the minimal model such that the model together with the data is described minimally. The model should capture all regularities and nothing more.

For causal inference, the set of Bayesian networks is used as a model class. The DAG of a Bayesian network gives a minimal description of the conditional independencies following from a causal structure. A system can, however, contain other regularities. Then, the assumptions and implications of causal model theory, such as faithfulness, modularity and the correctness of causal inference, may become invalid. It can give rise to other independencies so that the DAG becomes unfaithful. We will show that the presence of other regularities cannot be ignored.

In Section 27.2, we will introduce the concept of KMSS. In Section 27.3, we will give a survey of causal model theory and the learning algorithms. Section 27.4 discusses related work. In Section 27.5 we apply the principle of KMSS to inductive inference and show that a Bayesian network captures dependencies between variables. Section 27.6 establishes the link between minimality of Bayesian networks, compressibility and faithfulness. In Section 27.7 we will argue that causal inference is plausible if the minimal Bayesian network is the KMSS. Section 27.8 discusses various cases in which the minimal Bayesian network does not provide the minimal description.

27.2 Meaningful information

The *Kolmogorov complexity* of a string x is defined to be the length of the shortest computer program that prints the string and then halts (Li and Vitányi, 1997):

$$K(x) = \min_{p:\mathcal{U}(p)=x} l(p) \tag{27.1}$$

with $\mathcal{U}$ a universal computer and $l(p)$ the size in bits of program p. Patterns in the string allow for its compression, i.e. to describe the data using fewer symbols than the number of symbols to describe the data literally.

The string '00010001000100010001000100010001000100010001' can be described shorter by program `REPEAT 11 TIMES '0001'`. But not all bits of this program can be regarded as containing *meaningful information*. We consider meaningful information as the properties of the string that allow for its compression (Vitányi, 2002). Such properties are called patterns or *regularities*. The regularity of the example string is the repetition. The number of repetitions (11) or the substring '0001' is random information. A random string, which is incompressible, has no meaningful information at all.

For inductive inference, we will look for a minimal description in two parts, one containing the regularities or patterns of the data, which we put in the model, and one part containing the remaining random noise. Such a description is called a *two-part code*. This results in a generic approach for inductive inference, called *minimum description length* (MDL), according to which we have to pick out the model M_{mdl} from model class $\mathcal{M}$ where M_{mdl} is the model which minimizes the sum of the description length of M and of the data D encoded with the help of M (Grünwald, 1998):

$$M_{\mathrm{mdl}} = \arg\ \min_{M \in \mathcal{M}} \{L(M) + L(D \mid M)\} \tag{27.2}$$

with $L(.)$ the description length.

The MDL approach relies on the a priori chosen model class. It does not tell us how to make sure the models capture all regularities of the data. The KMSS provides a formal separation of meaningful and meaningless information. We limit the introduction of KMSS to models that can be related to a finite set of objects, called the *model set*. In the context of learning, we are interested in a model set S that contains string x and the objects that share x's regularities. All elements of a set S can be enumerated with a binary index of length $\log_2 |S|$ with $|S|$ the size of set S. We therefore say that x is *typical* for S if

$$K(x \mid p_S) \geq \log_2 |S| - \beta \tag{27.3}$$

with $K(x \mid p_S)$ the conditional Kolmogorov complexity. p_S denotes the shortest program that describes S and β an agreed upon constant. Given set S, x cannot be described shorter than by the set's index. Atypical elements have regularities that are not shared by most of the set's members and can therefore be described by a shorter description. Note that most elements of a set are typical, since, by counting arguments, only a small portion of it can be described shorter than $\log_2 |S|$.

The construction of S can be understood with the *Kolmogorov structure function* KSF. KSF(k, x) of x is defined as the $\log_2$-size of the smallest set including x which can be described with no more than k bits (Cover and Thomas, 1991):

$$\mathrm{KSF}(k, x) = \min_{\substack{p: l(p) \leq k \\ \mathcal{U}(p) = S x \in S}} \log_2 |S| \tag{27.4}$$

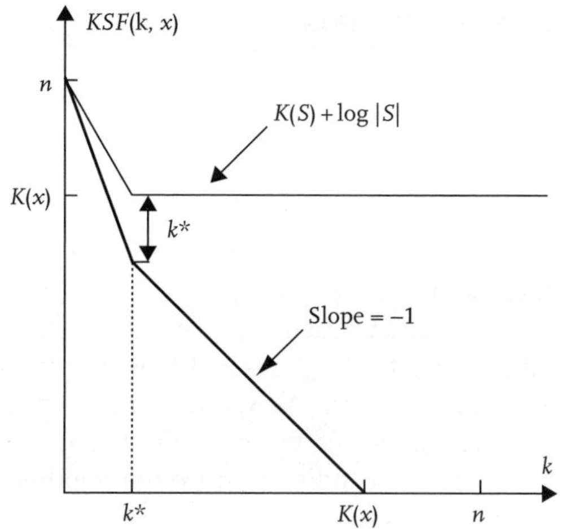

Fig. 27.1 Kolmogorov structure function for n-bit string x, k^* is the KMSS of x.

A typical graph of the structure function is illustrated in Figure 27.1. By taking $k = 0$, the only set that can be described is the entire set $\{0, 1\}^n$ containing 2^n elements, so that the corresponding log set size is n. By increasing k, the model can take advantage of the regularities of x in such way that each bit reduces the set's size more than halving it. The slope of the curve is smaller than -1. When k reaches k^*, all regularities are exploited. There are no more patterns in the data that allow for further compression. From then on each additional bit of k reduces the set by half. We proceed along the line of slope -1 until $k = K(x)$ and the smallest set that can be described is the singleton $\{x\}$. The curve $K(S) + \log_2 |S|$ is also shown on the graph. It represents the descriptive complexity of x by using the two-part code. With $k = k^*$ it reaches its minimum and equals to $K(x)$. When $k < k^*$, S is too general and is not a typical set for x. x is only typical for S if $k \geq k^*$. For random strings the curve starts at $\log_2 |S| = n$ for $k = 0$ and drops with a slope of -1 until reaching the x-axis at $k = n$. Each bit reveals one of the bits of x, and halves the model set.

The *Kolmogorov minimal sufficient statistic* (KMSS) of x is defined as the shortest program p^* which describes the smallest set S^* such that the two-stage description of x is as good as the minimal single–stage description of x (Gács *et al.*, 2001; Vitányi, 2002):

$$p^* = \arg\min_p \{l(p) \mid \mathcal{U}(p) = S^*,\ x \in S^*,\ K(S^*) + \log_2 |S^*| \leq K(x)\}. \quad (27.5)$$

The descriptive complexity of S^* is then k^*. Program p^* minimally describes the meaningful information present in x and nothing else. The definition ensures that x is a typical element of S^*.

27.3 Graphical causal models

This chapter will introduce graphical causal models and the accompanying learning algorithms (Pearl, 2000; Spirtes *et al.*, 1993).

27.3.1 Representation of causal relations

Graphical causal models intend to describe with a directed acyclic graph (DAG) the structure of the underlying physical mechanisms governing a system under study. The state of each variable, represented by a node in the graph, is generated by a stochastic process that is determined by the values of its parent variables in the graph. All variables that influence the outcome of the process are called *causes* of the outcome variable. An *indirect cause* produces the state of the effect indirectly, through another variable. If there is no intermediate variable among the known variables, the cause is said to be a *direct cause.*

Each process represents a physical mechanism. In its most general form it can be described by a conditional probability distribution (CPD) $P(X \mid Pa(X))$, where $Pa(X)$ is the set of parent nodes of X in the graph and constitute the direct causes of the variable. A causal model consists of a DAG over all variables and a CPD for each variable. The combination of the CPDs results in a joint probability distribution:

$$P(X_1, \ldots, X_n) = \prod_{i=1}^{n} P(X_i \mid Pa(X_i)). \tag{27.6}$$

For a discrete variable, the CPD is encoded by means of a tabular representation: for each possible assignment of values to the parents of X_i, we need to specify a distribution over the values that X_i can take. This is called a conditional probability table. For continuous variables, one often relies on prior knowledge or assumptions about the structure of the distribution. If one assumes linearly-related variables, the CPDs can be described by the following *structural equations*:

$$P(X_i \mid Pa(X_i)) = \sum_{X_j \in Pa(X_i)} a_{i,j}.X_j + U_i + c_i \tag{27.7}$$

where U_i represent the stochastic variations which cannot be explained by the model and c_i a constant term. One often assumes that U_i is normally distributed.

27.3.2 Modularity and the effect of changes to the system

A causal model represents a collection of processes that could account for the generation of the observed data. Each process is a stable and autonomous physical mechanism. It is then conceivable to change one such relationship without changing the others. This *modularity* permits one to predict the effect of external interventions or local reconfigurations of the mechanisms (Pearl, 2000). An *intervention* is defined as an atomic operation that fixates a variable to a given state and eliminates the corresponding factor (CPD) from the factorization (Eq. 27.6) (Pearl, 2000). Applied on a causal graph, an intervention on variable X sets the value of X and breaks all of the edges in the graph directed into X and preserves all other edges in the graph, including all edges directed out of X. This is called the Manipulation Theorem (Spirtes *et al.* 1993, p. 51). Intervening on a variable only affects its effects. Causes have to be regarded as if they were levers which can be used to manipulate their effects.

This approach does not directly define causality, but defines the implications of having a thorough knowledge of the mechanisms that make up a system. Manipulability puts a constraint of independentness on the mechanisms. The accuracy of the mutilated model relies on autonomy or modularity; a mechanism can be replaced by another without affecting the rest of the system. It is defined by Hausman and Woodward (1999, p. 545) as follows. They relate each CPD to a structural equation (Eq. 27.7).

Definition 27.1 (Modularity) For all subsets $\boldsymbol{Z}$ of the variable set $\boldsymbol{V}$, there is some non-empty range $\boldsymbol{R}$ of values of members of $\boldsymbol{Z}$ such that if one intervenes and sets the value of the members of $\boldsymbol{Z}$ within $\boldsymbol{R}$, then all equations except those with a member of $\boldsymbol{Z}$ as a dependent variable (if there is one) remain invariant.

27.3.3 Representation of independencies

The key for causal inference is the conditional independencies entailed by the system's causal structure. They are based on the property of Markov chains and v-structures. If X is affected by Y and Z, then we do not expect that X is independent of Y conditional on Z, except if Y affects X *via* Z. This is represented by a Markov chain. Random variables X, Z, Y are said to form a *Markov chain* in that order, denoted by $X \rightarrow Z \rightarrow Y$, if the joint probability mass function can be written as

$$P(X, Z, Y) = P(X).P(Z \mid X).P(Y \mid Z) \tag{27.8}$$

which is equivalent to the conditional independence of X and Y given Z. *Conditional independence* of X and Y given Z, written as $X \perp\!\!\!\perp Y \mid Z$, is defined as

$$P(X, Y \mid Z) = P(X \mid Z).P(Y \mid Z). \tag{27.9}$$

The conditional independence expresses that learning the value of X does not provide additional information about Y once the state of Z is known. We say that Z 'screens off' X from Y. Once the state of Z is observed, the state of Y no longer depends on that of X. For a *v-structure* on the other hand, for example $X \rightarrow Z \leftarrow Y$, X and Y are independent, but become dependent when conditioned on Z.

For a causal model, the *causal Markov condition* gives us the independencies that follow from the causal structure: each variable is probabilistically independent of its non-effects conditional on its direct causes. This condition is defined by Spirtes *et al.* (1993) as follows:

Definition 27.2 (Causal Markov condition) Let G be a causal graph with vertex set V and P be a probability distribution over the vertices in V generated by the causal structure represented by G. G and P satisfy the causal Markov condition if and only if for every W in V, W is independent of $V \setminus \boldsymbol{Descendants}(W) \setminus \boldsymbol{Parents}(W)$ given $\boldsymbol{Parents}(W)$.

These independencies are irrespective of the nature of the mechanisms, of the exact parameterization of the conditional probability distributions $P(X_i \mid Pa(X_i))$. Pearl and Verma constructed a graphical criterion, called d-separation, for retrieving, from the causal graph, all independencies following from the Causal Markov Condition.

A graph is called *faithful* to a distribution if all conditional independencies of the distribution correspond to a d-separation in the graph and vice versa. In other words, faithfulness means that if a graph represents a causal structure, all conditional independencies follow from the system's causal structure.

27.3.4 Correspondence with Bayesian networks

Graphical causal models provide a probabilistic account of causality (Spohn, 2001). This resulted in a close correspondence with Bayesian networks. In contrast to causal models, Bayesian networks are only concerned with offering a dense and manageable representation of joint distributions. A joint distribution over n variables can be *factorized* relative to a variable ordering $(X_1, \ldots, X_n)$:

$$P(X_1, \ldots, X_n) = \prod_{i}^{n} P(X_i \mid X_1, \ldots, X_{i-1}). \tag{27.10}$$

Variable X_j can be removed from the conditioning set of variable X_i if it becomes conditionally independent from X_i by conditioning on the rest of the set:

$$P(X_i \mid X_1 \ldots X_{i-1}) = P(X_i \mid X_1 \ldots X_{j-1}, X_{j+1} \ldots X_{i-1}). \tag{27.11}$$

Such conditional independencies reduce the complexity of the factors in the factorization. The conditioning sets of the factors can be described by a directed acyclic graph (DAG), in which each node represents a variable and has incoming edges from all variables of the conditioning set of its factor. The joint distribution is then described by the DAG and the conditional probability distributions (CPDs) of the variables conditional on their parents. A *Bayesian network* is a factorization that is edge-minimal, in the sense that no edge can be deleted without destroying the correctness of the factorization.

Although edge-minimality of a Bayesian network, the graph depends on the chosen variable ordering. Some orderings lead to the same networks, while others result in different topologies. Take five stochastic variables A, B, C, D and E. Figure 27.2(a) shows the graph that was constructed by simplifying the factorization based on variable ordering (A, B, C, D, E) by the three given conditional independencies. However, the Bayesian network, describing the same distribution, but based on ordering (A, B, C, E, D), depicted in Figure 27.2(b) contains two edges less because of five useful independencies. Both networks represent the probabilities just as well, except that the first one is more complex. We call the *minimal factorization* as the factorization which has the least total number of variables in the conditioning sets. The corresponding Bayesian network is called the *minimal Bayesian network* of a probability distribution.

Analogous to the causal Markov condition, the Markov condition gives the conditional independencies that follow from the structure of a Bayesian network: each variable is independent from all its non-descendants by conditioning on its parents in the graph. The equivalence of the Markov condition and factorizability can be proven (Hausman and Woodward, 1999, p. 532). This ensures the correspondence: causal models are also Bayesian networks. The difference lies in the causal component; causal models attribute a causal interpretation to the edges of the graph and are therefore called *causally interpreted Bayesian networks*.

27.3.5 Causal inference

The goal of causal inference is to learn the causal structure of a system based on observational data. Causal structure learning algorithms fall apart in two categories: scoring-based and constraint-based algorithms.

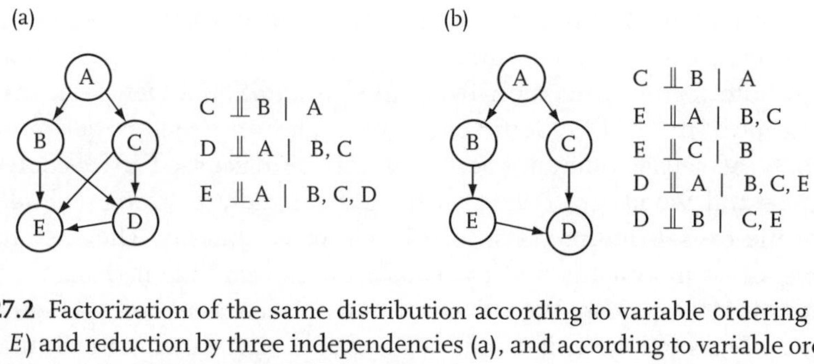

Fig. 27.2 Factorization of the same distribution according to variable ordering (A, B, C, D, E) and reduction by three independencies (a), and according to variable ordering (A, B, C, E, D) and five independencies (b).

Scoring-based algorithms are based on an optimized search through the set of all possible models, which tries to find the minimal model that best describes the data. Each model is given a score that is a trade-off between model complexity and goodness-of-fit. Different scoring criteria have been applied in these algorithms, such as a Bayesian scoring method (Cooper and Herskovits, 1992; Heckerman *et al.*, 1994), an entropy based method (Herskovits, 1991), a minimum message length (MML) method (Oliver *et al.*, 1992), and one based on the minimum description length (MDL) (Suzuki, 1996). As explained in the introduction, we are not investigating how to select the minimal model from the a priori chosen model class, but the model class which should be considered.

Constraint-based learning algorithms rely on the conditional independencies detected that follow from the system's causal structure. It is a kind of evidence-based construction, the decisions to include an edge and on the edge's orientation is based on the presence or absence of certain independencies. The algorithms assume the existence of a faithful graph, i.e. that all independencies follow from the causal structure. They also assume that the correct model is the minimal model. Minimality, faithfulness and the causal Markov condition give the three assumptions that ensure correct learning (Spirtes *et al.*, 1993). The minimality condition is an edge-minimal condition on the true causal graph.

It must be noted that some algorithms, such as the PC algorithm, also require *causal sufficiency*, i.e. that all common causes should be known: variables that are the direct cause of at least two variables. More sophisticated learning algorithms exist that are capable of detecting latent common causes. For now we will not take the presence of latent variables into consideration and discuss the consequences of this in Section 27.8.

27.4 Related work

The causal interpretation of a Bayesian network and the validity of faithfulness are often criticized (Freedman and Humphreys, 1999; Cartwright, 2001; Williamson, 2005; Hausman and Woodward, 1999). This paper would like to contribute to the discussion by giving an additional viewpoint through the concept of the KMSS. Some of the examples on which criticism on the possibility of causal inference is based will be discussed in Section 27.8. Hausman and Woodward (1999) on the other hand are strong defenders of linking the causal interpretation of models to modularity. They defend the equivalence of modularity and the causal Markov condition (Hausman and Woodward, 1999, p. 554). We will contribute to the discussion by motivating why and when modularity is a valid assumption, and showing the limitations of assuming faithfulness.

Pearl and others use *stability* as the main motivation for the faithfulness of causal models (Pearl, 2000, p. 48). Consider the model of Figure 27.2(b). In general, one expects A to depend upon D. A and D are independent only if the stochastic parameterization is such that the influences via paths $A \rightarrow B \rightarrow E \rightarrow D$ and $A \rightarrow C \rightarrow D$ cancel out exactly. This system is called unstable because a small change in the parameterization results in a dependency. The unhappy balancing act is a measure zero event, the probability of such a coincidence can therefore be regarded as zero. Hence, the majority of distributions compliant with a DAG are faithful (Pearl, 2000, p. 18). We argue that indeed typical distributions are faithful, but that nonetheless, unfaithful distributions appear.

Milan Studeny was one of the first to point out that the Bayesian networks cannot represent all possible sets of independencies. He constructed a different framework, called *imsets* (Studeny, 2001), which is capable of representing broader sets of independencies. We advocate a different approach. We will not look for a different representation of conditional independencies, but stick to Bayesian networks. Yet, we will try to find explanations (referring to regularities) for the presence of conditional independencies not coming from the system's causal structure.

27.5 Minimal description of distributions

In this section we start the analysis of causal inference by applying the KMSS principle on observed data of a collection of independent and identically distributed random variables. A minimal description for the data corresponds to the construction of a efficient code which on its turn corresponds to the description of a probability distribution (Grünwald *et al.*, 2005, Chapter 2). We thus have to investigate how distributions can be described compactly.

From the theory of Bayesian networks (Section 27.3.4), we know that a joint distribution can be described shorter by a factorization that is reduced by conditional independencies of the form of Eq. (27.11). The minimal factorization leads to $P(X_1, \ldots, X_n) = \prod \mathrm{CPD}_i$, with CPD_i the CPD of variable X_i. The descriptive size of the CPDs is determined by the number of variables in the conditioning sets. A two-part description of a joint distribution is then:

$$\begin{aligned}\mathrm{descr}(P(X_1 \ldots X_n)) &= \mathrm{descr}(\{Pa(X_1), \ldots, Pa(X_n)\}) \\ &\quad + \mathrm{descr}(\mathrm{CPD}_1) + \cdots + \mathrm{descr}(\mathrm{CPD}_n). \qquad (27.12)\end{aligned}$$

With $\mathrm{descr}(x)$ the description of x. Note that the parents' lists are described very compact by a DAG. If the description according to Eq. 27.12 is shorter than the literal description of the joint distribution, than the reduction of the factorization contains meaningful information. This meaningful information is described by the parents' lists or the DAG of the Bayesian network.

Theorem 27.3 *If the two-part code description of a probability distribution, given by Eq. 27.12 in which the CPDs are described literally, results in an incompressible string which is shorter than the literal description of the joint probability distribution, the first part is the Kolmogorov minimal sufficient statistic.*

Proof 1 The CPDs do not contain meaningful information (regularities), since they are literally described and they are incompressible. This last follows from the incompressibility of the total description. Since the total description is shorter than the literal description, the reduction of the factorization outweighs the description of the parents' lists. The parents' lists therefore contain meaningful information. Their incompressibility ensures that it is the KMSS. □

Concluding, we end up with a three-part code for the description of the observations:

$$\text{descr}(data) = \text{descr}(DAG) + \text{descr}(CPD_1) + \cdots + \text{descr}(CPD_n)$$
$$+\text{descr}(data \mid distribution). \qquad (27.13)$$

The data is described with the help of a probability distribution, which on its turn is described by a DAG and a list of CPDs. The regularities that allow the compact description of the data are the dependencies among the variables; knowing one variable gives information about the state of another variable. Conditional independencies, on the other hand, reduce the model's complexity. They reduce the number of variables to consider when describing the dependencies among the variables.

27.6 Minimality of Bayesian networks

The following two theorems show that the Bayesian network corresponding to the minimal factorization is the KMSS and faithful if its DAG and CPDs are random and incompressible.

Theorem 27.4 *If a faithful Bayesian network exists for a distribution, it is the minimal Bayesian network, i.e. the Bayesian network with the minimal number of edges.*

Proof 2 Recall that the absence of an edge between two variables X and Y in a Bayesian network implies that there exists a set of variables S not containing X and Y that makes X and Y conditionally independent: $X \perp\!\!\!\perp Y \mid S$. In case of faithfulness, the presence of an edge forbids the existence of such a set. Let A be a graph that has fewer edges then the faithful graph B. It follows that B contains an edge between two variables X and Y that A does not contain. The absence of the edge in A implies that X and Y become independent by conditioning on some set of the other variables. But this contradicts with the faithfulness of B which implies that X and Y cannot become independent. □

Neapolitan (2003, p. 107) provides a proof for edge-minimality, while here minimality in the global sense is considered.

The DAG of a Bayesian network corresponds to a set of conditional independencies. Intuitively we would expect that two variables are dependent if they are not d-separated. When this is true, the DAG is faithful to the probability distribution. The next theorem proves that two variables that are not d-separated can only be independent if there is a constraint between the probabilities. To illustrate the theorem, consider the model of Figure 27.2(b) and the set of distributions compatible with the DAG. For typical distributions, dependencies $D \not\perp\!\!\!\perp E$ and $A \not\perp\!\!\!\perp E$ hold. There are, however, specific parameterizations which lead to independencies $D \perp\!\!\!\perp E$ or $A \perp\!\!\!\perp E$. Such independencies only follow if specific equations between the free parameters are satisfied.

Theorem 27.5 *A Bayesian network for which the concatenation of the descriptions of the conditional probability distributions (CPDs) is incompressible, is faithful.*

Proof 3 Recall that a Bayesian network is a factorization that is edge-minimal. This means that for each parent $pa_{i,j}$ of variable X_i:

$$P(X_i \mid pa_{i,1}, \ldots pa_{i,j}, \ldots pa_{i,k}) \neq P(X_i \mid pa_{i,1}, \ldots pa_{i,j-1}, pa_{i,j+1}, \ldots pa_{i,k}). \tag{27.14}$$

Variables cannot be eliminated from the factors of the factorization. The proof will show that any two variables that are not d-separated are dependent, unless the probabilities of the CPDs are 'related', in the sense that some probabilities can be calculated from others and the set of CPDs is compressible. We derive the relations for discrete variables. For continuous variables, the analysis results in relations among the free parameters of the CPDs.

We have to consider the following possibilities. The two variables can be adjacent (a), related by a Markov chain (b),[1] a v-structure (c), a combination of both or connected by multiple paths (d).

First we prove that a variable marginally depends on each of its adjacent variables (a). Consider adjacent nodes D and E of the Bayesian network of Figure 27.2(b). We will demonstrate that $P(D \mid E) = P(D)$ results in a regularity. We expand the first term with all other parents of D:

$$P(D \mid E) = \sum_{c \in C_{\text{dom}}} P(D \mid E, c).P(c \mid E). \tag{27.15}$$

[1] Recall that a Markov chain is a path not containing v-structures.

C is also a parent of D, thus, by Eq. (27.14), there are at least two values of C_{dom} for which $P(D \mid E, c) \neq P(D \mid E)$.[2] Take c_1 and c_2 being such values for which

$$P(D \mid E, c_1) \neq P(D \mid E, c_2). \tag{27.16}$$

There are also at least two such values of E_{dom}, take e_1 and e_2. Equation (27.15) should hold for all values of E and equal to $P(D)$ to get an independency. This results in the following relation among the probabilities:

$$\begin{aligned} &P(D \mid e_1, c_1).P(c_1 \mid e_1) + P(D \mid e_1, c_2).P(c_2 \mid e_1) \\ = \; &P(D \mid e_2, c_1).P(c_1 \mid e_1) + P(D \mid e_2, c_2).P(c_2 \mid e_1). \end{aligned} \tag{27.17}$$

Note that the equation cannot be algebraically simplified: the conditional probabilities are not equal to $P(D)$ (Eq. 27.14) nor to each other (Eq. 27.16). The proof can easily be generalized for variables having more parents.

Next, by the same arguments it can be proved that variables connected by a Markov chain are by default dependent (b). Take $A \rightarrow B \rightarrow E$ in Figure 27.2(b), independence of A and E requires that

$$P(E \mid a) = \sum_{b \in B_{dom}} P(E \mid b).P(b \mid a) = P(E) \quad \forall a \in A, \tag{27.18}$$

and this would also result in a regularity among the CPDs.

In a v-structure, both causes are dependent when conditioned on their common effect (c), for $C \rightarrow D \leftarrow E$, $P(D \mid C, E) \neq P(D \mid E)$ is true by Eq. (27.14). Finally, if there are multiple unblocked paths connecting two variables, then independence of both variables implies a regularity as well (d). Take A and D in Figure 27.2(b):

$$P(D \mid A) = \sum_{b \in B_{dom}} \sum_{c \in C_{dom}} \sum_{e \in E_{dom}} P(D \mid c, e).P(c \mid A).P(e \mid b).P(b \mid A).$$

Note that $P(c, e \mid A) = P(c \mid A).P(e \mid A)$ follows from the independence of C and E given A. All factors from the equation satisfy Eq. (27.14), so that, again, the equation only would equal to $P(D)$ if there is a relation among the probabilities. □

From the theorem it follows that a Bayesian network with random CPDs is the minimal factorization. Bayesian networks not based on a minimal factorization, such as the one of Figure 27.2, are compressible, namely by the regularities among the CPDs that follow from the independencies not represented by the graph. Pearl hypothesizes that there is no bounded set of

[2] $P(D \mid E)$ is a weighted average of $P(D \mid E, C)$. If one probability $P(D \mid E, c_1)$ is different than this average, let's say higher, than there must be at least one value lower than the average, thus different.

conditions that would ensure the existence of a faithful graph (Pearl, 1988, p. 131). Indeed, as shown by the theorem, every dependence can be turned into an independence by a balanced parameterization of some CPDs.

It must be noted that if there exists a faithful Bayesian network, it is not necessarily unique. Multiple faithful models can exist for a distribution. These models represent the same set of independencies and are therefore statistically indistinguishable. They define a *Markov-equivalence class*. It is proved that they share the same skeleton and v-structures. They only differ in the orientation of some edges (Pearl, 2000). This set can be represented by a partially directed acyclic graph in which some of the edges are not oriented. The corresponding factorizations have the same number of conditioning variables and thus all models of a Markov-equivalence class have the same complexity.

27.7 When the minimal Bayesian network is the KMSS

In this section we will discuss the case in which there is exactly one minimal Bayesian network which is also the minimal description. This means that there are no other regularities and no other independencies than the conditional independencies represented by the model. The DAG is then the KMSS and minimally represents all regularities. It is also faithful.

The minimal Bayesian network decomposes the description of a joint distribution into a list of CPDs. This means that the minimal description of the system is a concatenation of descriptions, namely the description of the individual CPDs. In other words, we have found a unique and minimal decomposition of the model. This brings us to modularity and manipulability. We have discovered that the minimal description is a concatenation of unrelated components. The CPDs are independent; the concatenation of their descriptions cannot be compressed. Then, among all possible explanations, the simplest is that *each CPD corresponds to an independent part of reality*. Thus, following Occam's razor, modularity is the most likely hypothesis about the system under study. The correctness of Occam's razor cannot be proven, the principle must be interpreted as the most effective *strategy* for deciding among competing explanations (Grünwald, 1998). Modularity of the minimal Bayesian network must be regarded as the top-ranked hypothesis, which can be verified with background knowledge or experiments with interventions. Thus, the three conditions for causal inference are valid (Section 27.3.5): minimality and faithfulness are fulfilled, and the causal Markov condition follows from modularity. Description minimality is linked to causality through modularity.

Occam's razor is contradicted when the real system is more complex than suggested by the complexity of the observations. Take the impact of $Tax\ rate$ increase on $Tax\ revenue$ as shown in Figure 27.3(a). A $Tax\ rate$ increase has a negative effect on the *Economy* which could neutralize the increase of the tax revenues, such that *Tax rate* $\perp\!\!\!\perp$ *Tax revenue*. If so, the system is minimally described by the model of Figure 27.3(b). This model is faithful, incompressible and simpler than the true model. From observations alone, one cannot find indications for the more complex true model. Although not minimal in the global sense, the model of Figure 27.3(a) is edge-minimal: no edge can be removed without destroying the correctness of the model.

The CPD of a variable is also called the variable's *causal Markov kernel*. Note that by representing a causal model with a graph, the representation suggests that the edges – instead of the CPDs – are the basic components. This is however not true. A graphical model can therefore be misleading. A better representation is shown in Figure 27.4. It represents the same system as the causal model of Figure 27.2(b), but emphasizes that CPDs are the basic components.

Decomposition and thus also causality matches with a *reductionist* view, according to which the world can be studied in parts. Indeed, if the system cannot be decomposed, if there are no conditional independencies that simplify a factorization, then the DAG does not contain meaningful information. We end up with a Holist system in which everything depends on everything.

Note that uniqueness of the minimal Bayesian network is not essential. As discussed in the previous section, if the minimal Bayesian network is not unique, the Markov-equivalence class indicates exactly which parts are undecided (the orientation of some edges). So, we know exactly for which parts of the model we do not have enough information to decide upon the decomposition.

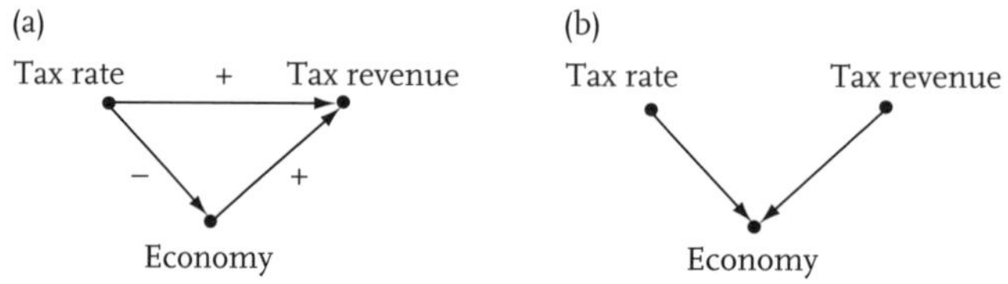

Fig. 27.3 Model in which the impact of *Tax rate* increase on *Tax revenue* is neutralized by the negative effect on the *Economy* (a). The minimal Bayesian network describing the system (b).

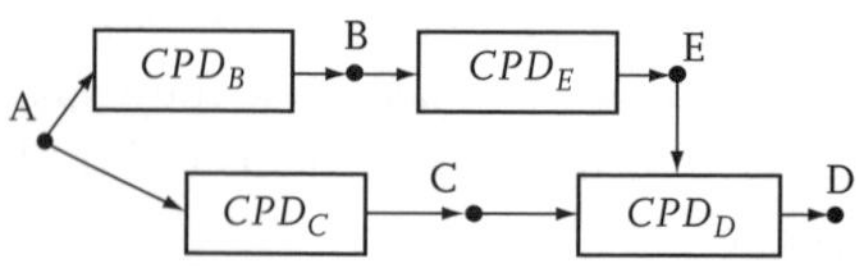

Fig. 27.4 Decomposition of the system represented by the causal model of Fig. 27.2(b) into independent components.

27.8 When the minimal Bayesian network is not the KMSS

To study the validity of faithfulness and the modularity property, we will in this section not assume incompressibility of the minimal Bayesian networks. They are denoted with BN_{min}. Instead, we will study a wide variety of cases, appearing throughout literature, in which regularities appear that are not described by a Bayesian network. We will analyse the properties of the True Causal Model (CM_{true}) and those of the BN_{min}.

Table 27.1 gives an overview of the answers for the next questions, which will be discussed in the following.

- Is the CM_{true} compressible? If so, is the compressibility due to the compressibility of the description of a single CPD or the compressibility of the concatenation of the descriptions of multiple CPDs?
- Is the compressibility of the minimal Bayesian networks due to the compressibility of the description of a single CPD or the compressibility of the concatenation of the descriptions of multiple CPDs?
- Is the CM_{true} present in BN_{min}? The answer to this and the next question determines the feasibility of causal inference.
- Is there a unique BN_{min}? Are the regularities under consideration responsible for the presence of multiple minimal Bayesian networks?
- Is the true causal model faithful to the system?
- Are the minimal Bayesian networks faithful to the system?
- Does modularity holds for the true causal model?

27.8.1 Compressibility of a single CPD

First we consider cases in which the description of an individual CPD is compressible. Faithfulness and the uniqueness of the minimal Bayesian network

Table 27.1 Answers to questions for the different case studies. A † indicates that Markov networks are considered. A ‡ indicates that Bayesian networks with latent variables are considered.

	Compress. CM_{true}	Compress. BN_{min}	$CM_{\text{true}} \in BN_{\text{min}}$	Unique BN_{min}	CM_{true} faithful	BN_{min} faithful	Modular CM_{true}
1. Local	single	single	Yes	Yes	Yes	Yes	Yes
2. PIM	single	single	Yes	No	No	No	Yes
3. Determ	single	single	Yes	No	No	No	Yes
4. Unfaithf	concat.	concat.	Yes	Yes	No	No	No/Yes
5. Markov	No†	concat.	No	No	Yes†	No	-
6. Latent	No‡	concat.	No	No	Yes‡	No	Yes‡
7. OO-nets	concat.	concat.	Yes	Yes	Yes	Yes	No/Yes

are not guaranteed, but the cases show that the modularity assumption still holds. The CPDs are independent.

Case 1. When individual CPDs can be compressed, we call this type of regularity *local structure* (Friedman and Goldszmidt, 1996). For discrete variables, the conditional probability tables are exponential in the number of parents of a variable X: for each possible assignment of values to the parents of X, we need to specify a distribution over the values X can take. When regularities among the probabilities appear these tables can be described more compactly, for example by decision trees. The regularities to construct the tree are called context-specific independencies (Boutilier *et al.*, 1996). On top of the independencies following from the causal structure, the system exhibits additional regularities. But the model remains faithful and the decomposition is correct.

Case 2. Variables in *pseudo-independent models* are pairwise independent but collectively dependent (Xiang *et al.*, 1996). For example, consider a binary variable X_3 that is determined by two other binary variables X_1 and X_2 by an *exclusive or* relation: $X_3 = X_1 \texttt{ EXOR } X_2$. This system can be represented by causal model $X_1 \rightarrow X_3 \leftarrow X_2$. Because of the pairwise independencies $X_3 \perp\!\!\!\perp X_1$ and $X_3 \perp\!\!\!\perp X_2$, the model is not faithful. There are three minimal Bayesian networks: besides the correct $X_1 \rightarrow X_3 \leftarrow X_2$, also $X_1 \rightarrow X_2 \leftarrow X_3$ and $X_2 \rightarrow X_1 \leftarrow X_3$. The CPD $P(X_3 \mid X_1, X_2)$ exhibits a strict regularity. Yet, pseudo-independent models fit in the reductionist approach of causal models. The only problem is that the conditional independencies do not provide enough information to conclude about the causal connections.

Case 3. Deterministic or functional relations among variables result in CPDs with a very specific form. Distributions with deterministic relations cannot be represented by a faithful graph (Spirtes *et al.*, 1993). Consider the system $X \rightarrow Y \rightarrow Z$ in which Y is a function of X: $Y = f(X)$. From the model (Markov chain) it follows that $X \perp\!\!\!\perp Z \mid Y$. By the functional relation, variable X got all information about Y, which implies $Y \perp\!\!\!\perp Z \mid X$. Both independencies imply a violation of the *intersection condition,* one of the conditions that Pearl imposes on a distribution in the elaboration of causal theory and its algorithms (Pearl, 1988). In (Lemeire, 2007) we call X and Y *information equivalent* with respect to Z, both variables have in some sense the same information about Z. Then, the set of minimal Bayesian networks contains graphs that connect X with Z and graphs that connect Y with Z. From the information about the conditional independencies alone we cannot decide upon which variable, X or Y, directly relates to Z. The solution we proposed for causal inference is to connect the variables that have the simplest relation (Lemeire, 2007). We defined an augmented causal model which also incorporates information of deterministic relations.

27.8.2 Compressibility of a set of CPDs

When the description of some CPDs taken together can be compressed, the CPDs are in some way related.

Case 4. The most-known example of unfaithfulness is when in the model of Figure 27.5(a), A and D appear to be independent (Spirtes *et al.*, 1993). This happens when the influences along the paths $A \to B \to D$ and $A \to C \to D$ exactly balance, so that they cancel each other out and the net effect results in an independence. For continuous variables this happens when an exact correspondence of the free parameters is fulfilled. The model is not faithful. This balancing act can give an indication of a global mechanism or *meta-mechanism*, such as evolution (Korb and Nyberg, 2006), controlling the mechanisms such that the parameters are calibrated until they neutralize. Modularity and autonomy of the CPDs depends on the meta-mechanism. Evolution works on the long-term, so modularity holds for a limited time period. For meta-mechanisms controlling the mechanisms instantly, the CPDs cannot be considered as being independent.

Case 5. Consider a system that is minimally described by a Markov network, as shown in Figure 27.5(b). Variables which are connected by a path in the network are dependent, unless each path is blocked by the conditioning variables. So is $B \not\perp\!\!\!\perp C \mid A$, but $B \perp\!\!\!\perp C \mid \{A, D\}$. For describing the same network with a DAG, we have to orient the edges of the network. For acyclicity, we have to create at least one v-structure. We can choose for example $B - D - C$. But then, for keeping the same dependencies, we have to add an edge, as shown in Figure 27.5(c). Without $B \to C$ we would have $B \perp\!\!\!\perp C \mid A$. Clearly, this Bayesian network is not minimal; the description is longer than that of a Markov network. The parameterizations of the CPDs contain redundancies. In the model of Figure 27.5(c), the parameterizations must ensure that $B \perp\!\!\!\perp C \mid \{A, D\}$, an independency which is not captured by the DAG. The causal interpretation of the CPDs (modularity) is not correct for the minimal Bayesian networks. It's unclear how and if a Markov network can be interpreted causally, so we leave the question open.

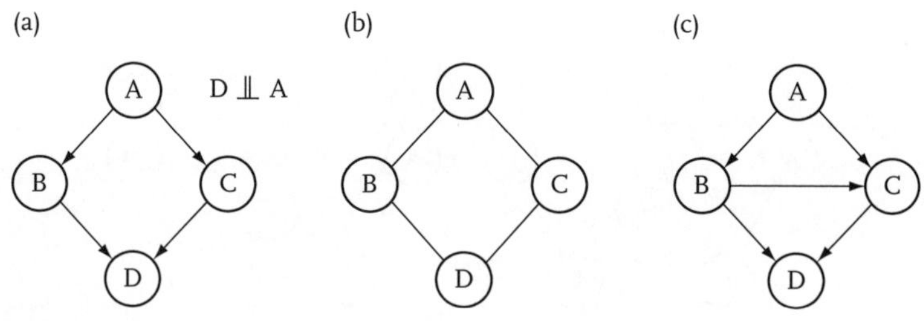

Fig. 27.5 O-structure in which A is independent from D (a). A Markov network (b) and one of the minimal Bayesian networks describing the same system (c).

Case 6. Causal sufficiency, the knowledge of all common causes, is an important property for correct causal learning. Take the system depicted in Figure 27.6(a) in which L is an unknown variable which is the cause of B and C. This gives rise to multiple minimal Bayesian networks, none of which models the system correctly. One of them is depicted in Figure 27.6(b). B and C are correlated, but none of the other known variables is the cause of both, so either B should be oriented towards C or vice versa. A should be connected to C to reflect dependency $A \not\perp\!\!\!\perp C \mid B$. But $A \perp\!\!\!\perp C$, thus there is a dependency between $P(B \mid A)$ and $P(C \mid A, B, D)$. The Bayesian network is therefore compressible and not faithful ($A \perp\!\!\!\perp C$ is not represented). The solution is to look for an alternative model class. Spirtes *et al.* (1995) propose the use of a *partially-oriented acyclic graph* (PAG) by which one can express the possibility of latent variables.

Case 7. Another regularity is the repetition of similar mechanisms in a system. This results in a causal model in which identical CPDs appear. The model is therefore compressible. The compressibility does not necessarily result in a dependence of the CPDs in terms of manipulability. As for case 4, it depends on the meta-mechanism responsible for the regularities in the system. The system could, for example, be designed by an engineer, such as a digital circuit. Then, modularity holds; one mechanism can be replaced by another without affecting the rest of the model. *Object-oriented nets* provide a representation format that explicitly capture similarities of mechanisms (Koller and Pfeffer, 1997).

27.9 Conclusions

A Bayesian network decomposes the description of a joint probability distribution into conditional probability distributions (CPDs). If a Bayesian network provides the Kolmogorov minimal sufficient statistic (KMSS) of a system, it gives the most plausible hypothesis about the causal structure of the system.

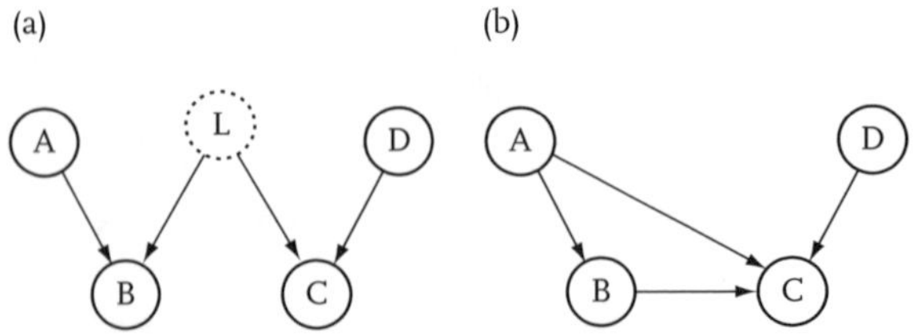

Fig. 27.6 Learning the model of a system with a latent variable L (a) and one of the minimal Bayesian networks (b).

The CPDs can be matched up with mechanisms of the underlying system. Decomposition reflects the causal component of graphical causal models.

Causal model theory expresses what typically can be expected from a causal structure. Typical distributions that are compatible with a causal structure are faithful. However, atypical distributions contain additional regularities and may invalidate the above conclusions. The minimal Bayesian networks of a probability distribution are then compressible and do not represent the KMSS.

If the description of a single CPD is compressible, this can result in unfaithfulness of the causal model. Causal inference is still possible, since the true model is an element of the set of minimal Bayesian networks and modularity is plausible. If on the other hand the concatenation of the CPDs is compressible, then the CPDs are no longer independent and the mapping of CPDs onto independent mechanisms becomes invalid. This can be due to a kind of meta-mechanism governing other mechanisms, or the incorrectness of considering the set of Bayesian networks as model class.

References

Craig Boutilier, Nir Friedman, Moises Goldszmidt, and Daphne Koller. Context-specific independence in Bayesian networks. In *Uncertainty in Artificial Intelligence*, pages 115–123, 1996.

Nancy Cartwright. What is wrong with Bayes nets? *The Monist*, pages 242–264, 2001.

Gregory F. Cooper and Edward Herskovits. A Bayesian method for the induction of probabilistic networks from data. *Machine Learning*, 9: 309–347, 1992.

Thomas M. Cover and Joy A. Thomas. *Elements of Information Theory*. John Wiley, 1991.

David A. Freedman and Paul Humphreys. Are there algorithms that discover causal structure? *Synthese*, 121: 29–54, 1999.

Nir Friedman and Moises Goldszmidt. Learning Bayesian networks with local structure. In *Proceedings of the 12th Conference on Uncertainty in Artificial Intelligence*, 1996.

Péter Gács, John Tromp, and Paul M. B. Vitányi. Algorithmic statistics. *IEEE Trans. Inform. Theory*, 47(6): 2443–2463, 2001.

Peter Grünwald. *The Minimum Description Length Principle and Reasoning under Uncertainty, ILLC Dissertation series 1998–03*. PhD thesis, University of Amsterdam, 1998.

Peter Grünwald, In-Jae Myung, and Mark Pitt. *Advances in Minimum Description Length Principle. Theory and Applications*. MIT Press, 2005.

Daniel M. Hausman and James Woodward. Independence, invariance and the causal Markov condition. *Bristish Journal For the Philosophy Of Science*, 50(4): 521–583, 1999.

David Heckerman, Dan Geiger, and David Chickering. Learning Bayesian networks: the combination of knowledge and statistical data. Technical Report MSR-TR-94-09, Microsoft Research, 1994.

Edward Herskovits. *Computer-Based Probabilistic Network Construction*. PhD thesis, Medical Information sciences, Stanford University, CA, 1991.

Daphne Koller and Avi Pfeffer. Object-oriented Bayesian networks. In *Proceedings of the Thirteenth Conference on Uncertainty in Artificial Intelligence (UAI-97)*, pages 302–313, 1997.

Kevin B. Korb and Erik Nyberg. The power of intervention. *Minds and Machines*, 16(3): 289–302, 2006.

Jan Lemeire. *Learning Causal Models of Multivariate Systems and the Value of it for the Performance Modeling of Computer Programs*. PhD thesis, Vrije Universiteit Brussel, 2007.

Ming Li and Paul M. B. Vitányi. *An Introduction to Kolmogorov Complexity and Its Applications*. Springer Verlag, 1997.

Richard E. Neapolitan. *Learning Bayesian Networks*. Prentice Hall, 2003.

Jonathan J. Oliver, David L. Dowe, and Chris S. Wallace. Inferring decision graphs using the minimum message length principle. In *Proceedings of the fifth Australian Joint Conference on Artificial Intelligence*, Tasmania, Australia, pages 361–367, 1992.

Judea Pearl. *Probabilistic Reasoning in Intelligent Systems: Networks of Plausible Inference*. San Mateo, CA, Morgan Kaufman Publishers, 1988.

Judea Pearl. *Causality. Models, Reasoning, and Inference*. Cambridge University Press, 2000.

Peter Spirtes, Christopher Meek, and Thomas Richardson. Causal inference in the presence of latent variables and selection bias. In *Proceedings of the 11th Conference on Uncertainty in Artificial Intelligence*, ed P. Besnard and S. Hanks, pages 499–506. Morgan Kaufmann, 1995.

Peter Spirtes, Clark Glymour, and Richard Scheines. *Causation, Prediction, and Search*, 2nd edition, Springer, Verlag, 1993.

Wolfgang Spohn. Bayesian nets are all there is to causal dependence. In *In Stochastic Causality, Maria Carla Galaviotti, Eds.* CSLI Lecture Notes, Springer, 2001.

Milan Studeny. On non-graphical description of models of conditional independence structure. In *HSSS Workshop on Stochastic Systems for Individual Behaviours*, Louvain la Neuve, Belgium, January 2001.

Joe Suzuki. Learning Bayesian belief networks based on the MDL principle: An efficient algorithm using the branch and bound technique. In *Proceedings of the International Conference on Machine Learning*, Bally, Italy, 1996.

Paul M. B. Vitányi. Meaningful information. In Prosenjit Bose and Pat Morin, editors, *ISAAC*, volume 2518 of *Lecture Notes in Computer Science*, pages 588–599. Springer, 2002.

Jon Williamson. *Bayesian Nets And Causality: Philosophical And Computational Foundations*. Oxford University Press, 2005.

Yang Xiang, S. K. Wong, and N. Cercone. Critical remarks on single link search in learning belief networks. In *Proceedings of the 12th Annual Conference on Uncertainty in Artificial Intelligence (UAI-96)*, pages 564–571. San Francisco, CA, Morgan Kaufmann Publishers, 1996.

·28·

Why making Bayesian networks objectively Bayesian makes sense

Dawn E. Holmes

Abstract

It is well-known that Bayesian networks are so-called because of their use of Bayes theorem for probabilistic inference. However, since Bayesian networks commonly use frequentist probabilities exclusively, is this sense they are not Bayesian. In this chapter it is argued that Bayesian networks that are objectively Bayesian, in other words those whose prior distribution is based on all and only the available information, have certain desirable properties and strengths over and above those based solely on the frequentist approach to probability. It is demonstrated, through an example, that these specially constructed graphical models may be used in otherwise intractable situations where data is unavailable or scarce and decisions need to be made.

28.1 Introduction

It is well-known that Bayesian networks are so-called, at least in part, because of their use of Bayes' theorem for probabilistic inference. However, such networks are not generally Bayesian as the term is used in Bayesian statistics because of the nature of their prior distributions, which use frequentist probabilities. Bayesian networks that are objectively Bayesian, in other words those whose prior distribution is derived from both subjective and frequentist probabilities, themselves based on all and only the available information and knowledge pertinent to the domain of application, have certain desirable properties and strengths over and above those based solely on the frequentist approach to probability. Jaynes, among others, supported the objective Bayesian interpretation of probability; in his view, probabilities should satisfy those constraints imposed on them by prior knowledge, whilst remaining as uncertain as possible. These specially constructed Bayesian networks may be used in otherwise intractable situations where data is unavailable or scarce and decisions need to be made. The crucial role of defining a prior distribution is the main topic for discussion, providing as it does the foundation upon which all decisions using a Bayesian network are based. Before focusing on the prior distributions of Bayesian networks, we discuss briefly the motivation for an objective Bayesian approach in general.

28.2 Objective Bayesianism

The paradoxes of probability emerged as part of the broader foundational crisis mathematics was experiencing in the early years of the twentieth century; indeed Keynes's Treatise on Probability (1921) was, by his own account, directly influenced by Russell's programme to reduce mathematics to logic. Whilst Keynes's view contrasts sharply with Venn's view that probabilities exist as a physical property of the world, he was determined to maintain the desire of the day for scientific objectivity and the supposed central role of logic as the all-pervading foundation. So, for Keynes, although probabilities were to be interpreted as rational degrees of belief, these were seen as leading to an objective or consensual belief. Since Keynes's foundational work is largely concerned with economics, it is hardly surprising that he was interested in the psychology of 'where do the numbers come from' and recognized that it was not always possible to assign numerical values to probabilities. However, in those cases where numerical values were deemed appropriate, he appealed to Bernoulli's principle of indifference (also known as the principle of insufficient reason). For Keynes, this was necessarily a logical principle and although some of the paradoxes it resulted in were eventually resolved by Keynes's modification of the principle, others proved intractable. See Ramsey (1931). Interestingly, but perhaps not surprisingly, just as Russell's programme led to paradox, so did Keynes's logical theory of probability.

The subjective theory of probability, proposed by both De Finetti (1974) and Ramsey (1931), came about as a response to the problems inherent in the logical theory that had become apparent by the 1920s. The crux of subjective theory is that probabilities are to be interpreted as an *individual's* rational degree of belief, in contrast with the *consensual* rational degrees of belief suggested by Keynes. A useful consequence of this interpretation is that repeated experimental trials are not necessary in order to assign probabilities; a feature which, as we will see, is particularly germane to the arguments presented below. Although there was initial criticism of this interpretation Ramsey (1931), Cox (1946), De Finetti (1974) and others successfully countered these arguments by showing that the axioms of probability may be justified in terms of their compliance with a number of rational, common-sense, intuitive or judicious properties. The choice of term is largely determined by the various authors' feel for the philosophical and psychological connotations attached to it, for example, Tribus (1969) favours 'rational'. This connection, between the formalization of probability as a set of axioms and the subjective interpretation of probability as a measure of belief in a proposition, has been made in different ways by various authors, notably Cox (1946), Schrodinger (1947a, b), Reichenbach (1949) and Tribus (1969). Each has shown that the axioms of probability are a necessary consequence of a set of properties, which we intuitively feel a measure of belief should have.

The question immediately arises as to how we determine these individual rational degrees of belief. Prior knowledge from relevant experience is the key. For example, in a medical scenario, a medical expert's individual rational degree of belief would quite clearly carry more weight than that of a non-medically trained person. As Suppes (2007) points out, there are thus many possible priors. So, although we may easily narrow our choice to domain-specific experts, it is not quite so simple to determine the optimal priors. On a practical level, subjective assessment is always possible and we can also perform sensitivity analysis subsequently. Clearly, we can start from a position in which total ignorance reigns, in which case we might consider it rational to assign equal probability to all possible outcomes; a position known as the principle of insufficient reason. Much of the work in this chapter relies on the application of this principle, which is justified initially by Jaynes's arguments supporting it as a useful tool in physics. See Jaynes (2003). For a detailed discussion of the psychological mechanisms at work in the formation of Bayesian priors, see Suppes (2007).

Jaynes developed a theory, based on Shannon and Weaver's (1948) information model, in which he propounded that an individual's probability should satisfy all the constraints known to be imposed on the system and further that from the many that fulfil this requirement, we should choose the one that maximizes entropy. Jaynes's interpretation of probability, known as the maximum entropy principle, is the fundamental principle of objective Bayesianism. See Jaynes (2003), Rosenkrantz (1977).

The theoretical work on which the current author's philosophical position is based is referred to in the foundational work of Williamson (2004, 2005) who is also very much in favour of applying maximum entropy to Bayesian networks. As noted in Rhodes *et al.* (1998) and Wiiliamson (2005) the main stumbling block in applying maximum entropy to Bayesian networks is that of computational feasibility. However, for the case under consideration in this chapter, that of multivalued, multiway trees, this did not prove to be an intractable problem; the linearity of the constraints leads to $O(n)$ processes and thus maximum entropy is applicable. In all other cases, non-linear constraints are unavoidable but even so closed-form solutions have been derived and the resulting algorithms, at least for inverted trees, lead to reliable approximations.

28.3 Bayesian networks

Bayesian networks, as first introduced by Pearl, grew out of a desire to represent multinomial probability distributions efficiently. Informally, a Bayesian network is a directed acyclic graph with nodes representing variables from a probability space. Each variable is independent of its non-descendents, given its parents and so the edges of the graph represent conditional probabilities. In order to construct a Bayesian network we must identify the information that influences our belief in a proposition as well as all the dependencies and independencies implied by this knowledge. Of course, we need to be sure that our graphical structure identifies just those independencies and dependencies that we have identified as pertinent. In order to do this, we use a result from graph theory, which provides a means of identifying the independencies under which any such system is constrained. *D*-separation, so-called in contrast to the analogous separation property of undirected graphs, comprises a set of rules devised by Pearl (1988) using which it is possible to decide for any set of variables in a network, whether they are independent from any other set of variables. *D*-separation is fundamental to the formal definition of a Bayesian network and also to the proof of Verma and Pearl's theorem, which determines all the conditional independencies implied by a directed acyclic graph in a Bayesian network. To define *d*-separation, let $\mathbf{G} = \langle \boldsymbol{V}, \boldsymbol{B} \rangle$ be a directed acyclic graph with $\boldsymbol{W} \subseteq \boldsymbol{V}$ and vertices u,v in $V - W$. Then u,v are said to be *d-separated* by $\boldsymbol{W}$ if every chain between u,v is blocked by $\boldsymbol{W}$. By 'chain' we mean a sequence of edges in the graph. If a chain is 'blocked' then there is a pair of arrows, somewhere on the chain, that meet either head-to-head, head-to-tail or tail-to-tail. Having defined *d*-separation, we may now define a Bayesian network. This particular form of definition is chosen so as to be useful later when maximum entropy is used, and because of this it contains certain redundancies; indeed, only constraint (2iii), those independence relationships implied by *d*-separation in the directed acyclic graph, is actually required, but it will be helpful to be able to refer to the constraints (2i) and (2ii) explicitly, in what follows.

Let:

(i) $\boldsymbol{V}$ be a finite set of vertices;

(ii) $\boldsymbol{B}$ be a set of directed edges between vertices with no feedback loops. The vertices together with the directed edges form a directed acyclic graph $\mathbf{G} = \langle \boldsymbol{V}, \boldsymbol{B} \rangle$;

(iii) a set of events be represented by the vertices of $\mathbf{G}$ and hence also represented by V, each event having a finite set of mutually exclusive outcomes;

(iv) E_i be a variable which can take any of the outcomes e_i^j of the event i, $j = 1, \ldots, n_i$;

(v) $\boldsymbol{P}$ be a probability distribution over the combinations of events, i.e. $\boldsymbol{P}$ consists of all possible $P\left(\bigcap_{i \in V} E_i\right)$.

Let **C** be the following set of constraints:

(2i) The elements of $\boldsymbol{P}$ sum to unity. (Although this constraint is already fulfilled by (v) above, it will be useful in what follows, to never to the sum to unity explicitly.)

(2ii) For each event i with a set of parents M_i there are associated conditional probabilities $P(E_i | \bigcap_{j \in M_i} E_j)$for each possible outcome that can be assigned to E_i and E_j. Probabilities are assumed non-zero.

(2iii) Those independence relationships implied by d-separation in the directed acyclic graph.

Then $\boldsymbol{N} = \langle \boldsymbol{G}, \boldsymbol{P}, \mathbf{C} \rangle$ is a Bayesian network if $\boldsymbol{P}$ satisfies **C**.

When the tree is represented by a graph, the nodes of the tree are the vertices of the graph and the branches of the tree are the edges of the graph. In this chapter, the topology of the graph will be restricted to the special case of a multivalued tree. The information in (2ii) above is therefore always of the form $\boldsymbol{P}(\boldsymbol{E}_{c_1} | \boldsymbol{M}_{c_1})$ except at the source nodes s, where it is of the form $P(E_s)$. The information required to specify the tree is thus given by:

(3i) the marginal probabilities of the sources nodes.

$$\boldsymbol{P}(e_i^j) = a_i^j; j = 1, \ldots, (n_{c_i} - 1) \text{ when } \boldsymbol{X}_i = \{\};$$

(3ii) the conditional probability of each non-source node c_i given the state of its parents

$$P(e_{c_i}^m | M_{c_1}) = \beta(c_i^m, M_{c_1})$$

for $m = 1, \ldots, (n_{c_i} - 1)$ when $\boldsymbol{X}_{c_1} \neq \{\}$ where α_i^j and $\beta(c_i^m, M_{c_1})$ are constants.

The general state S of the network is given by the conjunction $\bigcap_{i \in V} E_i$. A particular state is obtained by assigning some e_i^j to each E_i. Subtrees can have their states defined in a similar manner. Hence, any tree or subtree can only be in one of a finite set of states and the state of the tree is the conjunction of the states of any set of subtrees which constitutes a partition of the tree.

As mentioned previously, all independencies in a Bayesian network can be identified through Verma and Pearl's theorem. The reader is referred to Neapolitan (1990) for a proof of this theorem, which will be used later. We note that d-separation finds all independencies in a system and not a minimally sufficient set. When it comes to estimating missing or unknown

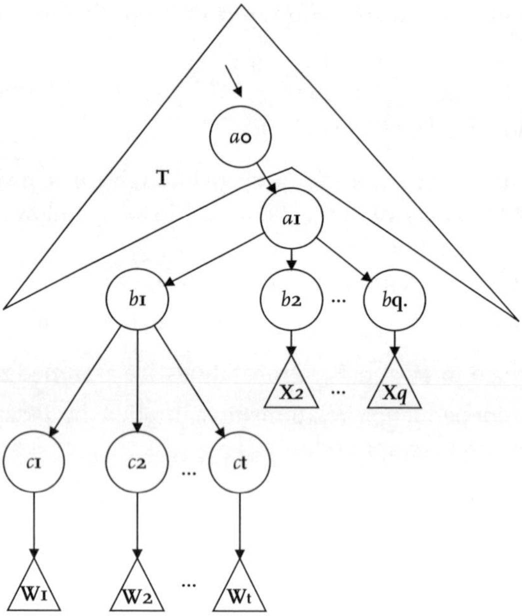

Fig. 28.1 A multiway Bayesian tree with a small part shown in detail.

probabilities in a Bayesian network using maximum entropy, we find the smallest set of independencies that is algebraically equivalent to the set determined by d-separation. The chain rule is used then in conjunction with d-separation to determine the joint probability distribution of the system being considered.

Figure 28.1 represents part of the induced graph of a Bayesian tree **N**. The nodes $a_0, a_1, b_1, \ldots, b_q, c_1, \ldots, c_t$ etc. are vertices of **V**. Triangles represent all the subtrees below a node and are labelled with bold capital letters. The polygon labeled **T** indicates the entire graph except for the subtree rooted at $a_1$and is itself a multiway Bayesian tree. A state S_i of the Bayesian tree is an assignment of some outcome e_i^j, to each variable E_i. For notational convenience, we note that $S_i = \bigcap_{j \in V} e_j^{\mathrm{sel}_j(i)}$ where sel_j is a solution function which chooses the outcome j that corresponds to the ith state. Subtrees can have their states defined in a similar manner and hence any tree or subtree can only be in one of a finite set of states and the state of the tree is the intersection of the states of its subtrees.

Let $\boldsymbol{T}$ denote a state of **T** in Figure 28.1, $\mathbf{X}_2 \ldots \mathbf{X}_q$ and $\mathbf{W}_1 \ldots \mathbf{W}_t$ denote the descendants of $b_2 \ldots b_q$ and $c_1 \ldots c_t$ respectively, and $X_2 \ldots X_q$ and $W_1 \ldots W_t$ denote a state of $\mathbf{X}_2 \ldots \mathbf{X}_q$ and $\mathbf{W}_1 \ldots \mathbf{W}_t$ respectively. Then the probability of any state S of **N** is given by:

$$P(S) = P(T, E_{a_1}, E_{b_1} \dots E_{b_q}, E_{c_1} \dots E_{c_t}, X_2 \dots X_q, W_1 \dots W_t). \tag{28.1}$$

Expanding (28.1) using the chain rule and simplifying using those independence relationships implied by d-separation we get:

$$\begin{aligned} P(S) = {} & P(T)\,P(E_{a_1} \mid E_{a_0})_{a_0 \in T}\,P(E_{b_1} \mid E_{a_1})\,P(E_{b_2} \mid E_{a_1}) \dots P(E_{b_q} \mid E_{a_1}) \\ & P(X_2 \mid E_{b_2}) \dots P(X_q \mid E_{b_q})\,P(E_{c_1} \mid E_{b_1})\,P(E_{c_2} \mid E_{b_1}) \dots P(E_{c_t} \mid E_{b_1}) \\ & P(W_1 \mid E_{c_1})\,P(W_2 \mid E_{c_2}) \dots P(W_t \mid E_{c_t}). \end{aligned} \tag{28.2}$$

Consider next the state of the Bayesian tree **N** where all variables are assigned arbitrary values; event a_1 is instantiated by $e^i_{a_1}$, event b_1 is instantiated by $e^j_{b_1}$, and so on. The instantiation of events in **T**by arbitrary values is denoted by T^u, those in **X** by X^x and those in **W** by W^w. Denote this state by $S_{b^j_1, a^i_1}$. Then,

$$P(S_{b^j_1, a^i_1}) = P(T^u, e^i_{a_1}, e^j_{b_1}, e^{i_2}_{b_2} \dots e^{i_q}_{b_q}, e^{j_1}_{c_1}, e^{j_2}_{c_2} \dots e^{j_t}_{c_t}, X^{x_2}_2 \dots X^{x_q}_q, W^{w_1}_1 \dots W^{w_t}_t). \tag{28.3}$$

Finally, consider the state $S_{b^n_1, a^i_1}$ where the eventb_1 is instantiated by its final or n^{th} value$e^n_{b_1}$, and all other events take on the same values as before. Thus:

$$P(S_{b^n_1, a^i_1}) = P(T^u, e^i_{a_1}, e^n_{b_1}, e^{i_2}_{b_2} \dots e^{i_q}_{b_q}, e^{j_1}_{c_1}, e^{j_2}_{c_2} \dots e^{j_t}_{c_t}, X^{x_2}_2 \dots X^{x_q}_q, W^{w_1}_1 \dots W^{w_t}_t). \tag{28.4}$$

Dividing (28.3) by (28.4) after first expanding both equations results in:

$$\frac{P\left(S_{b^j_1, a^i_1}\right)}{P\left(S_{b^n_1, a^i_1}\right)} = \frac{\beta(b^j_1, a^i_1) \prod_{k=1}^{k=t} \beta(c^{j_k}_k, b^j_1)}{\beta(b^n_1, a^i_1) \prod_{k=1}^{k=t} \beta(c^{j_k}_k, b^n_1)}. \tag{28.5}$$

This ratio is of particular interest because so many of the terms have cancelled and we will return to it subsequently.

28.4 The principle of maximum entropy and Bayesian networks

Bayesian networks can only be used for inference when all the information in 3(i) and 3(ii) has been provided; that is, when a prior distribution has been specified. Prior information, knowledge or experience, the modelling of which is an integral part of problem solving using Bayesian networks, is ignored by the classical paradigm. Objective Bayesianism contends that each problem as uniquely defined by its context and the experience of the analyst; this experience being expressed by the prior distribution. So the prior distribution incorporates what is known, both from experimentation relevant as it applies to a particular problem in the form of frequentist probabilities and as expert knowledge.

Although it is a necessary condition that each person's degrees of belief are consistent, i.e. they must obey the laws of probability, different people may have different degrees of belief in the same proposition because they have different information, knowledge or experience. Using frequentist probabilities alone results in a rigid network since new knowledge regarding prior probabilities cannot be accommodated once the network is built and clearly the results of Bayesian networks are only as good as their prior distribution. The term 'expert system' became unfashionable and was replaced by the term 'intelligent system' but the former seems more apposite to the present discussion since these systems rely on experts for their accuracy and usefulness, rather than on data exclusively.

Jensen (1996) points out that in a given domain there may be no sound theoretical method for determining all the required probabilities, and gives examples of how they are ascertained in practice. Sometimes they are guessed; sometimes a complex heuristic procedure is devised in order to produce an approximate and necessarily biased value. When multivalued events are to be modelled, the situation becomes complex. In some situations, there are inevitably too many conditional probabilities for an expert to reliably estimate. Thus, the need for a theoretically sound technique for estimating them in a minimally prejudiced fashion becomes apparent. The maximum entropy formalism, which grew out of the principle of insufficient reason, first formulated by Bernoulli, provides just such a technique.

The principle of insufficient reason states that if an event has many possible outcomes and there is insufficient reason for doing otherwise, equal values should be assigned to the probability of each possible outcome. A generalization of this principle, the principle of maximum entropy, is capable of determining a probability distribution for *any* combination of partial knowledge and partial ignorance; further, it has been shown by Jaynes (2003) to give the minimally prejudiced distribution. When ignorance is complete, to say that a distribution is minimally prejudiced is simply to say that it complies with the principle of insufficient reason. Hence, if we are completely ignorant about a situation, and have no evidence relevant to the probability distribution, we require the principle of maximum entropy to choose the uniform distribution and be equivalent to the principle of indifference and indeed, it has been proved that the uniform distribution has maximum entropy among all distributions supported on an interval.

The maximum entropy solution to a given problem takes into account ignorance of a situation, whilst conforming to the current state of knowledge and thus takes into account that ignorance may be reduced when more information becomes available. In the case where our ignorance is not total, we require that our inferences from Bayesian networks are based on all and only the information or knowledge we have available. To this end, we need to eliminate all those probability distributions that are in conflict with this knowledge and choose the one that maximizes entropy. To do this, we solve the optimization problem in which we find the maximum entropy distribution subject to those constraints that are imposed by our existing knowledge. Maximum entropy thus provides a technique for eliciting knowledge from incomplete information, without making any groundless assumptions. Since entropy can be interpreted as a mathematical measure of ignorance or partial knowledge, where higher entropy corresponds to greater ignorance, we are able to obtain knowledge that is consistent with the available information, but otherwise has maximum entropy.

There have been many justifications of the use of entropy as a measure of uncertainty. Those based on Shannon's (1948) original argument include Jaynes (2003) and Tribus (1969). Each has its own attractions. Wallis (2003) presents an argument that does not rely on any connection between probability and frequency. Hence, it does not assume that prior distributions result from repeatable experiments and this is highly attractive to those committed to Bayesian techniques. The many arguments supporting the use of entropy as a measure of uncertainty suggests that none is definitive and indeed, mathematical derivations continue to proliferate. Following a conjecture of Jaynes (2003), Shore and Johnson (1980) proved that any information measure other than entropy would eventually lead to inconsistencies. Shore and Johnson have also shown that the principle of maximum entropy can be derived from a set of consistency axioms for uncertain reasoning. Indeed, Klir and Folger (1995) note that any one of a set of requirements is sufficient for the characterization of Shannon entropy.

Whilst this is in itself cheering, it does not provide a link to real-world reasoning that the non-mathematician is likely to be convinced by. Much of the reasoning familiar to those who might wish to use a decision-support system or expert system, uses what is typically thought of as common-sense. In an attempt to fill this gap Paris and Vencovska (1996a) have evolved a set of desiderata which characterize common-sense reasoning. Paris and Vencovska (1997b) have proved that only the maximum entropy inference process conforms to the tenets of common-sense reasoning thus defined. These common-sense principles provide a means of convincing the end-user of an expert system that the support it gives is reliable or at least explicable.

When the prior information that individuals have concerning a problem differs, this results in different probability distributions, hence the results provided by maximum entropy are dependent on the person providing the prior information. Maximum entropy provides a rule for inductive reasoning, which is attractive to researchers in many fields. Most of the research in maximum entropy has focused on problems in physics. Application of the principle to areas as diverse as town planning and crystallography, as well as physics, is now being made. However, image processing is one of the current dominant areas of activity and Cheeseman has applied the principle of maximum entropy to this problem in the LandSat project. In particular, Cheeseman's work has made extensive use, at different times, of both Bayesian techniques and maximum entropy.

Using maximum entropy, we can thus accurately model subtle dependencies among variables in a way that is simply not possible with traditional predictive modelling techniques. This is particularly useful when modelling real-world problems, because nearly all of these involve prior information that is not reflected in other techniques. Traditional predictive models, such as decision trees, logistic regression, and neural networks, make assumptions about their data. However, in the maximum entropy model, all the available information is used to form the constraints for the optimization problem *and nothing else*. Hence, maximum entropy does not make any assumptions about its data and thus provides an unprejudiced distribution. Furthermore, of all the possible probability distributions that fit any available data, we are required to choose the one that maximizes entropy within the constraints imposed. If we were to choose any other it would necessarily have lower entropy and this would imply that information or knowledge other than that available had been incorporated.

A further objection is that since the probabilities obtained using maximum entropy have no experimental basis, they cannot give rise to physical predictions. In response to this objection, Jaynes (2003) points out that maximum entropy is applicable both in situations where a problem is represented by a single situation, where a repeatable experiment is not possible and thus the probabilities have no frequency connection and also, in situations where observed frequencies are available, in which case he demonstrates that the maximum entropy probabilities have a formal connection with frequencies. Much work has been done in this area and the reader is referred to the annual *Bayesian Inference and Maximum Entropy Methods in Science and Engineering* conference proceedings. Indeed, one important aspect of this work that is not covered here is that we must prove that the maximum entropy probability distribution preserves the conditional independence relationships implied by *d*-separation in the Bayesian model. See Holmes (1998). These conditional independence relationships are, in fact, only preserved in the case of a tree; in all other cases, the maximum entropy formalism does not preserve them and they must, therefore, be included as explicit nonlinear constraints.

In the case under discussion, that of a Bayesian tree, the constraints are all linear. However, when an inverted tree or a generalized Bayesian network is considered, nonlinear constraints must also be taken into consideration. For details of these cases the reader is referred to Holmes (2005). The current paper deals only with the case of discrete random variables with linear constraints. A further development by Holmes (2006), is the incorporation of inequality constraints. However, several authors have expressed concern about the more general use of the maximum entropy principle; in particular, Neapolitan (1991) offers an alternative to maximum entropy where interval constraints are concerned. In the case of a Bayesian network, any such constraint must be stated explicitly prior to maximization and this work has yet to be completed. However, following Neapolitan, suppose that we have a three-sided die, with sides labelled 1, 2 and 3, then in the case of total ignorance, the maximum entropy formalism assigns equal values as estimates of the unknown probabilities, as does Neapolitan's principle of interval constraints. We can further add that, should information become available, for example, $P(2) = 0.5$ then the maximum entropy formalism will assign $P(1) = 0.25$ and $P(3) = 0.25$, again the same as Neapolitan's generalization of the symmetric Dirichlet distribution. Since both Neapolitan's principle of interval constraints and the principle of maximum entropy are based on the principle of indifference, it is not surprising that both methods yield the same results in this case.

It can be shown that maximizing entropy is computationally infeasible in general. In a naive implementation, the time required to make a prediction is proportional to the number of possible outcomes, and the time required to optimize the model is proportional to the number of logically possible events. In real-world prediction problems, the number of logically possible events is infeasibly large. For such problems, a naïve implementation would require more computation than is available on all the world's computers. However, for problems that can be modeled using two-valued singly connected causal networks, it has been shown computationally feasible. We thus use the maximum entropy formalism to optimally estimate the prior distribution of a Bayesian network, using all and only the information available.

28.5 Maximum entropy in Bayesian networks

Consider again the knowledge domain represented by the multiway tree in Figure 28.1. The number of states N_S in the Bayesian tree **N** is given by $N_S = \prod_{i \in V} n_i$ where n_i is the number of values taken by the ith event. Denoting states by S_i where $i = 1, \ldots, N_S$, the maximum entropy formalism requires us to maximize $H = -\sum_{i=1}^{N_s} P(S_i) \ln P(S_i)$ whilst conforming to the constraints 2(i)–(iii). The requirement that the probability distribution sums to unity, referred to as the zeroth constraint is given by $\sum_{i=1}^{N_x} P(S_i) = 1$. Using the Lagrange multiplier technique to maximize H, Jaynes (2003) showed that the general expression for the probability of a state given a set of linear constraints is given by $P(S_i) = \prod_{j=0}^{N_C} \exp(-\lambda_j \sigma_{i,j})$ with $i = 1, \ldots, N_S$ where $\sigma_{i,j}$ is the coefficient of the ith state in the jth constraint and N_c is the number of required constraints excluding the sum to unity. Holmes (1999) has extended these results and applied them to Bayesian networks to find an expression for the ratio of probabilities of states found in (28.5).

Let $C_T, C_{X_2}, \ldots, C_{X_q}, C_{W_1}, \ldots, C_{W_t}$ be the constraints governing states $\boldsymbol{T}$, $X_2, \ldots, \boldsymbol{X}_q, \boldsymbol{W}_1, \ldots, W_t$ respectively. Then the ratio of the probabilities of any two states in which the probabilities of these states remain the same, will not contain any terms arising from $C_T, C_{X_2}, \ldots, C_{X_q}, C_{W_1}, \ldots, C_{W_t}$ since they will cancel. Similarly, the Lagrangian terms arising from the edges $\langle a_1, b_2\rangle, \ldots, \langle a_1, b_q\rangle$ will cancel if the outcomes of $a_1, b_1, \ldots, b_q$ are the same in both numerator and denominator. However, Lagrangian terms associated with the edges $\langle a_1, b_1\rangle\langle b_1, c_1\rangle, \ldots, \langle b_1, c_t\rangle$ will not cancel. Consider the constraints arising from $\langle a_1, b_1\rangle$, which will be denoted as follows:

$$P(e_{b_1}^j | e_{a_1}^i) = \beta(b_1^j, a_1^i).$$

Holmes (1999) has shown that by first expressing these constraints in terms of marginal probabilities, it is possible to re-express them in terms of state probabilities thus:

$$(1 - \beta(b_1^j, a_1^i)) \sum_{x \in X} P(S_x) - \beta(b_1^j, a_1^i) \sum_{y \in Y} P(S_y) = 0. \tag{28.6}$$

where

$$X = \left\{ x | \sum_x P(S_x) = P(e_{a_1}^i e_{b_1}^j) \right\} \text{ and } Y = \left\{ y | \sum_y P(S_y) = \sum_{\substack{k=1 \\ k \neq j}}^{k=n} P(e_{a_1}^i e_{b_1}^k). \right\}$$

Equation (28.6) determines a family of constraint equations for the edge $\langle a_1, b_1\rangle$. When information is complete, that is when all the probabilities in 2(i)–2(iii) are known, every edge will have such a family of constraint equations associated with it. With each constraint we associate a Lagrange multiplier λ. For the edge $\langle a_1, b_1\rangle$ the Lagrange multipliers are $\lambda(b_1^1, a_1^1), \lambda(b_1^1, a_1^2), \ldots, \lambda(b_1^n, a_1^p)$ where event a_1 has p outcomes and event b_1 has n outcomes. λ_0 denotes the Lagrange multiplier associated with the zeroth constraint. Partially differentiating constraints expressed in the form of state probabilities, an expression analogous to that found in (28.5) for the probability of states is found:

$$\frac{P(S_{b_1^j,a_1^i})}{P(S_{b_1^n,a_1^i})} =$$

$$\exp\left(-\lambda\left(b_1^j a_1^i\right)\right)\prod_{x=1}^{t}\left(\frac{\exp\left(\left(-\lambda\left(c_x^{j_x} b_1^j\right)\right)\right)}{\exp\left(\left(-\lambda\left(c_x^{j_x} b_1^n\right)\right)\right)}\prod_{k=1}^{n_{cx}}\frac{\exp\left(\left(-\lambda\left(c_x^k b_1^j\right)\right)\left(-\beta\left(c_x^k b_1^j\right)\right)\right)}{\exp\left(\left(-\lambda\left(c_x^k b_1^n\right)\right)\left(-\beta\left(c_x^k b_1^n\right)\right)\right)}\right). \tag{28.7}$$

Equating (28.5) and (28.7) leads to the following expression for missing information in the Bayesian network in Figure 28.1.

$$\exp\left(-\lambda\left(b_1^j a_1^i\right)\right) = \frac{\beta\left(b_1^j, a_1^i\right)}{\beta\left(b_1^m, a_1^i\right)}\prod_{k=1}^{t}\frac{\beta\left(c_k^{j_k}, b_1^j\right)}{\beta\left(c_k^{j_k}, b_1^m\right)}\prod_{x=1}^{t}$$

$$\times\left(\frac{\exp\left(\left(-\lambda\left(c_x^{j_x} b_1^m\right)\right)\right)}{\exp\left(\left(-\lambda\left(c_x^{j_x} b_1^j\right)\right)\right)}\prod_{k=1}^{n_{cx}}\frac{\exp\left(\left(-\lambda\left(c_1^k b_1^m\right)\right)\left(-\beta\left(c_1^k b_1^m\right)\right)\right)}{\exp\left(\left(-\lambda\left(c_1^k b_1^j\right)\right)\left(-\beta\left(c_1^k b_1^j\right)\right)\right)}\right). \tag{28.8}$$

Thus, all Lagrange multipliers on the edge between a given node and its children can be found in terms of known probabilities and the Lagrange multipliers associated with the edges between the children and the grandchildren. We are particularly interested in the situation where one or more of the prior probabilities is unknown. By removing those constraint equations for which there is no information given, it is possible to estimate the missing information. Suppose that $\beta(b_1^j, a_1^i)$ is missing, in which case the maximum entropy solution no longer generates the term $\exp(-\lambda(b_1^j, a_1^i))$ and (28.8) becomes

$$\frac{{}_{mp}\beta\left(b_1^j, a_1^i\right)}{\beta\left(b_1^m, a_1^i\right)}\prod_{k=1}^{t}\frac{\beta\left(c_k^{j_k}, b_1^j\right)}{\beta\left(c_k^{j_k}, b_1^m\right)} =$$

$$\prod_{x=1}^{t}\left(\frac{\exp\left[\left(-\lambda\left(c_x^{j_x}, b_1^j\right)\right)\right]}{\exp\left[\left(-\lambda\left(c_x^{j_x}, b_1^m\right)\right)\right]}\prod_{k=1}^{n_{cx}}\frac{\exp\left[\left(-\lambda\left(c_x^k, b_1^j\right)\right)\left(-\beta\left(c_x^k, b_1^j\right)\right)\right]}{\exp\left[\left(-\lambda\left(c_x^k, b_1^m\right)\right)\left(-\beta\left(c_x^k, b_1^m\right)\right)\right]}\right)$$

where ${}_{mp}\beta(b_1^j, a_1^i)$ denotes a minimally prejudiced estimate of $\beta(b_1^j, a_1^i)$. For example, if b_1 is a leaf node and all information is missing equal probabilities are assigned by this method.

Since estimating the probabilities at the leaves is always possible, all the missing information can be estimated by propagating up the tree, using a post-order traversal, to estimate missing information as required. It is then possible to use a pre-order traversal of the tree to propagate the information down the tree calculating values for marginals as it proceeds. Both of these are $O(n)$ operations, so the marginals can be found with an $O(n)$ process when the maximum entropy method has been used to estimate missing information. For details of special cases see Holmes and Rhodes (1998).

The conceptual justification of the formula (28.8) is that it incorporates maximum uncertainty. In failing to model ignorance, an unnecessarily biased prior distribution would have been used, leading to untenable conclusions. One of the advantages of the maximum entropy formalism is that it may lead to a very broad probability distribution, the interpretation of which is that there is insufficient information upon which to base a prediction. The distribution can always be interpreted in terms of what we are able to predict, based on the information given as a system of constraints.

In the case under discussion, that of a Bayesian tree, the constraints are all linear. However, when an inverted tree or a general Bayesian network are considered, non-linear constraints must also be taken into consideration. For details of these cases the reader is referred to Holmes (2005). A further development is the incorporation of inequality constraints (Holmes (2006)). This work is, however, far from complete.

28.6 Applications

Bayesian networks are being used in many applications; from ecology and medicine to printer trouble-shooting and educational testing. For an example of their use in ecological modeling, see McCann (0000). Sometimes reliable data is available and within an objective Bayesian interpretation, there is no problem in using it; however, it still presents problems for representation by a Bayesian network since, as well as the inflexibility of the prior distribution, it requires the closed-world assumption for success. In any situation where an expert has provided the underlying graphical structure, he will undoubtedly be able to estimate some of the information required by the Bayesian network. However, it is almost impossible for a person to provide an actual probability distribution. For example, in cases of rare diseases, even the most experienced expert may have no usable prior knowledge. The National Institute of Health classifies a rare disease as one with fewer than 200,000 cases in the country and so diseases exist that even he most experienced specialist may not see in an entire career. On a practical level, to conclude that a useful prior probability of any rare disease is $< 200,000/N$ is a useful prior, although strictly accurate, is not a directly usable piece of prior knowledge in a Bayesian network. Using the maximum entropy approach provides the option of distinguishing between $P(\text{disease}) < 200,000/N$ and $P(\text{disease}) = 200,000/N$; we can set the probability of the disease as missing and find the optimal value for it based on $P(\text{disease}) < 200,000/N$ and all the other information in the network. An experienced, expert health-care professional can provide a better estimate of the probability of a rare disease than that given by this definition and it is this valuable subjective knowledge that we wish to use.

The maximum entropy method would be particularly useful in models when empirical data is missing or where we wish to test the effect of hypothesized interactions between variables.

28.7 Conclusion

We have discussed objective Bayesianism as it relates to Bayesian networks. It has been argued that only one of the many possible prior distributions is the correct one to use in a given knowledge domain and a method was described for constructing such a minimally prejudiced prior distribution. Bayesian networks designed to assist diagnosis must provide theoretically sound priors for very low probability outcomes and this is precisely what maximum entropy affords.

References

Cox R. (1946). Probability, frequency and reasonable expectation. *American Journal of Physics*, **14**, 1–13.

De Finetti, B. (1974). *Theory of Probability* (translation of 1970 book), two volumes. New York: Wiley.

Holmes, D.E. Independence relationships implied by D-separation in the Bayesian model of a causal tree are preserved by the maximum entropy model, *Bayesian Inference and Maximum Entropy Methods in Science and Engineering*, AIP Conference Proceedings, Melville New York 567, edited by Joshua T. Rychert, Gary J. Erickson and C. Ray Smith.

Holmes, D.E. (1999). Efficient estimation of missing information in multivalued singly connected networks using maximum entropy. In *Maximum Entropy and Bayesian Methods*. pp. 289–300. W. von der Linden, V. Dose, R. Fischer and R. Preuss. (eds.) Dordrecht: Kluwer Academic.

Holmes, D.E. (2005). Optimizing inequality constrained priors in bayesian networks. *Bayesian Inference and Maximum Entropy Methods in Science and Engineering*. American Institute of Physics 25th Conference Proceedings, American Institute of Physics.

Holmes, D.E. (2006). Toward a generalized Bayesian network. *Bayesian Inference and Maximum Entropy Methods in Science and Engineering*. 26th Conference Proceedings, American Institute of Physics.

Holmes, D.E. (2006). Toward a generalized Bayesian network. *Bayesian Inference and Maximum Entropy Methods in Science and Engineering.* 26th Conference Proceedings, American Institute of Physics.

Holmes, D.E. and Rhodes, P.C. (1998). Reasoning with incomplete information in a multivalued causal tree using the maximum entropy formalism. *International Journal of Intelligent Systems.* Vol. 13 No.9 September 1998 pp. 841–859.

Holmes, D.E., Rhodes, P.C. and Garside, G.R. (1998). Efficient computation of marginal probabilities in multivalued causal inverted multiway trees given incomplete information.' *International Journal of Intelligent Systems.*

Jaynes, E. T. (2003). *Probability Theory: The Logic of Science,* Cambridge University Press.

Jensen Finn, V. (1996). *An Introduction to Bayesian Networks.* London: UCL Press.

Keyness, J.M. (1921). *Treatise on Probability.* London: Macmillan.

Klir, G.J. and Folger, T.A. (1995). *Fuzzy Sets, Uncertainty and Information.* Englewood Clips: Prentice-Hall.

McCann, R. *et al.* Bayesian belief networks: Applications in resource management. *Canadian Journal of Forest Research,* **36**, 3053–3062.

Neapolitan, R.E. (1990). *Probabilistic Reasoning in Expert Systems.* New York: John Wiley.

Neapolitan, R.E. (1991). The principle of interval constraints, a generalization of the symmetric Dirichlet distribution, *Mathematical Biosciences.*

Paris, J.B. and Vencovska, A. (1996a). Some Observations on the Maximum Entropy Inference Process. *Technical Report L1-96.* Department of Mathematics, University of Manchester, UK.

Paris, J.B. and Vencovska, A. (1996b). Some observations on the maximum entropy inference process. *Technical Report L1-96.* Department of Mathematics, University of Manchester, UK.

Pearl, J. (1988). *Probabilistic Reasoning in Intelligent Systems. Networks of Plausible Inference.* Morgan Kaufmann.

Ramsey, F.P. (1931). *The Foundations of Mathematics and other Logical Essays.* London: Routledge and Kegan Paul.

Reichenbach, H (1949). *Theory of Probability* Berkeley. University of California Press.

Rosenkrantz, R. (1977). *Inference, Method and Decision: Towards a Bayesian Philosophy of Science,* Dordrecht: D. Reidel.

Schrodinger, E. (1947). The foundation of the theory of probability I. *Proceedings of the Royal Irish Academy,* Series A, **51**, pp. 51–66.

Schrodinger, E. (1947). *The Foundation of the theory of Probability II. Proceedings of the Royal Irish Academy,* Series A, **51**, pp. 141–146.

Shannon, C.E. (1948). *A mathematical theory of communication.* The Bell System Technical Journal, **27**, pp. 379–423, 623–656.

Shannon, C.E. and Weaver, W. (1948). *The Mathematical Theory of Communication.* University of Illinois Press.

Shore, J.E. and Johnson R.W. (1980). Axiomatic derivation of the principle of maximum entropy and the principle of minimum cross-entropy. *IEEE Transactions On Information Theory. IT*-26 pp. 26–37.

Suppes, P (2007). Where do Bayesian Priors come from? *Synthese,* 2007 **156**, 441–471.

Tribus, M. (1969). *Rational Descriptions, Decisions and Designs.*Elmsford NY. Pergamon.

Williamson, J. (2004). 'Objective Bayesian nets', in: S. Artemov, H. Barringer, A.S. d'Avila Garcez, L.C. Lamb and J. Woods (eds.), *We Will Show Them: Essays in Honour of Dov Gabbay*, Vol. 2, pp. 713–730. London: College Publications.
Williamson, J. (2005). *Bayesian Nets and Causality: Philosophical and Computational Foundations*. Oxford University Press.
Wallis, G. in Jaynes, E.T. (2003). *Probability Theory: The Logic of Science*. Cambridge University Press.

·29·

Probabilistic measures of causal strength

Branden Fitelson and Christopher Hitchcock

Abstract

A number of theories of causation posit that causes raise the probability of their effects. In this chapter, we survey a number of proposals for analysing causal strength in terms of probabilities. We attempt to characterize just what each one measures, discuss the relationships between the measures, and discuss a number of properties of each measure.

One encounters the notion of 'causal strength' in many contexts. In linear causal models with continuous variables, the regression coefficients (or perhaps the standardized coefficients) are naturally interpreted as causal strengths. In Newtonian mechanics, the total force acting on a body can be decomposed into component forces due to different sources. Connectionist networks are governed by a system of 'synaptic weights' that are naturally interpreted as causal strengths. And in Lewis's account of 'causation as influence' (Lewis 2000), he claims that the extent to which we regard one event as a cause of another depends upon the degree to which one event 'influences' the other. In this chapter, we examine the concept of causal strength as it arises within probabilistic approaches to causation. In particular, we are interested in attempts to measure the causal strength of one binary variable for another in probabilistic terms. Our discussion parallels similar discussions in confirmation theory, in which a number of probabilistic measures of degree of confirmational support have been proposed. Fitelson (1999) and Joyce (MS) are two recent surveys of such measures.

29.1 Causation as probability-raising

The idea that causes raise the probabilities of their effects is found in many different approaches to causation. In probabilistic theories of causation, of the sort developed by Reichenbach (1956), Suppes (1970), Cartwright (1979), Skyrms (1980), and Eells (1991), C is a cause of E if C raises the probability of E in fixed background contexts. We form a partition $\{A_1, A_2, A_3, \ldots, A_n\}$, where each A_i is a background context. Then C is a cause of E in context A_i just in case $P(E|C \wedge A_i) > P(E| \sim C \wedge A_i)$, or equivalently, just in case $P(E|C \wedge A_i) > P(E|A_i)$.[1] The idea is that each background context controls

[1] Note that both inequalities fail, albeit for different reasons, if $P(\sim C|A_i) = 0$.

for confounding causes of E, so that any correlation that remains between C and E is not spurious. According to Cartwright (1979), each background context should hold fixed (either as being present, or as being absent), every cause of E that is not itself caused by C. Eells (1991) has a similar proposal. If we construct the background contexts in this way, we would expect the conditional probabilities of the form $P(E|C \wedge A_i)$ and $P(E| \sim C \wedge A_i)$ to take values of 0 or 1 if E is caused deterministically. However, as Dupré (1984) points out, this carves up the background conditions more finely than is needed if the goal is simply to avoid confounding. For this purpose, it suffices to hold fixed the common causes of C and E. If we construct the more coarse-grained partition in this way, the conditional probabilities $P(E|C \wedge A_i)$ and $P(E| \sim C \wedge A_i)$ might take intermediate values even if determinism is true. An issue remains about what it means to say that C causes E *simpliciter*: whether it requires that C raise the probability of E in *all* background contexts (the proposal of Cartwright 1979 and Eells 1991), whether it must raise the probability of E in some contexts and lower it in none (in analogy with Pareto-dominance, the proposal of Skyrms 1980), or whether C should raise the probability of E in a weighted average of background contexts (this is, essentially, the proposal of Dupré 1984; see Hitchcock 2003 for further discussion). We will avoid this issue by confining our discussion to the case of a single background context.

In his paper (1986), Lewis offers a probabilistic version of his counterfactual theory of causation. Lewis says that E *causally depends upon* C just in case (i) C and E both occur, (ii) they are suitably distinct from one another, (iii) the probability that E would occur at the time C occurred was x, and (iv) the following counterfactual is true: if C had not occurred, the probability that E would occur would have been substantially less than x. Lewis takes causal dependence to be sufficient, but not necessary, for causation proper. In cases of preemption or overdetermination, there can be causation without causal dependence. We will largely ignore this complication here. The reliance on counterfactuals is supposed to eliminate any spurious correlation between C and E. The idea is that we evaluate the counterfactual 'if C had not occurred . . . ' by going to the nearest possible world in which C does not occur. Such a world will be one where the same background conditions obtain. So common causes of C and E get held constant on the counterfactual approach, much as they do in probabilistic theories of causation.

The interventionist approach to causation developed by Woodward (2003) can also be naturally extended to account for probabilistic causation. The idea would be that interventions that determine whether or not C occurs result in different probabilities for the occurrence of E, with interventions that make C occur leading to higher probabilities for E than interventions that prevent C from occurring. The key idea here is that interventions are exogenous, independent causal processes that override the ordinary causes of

C. Thus even if C and E normally share a common cause, an intervention that determines whether or not C occurs disrupts this normal causal structure and brings C or $\sim C$ about by some independent means.

29.2 Assumptions

We will remain neutral about the metaphysics of causation, and about the best theoretical approach to adopt. For definiteness, we will work within the mathematical framework of probabilistic theories of causation. Conditional probabilities are simpler and more familiar than probabilities involving counterfactuals or interventions, although the latter are certainly mathematically tractable (e.g. in the framework of Pearl 2000). We will assume that we are working within one particular background context A_i. Within this context, C and E will be correlated only if C is causally relevant to E. We will leave open the possibility that the context is not specified in sufficient detail to ensure that the conditional probabilities $P(E|C \wedge A_i)$ and $P(E| \sim C \wedge A_i)$ take extreme values if determinism is true. To keep the notation simple, however, we will suppress explicit reference to this background context. Moreover, when we are considering more than one cause of E, C_1 and C_2, we will assume that the background condition also fixes any common causes of C_1 and C_2. In addition, we shall assume that C_1 and C_2 are probabilistically independent in this background context. This means that we are ignoring the case where C_1 causes C_2 or vice versa.

In all of our examples, we will assume binary cause and effect variables, X_C and X_E, respectively. These can take the values 1 and 0, representing the occurrence or non-occurrence of the corresponding events. We will also write C as shorthand for $X_C = 1$, and $\sim C$ as shorthand for $X_C = 0$, and analogously for X_E. We will have a probability function P defined over the algebra generated by X_C and X_E, and also including at a minimum the relevant background context. P represents some type of objective probability. We do not assume that this objective probability is irreducible. For instance, it may be possible to assign probabilities to the outcomes of games of chance, even if the underlying dynamics are deterministic. We leave it open that it may be fruitful to understand causation in such systems probabilistically.

It will often be useful to make reference to a population of individuals, trials, situations, or instances in which C and E are either present or absent. For instance, in a clinical drug trial, the population is the pool of subjects, and each subject either receives the drug or not. In other kinds of experiments, we may have a series of trials in which C is either introduced or not. Eells (1991, Chapter 1) has a detailed discussion of such populations. We will call the members of such populations 'individuals', even though they may not be people or even objects, but trials, situations, and so on. $P(C)$ is then understood as the probability that C is present for an individual in the population, and likewise for other events in the algebra on which P is defined. This probability is approximated by the frequency of C in the population, although we do not assume that the probability is identical to any actual frequency.

When we discuss counterfactuals, these are to be understood as *non-backtracking* counterfactuals, in the sense of Lewis (1979). The antecedents of these counterfactuals are to be thought of as brought about by small 'miracles' (Lewis 1979) or exogenous interventions (Woodward 2003). We will abbreviate the counterfactual 'if A had occurred, then B would have occurred' by $A > B$. In some cases, we will want to explore the consequences of assuming *counterfactual definiteness*. Counterfactual definiteness is an assumption similar to determinism. It requires that for every individual in a population, either $C > E$ or $C > \sim E$ is true, and either $\sim C > E$ or $\sim C > \sim E$. (This assumption is also called *conditional excluded middle*, and it implies that counterfactuals obey the logic of Stalnaker (1968) rather than Lewis (1973).) If counterfactual definiteness is true, we will assume that holding the relevant background condition fixed suffices to ensure that $P(E|C) = P(C > E)$ and $P(E|\sim C) = P(\sim C > E)$.[2] We will not, however, assume that counterfactual definiteness is true in general. In particular, counterfactual definiteness seems implausible if determinism does not hold. If counterfactual definiteness is not true, we will assume that holding the relevant background condition fixed ensures that $C > P(E) = p$, where $p = P(E|C)$, and likewise for $\sim C$. In other words, if C were the case, then the probability of E would have been p, where p is the actual conditional probability $P(E|C)$.

We are interested in measures of the causal strength of C for E. We will write generically $\mathrm{CS}(E, C)$ for this causal strength. Specific measures to be discussed will be denoted by appending subscripts to the function CS. These measures are to be characterized in terms of formulas involving probabilities such as $P(E|C)$, $P(E|\sim C)$, and perhaps others as well. It will be convenient to write $\mathrm{CS}(E, C)$ to represent the result of applying the mathematical formula to C and E, even if this cannot naturally be interpreted as a causal strength (for example, if C does not raise the probability of E).

When we are considering multiple causes, we will represent the causal strength of C_1 for E in the presence of C_2 as $\mathrm{CS}(E, C_1; C_2)$. This will be defined in the same way as CS, but using the conditional probability $P(\bullet|C_2)$ instead of $P(\bullet)$.

We will also be interested in measures of preventative strength, which we will denote $\mathrm{PS}(E, C)$. We define the preventative strength of C for E in the following way:

[2] Note that we are assuming that C and E do not themselves include counterfactuals. As Lewis (1976) shows, if we allow embeddings, we cannot equate probabilities of conditionals with conditional probabilities under pain of triviality.

$$\text{PS}(E, C) = -\text{CS}(\sim E, C).$$

That is, the preventative strength of C for E is just the causal strength of C for $\sim E$, with a change in sign.[3]

We will consider a variety of candidate measures of causal strength. Some of these have been explicitly proposed as measures of causal strength; others are naturally suggested by various probabilistic approaches to causation. We will discuss the properties of each measure, and try to give an informal explanation of what each one is measuring. Although our overall approach is pluralistic, we will make a few remarks regarding what we take to be the merits and demerits of each measure. We will also discuss the relationships between the measures.

For purposes of comparing measures, we will convert all measures to a unit scale. That is, we will adopt the following two scaling conventions for all measures of causal strength (CS) and preventative strength (PS):

$$\text{If } C \text{ causes } E, \text{ then } \text{CS}(E, C) \in (0, 1].$$

$$\text{If } C \text{ prevents } E, \text{ then } \text{PS}(E, C) = -\text{CS}(\sim E, C) \in [-1, 0).$$

Measures that are based on *differences* in probabilities will typically already be defined on a [–1,1] scale. But, measures that are based on *ratios* of probabilities will generally need to be rescaled. We adopt two *desiderata* for any such rescaling: (a) that it map the original measure onto the interval [–1,1], as described above, and (b) that it yields a measure that is *ordinally equivalent* to the original measure, where $\text{CS}_1(E, C)$ and $\text{CS}_2(E, C)$ are *ordinally equivalent* iff

$$\text{For all } C, E, C' \text{and} E': \text{CS}_1(E, C) \geq \text{CS}_1(E', C') \text{ iff } \text{CS}_2(E, C) \geq \text{CS}_2(E', C').$$

There are many ways to rescale a (probabilistic relevance) ratio measure of the form p/q, in accordance with these two rescaling *desiderata*. Here is a general (parametric) class of such rescalings, where $\lambda \geq 0$, and $p > q$[4]

$$p/q \rightarrow (p - q)/(p + \lambda q).$$

When $\lambda = 0$, we get:

$$p/q \rightarrow (p - q)/p$$

[3] This definition assumes that each measure $\text{CS}(E, C)$ has a corresponding measure of preventative strength $\text{PS}(E, C)$ *with the same functional form* (although replacing E with $\sim E$ in the formula will sometimes result in different terms appearing in the expressions for $\text{CS}(E, C)$ and PS(E, C) – see the discussion of continuity properties below). In the recent literature on measures of *confirmational* strength, some authors have proposed that confirmation and disconfirmation should be measured using *different functional forms* (Crupi *et al.* 2007). We will not discuss any such 'piecewise' measures of causal/preventative strength here, but this is an interesting (possible) class of measures that deserves further scrutiny.

[4] We thank Kenny Easwaran for suggesting this general parametric way of representing rescalings of measures.

and, when $\lambda = 1$, we have:

$$p/q \rightarrow (p - q)/(p + q).$$

We will discuss several applications of each of these two kinds of rescalings, below.

29.3 The measures

Although we will spend much of the chapter introducing the measures in leisurely fashion, we will begin by presenting all of the measures that we will discuss in tabular form. These are shown in Table 29.1. For example, the Eells measure will be represented with a subscript e, and defined as the difference in conditional probabilities: $CS_e(E, C) = P(E|C) - P(E| \sim C)$.

29.4 Venn and Boolean representations

In presenting and discussing the various measures, it will be helpful to represent the probabilities pictorially using Venn diagrams. These will facilitate gaining an intuitive understanding of each measure. Figure 29.1 represents a situation in which C raises the probability of E. The square has an area of one unit. It represents the entire space of possibilities. This space is divided into six cells. The right side of the rectangle corresponds to the occurrence of C, the left half to $\sim C$. The shaded region corresponds to the occurrence of E. The height of the shaded region on the right-hand side corresponds to the conditional probability $P(E|C)$, and the shaded column on the left side corresponds to $P(E| \sim C)$. The two dotted lines are the result of extending the top of each shaded column all the way across the diagram. They are a

Table 29.1 Measures of causal strength.

Eells:	$CS_e(E, C) = P(E\|C) - P(E\| \sim C)$
Suppes:	$CS_s(E, C) = P(E\|C) - P(E)$
Galton:	$CS_g(E, C) = 4P(C)P(\sim C)[P(E\|C) - P(E\| \sim C)]$
Cheng:	$CS_c(E, C) = (P(E\|C) - P(E\| \sim C))/P(\sim E\| \sim C)$
Lewis ratio:	$CS_{lr}(E, C) = P(E\|C)/P(E\| \sim C)$
	$CS_{lr1}(E, C) = [P(E\|C) - P(E\| \sim C)]/[P(E\|C) + P(E\| \sim C)]$
	$CS_{lr2}(E, C) = [P(E\|C) - P(E\| \sim C)]/P(E\|C)$
Good:	$CS_{ij}(E, C) = P(\sim E\| \sim C)/P(\sim E\|C)$
	$CS_{ij1}(E, C) = [P(\sim E\| \sim C) - P(\sim E\|C)]/[P(\sim E\| \sim C) + P(\sim E\|C)]$
	$CS_{ij2}(E, C) = [P(\sim E\| \sim C) - P(\sim E\|C)]/P(\sim E\| \sim C)$

Table 29.2 Pictorial representations.

Eells:	$CS_e(E, C) = c + d$
Suppes:	$CS_s(E, C) = c$
Galton:	$CS_g(E, C) = 4cd$
Cheng:	$CS_c(E, C) = d/(b + d)$
Lewis ratio:	$CS_{lr}(E, C) = (d + f)/f$
	$CS_{lr1}(E, C) = d/(d + e + f)$
	$CS_{lr2}(E, C) = d/(d + f)$
Good:	$CS_{ij}(E, C) = (b + d)/d$
	$CS_{ij1}(E, C) = d/(2b + d)$
	$CS_{ij2}(E, C) = d/(b + d)$

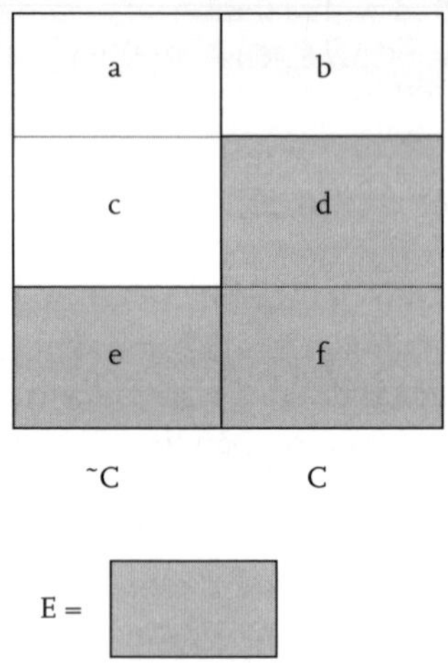

Fig. 29.1

mathematical convenience: they don't necessarily correspond to any events that are well-defined in the probability space. We will use the lower case letters a through f to denote the six regions in the diagram, and also to represent the areas of the regions. The ratios $a : c : e$ are identical to the ratios $b : d : f$. With this diagram, we can write, for example: $P(C) = b + d + f$; $P(E| \sim C) = e + f$; $P(E|C) - P(E| \sim C) = c + d$; and so on. The representations of the measures in terms of this figure are summarized in Table 29.2.

Additionally, several of the measures we will discuss can be given simple *Boolean representations*. A Boolean representation for $CS(E, C)$ is a probability space that has the following features:

(a) it includes as events C and E, and two additional events A and Q;

(b) E can be expressed as a Boolean function of the other three events: specifically, $E \equiv A \vee (Q \wedge C)$;

(c) the probabilities on the algebra generated by C and E are the same as the objective probabilities figuring in the measures of causal strength;

(d) $CS(E, C)$ is the (conditional or unconditional) probability of some event in the space involving Q.

Condition (b) is reminiscent of Mackie's definition of an INUS condition (Mackie 1974). C is an INUS condition for E just in case it is an *insufficient* but *non-redundant* part of an *unnecessary* but *sufficient* condition for E. In the expression $E \equiv A \vee (Q \wedge C)$, C is insufficient for E, since Q must also be present. $Q \wedge C$ is a sufficient condition for E, and C is not redundant: Q alone is insufficient. C is not necessary for E, since A may produce E even in the absence of C. Roughly, we may think of A as the proposition that conditions are right for E to occur in the absence of C, and we may think of Q as the proposition that conditions are right for C to cause E. If determinism is true, we may think of A as representing other causes that are sufficient for E, and of Q as representing the other background conditions that are necessary for C to be a cause of E. However, if there is genuine indeterminism, A and Q will not correspond to any physically real events, but are rather just mathematical conveniences; they may be thought of metaphorically as the results of God's dice rolls. The disjunctive form of the representation for E in (b), together with its probabilistic nature, has given it the name of a 'noisy or' representation.

We will give Boolean representations for four of our measures. These representations differ along two dimensions. First, they differ in the assumptions they make about the probabilistic relations that the new events A and Q bear to C and E and to each other. Second, they identify causal strength with the probabilities of different events, or with probabilities conditional upon different events. The Boolean representations are often helpful for giving an intuitive feel for just what the measures are measuring.

29.5 The Eells measure

Eells (1991) offers a probabilistic theory of causation according to which C is a (positive) cause of E just in case $P(E|C \wedge A_i) > P(E| \sim C \wedge A_i)$ for every background context A_i.[5] He then defined the 'average degree of causal significance' of C for E as: $ADCS(E, C) = \Sigma_i[P(E|C \wedge A_i) - P(E| \sim C \wedge A_i)]P(A_i)$.[6]

[5] In Eells' theory, causal claims are relativized to a population and a population type. We ignore this complication here.

[6] The proposal of Dupré (1984) that we should count C as a cause of E if it raises the probability of E in a 'fair sample' amounts to the claim that C is a cause of E just in case $ADCS(E, C) > 0$. Interestingly, Eells seems not to have understood this proposal. He was adamantly opposed to Dupré's suggestion and even suggests that it is conceptually confused. In particular, he seems

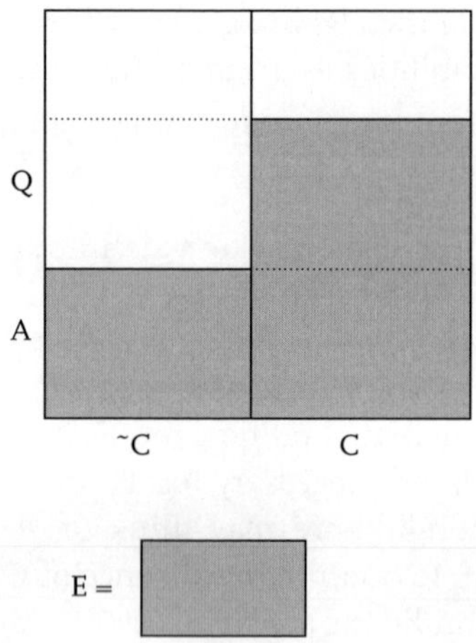

Fig. 29.2

This naturally suggests that when we confine ourselves to a single background context, we define causal strength as:

$$CS_e(E, C) = P(E|C) - P(E| \sim C).$$

This is equal to the area $c + d$ in Figure 29.1. Equivalently, it is the difference between the heights of the two shaded columns. The Eells measure is identical to what psychologists call the *probability contrast* – PC or ΔP for short (see e.g. Cheng and Novick 1990).

The Eells measure may be given a simple Boolean representation. We make the following assumptions about the new events A and Q:

(i) A and Q are mutually exclusive;

(ii) A and C are probabilistically independent; and

(iii) Q and C are probabilistically independent.

As is standard, we identify E with $A \vee (Q \wedge C)$. These assumptions are all shown diagrammatically in Figure 29.2. Given these assumptions, we have:

$$CS_e(E, C) = P(Q).^7$$

to interpret Dupré's call for averaging over background contexts – which is clearly done in the formula for ADCS – as equivalent to saying that C causes E just in case $P(E|C) > P(E| \sim C)$, where we do not control for confounding factors.

[7] All of the mathematical claims that appear in this chapter are verified in a companion *Mathematica* notebook, which can be downloaded from the following URL: http://fitelson.org/pmcs.nb [a PDF version of this notebook is available at http://fitelson.org/pmcs.nb.pdf]. The companion *Mathematica* notebook makes use of the PrSAT *Mathematica* package (Fitelson 2008), which can be downloaded from the following URL: http://fitelson.org/PrSAT/.

Intuitively, the Eells measure measures the difference that C's presence makes to the probability of E. If we had a population of individuals who all belonged to the relevant background context, and conducted a controlled experiment in which C is present for some individuals, and absent in others, the Eells measure would be an estimate of the difference between the relative frequencies of E in the two groups.

The Eells measure is related to a concept that statisticians call *causal effect*. Assume counterfactual definiteness, and let X and Y be two quantitative variables. Let x and x' be two possible values of X, and let i be an individual in the population. The causal effect of $X = x$ vs. $X = x'$ on Y for i (abbreviated $\text{CE}(Y, X = x, X = x', i)$) is the difference between the value Y would take if X were x and the value Y would take if X were x' for individual i. That is, $\text{CE}(Y, X = x, X = x', i) = y - y'$, where $X = x > Y = y$ and $X = x' > Y = y'$ are both true for i. Intuitively, the causal effect is the difference that a hypothetical change from $X = x'$ to $X = x$ would make for the value of Y. Assuming counterfactual definiteness, the Eells measure is the expectation of the causal effect of C vs. $\sim C$ on X_E: $\text{CS}_e(E, C) = E[\text{CE}(X_E, C, \sim C)]$. For example, if an individual i is such that $C > E$ and $\sim C > \sim E$, then for that individual, the causal effect of C vs. $\sim C$ on E is 1. The Eells measure corresponds to the expectation of this quantity. On the other hand, suppose that counterfactual definiteness is false. Then the Eells measure is equal to the causal effect of C vs. $\sim C$ on the probability of E, or equivalently, the expectation of X_E. Note that while the Eells measure itself is indifferent as to whether counterfactual definiteness is true or false, its interpretation in terms of causal effect is different in the two cases.

The Eells measure is also closely related to what Pearl (2000) calls the *probability of necessity and sufficiency* or PNS. Pearl assumes counterfactual definiteness, and defines $\text{PNS}(E, C) = P(C > E \wedge \sim C > \sim E)$. Intuitively, $\text{PNS}(E, C)$ is the probability that C is both necessary and sufficient for E, where necessity and sufficiency are understood counterfactually. *Monotonicity* is the assumption that $P(C > \sim E \wedge \sim C > E) = 0$. Intuitively, this means that there are no individuals that would have E if they lacked C, and also would have $\sim E$ if they had C. Under the assumption of monotonicity, $\text{CS}_e(E, C) = \text{PNS}(E, C)$. This is most easily seen by referring to Figure 29.1. Monotonicity is the assumption that no individuals in cell e are such that if they had C, they would be in cell b; and no individuals in cell b are such that if they lacked C, they would be in cell e. Then we can interpret the figure in the following way: e and f comprise the individuals for which $C > E$ and $\sim C > E$; a and b comprise the individuals for which $C > \sim E$ and $\sim C > \sim E$; and c and d comprise the individuals for which $C > E$ and $\sim C > \sim E$. The Eells measure is then the probability that an individual is in the last group. In other words, it is the proportion of the population for which C would make the difference between E and $\sim E$. We reiterate, however, that this interpretation assumes both counterfactual definiteness and monotonicity. In particular, if counterfactual definiteness fails, the Eells measure can continue to take positive values, while PNS is identically zero.

The Eells measure exhibits what we might call 'floor effects'.[7] If the background context A_i is one in which E is likely to occur even without C, then this will limit the size of $CS_e(E, C)$: there is only so much difference that C can make. In our Boolean representation, this is reflected in the assumption that A and Q are exclusive. If A is large, then Q must be small. This seems appropriate if we think of causal strength in terms of capacity to make a difference. On the other hand, if we think that the causal strength of C for E should be thought of as the intrinsic power of C to produce E, then it might seem strange that the causal strength should be limited by how prevalent E is in the absence of C.

29.6 The Suppes measure

Suppes (1970) required that for C to cause E, $P(E|C) > P(E)$. As we noted above, this is equivalent to the inequality $P(E|C) > P(E|\sim C)$. However, the two inequalities suggest different measures of causal strength. Thus we define the Suppes measure as

$$CS_s(E, C) = P(E|C) - P(E).$$

This quantity is equal to the area of region c in Figure 29.1.

The Suppes measure can be given a simple Boolean representation. Under the same assumptions as those made for the Eells measure, shown in Figure 29.2, we have

$$CS_s(E, C) = P(Q \wedge \sim C).$$

The Suppes measure is related to the Eells measure as follows:

$$CS_s(E, C) = P(\sim C) CS_e(E, C)$$

Table 29.3 provides a summary of all the mathematical inter-definitions. Note that we will only explicitly give the expression of a measure in terms of measures that have been previously introduced. The expression of the Suppes measure in terms of, e.g. the Galton measure can be derived simply by taking the appropriate inverse: e.g. $CS_s(E, C) = CS_g(E, C)/4P(C)$.

The Suppes measure may be understood operationally in the following way: it is the amount by which the frequency of E would increase if C were present for all individuals in the population. Indeed Giere (1979) offers a

[7] This terminology is slightly non-standard, since we are describing an upper bound on CS_e rather than a lower bound. However, looking at Figure 29.1, the bound results not from a ceiling that is low, but rather from a floor that is high.

Table 29.3 Inter-definability of the measures.

Suppes:	$CS_s(E, C)$	$= P(\sim C)CS_e(E, C)$
Galton:	$CS_g(E, C)$	$= 4P(C)P(\sim C)CS_e(E, C)$
		$= 4P(C)CS_s(E, C)$
Cheng:	$CS_c(E, C)$	$= CS_e(E, C)/P(\sim E \mid \sim C)$
		$= CS_s(E, C)/P(\sim E \wedge \sim C)$
		$= CS_g(E, C)/4P(C)P(\sim E \wedge \sim C)$
Lewis ratio:	$CS_{lr1}(E, C)$	$= [CS_{lr}(E, C) - 1]/[CS_{lr}(E, C) + 1]$
		$= CS_e(E, C)/[P(E \mid C) + P(E \mid \sim C)]$
	$CS_{lr2}(E, C)$	$= 1 - 1/CS_{lr}(E, C)$
		$= CS_e(E, C)/P(E \mid C)$
		$= CS_s(E, C)/P(E \mid C)P(\sim C)$
		$= CS_g(E, C)/4P(E \wedge C)P(\sim C)$
		$= CS_c(E, C)[P(\sim E \mid \sim C)/P(E \mid C)]$
	$CS_{ij}(E, C)$	$= CS_{lr}(\sim E, \sim C)$
Good:	$CS_{ij1}(E, C)$	$= [CS_{ij}(E, C) - 1]/[CS_{ij}(E, C) + 1]$
		$= CS_{lr1}(\sim E, \sim C)$
		$= CS_e(E, C)/[P(\sim E \mid C) + P(\sim E \mid \sim C)]$
	$CS_{ij2}(E, C)$	$= 1 - 1/CS_{ij}(E, C) = CS_c(E, C)$
		$= CS_e(E, C)/P(\sim E \mid \sim C)$
		$= CS_s(E, C)/P(\sim E \wedge \sim C)$
		$= CS_g(E, C)/4P(C)P(\sim E \wedge \sim C)$
		$= CS_{lr2}(\sim E, \sim C)$

probabilistic theory of causation in which causation is defined in just this way. This way of understanding the Suppes measure is only correct, however, if there is no frequency-dependent causation or inter-unit causation. In biology, mimicry is an example of frequency-dependent causation. For example, the tasty viceroy butterfly protects itself by mimicking the colour patterns of the unpalatable monarch butterfly. But the more prevalent the viceroys become, the less effective this ruse will become. So it may be that among butterflies, mimicking the monarch does in fact raise the probability of survival, but if all butterflies did it, the rate of survival would not go up. For an example of inter-unit causation, consider the effects of second-hand smoke. If everyone were to smoke, lung cancer rates would go up, in part because there would be more smokers, but also because at least some people would be exposed to greater amounts of second-hand smoke. In this case, the Suppes measure would underestimate the amount by which lung cancer would increase. Intuitively, what is going on in each of these cases is that the Suppes measure predicts the amount by which the prevalence of E will change *within a fixed background context*. However, when we increase the prevalence of C in the population, we also change the background context to which at least some members of the population belong. This will have an impact on the prevalence of E that goes beyond that predicted by the Suppes measure within a fixed background context.

The Suppes measure will exhibit floor effects in much the same way the Eells measure does. The Suppes measure is also sensitive to the unconditional value of $P(C)$: for fixed values of $P(E|C)$ and $P(E|\sim C)$, $CS_s(E, C)$ decreases as $P(C)$ increases. The feature seems *prima facie* undesirable if we construe causal strength as a measure of the intrinsic tendency or capacity of C to cause E. Such an intrinsic capacity should be independent of the prevalence of C.

29.7 The Galton measure

We name this measure after Francis Galton. With quantitative variables X and Y, we often evaluate the relationship between them in terms of the *covariance* or *correlation*. The covariance of two variables is defined as follows:

$$\text{Cov}(X, Y) = E(XY) - E(X)E(Y).$$

When X and Y are replaced by the indicator functions X_C and X_E, a little calculation gives us

$$\text{Cov}(X_E, X_C) = P(C)P(\sim C)[P(E|C) - P(E|\sim C)].$$

The multiplier $P(C)P(\sim C)$ takes a maximum value of ¼ when $P(C)$=0.5, so if we want to convert this measure to a unit scale we will need to normalize. One way to do this is to divide by the standard deviations of X_C and X_E, yielding the *correlation*. We will adopt the simpler expedient of multiplying by 4. Thus:

$$CS_g(E, C) = 4P(C)P(\sim C)[P(E|C) - P(E|\sim C)].$$

This is equal to 4 times the product of c and d in Figure 29.1. The Galton measure is related the Eells and Suppes measures as follows:

$$\begin{aligned} CS_g(E, C) &= 4P(C)P(\sim C)CS_e(E, C) \\ &= 4P(C)CS_s(E, C). \end{aligned}$$

Like the Suppes measure, the Galton measure will exhibit floor effects, and it will be sensitive to the unconditional probability of C. The Galton measure intuitively measures the degree to which there is variation in whether or not E occurs that is due to variation in whether or not C occurs. $CS_g(E, C)$ will take its maximum value when $P(E|C)$ is close to 1, $P(E|\sim C)$ is close to 0, and

$P(C)$ is close to 0.5. In these circumstances, $P(E)$ will be close to 0.5, so there is a lot of variation in the occurrence of E – sometimes it happens, sometimes it doesn't. When C occurs, there is very little variation: E almost always occurs; and when C doesn't occur, E almost never occurs. So there is a lot of variation in whether or not E occurs precisely because there is variation in whether or not C occurs. By contrast, suppose that $P(C)$ is close to 1. Then any variation in whether or not E occurs will almost all be due to the fact that $P(E|C)$ is non-extreme: E sometimes happens in the presence of C, and sometimes it doesn't. Likewise if $P(C)$ is close to 0. For example, it might be natural to say that smallpox is lethal: it is a potent cause of death. So we might think that the causal strength of smallpox for death is high. But the Galton measure would give it a low rating, perhaps even 0, since none of the actual variation in who lives and who dies during a given period is due to variation in who is exposed to smallpox: thankfully, no one is any more.

Note that the standard measure of *heritability* used in genetics and evolutionary biology is essentially a measure of correlation, and behaves much like the Galton measure. Because of the sensitivity of the heritability measure to the absolute level of variation in some trait among the parents in a population, heritability is a poor measure of the intrinsic tendency of parents to produce offspring that resemble them with respect to the trait in question.

29.8 The Cheng measure

The psychologist Patricia Cheng proposed that we have a concept of 'causal power', and that this explains various aspects of our causal reasoning (1997). Under the special assumptions we have made, causal power reduces to the following formula:

$$CS_c(E, C) = [P(E|C) - P(E|\sim C)]/P(\sim E|\sim C).$$

In our pictorial representation (Figure 29.1), this is equal to the ratio $d/(b + d)$.

It is well-known that the Cheng measure has a 'noisy or' representation (see, e.g. Glymour 1998). We make the following assumption:

A, Q, and C are both pairwise and jointly independent.

As always, E is identified with $A \vee (Q \wedge C)$. These assumptions are shown schematically in Figure 29.3. Then we can identify

$$CS_c(E, C) = P(Q).$$

Note that while both CS_e and CS_c are identified with $P(Q)$, the probabilistic assumptions underlying the two representations are different.

The Cheng measure is related to our other measures by the following formulae:

$$\begin{aligned} CS_c(E, C) &= CS_e(E, C)/P(\sim E|\sim C) \\ &= CS_s(E, C)/P(\sim E \wedge \sim C) \\ &= CS_g(E, C)/4P(C)P(\sim E \wedge \sim C). \end{aligned}$$

Only the first of these is particularly intuitive. One way of thinking about the Cheng measure is that it is like the Eells measure in focusing on the difference $P(E|C) - P(E|\sim C)$, but eliminates floor effects by dividing by $P(\sim E|\sim C)$. The idea is that it is only within the space allowed by $P(\sim E|\sim C)$ that C has to opportunity to make a difference for the occurrence of E, so we should rate C's performance by how well it does within the space allowed it.

Cheng conceives of her causal power measure in the following way. Assume that E will occur just in case C occurs and 'works' to produce E, or some other cause of E is present and 'works' to produce E. In our Boolean representation, shown in Figure 29.3, Q corresponds to C's 'working', and A corresponds to some other cause's working. $CS_c(E, C)$ is then the probability that C 'works'. These 'workings' are not mutually exclusive: it is possible that C is present and 'works' to produce E, and that some other cause also 'works' to produce E. Thus Cheng's model is compatible with causal overdetermination. A high probability for E in the absence of C needn't indicate that C isn't working most of the time when it is present. But this is at best a heuristic for thinking about causal power. The nature of this 'working' is metaphysically mysterious. If the underlying physics is deterministic, then perhaps we can understand C's 'working' as the presence of conditions that render C sufficient for E (represented by Q in our Boolean representation). If the causal relationship is

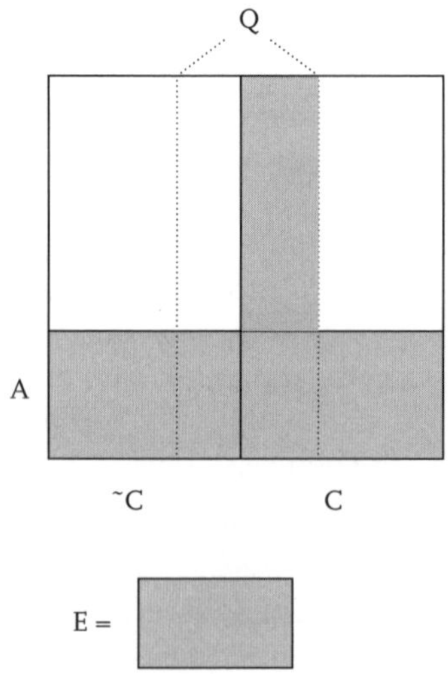

Fig. 29.3

indeterministic, however, it is hard to see what this 'working' could be. C and various other causes of E are present. In virtue of their presence E has a certain probability of occurring. On most conceptions of indeterministic causation, that is all there is to the story. (See, e.g. Lewis 1986 and Humphreys 1989, sections 29.10 and 29.11; Woodward (1990) challenges this conception. See also Hitchcock (2004) for discussion of the two different models.)

The Cheng measure is related to what Pearl (2000) calls the *probability of sufficiency* or POS. Assuming counterfactual definiteness, Pearl defines $\text{POS}(E, C) = P(C > E| \sim C \wedge \sim E)$. That is, in cases where neither C nor E occur, $\text{POS}(E, C)$ is the probability that E would occur if C were to occur. Conditioning on $\sim C \wedge \sim E$ means that we are in the rectangle occupied by a and c in Figure 29.1. Now assume monotonicity: that no individuals in region e would move to b if C were to occur, and no individuals in b would move to e if Cdid not occur. Then the result of hypothetically introducing C to the individuals in region a and c is to move them straight over to the right-hand side. So the proportion of individuals in regions a and c that will experience E when C is introduced is equal to $d/(b + d)$. So under the assumptions of counterfactual definiteness and monotonicity, $\text{CS}_c(E, C) = \text{POS}(E, C)$. If counterfactual definiteness does not hold, however, this interpretation cannot be employed. In this case, CS_g may still take positive values, while POS is identically zero.

The Cheng measure does not exhibit floor effects, and it is not sensitive to the absolute value of $P(C)$. For this reason it is a more plausible measure of the intrinsic capacity of C to produce E than any of the others we have discussed.

29.9 The Lewis ratio measure

In formulating the probabilistic extension of his counterfactual theory of causation, Lewis (1986) required that in order for E to be causally dependent upon C, the probability that E would occur if C had not occurred had to be *substantially less* than the actual probability of E. Lewis then remarks that the size of the decrease is measured by the *ratio* of the quantities, rather than their difference. This naturally suggests the following measure:

$$\text{CS}_{\text{lr}}(E, C) = P(E|C)/P(E| \sim C).$$

This is the ratio $(d + f)/f$ in Figure 29.1. The Lewis ratio measure is equivalent to the quantity called 'relative risk' in epidemiology and tort law: it is the risk of experiencing E in the presence of C, relative to the risk of E in the absence of C (see Parascandola 1996 for a philosophically sensitive discussion of these topics).

The Lewis ratio measure rates causes on a scale from one to infinity (and it gives numbers between zero and one when $P(E|C) < P(E|\sim C)$). Thus if we want to compare it directly with our other measures we will need to convert it to a unit scale. As discussed above, there are a number of ways of doing this. We will consider two. The first, corresponding to setting $\lambda = 1$ in our parametric rescaling formula above, is:

$$\mathrm{CS}_{\mathrm{lr1}}(E, C) = [P(E|C) - P(E|\sim C)]/[P(E|C) + P(E|\sim C)].$$

This is equal to $d/(d + e + f)$ in Figure 29.1. This re-scaling of the Lewis ratio measure is related to the Eells measure as follows:

$$\mathrm{CS}_{\mathrm{lr1}}(E, C) = \mathrm{CS}_e(E, C)/[P(E|C) + P(E|\sim C)].$$

Its mathematical relationship to the other measures is insufficiently elegant to be illuminating.

The second rescaling corresponds to setting $\lambda = 0$:

$$\mathrm{CS}_{\mathrm{lr2}}(E, C) = [P(E|C) - P(E|\sim C)]/P(E|C).$$

This is the ratio $d/(d + f)$ in Figure 29.1. This rescaling of the Lewis measure can be given a Boolean representation, using the same probabilistic assumptions as those used for the Eells and Suppes measures (shown in Figure 29.2). Then we have:

$$\mathrm{CS}_{\mathrm{lr2}}(E, C) = P(Q|C \wedge E).$$

This rescaling is related to our other measures via the following formulae:

$$\begin{aligned}\mathrm{CS}_{\mathrm{lr2}}(E, C) &= \mathrm{CS}_e(E, C)/P(E|C)\\ &= \mathrm{CS}_s(E, C)/P(E|C)P(\sim C)\\ &= \mathrm{CS}_g(E, C)/P(E \wedge C)P(\sim C)\\ &= \mathrm{CS}_c(E, C)[P(\sim E|\sim C)/P(E|C)].\end{aligned}$$

$\mathrm{CS}_{\mathrm{lr2}}(E, C)$ is equivalent to the quantity called the *probability of causation* in epidemiology and tort law. It is also related to what Pearl (2000) calls the *probability of necessity*, or PN. It will be helpful to consider the latter connection first. Assuming counterfactual definiteness, Pearl defines $\mathrm{PN}(E, C) = P(\sim C > \sim E|C \wedge E)$. That is, given that C and E both occurred, $\mathrm{PN}(E, C)$ is the probability that C is necessary for E, where necessity is understood counterfactually. If we assume monotonicity, then $\mathrm{PN}(E, C) = \mathrm{CS}_{\mathrm{lr2}}(E, C)$. The idea is if C and E both occur, we are in the region $d \cup f$ in Figure 29.1. Under the assumption of monotonicity, the effect of hypothetically removing C will be to shift individuals straight to the left. Thus the proportion of those in region $d \cup f$ that would no longer experience E if C did not occur would be $c/(c + e) = d/(d + f)$. If we define causation directly in terms of (definite) counterfactual dependence, as is done in the law, then $CS_{lr2}(E, C)$ is the probability that C caused E, given that C and E both occurred: hence the name 'probability of causation'. In our Boolean representation, Q can be thought of as C's being necessary for E, or C's causing E. 'Probability of causation' is important in tort law. In civil liability cases, the standard of evidence is 'more probable than not'. Thus if a plaintiff has been exposed to C, and suffers adverse reaction E, in order to receive a settlement she must establish that the probability is greater than one-half that C caused E. This is often interpreted as requiring that the 'probability of causation' is greater than 0.5.

It is worth remembering, however, that the interpretation of $CS_{lr2}(E, C)$ as the probability that C caused E depends upon three assumptions. The first is that counterfactual dependence is necessary for causation. This assumption fails in cases of preemption and overdetermination. We have chosen to ignore these particular problems, although as we have seen, the Cheng measure seems to be compatible with causal overdetermination. The second assumption is monotonicity. The third, and most important, is counterfactual definiteness. If counterfactual definiteness fails, then all we can say about those individuals that experience both C and E is that if C had not occurred, the probability of E would have been p, where $p$is $P(E| \sim C)$. Thus it is true for *all* the individuals that experience both C and E that the probability of E would have been lower if C had not occurred. So to the extent that there is a 'probability of causation', that probability is 1: for all the individuals that experience both C and E, C was *a* cause of E (although there may be other causes as well). This is how Lewis himself interprets indeterministic causation (Lewis 1986).[8]

Like the Eells, Suppes, and Galton measures, the Lewis ratio measure and its rescalings will exhibit floor effects. Like the Eells and Cheng measures, the Lewis ratio measures and its rescalings are not sensitive to the unconditional probability of C.

29.10 The Good measure

Good (1961–2) sought to define a measure $Q(E, C)$ of the *tendency of C to cause E*. The measure he ultimately proposed was $Q(E,C) = \log[P(\sim E|\sim C)/P(\sim E|C)]$. We propose to simplify this formula (in a way that does not affect its ordinal scale) by not taking the log (or equivalently, raising the base (e or 10) to the power of Q). Since we have already used the subscript 'g' for the Galton measure, we will use Good's well-known first initials '*ij*'.

$$CS_{ij}(E, C) = P(\sim E| \sim C)/P(\sim E|C).$$

[8] See also the discussion in Parascandola (1996) and Hitchcock (2004).

This is equal to the ratio $(b + d)/d$ in Figure 29.1. The Good measure is related to the Lewis ratio measure via the formula:

$$\text{CS}_{\text{ij}}(E, C) = \text{CS}_{\text{lr}}(\sim E, \sim C).$$

Like the Lewis ratio measure, the Good measure yields a scale from one to infinity when $P(E|C) > P(E|\sim C)$, and from zero to one otherwise. So we will consider two rescalings.

$$\text{CS}_{\text{ij1}}(E, C) = [P(\sim E|\sim C) - P(\sim E|C)]/[P(\sim E|\sim C) + P(\sim E|C)].$$

This is equal to the ratio $d/(2b + d)$ in Figure 29.1. This rescaling is related to other measures via the following formulae:

$$\begin{aligned}\text{CS}_{\text{ij1}}(E, C) &= \text{CS}_{\text{lr1}}(\sim E, \sim C)\\ &= \text{CS}_e(E, C)/[P(\sim E|C) + P(\sim E|\sim C)].\end{aligned}$$

It mathematical relationship to the other measures is insufficiently elegant to be illuminating. The second rescaling is:

$$\text{CS}_{\text{ij2}}(E, C) = [P(\sim E|\sim C) - P(\sim E|C)]/P(\sim E|\sim C)$$

which is equal to $d/(b + d)$. Interestingly, this second rescaling of the Good measure is identical to Cheng measure. Obviously, then, this rescaling will have the same properties, and be susceptible to the same interpretations, as the Cheng measure. Since the original Good measure and the first rescaling are ordinally equivalent to the second rescaling, they will be ordinally equivalent to the Cheng measure and also share many of its properties. Here are some other equivalences involving the second rescaling of the Good measure:

$$\begin{aligned}\text{CS}_{\text{ij2}}(E, C) &= \text{CS}_c(E, C)\\ &= \text{CS}_e(E, C)/P(\sim E|\sim C)\\ &= \text{CS}_s(E, C)/P(\sim E \wedge \sim C)\\ &= \text{CS}_g(E, C)/4P(C)P(\sim E \wedge \sim C)\\ &= \text{CS}_{\text{lr2}}(\sim E, \sim C).\end{aligned}$$

29.11 Other measures

It is fairly easy to generate other candidate measures. One would be the difference between the Eells and the Suppes measures, namely:

$$\text{CS}(E, C) = P(E) - P(E|\sim C).$$

This could be understood operationally as the amount by which the frequency of E would decline if C were completely eliminated (modulo worries about frequency dependent and inter-unit causation). We might think of this as the extent to which C is in fact causing E. Noting that the Lewis ratio measure is simply the ratio of the two quantities whose difference is the Eells measure, we could define a measure that is the ratio of the two quantities whose difference is the Suppes measure:

$$CS(E, C) = P(E|C)/P(E).$$

And of course we could then take different rescalings of this measure to convert it to a unit scale. We could also construct an analog of the Cheng measure that makes use of the difference that figures in the Suppes measure:

$$CS(E, C) = [P(E|C) - P(E)]/P(\sim E).$$

And so on. Since the measures that we have already discussed are more than enough to keep us busy, we will leave an exploration of the properties of these new measures as an exercise for the reader.[9]

29.12 Properties and comparisons

In the remaining sections, we will explore some further properties of the measures that we have introduced, and examine some relationships between them. First, we will consider whether any of our measures are ordinarily equivalent, or partially ordinally equivalent. Second, we will examine a number of continuity properties of measures – these involve the behaviours of the measures as $P(E|C)$ decreases from a value greater than $P(E|\sim C)$ to a value less than $P(E|\sim C)$. Finally, we will examine what the measures tell us about causal independence, and compare the independence judgments of the various measures.

29.13 Ordinal relationships between measures

Our two rescalings of the Lewis ratio measure are, by design, ordinally equivalent to the original Lewis ratio measure, and to each other. Likewise for the rescalings of the Good measure. Moreover, as we have already seen, one of our rescalings of Good's measure is *numerically identical* to Cheng's measure.

$$CS_{ij2}(E, C) = CS_c(E, C).$$

[9] The computational tools developed in the companion *Mathematica* notebook (see footnote 7) are quite general, and can be applied to various other possible measures of causal strength, and various other properties of measures as well.

Table 29.4 Ordinal equivalences between measures.

	Eells	Suppes	Galton	Cheng	Lewis ratio	Good
Eells	G-E	II-E	II-E	None	None	None
Suppes	II-E	G-E	II-E	None	None	None
Galton	II-E	II-E	G-E	None	None	None
Cheng	None	None	None	G-E	None	G-E
Lewis ratio	None	None	None	None	G-E	None
Good	None	None	None	G-E	None	G-E

Apart from these cases, no other pair of measures we're discussing here are numerically equivalent. Indeed, it turns out that no other pair of measures we're discussing here are *ordinally* equivalent (in general). But, some other pairs of measures are ordinally equivalent *in special types of cases.* Consider the following two special types of cases:

I. Cases involving a single effect (E) and two causes (C_1 and C_2).
II. Cases involving a single cause (C) and two effects (E_1 and E_2).

If two measures (CS_1 and CS_2) are such that, for all E, C_1 and C_2:

$$\text{CS}_1(E, C_1) \geq \text{CS}_1(E, C_2) \text{ iff } \text{CS}_2(E, C_1) \geq \text{CS}_2(E, C_2)$$

then CS_1 and CS_2 are ordinally equivalent *in all cases of Type I* (or 'I-equivalent', for short). And, if CS_1 and CS_2 are such that, for all C, E_1 and E_2:

$$\text{CS}_1(E_1, C) \geq \text{CS}_1(E_2, C) \text{ iff } \text{CS}_1(E_1, C) \geq \text{CS}_1(E_2, C)$$

then CS_1 and CS_2 are ordinally equivalent *in all cases of Type II* (or 'II-equivalent', for short). Various pairs of measures (which are not ordinally equivalent in general) turn out to be either I-equivalent or II-equivalent. For example, the Eells, Suppes, and Galton measures are all II-equivalent. This can be seen readily by examing the identities in Table 29.3. For a fixed C, the Eells, Suppes, and Galton measures are all fixed multiples of one another. Thus, for a fixed C, they will agree on comparative judgments of causal strength. Table 29.4 summarizes all ordinal relationships between measures (a 'G-E' in a cell of Table 29.4 means that the two measures intersecting on that cell are *generally* ordinally equivalent, a 'I-E' means they are I-equivalent, and a 'II-E' means they are II-equivalent).

29.14 Continuity properties of measures

Some of our measures exhibit the following *continuity* between causation and prevention ('Causation-Prevention Continuity'):

$$\text{(CPC)} \qquad \text{CS}(E, C) = -\text{CS}(\sim E, C).$$

Recall that we are *defining* $\text{PS}(E, C)$ as $-\text{CS}(\sim E, C)$. As such, we can also express (CPC) as asserting that the *absolute value* of $\text{CS}(E, C)$ is the same as the *absolute value* of $\text{PS}(E, C)$. If a measure satisfies (CPC), then we can plug probabilities into the measure without regard to whether C causes E or prevents E. If the measure yields a positive value, that is the causal strength of C for E; if it yields a negative value, that is the preventative strength of C for E. By contrast, if a measure does not satisfy (CPC), then we must first determine whether C causes E or prevents E before we know which probabilities to plug into the formula. If a measure violates (CPC), it would suggest that causation and prevention are somehow conceptually different – there is a 'discontinuity' where $P(E|C) = P(E|\sim C)$.[10] For example, the Eells measure is simply the difference between $P(E|C)$ and $P(E|\sim C)$. The effect of switching E and $\sim E$ is simply to reverse the sign. We can continue to use the same formula regardless of whether $P(E|C) > P(E|\sim C)$ or $P(E|C) < P(E|\sim C)$. The Suppes and Galton measures similarly obey (CPC). By contrast, the Cheng measure of the causal strength of C for E includes the term $P(\sim E|\sim C)$ in its denominator. Thus if C prevents E, and we want to assess $\text{PS}_c(E, C) = -\text{CS}_c(\sim E, C)$, we will need to replace $P(\sim E|\sim C)$ in the denominator with $P(E|\sim C)$, as well as merely changing the sign. So except for the special case where $P(\sim E|\sim C) = 0.5$, we will need to know whether C causes E or prevents E in order to know how to use the formula correctly.

Some measures exhibit the following continuity between causation and omission ('Causation-Omission Continuity'):

$$\text{(COC)} \qquad \text{CS}(E, C) = -\text{CS}(E, \sim C).$$

$\text{CS}(E, \sim C)$ may be thought of as the causal strength with which the omission or absence of C causes E. If a measure satisfies (COC), then, when C prevents E, $\text{CS}(E, C)$ will give us a measure of the extent to which the absence of C causes E (with the sign reversed).Thus such a measure may be thought to treat causation and causation by omission as on a par. For example, the Eells measure satisfies (COC): swapping $\sim C$ for C has the effect of switching the two terms, resulting in a change of sign. The Galton measure also satisfies (COC).

[10] We do not mean a literal discontinuity. All of our measures will take the value 0 when $P(E|C) = P(E|\sim C)$, and will approach this value from below and above.

Table 29.5 Continuity properties of measures.

	(CPC)	(COC)	(CPO)
Eells	Yes	Yes	Yes
Suppes	Yes	No	No
Galton	Yes	Yes	Yes
Cheng	No	No	No
Lewis Ratio rescaling #1 CS_{lr1}	No	Yes	No
Lewis ratio rescaling #2 CS_{lr2}	No	No	No
Good rescaling #1 CS_{ij1}	No	Yes	No
Good rescaling #2 CS_{ij2}	No	No	No

Interestingly, one of our rescalings of the Lewis ratio measure satisfies (COC) while the other does not; similarly for the Good measure. This suggests that the choice of rescaling will make a substantive difference to how the measures treat causation by omission. It also suggests that there is more to rescaling than simply preserving ordinal equivalence.

Finally, some measures exhibit the following continuity between causation, prevention, and omission ('Causation = Prevention by Omission'):

(CPO) $$CS(E, C) = CS(\sim E, \sim C).$$

Given our definition of PS, (CPO) says that the causal strength of C for E is equal in magnitude and opposite in sign to the preventative strength of $\sim C$ for E. It is easy to see that (CPO) is a logical consequence of the conjunction of (CPC) and (COC). So, any measure that satisfies both (CPC) and (COC) must also satisfy (CPO). But, the converse does not hold. That is, (CPO) is strictly weaker than (CPC) & (COC).[11] As reported in Table 29.5, the Eells and Galton measures satisfy both (CPC) and (COC). As a result, they both satisfy (CPO) as well. None of our other measures satisfy (CPO). Table 29.5 summarizes the behaviour of our measures of causal strength, with respect to these three continuity properties (see Section 29.5 of Eells and Fitelson 2002 for a formally similar table).

29.15 Causal independence

Causes sometimes operate *independently* of one another, and sometimes they do not. In this section, we will introduce a notion of causal independence and discuss some of its properties (vis-à-vis the measures of causal strength we are studying). First, we need a way of characterizing when two causes C_1 and C_2 of an effect E operate independently of one another (regarding E). The

[11] See (Eells and Fitelson 2002) for a discussion of these (and other) formal continuity properties of probabilistic relevance measures (in the context of *confirmation*).

intuitive idea behind our formal definition of causal independence is that C_1 and C_2 are *independent in causing* E just in case the causal strength of C_1 for E does not depend on whether or not C_2 is also present, and vice versa. This is *not* to say that C_1 and C_2 are (probabilistically) independent *of each other*.[12] Formally, this intuitive idea is best captured by the following definition:

C_1 and C_2 are *independent in causing* E, according to a measure of causal strength CS iff $\mathrm{CS}(E, C_1; C_2) = \mathrm{CS}(E, C_1; {\sim}C_2)$.

We will abbreviate this relation $\mathrm{I}_{\mathrm{CS}}(E, C_1, C_2)$. To avoid embedded subscripts, we will use I_n to label the independence relation generated by CS_n. Because we are assuming that C_1 and C_2 are probabilistically independent (given the background condition), the following two basic facts can be shown to hold – for *all* of our measures of causal strength CS (assuming each of C_1, C_2 causes E):

- $\mathrm{I}_{\mathrm{CS}}(E, C_1, C_2)$ iff $\mathrm{I}_{\mathrm{CS}}(E, C_2, C_1)$. [$\mathrm{I}_{\mathrm{CS}}$ *is symmetric in* C_1, C_2.]
- $\mathrm{I}_{\mathrm{CS}}(E, C_1, C_2)$ iff $\mathrm{CS}(E, C_1; C_2)$. $= \mathrm{CS}(E, C_1)$ [I_{CS} can be defined in terms of the *absence* of C_2, or just in terms of conditional *vs* unconditional CS-values.]

While all of our measures converge on these two fundamental properties of I_{CS}, there are also some important *divergences* between our CS-measures, when it comes to I_{CS}.

First, we will consider whether it is possible for various pairs of distinct CS-measures to *agree* on judgments of causal independence. That is, for which pairs of measures CS_1, CS_2 can we have *both* $\mathrm{I}_{\mathrm{CS}_1}(E, C_1, C_2)$ *and* $\mathrm{I}_{\mathrm{CS}_2}(E, C_1, C_2)$? It should be apparent that ordinal equivalence is sufficient for agreement in independence judgments, although it is not necessary. It follows that the different rescalings of the Lewis ratio measure will always agree on their independence judgments, as will the different rescalings of the Good measure. Moreover, the Good measure and its rescalings yield all the same independence judgments as the Cheng measure. Interestingly, among all the measures we're discussing here, not all pairs *can* agree on I_{CS}-judgments (apart from the trivial cases where one of C_1 or C_2 is not a cause of E). And, those pairs of measures that *can* agree on *some* I_{CS}-judgments, *must* agree on *all* I_{CS}-judgments. Table 29.6 summarizes these I_{CS}-agreement results.

[12] It is true that we are assuming (for simplicity) that C_1 and C_2 are probabilistically independent, relative to the background context. But, conceptually, this assumption is distinct from the assumption of the causal independence of C_1 and C_2 vis-à-vis E. A similar distinction needs to be made in the context of *confirmational* independence of two pieces of *evidence*, regarding a *hypothesis*. Various accounts of confirmational independence mistakenly conflate these two notions. See (Fitelson 2001, chapter 3).

Table 29.6 Do measures C_1 and C_2 agree on all, some, or none of their I_{CS}-judgments?

	Eells	Suppes	Galton	Cheng	Lewis ratio	Good
Eells	All	All	All	None	None	None
Suppes	All	All	All	None	None	None
Galton	All	All	All	None	None	None
Cheng	None	None	None	All	None	All
Lewis ratio	None	None	None	None	All	None
Good	None	None	None	All	None	All

Second, we will consider whether a measure CS's judging that $I_{CS}(E, C_2, C_1)$ places *substantive constraints* on the *individual* causal strengths $CS(E, C_1)$, $CS(E, C_2)$. Interestingly, some measures CS are such that $I_{CS}(E, C_2, C_1)$ *does* impose substantive constraints on the values of $CS(E, C_1)$, $CS(E, C_2)$. Specifically, the Eells, Suppes, and Galton measures all have the following property:

$$(\dagger)\ \text{If } I_{CS}(E, C_2, C_1),\ \text{then } CS(E, C_1) + CS(E, C_2) \leq 1.$$

Moreover, *only* the Eells, Suppes, and Galton measures have property (†). None of the other measures studied here are such that $I_{CS}(E, C_2, C_1)$ places such a substantive constraint on the values of $CS(E, C_1)$, $CS(E, C_2)$ for independent causes. (†) Strikes us as an undesirable property: it seems to indicate that there are a priori restrictions on which kinds of causes can act independently of one another.

Finally, we ask whether 'the conjunction of two independent causes is better than one'. More precisely, we consider the following question: which of our measures satisfy the following property for conjunctions of independent causes:

$$(S)\ \text{If } I_{CS}(E, C_2, C_1),\ \text{then } CS(E, C_1 \wedge C_2) > CS(E, C_i),\ \text{for both } i = 1 \text{ and } i = 2.$$

The intuition behind (S) is that if C_1 and C_2 are independent causes of E, then their conjunction should be a stronger cause of E than either individual cause C_1 or C_2. It is interesting to note that some of our measures *appear* to violate (S).[13] That is, if we think of (S) in *formal* terms, then measures like Eells and Cheng appear to violate (S). The problem here lies with the proper way to unpack. '$CS(E, C_1 \wedge C_2)$' for measures like Eells and Cheng, which compare $P(E|C)$ *and* $P(E|\sim C)$. When calculating $CS(E, C_1 \wedge C_2)$ for such measures, we should not simply compare $P(E|C_1 \wedge C_2)$ and $P(E|\sim(C_1 \wedge C_2))$,

[13] It is important to note here that *all* probabilistic relevance measures of degree of causal strength *must* satisfy the following, *weaker*, *qualitative* variant of (S):(S_0) If $I_{CS}(E, C_2, C_1)$, then $CS(E, C_1 \wedge C_2) > 0$ [i.e. $C_1 \wedge C_2$ *is a* cause of E]. And, this will be true on *either* way of unpacking '$CS(E, C_1 \wedge C_2)$' discussed below.

since that involves *averaging* over different possible *instantiations* of causal factors that might undergird the truth of '$\sim(C_1 \wedge C_2)$'. Rather, we should compare $P(E|C_1 \wedge C_2)$ and $P(E| \sim C_1 \wedge \sim C_2)$. Thus, for example, for the Eells measure, we would have $CS_e(E, C_1 \wedge C_2) = P(E|C_1 \wedge C_2) - P(E| \sim C_1 \wedge \sim C_2)$. Once we correct for this misleading way of unpacking '$CS(E, C_1 \& C_2)$' in (S), then it follows that *almost*[14] *all* of our measures of causal strength satisfy (S).

Note that if we redefine $CS(E, C_1 \wedge C_2)$ in this way, then some of the identities in Table 29.3 will not hold for conjunctive causes. For instance, the identity $CS_s(E, C) = P(\sim C) CS_e(E, C)$ relating the Eells and the Suppes measure for atomic causes, is not preserved. That is, it will not be the case that either $CS_s(E, C_1 \wedge C_2) = P(\sim(C_1 \wedge C_2))\ CS_e(E, C_1 \wedge C_2)$ or $CS_s(E, C_1 \wedge C_2) = P(\sim C_1 \wedge \sim C_2) CS_e(E, C_1 \wedge C_2)$ in general. Moreover, the redefinition of $CS(E, C_1 \wedge C_2)$ entails that in order to calculate causal strengths, we must identify the appropriate level of atomic causes. Most of the results in this chapter have to do only with such atomic (or fundamental/primitive) causal factors (and that is the intended domain for Table 29.3). The general problem of *combining* atomic causal factors into complex causal factors is a subtle one, which is beyond the scope of the present discussion.

Finally, we note that with this new definition of $CS(E, C_1 \wedge C_2)$, several of our measures yield fairly simple expressions for $CS(E, C_1 \wedge C_2)$ in terms of $CS(E, C_1)$ and $CS(E, C_2)$ in the case of independence:

$$I_e(E, C_1, C_2) \text{ implies } CS_e(E, C_1 \wedge C_2) = CS_e(E, C_1) + CS_e(E, C_2)$$
$$I_s(E, C_1, C_2) \text{ implies } CS_s(E, C_1 \wedge C_2) = CS_s(E, C_1) + CS_s(E, C_2)$$
$$I_c(E, C_1, C_2) \text{ implies } CS_c(E, C_1 \wedge C_2) = 1 - (1 - CS_c(E, C_1))(1 - CS_c(E, C_2))$$
$$I_{lr}(E, C_1, C_2) \text{ implies } CS_{lr}(E, C_1 \wedge C_2) = CS_{lr}(E, C_1) CS_{\mathrm{lr}}(E, C_2)$$
$$I_{lr2}(E, C_1, C_2) \text{ implies } CS_{lr2}(E, C_1 \wedge C_2) = 1 - (1 - CS_{\mathrm{lr2}}(E, C_1))(1 - CS_{\mathrm{lr2}}(E, C_2))$$
$$I_{ij}(E, C_1, C_2) \text{ implies } CS_{ij}(E, C_1 \wedge C_2) = 1 - (1 - CS_{\mathrm{ij}}(E, C_1))(1 - CS_{\mathrm{lr2}}(E, C_2)).$$

It bears remembering, however, that the antecedents are not all mutually satisfiable.[15]

[14] This question is particularly difficult to analyse for the Galton measure. We haven't been able to find any plausible redefinition of $CS_g(E, C_1 \wedge C_2)$ which ensures the satisfaction of (S) for the Galton measure. We suspect that the anomalous result occurs for CS_g because of the way we are trying to force what is essentially a *covariation* measure into a measure designed for binary random variables. Intuitively, from a perspecitive of covariation, it makes more sense to somehow think of '$C_1 \wedge C_2$' as a four-valued random variable. Considered just as a binary variable, it stands to reason that sometimes variation in whether or not '$C_1 \wedge C_2$' occurs won't capture some of the variation in whether E occurs, since some of the latter is due to variation in the different ways $\sim(C_1 \wedge C_2)$ can occur. This is a nice illustration of the subtlety of combining the causal strengths of individual ('atomic') causal factors.

[15] For more detailed treatment of the properties of conjunctive causes, see the accompanying notebook at http://fitelson.org/pmcs.nb or http://fitelson.org/pmcs.nb.pdf, pp. 22–30.

Acknowledgements

We would like to thank Jim Woodward, the audience at the Second Annual Formal Epistemology Festival (FEF2), and two anonymous referees for useful comments and discussion.

References

Cartwright, N. (1979). Causal laws and effective strategies, *Noûs* **13**: 419–437.

Cheng, P. (1997). From covariation to causation: A causal power theory, *Psychological Review* 104: 367–405.

Cheng, P. and L. Novick (1990). A probabilistic contrast model of causal induction, *Journal of Personality and Social Psychology* 58: 545–567.

Crupi, V., *et al.* (2007). On Bayesian measures of evidential support: Theoretical and empirical issues, *Philosophy of Science* 74(2): 229–252.

Dupré, J. (1984). Probabilistic causality emancipated, in Peter French, Theodore Uehling, Jr., and Howard Wettstein (eds.), *Midwest Studies in Philosophy IX* (Minneapolis: University of Minnesota Press), pp. 169–175.

Eells, E. (1991). *Probabilistic Causality*. Cambridge,: Cambridge University Press.

Eells, E. and Fitelson, B. (2002). Symmetries and assymmetries in evidential support, *Philosophical Studies* 107: 129–142 <http://fitelson.org/symmetry.pdf>.

Fitelson, B. (1999). On the plurality of Bayesian measures of confirmation and the problem of measure sensitivity, *Philosophy of Science* 66 (supplement): S362–S378 <http://fitelson.org/psa.pdf>.

Fitelson, B. (2001). *Studies in Bayesian Confirmation Theory*. PhD. Dissertation, University of Wisconsin–Madison <http://fitelson.org/thesis.pdf>.

Fitelson, B. (2008). A decision procedure for probability calculus with applications, *Review of Symbolic Logic* 1: 111–125 <http://fitelson.org/pm.pdf>.

Giere, R. (1979). *Understanding Scientific Reasoning*. New York: Holt, Rinehart and Winston.

Glymour, C. (1998). Learning causes: Psychological explanations of causal explanation 1, *Minds and Machines* 8(1): 39–60.

Good, I.J. (1961). A causal calculus I, *British Journal for the Philosophy of Science* **11**: 305–18.

Good, I.J. (1962). A causal calculus II, *British Journal for the Philosophy of Science* **12**: 43–51.

Hitchcock, C. (2003). Causal generalizations and good advice, in H. Kyburg and M. Thalos (eds.) *Probability is the Very Guide of Life* (Chicago: Open Court), pp. 205–232.

Hitchcock, C. (2004). Do all and only causes raise the probabilities of effects? in John, Collins, Ned Hall, and L.A. Paul, (eds.) *Causation and Counterfactuals* (Cambridge MA: MIT Press), pp. 403–418.

Humphreys, P. (1989). *The Chances of Explanation: Causal Explanations in the Social, Medical, and Physical Sciences*. Princeton: Princeton University Press.

Joyce, J. MS. On the plurality of probabilistic measures of evidential relevance, unpublished manuscript.
Lewis, D. (1973). *Counterfactuals.* Cambridge MA: Harvard University Press.
Lewis, D. (1976). Probabilities of conditionals and conditional probabilities, *Philosophical Review* 85: 297–315.
Lewis, D. (1979). Counterfactual dependence and time's arrow, *Noûs* 13: 455–76.
Lewis, D. (1986). Postscripts to 'Causation', in *Philosophical Papers, Volume II* (Oxford: Oxford University Press), pp. 173–213.
Lewis, D. (2000). Causation as influence, *Journal of Philosophy* 97: 182–197.
Mackie. J. (1974). *The Cement of the Universe.* Oxford: Oxford University Press.
Parascondola, M. (1996). Evidence and association: Epistemic confusion in toxic tort law. *Philosophy of Science* 63 (supplement): S168–S176.
Pearl, J. (2000). *Causality: Models, Reasoning, and Inference.* Cambridge: Cambridge University Press.
Reichenbach, H. (1956). *The Direction of Time.* Berkeley and Los Angeles: University of California Press.
Skyrms, B. (1980). *Causal Necessity.* New Haven and London: Yale University Press.
Stalnaker, R. (1968). A theory of conditionals, in N. Rescher, ed., *Studies in Logical Theory.* Oxford: Blackwell.
Suppes, P. (1970). *A Probabilistic Theory of Causality.* Amsterdam: North-Holland.
Woodward, J. (1990). Supervenience and singular causal claims, In D. Knowles (ed.) *Explanation and its Limits.* Cambridge: Cambridge University Press, pp. 211–246.
Woodward, J. (2003). *Making Things Happen: A Theory of Causal Explanation.* Oxford: Oxford University Press.

·30·

A new causal power theory

Kevin B. Korb, Erik P. Nyberg, and Lucas Hope

Abstract

The *causal power* of C over E is (roughly) the degree to which changes in C cause changes in E. A formal measure of causal power would be very useful, as an aid to understanding and modelling complex stochastic systems. Previous attempts to measure causal power, such as those of Good (1961), Cheng (1997), and Glymour (2001), while useful, suffer from one fundamental flaw: they only give sensible results when applied to very restricted types of causal system, all of which exhibit causal transitivity. Causal Bayesian networks, however, are not in general transitive. We develop an information-theoretic alternative, *causal information,* which applies to any kind of causal Bayesian network. Causal information is based upon three ideas. First, we assume that the system can be represented causally as a Bayesian network. Second, we use hypothetical interventions to select the causal from the non-causal paths connecting C to E. Third, we use a variation on the information-theoretic measure *mutual information* to summarize the total causal influence of C on E. Our measure gives sensible results for a much wider variety of complex stochastic systems than previous attempts and promises to simplify the interpretation and application of Bayesian networks.

30.1 Theories of causal power

Intuitively, causal power is the strength of the connection from a cause to an effect: the power of a drug to palliate a patient; the power of a medicine to cure a patient; the power of a carcinogen to kill a patient. Probably everyone concerned with understanding causal relationships would prefer to replace our intuitions about these powers with a formal measure of causal power. AI researchers would like a tool which could explain observed effects in terms of the most important observed causes in a causal Bayesian network. Cognitive psychologists studying human causal reasoning want a well-founded account of causal power in order to better assess human judgment and the effectiveness of training in causal reasoning. Philosophers of science would like a criterion that could help with theories of causation (both type and token). Perhaps, eventually, we can hope for tools to help us assess moral and legal responsibility. While these aims are all distinct, they are also all related.

Over the last century there have been several notable attempts at producing such an analysis of causal power. We begin by briefly reviewing these.

30.1.1 Wright's theory

The first such theory can fairly be attributed to Sewall Wright (1934). He developed the first formal theory of graphical causal models, namely linear path models, which gave rise to the structural equation modelling which has dominated causal analysis in the social and biological sciences ever since. Path models are standardized linear models, where every variable takes the unit normal distribution $N(0, 1)$, directed arcs between variables indicate direct causal connections (e.g. $C \to E$), and each arc is assigned a path coefficient ρ_{EC} relating its cause C (or **parent**) to its target E (or **child**). Wright demonstrated a strict relation between the path coefficients and linear correlations, allowing path coefficients to be calculated from observed correlations and vice versa. Between any two variables X and Z:

1. Each active path contributes a correlation equal to the product of the path coefficients along it (e.g. $X \to Y \to Z$ contributes $\rho_{YX}\rho_{ZY}$).
2. The total correlation is the sum of the contributions of all active paths.

Φ_k is an **active path** from X to Z if and only if it is a sequence of arcs connecting X to Z where no arc points backwards *after* an arc has pointed forwards. Active paths, therefore, are either chains, forwards or backwards, or paths that relate two variables via a common ancestor. They cannot include collisions (e.g. $X \to Y \leftarrow Z$).[1]

Wright did not explicitly attempt to characterize causal power. However, one straightforward way of doing so with path models would be to select only those paths that are forward chains from X to Y, and calculate the amount of correlation due to these paths. Another possibility is to apply the concept of intervention. **Interventions** can be represented by variables new to the modelled system, intentionally introduced to influence the value of one or more system variables. What we can call **perfect interventions** are those which successfully target a single variable, setting it to a particular distribution in a deterministic way, without regard for the original parents. While in reality perfect interventions are rare (Korb *et al.*, 2004), they can be very useful in developing theory; for one thing, graphically they can be represented simply by setting the target variable to its intended distribution and cutting all in-bound arcs to it. Suppose we apply such an intervention to C in a path model, imposing the unit normal distribution $N(0, 1)$. Only the forward chains from C to E will transmit the results of our intervention, and

[1] So, Wright's rule for identifying paths that contribute to the correlation between pairs of variables is essentially the definition of d-connection (Pearl, 1988), although Wright did not consider conditioning upon a collider, which will activate the path through it.

thus the resultant correlation with E will be equivalent to picking out these paths by hand. Whichever way we apply Wright's theory for analysing causal power, the resultant formula is:

Definition 30.1 (Wright's (implicit) causal power measure)
The causal power of C for E is:

$$\text{CP}(C, E) = \sum_k \left(\prod_{lm} \rho_{lm} \right)$$

for all forward chains $\Phi_k = C \rightarrow \ldots \rightarrow E$ and for all $X_m \rightarrow X_l \in \Phi_k$.

We believe this is a perfectly fine theory of causal power, as far as it goes. It is limited to recursive models, since non-recursive models lack directionality for some of their arcs. This can be interpreted as ignorance, either about arc orientation or about the possibility of unknown common causes relating the correlated variables. In either case, recursive models can be viewed as representing the underlying reality, with causal power being unknown until that reality is better revealed. So this limitation is not a defect in Wrightean power theory; it is a feature that all causal power theories ought to share. The problem is that Wright's theory is limited to linear Gaussian models, and many systems are nonlinear.

30.1.2 Good's theory

Good (1961) made the earliest explicit proposal for a causal power measure. He intended it to be more generally applicable than Wright's measure, by encompassing all kinds of multinomial networks (i.e. ones with discrete variables). Good made some general assumptions about the nature of a putative causal power measure, and thereby derived something equivalent to his Bayesian 'weight of evidence' formula (and vaguely analogous to electrical conductivity and resistance):

Definition 30.2 (Good's basic causal power measure)
The causal power of $C = c$ to produce $E = e$ is:

$$Q(e : c) = \log \frac{P(\neg e|\neg c)}{P(\neg e|c)}$$

provided that any dependency is entirely due to C affecting E.

$Q(e : c)$ plays a similar role to $\text{CP}(C, E)$ in Wright's theory. Like Wright, Good suggests a formula for calculating the causal power of a chain by combining the causal power of component arcs and also for additively calculating the causal power of multiple causal chains.

Unfortunately, Good's general assumptions do not hold in many multinomial networks.

- He assumes he can treat all variables as if they are binary, e.g. comparing c to $\neg c$ while ignoring any differences between sub-states of $\neg c$.
- Hence, his method of calculating the power of a causal chain, while it may hold for genuine binary variables, fails in general for discrete networks. For example, it entails causal transitivity: positive causal power from c to d and d to e implies positive causal power from c to e. This is inevitable for linear causal models, and for some others discussed below, but not for all multinomial networks. His formula can also give different answers depending upon the precise path, even if the end-to-end dependency is the same (Salmon, 1980), which contradicts his own assumptions.
- Good assumes that the power of multiple chains can simply be added, a method that fails wherever there is any causal interaction (even in binary networks).
- Good also provides no way of distinguishing causal from non-causal dependency paths.

Clearly, then, Good's theory is unsatisfactory as a general account of causal power.

30.1.3 Cheng's theory

Patricia Cheng (1997) developed her 'power PC' theory as an improvement over Rescorla and Wagner (1972), and it has been further developed by Glymour and Cheng (1998), Novick and Cheng (2004), and Glymour (2001). Cheng begins with a measure of positive statistical relevance, or 'positive probabilistic contrast'

$$\Delta P_c = P(e|c) - P(e|\neg c) > 0$$

which indicates 'candidate generative causation', echoing Suppes (1970), who called this *prima facie* causation. c is only a *prima facie* cause because the probabilistic contrast may actually be caused by a common ancestor that raises the probability of c and e occurring together. Suppes went on to lay down temporal and statistical conditions aiming to rule such cases out. These efforts have now been subsumed by developments in Bayesian network technology (cf. Twardy and Korb, 2004). Cheng rules these cases out by laying down some very stringent requirements for the causal relationships permitted in her models. First, the occurrence of c must be independent of all other parents of E. This implies either a limited causal structure in which there are no causal paths between C and these parents, or that the effect of these paths can be cancelled by fixing some background variables (which is not possible in some graphs). Second, the dependency between c and e must be independent of the dependency between e and any other parent, implying that there can be no causal interaction between C and any other parents of E.

Given these restrictions, it is clear that the probabilistic contrast must be caused by c. In other words, the occurrences of c must be 'generating' the additional occurrences of e. Cheng now defines the causal power of c for e as *the probability that any given occurrence of* c *will generate* e. This causal power of c is labelled p_c, leaving e implicit. Her basic insight is that ΔP is not a fair measure of p_c. There is a specific background rate at which e occurs even without c, namely $P(e|\neg c)$. This means that we can only detect the impact of c on the *remaining* instances of E, by measuring how many background occurrences of $\neg e$ are converted to e. $\neg e$ occurs with a background frequency of $1 - P(e|\neg c)$; it is converted with a frequency of ΔP; and therefore, the success rate of c must be the ratio of these two quantities.[2] Hence:

$$p_c = \frac{\Delta P_c}{1 - P(e|\neg c)}. \tag{30.1}$$

In contrast, a negative ΔP indicates 'candidate preventive causation', in which c appears to prevent e from occurring. To analyse this, Cheng places the same stringent restrictions on the parental relationships. She then defines the causal power of c to prevent e in an analogous way, as *the probability that any given occurrence of* c *will prevent* e. To distinguish prevention from generation, we write preventive powers as $\overline{p_c}$. By similar reasoning, we can only detect the success rate of c against the background rate of e, namely $P(e|\neg c)$. Thus:

$$\overline{p_c} = \frac{-\Delta P_c}{P(e|\neg c)}. \tag{30.2}$$

Cheng claims that these formulae are a significant improvement on previous theories, such as that of Rescorla and Wagner (1972), because (among other reasons) the formula for p_c provides the correct answer when e always occurs. If e always occurs, then the value for p_c is undefined, rather than a power of zero, as Rescorla and Wagner had suggested. Cheng deems leaving p_c unspecified to be correct because we should be unable to statistically assess the candidate causes of a universal event. Similarly, the value of $\overline{p_c}$ is undefined when e never occurs. However, we do not see this feature of her theory as a significant advantage. Rescorla and Wagner might well reply that no candidate cause could demonstrate any statistical power over a universal event, and therefore in such cases zero is a reasonable statistical assessment of causal power.

[2] Strictly speaking, this only gives us the success rate where $\neg e$ would otherwise have occurred. The success rate of c where e would have occurred anyway is a moot point. Cheng can either assume that it is the same, or else ignore these cases altogether, as being unimportant for assessing causal power.

The fundamental problem with Cheng's measure is that it has an extremely limited range of application.

- Like Good's theory, it is only applicable to questions about causal relations between values, as opposed to the variables themselves (which were addressed by Wright's theory).
- Like Good, Cheng treats all variables as if they are binary. Admittedly, this does not lead to the same contradictions in calculating the causal power of chains and networks with complex relations or structures. But then Cheng does not offer *any* way to calculate causal power in such cases.
- The structural independence restrictions upon parents are very severe, and will not be met by many Bayesian networks. This just leaves causal power undefined, despite the fact that C is clearly affecting E.
- Cheng's blanket ban on any causal interactions between parent variables are necessary to make her derivations of (30.1) and (30.2) work, but as Glymour (2001) has shown, it limits Cheng's theory to noisy-OR Bayesian networks. Novick and Cheng (2004) relax this last restriction by combining interactive parents in new variables. However, that ad hoc solution is, on the one hand, computationally infeasible when many parents of large arity are involved, and, on the other hand, leaves the restriction to non-interaction between all remaining parents untouched.[3]
- A notable consequence of the restriction to noisy-OR networks is causal transitivity. This, in fact, is a property of all the accounts discussed so far. Yet any account of causal power that entails transitivity is misleading, since causation in general is not transitive – a fact which is reflected in other types of Bayesian network. Take, for example, Richard Neapolitan's case of finesteride (Neapolitan, 2003). Finesteride reduces testosterone levels; lowered testosterone levels can lead to erectile dysfunction. However, finesteride fails to reduce testosterone levels *sufficiently* to cause erectile dysfunction. Such threshold effects do not occur in linear or noisy-OR networks, but they are common elsewhere.

Cheng's theory was intended, in part, to provide a psychological model for the causal attributions made by ordinary folk. Whatever its merits may be for this purpose, it lacks the generality that we would like from a sophisticated causal power measure.

[3] Luhmann and Ahn (2005) make the curious objection to power PC theory that it implies that all causes have powers of 0 or 1 to bring about their effects. They claim this follows from Cheng's assumptions 'unless for some inexplicable reason, the causal link between c and e is intrinsically indeterminate' (Luhmann and Ahn, 2005, p. 686). It is true that if all other causes of e are included in the model, and the model is deterministic, then powers of 0 or 1 will result. But this would not be unreasonable, nor is it a special problem for Cheng's account. The complaint amounts to the observation that Cheng plus determinism implies determinism!

30.1.4 Desiderata for causal power

Having briefly reviewed prior accounts of causal power, we can invert the list of their several or collective drawbacks to generate a list of features that would be desirable in a new account:

1. Wright's implicit causal power theory for linear models appears to be fine within its domain. Hence, if any new theory is applied to linear models, we should like it to attribute powers that directly correspond to Wright's. Specifically, causal powers between variables in linear networks should be ranked in the same order.
2. The measure should additionally be more general than Wright's theory by being applicable to all kinds of Bayesian network: even those with complex variables, structures, and dependencies.
3. The measure should not entail transitivity – simply because causation is not, in general, transitive. Of course, the measure needs to reflect transitivity when it appears.
4. The measure should be compatible with intervention. It should support the fundamental idea, illustrated above with Wright's theory, that interventions test causal power.
5. The measure should have an information-theoretic interpretation. This desideratum is not motivated by prior considerations, but we adopt it since causality gives rise to probabilistic relationships, which then ought to be interpretable using Shannon's information measure.[4]

Prior measures fulfil some of these requirements, but none of them successfully fulfils them all.

30.2 Causal Bayesian networks

There is a new paradigm for modelling probabilistic causal systems, arising from new technology, namely *causal Bayesian networks*. Such networks offer a powerful and general way to represent all kinds of stochastic causal relationships, and are being deployed in both theoretical and practical applications across a wide range of disciplines. Our own proposal for analysing causal power is (simultaneously) a proposal for reading the causal stories implicit within these networks, even if the entire network is too complex to fully comprehend.

Bayesian networks, popularized by Pearl (1988), use directed acyclic graphs to represent probabilistic relationships between random variables, e.g. $C \rightarrow D \rightarrow E$. There is an elementary conditional probability function $P(D|\pi_D)$

[4] For an excellent introduction to information theory, see Cover and Thomas (1991).

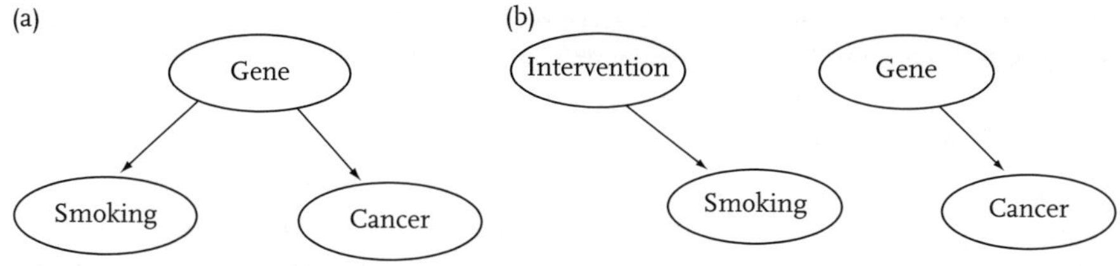

Fig. 30.1 (a) Fisher's smoking model; (b) Fisher's model with intervention.

associated with each node, which specifies a probability distribution for its variable, D, that depends only upon its parents, π_D (here only C), and not upon other variables (such as E). The linear models of Wright and the noisy-OR networks of Cheng are special cases. But in general the conditional probability functions are unrestricted and can include nonlinear interactions such as XOR relationships or threshold effects.

Bayesian networks were not originally intended to be interpreted *causally*: they were simply maps of probabilistic dependence, in which the arcs might be oriented in an anti-causal direction (e.g. $C \leftarrow D$). But in a causal Bayesian network the arcs are also supposed to reflect the direction of causation, and this interpretation has become increasingly important. Many causal discovery algorithms have been developed to learn causal Bayesian networks from data, and they have been quite successful (e.g. Verma and Pearl, 1990; Spirtes, Glymour, and Scheines, 2000; Neapolitan, 2004; Korb and Nicholson, 2004).

Given an explicitly causal Bayesian network, it becomes possible to model the implied effects of interventions – and this can be crucial for determining the causal story. For example, the fact that smoking is correlated to cancer is usually explained by the model *Smoking* → *Cancer*. But Sir Ronald Fisher proposed Figure 30.1(a) as an alternative model. His point was two-fold: (1) observational, correlational data alone cannot distinguish between the two models; (2) interventional data can do so. If his model were right and we intervened to force people to smoke, as in Figure 30.1(b), then there would be no resulting increase in cancers – whereas the usual model implies just such an increase. Thus, interventions can't be modeled without causation, and conversely, interventions can expose the difference between correlation and causation.

30.3 Causal information

Given these tools, we can now present our solution to the problem of measuring causal power: causal information.[5] Given some causal Bayesian network,

[5] Causal information was first introduced by Hope and Korb (2005); here we further develop the account and compare it to earlier theories. Ay and Polani (2008) have subsequently developed a similar information-theoretic approach.

the problem is to state the causal power of one variable, C, over another variable, E, implied by that network.[6]

30.3.1 Background conditions: ψ_h

Causal questions are always put relative to some background conditions. The propensity of smoking to induce lung cancer, for example, is likely to be an issue for a large class of adult humans, but not, perhaps, for humans who already have cancer. A full account of how background context should be modelled is a difficult and unsolved problem and one which surely must involve a treatment of conversational implicature and psychological theory. What is counted as an appropriate context depends upon our interests, perhaps varying from moment to moment as we shift from a historical query to a counterfactual query. Without attempting to provide an analysis of such complex issues, we simply point out that Bayesian networks offer some useful resources for representing context. Here we will simply represent such conditions by identifying a set of network variables whose values should be fixed, $\Psi = \psi_h$.[7] Thus, all the probabilities discussed in the following sections will be conditional probabilities of the form $P(\cdot|\psi_h)$, but for brevity we will omit this condition in our formulae for causal power.

30.3.2 Hypothetical interventions: $P^*(C)$

As we have seen, intervention upon C provides a straightforward way to distinguish between non-causal paths to E, e.g. those that run through common ancestors, and causal paths to E, i.e. forward chains. For this purpose we will apply hypothetical interventions that are targeted strictly at C, independent of any other parents of C and overwhelm their effects, and which impose a specific distribution on C.[8] So we augment the model M to M^*, with just one new intervention node and arc $I_C \to C$, and just one new elementary conditional probability function $P^*(C|\pi_C, I_C)$ over C, replacing the original $P(C|\pi_C)$. Since the intervention is overwhelming, when $I_C = Yes$ all inferential paths that begin backwards from C are cut. Since the intervention is stochastic, C still varies, and therefore dependency can still be transmitted by any path that begins forwards from C. For brevity, we will assume that $I_C = Yes$ has been added to ψ_h whenever we refer to $P^*(\cdot)$.

[6] Thus, we are here concerned with accounting for the total power of one variable to influence another, under various background conditions, across all causal paths connecting them. It would also be of interest to account for the causal powers of individual paths and relate them to the total causal power of all paths; that is a matter of current research.

[7] We will assume that Ψ does not include any common effect variable that activates a non-causal path from X to Y.

[8] We emphasize that these interventions are only hypothetical. Their purpose is simply to reveal features already implicit in the given causal Bayesian network. So it is not necessary that they be practical or even physically possible; it is sufficient that they can be modeled by augmentation.

But what intervention distribution should be imposed upon C? There are three alternative choices that strike us as reasonable, each serving a slightly different purpose.

Original $P^*(C)$

The first option is to reassert the original distribution over C. The new model M^* will still differ from M; however, by reimposing the original distribution on C we minimize those differences. Not only are all the causal paths between C and E preserved, but also the variation in C itself. The similarity between M and M^* means that the causal power of C over E in M^* reflects the original situation in M as closely as possible. For example, we can use M^* to consider, 'Given the variation in blood pressure among the general population, how much is this variable affecting heart attack outcomes?' We should note, however, that even if we impose the original distribution upon C, the resulting distribution upon E may still be considerably different in M^* than it was in M, simply because (as intended) C is no longer dependent on its original parents.

Uniform $P^*(C)$

We may not always wish to measure causal power relative to the original distribution over C. For example, in some subpopulation which regularly exercises and mostly eats fish, there may be very little natural variation in blood pressure. In consequence, the connection between blood pressure and heart attack outcomes will be concealed by this healthy lifestyle. So one way to bring out this latent feature of M is to consider a different intervention distribution over C, even though it is not the naturally occurring distribution in M. Any investigation into the power of blood pressure in M would certainly not randomize its subjects so that they all fell into the low blood pressure group; instead, it might mimic randomized experimental design by distributing C uniformly, with equal numbers across blood pressure categories. Thus, one plausible choice is a uniform distribution on C, so that there are equal numbers of subjects in every blood pressure group. In comparing the effects of different blood pressures, this provides a 'level playing field' in which the results are not biased by different actual frequencies for these blood pressures. Similarly, in comparing the influence of variables, it provides a standard distribution for comparison.

To be most exact, we would impose some intervention distribution $P^*(C)$ such that after we take into account the background conditions, the resulting distribution $P^*(C|\psi_h)$ is uniform. That is, $P^*(c_i|\psi_h) = \frac{1}{C}$ for each c_i. We note that to achieve this, $P^*(C)$ itself will not always be uniform.

Maximizing $P^*(C)$

Another reasonable question to ask is: what is the maximum impact that C could possibly have on E, according to M? To be precise, we can search the space of possible intervention distributions $P^*(C)$, to find those that maximize our causal power measure given the background conditions.

Note that this will not always be the uniform distribution considered above, even though for unbiased channels the uniform distribution maximizes mutual information. The 'channel' here – of causal power – will frequently be biased. Suppose, for example, that there are only three blood pressure categories, and while both low and medium blood pressures result in a similar risk of heart attack, high blood pressure results in a much higher risk. Then the maximum probabilistic dependence between *Blood Pressure* and *Heart Attack* will result from a distribution in which nearly 50% of subjects have high blood pressure, rather than 33%.

These three possible intervention distributions are complementary, attempting to measure three different forms of causal power of C over E: the original causal power, a uniform causal power, and the maximum causal power. Either of the latter two standardize causal power comparisons between models in that they eliminate the influence of diversity in prior distributions over C. In the formulae that follow we leave open the choice of intervention distribution, which is simply denoted $P^*(C)$. But to make illustrative computations, we will imagine that the original distribution has been imposed upon *Blood Pressure* to measure its causal power over *Heart Attack*, as in Figure 30.2.

30.3.3 Causal information formulae

Two values: c and e

We begin with the simplest formula and then work our way through the more complicated ones. Each formula answers a slightly different causal question.

In particular, we begin with the question: what is the causal power of one value, c, to affect another value, e? Value-to-value questions such as 'If I have high blood pressure, then how much does this affect my risk of having a heart attack?' are quite common. The causal information answer is:

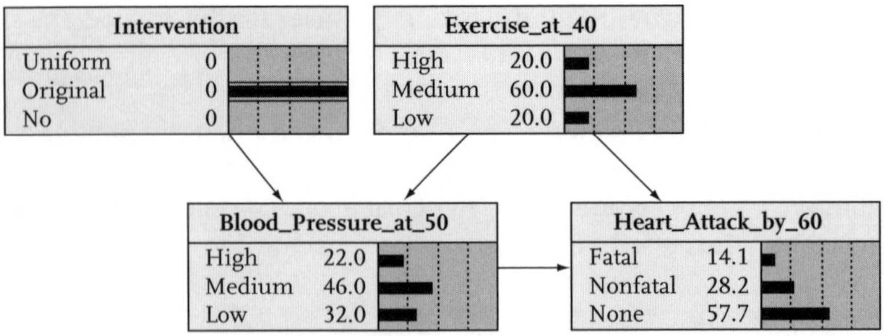

Fig. 30.2 An original intervention on blood pressure, to measure its causal power.

$$\mathrm{CI}(c,e) = P^*(e|c)\log\frac{P^*(e|c)}{P^*(e)}. \tag{30.3}$$

In information theory, this formula gives the information about e that is provided by the discovery that $C = c$ compared to knowing the prior distribution $P^*(C)$. Given that only causal paths are active, we suggest that this formula can also serve as a good measure of the causal power of $C = c$ to affect the probability of $E = e$. For example, suppose we observe that a patient has high blood pressure, c. This increases the probability of having a fatal heart attack, e, from the average probability $P^*(e) = 0.14$ to $P^*(e|c) = 0.23$. So $P^*(e|c)/P^*(e) = 1.63$. This is converted to a logarithm base 2,[9] which takes the positive value 0.71. It is multiplied by the probability of having a heart attack given high blood pressure, 0.23. So the causal power of high blood pressure to promote heart attack is 0.16.

Causal information compares $P^*(e|c)$ to the marginal probability $P^*(e)$, rather than the complementary probability $P^*(e|\neg c)$. This is similar to the standard formula for statistical relevance (SR) used in philosophy, rather than the standard formula used in psychology (Δp). Causal information compares these two probabilities as a ratio rather than a difference, thus measuring the proportional change (like the Bayes factor in confirmation theory) rather than the absolute change (like SR). This proportion is converted to a logarithm, which is usual in information theory. This logarithm is then weighted by the probability $P^*(e|c)$. The CI(c,e) measure is positive for promoting causes and negative for preventive causes, just like Δp.

Note that in this formula the prior probability of high blood pressure, $P^*(c)$, does not feature as a weighting factor. $C = c$ is treated as a given, as in the example question 'If I have high blood pressure, ...', so we set $P^*(c) = 1$.

Variable causes: C and e

The next formula addresses the question: what is the causal power of one variable, C, to affect a particular value, e?

$$\mathrm{CI}(C,e) = \sum_{c\in C} P^*(c)\,P^*(e|c)\log\frac{P^*(e|c)}{P^*(e)}. \tag{30.4}$$

This formula gives the expected information about e that will be provided by discovering the value of C, whatever that turns out to be, compared to knowing the distribution $P^*(C)$. The difference between this and Equation 30.3 is that the value of C is no longer treated as a given. Instead, we take the information (or power) from each individual value c_i, and weight this by the probability $P^*(c_i)$, to calculate the expected value. We suggest that this formula can also serve as a good measure of the causal power of C to affect the probability of

[9] The base is unimportant, and frequently natural logs are used. We use base 2 here to simplify the interpretation of causal information as code length.

$E = e$. For example, 'How much does variation in blood pressure affect the risk of having a heart attack?' is a variable-to-value question.

Note that some of the individual figures for causal power will be positive, and other figures will be negative. If we took a weighted average of the absolute magnitudes, then this would be the expected magnitude of the causal power exerted when C takes a specific value. However, the information-theoretic formula given above does not use absolute magnitudes, and negative individual powers will partially offset the positive ones. Therefore, CI(C,e), while useful for comparing alternative models, cannot be directly compared to the magnitude of CI(c,e). The CI(C,e) measure will always be positive, provided that C has some effect, i.e. $\exists i, j : \; P^*(e|c_i) \neq P^*(e|c_j)$, and otherwise it will be zero (see Appendix B).

Variable effects: c and E

What is the causal power of one particular value, c, to affect a variable, E?

$$\mathrm{CI}(c,\, E) = \sum_{e \in E} P^*(e|c) \log \frac{P^*(e|c)}{P^*(e)}. \tag{30.5}$$

This formula gives the total information about E that is provided by the discovery that $C = c$ compared to knowing the distribution $P^*(C)$. The difference between this and Equation 30.3 is that we are interested in all the values of E, not just one e. So we take the information from c for each individual value e_i, and add them to calculate the total value. We suggest that this formula can also serve as a good measure of the total causal power of c to affect the probability of E. For example, 'How much does having high blood pressure affect heart attack outcomes?' is a value-to-variable question. Again, note that our information-theoretic formula does not use absolute magnitudes, and the negative individual powers will partially offset the positive ones. The CI(c,E) measure is equivalent to the Kullback–Leibler divergence between $P^*(E|c)$ and $P^*(E)$, which is always positive, provided that there is some difference between the distributions.

Two variables: C and E

What is the causal power of one variable, C, to affect another variable, E?

$$\mathrm{CI}(C,\, E) = \sum_{c \in C, e \in E} P^*(c)\, P^*(e|c) \log \frac{P^*(e|c)}{P^*(e)}. \tag{30.6}$$

This formula gives the expected information about E that will be provided by discovering the value of C, whatever that turns out to be. It uses both the weighted average over the values of C and the sum over the values of E. We suggest that this formula can also serve as a good measure of the total causal power of C to affect the probability of E. For example, 'How much does variation in blood pressure affect heart attack outcomes?' is a variable-to-variable question. Again, the negative individual powers will partially offset the positive ones, but the CI(C,E) measure will always be positive, provided that C has some effect on E.

The number of alternative formulae reflect the fact that there are several related questions about the causal power of C over E. So it is important to disambiguate informal queries such as 'How much does blood pressure affect heart attacks?' before attempting to find an answer.

30.3.4 Mutual information

This last variable-to-variable equation can be transformed as follows:

$$\mathrm{CI}(C, E) = \sum_{c\in C, e\in E} P^*(c)\, P^*(e|c) \log \frac{P^*(e|c)}{P^*(e)} \tag{30.7a}$$

$$= \sum_{c\in C, e\in E} P^*(c, e) \log \frac{P^*(e|c)}{P^*(e)} \tag{30.7b}$$

$$= \sum_{c\in C, e\in E} P^*(c, e) \log \frac{P^*(c, e)}{P^*(c)\, P^*(e)} \tag{30.7c}$$

$$= \mathrm{MI}(C, E). \tag{30.7d}$$

This shows that causal information is identical to the information-theoretic quantity *mutual information* (MI), when applied to the two variables C and E, given the intervention upon C. The mutual information formula looks a little different. It compares the probability that c and e will occur together, $P^*(c, e)$, to the probability that they *would* occur together *if the two variables were independent*, $P^*(c)\,P^*(e)$. Thus, it measures the amount of dependency that exists between each pair of variable values. The accumulated dependency for the two variables is obtained by weighting these ratios according to the probability that this pair of values will actually arise, $P^*(c, e)$. In fact, the causal information formula does the same thing, but it has been expressed in an asymmetrical fashion to suit the asymmetry between cause and effect.

By definition, mutual information is the expected amount of information that one variable provides about another (or the loss of information that arises by falsely assuming that they are independent).[10] But, as above, it can also be interpreted as the amount of dependency between them. Therefore, it would be a good measure of causal power – except that some of this dependency can arise from non-causal links. Causal information corrects this defect.

[10] From Shannon (1948), the negative log of the probability of an event is the optimal code length to describe that event. Hence, mutual information can also be interpreted as the expected excess code length involved in recording the values of X and Y while wrongly assuming that they are independent.

30.3.5 Entropy

Mutual information is also closely related to the entropy measure of randomness. The information entropy on the variable E is defined as follows (Cover and Thomas, 1991):[11]

$$H(E) = -\sum_{e \in E} P(e) \log P(e). \tag{30.8}$$

Entropy is zero when $P(E = e_i) = 1$ for some value e_i, when there is no uncertainty about the value of E. It is maximized when $P(E)$ is uniform across all the possible values of E, when uncertainty is highest.

Similarly, conditional entropy measures the randomness of one variable given knowledge of another:

$$H(E|C) = -\sum_{c \in C} \sum_{e \in E} P(c, e) \log P(e|c). \tag{30.9}$$

Thus:

$$\mathrm{MI}(C, E) = H(E) - H(E|C). \tag{30.10}$$

This supports the interpretation of mutual information as the reduction in the uncertainty of E due to the knowledge of C.

30.4 Comparisons

Causal information has some clear advantages over the rival measures of causal power.

- Causal information is well defined for all causal Bayesian networks. This includes all the restricted classes of network for which other measures were designed: linear models, Cheng's binary models and their extensions, and whatever models Good had in mind. But it also includes classes of network for which these rival measures are not well-defined: e.g. ones with interactive causes, intransitivity, or multinomial (discrete) variables.
- Causal information is well defined for a wider variety of questions. It relates any causal variable or value (either observed or observable) to any effect variable or value. It does so with a uniform approach, unlike Cheng's measure (for example), which uses a different formula for promoting and preventive causes.
- Causal information yields appropriate results in all the restricted classes of network, where it mirrors the local properties. For example, in any

[11] Entropy is defined subject to the common assumption that $0 \log 0 = 0$, which is justified by continuity arguments.

network that exhibits causal transitivity, $C \to D \to E$ implies that E is dependent upon C. But it follows immediately that $\mathrm{CI}(C, D) \neq 0$, $\mathrm{CI}(D, E) \neq 0$, and $\mathrm{CI}(C, E) \neq 0$. So causal information itself exhibits causal transitivity, simply by accurately summarizing the true amount of dependency. Similarly, in linear path models, causal information is a monotonically increasing function of the magnitude of correlation (Hope, 2008, Chap 6). Therefore, the fact that other measures are *necessarily* transitive offers no advantage, even when they are applied to their own preferred class of network.

- Causal information yields appropriate results in the other classes of network, where it does not impose inappropriate properties. For example, in any network that exhibits causal intransitivity or interaction, causal information itself exhibits intransitivity or interaction, since it is based directly upon the corresponding probability distributions. It follows that causal information can be applied uniformly, without making restrictive assumptions about the structure of the network or its dependencies.

30.4.1 Quantitative comparisons

We will now provide some quantitative comparisons of the competing causal power measures, by applying them to appropriate graphs.

30.4.2 CI versus Wright

Since Wright's coefficients are only defined for linear models, we have constructed a linear approximation of our discrete blood pressure graph, depicted in Figure 30.3. This includes just the variables *Exercise, Blood Pressure,* and *Heart Attack*. We have assumed that these variables have been converted to scalar quantities that are distributed according to the standardized Gaussian distribution $N(0, 1)$. Positive numbers therefore represent higher than average levels, and negative numbers represent lower than average levels.

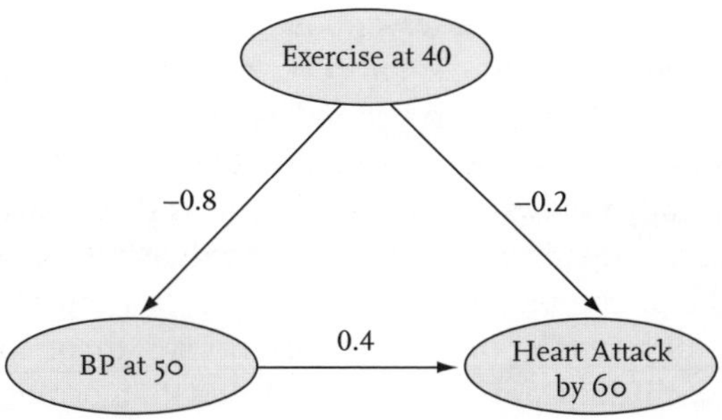

Fig. 30.3 Linear approximation to the *Blood Pressure* graph.

The size and direction of the dependencies in the discrete graph have been converted to similar path coefficients: −0.8 between *Exercise* and *Blood Pressure*; −0.2 between *Exercise* and *Heart Attack*; and +0.4 between *Blood Pressure* and *Heart Attack*.

The challenge is to measure the causal power of *Blood Pressure* over *Heart Attack*, despite their common cause *Exercise*, which forms the backpath

$$\textit{Blood Pressure} \leftarrow \textit{Exercise} \rightarrow \textit{Heart Attack}$$

Since there are two paths between *Blood Pressure* and *Heart Attack*, Wright's equations tell us that the total correlation between them is the sum of the correlations due to each of these paths:

$$r_{HB} = \rho_{HB} + \rho_{HX}\rho_{BX} = +0.4 + (-0.2 \times -0.8) = +0.56. \qquad (30.11)$$

This figure obviously depends upon the strength of the backpath. Yet the arc directions imply that *Blood Pressure* does not actually exert any causal influence by this path. In fact, since this is a linear graph, either changing the path coefficients on this path or conditioning upon *Exercise* can have no effect upon the causal dependency between *Blood Pressure* and *Heart Attack*. So, it follows that any measure that is affected by the backpath is incorrect.

To convert Wright's approach to a measure of causal power, we can pick out just the directed causal paths from *Blood Pressure* to *Heart Attack*, and calculate their contribution alone:

$$r_{HB} = \rho_{HB} = +0.4. \qquad (30.12)$$

This is a plausible measure since it only depends upon the causal path and increases with the size of the correlation that it induces. Ignoring causal direction in this case would result in a 40% overestimate of causal power (as measured by correlation).

To apply the CI measure, we can simulate an overwhelming intervention upon *Blood Pressure* with an intervention distribution that is the original marginal distribution, i.e. $N(0, 1)$. This overwhelms the arc between *Exercise* and *Blood Pressure*. Since the intervention has removed the backpath, the total correlation is now equal to the causal correlation. As noted above, CI is a strictly monotonically increasing function of this restricted causal correlation; i.e. the CI approach is equivalent to Wright's approach for determining which paths are causal and combining them to form a total causal correlation. The only substantial differences are (a) the information-theoretic rescaling, and (b) the CI approach does not need any additional algorithm to identify the paths or make the calculations – once the intervention is added, any standard Bayesian updating algorithm can do that work.

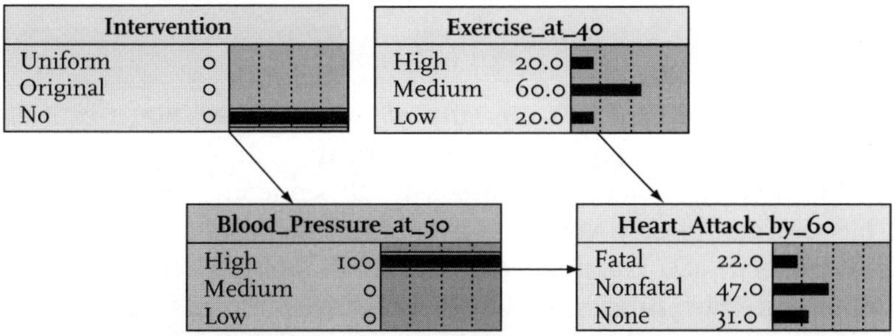

Fig. 30.4 Noisy-OR approximation to the blood pressure graph.

30.4.3 CI versus Cheng

Cheng's measure can be applied to any discrete graph, but, unlike CI, it will not generally be valid if her strict assumptions are not met. To illustrate, we have constructed an alternative version of our discrete blood pressure graph, which has been simplified to accommodate Cheng's restrictions (Figure 30.4):

1. We have removed the connection between *Exercise* and *Blood Pressure*, so that they are not marginally dependent.
2. Cheng's formulae are defined for binary variables. We can only apply the power PC theory to our ternary variables by partitioning the states into two mutually exclusive subsets. We shall let the causal event c be high blood pressure, so that $\neg c$ is either medium or low blood pressure. We assume that the relative marginal probabilities of these two sub-states of $\neg c$ remain the same. We shall let the effect event e be a fatal heart attack, and the other possible cause of this event be low exercise.
3. We have altered the effects of low exercise and high blood pressure on fatal heart attacks so that their true interaction is replaced by a noisy-OR approximation. Specifically, there is a 20% chance that high blood pressure alone will result in a fatal attack; 12.5% for low exercise alone; 30% for both factors together; and 0% where neither factor is present. All necessary adjustments to the rates of fatal heart attack have been balanced by changes to the rates of non-fatal heart attack.

These simplifications are not very realistic, but these or similar changes are necessary in order to satisfy Cheng's restrictions.

We can now apply Cheng's formula to assess the power of high blood pressure to promote fatal heart attacks. We are supposed to ignore *Exercise*, since it is assumed that this does not affect the causal power of *Blood Pressure*. So we shall just assume the marginal distribution over this variable. The result is (see Appendix A for computation details):

$$p_c = \frac{\Delta p_c}{1 - P(e|\neg c)} = 0.20.$$

Cheng's figure of 0.20 is a measure of causal power, but it can also be given a straightforward probabilistic interpretation. It is supposed to be the probability that high blood pressure will kill someone, given that without high blood pressure they would survive. Since this is a noisy-OR graph, this figure is perfectly accurate relative to that graph and is a perfectly reasonable way to measure how much high blood pressure promotes fatal heart attacks. Nevertheless, even in this case, CI does interestingly differ from Cheng's measure, beyond its different scaling.

The CI formula, where we assume an intervention upon *Blood Pressure* that imposes the original marginal distribution, reports the causal power:

$$\text{CI}(c, e) = P^*(e|c) \log \frac{P^*(e|c)}{P^*(e)} = 0.37.$$

The CI figure of 0.37 is the expected number of bits saved by learning that $C = c$, when efficiently encoding the fatal outcome $E = e$. Whereas PC reflects only the proportional increase in risk, CI reflects the absolute increase in risk, through the weighting by $P^*(e|c)$: the absolute probability of a fatal heart attack given high blood pressure. In consequence, the CI measure, unlike power PC, can report different amounts of causal power depending upon the state of *Exercise*. For example, suppose that one does some exercise, so the chance of dying is low. In that case, high blood pressure would result in bigger absolute increase to the chance of dying, and accordingly, $\text{CI}(c, e) = 0.44$. On the other hand, if one does low exercise, then the chance dying is higher anyway, so having high blood pressure would result in a smaller absolute increase, and accordingly, $\text{CI}(c, e) = 0.26$. Clearly, there is a sense in which *Exercise* does affect the power of high blood pressure for fatal heart attacks. CI reflects this, whereas Cheng's measure does not.

So far we have found some differences between CI and power PC, without establishing any advantage over the latter. Suppose now that we reintroduce the original connection between *Exercise* and *Blood Pressure*. By violating Cheng's assumptions, this will increase the divergence between the CI and PC analyses. The power PC theory's assessment of causal power becomes:

$$p_c = \frac{\Delta p_c}{1 - P(e|\neg c)} = 0.22.$$

The backpath has caused Cheng's causal power estimate to increase by 10%. But in fact the noisy-OR interaction has not changed, so the true probability that high blood pressure will kill has not changed. Regardless of one's level of exercise, high blood pressure still reduces the chance of survival by 20%. The 10% difference therefore represents an error in Cheng's estimate of

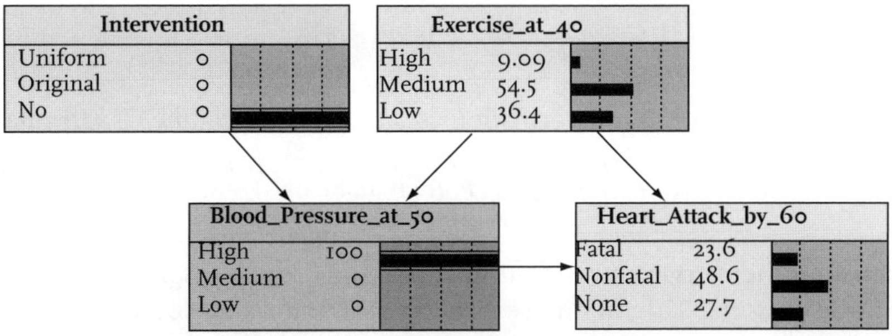

Fig. 30.5 The original *Blood Pressure* graph with messy interaction and backpath.

causal power, overstating the power of high blood pressure. In contrast, the CI measure is unchanged, since the reintroduced arc is overwhelmed by the intervention distribution. This is a clear advantage of CI.

Now let us revert to our original graph, relating *Exercise, Blood Pressure,* and *Heart Attack,* as depicted in Figure 30.5. The PC calculation for it yields:

$$p_c = \frac{\Delta p_c}{1 - P(e|\neg c)} = 0.15.$$

While Cheng's measure can again be applied, it is now vulnerable to two sources of error: the backpath and the messy interaction. The interaction between *Exercise* and *Blood Pressure* creates an additional kind of problem for PC. Cheng's theory suggests that we can afford to ignore the state of *Exercise* and gives us only one figure for the power of *Blood Pressure.* But unlike the noisy-OR graph, the true probability may differ depending upon one's specific level of *Exercise* – for example, low exercise and high blood pressure may have a synergistic effect in promoting fatal heart attacks. Given any such interaction, Cheng's measure will be unreliable for specific states of *Exercise.* The CI account acknowledges such possibilities and avoids errors due either to backpaths or to interaction effects.

Note also that for Figure 30.5 Cheng's causal power estimate has decreased from 0.20 to 0.15, a difference of 25%. The CI formula calculates causal power as $\text{CI}(c, e) = 0.16$, which is less than 50% of the 0.37 obtained from the noisy-OR graph. This shows that on either measure, major errors resulted from using the noisy-OR approximation – and so we are much better off using a more general measure applicable to arbitrary causal networks.

30.4.4 CI versus MI

Finally, we will contrast causal information with mutual information, again using the original graph in Figure 30.5. The MI formula applied to *Blood Pressure* and *Heart Attack* (assuming the original marginal distributions) is equivalent to:

$$\mathrm{MI}(C, E) = \sum_{c \in C, e \in E} P(c)\, P(e|c) \log \frac{P(e|c)}{P(e)} \tag{30.13}$$

which reports the mutual information as 0.28. In contrast, the CI formula is:

$$\mathrm{CI}(C, E) = \sum_{c \in C, e \in E} P^*(c)\, P^*(e|c) \log \frac{P^*(e|c)}{P^*(e)} \tag{30.14}$$

calculating the causal power as 0.13, if we impose the original distribution on *Blood Pressure*. The difference between MI and CI is not due to scaling; it is because MI is affected by the backpath *Blood Pressure* ← *Exercise* → *Heart Attack*, while CI is not. So MI would be very misleading as a measure of causal power here: it *doubles* the realistic figure for the power of *Blood Pressure* over *Heart Attack*. It could lead to over-prescription of blood pressure medication, rather than recommending lifelong exercise!

30.5 Conclusions

Causal information, our new measure of causal power, is theoretically well-founded. Causal Bayesian networks provide a very general and powerful way to represent complex stochastic systems. Hypothetical interventions, when properly modelled in causal Bayesian networks, provide a clear separation of causal from non-causal paths. In mutual information, information theory provides an appropriate summary measure for cumulative causal influence, which applies to all sorts of networks and interventions, and can be tailored to specific purposes. The combination of the two, interventions and mutual information, yields causal information.

The result is a measure of causal power that has much wider application than previous accounts. Causal information can be applied to a wider variety of systems, including those with nonlinear probabilistic influences and intricate structural relationships between variables. In such cases it still yields sensible results, unlike the alternative measures put forward by Cheng (1997), Glymour (2001), and Good (1961). These alternative measures were designed for simpler cases, such as noisy-OR networks that exhibit causal transitivity. But in these cases, too, our measure still yields appropriate results. And causal information is the only measure that is well defined for relating any combination of values and variables.

We look forward to applying causal information to theoretical problems in philosophy and AI. Causal information is also a promising measure for summarizing explanatory information encoded in a Bayesian network and so offers new means for simplifying the interpretation of complex Bayesian networks.

Acknowledgements

We are grateful for the useful comments by the referees. This work was supported in part by a grant from Monash University.

References

Ay, N. and D. Polani (2008). Information flows in causal networks. *Advances in Complex Systems 11*, 17–41.

Cheng, P. W. (1997). From covariation to causation: A causal power theory. *Psychological Review 104*(2), 367–405.

Cover, T. M. and J. A. Thomas (1991). *Elements of Information Theory*. New York: John Wiley.

Glymour, C. (2001). *The Mind's Arrows: Bayes Nets and Graphical Causal Models in Psychology*. MIT: MIT Press.

Glymour, C. and P. Cheng (1998). Causal mechanism and probability: a normative approach. In M. Oaksford and N. Chater (Eds.), *Rational Models of Cognition*. Oxford: Oxford University Press.

Good, I. J. (1961). A causal calculus. *British Journal for the Philosophy of Science 11*, 305–318.

Hope, L. (2008). *Information Measures for Causal Explanation and Prediction*. PhD thesis, Monash University.

Hope, L. R. and K. B. Korb (2005). An information-theoretic causal power theory. In *Lecture Notes in Artificial Intelligence, Volume 3809*, pp. 805–811. Berlin: Springer-Verlag.

Korb, K. B., L. R. Hope, A. E. Nicholson, and K. Axnick (2004). Varieties of causal intervention. In *PRICAI'04—Proceedings of the 8th Pacific Rim International Conference on Artificial Intelligence*, Auckland, New Zealand, pp. 322–331.

Korb, K. B. and A. E. Nicholson (2004). *Bayesian Artificial Intelligence*. Boca Raton: Chapman & Hall / CRC.

Luhmann, C. C. and W. Ahn (2005). The meaning and computation of causal power. *Psychological Review 112*, 685–692.

Neapolitan, R. E. (2003). Stochastic causality. In *International Conference on Cognitive Science, Sydney, Australia*.

Neapolitan, R. E. (2004). *Learning Bayesian Networks*. Englewood Cliffs: Prentice Hall.

Novick, L. and P. Cheng (2004). Assessing interactive causal influence. *Psychological Review 111*, 455–485.

Pearl, J. (1988). *Probabilistic Reasoning in Intelligent Systems*. San Mateo, CA: Morgan Kaufmann.

Rescorla, R. A. and A. R. Wagner (1972). A theory of Pavlovian conditioning. In A. H. Black and W. Prokasy (Eds.), *Classical Conditioning II: Current Theory and Research*, pp. 64–99. Appleton-Century-Crofts.

Salmon, W. C. (1980). Probabilistic causality. *Pacific Philosophical Quarterly 61*, 50–74.

Spirtes, P., C. Glymour, and R. Scheines (2000). *Causation, Prediction and Search: Second Edition*. Cambridge, MA: The MIT Press.

Suppes, P. (1970). *A Probabilistic Theory of Causality*. Amsterdam: North-Holland.

Twardy, C. R. and K. B. Korb (2004). A criterion of probabilistic causality. *Philosophy of Science 71*, 241–62.

Verma, T. and J. Pearl (1990). Equivalence and synthesis of causal models. In *Proceedings of the sixth conference on uncertainty in artificial intelligence*, San Francisco, pp. 462–470. UAI: Morgan Kaufmann.

Wright, S. (1934). The method of path coefficients. *Annals of Mathematical Statistics 5*, 161–215.

Appendix A: CI versus Cheng computations

The following computations were made for the examples in Section 30.4.3.
The first Cheng computation:

$$
\begin{aligned}
P(e|c) &= 0.22 \\
P(e|BP = Low) &= 0.025 \\
P(e|BP = Med) &= 0.025 \\
P(e|\neg c) &= P(e|BP = Low)[0.32/0.78] + P(e|BP = Med)[0.46/0.78] \\
&= 0.025 \times 0.41 + 0.025 \times 0.59 \\
&= 0.025 \\
\Delta p_c &= P(e|c) - P(e|\neg c) \\
&= 0.22 - 0.025 \\
&= 0.195 \\
p_c &= \Delta p_c/[1 - P(e|\neg c)] \\
&= 0.195/[1 - 0.025] \\
&= 0.20.
\end{aligned}
$$

The first CI computation:

$$
\begin{aligned}
P^*(e|c) &= 0.22 \\
P^*(e) &= 0.068 \\
\mathrm{CI}(c, e) &= P^*(e|c)\log_2[P^*(e|c)/P^*(e)] \\
&= 0.22\log_2[0.22/0.068] \\
&= 0.37.
\end{aligned}
$$

The CI computation for high and medium exercisers:

$$
\begin{aligned}
P^*(e|c) &= 0.20 \\
P^*(e) &= 0.0440 \\
\mathrm{CI}(c, e) &= P^*(e|c)\log_2[P^*(e|c)/P^*(e)] \\
&= 0.20\log_2[0.20/0.0440] \\
&= 0.44.
\end{aligned}
$$

The CI computation for low exercisers:

$$\begin{aligned} P^*(e|c) &= 0.30 \\ P^*(e) &= 0.0164 \\ \mathrm{CI}(c, e) &= P^*(e|c)\log_2[P^*(e|c)/P^*(e)] \\ &= 0.30\log_2[0.30/0.0164] \\ &= 0.26. \end{aligned}$$

The Cheng computation with the backpath reintroduced:

$$\begin{aligned} P(e|c) &= 0.24 \\ P(e|BP = Low) &= 0.0078 \\ P(e|BP = Med) &= 0.027 \\ P(e|\neg c) &= P(e|BP = Low)[0.32/0.78] + P(e|BP = Med)[0.46/0.78] \\ &= 0.0078 \times 0.41 + 0.027 \times 0.59 \\ &= 0.0032 + 0.016 \\ &= 0.019 \\ \Delta p_c &= P(e|c) - P(e|\neg c) \\ &= 0.24 - 0.019 \\ &= 0.22 \\ p_c &= \Delta p_c/[1 - P(e|\neg c)] \\ &= 0.22/[1 - 0.019] \\ &= 0.22. \end{aligned}$$

The Cheng computation with both the backpath and interaction reintroduced:

$$\begin{aligned} P(e|c) &= 0.24 \\ P(e|BP = Low) &= 0.038 \\ P(e|BP = Med) &= 0.16 \\ P(e|\neg c) &= P(e|BP = Low)[0.32/0.78] + P(e|BP = Med)[0.46/0.78] \\ &= 0.038 \times 0.41 + 0.16 \times 0.59 \\ &= 0.016 + 0.095 \\ &= 0.11 \\ \Delta p_c &= P(e|c) - P(e|\neg c) \\ &= 0.24 - 0.11 \\ &= 0.13 \\ p_c &= \Delta p_c/[1 - P(e|\neg c)] \\ &= 0.13/[1 - 0.11] \\ &= 0.15. \end{aligned}$$

Appendix B: CI for variables is non-negative

Theorem 30.1. $\mathrm{CI}(C, E)$, $\mathrm{CI}(c, E)$ *and* $\mathrm{CI}(C, e)$ *are always non-negative.*

Proof.
(a) $\mathrm{CI}(C, E)$ is equivalent to the mutual information $\mathrm{MI}(C, E)$ under the specified interventions, as shown in Section 30.3.4. Mutual information is always non-negative (Cover and Thomas, 1991).

(b) $\mathrm{CI}(c, E)$ is equivalent to Kullback–Leibler divergence between $P^*(E|c)$ and $P^*(E)$. Kullback–Leibler divergence is always non-negative (Cover and Thomas, 1991).

(c) The Kullback–Leibler divergence between $P^*(C|e)$ and $P^*(C)$ is

$$\begin{aligned}\sum_c P^*(c|e)\log\frac{P^*(c|e)}{P^*(c)} &= \sum_c \frac{P^*(c,e)}{P^*(e)}\log\frac{P^*(c,e)}{P^*(c)\,P^*(e)} \\ &= \frac{1}{P^*(e)}\sum_c P^*(c,e)\log\frac{P^*(c,e)}{P^*(c)\,P^*(e)} \\ &= \frac{1}{P^*(e)}\sum_c P^*(c)\,P^*(e|c)\log\frac{P^*(e|c)}{P^*(e)} \\ &= \frac{\mathrm{CI}(C,e)}{P^*(e)}.\end{aligned}$$

Since $P^*(e)$ and the Kullback–Leibler divergence are always non-negative, so is $\mathrm{CI}(C, e)$. □

·31·

Multiple testing of causal hypotheses

Samantha Kleinberg and Bud Mishra

Abstract

A primary problem in causal inference is the following: From a set of time course data, such as that generated by gene expression microarrays, is it possible to infer all significant causal relationships between the elements described by this data? In prior work (Kleinberg and Mishra, 2009), we have proposed a framework that combines notions of causality in philosophy, with algorithmic approaches built on model checking and statistical techniques for significance testing. The causal relationships can then be described in terms of temporal logic formulæ, thus reframing the problem in terms of model checking. The logic used, PCTL, allows description of both the time between cause and effect and the probability of this relationship being observed. Borrowing from philosophy, we define *prima facie* causes in terms of probability raising, and then determine whether a causal relationship is significant by computing the average difference a *prima facie* cause makes to the occurrence of its effect, given each of the other *prima facie* causes of that effect. However, this method faces many interesting issues confronted in statistical theories of hypothesis testing, namely, given these causal formulæ with their associated probabilities and our average computed differences, instead of choosing an arbitrary threshold, how do we decide which are 'significant'? To address this problem rigorously, we use the concepts of multiple hypothesis testing (treating each causal relationship as a hypothesis), and false discovery control. In particular, we apply the empirical Bayesian formulation proposed by Efron (2004). This method uses an empirical rather than theoretical null, which has been shown to be better equipped for cases where the test statistics are dependent – as may be true in the case of complex causal structures. The general approach may be used with many of the traditional philosophical theories where thresholds for significance must be identified.

31.1 Introduction

Large temporal data sets such as gene expression microarrays, neural spike trains, and stock price movements are prevalent and ripe for causal inference, as one may hypothesize that there are underlying causal structures or rules governing the behaviours of these systems. One neuron *causes* another to fire; two genes being active for some period of time until a third joins them *cause* another to become active; a fluctuation in a key economic factor may *cause* two stock prices to move against each other. However, in the case of these large scale experiments involving thousands of 'elements', there are astronomically many possible causal hypotheses to consider. Starting with the simple idea of 'a causes b' (disregarding the time between a and b and other possibly relevant factors), and without any preconceived ideas to guide our testing, we already face a minimum of $O(|N|^2)$ (where $|N|$ is the number of genes or neurons, etc.) hypotheses to test. In these experiments $|N|$ can be in the thousands, so it is not possible to carry out a manual analysis to evaluate the merits of each hypothesis on its own. However, it is still possible to make valuable causal inferences assuming that we can devise a systematic way of determining which of these should be accepted and which rejected. Since there is always the possibility of making errors in how we accept and reject, it would be hopeless to require that we make these types of inferences in a way that guarantees completely error-free results.

In this chapter we focus on how multiple hypothesis testing and false discovery control may be applied to identify causal hypotheses that are statistically interesting. We are aware that the approach is a general one that can be utilized with many probabilistic definitions of causality. Nevertheless, we illustrate its use primarily in conjunction with an efficient computational method of causal inference, which is based on temporal logic and model checking. The basic premise of the approach is the following: we wish to study a system, for which a lot of data (usually time-course) is available but without any a priori hypotheses about the underlying causal structure; we want our attention drawn to the most promising relationships concealed in these vast data sets. For example, in the case of microarray gene expression data, we will not necessarily uncover the complete and correct story of a biological process with this method, but we can direct biologists' attention to a relevant subset and suggest future experiments that may refute or validate certain hypotheses suggested by the data analysis.

31.2 Causal hypotheses

Probabilistic definitions of causality – based generally on the idea that causes raise the probability of their effects – lead to many events erroneously being labelled as causes. In one classic example, a falling barometer may seem to raise the probability of rain. However, once we find that the air pressure is decreasing, this accounts for both the falling barometer and the rain. That is, the barometer is no longer informative once we know about the decreasing air pressure. To address this difficulty, much work has centered on distinguishing between spurious and genuine causes using probabilities and time of occurrence of the events. These include calling the earliest cause that can account for an effect genuine and all others spurious (Suppes, 1970), looking for earlier common causes that screen effects off from one another (Reichenbach, 2000), extending this to inference from graphs by enforcing these properties (Spirtes *et al.*, 2000), and holding fixed sets of situations and requiring causes to raise the probability of their effects in all such situations (Cartwright, 1979; Eells, 1991).

With any method that leads to a numerical assessment – be it a probability, an average of some tests, etc. – of a cause's 'value', we must decide which values correspond to genuine causation. For example, we may compute an average degree of causal significance as Eells (1991) suggests, but then we are confronted with many unanswered questions: e.g. what average values are acceptable? In a small-scale experiment, we can look at the computed values for all possible causes of a particular effect on a case-by-case basis. However, with many hypotheses and no priors, we need a method of determining where the cutoff should be. Fortunately, we can use the idea that, when testing a large number of possible causes, we expect only a small portion to be genuine. For example, we would not expect every gene to be causally related to every other gene.

When determining the following: (1) how we plan to infer causal relationships and (2) what constitutes a genuine (or, more accurately in our case, significant) causal relationship, we must keep in mind the types of data being analysed and how we plan to use the results. In the wide range of areas where we seek to apply these ideas – biology, neuro-science, finance, politics, to name a few – finding simply the earliest or most likely cause will not do. For instance, there may be known constraints on the problem. In the case of synthetic data simulating the activity of neurons, there was known to be a window of time after one neuron fired before it could cause another to fire. In that case, the time is restricted such that if we only looked for the earliest cause, we would not correctly infer the desired network. In other cases, such as that of microarray gene expression data, there may be a large number of genes being tested and a large number of spurious associations. Rather than defining some arbitrary threshold for the probability of a causal relationship, or testing against a standard null hypothesis, we must exploit the structure of the large quantity of data to aid our testing. That is, we can use an empirical null hypothesis that would be suggested by an empirical Bayes approach.

Finally, we must address the question of which causal hypotheses should be tested. While most methods test for conditional independence between events or types of events (Pearl, 2000), this may not always give us the whole story. This method tells us nothing about the time between cause and effect, which, as seen in the examples above, can be quite important. Were we to find causal relationships in a biological system, but have no information on the time between the cause and effect, we could miss an opportunity for intervention via a drug or vaccine, which must be used in a time- and dosage-restricted manner to have their intended effect. Further, some causal relationships are more complicated than the ones suggested by the form 'A causes B'. A may be a conjunction of events that must act in concert for some period of time until they are joined by an intermediate event D, before ultimately causing B. These types of relationships can be arbitrarily complex and the only limitations on their inference are computing power and the available data. While it is not currently practical to generate vast amounts of such hypotheses or perform microarray experiments with hundreds of time-points, that will not always be the case. We foresee a time, in not too distant a future, when we will be able to routinely generate and test such hypotheses as better bio- and nano-technologies revolutionize the experimental methods. Thus we ignore objections concerning the data sizes needed to infer causality reliably, and simply focus our discussion on a description of how hypotheses are represented and later, expand it with illustrative and realistic examples of how they are tested.

31.2.1 Using temporal logic

In order to represent complex causal hypotheses, we propose a framework that defines causal relationships in terms of a probabilistic temporal logic and then treats the problem of inference as one of model checking. We will briefly describe the method and then discuss how it is enhanced through the use of multiple testing methods. Further details can be found in Kleinberg and Mishra (2009).

Suppose we are given a series of data describing events over time, with some labels for the types of events. Then, at each instant in time, we know which events are true. That is, we can view each time instant as a *state* (that is labelled with the properties, or events, true within it). So, when we view the progression of the system over time, we have seen one possible run of the system: a series of transitions between these states. If the run is sufficiently long, then we have seen almost all of the states it can occupy as well as their occupation probabilities (estimated by counting how frequently the system occupies those states). From that observation, we want to find the underlying structure of the system. For example, we can imagine watching a game of checkers and trying to determine the rules of the game from just that observation.

When we attempt to determine if some 'cause', c (possibly a complex logical formula) causes some 'effect', e, we wish to answer the following: if we are in a state where c is true, how likely is it that we will transition to a state where e is true? As long as we can represent c as a logical formula, we can test it as a causal hypothesis. The main steps of the procedure are representing our causal hypotheses as logical formulæ, checking whether the system satisfies the formulæ, and determining which of the satisfied formulæ are significantly causal.

Our approach is founded upon two premises: (i) causes raise the probability of their effects, and (ii) causes occur temporally prior to their effects.[1] Then, using the logic described by Hansson and Jonsson (1994), which is a probabilistic variant of Computation Tree Logic (CTL) called PCTL, we define possible, or *prima facie*, causes as follows.

Definition 31.1. c is a *prima facie* cause of e iff:

1. $F^{\leq\infty}_{>0} c$,
2. $c \leadsto^{\geq 1, \leq\infty}_{\geq p} e$, and
3. $F^{\leq\infty}_{<p} e$.

This means that (1) the system is eventually in a state where c is true with non-zero probability, and (3) the probability of eventually being in a state where e is true is (2) less than the probability of being in a state where e is true after being in a state where c is true, with at least one transition between the c and e state. Next, we need to decide if such a pattern may have been satisfied by happenstance, and the causal relation inferred is likely to be insignificant. To test whether a *prima facie* cause is insignificant, we take our cues from the methods that have been proposed in the philosophical literature (Suppes, 1970; Eells, 1991). More specifically, to determine whether a particular c as a cause of e is insignificant, we take the set X of *prima facie* causes of e and for each $x \in X\backslash c$, compute the difference c makes to the probability of e in relation to x. That is, this process estimates c's average value as a cause, by examining the the probability of e after $c \wedge x$ in comparison to that after $\bar{c} \wedge x$.

With

$$\varepsilon_x(c, e) = P(e|c \wedge x) - P(e|\bar{c} \wedge x) \tag{31.1}$$

we compute:

$$\varepsilon_{\text{avg}}(c, e) = \frac{\sum\limits_{x \in X\backslash c} \varepsilon_x(c, c)}{|X \setminus c|}. \tag{31.2}$$

This ε_{avg} is then used to determine whether c is a significant cause of e. Note that c and x lead to e with certain time bounds, but these have been omitted here. We define that:

Definition 31.2. A cause, c, is an *ε-insignificant cause* of an effect, e, iff: c is a *prima facie* cause of e and $\varepsilon_{\text{avg}} < \varepsilon$.

[1] There is no inherent reason that we should not allow causes to occur simultaneously with their effects, though this has not been relevant to our applications.

Definition 31.3. A *prima facie* cause, c, of an effect, e, is an *ε-significant cause* of e, iff it is not an ε-insignificant cause of e.

To determine what value of ε is appropriate, one may use background knowledge of the problem or simulations. Another method is to determine this value statistically, using multiple hypothesis testing methods.

31.3 Multiple hypothesis testing and the false discovery rate

When testing a single hypothesis, we can make the decision about whether to accept or reject the null hypothesis based on the probability that the result would occur if the null hypothesis were true. However, when we are testing multiple hypotheses at once, the probability that we will get such results – even under the null hypothesis – increases and we must account for this. For example, we may test the fairness of a coin by flipping it 10 times and seeing how many times it comes up heads and how many times it comes up tails. If there were nine heads, we would likely say that it is biased, as the probability of this happening when the coin is fair is 0.01, and the p-value 0.022. Frequently, a significance level of $p \leq \alpha = 0.05$ is sufficient to reject the null hypothesis, and thus we would call the coin unfair. In the case where we are testing 100 fair coins, we may incorrectly deem 5 unfair ($n_{\text{tests}} \times \alpha$). In fact, the probability of doing so is over 99%. Thus it is necessary to account for the fact that we are performing many tests simultaneously, increasing our chances of seeing unlikely or anomalous behaviour (Storey and Tibshirani, 2003; Efron, 2004; Benjamini and Yekutieli, 2001).

31.3.1 Basic definitions

First, we define two types of error. *Type I errors*(α) are those where we reject a true null hypothesis. The *per-comparison error rate* is the probability of making such an error during each significance test. *Type II errors*(β) are those where the null hypothesis is not rejected when it should be. Whereas Type I errors mean we have made a false discovery (false positive), Type II errors mean we have missed an opportunity for discovery (false negative). While it is desirable to reduce both types of error, it may only be possible to trade one kind off against the other. The best trade-offs are judged in terms of the relative costs of these errors in a particular domain of application.

Thus, we define next the error rates over all the hypotheses being tested. The *familywise error* (FWE) rate is the probability of rejecting one or more true null hypotheses (i.e. the probability of having at least one Type I error), during all tests. For the FWE to approach a desired bound of $\alpha \ll 1$ we need each of the, say, n tests to be conducted with an even stricter bound, such as $\frac{\alpha}{n}$, as required by the so-called Bonferroni correction (Benjamini and Yekutieli,

2001). However, the FWE has low power, meaning that we have a good chance of making a Type II error (Benjamini and Hochberg, 1995). Another measure, called the False Discovery Rate (FDR), estimates the proportion of Type I errors among all rejected hypotheses (that is, the number of false discoveries divided by the total number of discoveries). This measure results in more power than the FWER while still bounding the error. The main idea is that, if we are rejecting only a few null hypotheses, then each false discovery we make in that case is more significant than rejecting a large number of null hypotheses and making more false discoveries. That is, in the first case, the false discoveries are a larger percentage of the overall number of discoveries than they are in the latter case. In this work we concentrate on applying the FDR to partition all *prima facie* causes into significant and insignificant causes.

31.3.2 Controlling the FDR

The introduction of the FDR and procedures for controlling it are described by Benjamini and Hochberg (1995). The procedure is as follows. Using a method similar to that of Bonferonni, when testing m hypotheses, order the p-values $P_{(1)} \leq P_{(2)} \cdots \leq P_{(m)}$. Then with k selected as the largest i such that:

$$P_{(i)} \leq \frac{i}{m}\alpha, \tag{31.3}$$

we reject all $H_{(i)}$, $i = 1, 2, \ldots, k$. In the case when all hypotheses are true this controls the FWE, and otherwise controls the proportion of erroneous rejections. For *independent* test statistics, this procedure controls the FDR at rate α. However, it was later shown that this also holds for positively dependent test statistics and can be modified to control the FDR in other cases (Benjamini and Yekutieli, 2001).

31.3.3 Using an empirical null hypothesis

In the methods described so far, it was necessary to use a theoretical null hypothesis, namely, that values have a standard normal distribution. However, this may not be appropriate for all data. It is possible, then, to take advantage of the multitude of hypotheses being tested and to determine the correct null hypotheses from the data. The use of an empirical null hypothesis was described by Efron (2004), and provides a novel empirical Bayesian solution to the problem. In that work, Efron described how one may estimate the empirical null distribution and how the choice of null hypothesis has a large impact on the discoveries made. Take, for example, Figures 31.1, 31.2, and 31.4 below showing the results of analysis described in Section 31.4. Each of these data sets has a different underlying distribution and thus their empirical nulls vary from the theoretical null in different ways. In the example shown in Figure 31.1, using the theoretical null would lead to no null hypotheses being rejected, while many would be rejected under the empirical null. This is an extreme example, as the tests are highly dependent, but it illustrates the importance of selecting a proper null hypothesis as well as the importance of this choice when examining causal inferences from data, where we cannot avoid assuming at least some level of dependence. In practice, most methods for inferring the null hypothesis empirically attempt to fit to the central peak of the data.

31.3.4 Computing the FDR

Here, we use *local false discovery rate* (fdr) calculations, which use densities, as our N's are large, though these methods may also be used with standard tail-area FDR methods such as that described in Section 31.3.2 (Efron, 2007). We follow the formulation described by Efron (2004).

With N hypotheses $H_1, H_2, \ldots, H_N$ we have the corresponding z-values $z_1, z_2, \ldots, z_N$. These values, also called the standard score, are the number of standard deviations by which a result deviates from the mean. In the case of our causal analyses, these z-values are computed from the ε_{avg}s. We begin by assuming the N cases fall into two classes: one where the effects are either spurious or not large enough to be interesting (and thus where we accept the null causal hypotheses), and another where the effects are large enough to be interesting (and where we will accept the non-null hypotheses as true). We also assume the proportion of non-null cases are small relative to N, say, around 10%. Then, p_0 and $p_1 = 1 - p_0$ are the prior probabilities of a case (here, a causal hypothesis) being in the 'uninteresting' or 'interesting' classes respectively. The densities, $f_0(z)$ and $f_1(z)$, of each class describe the distribution of these probabilities. When using a theoretical null, $f_0(z)$ is the standard $N(0, 1)$ density. Note that we need not know $f_1(z)$, though we must estimate p_0 (usually $p_0 \geq 0.9$). We define the mixture density:

$$f(z) = p_0 f_0(z) + p_1 f_1(z), \tag{31.4}$$

then the posterior probability of a case being uninteresting given z is

$$Pr\{null|z\} = p_0 f_0(z)/f(z), \tag{31.5}$$

and the *local false discovery rate*, is:

$$fdr(z) \equiv f_0(z)/f(z). \tag{31.6}$$

Note that, in this formulation, the p_0 factor is ignored, yielding an upper bound on $fdr(z)$. Assuming that p_0 is large (close to 1), this simplification does not lead to massive overestimation of $fdr(z)$. One may also choose to estimate p_0 and thus include it in the FDR calculation, making $fdr(z) = Pr\{null|z\}$. The procedure is then:

1. Estimate $f(z)$ from the observed z-values;
2. Define the null density $f_0(z)$ either from the data or using the theoretical null;
3. Calculate $fdr(z)$ using equation (31.6);
4. Label H_i where $fdr(z_i)$ is less than a threshold (say, 0.10) as interesting, or in our case, causally significant.

Overall the procedure is to enumerate a set of causal hypotheses (represented by logical formulæ), test these in the data to see which satisfy the conditions for *prima facie* causality, then for each identified *prima facie* cause, compute its associated ε_{avg}. The causes where ε_{avg} correspond to z-values with fdr less than a small threshold are called ε-significant, and the rest ε-insignificant.

31.4 Examples

We illustrate the proposed method on three different types of data: biological microarrays, political speeches and favorability ratings, and neural spike trains. For all examples, the empirical null was calculated using the method of Jin and Cai (2006) and the R program they have made publicly available.

31.4.1 Microarrays

The data is a set of time-course gene-expression microarrays covering the 48 hour Intraerythrocytic Developmental Cycle(IDC) of *Plasmodium falciparum* (Bozdech *et al.*, 2003). Microarray data, where expression levels may be measured for thousands of genes at a time, have been the subject of many studies on both multiple hypothesis testing (Efron and Tibshirani, 2002; Dudoit *et al.*, 2003) as well as causal inference (Friedman *et al.*, 2000; Spirtes *et al.*, 2001; Opgen-Rhein and Strimmer, 2007; Murphy and Mian, 1999).

In this example, we want to find causal relationships between genes that may enlighten the complex mechanisms underlying this cycle. All *P. falciparum* genes are active at some point during the IDC, forming a so-called 'cascade' of activity. We look at relationships over the entire time course without taking into account the structures imposed by individual IDC stages, but hope to address the finer causal properties by considering these stages separately. The relationships we wished to test were generated, using the PCTL formulation, by taking all pairs of genes where the influence occurs at the next unit of time. In other words, we have considered all formulæ of the form: $c \leadsto^{\geq 1, \leq 1}_{\geq p} e$, where c and e represent the under- or over-expression of particular genes.

After restricting our data set such that only genes that are known to be involved in protein–protein interactions were tested, we were left with $N = 2846$ unique genes. To estimate $f(z)$, we used a spline fit to the histogram.

In Figure 31.1, we notice that the data falls mostly within the plotted theoretical null $N(0, 1)$. Were we to use that as our null hypothesis, we would hardly find any interesting cases. However, a visual examination of the data indicates that the data follow almost a normal distribution, but with a positive skew (longer right tail). The empirical null $N(-1.00, 0.89)$ takes this into account and is thus shifted much further over than the theoretical null. Note that in this case, we have thousands of *prima facie* causes where $f(z_i) < 0.1$. There are a few ways to explain this structure of the null distribution, which convinced us that the empirical null is correct and we should indeed detect a multitude of non-insignificant causes. As is usually believed, biological systems are, by necessity, quite robust (Kitano, 2004), giving rise to a large number of correlated cause–effect relationships that orchestrate the system's dynamics in a fail-safe manner. In this particular example, we have three main phases of the IDC, and during each the genes related to that phase act in concert producing the cascade. It is quite likely that the dependencies arise in two important manners: (1) many genes are causally related to many other genes that are active in the same IDC stage, and are organized in a complex network of interactions, and (2) there are many back-up mechanisms to allow the cascade to continue uninterrupted in case of some perturbation to the system. In this case we could use a lower value for the acceptable FDR or speculate two classes of causal relationships: primary causes, and backup causes. These and other such finer analyses are deferred to future research.

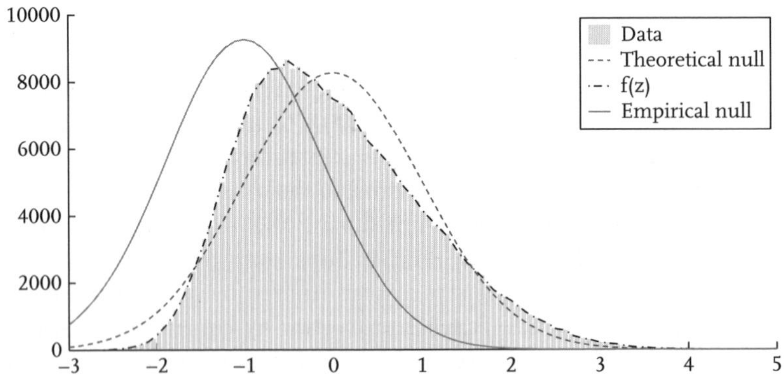

Fig. 31.1 *P. Falciparum* microarray example. Here we are testing causal relationships amongst all pairs of genes over the entire time-course. The histogram shows the number of prima facie causes with a given *z*-value.

31.4.2 Neural spike trains

In our next example, we consider another time-course data set from a different domain: neural-activity data. This data set consists of a series of synthetically generated patterns, and thus has the ability to eventually reveal the assumed true causal neural networks that were embedded in the simulations.[2] The data were created to mimic multi-neuronal electrode array experiments, in which neuron firings may be tracked over a period of time. We ran our inference algorithm on this set of data, each containing 100,000 firings of a set of neurons, each denoted by a character of the English alphabet. Each data set was embedded with a different causal network. At each time point a neuron can fire randomly (dependent on the noise level selected) or may be triggered to fire by one of its 'cause neurons'. Additional background knowledge was known and used by the inference algorithm: there is a 20 unit refractory period after which a neuron fires before it may trigger another, and then a 20 unit window of time when it may trigger the other to fire. Consequently, our algorithm only needed to search for relationships, where one neuron causes another to fire during a window of 20–40 time units after the causal neuron fires. Full results for all five structures and comparison with the PC algorithm of Spirtes *et al.* (2000) and Granger causality (Granger, 1969) are available at: `http://people.dbmi.columbia.edu/samantha/papers/tlcs.html`. We discuss one of the structures in detail here. In this example, the underlying structure was a binary tree of four levels. Our algorithm enumerated 642 *prima facie* causal hypotheses, satisfied by the data.

The results are shown in Figure 31.2, which indicate that there are far fewer significant causal relationships than in the previous example. The empirical null in this case is given by $N(-0.14, 0.39)$, so it is shifted slightly to the left of the theoretical null, and is significantly narrower. The tail of the distribution extends quite far to the right, continuing up to eight standard deviations away from the mean. A close up of this area is shown in Figure 31.3. The results obtained here are consistent with the known causal structures that were used to create the simulated data and agree with our earlier work on this data. The ten genuine causal relationships were the only hypotheses with z-values greater than three, though there were seven others that, like these, had an FDR of zero. With no prior knowledge, there are two methods for determining the actual causes. First, in a case where there are few causal relationships found, such as in this example, we can look at the individual hypotheses and manually filter the causal hypotheses. For instance, if there are two causes of an effect, say, one with z-value 7.2 and the other with a value 1.3, we may speculate that the former is more likely to be the genuine cause. If the data were experimental, we could do further testing to validate (or refute) this

[2] The data was provided as part of the Fourth KDD Workshop on Temporal Data Mining. It is publicly available at: `http://people.cs.vt.edu/~ramakris/kddtdm06/`.

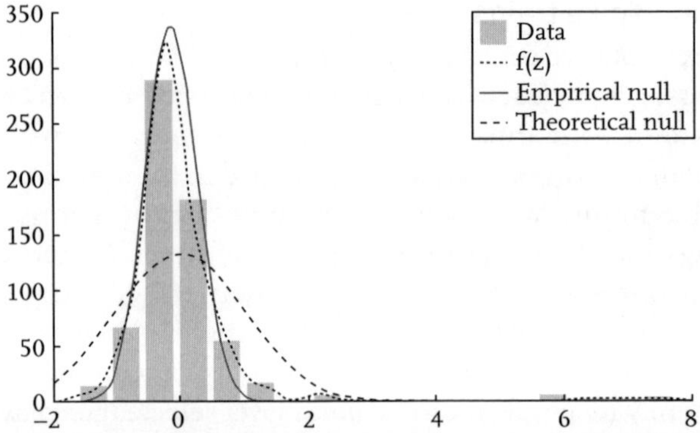

Fig. 31.2 Neural spike train example. In this example, we tested pairwise causal relationships, taking into account the known temporal constraints on the system.

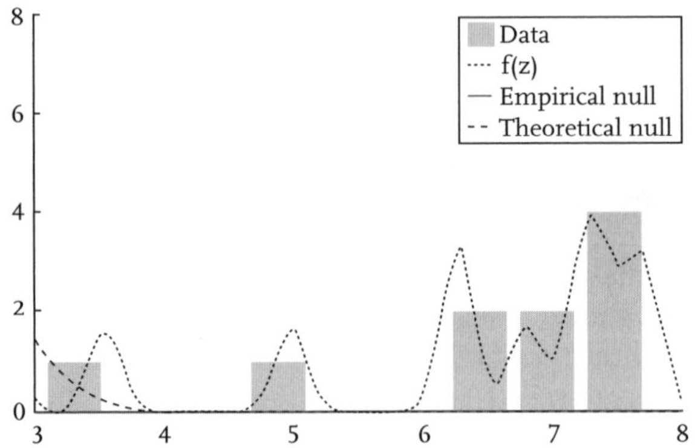

Fig. 31.3 Neural spike train example.

claim. Second, had there been a larger number of *prima facie* causes of each effect, we could treat each of those as a family of hypotheses, conducting the procedure after the computation of ε_{avg} on each of these families individually.

While, constrained by space, we relegate the full comparison to the website cited earlier, we will briefly summarize the results here. This synthetic dataset is the only one of those tested where the true structure is known, thus allowing us to assess our performance. We used the TETRAD IV implementation of the PC algorithm with the chi-square significance test (using default parameters $\alpha = 0.05$ and depth $= 0.01$), and the `granger.test` function in the MSBVAR R package (with a lag of 20 time units). For the Granger tests, we applied the same FDR control procedure as for our own tests. In the case of the PC results, unlike the other two algorithms, both directed and undirected edges are returned, but we did not include the undirected edges in computing the FDR

Table 31.1 Comparison of false discovery rate (FDR), false non-discovery rate (FNR) and intersection between runs. Results are over all synthetic MEA structures.

Method	FDR	FNR	Intersection
Kleinberg–Mishra	0.0093	0.0005	0.9583
Granger	0.5079	0.0026	0.7530
PC	0.9608	0.0159	0.0671

and FNR results. Over all five structures (ranging from chains of neurons, to a binary tree, to single, multiple and linked scatter–gather relationships) and both noise levels, the results are as follows.

The false discovery rate (FDR) is the number of false positives divided by all positives, and the false non-discovery (or false negative) rate (FNR) is the number of false negatives divided by all negatives.[3] As shown in Table 31.1 our FDR is quite low, at 0.0093, while for the Granger test more than half of all discoveries were erroneous, and nearly all PC discoveries were false. While the FNR is low for all algorithms, we note that after accounting for the usual trade off between false positives and negatives, our algorithm still achieved the lowest rate for both measures. Finally, since for each parameter setting (noise level and underlying structure), two datasets were generated, we compared how many of the discoveries were made in both runs for a particular setting. That is, this is the proportion of relationships found in both runs out of all relationships found. According to this measure, our results were also the most consistent.

31.4.3 Politics

The final example, considered next, consists of a set of political speeches and job approval ratings. We collected President Bush's weekly radio addresses for a seven year period along with his Gallup job approval polls for the same period. The speeches were processed into phrases (corresponding to either exact wordings in the text, manually defined synonyms, or categories containing more specific wordings), and their dates of use. We operated under the assumption that a phrase's effect on job approval rating (causing it to go either up or down), would be reflected in the immediately following poll, and not any later polls.

The results are shown in Figure 31.4. Unlike the other two examples, the underlying distribution appears to be bimodal. The empirical null, $N(0.39, 0.96)$ is very close to the theoretical null, shifted slightly to the right.

[3] Given that we are testing such a large number of hypotheses, the FNR is bound to be artificially low.

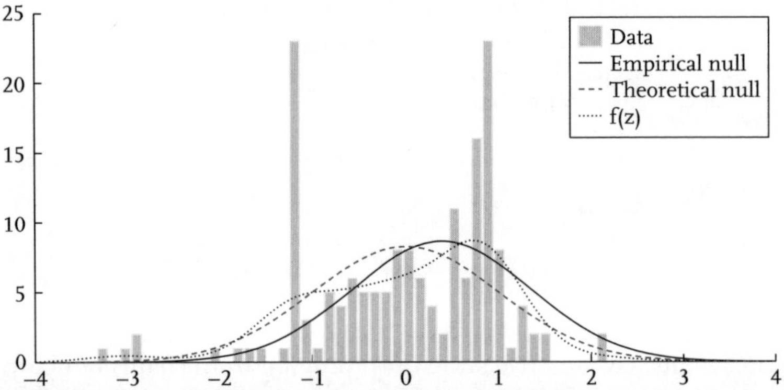

Fig. 31.4 Political example, where phrases were tested as potential causes of changes in job approval ratings.

Interestingly, if we look at the positive *z*-values, there appear to be few interesting hypotheses. In fact, no phrases were found to be significantly causal. However, when we look at the left side of the figure, the negative *z*-values, we see a spike around -3. There were three such phrases with $fdr < 0.1$. This is saying that the use of the phrases with those *z*-values does not have a causal influence on the changing speech ratings but rather this area of the graph represents causation by omission. That is, failure to use these phrases has an effect on job approval. In general, data sets of this nature are expected to be much harder to analyze, as in these cases there are many unmodeled states (for instance, the contexts created by knowledge and belief propagation among the voters that remain hidden and highly history-dependent).

31.5 Related work

Many methods exist for determining the significance of causal relationships. They include the purely philosophical – looking primarily at what it means to be a cause, as well as the computational – using mathematical properties to identify causes. We outline the main approaches and how they relate to the method we have described.

31.5.1 Philosophical

The closest methods to ours in the philosophical literature are those concentrating on probabilistic causality. In these methods, a cause C raises the probability of its effect E beyond the normal probability of E. That is, $P(E|C) > P(E)$. One problem for probabilistic theories of causality is that there may be cases where two events are the result of an earlier common cause. In one commonly used example, we may frequently see yellow stained fingers and lung cancer together. However, we cannot simply say that yellow stained fingers cause lung cancer, or that lung cancer causes yellow stained fingers. Using more information, we can find an earlier common cause of both – smoking. Here, smoking 'screens-off' lung cancer from yellow stained fingers. That is, when we hold fixed that someone is a smoker, the relationship between stained fingers and lung cancer disappears. This idea of earlier 'screening off' causes was introduced by Reichenbach (2000).

Later work, such as that by Suppes (1970) and Eells (1991), also focuses on finding other causes that better account for the effects in question. In Suppes' theory, a *prima facie* cause is one that raises the probability of its effect and occurs strictly earlier than the effect. *Spurious* causes are defined in a few ways by Suppes, but all look for strictly earlier events or classes of events that render a prima facie cause uninformative. One such definition is:

Definition 31.4. $B_{t'}$, a *prima facie* cause of A_t, is a *spurious cause* iff there is a partition, $\pi_{t''}$ where $t'' < t'$ and for every $C_{t''}$ in $\pi_{t''}$

1. $P(B_{t'} C_{t''}) > 0$
2. $P(A_t | B_{t'} C_{t''}) = \mathrm{P}(A_t | C_{t''})$.

Here, a partition, π_t may be of either the sample space or universe and consists of 'pairwise disjoint, nonempty sets whose union is the whole space (Suppes, 1970)'. Similarly, Suppes defines ε-spuriousness, with the difference being that

$$|P(A_t | B_{t'} C_{t''}) - P(A_t | C_{t''})| < \varepsilon. \tag{31.7}$$

This definition raises the question: what is an appropriate value for ε? This is a place where no fixed threshold can be offered as an answer, especially when testing a multitude of causal hypotheses. Note that when we have many *prima facie* causes and do not necessarily wish to select the extreme cases (corresponding to only one or several genuine cause(s)), we can obtain a more flexible solution as follows: compute the differences shown in equation (31.7), convert those differences to *z*-values, then use the methods described here to determine what that cutoff ε should be.

Similarly, Eells (1991) provides probabilistic definitions of *positive causal factors* as well as measures of causal *relevance*. Eells states:

Definition 31.5. C is a *positive causal factor* for E iff for each i

$$P(E | K_i \wedge C) > P(E | K_i \wedge \neg C), \tag{31.8}$$

where the K_i's are causal background contexts. By causal background contexts, we mean that if there are n factors independently of C relevant to E there are 2^n ways of holding these fixed. Those that occur with nonzero probability in conjunction with C as well as $\neg C$ (i.e. $P(C \wedge K_i) > 0$ and $P(\neg C \wedge K_i) > 0$)

constitute a background context. For example, if we have three factors, say, x_1, x_2, and x_3, one possible background context would be $K_i = \overline{x_1} \wedge x_2 \wedge \overline{x_3}$. We may also define negative as well as neutral causal factors by changing the $>$ in equation 31.8 to $<$ and $=$ respectively.

A cause C, may also have *mixed* relevance for effect, E – it is not negative, positive or neutral. This situation corresponds to C's role varying depending on the context (as may have happened in our political examples). Eells defines that C is *causally relevant* to E if it has mixed, positive, or negative relevance for E, i.e. it is not causally neutral.

In addition to determining whether or not C is causally relevant to E, we may also wish to describe *how* relevant C is to E. As in Suppes' theory, it is possible to have causally relevant factors with small roles. One method Eells gives for measuring the significance of a factor X to a factor Y is:

$$\sum_i \Pr(K_i)[\Pr(Y|K_i \wedge X) - \Pr(Y|K_i \wedge \neg X)]. \tag{31.9}$$

This approach once again leads us to the problem of determining what values we may deem significant. In this case, we could also convert these sums to z-values, and apply the multiple hypotheses testing procedure to determine whether to accept or reject each hypothesis.

31.5.2 Computational

In additional to the philosophical foundations for detecting causality, we need algorithmic methods that allow us to infer it automatically from data (whether experimental or observational). One of the primary efforts in this area, that of Spirtes *et al.* (2000) (SGS), uses Bayesian networks (BNs). In general, this approach as well as the work of Pearl and Verma (1991), uses graphical models where the causal structure is represented as a graph, with nodes representing variables and directed edges between nodes representing conditional dependence. In order to infer these models from data, which is not necessarily time-course, a number of assumptions must be made in order to find both the relationships and their direction. The result of the inference is one or a set of possible directed acyclic graphs (DAGs) where a directed edge between two nodes is given the interpretation that the node at the tail causes the node at the head. We will primarily review the work of SGS, but note that there are many variants on this general framework.

The first assumption, called the *causal Markov condition* (CMC), is that every node in the graph is independent of its non-descendants, given its parents. Graphs are also assumed to be complete, in that common causes of any two nodes are included and that all causal relationships among the variables are included in the graph. The second necessary assumption, *Faithfulness*, is that the independence relations in the graph are exactly those of distributions produced by the graph. That is, the independence relations obtained from the graph are due to the causal structure and not coincidence or latent variables (for example, a positive effect and indirect negative effect canceling out). Finally, the last assumption is *causal sufficiency*, which is that the set of measured variables includes all common causes of pairs on that set. When this does not hold, some methods will produce a set of graphs accounting for the observed distribution that include nodes representing unmeasured common causes.

One issue with this approach is that there is little discussion about the idea of conditional independence. It is highly unlikely that one will find exact independence from a finite data sample, so it is necessary to determine at what point pairs are considered independent. While most methods use standard significance tests (such as mutual information or chi-square) with a set threshold, it is possible to use a multiple hypothesis testing method such as the one described here in order to define a threshold for this value. Another issue is that in these models, there is no innate method of representing the time between the cause and effect or more complex relationships than simply one variable causing another.

Recently, extensions using dynamic Bayesian networks (DBNs) have been proposed to address the temporal component of the causal relationships. In their simplest form, DBNs begin with a prior distribution (described by a DAG) as well as two more DAGs: one that represents the system at time t and one that represents the system at $t+1$, where these relationships hold for any values of t. The two DAGs are connected to denote how a variable at t influences another at the next time point. Thus it is assumed that the structure and dependencies persist over time without change (since the relationships from t to $t+1$ are the same as those from $t+10$ to $t+11$). The DAGs have the same structure as those previously described, with each variable denoted by a node in the graph. There is still no representation of the length of time between cause and effect in this model, but recent work by Eichler and Didelez (2007) has focused on capturing that information. This work builds on prior work by Granger (1969) who defined a type of causality, applied primarily to economics, between time series. There, one time series Granger-causes another if lagged values of the first are informative about the second. Similarly, in the method of Eichler and Didelez, one time series is said to cause another if an intervention on the first alters the other at a later time. Thus the lag between cause and effect may have any arbitrary value, and this value is found as part of the inference process. One could potentially extend this approach to represent more complex relationships, but the framework does not lead to a general method for testing such relationships. Similarly, while DBNs can compactly represent distributions in sparse structures, it can be difficult to extend them in the case of large data sets (thousands or tens of thousands of variables) that are highly dependent.

While we can define variables in arbitrary ways, the difficulty comes in finding their probabilities and whether they are satisfied by the data. This process still requires model checking. Further, we want to be able to test new formulæ, as one can in structures built for model checking. Recent work by Langmead *et al.* (2006) has attempted to bridge the gap between temporal logic model checking and DBNs by translating the DBNs into structures that allow for such model checking. In this method, DBNs may be used for inferring relationships described by temporal logic formulæ. However, later work by Langmead (2008) showed that not all DBNs may be translated in this manner, so this approach is limited compared to one where we use a model that already allows for model checking.

31.6 Conclusion

Trying to understand and explain our natural and social worlds in terms of a complex and intertwined causal network has been at the core of the Western scientific tradition. In many cases, these ideas have been surprisingly successful, but also in few rare but important cases, the idea of causality has been a major source of bewilderment. While there are some who have even proposed abolishing the very notion of causality from our vocabulary, there are others (mainly philosophers) who have made clear progress in creating a more meaningful philosophical notion of causality that to a large extent coincides with its common-sense meaning while containing enough structure to permit the qualitative causal reasoning that remain necessary in understanding complex phenomena in biology, and social sciences, where our theories are still incomplete and fragmentary.

We noticed and exploited the fact that these ideas from philosophy naturally lend themselves to being expressed in a probabilistic temporal logic, thus allowing these logical formulæ to be checked by automated model-checking processes. Furthermore, since the data that are mined to excavate causal relations are noisy and non-deterministic, they frequently yield insignificant causal rules that by pure coincidence resemble the structural patterns of genuine causes. Fortunately, using an empirical Bayesian approach, one can, from the data itself, extract an empirical null-distribution for such insignificant causes and filter out putative genuine causes (which can then be tested by appropriate experiments). Because of these two important tools from computer science (an efficient algorithm for model checking) and statistics (an empirical method of false discovery control), it is now possible to consider the accepted views of causality from philosophy and systematically and agnostically verify them in many data-rich domains of discourse (e.g. biology, neuroscience, economics, social sciences, etc.)

Such an approach has a clear pay-off: if our ideas of causality are sound and effective, they will give us enough useful capabilities, for instance, to diagnose and cure diseases, to exploit arbitrages and transient inefficiencies in the market, to strategically manipulate elections, or to auction off worthless goods to unsuspecting virtual friends in our social networks. If on the other hand, there are serious shortcomings in these notions of causality, they will be exposed in their particular contexts, thus suggesting to philosophers how they may think about these ideas more rigorously. To illustrate this process, we mention two situations we have already observed. One occurs in the example from biology (in the study of malaria parasites) which makes it clear that the causal relations must respect a strict notion of locality in time (for instance, causal structures may vary from stage to stage in the parasite's life cycle), and need to incorporate the notion of a natural temporal scale at which causes and effects relate to each other within the system. By examining the empirical null-distributions in a multi-scale manner these time-scale properties can be automatically discovered. The other example occurs in the domain of politics where it becomes clear that there are many hidden variables, and few relationships of significance.

References

Benjamini, Y. and Hochberg, Y. (1995). Controlling the false discovery rate: A practical and powerful approach to multiple testing. *Journal of the Royal Statistical Society. Series B (Methodological)*, **57**(1), 289–300.

Benjamini, Y. and Yekutieli, D. (2001). The control of the false discovery rate in multiple testing under dependency. *Annals of Statistics*, **29**(4), 1165–1188.

Bozdech, Z., Llinás, M., Pulliam, BL, Wong, ED, Zhu, J. *et al.* (2003). The transcriptome of the intraerythrocytic developmental cycle of Plasmodium falciparum. *PLoS Biol*, **1**(1), e5.

Cartwright, N. (1979). Causal laws and effective strategies. *Nous*, **13**(4), 419–437.

Dudoit, S., Shaffer, J.P., and Boldrick, J.C. (2003). Multiple hypothesis testing in microarray experiments. *Statistical Science*, **18**(1), 71–103.

Eells, E. (1991). *Probabilistic Causality*. Cambridge University Press.

Efron, B. (2004). Large-scale simultaneous hypothesis testing: The choice of a null hypothesis. *Journal of the American Statistical Association*, **99**(465), 96–105.

Efron, B. (2007). Size, power, and false discovery rates. *Annals of Statistics*, **35**(4), 1351–1377.

Efron, B. and Tibshirani, R. (2002). Empirical Bayes methods and false discovery rates for microarrays. *Genetic Epidemiology*, **23**(1), 70–86.

Eichler, M. and Didelez, V. (2007). Causal reasoning in graphical time series models. In *Proceedings of the 23rd Annual Conference on Uncertainty in Artificial Intelligence*.

Friedman, N., Linial, M., Nachman, I., and Pe'er, D. (2000). Using Bayesian networks to analyze expression data. *Journal of Computational Biology*, **7**(3–4), 601–620.

Granger, C.W. (1969). Investigating causal relations by econometric models and cross-spectral methods. *Econometrica: Journal of the Econometric Society*, **37**(3), 424–438.

Hansson, H. and Jonsson, B. (1994). A logic for reasoning about time and reliability. *Formal Aspects of Computing*, **6**(5), 512–535.

Jin, J. and Cai, T.T. (2006). Estimating the null and the proportion of non-null effects in large-scale multiple comparisons. Journal of the American statistical Association, **102**, 495–506.

Kitano, H. (2004). Biological robustness. *Nature Reviews Genetics*, **5**(11), 826–837.

Kleinberg, S. and Mishra, B. (2009). The temporal logic of causal structures. In *Proceedings of the 25th Conference on Uncertainty in Artificial Intelligence (UAI)*, Montreal, Quebec.

Langmead, C.J., Jha, S.K., and Clarke, E.M. (2006). Temporal logics as query languages for dynamic Bayesian networks: Application to d. melanogaster embryo development. Technical Report CMU-CS-06-159, Carnegie Mellon University.

Langmead, C. J. (2008). Towards inference and learning in dynamic Bayesian networks using generalized evidence. Technical Report CMU-CS-08-151, Carnegie Mellon University.

Murphy, Kevin and Mian, Saira (1999). Modelling gene expression data using dynamic Bayesian networks. Technical report, University of California, Berkeley, CA.

Opgen-Rhein, R. and Strimmer, K. (2007). From correlation to causation networks: a simple approximate learning algorithm and its application to high-dimensional plant gene expression data. *BMC Systems Biology*, 1: 37.

Pearl, J. (2000). *Causality: Models, Reasoning, and Inference*. Cambridge University Press.

Pearl, J. and Verma, T. S. (1991). A theory of inferred causation. In *KR'91: Principles of Knowledge Representation and Reasoning* (ed. J. F. Allen, R. Fikes, and E. Sandewall), San Mateo, California, pp. 441–452. Morgan Kaufmann.

Reichenbach, H. (2000). *The Direction of Time*. New York: Dover Publications.

Spirtes, P., Glymour, C., and Scheines, R. (2000). *Causation, Prediction, and Search*. Cambridge MA: MIT Press.

Spirtes, P., Glymour, C., Scheines, R. *et al.* (2001). Constructing Bayesian network models of gene expression networks from microarray data. In *Proceedings of the Atlantic Symposium on Computational Biology, Genome Information Systems and Technology*.

Storey, J.D. and Tibshirani, R. (2003). Statistical significance for genomewide studies. *Proceedings of the National Academy of Sciences*, **100**(16), 9440.

Suppes, P. (1970). *A Probabilistic Theory of Causality*. Amsterdam: North-Holland.

·32·

Measuring latent causal structure

Ricardo Silva

Abstract

The presence of latent variables makes the task of estimating causal effects difficult. In particular, it might not even be possible to record important variables without measurement error, a common fact in fields such as psychology and social sciences. A fair amount of theory is often used to design instruments to indirectly measure such latent variables, such that one obtains estimates of measurement error. If the measurement error is known, then causal effects can be identified in a variety of scenarios. Unfortunately, a strictly theoretical approach for formalizing a measurement model is error prone and does not provide alternative models that could equally or better explain the data. We introduce an algorithmic approach that, given a set of observed indicators of latent phenomena of interest and common assumptions about the causal structure of the world, provides a set of measurement models compatible with the observed data. This approach extends previous results in the literature which would select an observed variable only if it measured a single latent variable. Our extensions cover cases where some variables are allowed to be indicators of more than one hidden common cause.

32.1 Introduction

Discovering latent representations of the observed world has become increasingly more relevant in the artificial intelligence literature (Hinton and Salakhutdinov, 2006; Bengio and Cun, 2007). Much of the effort concentrates on building latent variables which can be used in prediction problems, such as classification and regression.

A related goal of learning latent structure from data is that of identifying which hidden common causes generate the observations. This becomes relevant in applications that require predicting the effect of policies.

As an example, consider the problem of identifying the effects of the 'industrialization level' of a country on its 'democratization level' across two different time points. Democratization levels and industrialization levels are not directly observable: they are hidden common causes of observable *indicators* which can be recorded and analysed. For instance, gross national product (GNP) is an indicator of industrialization level, while expert assessments of freedom of the press can be used as indicators of democratization. Extended discussions on the distinction between indicators and the latent variables they measure can be found in the literature of structural equation models (Bollen, 1989) and error-in-variables regression (Carroll, Ruppert and Stefanski, 1995).

Causal networks can be used as a language to represent this information. We postulate a graphical encoding of causal relationships among random variables, where vertices in the graph representing random variables and directed edges $V_i \rightarrow V_j$ represent the notion that V_i is a direct cause of V_j. Formal definitions of direct causation and causal networks are given by Spirtes, Glymour and Scheines (2000) and Pearl (2000).

In our setup, we explicitly represent latent variables of interest as vertices in the graph. For example, in Figure 32.1 we have a network representation for the problem of causation between industrialization and democratization levels. This model makes assumptions about the connections among latent variables themselves: e.g. industrialization causes democratization, and the possibility of unmeasured confounding between industrialization and democratization is not taken into account (which, of course, can be criticized and refined).

Following the mixed graph notation (Richardson and Spirtes, 2002; Spirtes *et al.*, 2000; Pearl, 2000), we also use bi-directed edges $V_i \leftrightarrow V_j$ to denote implicit paths due to latent common causes. That is, $V_i \leftrightarrow V_j$ denotes a set of causal paths (e.g. $V_i \leftarrow X \rightarrow V_j$) that originate from common causes that have been marginalized (such as X in the previous example), as discussed in full detail by Richardson and Spirtes (2002). The distinction between 'explicit' and 'implicit' latent variables is problem dependent: if we do not wish to establish causal effects for some hidden variables, then they can be marginalized.

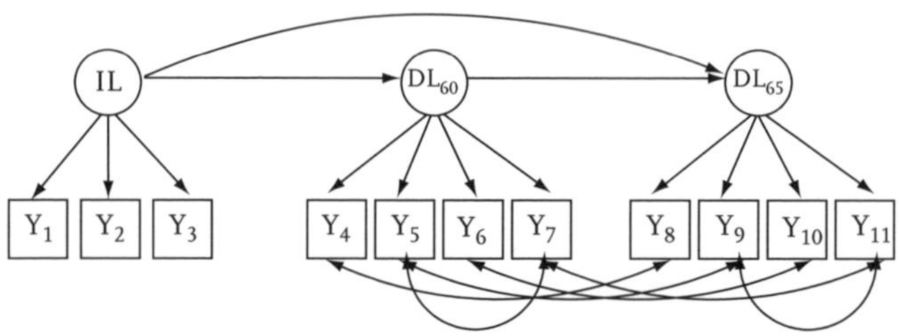

Fig. 32.1 A causal diagram connecting industrialization levels (IL) of a country, in 1960, to democratization levels in 1960 and 1965 (DL_{60} and DL_{65}, respectively). In our diagrams, we follow the standard structural equation modelling notation and use square vertices represent observable variables, circles represent latent variables (Bollen, 1989). The industrialization indicators are: Y_1, gross national product; Y_2, energy consumption per capita; Y_3, percentage of labor force in industry. The democratization indicators are: Y_4/Y_8, freedom of press; Y_5/Y_9, freedom of opposition; Y_6/Y_{10}, fairness of elections; Y_7/Y_{11}, elective nature of legislative body. Details are given by Bollen (1989).

Establishing the causal connections among latent variables is an important causal question, but it is only meaningful if such hidden variables are connected to our observations. A complementary and perhaps even more fundamental problem is that of finding which latent variables exist, and how they cause the observed measures.

This will be the main problem tackled in our contribution: given a dataset of indicators assumed to be generated by *unknown* and *unmeasured* common causes, we wish to discover which hidden common causes are those, and how they generate our data. Using a definition from the structural equation modelling literature, we say we are interested in learning the *measurement model* of our problem (Bollen, 1989). Our contribution generalizes the approach introduced by Silva *et al.* (2006).

In the context of the example of Figure 32.1, suppose we are given a dataset with 11 indicators, and wish to discover the respective latent common causes and measurement model. Assuming Figure 32.1 as the unknown gold standard, we are successful if we predict that $\{Y_1, Y_2, Y_3\}$ are generated by a particular hidden common cause, $\{Y_4, \ldots, Y_7\}$ are generated by another hidden common cause and so on.

The approach of Silva *et al.* (2006) works as follows:

1. for each pair of observed variables Y_i and Y_j, search for constraints in the observed covariance matrix that support the fact that Y_i and Y_j do not have a common parent in the true graph;
2. for each pair of observed variables Y_i and Y_j, search for constraints in the observed covariance matrix that support the fact that Y_i and Y_j are conditionally independent given some hidden variable in the true graph.

These two pieces of evidence allow for the identification of *substructures*: notice that in order to find a measurement model where each observed has a single parent, we can use the second condition above. To identify that two or more latent variables are different variables in the true graph, we use the first condition. The limitation of this procedure, however, is that it discards any observed variable that has more than one latent parent. We will introduce a method that generalizes Silva *et al.* (2006) by allowing for indicators with multiple hidden causes.

The solution to this problem lies at the intersection of artificial intelligence techniques to infer causal structures, statistical models and the exploitation of assumptions commonly made in some applied sciences such as psychology and social sciences.

Success will depend on how structured the real-world causal network is and how valid our assumptions are. If the postulated true network that generated our data is not sparse, for instance, then there will be so many models compatible with the observed data that no useful conclusion can be made. This situation, however, is not different from the limitations of standard causal network discovery procedures (with no latent variables) (Spirtes *et al.*, 2000), which rely on the existence of many conditional independence constraints. Even if we learn only some partial information about the measurement model, in principle it is still possible to infer some of the causal structure among the latent variables – the ultimate goal in many applications of latent variable models (Silva *et al.*, 2006; Spirtes *et al.*, 2000).

We describe our assumptions and a formal problem statement in full detail in Section 32.2. An algorithm to tackle the stated problem is provided in Section 32.3. Experiments are show in Section 32.4, followed by a Conclusion. Before that, however, we discuss what is the current common practice for unveiling the causal measurement structure of the world, and why they fall short on providing a reasonable solution.

A motivating example

Exploratory factor analysis is still the method of choice for suggesting hidden common cause models in the sciences. A detailed description of the method within the context of psychology and social sciences is given by Bartholomew and Knott (1999). In this section, we will illustrate the weaknesses of factor analysis. This motivates the need for more advanced methods resulting from artificial intelligence techniques in causal discovery.

In a nutshell, the main assumption of factor analysis states that each observed variable Y_i should be the effect of a set of latent factors $\mathbf{X} \equiv \{X_1, \ldots, X_L\}$ plus some independent error term ϵ_i:

$$Y_i = \sum_{j=1}^{L} \lambda_{ij} X_j + \epsilon_i. \tag{32.1}$$

Variables are assumed to be jointly Gaussian, although this is not strictly necessary. The measurement model is given by the coefficients $\{\lambda_{ij}\}$ and variances of the error terms $\{\epsilon_i\}$. Learning the measurement model is the key task, which is required in order to understand what the hidden common causes should represent in the real world. The factor analysis model is agnostic with respect to the causal structure of $\mathbf{X}$, but knowing the measurement model would also help us to learn the causal structure among latent variables (Spirtes *et al.*, 2000). In the following discussion, we will assume that we know how many latent variables exist, and then illustrate how such a widely used method is unreliable even under this highly favourable circumstance.

Given the observed covariance matrix of $\mathbf{Y} \equiv \{Y_1, \ldots, Y_p\}$, it is possible to infer the coefficients λ_{ij} and the covariance matrix of $\mathbf{X}$, but not in an unique way. Without going into details, there are ways of choosing a solution among this equivalence class such that the measurement structure is as simple 'as possible' (within the selection criterion of choice) (Bartholomew and Knott,

1999). Simplicity here means having many coefficients $\{\lambda_{ij}\}$ set to zero, indicating that each observed variable measures only a few of the latent variables. Getting the correct sparse structure is essential in order to interpret what the hidden common causes are. Notice that this corresponds to a directed causal network, where non-zero coefficients are encoded as directed edges in the graph.

Such methods will work when the true model that generated the data is in fact a 'simple structure', or a 'pure measurement model', in the sense that each observed variable has a single parent in the corresponding causal network. However, any deviance from this simple structure will strongly compromise the result.

We provide an example in Figure 32.2. We generated data from a linear causal model that follows the causal diagram of Figure 32.2(a).[1] Given data for the observed variables $Y_1, \ldots, Y_6$, we ideally would like to get a structure such as the one in Figure 32.2(b), where the question marks emphasize that labels for the latent variables should be provided by background knowledge.

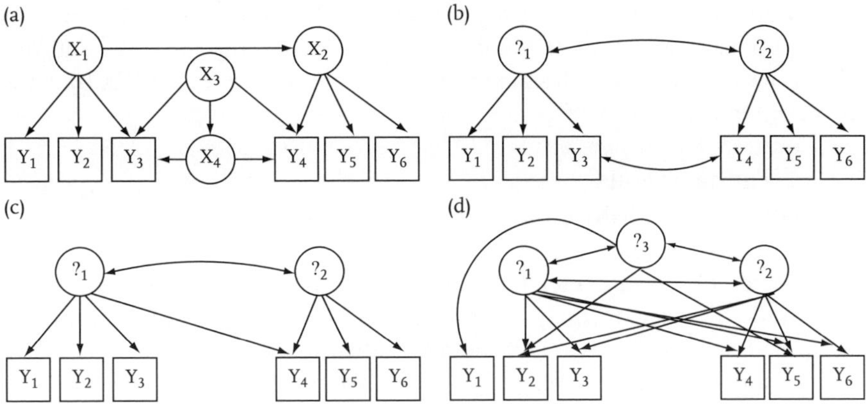

Fig. 32.2 In (a), we show a synthetic structure from which we generated 200 data points. Our algorithm is able to reconstruct the causal graphical structure in (b), which captures several features of the original model. Bi-directed edges represent conditional association and the possibility of some unidentified set of hidden common causes of the corresponding vertices. Inferred latent variables are not labelled, but can then be interpreted from the resulting structure and the explicit assumptions discussed in Section 32.2. In (c), the resulting structure obtained with exploratory factor analysis using two latent variables and the promax rotation technique. One implication of this structure is that Y_3 and Y_4 are now independent given two latent variables, contrary to (a). This extra constraint leads to biases in the estimation of the model. In (d), the factor analysis result we got when attempting to fit three latent variables.

[1] Coefficients were generated uniformly at random on the inverval $[-1.5, -0.5] \cup [0.5, 1.5]$ while variances of error terms were generated uniformly in $[0, 0.5]$.

Notice that in this contribution our aim is not to learn the structure connecting the latent variables, and the bi-directed edge in this case denotes an arbitrary causal connection.

Exploratory factor analysis fails to provide sensible answers.[2] In Figure 32.2, we show results obtained with different numbers of latent variables. Figure 32.2(c) shows a common outcome when we indicate that the model should have two hidden common causes. There exists no theory that provides a clear interpretation for these edges. Even worse, results can easily become meaningless. In Figure 32.2(d), we depict the result of exactly the same procedure, but where now we allow for three hidden common causes. The method we describe in our contribution is able to recover Figure 32.2(b).

32.2 Problem statement and assumptions

We start this section with a general view of the problem, and how it relates to previous work and other issues of scientific discovery. This is followed by a formal characterization of the problem and its assumptions. We end this section by describing which kind of results we can and cannot obtain by using by our methods.

32.2.1 Fundamentals

Exploratory factor analysis has been an important tool in applied sciences (Loehlin, 2004), when the goal is to identify hidden common causes responsible for the observed associations among recorded variables. However, it is well-known that factor analysis provides no basis for choosing among different causal hypotheses that generate undistinguishable probability distributions among the observed variables. As briefly discussed in the previous section, and demonstrated in detail by Silva *et al.* (2006) through several experiments, exploratory factor analysis is unreliable.

Our work fully embraces the framework of Spirtes *et al.* (2000) and Pearl (2000): it assumes that *there is* a causal structure that generates the observed data, and that such causal structure can be formally *represented* as a directed acyclic graph (DAG) using the axioms discussed by Spirtes *et al.* (2000). The consequence is that different DAG structures will imply different constraints in the observed distribution, which can be tested from data. The view of this work is that assuming the existence of a true structure (be it a DAG or some

[2] The number of latent variables can be chosen by a variety of standard techniques (Loehlin, 2004). For instance: by maximizing a score function that trades-off the complexity of the model against its fitness, or by choosing the minimal number such that the model passes a statistical significance test. In our example, a model with three latent variables passed a chi-square test at a 0.05 significance level.

other mathematical object) is unavoidable for both theoretical and practical reasons.

The postulated DAG structure contains both observed and unobserved (hidden, latent) variables as vertices. Such hidden variables are unknown unknowns: we assume they represent real entities, but we do not know which ones exist and how they are related to the observations. We regard this practice as fundamentally different from some typical applications of factor analysis: the resulting structures that we obtain from our analysis, as well as the generation of hidden common causes, follow from our data and assumptions, not from a post-hoc reification of equivalent models fit to the data.

Interpreting the resulting latent variables and linking them to real entities and possible interventions still requires knowledge of the domain. However, such latent variables are a consequence of the core assumptions discussed in the next section. If assumptions are not to be believed, then conclusions should not be warranted. However, if for a particular domain assumptions are deemed reasonable, we can provide some guarantees concerning the set of models that is compatible with the data. Although the result is underdetermined, it can still report valuable information, as we will see through several examples.

Notice that the DAG structure can also have a non-causal meaning and the results that follow could in principle refer to models of independence constraints only, as in the applications described by Hinton and Salakhutdinov (2006). However, this does not change the fact that DAGs provide a formal language for encoding causal statements. Since it is commonly accepted that causation and probabilistic independence are partially related, this is not coincidental.[3] As a matter of fact, exploiting independence constraints is indeed the source of underdetermination, because the concepts of causation and conditional independence are not the same despite sharing a common notation.

Underdetermination does not imply that the problem leads to no solution, but that solutions should deal with underdetermination. As discussed by Leplin (2004), even postulated unobserved entities later discarded by further evidence (such as phlogiston) can have identifiable successors in later theories (such as oxygen): new evidence narrows down properties of latent variables discovered from previous data, but partial knowledge about such hidden variables and their measurements can still be used to formulate useful scientific theories (such as the link between combustion and rust).

Our main motivation is an extension of the methods of Silva *et al.* (2006). Those methods allows for the identification of causal structures indicating hidden common causes of observed variables. The main idea is to exploit

[3] For instance, if an intervention on variable X at its different levels always results in the same distribution for a variable Y, X is not considered a cause of Y, other things being equal.

constraints in the covariance matrix of such variables (which can be estimated from data) to rule out causal structures that fail to satisfy such constraints or, that support constraints that are not in the covariance matrix. The key advantage of such a method is that it allows for the identification of substructures that are common to all possible DAGs compatible with said constraints. However, this is a conservative procedure that can discard many variables for which no causal information can be discovered. The main goal of the contribution is to generalize the procedure, so that other latent variables and more of their causal relationships to observed variables can be identified.

32.2.2 Formal problem statement

Assume that that our data follows a distribution $\mathcal{P}$ generated according to a directed acyclic causal graph (DAG) $\mathcal{G}$ (Spirtes *et al.*, 2000; Pearl, 2000) with observed nodes **Y** and hidden nodes **X**. We also assume that the resulting distribution $\mathcal{P}$ is *faithful* to $\mathcal{G}$ (Spirtes *et al.*, 2000), that is, a conditional independence constraint holds in $\mathcal{P}$ if and only if it holds in $\mathcal{G}$ (using the common criterion of *d-separation* – please refer to Pearl (2000) for a definition and examples). These are all standard assumptions from the causal discovery literature.[4]

Our more particular assumptions are

- no observed node $Y \in \mathbf{Y}$ is a parent in $\mathcal{G}$ of any hidden node $X \in \mathbf{X}$;
- each random variable in $\mathbf{Y} \cup \mathbf{X}$ is a linear combination of its parents, plus additive noise, as in Equation (32.1).

The first assumption is motivated by applications in structural equation modelling (Bollen, 1989), where prior knowledge is used to distinguish between standard indicators and 'causal indicators', which are causes of the latent variables of interest. Both of these assumptions can be relaxed to some extent, although any claims concerning the resulting causal structures learned from data will be weaker. Silva *et al.* (2006) discuss the details.

For the purposes of simplifying the presentation of this chapter, we also introduce the following two assumptions:

- no observed node $Y \in \mathbf{Y}$ is an ancestor in $\mathcal{G}$ of any other observed node;
- every pair of observed nodes in **Y** has a common latent ancestor[5] in $\mathcal{G}$.

These two assumptions can be dropped without any loss of generality (Silva *et al.*, 2006), but they will be useful for presentation purposes. Notice that the latter assumption implies that there are no conditional independence constraints in the marginal distribution of **Y**. As such, a standard causal

[4] More precisely, we assume a stronger version of faithfulness in which the constraints we describe later in this section are *linearly implied* by $\mathcal{G}$. See Chapter 6 of Spirtes *et al.* (2000).

[5] As a reminder, this is not the same as having parents in common.

inference algorithm such as the PC algorithm (Spirtes *et al.*, 2000) cannot provide any information.

Notice also that we do not assume any other form of background knowledge concerning the number of latent variables or particular information concerning which observed variables have common hidden parents.

Having clarified all assumptions on which our methods rely, the problem we want to solve can be formalized. Let the *measurement model* $\mathcal{M}$ of $\mathcal{G}$ be a graph given by all vertices of $\mathcal{G}$, and the edges of $\mathcal{G}$ that connect latent variables to observed variables. In order to be agnostic with respect to the causal structure among latent variables, we connect each pair of latent variables by a bi-directed edge as a general symmetric representation of dependency. Ideally, given the distribution $\mathcal{P}$ over the observed variables and that our assumptions hold, we would like to reconstruct $\mathcal{M}$. Since $\mathcal{P}$ has to be estimated from the data, it is of practical interest to use only features of $\mathcal{P}$ that can be easily estimated. As such, we rephrase our problem as learning $\mathcal{M}$ given Σ, the covariance matrix of $\mathbf{Y}$.

However, in general this is only possible if the true model entails that Σ is constrained in ways that cannot be explained by other models. For instance, if there are more latent variables than observed variables, and each latent variable is a parent of all elements of $\mathbf{Y}$, then Σ has no constraints and an infinite number of models will be compatible with the data.

Silva *et al.* (2006) formalize the problem by extracting only *pure measurement submodels* of the true model, subgraphs of $\mathcal{M}$ where each observed variable Y has a single parent, and where this parent d-separates Y from all other vertices of the submodel in $\mathcal{G}$. Such single-parent vertices are also called *pure indicators*. Moreover, the output of the procedure described by Silva *et al.* (2006) only generates submodels where each latent variable has at least three pure indicators. If such models exist, they can be discovered given Σ. The scientific motivation is that many datasets studied through structural model analysis and factor analysis support the existence of pure measurement submodels. As we mentioned in the previous section, methods for providing 'simple structures' in factor analysis are hard to justify unless some pure measurement submodel exists. Therefore, it would be hard to justify factor analysis as a more flexible approach, since its output would be unreliable anyway. An important advantage of the causal discovery approach discussed here is that it knows its limitations: if there is no pure measurement submodel for all latent variables in the true model, it will report a model for a subset of the variables only. This also means that an empty structure might be reported if no pure submodel exists.

Our contribution is to extend the work of Silva *et al.* by allowing several 'impurities' in the output of our new procedure. To give an example where this is necessary, consider Figure 32.2(a) again. It is not possible to include both latent variables using the procedure of Silva *et al.*: if latent variables X_1

and X_2, and their respective three indicators, are included, it turns out Y_3 is not d-separated from Y_4 by either X_1 or X_2. The best Silva *et al.* (2006) can do is to include, say, X_1 and its indicators, plus one of its descendants as an indirect indicator which does not violate the separations in the true model. For instance, the model with edges $X_1 \rightarrow Y_i$, for $i \in \{1, 2, 3, 5\}$ and no other variable, satisfies this condition. In contrast, the new procedure described here is able to generate Figure 32.2(b).

In practice, Σ has to be estimated from data. In the discussion that follows, we assume that we know Σ so that we can concentrate on the theory and the main ideas. Section 32.4 provides methods to deal with an estimate of Σ and which practical issues arise in this case.

32.2.3 Description of output

Our output is a *measurement pattern* $\mathcal{M}_P$ which, under the above specified assumptions and given the population matrix Σ of a set of observed variables **Y**, provides provably correct causal claims concerning the true structure $\mathcal{M}$. The measurement pattern is a directed mixed graph with labeled edges (as explained below), with hidden nodes $\{L_i\}$ and observed nodes that form a subset of **Y**. $\mathcal{M}_P$ includes directed edges from latent variables to observed variables, and bi-directed edges between observed variables.

Before introducing the new procedure in Section 32.3, we formalize the causal claims that a measurement pattern $\mathcal{M}_P$ provides:

1. each hidden variable L_i in $\mathcal{M}_P$ corresponds to some hidden variable X_j in $\mathcal{G}$. In the items below, we call this variable $X(L_i)$. Moreover, for two different latent variables L_i and L_j in $\mathcal{M}_P$, $X(L_i) \neq X(L_j)$ in any possible mapping from hidden variables in $\mathcal{M}_P$ to hidden variables in $\mathcal{G}$;
2. if Y_i is in $\mathcal{M}_P$ but it is not a child of latent variable L_j, then Y_i is independent of $X(L_j)$ in $\mathcal{G}$ given its parents in $\mathcal{M}_P$;
3. given *any* pure measurement submodel of $\mathcal{M}_P$ with at least three indicators per latent variable, and a total of at least four observed variables, then *at most one* of the latent-to-indicator edges $L_i \rightarrow Y_j$ does not correspond to the true causal relationship in $\mathcal{G}$. That is, it is possible that for one pair, $X(L_i)$ is not a cause of Y_j and/or the relationship is confounded;

The last item needs to be clarified with an example, since it is not intuitive. Let Figure 32.3(a) be a true causal structure from which we can measure the covariance matrix of $Y_1, \ldots, Y_6$. The structure reported by our procedure is the one in Figure 32.3(b). Five out of Six edges correspond to the correct causal statement, except $L_2 \rightarrow Y_6$ (which should be confounded). We cannot know which one, but at least we know this is the case. As in any causal discovery algorithm (Spirtes *et al.*, 2000; Pearl, 2000), background knowledge is necessary to refine the information given by an equivalence class of graphical

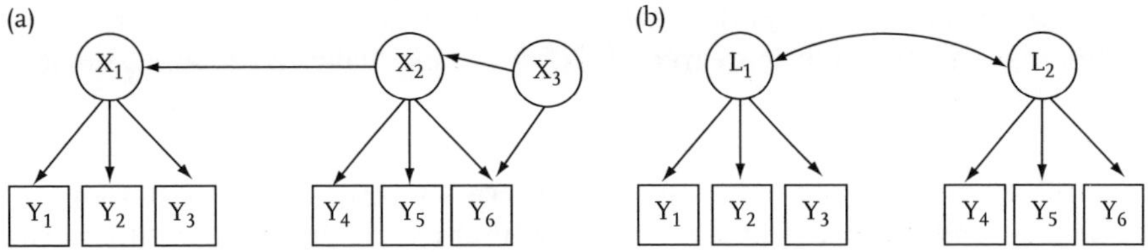

Fig. 32.3 A covariance matrix obtained from the true model in (a) will result in the structure shown in (b). This structure should be interpreted as making claims about an equivalence class of models: in this case, at most one of the directed edges does not correspond to the exact claim that there is an unconfounded causal relationship between the latent variable and its respective child. But notice that there are only Seven models compatible with this claim, instead of the $2^6 = 32$ possibilities (each of the six relations $L_i \rightarrow Y_j$ being confounded or not) that the same adjacencies could provide.

structures. In our case, such models are equivalent in the sense they imply the same testable constraints.

Finally, edges are labelled as 'confirmed' (they do correspond to actual paths in the true graph) or 'unconfirmed' (we cannot decide whether a corresponding path exists in the true graph). In the next section we clarify how 'unconfirmed' edges appear. Unless otherwise stated, all other edges are 'confirmed' edges.

32.3 An algorithm for inferring an impure measurement model

Let a *one-factor submodel* of $\mathcal{G}$ be a set composed of one hidden variable X and four observed variables $\{Y_A, Y_B, Y_C, Y_D\}$, such that each pair of distinct variables is d-separated by X.

One-factor submodels play an important role in our procedure. A vertex Y_i will be included in our output measurement pattern $\mathcal{M}_P$ if and only if it belongs to some one-factor submodel of $\mathcal{G}$. Also, X will correspond to some output latent if and only if it belongs to some one-factor submodel. Figure 32.2(a) illustrates the concept: the sets $\{X_1, Y_1, Y_2, Y_3, Y_5\}$ and $\{X_2, Y_4, Y_5, Y_6, Y_1\}$ are one-factor submodels. No one-factor submodels exist for X_3 and X_4.

This fact should not be surprising. It is well-known in the structural equation modelling literature that the folllowing model is testable:

$$Y_i = \lambda_i X + \epsilon_i \tag{32.2}$$

where $i \in \{1, 2, 3, 4\}$, and X and $\{\epsilon_i\}$ are mutually independent Gaussian variables of zero mean. This corresponds to a Gaussian causal network with corresponding edges $X \to Y_i$. Adding an extra edge, and hence a new parameter, would remove one degree of freedom and make the model undistinguishable from models with two latent variables (Silva *et al.*, 2006).

One way to characterize which constraints are entailed by this model is by writing down the *tetrad constraints* of this structure. If σ_{ij} is the covariance of Y_i and Y_j, and σ_X^2 is the variance of X, then for the model (32.2) the following identify holds:

$$\sigma_{12}\sigma_{34} = \lambda_1\lambda_2\sigma_X^2 \times \lambda_3\lambda_4\sigma_X^2 = \lambda_1\lambda_3\sigma_X^2 \times \lambda_2\lambda_4\sigma_X^2 = \sigma_{13}\sigma_{34}. \qquad (32.3)$$

Similarly, $\sigma_{12}\sigma_{34} = \sigma_{14}\sigma_{23}$. For a set of four variables $\{Y_A, Y_B, Y_C, Y_D\}$, we represent the statement $\sigma_{AB}\sigma_{CD} = \sigma_{AC}\sigma_{BD} = \sigma_{AD}\sigma_{BC}$ by the predicate $\mathcal{T}(ABCD)$. Notice this is entailed by the graphical structure, since the relationship does not depend on the precise values of $\{\lambda_i\}$ or σ_X^2.

For the causal discovery goal, however, the relevant concept is the converse: given observable constraints that can be tested, which causal structures are compatible with them? Concerning one-factor submodels, the converse has been proved[6] by Silva *et al.* (2006):

Lemma 1. *If $\mathcal{T}(ABCD)$ is true, then there is a latent variable in $\mathcal{G}$ that d-separates $\{Y_A, Y_B, Y_C, Y_D\}$.*

For example, in Figure 32.3(a), X_1 d-separates each pair of distinct variables in $\{Y_1, Y_2, Y_3, Y_4\}$, although it is not a cause of Y_4. A result such as Lemma 1 is important for discovering latent variables, but it is of limited use unless there are ways of ruling out the possibility that some latent variables are causes of some indicators. It turns out that the $\mathcal{T}(\cdot)$ constraint can also be used for this purpose.

Consider Figure 32.3(a) again. If we pick all three indicators of one latent variables along with some indicator of the other latent variables, we have a one-factor model that passes the conditions of Lemma 1. One possibility is that all six indicators are pure indicators of a single latent cause: after all, each pair $\{Y_A, Y_B\}$ is d-separated by some single latent variable. However, this does not tell us whether the latent variable that separates one group is the same as the one that separates another group. This is clear from Figure 32.3(a): X_1 d-separates any pair in $\{Y_1, Y_2, Y_3\} \times \{Y_4, Y_5, Y_6\}$. However, it does not d-separate any pair in $\{Y_4, Y_5, Y_6\} \times \{Y_4, Y_5, Y_6\}$. We have to deduce this information without looking at the true graph, but only at the marginal covariance matrix of $\mathbf{Y}$.

[6] To avoid unnecessary repetition, from now on we establish the convention that all results use the assumptions of Section 32.2, without explicitly mentioning them in the theoretical development.

One way of discarding connections from latents to indicators, and deducing that two unobserved variables are not the same, is given by the following result:

Lemma 2. *Consider the observed variables* $\{Y_A, Y_B, Y_C, Y_D, Y_E, Y_F\}$. *If both* $\mathcal{T}(ABCD)$ *and* $T(ADEF)$ *are true, but* $\sigma_{AB}\sigma_{DE} \neq \sigma_{AD}\sigma_{BE}$, *then* Y_A *and* Y_D *cannot have any common parent in* $\mathcal{G}$.

A detailed proof is given by Silva *et al.* (2006). The intuitive explanation is that, if Y_A and Y_D did have a common parent (say, X_{AD}), then this latent variable would be precisely the one, and only one, responsible for both constraints $\mathcal{T}(ABCD)$ and $\mathcal{T}(ADEF)$. It would not be hard to show that this would imply $\sigma_{AB}\sigma_{DE} = \sigma_{AD}\sigma_{BE}$, contrary to the assumption.

Notice that these two results are already enough to find a pure measurement submodel. The general skeleton of the procedure is to find a partition $\{\mathbf{M}_1, \ldots, \mathbf{M}_C\}$ such that $\cup_{i=1}^{C}\mathbf{M}_i \subseteq \mathbf{Y}$ and

1. elements in $\mathbf{M}_i$ are d-separated by some hidden variable (using Lemma 1);
2. elements in $\mathbf{M}_i$ and $\mathbf{M}_j$ cannot have common parents (using Lemma 2).

Many more details need to be described in order to provide an equivalence class of pure measurement models with three indicators per latent variable, but this is the general idea. What is missing from this procedure is a way of coping with impure measurement models so that a structure such as the one in Figure 32.2(b) can be obtained. We now introduce the first theoretical results that accomplish that.

32.3.1 Finding impure indicators

Consider what can happen if we observe the covariance matrix generated by the model of Figure 32.2(a). We know that there is no single latent variable that d-separates (say) $\{Y_1, Y_2, Y_3, Y_4\}$. However, we know that there is some hidden L that d-separates $\{Y_1, Y_2, Y_3, Y_5\}$, as well as some hidden L' that d-separates $\{Y_1, Y_4, Y_5, Y_6\}$. So far, we could infer an approximate graph such as:

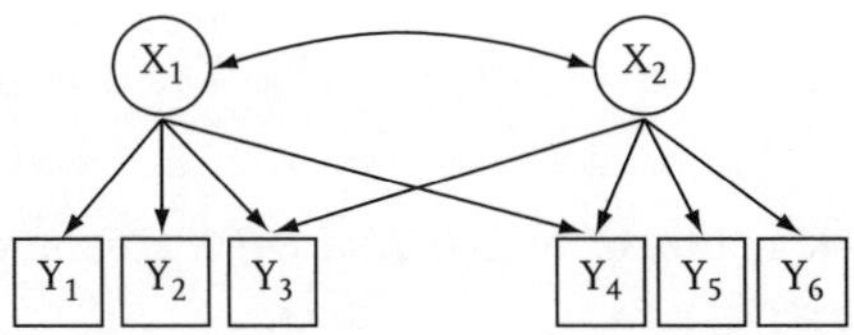

However, we cannot stop here and report this as a possible solution: we will get an inconsistent estimate for the covariance of the latent variables, which can lead to wrong conclusions about the causal structure of the latents. We would like to account for the possibility that the impurities arise not from our identified latents, but from some other source. This is the result summarized by Lemma 3:

Lemma 3. *Consider the observed variables* $\{Y_A, Y_B, Y_C, Y_D, Y_E, Y_F\}$. *If the following predicates are true:*

$$\mathcal{T}(ABCE), \mathcal{T}(ABCF), \mathcal{T}(ADEF), \mathcal{T}(BDEF)$$

and the following predicates are false

$$\mathcal{T}(ABEF), \mathcal{T}(ABCD), \mathcal{T}(CDEF)$$

then in the corresponding causal graph $\mathcal{G}$, *we have that:*

- $\mathcal{G}$ *contains at least two different latent variables,* L_1 *and* L_2;
- L_1 *d-separates all pairs in* $\{Y_A, Y_B, Y_C\} \times \{Y_D, Y_E, Y_F, L_2\}$, *except* $Y_C \times Y_D$;
- L_2 *d-separates all pairs in* $\{Y_A, Y_B, Y_C, L_1\} \times \{Y_D, Y_E, Y_F\}$, *except* $Y_C \times Y_D$;
- Y_C *and* Y_D *have extra hidden common causes not in* $\{L_1, L_2\}$.

A formal proof of a slightly more general result is given by Silva (2006). The core argument is as follows. The existence of L_1 and L_2 follows from Lemma 1 and the constraints $\mathcal{T}(ABCE)$ and $\mathcal{T}(ADEF)$. That $L_1 \neq L_2$ follows from Lemma 2 and the fact that $\mathcal{T}(ABEF)$ is false. The other d-separations follow from Lemma 1 and the given tetrad constraints. Finally, if Y_C and Y_D did not have any other hidden common cause, we could not have both $\mathcal{T}(ABCD)$ and $\mathcal{T}(CDEF)$ falsified at the same time, contrary to our hypothesis.

In our diagrams, we represent such extra hidden common causes by bi-directed edges $Y_C \leftrightarrow Y_D$. That is, we do not specify how many hidden common causes for this pair exist or how they are connected to other latent variables. Notice that we never claim that the implicit latent variables represented by bi-directed edges are independent of the discovered latent variables. Figure 32.4 illustrates a case.

The second type of impurity we will account for nodes that have more than one represented latent parent.

Lemma 4. *Consider the observed variables* $\{Y_A, Y_B, Y_C, Y_D, Y_E, Y_F, Y_G\}$. *If the folllowing predicates are true:*

$$\mathcal{T}(ABCK), \textit{for } K \in \{D, E, F, G\}; \mathcal{T}(KEFG), \textit{for } K \in \{A, B, C, D\};$$

and the following predicates are false

$$\mathcal{T}(K_1K_2K_3K_4), \textit{for } \{K_1, K_2\} \subset \{A, B, C\}, \{K_3, K_4\} \subset \{E, F, G\};$$
$$\mathcal{T}(ADEF), \mathcal{T}(ABDE)$$

then in the corresponding causal graph $\mathcal{G}$, *we have that:*

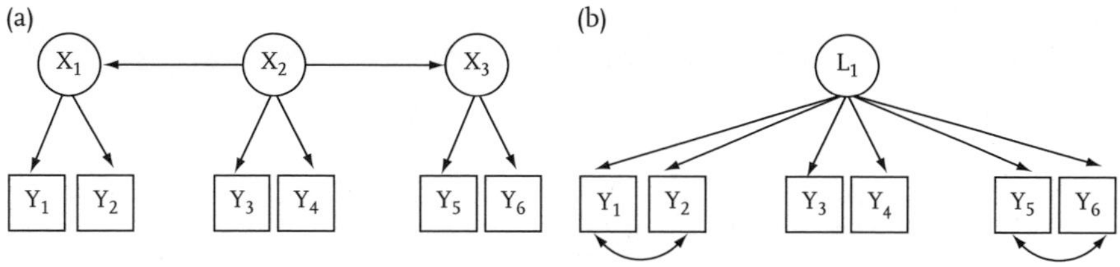

Fig. 32.4 A model such as the one in (a) generates the measurement pattern in (b). Notice that the indication of extra hidden common causes, as represented by, e.g. $Y_1 \leftrightarrow Y_2$, does not imply that these unrepresented causes are independent of the represented ones. Moreover, notice that if the edge $X_1 \leftarrow X_2$ is *switched* in (a) to $X_1 \rightarrow X_2$ instead, the corresponding pattern would still be the one in (b). In this case, it is clear that both edges $L_1 \rightarrow Y_1$ and $L_1 \rightarrow Y_2$ are not representing the actual causal directions of the true graph: in the modified (a), we now have the path $Y_1 \leftarrow X_1 \rightarrow X_2$. Since L_1 is mapped to X_2, edge $L_1 \rightarrow Y_1$ should not be interpreted as X_2 being a cause of Y_1. This does not contradict our previous claims: the corresponding measurement pattern claim is about pure submodels. In this case, a pure submodel could be the one containing Y_1, Y_3, Y_4 and Y_5 only. One edge, $L_1 \rightarrow Y_1$, still does not explicitly indicate the confounding given by X_1, but this is compatible with the measurement pattern description.

- *$\mathcal{G}$ contains at least two different latent variables, L_1 and L_2;*
- *L_1 d-separates all pairs in $\{Y_A, Y_B, Y_C, Y_D\}$;*
- *L_2 d-separates all pairs in $\{Y_D, Y_E, Y_F, Y_G\}$;*
- *L_1 d-separates all pairs in $\{Y_A, Y_B, Y_C\} \times \{Y_E, Y_F, Y_G, L_2\}$, but not $Y_D \times L_2$;*
- *L_2 d-separates all pairs in $\{Y_A, Y_B, Y_C, L_1\} \times \{Y_E, Y_F, Y_G\}$, but not $Y_D \times L_1$.*

The nature of this result complements the previous one: instead of searching for evidence to remove edges from latents into indicators, this result provides identification of edges that *cannot* be removed. That is, if no third identifiable latent variable L_i can separate Y_D from L_1 and L_2, then edges $L_1 \rightarrow Y_D$ and $L_2 \rightarrow Y_D$ cannot be removed. Doing so would imply d-separations (e.g. Y_D and L_2 given L_1) contrary to the implications of our assumptions.

The argument again exploits Lemmas 1 and 2. A more detailed proof is given by Silva (2006). Notice the need for extra indicators in this case: this is another illustration of the need for one-factor models for each latent variable. Without Y_7 in the example of Figure 32.5(a), the result would be the measurement pattern of Figure 32.5(b).

Notice that if there are indicators that share more than one common parent in $\mathcal{G}$ then, by using tetrad constraints only, we cannot separate them (i.e. avoid a bi-directed edge) even if their parents are identified in the model. Figure 32.6 illustrates what the measurement pattern should report. Using higher-order constraints than tetrad constraints might be of help in this situation

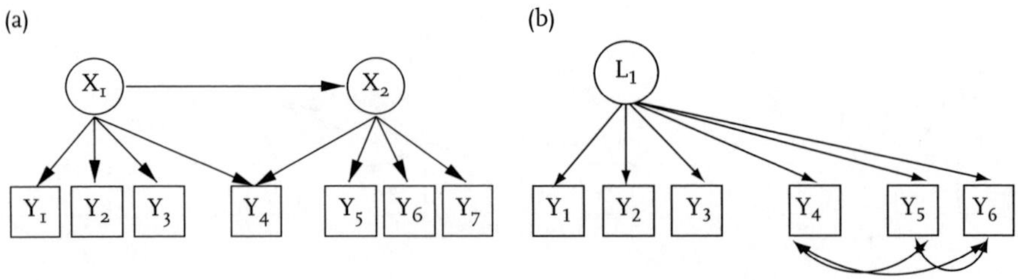

Fig. 32.5 The graph in (a) can be rebuilt exactly. However, without a fourth indicator of X_2 (e.g. Y_7), this latent variable can not be detected and the result would be the graph in (b).

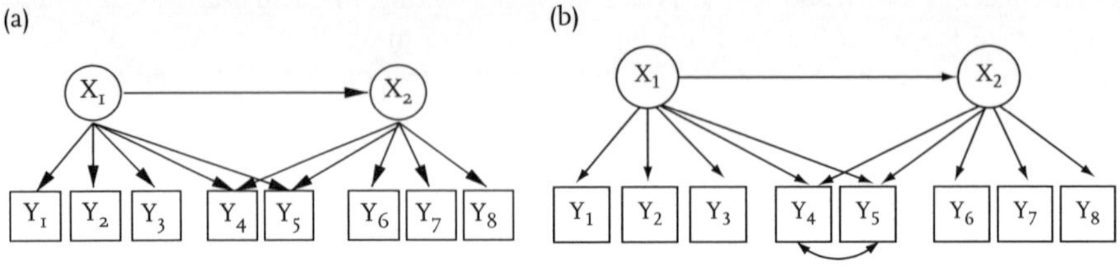

Fig. 32.6 The graph in (a) is not fully identifiable with tetrad constraints only. A conservative measurement pattern needs to report the structure in (b).

(Sullivant and Talaska, 2008), but this is out of the scope of the current contribution.

To summarize:

- Lemma 1 provides the evidence to include latent variables;
- Lemma 2 provides the evidence to distinguish between different latent variables;
- Lemma 3 allows for the removal of extra edges from latents into indicators and proves the necessity of some bi-directed edges;
- Lemma 4 proves the necessity of some edges from latents into indicators, but does not prove the necessity of adding some bi-directed edges.

32.3.2 Putting the pieces together

So far, we have described how to identify particular pieces of information about the underlying causal graph. While those results allow us to identify isolated latent variables and to remove or confirm particular connections, we need to combine such pieces within a measurement pattern. Unlike the procedure of Silva *et al.* (2006), this pattern should be able to represent several pure measurement submodels within a single graphical object and to possibly include more latent variables than any pure model.

In this section, we assume that we have the population covariance matrix Σ. Important practical issues arising from the need of estimating the covariance matrix from data are discussed in the next section. We start by finding groups of variables that are potential indicators of a single latent variable. We first build an auxiliary undirected graph $\mathcal{H}$ as follows:

InitialPass: this procedure returns an undirected graph $\mathcal{H}$.

1. let $\mathcal{H}$ be a fully connected undirected graph with nodes $\mathbf{Y}$;
2. for all groups of six variables $\{Y_A, Y_B, Y_C, Y_D, Y_E, Y_F\}$ that form a clique in $\mathcal{H}$, if $\mathcal{T}(ABCD)$ and $\mathcal{T}(ADEF)$ are true but $\sigma_{AB}\sigma_{DE} \neq \sigma_{AD}\sigma_{BE}$, remove the edge $Y_A - Y_D$;
3. if for a given Y_A in $\mathcal{H}$ there is no triplet $\{Y_B, Y_C, Y_D\}$ such that $\mathcal{T}(ABCD)$ holds, then remove Y_A from $\mathcal{H}$, since there will be no one-factor model including Y_A;
4. return $\mathcal{H}$.

Notice that if two vertices are not adjacent in $\mathcal{H}$, they cannot possibly be children of the same latent variable (it follows from Lemma 2). This motivates us to look for one-factor models within cliques of $\mathcal{H}$ only. In Step 3, we discard variables not in one-factor models, since nothing informative can be claimed about them using our methods.

In the next step, we obtain a set of tentative subgraphs, where each subgraph contains a single latent variable and its indicators:

SingleLatents: given $\mathcal{H}$, this procedure returns a set $\mathcal{S}$ of graphs with a single latent variable each.

1. initialize $\mathcal{S}$ as the empty set;
2. for each clique $\mathcal{C}$ in $\mathcal{H}$
3. if there is no $\{Y_A, Y_B, Y_C\} \subset \mathcal{C}$ and $Y_D \in \mathbf{Y}$ such that $\mathcal{T}(ABCD)$ holds, continue to next clique;
4. create a graph $\mathcal{G}_i$ with latent vertex L_i, $i = |\mathcal{S}| + 1$, and with children given by $\mathcal{C}$;
5. for each $\{Y_A, Y_B\} \subset \mathcal{C}$, if there is no $Y_C \in \mathcal{C}$ and $Y_D \in \mathbf{Y}$ such that $\mathcal{T}(ABCD)$ is true, then add edge $Y_A \leftrightarrow Y_B$ to $\mathcal{G}_i$. Mark this edge as 'unconfirmed';
6. add the new graph to $\mathcal{S}$;
7. return $\mathcal{S}$.

By Lemma 1, every single latent variable created in this procedure corresponds to at least one possible latent variable in $\mathcal{G}$. The rationale for Step 5 is that L_i does not d-separate Y_A and Y_B. It is possible to confirm many such edges using an argument similar to Lemma 4, but we leave out a detailed analysis to simplify the presentation.[7]

Finally, all single graphs are unified into a coherent measurement pattern:

FIND MEASUREMENT PATTERN: returns a measurement pattern $\mathcal{M}_P$ given $\mathcal{S}$.

1. let $\mathcal{M}_P$ be the union of all graphs in $\mathcal{S}$, where all latents are connected by bi-directed edges;
2. for every pair $\{\mathcal{S}_i, \mathcal{S}_j\} \subset \mathcal{S}$ do
3. consider all triplets $\{Y_A, Y_B, Y_C\} \subset \mathcal{S}_i \cup \mathcal{S}_j$ such that $\mathcal{T}(ABCD)$ holds for some Y_D. If such triplets are also in $\mathcal{S}_i \cap \mathcal{S}_j$, set the children of L_j to be children of L_i and discard L_j. Set all $L_i \rightarrow Y_k$ to be 'unconfirmed' if Y_k is not in $\mathcal{S}_i \cap \mathcal{S}_j$. Continue to next pair;
4. for every pair $\{Y_C, Y_D\} \subset \mathcal{S}_i \cap S_j$, add 'unconfirmed' edge $\mathcal{Y}_C \leftrightarrow Y_D$ to $\mathcal{M}_P$. If Lemma 3 can be applied to $\{Y_C, Y_D\}$ where $\{Y_A, Y_B\} \subset S_i$ and $\{Y_E, Y_F\} \subset S_j$, then remove edges $L_j \rightarrow Y_C$ and $L_i \rightarrow Y_D$ and mark $\mathcal{Y}_C \leftrightarrow Y_D$ as 'confirmed';
5. if Y_j has more than one parent, mark all directed edges $L_i \rightarrow Y_j$ unsupported by Lemma 4 as 'unconfirmed';
6. return M_P.

The justification for most steps follows directly from our previous results.[8] To understand Step 3, however, we need an example. In Figure 32.7(a), we have a true model. We can separate Y_4 from Y_8 using Lemma 2. The result of FIRST PASS is the graph $\mathcal{H}$ shown in Figure 32.7(b). Sets $\{Y_4, Y_5, Y_6, Y_7\}$ and $\{Y_5, Y_6, Y_7, Y_8\}$ are cliques in $\mathcal{H}$, but they refer to the same latent variable X_2. There will be edges $L_2 \rightarrow Y_4$ and $L_2 \rightarrow Y_8$ in the measurement pattern, but they will not be confirmed edges. Notice that there might be ways of removing $L_2 \rightarrow Y_4$ and $L_2 \rightarrow Y_8$, but they are out of the scope of this chapter. Our goal is not to provide complete identification methods, but to show the main tools and the difficulties of learning impure measurement models.

[7] An example: in Figure 32.5(b), all bi-directed edges can be confirmed, because each of $\{Y_4, Y_5, Y_6\}$ is separated from $\{Y_1, Y_2, Y_3\}$ by L_1. We can therefore isolate the failure of having a one-factor model composed of $\{L_1, Y_1, Y_2, Y_4, Y_5\}$ down to the $Y_4 \leftrightarrow Y_5$ edge.

[8] Notice also that, among all confirmed edges, it is still the case that the directionality is not individually determined, as stated in Section 32.2.3: within any given measurement submodel, at most one edge $L_i \rightarrow Y_j$ might not correspond to a direct, unconfounded, causal relationship.

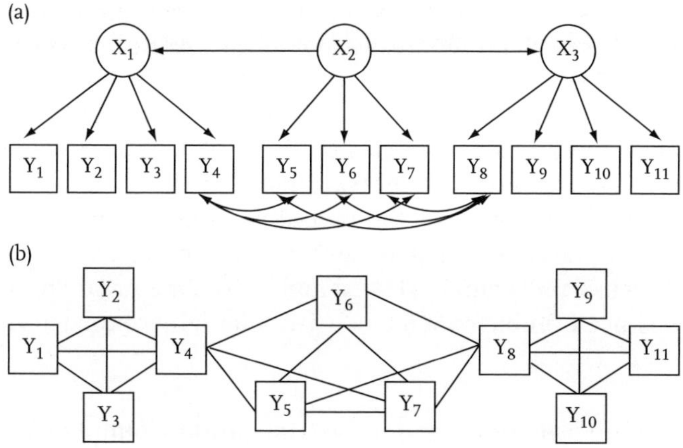

Fig. 32.7 With the true graph being (a), we obtain two cliques of variables $\{Y_4, Y_5, Y_6, Y_7\}$ and $\{Y_5, Y_6, Y_7, Y_8\}$ in $\mathcal{H}$ (b), since we can discover that Y_4 and Y_8 are indicators of different variables. However, these two cliques are related to the same true latent X_2 and have to be merged. The side-effect is that we cannot confirm the edges $L_2 \rightarrow Y_4$ and $L_2 \rightarrow Y_8$, although we know both cannot possibly exist at the same time.

To summarize:

- different latent variables in the output cannot be mapped to the same latent variable in the true graph;
- lack of edges in the measurement pattern correspond to conditional independence constraints in the true graph;
- the two main sources of causal indeterminacy are: some edges are labelled as unconfirmed, in the sense the corresponding causal path might not exist in the true graph; some of the causal relationships indicated by edges $X_i \rightarrow Y_j$ might be confounded, but no more than one is confounded within any pure measurement submodel of the output measurement pattern.

32.4 Experiments

In this section, we illustrate how the theory can be applied by analysing two simple datasets.

In practice, we will not know Σ, but only an estimate obtained from a sample. Robust statistical procedures to score models and test constraints from finite samples are described at length by Silva *et al.* (2006).

In the following experiments, we assume data are multivariate Gaussian. Wishart's tetrad test can be used to evaluate $\mathcal{T}(\cdot)$, which we accept as true if the p-value for the test is greater or equal to 0.05 (Silva *et al.*, 2006). In the

SINGLELATENTS procedure, for each clique C we add extra bi-directed edges to $\mathcal{G}_i$ by a greedy search procedure: we look at each pair of variables and evaluate the Bayesian information criterion (BIC, Schwarz (1978)) for the model with the added edge. If the best model is better than the current one, we keep the edge. Otherwise, we stop modifying $\mathcal{G}_i$. An analogous procedure is performed to add bi-directed edges in FINDMEASUREMENTPATTERN.

In the worst-case scenario, the procedure scales at an exponential rate in the number of variables due to the necessity of finding cliques in a graph (the SINGLELATENTS procedure). The examples are small and sparse enough so that this is not a problem. Some heuristics for larger problems are described by Silva *et al.* (2006).

32.4.1 Democratization and industrialization example

This is the study described at the beginning of Section 34.1 and discussed by Bollen (1989). A sample of 75 countries was collected. We will discuss the outcome of our procedure and how it relates to the 'gold standard' of Figure 32.1.[9]

If the true model is indeed Figure 32.1 and if we had access to an oracle that could answer exactly which tetrad constraints hold and do not hold in the true model, then the result of our algorithm would be Figure 32.8(a). The result obtained with our implementation is shown in Figure 32.8(b). With only 75 samples, it is not surprising that the BIC score tends to produce models with fewer edges than expected. Still, the model reveals a lot of information present in the expected pattern. It also suggests ways of extending the procedure, such as allowing for the background knowledge that some variables have the same definition, but recorded over time. Recall that the resulting model was obtained without any extra information.

32.4.2 Depression example

The next dataset is a depression study with five indicators of self-steem ($SELF$), four indicators of depression ($DEPRES$) and three indicators of impulsiveness ($IMPULS$). This dataset is one of the examples that accompany the LISREL software for structural equation modeling. The depression data and the meaning of the corresponding variables can also be found at

- `http://www.ssicentral.com/lisrel/example1-2.html`

[9] Caveat emptor: in our setup, a gold standard means a theoretical model, one that might be wrong but that reflects substantive prior knowledge. Since indicators are built with the purpose of measuring particular latent variables, and are frequently used across several different studies, we believe that the existence of the chosen latent variables and edges connecting latent variables to indicators are to be trusted. Incorrect choices of bi-directed edges are not unlikely, however, as well as the the possibility of extra directed edges and latent variables not accounted by the theoretical model. We do not have an objective way of evaluating them.

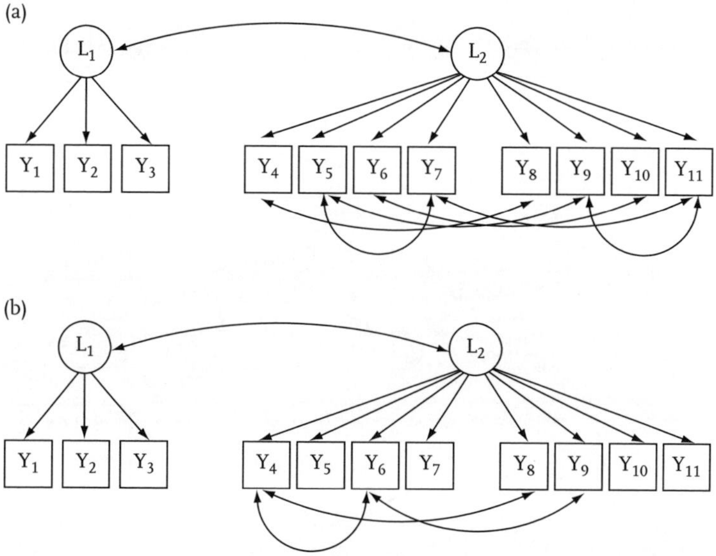

Fig. 32.8 In (a), the measurement pattern that corresponds to the gold standard. In (b), the result of our procedure. All directed edges are correct. With small sample sizes, the BIC score tends to produce models simpler than expected, so it is not surprising that the model lacks several of the bi-directed edges.

A theoretical gold standard is show in Figure 32.9(a). It is worth mentioning that, treated as a Gaussian model, this graphical structure does not fit the data: the chi-square score is 122.8 with 51 degrees of freedom. The sample size is 204.

Our result is shown in Figure 32.9(b). It was impossible to find a hidden common cause for the indicators of impulsiveness: the correlations of $IMPULS1$ and $IMPULS2$ with the other items were just too low, and those items had to be discarded. The only major difference against the gold standard was assigning $SELF5$ with the incorrect latent parent (the role of $IMPULS3$ in the solution is compatible with the properties of a measurement pattern). Given the number of bi-directed connections into $SELF5$, however, this indicator seems particularly problematic.

It is relevant to stress that in this study, the indicators are ordinal (in a 0 to 4 scale), not continuous. We were still able to provide relevant information despite using a Gaussian model. In future work, methods to deal with ordinal data will be developed. The theory for ordinal data is essentially identical, as discussed by Silva *et al.* (2006). However, non-Gaussian models need to be used, which increases the computational cost of the procedure considerably.

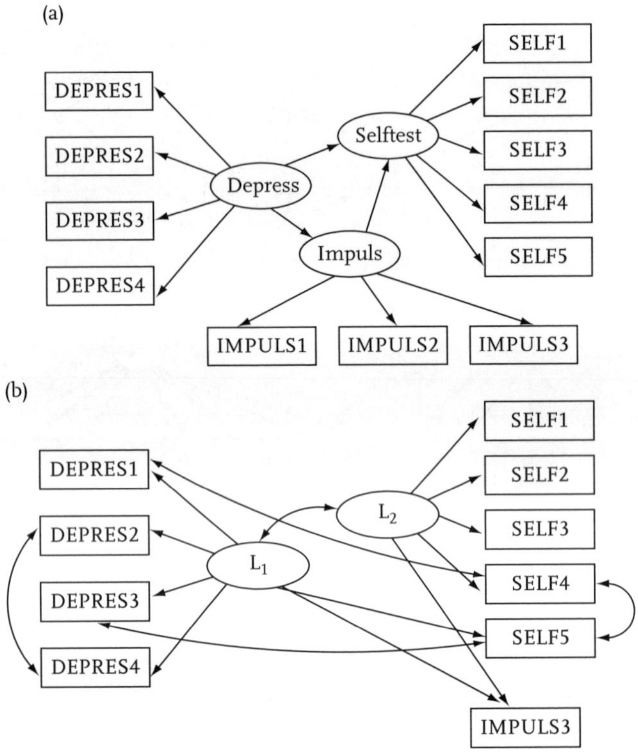

Fig. 32.9 In (a), the gold standard of the depression study. The measurement pattern is precisely the same (except for the latent connections). In (b), the result of our procedure. The inferred model cannot contain the impulsiveness latent variables, as it turns out the correlation of $IMPULS1$ and $IMPULS2$ with other variables are statistically too close to zero at a 0.10 level.

32.5 Conclusion

Learning measurement models is an important causal inference task in many applied sciences. Exploratory factor analysis is a popular tool to accomplish this task, but it can be unreliable and causal assumptions are often left unclear. Better approaches are needed. Loehlin (2004) argues that while there are several approaches to automatically learn causal structure, none can be seem as competitors of exploratory factor analysis. Procedures such as the one introduced by Silva *et al.* (2006) and extended here are important steps that fill this gap.

The new procedure introduced in this work allows for impure indicators. For the ultimate goal of estimating causal relationships among latent variables, this is sometimes essential: in the example of Figure 32.2, we are not able to keep both latent variables in our model if we use the method of Silva *et al.* (2006), which requires three pure indicators per latent variable. In other cases, the advantage is statistical: in the example of Figure 32.5, by keeping more indicators we have more data that can be used to better estimate the measurement model and the corresponding causal parameters connecting the latent variables. As future work, we plan to perform a full theoretical and practical study of the advantages of discovering impure measurement models as a way of obtaining better estimates of latent causal effects.

The inclusion of impure indicators is an important step to make such approaches more generally applicable. As hinted in our discussion, other identification results to confirm or remove edges can be further developed. Higher-order constraints in the covariance matrix, besides tetrad constraints, are yet to be exploited (Sullivant and Talaska, 2008). Exploring the higher-order moments of the observed distribution (i.e. not only the covariance matrix) has been a successful approach to identify the causal structure of linear models (Shimizu *et al.*, 2006), but how to adapt them to discover a measurement model is still unclear. Finally, some progress on allowing for nonlinearities has been made (Silva and Scheines, 2005), but more robust statistical procedures and further identification results are necessary.

Acknowledgements

The author would like to thank two anonymous reviewers for sharp and detailed comments that improved the content and presentation of this chapter.

References

Bartholomew, D. and Knott, M. (1999). *Latent Variable Models and Factor Analysis.* London: Arnold Publishers.

Bengio, Y. and Cun, Y. Le (2007). Scaling learning algorithms towards AI. *Large Scale Kernel Machines.* 321–329, MIT Press, Boston, MA.

Bollen, K. (1989). *Structural Equations with Latent Variables.* New John Wiley & Sons.

Carroll, R., Ruppert, D., and Stefanski, L. (1995). *Measurement Error in Nonlinear Models.* L. Chapman & Hall.

Hinton, G. and Salakhutdinov, R. (2006). Reducing the dimensionality of data with neural networks. *Science*, **313**(5786), 504–507.

Leplin, J. (2004). A theory's predictive success can warrant belief in the unobservable entities it postulates. In C. Hitchcock (Ed.), *Contemporary Debates in Philosophy of Science*, 117–132. Blackwell Publishing, Makden, MA.

Loehlin, J. (2004). *Latent Variable Models: An Introduction to Factor, Path and Structural Equation Analysis.* Lawrence Erlbaum.

Pearl, J. (2000). *Causality: Models, Reasoning and Inference.* Cambridge University Press.

Richardson, T. and Spirtes, P. (2002). Ancestral graph Markov models. *Annals of Statistics*, **30**, 962–1030.

Schwarz, G. (1978). Estimating the dimension of a model. *Annals of Statistics*, 6(2), 461–464.

Shimizu, S., Hoyer, P., Hyvärinen, A., and Kerminen, Antti (2006). A linear non-Gaussian acyclic model for causal discovery. *Journal of Machine Learning Research*, **7**, 2003–2030.

Silva, R. (2006). Principled selection of impure measures for consistent learning of linear latent variable models. *NIPS Workshop on Causality and Feature Selection*.

Silva, R. and Scheines, R. (2005). New d-separation identification results for learning continuous latent variable models. *Proceedings of the 22nd Interational Conference in Machine Learning*.

Silva, R., Scheines, R., Glymour, C., and Spirtes, P. (2006). Learning the structure of linear latent variable models. *Journal of Machine Learning Research*, **7**, 191–246.

Spirtes, P., Glymour, C., and Scheines, R. (2000). *Causation, Prediction and Search*. Cambridge University Press.

Sullivant, S. and Talaska, K. (2008). Trek separation for Gaussian graphical models. *arxiv::0812.1938*.

·33·

The structural theory of causation

Judea Pearl

Abstract

This chapter presents a general theory of causation based on the Structural Causal Model (SCM) described by Pearl (2000*a*). The theory subsumes and unifies current approaches to causation, including graphical, potential outcome, probabilistic, decision analytical, and structural equation models, and provides both a mathematical foundation and a friendly calculus for the analysis of causes and counterfactuals. In particular, the chapter demonstrates how the theory engenders a coherent methodology for inferring (from a combination of data and assumptions) answers to three types of causal queries: (1) queries about the effects of potential interventions, (2) queries about probabilities of counterfactuals, and (3) queries about direct and indirect effects.

33.1 Introduction

Twentieth-century science has witnessed a lingering tension between the questions that researchers wish to ask and the language in which they were trained – statistics.

The research questions that motivate most studies in the health, social and behavioural sciences are not statistical but causal in nature. For example, what is the efficacy of a given drug in a given population? Whether data can prove an employer guilty of hiring discrimination? What fraction of past crimes could have been avoided by a given policy? What was the cause of death of a given individual, in a specific incident? These are causal questions because they require some knowledge of the data-generating process; they cannot be computed from distributions alone.

Any conception of causation worthy of the title 'theory' must be able to (1) represent these questions in a formal language, (2) provide a precise language for communicating assumptions under which the questions need to be answered, (3) provide a systematic way of answering at least some of these questions and labelling others 'unanswerable', and (4) provide a method of determining what assumptions or new measurements would be needed to answer the 'unanswerable' questions.[1]

[1] For example, a theory may conclude that the information at hand is not sufficient for determining the efficacy of a drug unless certain assumptions are deemed plausible, or unless

A 'general theory' of causation should do more. In addition to embracing *all* questions judged to have causal character, a general theory must also *subsume* any other theory or method that scientists have found useful in exploring the various aspects of causation, be they epistemic, methodological or practical. In other words, any alternative theory need to evolve as a special case of the 'general theory' when restrictions are imposed on either the model, the type of assumptions admitted, or the language in which those assumptions are cast.

This paper presents a theory of causation that satisfies the criteria above. It is based on the Structural Causal Model (SCM) developed by Pearl (1995; 2000*a*) which combines features of the structural equation models (SEM) used in economics (Haavelmo, 1943) and social science (Duncan, 1975), the potential-outcome notation of Neyman (1923) and Rubin (1974), and the graphical models developed for probabilistic reasoning (Pearl, 1988; Lauritzen, 1996) and causal analysis (Spirtes *et al.*, 2000; Pearl, 2000*a*). The theory presented forms a coherent whole that supercedes the sum of its parts.

Although the basic elements of SCM were introduced in the mid-1990s (Pearl, 1995), and have been adapted warmly by epidemiologists (Greenland *et al.*, 1999; Glymour and Greenland, 2008), statisticians (Cox and Wermuth, 2004; Lauritzen, 2001), and social scientists (Morgan and Winship, 2007), its potentials as a comprehensive theory of causation are yet to be fully utilized. Some have congratulated the SCM for generalizing econometric models from linear to non-parametric analysis (Heckman, 2008), some have marvelled at the clarity and transparency of the graphical representation (Greenland and Brumback, 2002), others praised the flexibility of the $do(x)$ operator (Hitchcock, 2001; Lindley, 2002; Woodward, 2003) and, naturally, many have used the SCM to weed out myths and misconceptions from outdated traditions (Meek and Glymour, 1994; Greenland *et al.*, 1999; Cole and Hernán, 2002; Arah, 2008; Shrier, 2009; Pearl, 2009*b*) Still, the more profound contributions of SCM, those stemming from its role as a comprehensive theory of causation, have not been fully explicated. These include:

1. The unification of the graphical, potential outcome, structural equations, decision analytical (Dawid, 2002), interventional (Woodward, 2003), sufficient component (Rothman, 1976) and probabilistic approaches to causation; with each approach viewed as a restricted special aspect of the SCM.
2. The axiomatization and algorithmization of counterfactual expressions.
3. Defining and identifying joint probabilities of counterfactual statements.
4. Reducing the evaluation of actions and policies to algorithmic level of analysis.

data from a specific experimental study were made available. Such conclusion constitutes a valid 'solution', provided no better solution exists.

5. Solidifying the mathematical foundations of the potential-outcome model, and formulating the counterfactual foundations of structural equation models.
6. Demystifying enigmatic notions such as 'confounding', 'ignorability', 'exchangeability', 'superexogeneity' and others, which have emerged from 'black-box' approaches to causation.
7. Providing a transparent language for communicating causal assumptions and defining causal problems.

This chapter presents the main features of the structural theory by, first, contrasting causal analysis with standard statistical analysis (Section 33.2), second, presenting a friendly formalism for counterfactual analysis, within which most (if not all) causal questions can be formulated and resolved (Sections 33.3 and 33.4) and, finally, contrasting the structural theory with two other frameworks: probabilistic causation (Section 33.5) and the Neyman–Rubin potential-outcome model (Section 33.6). The analysis will be demonstrated by attending to three types of queries: (1) queries about the effect of potential interventions (Section 33.3.1 and 33.3.2), (2) queries about counterfactuals (Section 33.3.3) and (3) queries about direct and indirect effects (Section 33.4).

33.2 From statistical to causal analysis: Distinctions and barriers

33.2.1 The basic distinction: Coping with change

The aim of standard statistical analysis, typified by regression, estimation, and hypothesis testing techniques, is to assess parameters of a distribution from samples drawn of that distribution. With the help of such parameters, one can infer associations among variables, estimate the likelihood of past and future events, as well as update the likelihood of events in light of new evidence or new measurements. These tasks are managed well by standard statistical analysis so long as experimental conditions remain the same. Causal analysis goes one step further; its aim is to infer not only the likelihood of events under static conditions, but also the dynamics of events under *changing conditions*, for example, changes induced by treatments or external interventions.

This distinction implies that causal and statistical concepts do not mix. There is nothing in the joint distribution of symptoms and diseases to tell us that curing the former would or would not cure the latter. More generally, there is nothing in a distribution function to tell us how that distribution would differ if external conditions were to change – say from observational to experimental setup – because the laws of probability theory do not dictate how one property of a distribution ought to change when another property is modified. This information must be provided by causal assumptions which identify relationships that remain invariant when external conditions change.

These considerations imply that the slogan 'correlation does not imply causation' can be translated into a useful principle: one cannot substantiate causal claims from associations alone, even at the population level – behind every causal conclusion there must lie some causal assumption that is not testable in observational studies.[2]

33.2.2 Formulating the basic distinction

A useful demarcation line that makes the distinction between associational and causal concepts crisp and easy to apply, can be formulated as follows. An associational concept is any relationship that can be defined in terms of a joint distribution of observed variables, and a causal concept is any relationship that cannot be defined from the distribution alone. Examples of associational concepts are: correlation, regression, dependence, conditional independence, likelihood, collapsibility, propensity score, risk ratio, odd ratio, marginalization, conditionalization, 'controlling for', and so on. Examples of causal concepts are: randomization, influence, effect, confounding, 'holding constant', disturbance, spurious correlation, faithfulness/stability, instrumental variables, intervention, explanation, attribution, and so on. The former can, while the latter cannot be defined in term of distribution functions.

This demarcation line is extremely useful in causal analysis for it helps investigators to trace the assumptions that are needed for substantiating various types of scientific claims. Every claim invoking causal concepts must rely on some premises that invoke such concepts; it cannot be inferred from, or even defined in terms statistical associations alone.

33.2.3 Ramifications of the basic distinction

This principle has far reaching consequences that are not generally recognized in the standard statistical literature. Many researchers, for example, are still convinced that confounding is solidly founded in standard, frequentist statistics, and that it can be given an associational definition saying (roughly): 'U is a potential confounder for examining the effect of treatment X on outcome Y when both U and X and U and Y are not independent'. That this definition and all its many variants must fail (Pearl, 2000*a*, Section 6.2)[3] is obvious from the demarcation line above; if confounding were definable in terms

[2] The assumption of 'faithfulness' or 'stability' as defined in the 'causal discovery' literature (Spirtes, *et al.* 2000; Pearl, 2000*a*, chapter 2) is likewise a causal assumption, albeit a genetic one, for it restricts any causal model from generating data that hide the structure of the model (e.g. by cancellation).

[3] For example, any intermediate variable U on a causal path from X to Y satisfies this definition, without confounding the effect of X on Y.

of statistical associations, we would have been able to identify confounders from features of non-experimental data, adjust for those confounders and obtain unbiased estimates of causal effects. This would have violated our golden rule: behind any causal conclusion there must be some causal assumption, untested in observational studies. Hence the definition must be false. Therefore, to the bitter disappointment of generations of epidemiologist and social science researchers, confounding bias cannot be detected or corrected by statistical methods alone; one must make some judgmental assumptions regarding causal relationships in the problem before an adjustment (e.g. by stratification) can safely correct for confounding bias.

Another ramification of the sharp distinction between associational and causal concepts is that any mathematical approach to causal analysis must acquire new notation for expressing causal relations – probability calculus is insufficient. To illustrate, the syntax of probability calculus does not permit us to express the simple fact that 'symptoms do not cause diseases', let alone draw mathematical conclusions from such facts. All we can say is that two events are dependent – meaning that if we find one, we can expect to encounter the other, but we cannot distinguish statistical dependence, quantified by the conditional probability $P(disease|symptom)$ from causal dependence, for which we have no expression in standard probability calculus. Scientists seeking to express causal relationships must therefore supplement the language of probability with a vocabulary for causality, one in which the symbolic representation for the relation 'symptoms cause disease' is distinct from the symbolic representation of 'symptoms are associated with disease'.

33.2.4 Two mental barriers: Untested assumptions and new notation

The preceding two requirements: (1) to commence causal analysis with untested,[4] theoretically or judgmentally based assumptions, and (2) to extend the syntax of probability calculus, constitute the two main obstacles to the acceptance of causal analysis among statisticians and among professionals with traditional training in statistics.

Associational assumptions, even untested, are testable in principle, given sufficiently large sample and sufficiently fine measurements. Causal assumptions, in contrast, cannot be verified even in principle, unless one resorts to experimental control. This difference stands out in Bayesian analysis. Though the priors that Bayesians commonly assign to statistical parameters are untested quantities, the sensitivity to these priors tends to diminish with increasing sample size. In contrast, sensitivity to prior causal assumptions, say that treatment does not change gender, remains substantial regardless of sample size.

[4] By 'untested' I mean untested using frequency data in non-experimental studies.

This makes it doubly important that the notation we use for expressing causal assumptions be meaningful and unambiguous so that one can clearly judge the plausibility or inevitability of the assumptions articulated. Statisticians can no longer ignore the mental representation in which scientists store experiential knowledge, since it is this representation, and the language used to access it that determine the reliability of the judgments upon which the analysis so crucially depends.

How does one recognize causal expressions in the statistical literature? Those versed in the potential-outcome notation (Neyman, 1923; Rubin, 1974; Holland, 1988), can recognize such expressions through the subscripts that are attached to counterfactual events and variables, e.g. $Y_x(u)$ or Z_{xy}. (Some authors use parenthetical expressions, e.g. $Y(0)$, $Y(1)$, $Y(x, u)$ or $Z(x, y)$.) The expression $Y_x(u)$, for example, stands for the value that outcome Y would take in individual u, had treatment X been at level x. If u is chosen at random, Y_x is a random variable, and one can talk about the probability that Y_x would attain a value y in the population, written $P(Y_x = y)$ (see Section 33.6 for semantics). Alternatively, Pearl (1995) used expressions of the form $P(Y = y|set(X = x))$ or $P(Y = y|do(X = x))$ to denote the probability (or frequency) that event $(Y = y)$ would occur if treatment condition $X = x$ were enforced uniformly over the population.[5] Still a third notation that distinguishes causal expressions is provided by graphical models, where the arrows convey causal directionality.[6]

However, few have taken seriously the textbook requirement that any introduction of new notation must entail a systematic definition of the syntax and semantics that governs the notation. Moreover, in the bulk of the statistical literature before 2000, causal claims rarely appear in the mathematics. They surface only in the verbal interpretation that investigators occasionally attach to certain associations, and in the verbal description with which investigators justify assumptions. For example, the assumption that a covariate not be affected by a treatment, a necessary assumption for the control of confounding (Cox, 1958, p. 48), is expressed in plain English, not in a mathematical expression.

Remarkably, though the necessity of explicit causal notation is now recognized by many academic scholars, the use of such notation has remained

[5] Clearly, $P(Y = y|do(X = x))$ is equivalent to $P(Y_x = y)$, This is what we normally assess in a controlled experiment, with X randomized, in which the distribution of Y is estimated for each level x of X. Still, the former can be defined without resorting to counterfactual notation (Pearl, 2000*a*, pp. 23–4) to the delight of those who prefer to deny mathematical notation to any assertion that is not experimentally testable in isolation (Dawid, 2002).

[6] These notational clues should be useful for detecting inadequate definitions of causal concepts; any definition of confounding, randomization or instrumental variables that is cast in standard probability expressions, void of graphs, counterfactual subscripts or $do(*)$ operators, can safely be discarded as inadequate.

enigmatic to most rank and file researchers, and its potentials still lay grossly underutilized in the statistics based sciences. The reason for this, can be traced to the unfriendly semi-formal way in which causal analysis has been presented to the research community, resting primarily on the restricted paradigm of controlled randomized trials advanced by Rubin (1974).

The next section provides a conceptualization that overcomes these mental barriers; it offers both a friendly mathematical machinery for cause–effect analysis and a formal foundation for counterfactual analysis.

33.3 Structural Causal Models (SCM) and the language of diagrams

33.3.1 Semantics: Causal effects and counterfactuals

How can one express mathematically the common understanding that symptoms do not cause diseases? The earliest attempt to formulate such relationship mathematically was made in the 1920s by the geneticist Sewall Wright (1921), who used a combination of equations and graphs. For example, if X stands for a disease variable and Y stands for a certain symptom of the disease, Wright would write a linear equation:

$$y = \beta x + u_Y \qquad (33.1)$$

where x stands for the level (or severity) of the disease, y stands for the level (or severity) of the symptom, and u_Y stands for all factors, other than the disease in question, that could possibly affect Y. In interpreting this equation one should think of a physical process whereby Nature *examines* the values of x and u_Y and, accordingly, *assigns* variable Y the value $y = \beta x + u_Y$. Similarly, to 'explain' the occurrence of disease X, one could write $x = u_X$, where U_X stands for all factors affecting X.

To express the directionality inherent in this process, Wright augmented the equation with a diagram, later called a 'path diagram', in which arrows are drawn from (perceived) causes to their (perceived) effects and, more importantly, the absence of an arrow makes the empirical claim that Nature assigns values to one variable while ignoring the other. In Figure 33.1, for example, the absence of arrow from Y to X represent the claim that symptom Y is not among the factors U_X which affect disease X.

The variables U_X and U_Y are called 'exogenous'; they represent observed or unobserved background factors that the modeller decides to keep unexplained, that is, factors that influence but are not influenced by the other variables (called 'endogenous') in the model.

If correlation is judged possible between two exogenous variables, U_Y and U_X, it is customary to connect them by a dashed double arrow, as shown in Figure 33.1(b).

$$x = u_X$$
$$y = \beta x + u_Y$$

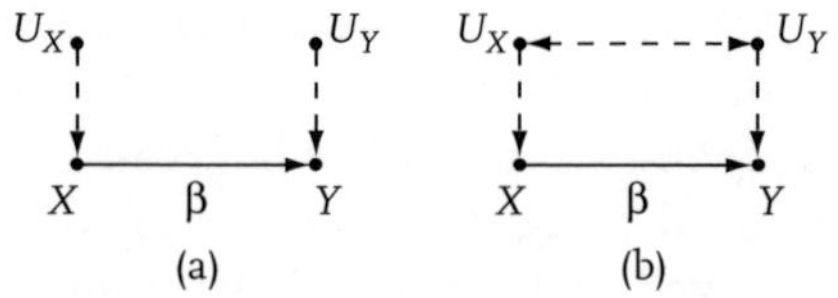

Fig. 33.1 A simple structural equation model, and its associated diagrams. Unobserved exogenous variables are connected by dashed arrows.

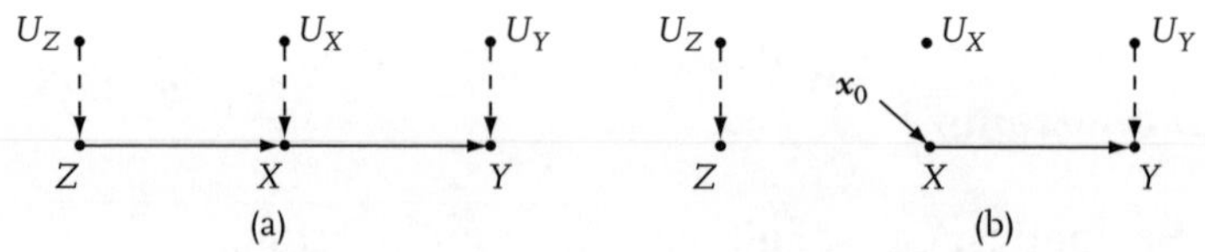

Fig. 33.2 (a) The diagram associated with the structural model of equation (33.2). (b) The diagram associated with the modified model, M_{x_0}, of equation (33.3), representing the intervention $do(X = x_0)$.

To summarize, path diagrams encode causal assumptions via missing arrows, representing claims of zero influence, and missing double arrows (e.g. between U_X and U_Y), representing the assumption $Cov(U_Y, U_X) = 0$. Note that, despite its innocent appearance in associational vocabulary, the latter assumption is causal, not statistical, for it cannot be confirmed or denied from the joint distribution of observed variables, in case the U's are unobservable.

The generalization to nonlinear systems of equations is straightforward. For example, the non-parametric interpretation of the diagram of Figure 33.2(a) corresponds to a set of three functions, each corresponding to one of the observed variables:

$$\begin{aligned} z &= f_Z(u_Z) \\ x &= f_X(z, u_X) \\ y &= f_Y(x, u_Y) \end{aligned} \tag{33.2}$$

where U_Z, U_X and U_Y are assumed to be jointly independent but, otherwise, arbitrarily distributed.

Remarkably, unknown to most economists and pre-2000 philosophers,[7] structural equation models provide a formal interpretation and symbolic machinery for analysing counterfactual relationships of the type: 'Y would

[7] Connections between structural equations and a restricted class of counterfactuals were recognized by Simon and Rescher (1966). These were generalized by Balke and Pearl (1995) who used modified models to permit counterfactual conditioning on dependent variables. This development seems to have escaped Collins *et al.* (2004).

be y had X been x in situation $U = u$', denoted $Y_x(u) = y$. Here U represents the vector of all exogenous variables.[8]

The key idea is to interpret the phrase 'had X been x_0' as an instruction to modify the original model and replace the equation for X by a constant x_0, yielding the sub-model.

$$\begin{aligned} z &= f_Z(u_Z) \\ x &= x_0 \\ y &= f_Y(x, u_Y) \end{aligned} \tag{33.3}$$

the graphical description of which is shown in Figure 33.2(b).

This replacement permits the constant x_0 to differ from the actual value of X (namely $f_X(z, u_X)$) without rendering the system of equations inconsistent, thus yielding a formal interpretation of counterfactuals in multi-stage models, where the dependent variable in one equation may be an independent variable in another (Balke and Pearl, 1994*ab*; Pearl, 2000*b*). For example, to compute $E(Y_{x_0})$, the expected effect of *setting* X to x_0 (also called the *average causal effect* of X on Y, denoted $E(Y|do(x_0))$ or, generically, $E(Y|do(x))$), we solve equation (33.3) for Y in terms of the exogenous variables, yielding $Y_{x_0} = f_Y(x_0, u_Y)$, and average over U_Y. It is easy to show that in this simple system, the answer can be obtained without knowing the form of the function $f_Y(x, u_Y)$ or the distribution $P(u_Y)$. The answer is given by:

$$E(Y_{x_0}) = E(Y|do(X = x_0) = E(Y|x_0)$$

which is computable from the distribution $P(x, y, z)$, hence estimable from observed samples of $P(x, y, z)$. This result hinges on the assumption that U_Z, U_X, and U_Y are mutually independent and on the topology of the graph (e.g. that there is no direct arrow from Z to Y).

In general, it can be shown (Pearl, 2000*a*, Chapter 3) that, whenever the graph is Markovian (i.e. acyclic with independent exogenous variables) the post-interventional distribution $P(Y = y|do(X = x))$ is given by the following expression:

$$P(Y = y|do(X = x)) = \sum_t P(y|t, x) P(t) \tag{33.4}$$

where T is the set of direct causes of X (also called 'parents') in the graph. Again, we see that all factors on the right-hand side are estimable from the distribution P of observed variables and, hence, the counterfactual probability $P(Y_x = y)$ is estimable with mere partial knowledge of the generating

[8] Because $U = u$ may contain detailed information about a situation or an individual, $Y_x(u)$ is related to what philosophers called 'token causation', while $P(Y_x = y | Z = z)$ characterizes 'Type causation', that is, the tendency of X to influence Y in a subpopulation characterized by $Z = z$.

process – the topology of the graph and independence of the exogenous variables is all that is needed.

When some variables in the graph (e.g. the parents of X) are unobserved, we may not be able to learn (or 'identify' as it is called) the post-intervention distribution $P(y|do(x))$ by simple conditioning, and more sophisticated methods would be required. Likewise, when the query of interest involves several hypothetical worlds simultaneously, e.g. $P(Y_x = y, Y_{x'} = y')$,[9] the Markovian assumption may not suffice for identification and additional assumptions, touching on the form of the data-generating functions (e.g. monotonicity) may need to be invoked. These issues will be discussed in Sections 33.3.3 and 33.6.

This interpretation of counterfactuals, cast as solutions to modified systems of equations, provides the conceptual and formal link between structural equation models, used in economics and social science and the Neyman–Rubin potential-outcome framework to be discussed in Section 33.6. But first we discuss two long-standing problems that have been completely resolved in purely graphical terms, without delving into algebraic techniques.

33.3.2 Confounding and causal effect estimation

The central target of most studies in the social and health sciences is the elucidation of cause–effect relationships among variables of interests, for example, treatments, policies, preconditions and outcomes. While good statisticians have always known that the elucidation of causal relationships from observational studies must be shaped by assumptions about how the data were generated, the relative roles of assumptions and data, and ways of using those assumptions to eliminate confounding bias have been a subject of much controversy. The structural framework of Section 33.3.1 puts these controversies to rest.

Covariate selection: The back-door criterion

Consider an observational study where we wish to find the effect of X on Y, for example, treatment on response, and assume that the factors deemed relevant to the problem are structured as in Figure 33.3; some are affecting the response, some are affecting the treatment and some are affecting both treatment and response. Some of these factors may be unmeasurable, such as genetic trait or life style, others are measurable, such as gender, age, and salary level. Our problem is to select a subset of these factors for measurement and adjustment, namely, that if we compare treated vs. untreated subjects having the same values of the selected factors, we get the correct treatment effect in that subpopulation of subjects. Such a set of factors is called a 'sufficient set' or a set 'appropriate for adjustment'. The problem of defining a sufficient set, let

[9] Read: The probability that Y would be y if X were x and y' if X were x'.

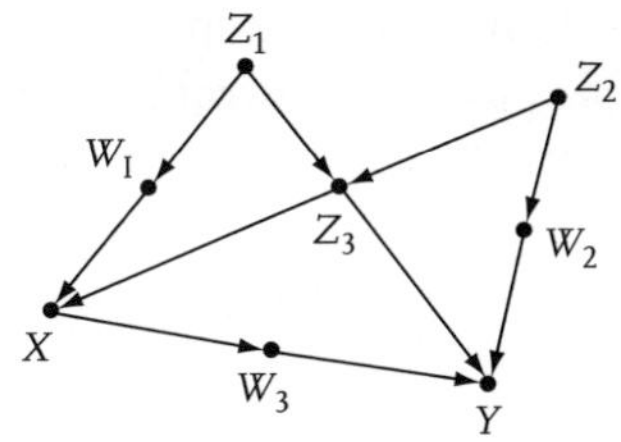

Fig. 33.3 Graphical model illustrating the back-door criterion for identifying the causal effect of X on Y. Error terms are not shown explicitly.

alone finding one, has baffled epidemiologists and social science for decades (see Greenland *et al.*, 1999; Pearl, 1998, 2003*b* for reviews).

The following criterion, named 'back-door' by Pearl (1993*a*), settles this problem by providing a graphical method of selecting a sufficient set of factors for adjustment. It states that a set S is appropriate for adjustment if two conditions hold:

1. No element of S is a descendant of X.
2. The elements of S 'block' all 'back-door' paths from X to Y, namely all paths that end with an arrow pointing to X.[10]

Based on this criterion we see, for example, that the sets $\{Z_1, Z_2, Z_3\}$, $\{Z_1, Z_3\}$, and $\{W_2, Z_3\}$, are each sufficient for adjustment, because each blocks all back-door paths between X and Y. The set $\{Z_3\}$, however, is not sufficient for adjustment because, as explained above, it does not block the path $X \leftarrow W_1 \leftarrow Z_1 \rightarrow Z_3 \leftarrow Z_2 \rightarrow W_2 \rightarrow Y$.

The implication of finding a sufficient set S is that, stratifying on S is guaranteed to remove all confounding bias relative the causal effect of X on Y. In other words, it renders the causal effect of X on Y estimable, via

$$P(Y = y|do(X = x)) = \sum_s P(Y = y|X = x, S = s)\,P(S = s). \qquad (33.5)$$

Since all factors on the right-hand side of the equation are estimable (e.g. by regression) from the pre-interventional data, the causal effect can likewise be estimated from such data without bias.

The back-door criterion allows us to write equation (33.5) directly, after selecting a sufficient set S from the diagram, without resorting to any algebraic manipulation. The selection criterion can be applied systematically to

[10] A set S of nodes is said to block a path p if either (i) p contains at least one arrow-emitting node that is in S, or (ii) p contains at least one collision node that is outside S and has no descendant in S. See (Pearl, 2000*a*, pp. 16–7). If S blocks *all* paths from X to Y it is said to 'd-separate X and Y' and, then, X and Y are independent given S.

diagrams of any size and shape, thus freeing analysts from judging whether 'X is conditionally ignorable given S', a formidable mental task required in the potential-outcome framework (Rosenbaum and Rubin, 1983). The criterion also enables the analyst to search for an optimal set of covariate – namely, a set S that minimizes measurement cost or sampling variability (Tian *et al.*, 1998).

General control of confounding
Adjusting for covariates is only one of many methods that permits us to estimate causal effects in non-experimental studies. A much more general identification criterion is provided by the following theorem:

Theorem 33.1. (*Tian and Pearl 2002*)
A sufficient condition for identifying the causal effect $P(y|do(x))$ is that every path between X and any of its children traces at least one arrow emanating from a measured variable.[11]

For example, if W_3 is the only observed covariate in the model of Figure 33.3, then there exists no sufficient set for adjustment (because no set of observed covariates can block the paths from X to Y through Z_3), yet $P(y|do(x))$ can nevertheless be estimated since every path from X to W_3 (the only child of X) traces either the arrow $X \rightarrow W_3$, or the arrow $W_3 \rightarrow Y$, both emanating from a measured variable (W_3). In this example, the variable W_3 acts as a 'mediating instrumental variable' (Pearl, 1993*b*; Chalak and White, 2006) and yields the estimand:

$$\begin{aligned} & P(Y = y|do(X = x)) \\ & = \sum_{w_3} P(W_3 = w_3|do(X = x))\, P(Y = y|do(W_3 = w_3)) \\ & = \sum_{w_3} P(w_3|x) \sum_{x'} P(y|w_3, x')\, P(x'). \end{aligned} \tag{33.6}$$

More recent results extend this theorem by (1) presenting a necessary and sufficient condition for identification (Shpitser and Pearl, 2006*a*), and (2) extending the condition from causal effects to any counterfactual expression (Shpitser and Pearl, 2007, 2009). The corresponding unbiased estimands for these causal quantities are readable directly from the diagram.

The mathematical derivation of causal effect estimands, like equations (33.5) and (33.6), is merely a first step toward computing quantitative estimates of those effects from finite samples, using the rich traditions of statistical estimation and machine learning. Although the estimands derived in (33.5) and (33.6) are non-parametric, this does not mean that one should

[11] Before applying this criterion, one may delete from the causal graph all nodes that are not ancestors of Y.

refrain from using parametric forms in the estimation phase of the study. For example, if the assumptions of Gaussian, zero-mean disturbances and additive interactions are deemed reasonable, then the estimand given in (33.6) can be converted to the product $E(Y|do(x)) = r_{W_3X} r_{YW_3 \cdot X} x$, where $r_{YZ \cdot X}$ is the (standardized) coefficient of Z in the regression of Y on Z and X. More sophisticated estimation techniques can be found in Rosenbaum and Rubin (1983), and Robins (1999). For example, the 'propensity score' method of Rosenbaum and Rubin (1983) was found to be quite useful when the dimensionality of the adjusted covariates is high and the data is sparse (see Pearl, 2009*a*, pp. 348–52).

It should be emphasized, however, that contrary to conventional wisdom (e.g. Rubin, 2009), propensity score methods are merely efficient estimators of the right-hand side of (33.5); they cannot be expected to reduce bias in case the set S does not satisfy the back-door criterion (Pearl, 2009*abc*).

33.3.3 Counterfactual analysis in structural models

Not all questions of causal character can be encoded in $P(y|do(x))$ type expressions, in much the same way that not all causal questions can be answered from experimental studies. For example, questions of *attribution* (also called 'causes of effects' (Dawid, 2000), e.g. I took an aspirin and my headache is gone, was it *due* to the aspirin?) or of *susceptibility* (e.g. I am a healthy nonsmoker, would I be as healthy had I been a smoker?) cannot be answered from experimental studies, and naturally, this kind of questions cannot be expressed in $P(y|do(x))$ notation.[12] To answer such questions, a probabilistic analysis of counterfactuals is required, one dedicated to the relation 'Y would be y had X been x in situation $U = u$', denoted $Y_x(u) = y$.

As noted in Section 33.3.1, the structural definition of counterfactuals involves modified models, like M_{x_0} of equation (33.3), formed by the intervention $do(X = x_0)$ (Figure 33.2b). Call the solution of Y in model M_x the *potential response* of Y to x, and denote it by the symbol $Y_x(u)$. In general, then, the formal definition of the counterfactual $Y_x(u)$ in SCM is given by (Pearl, 2000*a*, p. 98):

$$Y_x(u) \triangleq Y_{M_x}(u). \tag{33.7}$$

The quantity $Y_x(u)$ can be given experimental interpretation; it stands for the way an individual with characteristics (u) would respond, had the treatment

[12] The reason for this fundamental limitation is that no death case can be tested twice, with and without treatment. For example, if we measure equal proportions of deaths in the treatment and control groups, we cannot tell how many death cases are actually attributable to the treatment itself; it is quite possible that many of those who died under treatment would be alive if untreated and, simultaneously, many of those who survived with treatment would have died if not treated.

been x, rather than the treatment $x = f_X(u)$ actually received by that individual. In our example, since Y does not depend on v and w, we can write:

$$Y_{x_0}(u_Y, u_X, u_Z) = Y_{x_0}(u_Y) = f_Y(x_0, u_Y).$$

Clearly, the distribution $P(u_Y, u_X, u_Z)$ induces a well-defined probability on the counterfactual event $Y_{x_0} = y$, as well as on joint counterfactual events, such as '$Y_{x_0} = y$ AND $Y_{x_1} = y'$', which are, in principle, unobservable if $x_0 \neq x_1$. Thus, to answer attributional questions, such as whether Y would be y_1 if X were x_1, given that in fact Y is y_0 and X is x_0, we need to compute the conditional probability $P(Y_{x_1} = y_1 | Y = y_0, X = x_0)$ which is well defined once we know the forms of the structural equations and the distribution of the exogenous variables in the model. For example, assuming linear equations (as in Figure 33.1),

$$x = u_X, \qquad y = \beta x + u_Y,$$

the conditions $Y = y_0$ and $X = x_0$ yield $u_X = x_0$ and $u_Y = y_0 - \beta x_0$, and we can conclude that, with probability one, Y_{x_1} must take on the value: $Y_{x_1} = \beta x_1 + u_Y = \beta(x_1 - x_0) + y_0$. In other words, if X were x_1 instead of x_0, Y would increase by β times the difference $(x_1 - x_0)$. In nonlinear systems, the result would also depend on the distribution of U and, for that reason, attributional queries are generally not identifiable in nonparametric models (Pearl, 2000*a*, Chapter 9).

In general, if x and x' are incompatible then Y_x and $Y_{x'}$ cannot be measured simultaneously, and it may seem meaningless to attribute probability to the joint statement 'Y would be y if $X = x$ *and* Y would be y' if $X = x'$'.[13] Such concerns have been a source of objections to treating counterfactuals as jointly distributed random variables (Dawid, 2000). The definition of Y_x and $Y_{x'}$ in terms of two distinct submodels neutralizes these objections (Pearl, 2000*b*), since the contradictory joint statement is mapped into an ordinary event, one where the background variables satisfy both statements simultaneously, each in its own distinct submodel; such events have well defined probabilities.

The structural interpretation of counterfactuals also provides the conceptual and formal basis for the Neyman–Rubin potential-outcome framework, an approach to causation that takes a controlled randomized trial (CRT) as its starting paradigm, assuming that nothing is known to the experimenter about the science behind the data. This 'black-box' approach, which has thus far been denied the benefits of graphical or structural analyses, was developed by statisticians who found it difficult to cross the two mental barriers discussed in Section 33.2.4. Section 33.6 establishes the precise relationship between the

[13] For example, 'The probability is 80% that Joe belongs to the class of patients who will be cured if they take the drug and die otherwise.'

structural and potential-outcome paradigms, and outlines how the latter can benefit from the richer representational power of the former.

33.4 Mediation: Direct and indirect effects

33.4.1 Direct versus total effects

The causal effect we have analysed so far, $P(y|do(x))$, measures the *total* effect of a variable (or a set of variables) X on a response variable Y. In many cases, this quantity does not adequately represent the target of investigation and attention is focused instead on the direct effect of X on Y. The term 'direct effect' is meant to quantify an effect that is not mediated by other variables in the model or, more accurately, the sensitivity of Y to changes in X while all other factors in the analysis are held fixed. Naturally, holding those factors fixed would sever all causal paths from X to Y with the exception of the direct link $X \rightarrow Y$, which is not intercepted by any intermediaries.

A classical example of the ubiquity of direct effects involves legal disputes over race or sex discrimination in hiring. Here, neither the effect of sex or race on applicants' qualification nor the effect of qualification on hiring are targets of litigation. Rather, defendants must prove that sex and race do not *directly* influence hiring decisions, whatever indirect effects they might have on hiring by way of applicant qualification.

Another example concerns the identification of neural pathways in the brain or the structural features of protein-signalling networks in molecular biology (Brent and Lok, 2005). Here, the decomposition of effects into their direct and indirect components carries theoretical scientific importance, for it predicts behavior under a rich variety of hypothetical interventions.

In all such examples, the requirement of holding the mediating variables fixed must be interpreted as (hypothetically) setting the intermediate variables to constants by physical intervention, not by analytical means such as selection, conditioning, or adjustment. For example, it will not be sufficient to measure the association between gender (X) and hiring (Y) for a given level of qualification Z, because, by conditioning on the mediator Z, we may create spurious associations between X and Y even when there is no direct effect of X on Y. This can easily be illustrated in the model $X \rightarrow Z \leftarrow U \rightarrow Y$, where X has no direct effect on Y. Physically holding Z constant should eliminate the association between X and Y, as can be seen by deleting all arrows entering Z. But if we were to condition on Z, a spurious association would be created through U (unobserved) that might be construed as a direct effect of X on Y.

Using the $do(x)$ notation, and focusing on differences of expectations, this leads to a simple definition of *controlled direct effect*:

$$CDE \triangleq E(Y|do(x'), do(z)) - E(Y|do(x), do(z))$$

or, equivalently, using counterfactual notation:

$$CDE \triangleq E(Y_{x'z}) - E(Y_{xz})$$

where Z is any set of mediating variables that intercept all indirect paths between X and Y. Graphical identification conditions for expressions of the type $E(Y|do(x), do(z_1), do(z_2), \ldots, do(z_k))$ were derived by Pearl and Robins (1995) (see Pearl 2000*a*, Chapter 4) and invoke sequential application of the back-door conditions discussed in Section 33.3.2.

33.4.2 Natural direct effects

In linear systems, the direct effect is fully specified by the path coefficient attached to the link from X to Y; therefore, the direct effect is independent of the values at which we hold Z. In nonlinear systems, those values would, in general, modify the effect of X on Y and thus should be chosen carefully to represent the target policy under analysis. For example, it is not uncommon to find employers who prefer males for the high-paying jobs (i.e. high z) and females for low-paying jobs (low z).

When the direct effect is sensitive to the levels at which we hold Z, it is often meaningful to define the direct effect relative to some 'natural' baseline level that may vary from individual to individual and represents the level of Z just before the change in X. Conceptually, we can define the average direct effect $DE_{x,x'}(Y)$ as the expected change in Y induced by changing X from x to x' while keeping all mediating factors constant at whatever value they *would have obtained* under $do(x)$. This hypothetical change, which Robins and Greenland (1991) called 'pure' and Pearl (2001) called 'natural', mirrors what lawmakers instruct us to consider in race or sex discrimination cases: 'The central question in any employment-discrimination case is whether the employer would have taken the same action had the employee been of a different race (age, sex, religion, national origin, etc.) and everything else had been the same'. (In *Carson versus Bethlehem Steel Corp.*, 70 FEP Cases 921, 7th Cir. (1996)).

Extending the subscript notation to express nested counterfactuals (Pearl, 2001) gave the following definition for the 'natural direct effect':

$$DE_{x,x'}(Y) \triangleq E(Y_{x',Z_x}) - E(Y_x). \qquad (33.8)$$

Here, Y_{x',Z_x} represents the value that Y would attain under the operation of setting X to x' and, simultaneously, setting Z to whatever value it would have obtained under the original setting $X = x$. We see that $DE_{x,x'}(Y)$, the natural direct effect of the transition from x to x', involves probabilities of *nested counterfactuals* and cannot be written in terms of the $do(x)$ operator. Therefore, the natural direct effect cannot in general be identified, even with the help of ideal, controlled experiments, for we cannot rerun history and re-condition on an action actually taken (see footnote 12). Pearl (2001) has nevertheless shown that, if certain assumptions of 'no confounding' are deemed valid,[14] the natural direct effect can be reduced to

$$DE_{x,x'}(Y) = \sum_{z} [E(Y|do(x', z)) - E(Y|do(x, z))] P(z|do(x)). \tag{33.9}$$

The intuition is simple; the natural direct effect is the weighted average of the controlled direct effect, using the causal effect $P(z|do(x))$ as a weighing function.

In particular, expression (33.9) is both valid and identifiable in Markovian models, where each term on the right can be reduced to a 'do-free' expression using equation (33.4).

33.4.3 Natural indirect effects

Remarkably, the definition of the natural direct effect (33.8) can easily be turned around and provide an operational definition for the *indirect effect* – a concept shrouded in mystery and controversy, because it is impossible, using the $do(x)$ operator, to disable the direct link from X to Y so as to let X influence Y solely via indirect paths.

The natural indirect effect, IE, of the transition from x to x' is defined as the expected change in Y affected by holding X constant, at $X = x$, and changing Z to whatever value it would have attained had X been set to $X = x'$. Formally, this reads (Pearl, 2001):

$$IE_{x,x'}(Y) \triangleq E[(Y_{x,Z_{x'}}) - E(Y_x)], \tag{33.10}$$

which is almost identical to the direct effect (equation (33.8)) save for exchanging x and x'.

Indeed, it can be shown that, in general, the total effect TE of a transition is equal to the *difference* between the direct effect of that transition and the indirect effect of the reverse transition. Formally,

$$TE_{x,x'}(Y) \triangleq E(Y_{x'} - Y_x) = DE_{x,x'}(Y) - IE_{x',x}(Y). \tag{33.11}$$

In linear systems, where reversal of transitions amounts to negating the signs of their effects, we have the standard additive formula

$$TE_{x,x'}(Y) = DE_{x,x'}(Y) + IE_{x,x'}(Y). \tag{33.12}$$

Since each term above is based on an independent operational definition, this equality constitutes a formal justification for the additive formula used routinely in linear systems.

[14] One sufficient condition is that $Z_x \perp\!\!\!\perp Y_{x',z} | W$ holds for some set W of measured covariates. See details and graphical criteria in Pearl (2001, 2005) and in Petersen *et al.* (2006).

Note that, although it cannot be expressed in do-notation, the indirect effect has clear policy-making implications. For example: in the hiring discrimination context, a policy maker may be interested in predicting the gender mix in the work force if gender bias is eliminated and all applicants are treated equally – say, the same way that males are currently treated. This quantity will be given by the indirect effect of gender on hiring, mediated by factors such as education and aptitude, which may be gender-dependent.

More generally, a policy maker may be interested in the effect of issuing a directive to a select set of subordinate employees, or in carefully controlling the routing of messages in a network of interacting agents. Such applications motivate the analysis of *path-specific effects*, that is, the effect of X on Y through a selected set of paths (Avin *et al.*, 2005).

Note that in all these cases, the policy intervention invokes the selection of signals to be sensed, rather than variables to be fixed. Pearl (2001) has suggested therefore that *signal sensing* is more fundamental to the notion of causation than *manipulation*; the latter being but a crude way of stimulating the former in experimental setup. The mantra 'No causation without manipulation' must be rejected. (See Pearl, 2009*a* [Section 11.4.5.])

It is remarkable that counterfactual quantities like DE and ID that could not be expressed in terms of $do(x)$ operators, and appear therefore void of empirical content, can, under certain conditions be estimated from empirical studies. A general characterization of those conditions, including a complete identification of ETT, is given by Shpitser and Pearl (2007, 2009).

Additional examples of this 'marvel of formal analysis' are given by Pearl (2009*a*, Chapters 7, 9, and 11). It constitutes an unassailable argument in defence of counterfactual analysis, as expressed in Pearl (2000*b*) against the stance of Dawid (2000).

33.5 Structural versus probabilistic causality

Probabilistic causality (PC) is a branch of philosophy that has attempted, for the past several decades, to characterize the relationship between cause and effect using the tools of probability theory (Hitchcock, 2003; Williamson, ming). Our discussion of Section 33.2 rules out any such characterization and, not surprisingly, the PC program is known mainly for the difficulties it has encountered, rather than its achievements. This section explains the main obstacle that has kept PC at bay for over half a century, and demonstrates how the structural theory of causation clarifies relationships between probabilities and causes.

33.5.1 The 'probability raising' trap

The idea that causes raise the probability of their effects has been the engine behind most of PC explorations. It is a healthy idea, solidly ensconced in intuition. We say, for example, 'reckless driving causes accidents' or 'you will fail the course because of your laziness' (Suppes, 1970), knowing quite well that the antecedents merely tend to make the consequences more likely, not absolutely certain. One would expect, therefore, that probability raising should become the defining characteristic of the relationship between a cause (C) and its effect (E). Alas, though perfectly valid, this intuition cannot be expressed using the tools of probabilities; the relationship 'raises the probability of' is counterfactual (or manipulative) in nature, and cannot, therefore, be captured in the language of probability theory.

The way philosophers tried to capture this relationship, using inequalities such as[15]

$$P(E|C) > P(E) \tag{33.13}$$

was misguided from the start – counterfactual 'raising' cannot be reduced to evidential 'raising', or 'raising by conditioning'. The correct inequality, according to the structural theory of Section 33.3, should read:

$$P(E|do(C)) > P(E) \tag{33.14}$$

where $do(C)$ stands for an external intervention that compels the truth of C. The conditional probability $P(E|C)$, as we know from Section33.3 represents a probability resulting from a passive observation of C, and rarely coincides with $P(E|do(C))$. Indeed, observing the barometer falling increases the probability of a storm coming, but does not 'cause' the storm; if the act of manipulating the barometer were to change the probability of storms, the falling barometer would qualify as a cause of storms.

Reformulating the notion of 'probability raising' within the calculus of *do*-operators resolves the difficulties that PC has encountered in the past half-century.[16] Two such difficulties are worth noting here, for they can be resolved by the analysis of Section 33.3.

33.5.2 The mystery of 'background context'

Recognizing that the basic inequality $P(E|C) > P(E)$ may yield paradoxical results in the presence of confounding factors (e.g. the atmospheric pressure in the example above), philosophers have modified the inequality by conditioning on a background factor K, yielding the criterion: $P(E|C, K = k) > P(E|K = k)$ where K consists on a set of variables capable of creating spurious

[15] Some authors write $P(E|C) > P(E|\neg C)$, which is equivalent to (33.13); the latter is easier to generalize to the non-binary case.

[16] This chapter focuses on 'type causation' namely, the tendency of the cause to bring about the effect. Token causation, also known as 'actual causation' (Pearl, 2000*a*, Chapter 10) requires heavier counterfactual machinery.

dependencies between the cause and the effect. However, the question of what variables should enter K led to speculations, controversies and fallacies.[17]

Cartwright (1983), for example, states that a factor F should enter into K if and only if F is *causally relevant* to the effect, that is, F tends to either promote or prevent E. Eells (1991) on the other hand dropped the 'only if' part and insisted on the 'if.' The correct answer, as we know from our analysis of Section 33.3, is neither Cartwright's nor Eell's; K should merely satisfy the back-door criterion of Section 33.3.2, which may or may not include variables that are causally relevant to the effect E.

The background-context debate is symptomatic of the fundamental flaw of the probabilistic causality program; the program first misrepresented the causal relation $P(E|do(C))$ by a conditional probability surrogate $P(E|C)$, and then, to escape the wrath of spurious associations, attempted to patch-up the distortion by adding remedial conditionalizations, only to end up with a contested $P(E|C, K)$. The correct strategy should have been to define 'probability raising' directly in terms of the $do(x)$ operator (or counterfactual variables Y_x), which would have yielded general and coherent results with no need for remedies.[18]

33.5.3 The epistemology of causal relevance and probability raising

The introduction of a 'causal relevance' relation into the definition of 'cause' is of course circular, for it compromises the original goal of reducing causality to purely probabilistic relations. It gave rise however to an interesting epistemological problem whose aim is not reductive but interpretative: Given that humans store experience in the form of qualitative 'causal relevance' relationships, (with variable X being 'causally relevant' to Y whenever it can influence Y in some way), we ask whether this knowledge, together with a probability function P is sufficient for determining whether event $X = x$ is a cause of event $Y = y$ in the 'probability raising' sense.[19]

The problem is interesting because it connects judgments of three different types: judgments about 'causal relevance' (R), about probabilities (P), and about cause–effect relations (CE). There is little doubt that causal-relevance relationships form part of an agent epistemic state; such relationships are implied by people's understanding of mechanisms, and how mechanisms are put together in the world around them. It is also reasonable to assume that an

[17] Conditioning on *all* factors F preceding C (Good, 1961; Suppes, 1970) would lead to counter intuitive conclusions (Pearl, 2000*a*, p. 297).

[18] Lewis (1986) proposed indeed to treat probability raising in the context of his counterfactual theory. However, lacking structural semantics, PC advocates viewed Lewis's counterfactuals as resting on shaky formal foundation 'for which we have only the beginnings of a semantics (via the device of measures over possible worlds)' (Cartwright, 1983, p. 34).

[19] This is my interpretation of Eell's (1991) epistemic consistency problem (Pearl, 2000*a*, p. 252).

agent's epistemic state contains some representation of a probability function P that summarizes facts, observations, and associations acquired by the agent, either directly or indirectly (say through hearsay, or reading scientific reports). Finally, people usually reach consensus judging whether a given event $X = x$ 'causes' event $Y = y$, and generally agree with the 'probability raising' maxim.

The epistemic question above amounts to asking whether the three types of judgments, R, P, and CE, are compatible with each other. Put differently, the question we may ask is whether CE judgments are compatible with the pair $\langle R, P\rangle$ and the probability raising maxim given in (33.14). To answer such questions we must first determine whether the pair $\langle R, P\rangle$ is sufficient for deriving inequalities of the type given in (33.14).

The structural theory of causation gives a definitive solution to this problem which reads as follows:

Given: A graph G on a set V of variables, such that there is a directed path from X to Y in V iff X is judged to be 'causally relevant' to Y.
Also given: a probability measure $P(v)$ that is compatible with G.
Problem: Decide, for a given X and Y in V, whether the probability raising inequality (33.14) holds for $C : X = x$ and $E : Y = y$, namely whether the causal effect

$$CE = P(y|do(x)) - P(y) \tag{33.15}$$

is greater than zero, given G and P.

The solution follows immediately from the identification of causal effects in Markovian models, which permits the derivation of CE from G and P, for example, by the causal effect formula of equation (33.4).

The solution is less obvious when P is defined over a proper subset W of V, where $\{V - W\}$ represents the set of unmeasured variables. The problem then reduces to that of identifying CE in semi-Markovian models such as those addressed in Theorem 33.1. Fortunately, the completeness results of Tian and Pearl (2002) and Shpitser and Pearl (2006*b*) reduce this problem to algorithmic routine on the graph G and, furthermore, they provide a guarantee that, if the algorithm fails, then any algorithm would fail, namely the causal effect of x on y does not have a unique value, given R and P.

I venture to conjecture that *every* epistemic problem concerned with the relationship between causes and probabilities is now amenable to algorithmic solution, provided that one explicates formally what is assumed and what needs to be decided.

33.5.4 Is probabilistic causality subsumed by the structural theory?

In view of the difficulties described above, it is fair to ask whether PC should be regarded as a special case of the structural theory, or, for that matter, whether it should qualify as a theory of causation by the four criteria set forth in Section 33.1. The answer is that, although PC fails to satisfy these criteria, its aspirations were to provide a formal language for causal assertions of the 'probability raising' variety. While the notation chosen for the task was inadequate, the reasoning behind most PC investigations was clearly guided by structural considerations. The introduction of a 'causal relevance' relation into the theory attests to the structural nature of that reasoning. The structural theory now permits PC investigators to re-articulate philosophical and epistemological problems in an unambiguous formal language and derive, using the notational machinery provided by the SCM, answers to pending questions in this area of inquiry. Section 33.5.3 demonstrates the benefits of this machinery; similar benefits were demonstrated in problems posed by Woodward (Pearl, 2003*a*) and Cartwright (Pearl, 2009*a*, pp. 362–5).

33.6 Comparison to the potential-outcomes framework

The primitive object of analysis in the potential-outcome framework is the unit-based response variable, denoted $Y_x(u)$, read: 'the value that outcome Y would obtain in experimental unit u, had treatment X been x' (Neyman, 1923; Rubin, 1974). Here, *unit* may stand for an individual patient, an experimental subject, or an agricultural plot. In Section 33.3.3 we saw that this counterfactual entity has the natural interpretation as representing the solution for Y in a modified system of equations, where *unit* is interpreted a vector u of background factors that characterize an experimental unit. Each structural equation model thus carries a collection of assumptions about the behavior of hypothetical units, and these assumptions permit us to derive the counterfactual quantities of interest. In the potential-outcome framework, however, no equations are available for guidance and $Y_x(u)$ is taken as primitive, that is, an undefined quantity in terms of which other quantities are defined; not a quantity that can be derived from some model. In this sense the structural interpretation of $Y_x(u)$ given in (33.7) provides the formal basis for the potential-outcome approach; the formation of the submodel M_x explicates mathematically how the hypothetical condition 'had X been x' could be realized, and what the logical consequence are of such a condition.

33.6.1 The 'black-box' or 'missing-data' paradigm

The distinct characteristic of the potential-outcome approach is that, although investigators must think and communicate in terms of undefined, hypothetical quantities such as $Y_x(u)$, the analysis itself is conducted almost entirely within the axiomatic framework of probability theory. This is accomplished, by postulating a 'super' probability function on both hypothetical and real events. If U is treated as a random variable then the value of the counterfactual $Y_x(u)$ becomes a random variable as well, denoted as Y_x. The potential-outcome analysis proceeds by treating the observed distribution $P(x_1, \ldots, x_n)$ as the marginal distribution of an augmented probability function P^* defined over both observed and counterfactual variables. Queries about causal effects (written $P(y|do(x))$ in the structural analysis) are phrased as queries about the marginal distribution of the counterfactual variable of interest, written $P^*(Y_x = y)$. The new hypothetical entities Y_x are treated as ordinary random variables; for example, they are assumed to obey the axioms of probability calculus, the laws of conditioning, and the axioms of conditional independence.

Naturally, these hypothetical entities are not entirely whimsy. They are assumed to be connected to observed variables via consistency constraints (Robins, 1986) such as

$$X = x \Longrightarrow Y_x = Y, \tag{33.16}$$

which states that, for every u, if the actual value of X turns out to be x, then the value that Y would take on if 'X *were* x' is equal to the actual value of Y. For example, a person who chose treatment x and recovered, would also have recovered if given treatment x by design. Whether additional constraints should tie the observables to the unobservables is not a question that can be answered in the potential-outcome framework, which lacks an underlying model.

The main conceptual difference between the two approaches is that, whereas the structural approach views the intervention $do(x)$ as an operation that changes the distribution but keeps the variables the same, the potential-outcome approach views the variable Y under $do(x)$ to be a different variable, Y_x, loosely connected to Y through relations such as (33.16), but remaining unobserved whenever $X \neq x$. The problem of inferring probabilistic properties of Y_x, then becomes one of 'missing-data' for which estimation techniques have been developed in the statistical literature.

Pearl (2000*a*, Chapter 7) shows, using the structural interpretation of $Y_x(u)$, that it is indeed legitimate to treat counterfactuals as jointly distributed random variables in all respects, that consistency constraints like (33.16) are automatically satisfied in the structural interpretation and, moreover, that investigators need not be concerned about any additional constraints except the following two:

$$Y_{yz} = y \text{ for all } y, \text{ subsets } Z, \text{ and values } z \text{ for } Z \tag{33.17}$$

$$X_z = x \Rightarrow Y_{xz} = Y_z \text{ for all } x, \text{ subsets } Z, \text{ and values } z \text{ for } Z. \tag{33.18}$$

Equation (33.17) ensures that the interventions $do(Y = y)$ results in the condition $Y = y$, regardless of concurrent interventions, say $do(Z = z)$, that may be applied to variables other than Y. Equation (33.18) generalizes (33.16) to cases where Z is held fixed, at z.

33.6.2 Problem formulation and the demystification of 'ignorability'

The main drawback of this black-box approach surfaces in problem formulation, namely, the phase where a researcher begins to articulate the 'science' or 'causal assumptions' behind the problem at hand. Such knowledge, as we have seen in Section 33.1, must be articulated at the onset of every problem in causal analysis – causal conclusions are only as valid as the causal assumptions upon which they rest.

To communicate scientific knowledge, the potential-outcome analyst must express assumptions as constraints on P^*, usually in the form of conditional independence assertions involving counterfactual variables. For instance, in our example of Figure 33.2(a), to communicate the understanding that the Z is randomized (hence independent of U_X and U_Y), the potential-outcome analyst would use the independence constraint $Z \perp\!\!\!\perp \{Y_{z_1}, Y_{z_2}, \ldots, Y_{z_k}\}$.[20] To further formulate the understanding that Z does not affect Y directly, except through X, the analyst would write a, so called, 'exclusion restriction': $Y_{xz} = Y_x$.

A collection of constraints of this type might sometimes be sufficient to permit a unique solution to the query of interest. For example, if one can plausibly assume that, in Figure 33.3, a set Z of covariates satisfies the conditional independence

$$Y_x \perp\!\!\!\perp X | Z \tag{33.19}$$

(an assumption termed 'conditional ignorability' by Rosenbaum and Rubin (1983)), then the causal effect $P(y|do(x)) = P^*(Y_x = y)$ can readily be evaluated to yield

$$\begin{aligned} P^*(Y_x = y) &= \sum_z P^*(Y_x = y|z) P(z) \\ &= \sum_z P^*(Y_x = y|x, z) P(z) \quad \text{(using (33.19))} \\ &= \sum_z P^*(Y = y|x, z) P(z) \quad \text{(using (33.16))} \\ &= \sum_z P(y|x, z) P(z). \end{aligned} \tag{33.20}$$

The last expression contains no counterfactual quantities (thus permitting us to drop the asterisk from P^*) and coincides precisely with the standard covariate-adjustment formula of equation (33.5).

We see that the assumption of conditional ignorability (33.19) qualifies Z as a sufficient covariate for adjustment; it is entailed indeed by the 'back-door'

[20] The notation $Y \perp\!\!\!\perp X|Z$ stands for the conditional independence relationship $P(Y = y, X = x|Z = z) = P(Y = y|Z = z) P(X = x|Z = z)$ (Dawid, 1979).

criterion of Section 33.3.2, which qualifies such covariates by tracing paths in the causal diagram.

The derivation above may explain why the potential-outcome approach appeals to mathematical statisticians; instead of constructing new vocabulary (e.g. arrows), new operators ($do(x)$) and new logic for causal analysis, almost all mathematical operations in this framework are conducted within the safe confines of probability calculus. Save for an occasional application of rule (33.18) or (33.16), the analyst may forget that Y_x stands for a counterfactual quantity – it is treated as any other random variable, and the entire derivation follows the course of routine probability exercises.

However, this mathematical orthodoxy exacts a very high cost: all background knowledge pertaining to a given problem must first be translated into the language of counterfactuals (e.g. ignorability conditions) before analysis can commence. This translation may in fact be the hardest part of the problem. The reader may appreciate this aspect by attempting to judge whether the assumption of conditional ignorability (33.19), the key to the derivation of (33.20), holds in any familiar situation, say in the experimental setup of Figure 33.2(a). This assumption reads: 'the value that Y would obtain had X been x, is independent of X, given Z'. Even the most experienced potential-outcome expert would be unable to discern whether any subset Z of covariates in Figure 33.3 would satisfy this conditional independence condition.[21] Likewise, to derive equation (33.6) in the language of potential-outcome (see Pearl (2000*a*, p. 223)), one would need to convey the structure of the chain $X \rightarrow W_3 \rightarrow Y$ using the cryptic expression: $W_{3_x} \perp\!\!\!\perp \{Y_{w_3}, X\}$, read: 'the value that W_3 would obtain had X been x is independent of the value that Y would obtain had W_3 been w_3 jointly with the value of X'. Such assumptions are cast in a language so far removed from ordinary understanding of scientific theories that, for all practical purposes, they cannot be comprehended or ascertained by ordinary mortals. As a result, researchers in the graph-less potential-outcome camp rarely use 'conditional ignorability' (33.19) to guide the choice of covariates; they view this condition as a hoped-for miracle of nature rather than a target to be achieved by reasoned design.[22]

Replacing 'ignorability' with a simple condition (i.e. back-door) in a graphical model permits researchers to understand what conditions covariates must fulfil before they eliminate bias, what to watch for and what to think about

[21] Inquisitive readers are invited to guess whether $X_z \perp\!\!\!\perp Z | Y$ holds in Figure 33.2(a).

[22] The opaqueness of counterfactual independencies explains why many researchers within the potential-outcome camp are unaware of the fact that adding a covariate to the analysis (e.g. Z_3 in Figure 33.3) may actually *increase* confounding bias. Paul Rosenbaum, for example, writes: 'there is no reason to avoid adjustment for a variable describing subjects before treatment' (Rosenbaum, 2002, p. 76). Rubin (2009) goes as far as stating that refraining from conditioning on an available measurement is 'nonscientific ad hockery' for it goes against the tenets of Bayesian philosophy (see (Pearl, 2009*bc*) for a discussion of this fallacy).

when covariates are selected, and what experiments we can do to test, at least partially, if we have the knowledge needed for covariate selection.

Aside from offering no guidance in covariate selection, formulating a problem in the potential-outcome language encounters three additional hurdles. When counterfactual variables are not viewed as byproducts of a deeper, process-based model, it is hard to ascertain whether *all* relevant counterfactual independence judgments have been articulated, whether the judgments articulated are *redundant*, or whether those judgments are *self-consistent*. The need to express, defend, and manage formidable counterfactual relationships of this type explain the slow acceptance of causal analysis among health scientists and statisticians, and why economists and social scientists continue to use structural equation models instead of the potential-outcome alternatives advocated by Angrist *et al.* (1996); Holland (1988); Sobel (1998).

On the other hand, the algebraic machinery offered by the counterfactual notation, $Y_x(u)$, once a problem is properly formalized, can be extremely powerful in refining assumptions (Angrist *et al.*, 1996), deriving consistent estimands (Robins, 1986), bounding probabilities of necessary and sufficient causation (Tian and Pearl, 2000), and combining data from experimental and non-experimental studies (Pearl, 2000*a*). Pearl (2000*a*, p. 232) presents a way of combining the best features of the two approaches. It is based on encoding causal assumptions in the language of diagrams, translating these assumptions into counterfactual notation, performing the mathematics in the algebraic language of counterfactuals (using (33.16), (33.17), and (33.18)) and, finally, interpreting the result in plain causal language. The mediation formulas derived in Section 33.4 illustrate such symbiosis.

In comparison, when the mediation problem is approached from an orthodox potential-outcome viewpoint, void of the structural guidance of equation (33.7), pardoxical results ensue (Rubin, 2004). For example, the direct effect is definable only in units absent of indirect effects. This means that a grandfather would be deemed to have no direct effect on his grandson's behaviour in families where he has had some effect on the father. This leaves us mostly with odd families, absent of grandfathers or fathers. In linear systems, to take a sharper example, the direct effect would be undefined whenever indirect paths exist from the cause to its effect. Such paradoxical conclusions underscore the wisdom, if not necessity of a symbiotic analysis, in which the counterfactual notation $Y_x(u)$ is governed by the structural semantics of the SCM.

33.7 Conclusions

Theories of causation require two ingredients that are absent from probabilistic or logical theories; a science-friendly language for articulating causal knowledge, and a mathematical machinery for processing that knowledge, combining it with data and drawing new causal conclusions about a phenomenon. This chapter introduces a general theory of causation, based on non-parametric structural equation models, that supplements statistical methods with the needed ingredients. The algebraic component of the theory coincides with the potential-outcome framework, and its graphical component embraces Wright's method of path diagrams (in its non-parametric version). When unified and synthesized, the two components offer empirical investigators a powerful and comprehensive methodology for causal inference, and a general framework for viewing other, less general approaches to causation, including probabilistic causation (Section 33.5) and the potential-outcome model (33.6).

Acknowledgments

Portions of this chapter are based on my book *Causality* (Pearl, 2000, 2nd edition 2009), and have benefited appreciably from conversations with Chris Hitchcock. This research was supported in parts by an ONR grant #N000-14-09-1-0665, and NSF grant #IIS-0914211.

References

Angrist, J.D., Imbens, G.W., and Rubin, D.B. (1996). Identification of causal effects using instrumental variables (with comments). *Journal of the American Statistical Association*, **91**(434), 444–472.

Arah, O.A. (2008). The role of causal reasoning in understanding Simpson's paradox, Lord's paradox, and the suppression effect: Covariate selection in the analysis of observational studies. *Emerging Themes in Epidemiology*, **4**, doi:10.1186/1742–7622–5–5. Online at http://www.ete-online.com/content/5/1/5.

Avin, C., Shpitser, I., and Pearl, J. (2005). Identifiability of path-specific effects. In *Proceedings of the Nineteenth International Joint Conference on Artificial Intelligence IJCAI-05*, Edinburgh, UK, pp. 357–363. Morgan-Kaufmann Publishers.

Balke, A. and Pearl, J. (1994*a*). Counterfactual probabilities: Computational methods, bounds, and applications. In *Uncertainty in Artificial Intelligence 10* (ed. R. L. de Mantaras and D. Poole), pp. 46–54. Morgan Kaufmann, San Mateo, CA.

Balke, A. and Pearl, J. (1994*b*). Probabilistic evaluation of counterfactual queries. In *Proceedings of the Twelfth National Conference on Artificial Intelligence*, Volume I, pp. 230–237. MIT Press, Menlo Park, CA.

Balke, A. and Pearl, J. (1995). Counterfactuals and policy analysis in structural models. In *Uncertainty in Artificial Intelligence 11* (ed. P. Besnard and S. Hanks), pp. 11–18. Morgan Kaufmann, San Francisco.

Brent, R. and Lok, L. (2005). A fishing buddy for hypothesis generators. *Science*, **308**(5721), 523–529.

Cartwright, N. (1983). *How the Laws of Physics Lie*. Clarendon Press, Oxford.

Chalak, K. and White, H. (2006, July). An extended class of instrumental variables for the estimation of causal effects. Technical Report Discussion Paper, UCSD, Department of Economics.

Cole, S.R. and Hernán, M.A. (2002). Fallibility in estimating direct effects. *International Journal of Epidemiology*, **31**(1), 163–165.

Collins, J., Hall, N., and Paul, L.A. (eds.) (2004). *Causation and Counterfactuals*. MIT Press, Cambridge, MA.

Cox, D.R. (1958). *The Planning of Experiments*. John Wiley and Sons, New York.

Cox, D.R. and Wermuth, N. (2004). Causality: A statistical view. *International Statistical Review*, **72**(3), 285–305.

Dawid, A.P. (1979). Conditional independence in statistical theory. *Journal of the Royal Statistical Society, Series B*, **41**(1), 1–31.

Dawid, A.P. (2000). Causal inference without counterfactuals (with comments and rejoinder). *Journal of the American Statistical Association*, **95**(450), 407–448.

Dawid, A.P. (2002). Influence diagrams for causal modelling and inference. *International Statistical Review*, **70**, 161–189.

Duncan, O.D. (1975). *Introduction to Structural Equation Models*. Academic Press, New York.

Eells, E. (1991). *Probabilistic Causality*. Cambridge University Press, Cambridge, MA.

Glymour, M.M. and Greenland, S. (2008). Causal diagrams. In *Modern Epidemiology* (3rd edn) (ed. K. Rothman, S. Greenland, and T. Lash), pp. 183–209. Lippincott Williams & Wilkins, Philadelphia, PA.

Good, I.J. (1961). A causal calculus (I). *British Journal for the Philosophy of Science*, **11**, 305–318.

Greenland, S. and Brumback, B. (2002). An overview of relations among causal modelling methods. *International Journal of Epidemiology*, **31**, 1030–1037.

Greenland, S., Pearl, J., and Robins, J.M (1999). Causal diagrams for epidemiologic research. *Epidemiology*, **10**(1), 37–48.

Haavelmo, T. (1943). The statistical implications of a system of simultaneous equations. *Econometrica*, **11**, 1–12. Reprinted in D.F. Hendry and M.S. Morgan (Eds.), *The Foundations of Econometric Analysis*, Cambridge University Press, 477–490, 1995.

Heckman, J.J. (2008). Econometric causality. *International Statistical Review*, **76**(1), 1–27.

Hitchcock, C. (2001). Book reviews: Causality: Models, Reasoning and Inference. *The Philosophical Review*, **110**(4), 639–641.

Hitchcock, C.R. (2003). Probabilistic causation. In *Stanford Encyclopedia of Philosophy* (Winter 2003 Edition) (ed. E. Zalta). URL = http://plato.stanford.edu/entries/causation-probabilistic/, Stanford, CA.

Holland, P.W. (1988). Causal inference, path analysis, and recursive structural equations models. In *Sociological Methodology* (ed. C. Clogg), pp. 449–484. American Sociological Association, Washington, D.C.

Lauritzen, S.L. (1996). *Graphical Models*. Clarendon Press, Oxford.

Lauritzen, S.L. (2001). Causal inference from graphical models. In *Complex Stochastic Systems* (ed. D. Cox and C. Kluppelberg), pp. 63–107. Chapman and Hall/CRC Press, Boca Raton, FL.

Lewis, D. (1986). *Philosophical Papers*, Volume II. Oxford University Press, New York.

Lindley, D.V. (2002). Seeing and doing: The concept of causation. *International Statistical Review*, **70**, 191–214.

Meek, C. and Glymour, C.N. (1994). Conditioning and intervening. *British Journal of Philosophy Science*, **45**, 1001–1021.

Morgan, S.L. and Winship, C. (2007). *Counterfactuals and Causal Inference: Methods and Principles for Social Research (Analytical Methods for Social Research)*. Cambridge University Press, New York.

Neyman, J. (1923). On the application of probability theory to agricultural experiments. Essay on principles. Section 9. *Statistical Science*, **5**(4), 465–480.

Pearl, J. (1988). *Probabilistic Reasoning in Intelligent Systems*. Morgan Kaufmann, San Mateo, CA.

Pearl, J. (1993*a*). Comment: Graphical models, causality, and intervention. *Statistical Science*, **8**(3), 266–269.

Pearl, J. (1993*b*). Mediating instrumental variables. Technical Report TR-210, http://ftp.cs.ucla.edu/pub/stat_ser/R210.pdf, Department of Computer Science, University of California, Los Angeles.

Pearl, J. (1995). Causal diagrams for empirical research. *Biometrika*, **82**(4), 669–710.

Pearl, J. (1998). Graphs, causality, and structural equation models. *Sociological Methods and Research*, **27**(2), 226–284.

Pearl, J. (2000*a*). *Causality: Models, Reasoning, and Inference*. Second edn. 2009 Cambridge University Press, New York.

Pearl, J. (2000*b*). Comment on A.P. Dawid's, Causal inference without counterfactuals. *Journal of the American Statistical Association*, **95**(450), 428–431.

Pearl, J. (2001). Direct and indirect effects. In *Proceedings of the Seventeenth Conference on Uncertainty in Artificial Intelligence*, pp. 411–420. Morgan Kaufmann, San Francisco, CA.

Pearl, J. (2003*a*). Reply to Woodward. *Economics and Philosophy*, **19**, 341–344.

Pearl, J. (2003*b*, December). Statistics and causal inference: A review. *Test Journal*, **12**(2), 281–345.

Pearl, J. (2005). Direct and indirect effects. In *Proceedings of the American Statistical Association, Joint Statistical Meetings*, pp. 1572–1581. MIRA Digital Publishing, Minn., MN.

Pearl, J. (2009*a*). *Causality: Models, Reasoning, and Inference*, Second edn. Cambridge University Press, New York.

Pearl, J. (2009*b*). Letter to the editor: Remarks on the method of propensity scores. *Statistics in Medicine*, **28**, 1415–1416. http://ftp.cs.ucla.edu/pub/stat_ser/r345-sim.pdf.

Pearl, J. (2009*c*). Myth, confusion, and science in causal analysis. Technical Report R-348, University of California, Los Angeles, CA. http://ftp.cs.ucla.edu/pub/stat_ser/r348.pdf.

Pearl, J. and Robins, J.M. (1995). Probabilistic evaluation of sequential plans from causal models with hidden variables. In *Uncertainty in Artificial Intelligence 11* (ed. P. Besnard and S. Hanks), pp. 444–453. Morgan Kaufmann, San Francisco.

Petersen, M.L., Sinisi, S.E., and van der Laan, M.J. (2006). Estimation of direct causal effects. *Epidemiology*, **17**(3), 276–284.

Robins, J.M. (1986). A new approach to causal inference in mortality studies with a sustained exposure period – applications to control of the healthy workers survivor effect. *Mathematical Modeling*, **7**, 1393–1512.

Robins, J.M. (1999). Testing and estimation of directed effects by reparameterizing directed acyclic with structural nested models. In *Computation, Causation, and Discovery* (ed. C. Glymour and G. Cooper), pp. 349–405. AAAI/MIT Press, Cambridge, MA.

Robins, J.M. and Greenland, S. (1991). Estimability and estimation of expected years of life lost due to a hazardous exposure. *Statistics in Medicine*, **10**, 79–93.

Rosenbaum, P.R. (2002). *Observational Studies*, Second edn. Springer-Verlag, New York.

Rosenbaum, P. and Rubin, D. (1983). The central role of propensity score in observational studies for causal effects. *Biometrika*, **70**, 41–55.

Rothman, K.J. (1976). Causes. *American Journal of Epidemiology*, **104**, 587–592.

Rubin, D.B. (1974). Estimating causal effects of treatments in randomized and nonrandomized studies. *Journal of Educational Psychology*, **66**, 688–701.

Rubin, D.B. (2004). Direct and indirect causal effects via potential outcomes. *Scandinavian Journal of Statistics*, **31**, 161–170.

Rubin, D.B. (2009). Author's reply: Should observational studies be designed to allow lack of balance in covariate distributions across treatment group? *Statistics in Medicine*, **28**, 1420–1423.

Shpitser, I. and Pearl, J. (2006*a*). Identification of conditional interventional distributions. In *Proceedings of the Twenty-Second Conference on Uncertainty in Artificial Intelligence* (ed. R. Dechter and T. Richardson), pp. 437–444. AUAI Press, Corvallis, OR.

Shpitser, I. and Pearl, J. (2006*b*). Identification of joint interventional distributions in recursive semi-Markovian causal models. In *Proceedings of the Twenty-First National Conference on Artificial Intelligence*, pp. 1219–1226. AAAI Press, Menlo Park, CA.

Shpitser, I. and Pearl, J. (2007). What counterfactuals can be tested. In *Proceedings of the Twenty-Third Conference on Uncertainty in Artificial Intelligence*, pp. 352–359. AUAI Press, Corvallis, OR. Also, *Journal of Machine Learning Research*, 9: 1941–1979, 2008.

Shpitser, I. and Pearl, J. (2009). Effects of treatment on the treated: Identification and generalization. In *Proceedings of the Twenty-Fifth Conference on Uncertainty in Artificial Intelligence*, pp. 514–521. AUAI Press, Corvallis, OR.

Shrier, I. (2009). Letter to the editor: Propensity scores. *Statistics in Medicine*, **28**, 1317–1318. See also Pearl 2009 <http://ftp.cs.ucla.edu/pub/stat_ser/r348.pdf>.

Simon, H.A. and Rescher, N. (1966). Cause and counterfactual. *Philosophy and Science*, **33**, 323–340.

Sobel, M.E. (1998). Causal inference in statistical models of the process of socioeconomic achievement. *Sociological Methods & Research*, **27**(2), 318–348.

Spirtes, P., Glymour, C.N., and Scheines, R. (2000). *Causation, Prediction, and Search*, second edn. MIT Press, Cambridge, MA.

Suppes, P. (1970). *A Probabilistic Theory of Causality*. North-Holland, Amsterdam.

Tian, J., Paz, A., and Pearl, J. (1998). Finding minimal separating sets. Technical Report R-254, University of California, Los Angeles, CA.

Tian, J. and Pearl, J. (2000). Probabilities of causation: Bounds and identification. *Annals of Mathematics and Artificial Intelligence*, **28**, 287–313.

Tian, J. and Pearl, J. (2002). A general identification condition for causal effects. In *Proceedings of the Eighteenth National Conference on Artificial Intelligence*, pp. 567–573. AAAI Press/The MIT Press, Menlo Park, CA.

Williamson, J. (2010). Probabilistic theories of causality. In *The Oxford Handbook of Causation* (ed. H. Beebee, C. Hitchcock, and P. Peter). Oxford University Press, New York.

Woodward, J. (2003). *Making Things Happen*. Oxford University Press, New York, NY.

Wright, S. (1921). Correlation and causation. *Journal of Agricultural Research*, **20**, 557–585.

·34·

Defining and identifying the effect of treatment on the treated

Sara Geneletti and A. Philip Dawid

Abstract

The *effect of treatment on the treated* (ETT) is of interest to econometricians as a measure of the effectiveness of schemes (such as training programmes) that require voluntary participation from eligible members of the population; it is also of interest in epidemiologic and similar contexts in cases where treatment randomization is not possible.

ETT has usually been expressed and analysed in terms of potential responses. Here we describe a new approach to formulating and evaluating ETT, based on an alternative decision-theoretic framework for causal inference. We give simple conditions under which ETT is well-defined, and identifiable given data from both an observational study and a control group, and further conditions allowing identification of ETT from purely observational data, with the assistance of a suitable instrumental variable.

We further show that the potential response formulation can be treated as a special case of our decision-theoretic approach.

34.1 Introduction

One rôle of labour economics is to evaluate the impact of government initiatives, such as employment and education schemes, on economic indicators such as income, the purpose being to inform the introduction of future schemes and policy changes: Should more adult training programmes be funded? Should education be made compulsory until the age of 18? Evaluating the effect of such policies is far from straightforward, most especially on account of the fundamental problem of *self-selection* (Heckman, 1979). Because of self-selection it is typically unclear whether, and to what extent, changes in economic indicators can be attributed to government programmes where participation is voluntary, since individuals that take part in such programmes (i.e. the self-selected) tend to be more motivated and receive higher incomes, irrespective of participation.

A similar problem emerges in epidemiologic contexts. A recent ruling in the US (Okie, 2006) gives terminal cancer patients the right to be treated with experimental (Phase I) drugs. This means that patients can self-select themselves into the treatment group, without randomization. Data on response from this group of patients will not yield reliable estimates of the *average causal effect* (ACE) as estimates will be confounded by the patients' attitude and health.

We will use the following two examples, one from labour economics and another from epidemiology, where effect evaluation is hampered by unknown selection criteria, to illustrate aspects of the methodology we develop in this chapter.

Example 34.1 Training programme
As a local government initiative, a mathematics refresher course aimed at adults with no higher education is introduced into a community. After some time, the local government wants to know whether and to what extent the course has had an impact on the income of the participants, as it plans to introduce more refresher courses and make participation in such courses a requirement for job-seekers enrolled in employment schemes. The problem with evaluating the impact of the inititative is that it will be confounded by partially unobserved individual characteristics. Thus estimates of the effect are typically obtained by relying on additional – and generally untestable – assumptions (Heckman, 1979). □

Example 34.2 Invalid randomization
Consider an epidemiologic trial where drug treatment is not appropriately randomised to patients, for instance in clinical trials with invalid blinding schemes. This may be due to a faulty protocol or it may be that doctors involved in the trial are aware of the health status of the patients. In contrast to Example 34.1, it is the doctor in charge of administering treatment who does the 'selecting': if he believes that the drug works, he will tend to give it to patients he thinks will benefit the most, e.g. those that are younger or fitter. The doctor's 'hunch' will thus be a confounder for the effect of the drug, as it will both determine treatment assignment and be predictive of health outcomes. □

Although the situations described above are formally analogous, they differ in focus. In Example 34.1 the quantity of interest is the effect of participation for those who chose to participate. This is termed the *effect of treatment on the treated* (ETT). This is also sometimes referred to as the average treatment effect on the treated (ATET) (Hotz *et al.* 2006). In contrast, in Example 34.2 the quantity of real interest is the *average causal effect* (ACE), as this is what is required for FDA drug approval, for example. The average causal effect can not usually be identified from confounded observational data unless strong additional assumptions are made. However, with weaker assumptions it may still be possible to identify ETT. This may not be exactly what is really wanted, but can provide some useful information on treatment effects.

Potential responses

Current statistical approaches to defining and estimating ETT are almost exclusively based on the *potential response* (PR) framework (Rubin, 1974, 1978). Thus Heckman and Robb (1985) introduced ETT in the following terms:[1]

$$\text{ETT} := \text{E}(Y_1 - Y_0 \mid T = 1). \tag{34.1}$$

Here T is the treatment variable, with value 1 for a subject receiving active treatment and 0 for a control; while Y_1 and Y_0 are the putative 'potential responses' (Rubin, 1978) of a subject to each of these treatments. By definition, it is possible to observe at most one of the two potential responses for any given subject – the other then becoming *counterfactual*. Inference about counterfactuals is sensitive to arbitrary and necessarily untestable assumptions (Dawid, 2000).

Equation (34.1) can also be expressed as

$$\text{ETT} = \text{E}(Y_1 \mid T = 1) - \text{E}(Y_0 \mid T = 1). \tag{34.2}$$

It appears *prima facie* that, in order for the expectation in (34.1) to be meaningful, we must have a joint probability distribution for (Y_1, Y_0, T) – or at the very least, using (34.2), a conditional distribution for $(Y_1 \mid T = 1)$ and one for $(Y_0 \mid T = 1)$. However since we can never observe Y_0 when $T = 1$, learning the latter distribution from data – and, thus, learning ETT – appears, on the face of it, problematic.

Another approach

The above formulation employs counterfactual logic and assumptions. We consider that the current evaluation literature is unnecessarily complicated by the many typically untestable assumptions[2] needed to use counterfactuals.

Our principal aim in this chapter is to demonstrate how a different formalism, eschewing counterfactuals, can helpfully be used to interpret and identify ETT. Our approach is grounded on the *decision theoretic* (DT) framework for causal inference introduced by Dawid (2002, 2007). This supplies a formal language by means of which causal questions can be rigorously posed and analysed, using clear and meaningful assumptions; moreover, fewer such assumptions are typically required than for other approaches such as PR.

[1] Their general definition of ETT allows for further conditioning on a set of observed covariates X, as well as on T. For simplicity we shall omit X wherever this does not affect the thrust of our argument.

[2] Some of these are *ignorability* and the *stable unit-treatment value assumption* (Rubin, 1986). The former requires that the counterfactual outcomes be independent of the treatment received, the latter requires that the potential response to a treatment of one individual be well-defined, independently of the treatments assigned to other individuals. Another assumption usually invoked in counterfactual theory is *consistency* (Robins, 1986), which requires that the realized response, when treatment t is actually applied, be the same as the corresponding potential response.

In the DT framework, causal assumptions are expressed in terms of *conditional independence* statements, that can, in principle if not always in practice, be tested, since all quantities involved are jointly observable. Thus the DT framework provides a more concise, economical and justifiable approach to inference on treatment effects.

We shall present two alternative descriptions of ETT in decision-theoretic terms, and show that they are equivalent. In particular, we show that, contrary to first impressions, ETT is well-defined, being fully determined by the probabilistic behaviour of observable variables. We further show how the PR framework can be formally subsumed within the DT framework as a special case, and deduce that (again, contrary to first impressions) the PR formulation of ETT is itself well-defined in this sense.

Outline

The chapter is laid out as follows. Section 34.2 introduces the basic principles of the DT framework. In Section 34.3 we provide a DT definition of ETT in terms of a 'preference variable', governing treatment choice whenever that can be exercised freely, and show that ETT can be identified from observational and interventional data. In Section 34.4 we develop an alternative DT account based on a 'sufficient covariate', and prove that this leads to a unique definition, also agreeing with the earlier one. Moreover, the traditional PR account can be subsumed as a special case of this treatment, and so must deliver the identical answer. In Section 34.5 we provide a DT description of two approaches for estimating ETT, using randomised availability trials, or instrumental variables to estimate ETT from observational data alone. We make some concluding remarks in Section 34.6.

34.2 Decision-theoretic approach to causal inference

The decision-theoretic approach to causal inference (Dawid, 2002, 2003; Dawid and Didelez, 2010; Dawid, 2007) is grounded in the statistical theory of decision-making under uncertainty (Raiffa, 1968; Smith, 1988). Rather than split the response Y of a subject into several potential responses, we consider a variety of stochastic behaviours for the single variable Y (jointly with other relevant variables), under various different *regimes* that may be operating. Our principal purpose is to identify and compare the distributions of Y for a variety of contemplated *interventional* regimes. However data may only have been collected under some *observational* regime. From this standpoint, the major problems to be addressed are whether, when and how probabilistic information can usefully be transferred across regimes.

For simplicity we restrict attention here to a comparison of two treatments, 'active' and 'control', and the three corresponding regimes, one observational and two interventional. These represent respectively, the circumstances in which a particular observational study of interest is conducted, and those in which one or other of the treatments is administered to a subject. Within each regime, subjects are regarded as exchangeable. We can consequently regard their values on all relevant variables as being drawn, independently across subjects, from some fixed (though generally unknown) joint distribution – which will however generally differ across regimes. The real-world meaning of these regimes will necessarily be context-specific, and the plausibility of any assumptions that may be made about them must be assessed in relation to those real-world meanings.

It is the interventional regimes that are the objects of principal interest, and about which we should like to learn from data, since these will be of direct relevance for guiding future action or policy choice. Thus the government, in deciding whether or not to introduce a new initiative, would want to assess, and compare, the consequences both of action and of inaction. A patient, faced with the decision as to whether or not to take a treatment, needs to assess what the response might be if he did, or if he did not. In either case, knowledge of the distribution of the response under each proposed intervention is exactly what is required to support rational choice between the options. But these interventional distributions may be difficult to identify directly from data collected under observational conditions. One might, naïvely, regard the observational distribution of the response, among those patients who happened to get the treatment, as directly informative about a new patient's response, if he were to decide to take the treatment; but this would be valid only if he could consider himself exchangeable (on relevant pre-treatment variables) with that observational group. Likewise, for the control group data to be directly relevant for this patient, he would need to regard himself as exchangeable with the observational control group. However it will clearly be impossible to satisfy both conditions simultaneously if – as is common in observational studies – those two groups are not even exchangeable with each other. Then some more refined analysis, typically requiring extra assumptions to be imposed and justified, becomes essential.

34.2.1 Formal set-up

Denote the treatment variable by T, taking value 1 for active treatment, and 0 for control treatment. We introduce a further variable F, the *intervention variable* or *regime indicator*. The possible values for F are $\emptyset$, 0 and 1, indexing the regimes under consideration. When $F = \emptyset$, this indicates that variables are being generated under observational conditions, whereas $F = t$ $(t = 0, 1)$ indicates that they are generated under an intervention that sets $T = t$.

Whereas T and Y are chance variables, F is a *decision* or *parameter* variable, and has no uncertainty associated with it. In particular, this means that all probability statements made must be explicitly or implicitly conditional on F.

We denote the distribution [resp., expectation] of the chance variables under regime $F = \tau$ ($\tau = 0, 1, \emptyset$) by $p_\tau(\cdot)$ [resp., $\mathrm{E}_\tau(\cdot)$]. We note that, under our interpretation of F, we must have, for $t = 0, 1$:

$$F = t \Rightarrow T = t, \tag{34.3}$$

so that

$$p_t(T = t) = 1 \quad (t = 0, 1). \tag{34.4}$$

Average causal effect
The *average causal effect* (ACE) (of treatment $T = 1$, relative to treatment $T = 0$) on Y is defined as follows:

$$\mathrm{ACE} = \mathrm{E}_1(Y) - \mathrm{E}_0(Y). \tag{34.5}$$

This is a simple comparison of the expected value (implicity assumed to exist) of Y under intervention to apply treatment 1, with that under intervention to apply treatment 0. When utility is linear in the value of the outcome Y, a rational subject with no additional relevant information would prefer treatment 1 to treatment 0 if and only if ACE > 0.[3]

No confounding
In some cases (e.g. randomised trials) we might be prepared to assume that the following conditional independence relation (Dawid, 1979, 2000, 2002) holds:

$$Y \perp\!\!\!\perp F \mid T. \tag{34.6}$$

This says that, for $t = 0, 1$, $p_\emptyset(y \mid T = t) = p_t(y \mid T = t)$ $(= p_t(y)$, by (34.3)); i.e. given either treatment, the distribution of the response is assumed the same in both the observational regime and the relevant interventional regime. This is the case of 'no confounding', when, for the purpose of estimating the distributions of Y given T, we can treat the observational regime exactly as if it had been interventional. When (34.6) holds, $\mathrm{ACE} = \mathrm{E}_\emptyset(Y \mid T=1) - \mathrm{E}_\emptyset(Y \mid T=0)$, and so can be identified directly from observational data.

The conditional independence assumption (34.6) can be represented graphically by means of the *influence diagram* (Dawid, 2002) of Figure 34.1. This is a decision-theoretic version of a directed acyclic graph (DAG), with chance variables represented by round nodes, and decision variables by square nodes. Associated with the arrow from F to T is a specification of the distribution of T in each regime specified by F (in fact degenerate for $F = 0$

[3] We could, without adding any real complication, replace Y in (34.5) by some function of Y, e.g. a nonlinear measure of the utility of outcome Y. Still more generally, we could compare some other chosen feature, e.g. variance, of the distributions of Y under the two experimental regimes $F = 1$ and $F = 0$.

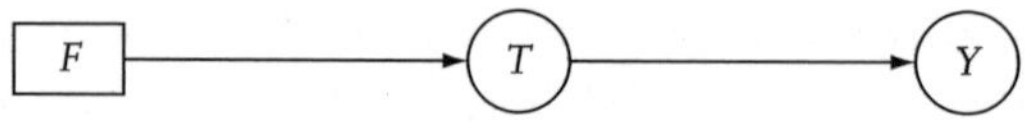

Fig. 34.1 Influence diagram representing the conditional independence assumption $Y \perp\!\!\!\perp F \mid T$.

or 1, though non-degenerate for $F = \emptyset$). Associated with that from T to Y is a specification of the conditional distribution of the response Y, given that treatment T has been administered. The *absence* of any arrow from F to Y encodes assumption (34.6): that this conditional distribution does not further depend on which regime is in operation. (Note however that the property (34.4) is not encoded in the graph, and has to be introduced explicitly when needed.) Finally we remark that, since F is a decision variable, no probability distribution is associated with it.

In most contexts (34.6) can *not* reasonably be assumed (the case of *confounding*). It is such cases that form the focus of this chapter.

34.3 Decision-theoretic formulation of ETT. I. Preference variable

34.3.1 Treatment allocation and treatment application

When we compare the responses in two or more treatment groups in an observational setting, what we actually see is a combination of two quite distinct effects:

Treatment effect
The specific power of the treatment to make a difference to the outcome of interest.

Selection effect
The fact that we are not observing random subsets of the population of interest.

In particular, even if there were no treatment effect whatsoever, the existence of a differential selection effect would typically lead to systematic differences between the outcomes in the different treatment groups, because we would not be comparing like with like: this is the essence of the problem of confounding.

We will find it helpful to elaborate our description and notation to make the above important distinction explicit. We introduce a *preference variable* D, describing the treatment which an individual would choose/be chosen to take if free to do so. Quite separately we have the *treatment variable* T, which indicates which treatment is actually taken or applied. In the observational setting the preferred treatment will be applied, $T = D$: thus D and T are completely confounded with each other. However, in an interventional setting we could override the initial preference, and so need not have $T = D$.

A similar construction was used by Robins *et al.* (2006). However their analysis involved an additional latent variable U. In Section 34.4, in an alternative to our approach above, we too introduce such a variable U – but as we never need to consider *both* U and D, our description and analysis are more straightforward.

Consider Example 34.1. We are interested in the effect of the programme on the income of potential participants, i.e. ETT. Each eligible individual, once made aware of the opportunity, makes a personal choice, D, whether to participate or not. We now consider two scenarios. In the first, participation is voluntary: this is the observational regime, $F = \emptyset$, in which $T = D$. The second is the interventional setting: e.g. a controlled trial run by the local government that randomizes all eligible adults to participate or not in the programme, irrespective of their personal preferences. This gives rise to two interventional regimes, $F = t$ $(t = 0, 1)$. These would also be relevant to the considerations a new subject who needs to decide whether or not to participate.

In the context of Example 34.2, the observational regime, $F = \emptyset$, describes the scenario in which the doctor has a treatment preference D, based perhaps on his hunches about the patient's likely recovery, and then actually gives the corresponding treatment. The interventional regimes $F = t$ $(t = 0, 1)$ refer to the case where the hospital management overrules the doctor's preference and administers treatment t, or to the treatment decision problem faced by a new patient. We will know the value of D in the observational regime, since it is then identical with the treatment T actually received; but we generally will not know it in an interventional regime, where this need not hold.

34.3.2 Assumptions

Certain variables can be classified as 'pre-treatment' variables: their values are supposed fully determined before the point at which treatment is actually applied. In particular this applies to a 'covariate', i.e. a permanent pre-existing attribute of a subject. Other variables will be 'post-treatment' variables, arising only after the point of treatment application. Although this is not a formal description, in most contexts it will be clear whether a variable belongs to one or other of these groups. Our models will involve only pre-treatment variables, together with a single post-treatment variable, the response Y – and, of course, the treatment variable T itself.[4]

We require the following assumptions:

[4] Thus we do not here consider situations such as that treated by Robins (1989), involving sequential decisions based on accruing time-varying information.

(i) D is a pre-treatment variable.

(ii) Pre-treatment variables have the same distribution in all regimes, interventional and observational.

(iii) A subject's response to a treatment does not depend on whether the treatment is self-selected or externally imposed.

Assumption (i) is usually plausible, since it will be reasonable to suppose that, for example:

- An individual knows whether he would wish to participate in a programme, before, and regardless of whether, he is forced to participate or not participate.
- A doctor can decide which treatment he would like to assign to whom, before, and regardless of, whatever the hospital management might decide to do.

At first glance assumption (ii) appears tantamount to assuming 'no backwards causation', which appears entirely reasonable. However, it can fail if, for example, the observational study was done in one population but the proposed interventions relate to a new population, or for a subject who can not be treated as exchangeable (on pre-treatment variables) with those in the study.

While little formal progress can be made without an assumption such as (iii), we warn that it may well be unreasonable in many contexts. Thus in Example 34.1 an individual might well respond differently to the training programme if forced to undertake it from how he would if he himself had chosen to do so; in Example 34.2, the placebo effect could lead to a patient who knows (or whose doctor knows) what treatment has been assigned to him responding differently from how he would if the same treatment had been applied in a double-blind clinical trial. (Note that these effects are quite distinct from those associated with confounding.) In such cases we should really regard the 'same' treatment applied in different settings as constituting several distinct treatments, and in such a case our analyis may simply not apply. In any case, whenever we talk about 'the effect of treatment' we should be very clear as to the nature of the treatment we are referring to, and of that with which we are comparing it.

34.3.3 Regimes

Suppose we have fully specified the preference variable D, and the joint distribution of all relevant observables in the two interventional regimes $F = t$ ($t = 0, 1$). By assumption (ii), these must agree when restricted to pre-treatment variables – including, by assumption (i), D. We can now explicitly construct the joint observational distribution, $F = \emptyset$, of all variables as follows. We first generate all relevant pre-treatment variables, including D, from their joint distribution (which, by (ii), is the same in both interventional regimes);

and then use the realized value of D to determine which treatment T to give.[5] Finally, if, e.g. $D = 1$, we assign to Y the distribution it would have in the active treatment regime, $F = 1$, *conditional on* $D = 1$ (here we use assumption (iii)).

Note that this process cannot be reversed: knowing only the observational distribution, in which necessarily $T = D$, we cannot in general identify, e.g. the distribution of Y given $D = 0$, $T = 1$, which is a necessary ingredient of the interventional regime $F = 1$.

It is easy to see that D has the following properties:

$$D \perp\!\!\!\perp F \tag{34.7}$$

$$Y \perp\!\!\!\perp F \mid (D, T). \tag{34.8}$$

Here (34.7) says that D has the same distribution in all regimes, be they interventional or observational; while (34.8) says that the distribution of Y given D and T is the same in all regimes, whenever it is meaningfully defined: because of the deterministic dependence of T on F and D, this means that, for $t = 0, 1$, the distribution of Y given $D = t$ is the same in the observational regime $F = \emptyset$ and the interventional regime $F = t$.

The *effect of treatment on the treated* is now defined as:

$$\text{ETT} := \text{E}_1(Y \mid D = 1) - \text{E}_0(Y \mid D = 1). \tag{34.9}$$

This is essentially ACE, as defined in (34.5), but calculated for a specific subpopulation of patients: those having $D = 1$, i.e. those who would choose/be chosen to receive treatment 1 – whether or not they actually receive it.

We remark that (34.9) displays a clear separation of 'treatment effect' from 'differential selection effect'. The former is effected by the comparison of expected responses under the two interventional regimes $F = 1$ and $F = 0$; the latter is excluded because we only compare interventional regimes, and condition on the identical property (namely preference for active treatment, $D = 1$) in both.

34.3.4 ETT is identifiable

In a randomised controlled trial, if we could record the preference variable D for all subjects, we could identify ETT straightforwardly. In practice, however, we will not usually be able to observe D in interventional regimes – although we can do so indirectly in observational regimes, since then we know $D = T$, and T is observed. It thus appears *a priori* that it would be impossible to identify the term $\text{E}_0(Y \mid D = 1)$ from available data, and thus impossible to identify ETT.

[5] Note that T is functionally determined by D and F: $T = t$ when $F = t$ $(t = 0, 1)$, and $T = D$ when $F = \emptyset$.

The following analysis shows that, contrary to this initial appearance, ETT *can* be identified – so long as we can gather data under both observational and (some) interventional circumstances. (We must also suppose $p_{\emptyset}(T = 1) > 0$.)

The first term in (34.9), $E_1(Y \mid D = 1)$, presents no difficulty. We have:

$$\begin{aligned} E_1(Y \mid D = 1) &= E_1(Y \mid D = 1, T = 1) \\ &= E_{\emptyset}(Y \mid D = 1, T = 1) \\ &= E_{\emptyset}(Y \mid T = 1) \end{aligned} \tag{34.10}$$

where the first equality holds because $F = 1 \Rightarrow T = 1$, the second from (34.8), and the third because $T = 1 \Rightarrow D = 1$ under $F = \emptyset$. Thus $E_1(Y \mid D = 1)$ is directly identifiable from observational data on (T, Y).

To get a handle on the problematic second term of (34.9), $E_0(Y \mid D = 1)$, we argue as follows. We have

$$E_0(Y) = E_0(Y \mid D = 0) \times p_0(D = 0) + E_0(Y \mid D = 1) \times p_0(D = 1). \tag{34.11}$$

From data gathered under 'control' conditions, $F = 0$, we can identify the left-hand side of (34.11), $E_0(Y)$.

From (34.8), $p_0(D = 0) = p_{\emptyset}(D = 0)$, and this in turn is $p_{\emptyset}(T = 0)$, since $D = T$ in the observational regime $F = \emptyset$. Likewise, $p_0(D = 1) = p_{\emptyset}(T = 1)$. So these terms can be identified from observational data.

Suppose for the moment (an extreme special case) that $p_{\emptyset}(T = 1) = 1$, whence $p_{\emptyset}(T = 0) = 0$. Then from (34.11) we deduce $E_0(Y \mid D = 1) = E_0(Y)$. Also, (34.10) becomes $E_{\emptyset}(Y)$. We thus have, from (34.9),

$$\text{ETT} = E_{\emptyset}(Y) - E_0(Y). \tag{34.12}$$

In this case the observational group behaves just like an interventional treatment group, $E_{\emptyset}(Y) = E_1(Y)$, and ETT = ACE.

Otherwise, we have, in parallel fashion to (34.10), $E_0(Y \mid D = 0) = E_{\emptyset}(Y \mid T = 0)$, which can be identified from observational data on (T, Y). The remaining term in equation (34.11), $E_0(Y \mid D = 1)$, can thus be solved for. Since we now have both $E_1(Y \mid D = 1)$ from (34.10) and $E_0(Y \mid D = 1)$ from (34.11), we can obtain ETT from (34.9). Doing the algebra, we obtain:

$$\text{ETT} = \frac{E_{\emptyset}(Y) - E_0(Y)}{p_{\emptyset}(T = 1)}, \tag{34.13}$$

a general form that also includes the special case (34.12).

It follows that ETT is fully identified by the distributions of the observables (T, Y) in the various regimes: indeed, we can identify ETT so long as, in addition to observational data on Y and T, we have data on Y from experimental subjects under control conditions.

Although based on different assumptions, formula (34.13) is essentially the same as formula (8.20) in Pearl (2009). In Section 34.4.2 we shall see why this must be.

34.4 Decision-theoretic formulation of ETT. II. Unobserved confounder

The above development of ETT relies on the existence and meaningfulness of the preference variable D in all regimes, observational and interventional. While this may be a reasonable assumption in, e.g. economic contexts, where agents may be supposed to form preferences in accordance with rational principles such as maximization of expected utility, in other contexts it may seem somewhat far-fetched.

We now present an alternative, more general, construction – which, as we shall see, is fully consistent with that described above based on the preference variable. We consider a (typically multivariate, typically unobserved) covariate U (i.e. a permanent attribute of a subject) that can be considered as determining treatment choice – typically only probabilistically – in the observational regime. For Example 34.1, U might comprise the personal characteristics of the individuals, their motivation, natural talent, confidence, etc. For Example 34.2, U could represent the attributes of the patient that determine the doctor's hunches as to who will benefit more from the treatment. Such an unobserved variable U, associated with treatment in the observational regime, will be a *confounder* if it is also predictive of outcome.

In contrast to our analysis in Section 34.3.1, we do not now directly construct the observational regime $F = \emptyset$ from the interventional regimes; rather, we regard it as having an entirely independent existence. Then to make progress we must make (and justify!) assumptions relating this to the interventional regimes. Our fundamental requirement is that U be a *sufficient covariate* (Dawid, 2002): that is, for $F \in \{0, 1, \emptyset\}$:

$$U \perp\!\!\!\perp F \tag{34.14}$$

$$Y \perp\!\!\!\perp F \mid (U, T). \tag{34.15}$$

Here (34.14) requires that U (being a pre-treatment variable) have the same distribution in all regimes; while (34.15) requires that, *if only we knew, and conditioned on,* U, the distribution of the response to an applied treatment would be the same, no matter whether that treatment had been applied under interventional or observational conditions.

The conditional independence relations (34.14) and (34.15) can be represented graphically by means of the influence diagram of Figure 34.2. Note that the arrows in Figure 34.2 represent stochastic dependence: in particular, T and U need not fully determine Y, but merely modify its distribution. This probabilistic interpretation of a 'causal model' may be contrasted with that of Pearl (2009, Section 7.1), which would treat U as an undefined exogenous ('error')

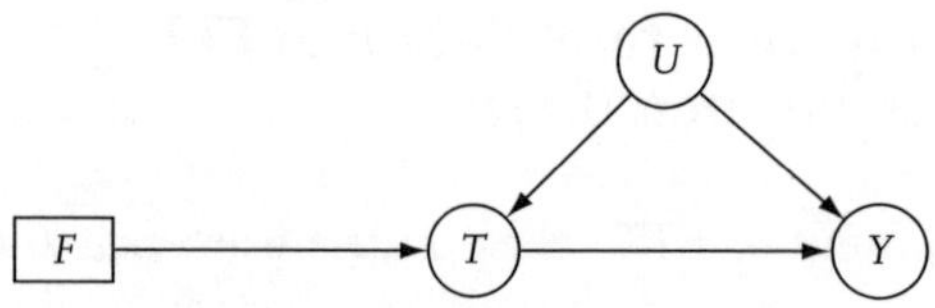

Fig. 34.2 A sufficient covariate U.

variable, and Y as functionally determined by T and U.[6] Our stochastic model is more general, and appropriate to our intended interpretation of U as a pre-existing real-world attribute of a subject, that could, in principle at least, be identified and measured. It also explicitly allows treatment choice, even in the interventional regime, to incorporate an element of randomization. However, so far as the mathematics is concerned, Pearl's functional interpretation can be treated as a special case of our model.

Specific causal effect
We now introduce the *specific causal effect* of treatment, relative to the specified sufficient covariate U. This is a a function of U, defined by

$$\mathrm{SCE}_U := \mathrm{E}_1(Y \mid U) - \mathrm{E}_0(Y \mid U), \tag{34.16}$$

i.e.

$$\mathrm{SCE}_U(u) := \mathrm{E}_1(Y \mid U = u) - \mathrm{E}_0(Y \mid U = u).$$

That is, $\mathrm{SCE}_U(u)$ is the average causal effect in the subpopulation of individuals having the specified value u for U.

By (34.15) we can also express

$$\mathrm{SCE}_U = \mathrm{E}_\emptyset(Y \mid T = 1, U) - \mathrm{E}_\emptyset(Y \mid T = 0, U). \tag{34.17}$$

Thus SCE_U could be identified from observational data if U were to be observed in addition to T and Y.[7] However typically this will not be the case.

We remark that, by (34.14), for $t = 0, 1$,

$$\begin{aligned}\mathrm{E}_\emptyset\{\mathrm{E}_t(Y \mid U)\} &= \mathrm{E}_t\{\mathrm{E}_t(Y \mid U)\} \\ &= \mathrm{E}_t(Y),\end{aligned}$$

whence

$$\begin{aligned}\mathrm{E}_\emptyset(\mathrm{SCE}_U) &= \mathrm{E}_1(Y) - \mathrm{E}_0(Y) \\ &= \mathrm{ACE}.\end{aligned} \tag{34.18}$$

[6] There appears to be a common perception that graphical models are somehow tied to such a deterministic interpretation. In fact the opposite is true: they are fundamentally tools for representing and handling probability distributions, not functional relationships.

[7] We also need that, for each value of U, both values of T are observed in the data – which in particular would disallow the choice $U = D$.

In particular, $\mathrm{E}_\emptyset(\mathrm{SCE}_U)$ cannot depend on the choice of sufficient covariate U. Formulas (34.17) and (34.18) enable us to identify ACE from an observational study whenever we can measure some sufficient covariate.[8]

34.4.1 Definition and uniqueness of ETT

After the above preliminaries, we are ready to define the *effect of treatment on the treated* in this setting:

$$\mathrm{ETT}_U := \mathrm{E}_\emptyset\{\mathrm{SCE}_U \mid T = 1\}. \tag{34.19}$$

That is, ETT_U is the average, in the observational regime, of the specific causal effect (relative to U), for those individuals who in fact receive the active treatment $T = 1$.[9] This could be identified if we had observational data on all three variables (T, U, Y). But in general U will not be observable, in any regime: a seemingly fatal handicap to identifying ETT_U.

When a sufficient covariate exists, it need not be unique. The above definition of ETT_U appears, *prima facie*, to depend on the specific choice of sufficient covariate U, and its probabilistic relationship with the observables (T, Y). However, Theorem 34.1 below (a generalization of the argument of Section 34.3.4) will show that this is not in fact the case: ETT_U does not depend on the choice of U, but only on the joint distributions of the observables in the various regimes. In particular, it can be identified from such data *even when we do not specify, or observe, any sufficient covariate.*

Theorem 34.1. *Suppose $p_\emptyset(T = 1) > 0$. Then, for any sufficient covariate U,*

$$\mathrm{ETT}_U = \frac{\mathrm{E}_\emptyset(Y) - \mathrm{E}_0(Y)}{p_\emptyset(T = 1)}. \tag{34.20}$$

Proof Suppose first $p_\emptyset(T = 0) > 0$. For $t = 0, 1$, define

$$k(t) := \mathrm{E}_\emptyset\{\mathrm{E}_0(Y \mid U) \mid T = t\} \tag{34.21}$$

$$= \mathrm{E}_\emptyset\{\mathrm{E}_\emptyset(Y \mid U, T = 0) \mid T = t\} \tag{34.22}$$

by (34.15). In particular, $k(0) = \mathrm{E}_\emptyset(Y \mid T = 0)$.

By (34.17) and conditional independence property (34.15), (34.19) is equal to

$$\mathrm{ETT}_U = \mathrm{E}_\emptyset(Y \mid T = 1) - k(1). \tag{34.23}$$

[8] Of course, if we could also measure U on a new individual, it would be SCE_U, rather than ACE, that would be relevant for his decision problem.

[9] For this to be meaningful we need to assume $p_\emptyset(T = 1) > 0$. Note also that (34.16) defines SCE_U only up to a set of probability 0 under the distribution (common to all regimes considered) of U; but since such a set also has probability 0 in the observational regime conditional on $T = 1$, ETT_U is well-defined.

Also,

$$\begin{aligned} E_0(Y) &= E_0\{E_0(Y \mid U)\} \\ &= E_0\{E_\emptyset(Y \mid U, T = 0)\} \\ &= E_\emptyset\{E_\emptyset(Y \mid U, T = 0)\} \end{aligned} \tag{34.24}$$

by (34.15) and (34.14). It follows that

$$\begin{aligned} E_0(Y) &= E_\emptyset\{k(T)\} \\ &= p_\emptyset(T = 0)\, k(0) + p_\emptyset(T = 1)\, k(1). \end{aligned}$$

Hence

$$k(1) = \frac{E_0(Y) - p_\emptyset(T = 0)\, E_\emptyset(Y \mid T = 0)}{p_\emptyset(T = 1)}. \tag{34.25}$$

Formula (34.20) now follows on substituting into (34.23).

Finally, the special case $p_\emptyset(T = 0) = 0$ can be handled by a similar (and simpler) argument. □

In the light of the above result, we no longer need to specify which sufficient covariate U is used to define the effect of treatment on the treated; consequently we can just use the notation ETT.

Comments

(i) Our analysis in Section 34.3.1 in terms of the preference variable D can be treated as a special case of that above, on identifying U with D.

(ii) Suppose that in fact there is no confounding. In that case $E_\emptyset(Y) = E_0(Y) \times p_\emptyset(T = 0) + E_1(Y) \times p_\emptyset(T = 1)$, and formula (34.20) reduces to ETT = ACE.

(iii) It is surprising, though of no ultimate significance, that to identify ETT we do not need data on subjects receiving the treatment by intervention.

(iv) The fact that ETT is well-defined in terms of observable distributions in the various regimes does *not* mean that, in Example 34.1, it would be the same for two communities with different population distributions and attitudes – since the relevant observational regimes would be different. Similarly in Example 34.2, different distributions of patients, or different behaviour of the doctors, would typically yield different values for ETT. It is thus a matter for careful consideration whether, and under what circumstances, ETT will be informative about subjects who can not be regarded as exchangeable with those in the study population.

(v) Although the value of ETT does not depend on which sufficient covariate U is being considered, its definition and interpretation require that some such variable U should exist. In some contexts we might not be willing to accept this assumption. Then – notwithstanding the fact that we could still calculate the right-hand side of (34.20) from knowledge of the distributions of observables in the observational and control regimes – we perhaps should not attempt to interpret this as 'the effect of treatment on the treated'.

34.4.2 Application to potential response framework

In the potential response framework we conceive of the existence of the pair of potential responses (Y_0, Y_1). These are implicitly supposed to have the same values, and the same joint distribution, no matter what regime operates. That is:

$$(Y_0, Y_1) \perp\!\!\!\perp F. \tag{34.26}$$

It is also assumed that, no matter what regime operates, the actual response Y is fully determined by the pair (Y_0, Y_1) and the treatment T applied: $Y = Y_T$. This functional dependence implies, in particular:

$$Y \perp\!\!\!\perp F \mid (Y_0, Y_1, T). \tag{34.27}$$

Comparing (34.26) and (34.27) with (34.14) and (34.15), we see that we can formally treat $U^* = (Y_0, Y_1)$ as a sufficient covariate. Since $E_t(Y \mid U^*) = Y_t$, the associated specific causal effect is $\text{SCE}^* = Y_1 - Y_0$, and hence the associated definition of ETT is $\text{ETT}^* = E_\emptyset(Y_1 - Y_0 \mid T = 1)$. This recovers the 'traditional' definition (1) of ETT in the potential response framework.

Now Theorem 34.1 shows that:

(i) The PR definition can be expressed as in (34.20).

(ii) It agrees with the definition of ETT given in Section 34.3.1, as well as with any of the variations, as in (34.19), in terms of an arbitrary sufficient covariate U.

34.5 Estimating ETT

34.5.1 Availability trials and policy analysis

It follows from (34.20) that we could readily identify ETT if, in addition to having data on T and Y from an observational study, we had data on Y from a group of subjects (randomly selected from the same population) who were made to take the control treatment.

One way of obtaining the required data is by performing a randomised *availability trial*: a study in which the treatment is made available to (but not forced on) a random subgroup of the population, while being denied to another such subgroup. Assuming that both the take-up and the response behaviour of the individuals in the first (availability) subgroup is unaffected by the conduct of the trial, we can use the data from that subgroup directly to estimate $E_\emptyset(Y)$ and $p_\emptyset(T = 1)$; while $E_0(Y)$ can be estimated from the other (control) subgroup. We can then substitute into formula (34.20) to estimate ETT.

Such a trial would be of obvious direct relevance to *policy analysis*, where the aim is to assess the benefit of the policy decision to make the treatment available. The numerator of (34.13) is exactly what is being estimated by the simple contrast between the average responses in the two arms of the availability trial: it directly measures the additional benefit, per individual in the population, of introducing this policy. This average benefit could then be balanced against the average cost per individual of introducing the policy, to aid the decision as to whether or not to make treatment available.

The further inclusion of the denominator in formula (34.20) has the effect of adjusting it to measure the benefit of the policy introduction *per individual taking up the treatment* – so supplying a simple and intuitive interpretation of formula (34.20). It could be argued that this is of less direct value for policy analysis than the straightforward use of the numerator without this adjustment. If, however, the cost of introduction were directly proportional to the number taking up the treatment, then it would indeed be appropriate to base the decision on ETT. In more realistic scenarios, one would need to combine, appropriately, the overall benefit (proportional to the numerator of (34.20) and population size), fixed set-up costs, costs pro rata to population size, and costs pro rata to the number taking up the treatment. While such an analysis would still use the ingredients of (34.20), it could do so in a more flexible manner.

34.5.2 Estimation using an instrumental variable

When no controlled randomization is possible, we can sometimes substitute pseudo-randomization devices, such as instrumental variables (Heckman and Navarro-Lozano, 2004; Heckman, 2005; Didelez and Sheehan, 2007).

Consider Example 34.2 with a twist. The doctor visits a group of patients and, based on his initial diagnosis, he prescribes some of them the new drug. After the doctor has completed this process, these patients are given a preliminary allergy test, and a subset group of them is identified as allergic to some of the components of the drug. As a consequence, the doctor's prescription is overruled for these patients and they are not administered the drug.

We assume:

(i) The presence of the allergy is independent of the doctor's treatment preference.

(ii) Conditional on the doctor's treatment preference, response to control treatment (in this case, not taking the drug) is independent of the presence of allergy.

Property (i) will be plausible when the doctor cannot tell which patients are allergic just by talking to them; it also requires that the patient's medical records do not contain information on allergies to the drug ingredients. This might be the case for instance, when the drug is new, or the ingredients it contains are not in commonly available medication and thus allergy has not been reported. Property (ii) will be plausible if the physiological systems responsible for the allergy and the disease under study are unconnected. Thus, for patients who do not receive the treatment (whether due to the doctor's treatment preference, or because they have the allergy), response will not be systematically different amongst those who have the allergy. When (i) and (ii) hold, the allergy status of a patient is an *instrumental variable.*

In this case we can use the allergic patients, from whom treatment is withheld, as a proxy for an experimental control group, and so estimate $\mathrm{E}_0(Y)$ from these; the remaining patients form, essentially, an observational study group, allowing us to estimate $\mathrm{E}_\emptyset(Y)$ and $p_\emptyset(T = 1)$. Hence we can identify ETT from (34.20).

Formal construction

Informal arguments such as the above can be valuable and revealing, but it is necessary to have a rigorous formal system to express and derive such results. To demonstrate how this is provided by our decision-theoretic framework, we present the formal construction below.

We frame our argument in the set-up of Section 34.3 (a similar argument could be applied for that of Section 34.4). We have variables F, D, T, Y as before – except now we only need to consider $F \in \{0, \emptyset\}$. In addition we have a binary pre-treatment variable A (indicating presence of allergy) such that $p(A = 0) > 0$, and *in the observational regime* $F = \emptyset$, whatever be the value of U:

$$A = 0 \Rightarrow T = 0. \tag{34.28}$$

In particular, when $A = 0$ this overrides the original treatment preference D. Clearly D, being a pre-treatment variable, continues to satisfy (34.7). However our original argument for (34.8) no longer stands. We replace it by

$$Y \perp\!\!\!\perp F \mid (D, T = 0), \tag{34.29}$$

which says that, conditional always on treatment preference, the observational distribution of response among those who receive the control treatment is the same as its distribution in the control interventional regime. In particular this will hold if D is a sufficient covariate.

We further assume that *in the observational regime* $F = \emptyset$:

$$A \perp\!\!\!\perp D \tag{34.30}$$

$$Y \perp\!\!\!\perp A \mid (D,\ T = 0). \tag{34.31}$$

Here (34.30) and (34.31) formalize (i) and (ii) above, respectively. Property (34.30) must in fact hold in all regimes, since A and D are pre-treatment variables. Condition (34.31) could have been imposed for active as well as control treatment, but this turns out not to be required for our analysis below.

Note that (34.28), (34.30) and (34.31) are analogous to (34.3) (for $t = 0$), (34.7) and (34.29), respectively, but with A replacing F.

Theorem 34.2. *Under the above conditions,* $\mathrm{E}_\emptyset(Y \mid A = 1) = \mathrm{E}_0(Y)$.

Proof

$$\begin{aligned}
\mathrm{E}_\emptyset(Y \mid A = 0) &= \mathrm{E}_\emptyset\{\mathrm{E}_\emptyset(Y \mid D,\ A = 0) \mid A = 0\} \\
&= \mathrm{E}_\emptyset\{\mathrm{E}_\emptyset(Y \mid D,\ A = 0)\} \\
&\qquad \text{by (34.30)} \\
&= \mathrm{E}_\emptyset\{\mathrm{E}_\emptyset(Y \mid D,\ A = 0,\ T = 0)\} \\
&\qquad \text{by (34.28)} \\
&= \mathrm{E}_0\{\mathrm{E}_\emptyset(Y \mid D,\ A = 0,\ T = 0)\} \\
&\qquad \text{by (34.7)} \\
&= \mathrm{E}_0\{\mathrm{E}_\emptyset(Y \mid D,\ T = 0)\} \\
&\qquad \text{by (34.31)} \\
&= \mathrm{E}_0\{\mathrm{E}_0(Y \mid D,\ T = 0)\} \\
&\qquad \text{by (34.29)} \\
&= \mathrm{E}_0\{\mathrm{E}_0(Y \mid D)\} \\
&\qquad \text{by (34.3)} \\
&= E_0(Y).
\end{aligned}$$

□

Using (34.13), the above result now enables us to identify ETT from data collected under purely observational conditions. The proof extends trivially to the whole distribution of Y, not merely its expectation, allowing us to consider alternative loss functions.

IV identification in practice

For the above identification of ETT to work, we only need use an IV to create a proxy for the response of an experimental control group. We are thus more likely to be able to find an appropriate instrument than when IV methods are used to identify ACE, requiring both experimental treatment and control proxies.

A disadvantage is that (34.31) cannot be empirically tested if, as would be common, we can not observe D when $A = 0$. We will often need to rely on bold and debatable arguments as to the suitability of a purported IV.

For example, Denmark has recently outlawed the use of trans-fats (trans fatty acids) in packaged foods (Stender *et al.* 2006). Thus, if we wanted to investigate the effect of trans-fats on some health outcomes in Nordic countries, where diets might plausibly be assumed similar, we might treat 'being Danish' as an IV, so using a random sample of the Danish population as a proxy for an experimental control group, and a random sample of the population of other Nordic countries as the non-experimental treatment group. The assumption of similar diets and lifestyles is debatable, and trans-fats are so widespread in packaged foods that it might be difficult to find a large enough sample of non-experimental untreated. However, it is a plausible approach that might provide valuable information.

34.5.3 Matching and control functions

Two other methods commonly used to identify ETT from observational data are *matching* and *control functions* (Heckman and Navarro-Lozano, 2004; Heckman and Vytlacil, 2005; Rubin, 2006), developed for use in the context of labour economics.

Matching essentially defines the problem away by assuming we have an observable sufficient covariate, so allowing identification of ETT by (34.19), or indeed of ACE by (34.18).

Control functions seek to relate the observed variables to unobserved variables *via* deterministic functions – in particular, in an econometric setting *personal utility functions*, which are used to measure the likelihood of self-selection, are of crucial importance. Although the method of control functions can be given a DT formulation, we do not consider it further here, since we seek to avoid strong assumptions about unobservable deterministic relations. We do however recognize that in particular contexts, such as the economic problems for which they were introduced, such strong assumptions may be acceptable.

34.6 Discussion

We have described two related ways in which the concept of 'effect of treatment on the treated' can be given a meaningful decision-theoretic interpretation. The first operates by distinguishing between 'selection for treatment' and 'receipt of treatment'. The second makes use of the existence of an unobserved 'sufficient covariate' U. We have shown that the latter approach yields a unique value for ETT, no matter what choice may be made for U, and that this agrees with the value delivered by the former approach. There is also a formal connexion with the approach based on potential responses, ensuring agreement with that too. We have further shown that ETT can be identified so long as we have data both from the observational regime and from an experimental control group.

We have also proposed two alternative ways of estimating ETT and by using the availability trial approach in Section 34.5.1 have given an alternative interpretation of ETT as effectiveness of making treatment available per individual taking up the treatment.

Identifying ETT in our DT framework relies only on the standard probabilistic machinery once the assumptions (i)–(iii) in Section 34.3.2 are deemed to hold. We do not need to impose additional assumptions, required in the potential response framework to construct counterfactuals, such as consistency (Robins, 1986) or stable unit-treatment value assumption (Rubin, 1974). The DT approach should be more acceptable to those who appreciate the overall stochastic emphasis of the enterprise of statistical science, since it does not demand a deterministic understanding of causality such as advocated by e.g. Heckman (2005).

Whether our assumptions (i)–(iii) in Section 34.3.2, or alternatively (34.14)–(34.15) in Section 34.4, are appropriate will depend on the context under consideration, and must be evaluated in the light of specific information. But since these assumptions relate only to the actual world, not to counterfactual parallel worlds, they are relatively straightforward to think about.

Acknowledgements

We are grateful to two anonymous referees for valuable feedback and suggestions.

References

Dawid, A. P. (1979). Conditional independence in statistical theory (with Discussion). *Journal of the Royal Statistical Society, Series B 41*, 1–31.

Dawid, A. P. (2000). Causal inference without counterfactuals (with Discussion). *Journal of the American Statistical Association 95*, 407–448.

Dawid, A. P. (2002). Influence diagrams for causal modelling and inference. *International Statistical Review 70*, 161–189. Corrigenda, *ibid.*, 437.

Dawid, A. P. (2003). Causal inference using influence diagrams: The problem of partial compliance (with Discussion). In P. J. Green, N. L. Hjort, and S. Richardson (Eds.), *Highly Structured Stochastic Systems*, pp. 45–81. Oxford University Press.

Dawid, A. P. (2007). Fundamentals of statistical causality. Research Report 279, Department of Statistical Science, University College London.

Dawid, A. P. and V. Didelez (2010). Identifying the consequences of dynamic treatment strategies: A decision-theoretic overview. *Statistics Surveys 4*, 184–231. DOI 10.1214/10-SS081.

Didelez, V. and N. A. Sheehan (2007). Mendelian randomisation as an instrumental variable approach to causal inference. *Statistical Methods in Medical Research 16*, 309–330.

Forcina, A. (2006). Causal effects in the presence of non-compliance: A latent variable interpretation (with Discussion). *Metron 64*, 275–302.

Heckman, J. (1979). Sample selection bias as a specification error. *Econometrica 47*, 153–161.

Heckman, J. (2005). The scientific model of causality. *Sociological Methodology 35*, 1–98.

Heckman, J. and S. Navarro-Lozano (2004). Using matching, instrumental variables, and control functions to estimate economic choice models. *The Review of Economics and Statistics 80*, 30–57.

Heckman, J. and R. Robb (1985). Alternative methods for estimating the impact of interventions. In J. Heckman and B. Singer (Eds.), *Longitudinal Analysis of Labor Market Data*, pp. 156–245. New York: Cambridge University Press.

Heckman, J. and E. Vytlacil (2005). Structural equations treatment effects and econometric policy evaluation. *Econometrica 73*, 669–738.

Hotz, J. V., G. W. Imbens, and J. A. Klerman (2006). Evaluating the differential effects of alternative welfare-to-work training components: A reanalysis of the California GAIN program. *Journal of Labor Economics 24*, 521–566.

Okie, S. (2006). Access before approval – a right to take experimental drugs? *New England Journal of Medicine 355*, 437–440.

Pearl, J. (2009). *Causality: Models, Reasoning and Inference*, Second ed. Cambridge: Cambridge University Press.

Raiffa, H. (1968). *Decision Analysis: Introductory Lectures on Choices under Uncertainty*. Reading, Massachusetts: Addison-Wesley.

Robins, J. M. (1986). A new approach to causal inference in mortality studies with a sustained exposure period – application to control of the healthy worker survivor effect. *Mathematical Modelling 7*, 1393–1512.

Robins, J. M. (1989). The analysis of randomized and nonrandomized AIDS treatment trials using a new approach to causal inference in longitudinal studies. In L. Sechrest, H. Freeman, and A. Mulley (Eds.), *Health Service Research Methodology: A Focus on AIDS*, pp. 113–159. NCSHR, US Public Health Service.

Robins, J. M., T. J. VanderWeele, and T. S. Richardson (2006). Comment on 'Causal effects in the presence of non compliance: A latent variable interpretation' by Antonio Forcina. *Metron 64*, 288–298.

Rubin, D. (1986). Comment: Statistics and causal inference. *Journal of the American Statistical Association 81*, 968–970.

Rubin, D. (2006). *Matched Sampling for Causal Effects*. Harvard University Press.

Rubin, D. B. (1974). Estimating causal effects of treatments in randomized and nonrandomized studies. *Journal of Educational Psychology 66*, 688–701.

Rubin, D. B. (1978). Bayesian inference for causal effects: The rôle of randomization. *Annals of Statistics 6*, 34–68.

Smith, J. Q. (1988). *Decision Analysis: A Bayesian Approach*. London: Chapman and Hall.

Stender, S., J. Dyerberg, A. Bysted, T. Leth, and A. Astru (2006). A trans world journey. *Atherosclerosis Supplements 7*, 47–52.

·35·

Predicting 'It will work for us': (Way) beyond statistics

Nancy Cartwright

Abstract

A great deal of attention in evidence-based policy and practice is directed to statistical studies – especially randomized controlled trials – that support causal conclusions, which this chapter dubs 'It-works-somewhere claims'. What's needed for policy and practice, however, are conclusions that the policy will work for us, as when and how we would implement it. Despite widespread recognition of the problem of external validity, it is all too easy to suppose that conclusions of the first sort provide strong evidence for those of the second sort. This chapter argues that this is not the case. Further, 'external validity' is the wrong way to characterize the problem. Usually the only reliable way to use an it-works-somewhere result as evidence for 'It will work for us' is via what J.S. Mill calls a 'tendency' claim (and I call a 'capacity' claim). This however points out how weak 'It works somewhere' is in support of 'It will work for us', for two reasons. (1) It takes a great deal of theory, observation and experiment, far beyond the statistical study itself, to establish a tendency/capacity claim; (2) Reliable prediction requires in addition a great deal of local knowledge supplied by neither the statistical study nor the capacity claim.

35.1 Introduction

The topic of this chapter is 'external validity' and its problems. The discussion will be confined to a special class of conclusions: causal conclusions drawn from statistical studies whose fundamental logic depends on J.S. Mill's method of difference. These include randomized control trials (RCTs), case control studies and cohort studies.

These kinds of studies aim to establish conclusions of the form 'Treatment T causes outcome O' by finding a difference in the probability (or mean value) of O between two groups, commonly called the 'treatment' and the 'control' groups.[1] Given the method-of-difference idea, in order for the causal

[1] Naturally only a difference in frequency is observed. There is thus a preliminary question of statistical inference: what probabilities to infer from the observed frequencies. I set this question aside here because I want to focus on the issue of *causal* inference.

conclusion to be justified the two groups must have the same distribution of causal factors for O except T itself and its downstream effects. The underlying supposition is that differences in probabilities require a causal explanation; if the distribution of causes in the two groups is the same but for T yet the probability of O differs between them, the only possible explanation is that T causes O. The studies differ by how they go about trying to ensure as best possible that the two study groups do have the same distribution for causal factors other than T. There are, as we know, heated debates about the importance of randomization in this regard but these debates are tangential to my topic.

I want to separate issues in order to focus on a question of *use*. Suppose, contrary to realistic fact, that we could be completely satisfied that the two groups had identical distributions for the other factors causally relevant to O. I shall call this an *ideal* Mill's method-of-difference study. What is the form of the conclusion that can be drawn from that and of what use is it? In particular of what use is it in predicting whether T will cause O, or produce an improvement in the probability or mean of O, 'for us' – in a population we are concerned with, implemented as it may be implemented there?

The basic problem is that the kinds of conclusions that are properly warranted by the method-of-difference design are conclusions confined to the population in the study. That is seldom, indeed almost never, the population that we want to know about.[2] A difference in the probability of the outcome in this kind of study can at best establish what I call 'it-works-somewhere' claims and the *somewhere* is never where we aim to make further predictions. We want to know, 'Will it work for us in our target population as it would be implemented there?' This questions often goes under the label of an 'effectiveness' claim. I call it more perspicuously an 'it-will-work-for-us' claim. The problem of how to move from an it-works-somewhere claim to an it-will-work-for-us claim usually goes under the label 'external validity' and is loosely expressed as the question 'Under what conditions can the conclusion established in a study be applied to other populations?'

In this chapter I shall argue for two claims: a negative claim that external validity is the wrong idea and a positive claim that what I call 'capacities' and Mill called 'tendencies' are almost always the only right idea. The currently popular solution to the problem of external validity from philosophers and statisticians alike is to study the 'invariance' characteristics of the probability distribution that describes the population in the study. I shall argue that external validity is the wrong way to express the problem and invariance is a poor strategy for fixing it. Probabilistic results are invariant under only the

[2] Even if the entire target population were enrolled in the study, predictions will be about future effectiveness where there may be no guarantee that this population stays the same over time with respect to the causally relevant factors.

narrowest conditions, almost never met. What's useful is to establish not the invariance of the probabilistic result but the invariance of the *contribution* the cause produces, where the concept of 'contribution' only applies where a 'tendency claim' is valid. Tendencies, I shall argue, are the primary conduit by which 'it-works-somewhere' claims can support that it will work for us.

This raises a serious problem that I want to stress: Reasoning involving capacities/tendencies requires a lot more evidence and evidence of far different kinds than we are generally instructed to consider and we lack good systematic accounts of what this evidence can or should look like.[3]

In particular I shall argue:

1. We need lots more than statistics to establish tendency claims.
2. The very way tendencies operate means that building a good model to predict effectiveness is a delicate, creative enterprise requiring a large variety of information, at different levels of generality, from different fields and of different types.
3. Correlatively we need a large amount of varied evidence to back up the information that informs the model.

35.2 What can Mill's method of difference establish, even in the ideal?

I should begin with a couple of caveats. My discussion takes 'ideal' seriously. What can be done in the real world is far from the ideal and I will not discuss how to handle that obvious fact. I want to stress problems that we have even where some reasonable adjustment for departures from the ideal is possible. The second caveat is that I discuss only inferences of a narrow kind, from 'T causes O somewhere' to 'T, as T will be implemented by us, will cause O for us'. For most practical policy purposes, inferences that start from 'T cause O somewhere' need to end up with conclusions of a different form from this, often at best at 'T' will cause O' for us' where T' and O' bear some usually not very well understood relation to T and O. I suppose here that the inferences made assume at least that T and O are fixed from premise to conclusion, though other causal factors may be changed as a result of our methods of implementation.[4] With these caveats in place, turn now to the meat of what I want to discuss.

[3] Consider as a smattering of examples the evidence use guideliness from the US Dept of Education (2003), the Scottish Intercollegiate Guideline Network (2008), Sackett *et al.* (2000), Atkins *et al.* (2004) or the Cabinet Office (2000).

[4] Exactly what counts as changing T versus changing additional factors that were in place in the study but are not in place in the target implementation is a little arbitrary. But drawing a rough distinction helps make clear what additional problems still face us even if T and O are entirely fixed. (Thanks to John Worrall for urging me to make these two caveats explicit. For more on

If the conclusion that we look for in answer to the question in the title of this section is to be a causal claim (as opposed to a merely probabilistic claim) about T and O, then here is at least one valid conclusion that can be drawn using Mill's methods, supposing them applied ideally (which of course we can only hope to do approximately and even then, we seldom are in a strong position to know whether we have succeeded):

> The treatment, T, administered as it is in the study, causes the outcome, O, in some individuals in the study population, X.

This conclusion depends on the assumption that if there are more cases of O in the subpopulation of X where T obtains (the 'treatment group') than in the subpopulation in which it does not (the 'control group'), then at least some individuals in the treatment group have been caused to be O by T.

Since this conclusion depends on taking causal notions seriously and in particular on taking the notion of singular causation[5] as already given, those who are suspicious about causation tend instead to look for mere probabilistic conclusions. The usual one to cite is *mean effect size*: the mean of O in the treatment group minus the mean of O in the control group ($\langle O\rangle_T - \langle O\rangle_C$).

What about the external validity of this conclusion?

> *ESEV* (effect size external validity): When will the mean difference be the same between the study population X and a target population θ?

- *ESEV Answer 1*: If T makes the same difference in O for every member of X and θ.

This, however, is a situation that we can expect to be very rare. Usually the effect of a cause will be relational, depending in particular on characteristics of the systems affected. Consider an uncontroversial case, well-known and well-understood. The effect of gravity or of electromagnetic attraction and repulsion on the force an object is subject to depends, for gravity, on the mass of that object, and for electromagnetism, on the magnetic or electric charge of the affected object.

A more widely applicable answer than ESEV Answer 1 is available wherever the *probabilistic theory of causation* holds. This theory supposes that the probability (in the sense of objective chance) of an effect O is the same for

both issues, I suggest looking at Worrall's many papers on these subjects. Cf. Worrall (2007) and references therein.)

[5] That is, that 'T causes O in individual *i*' is already understood. Alternatively, one could presuppose the probabilistic theory of causality in which T causes O in a population ϕ that is causally homogeneous but for T and its downstream effects just in case in ϕ, Prob(O/T) > Prob(O/−T). Then if Prob(O) in the experimental population with T > Prob(O) in the experimental population with −T, we can be assured that there is a subpopulation of X in which 'T causes O'. (But note that if the two probabilities are equal, we have no reason to judge that T causes O in no subpopulations rather than that its positive effects in some cancel its negative effect in others.)

any population that has all the same causes of O and for which the causes of O all take the same value; i.e. the probability is the same for all members of a causally homogeneous subclass.[6] Loosely, 'The probability of an effect is set once the values of all its causes are fixed'. The set of causes of O that are supposed fixed in this assumption are those characteristics that appear in the antecedent of a complete and correct causal law for O.[7] The probabilistic theory of causation then provides a second sufficient condition for effect size external validity.

- *ESEV Answer 2*: When X and θ are the same with respect to
    - (a) The causal laws affecting O AND
    - (b) Each 'causally homogeneous' subclass has the same probability in θ as in X.

Sufficiency follows from the probabilistic theory of causation. In addition, these two are also almost necessary. When they do not hold then ESEV is an accident of the numbers. This can be seen by constructing cases with different causal laws (hence different subclasses that are causally homogenous) or with different probabilities for the causally homogeneous subclasses (e.g. shifting weights between those subclasses in which T is causally positive for O and those for which it is causally negative or less strongly positive).[8]

These are strong conditions, and they are recognized as such by many scholars who try to be careful about external validity. One good example appears in a debate about the legitimacy of reanalysing the results from RCTs on the effects on families from disadvantaged neighbourhoods of moving to socioeconomically better neighbourhoods. In 'What Can We Learn about Neighborhood Effects from the Moving to Opportunity Experiment?'[9] Ludwig *et al.* take the purist position: They oppose taking away lessons that the study was not designed to teach. In a section titled '*Internal versus External Validity*', these authors further caution:

> ...MTO defined its eligible sample as...[see below]. Thus MTO data...are strictly informative only about this population subset – people residing in high-rise public housing in the mid-1990s, who were at least somewhat interested in moving and sufficiently organized to take note of the opportunity and complete an application. The MTO results should only be extrapolated to other populations if the other families, their residential environments, and their motivations for moving are similar to those of the MTO population.

[6] These probabilities will be zero or one where determinism holds but not in cases where causality can be purely probabilistic.

[7] What counts as 'complete' and correct here requires some care in defining; delving into this issue takes us too far from the main topic of this chapter.

[8] The constructions resemble those illustrating Simpson's paradox. Cf. Cartwright (1979); Salmon (1971).

[9] Ludwig *et al.* (2008).

The trouble here is that RCTs are urged in the first place because we do not know what the other causes of the outcome are, let alone knowing that they have the same distribution in the study population as in possible target populations. This is a fact the authors themselves make much of in insisting that only conclusions based on the full RCT design can be drawn. For instance, they explain

> The key problem facing nonexperimental approaches is classic omitted-variable bias.

and

> A second problem ... is our lack of knowledge of which neighborhood characteristics matter ... Suppose it is the poverty rate in a person's apartment building, and not in the rest of the census tract ... [BUT an experimental] mobility intervention changes an entire bundle of neighborhood characteristics, and the total impact of changing this entire bundle ... can be estimated even if the researcher does not know which neighborhood variables matter.

The overall lesson I want to urge from this is that effect size will seldom travel from the study population to target populations and even when it does, we seldom have enough background knowledge to be justified in assuming so.

Effect size is a very precise result however. Perhaps we would be happy with something weaker, for instance, the direction of the effect. So we should ask:

> *Effect direction external validity (EDEV)*: When will an increase (resp. decrease or no difference) in the probability or mean of O given T in a study population X be sufficient for an increase (resp. decrease or no difference) in a target population θ?

There are a variety of answers that can supply sufficient conditions, including

- *EDEV Answer 1*: If X and θ
    - (a) Have the same causal laws, AND
    - (b) *Unanimity:* T acts in the same direction with respect to O in all causally homogeneous subpopulations.
- *EDEV Answer 2*: If θ has 'the right' subpopulations in the 'right' proportions.

Both these answers are still very demanding. Clearly they require a great deal of background knowledge before we are warranted in assuming that they hold. In the end I shall argue that there is no substitute for knowing a lot, though there will be different kinds of things we need to know to follow the alternative route I propose – that of exporting facts about the contributions of stable tendencies. The tendency route is often no more epistemically demanding[10] than

[10] Nor, sadly, do I think we can hope for answers that are less demanding epistemically if we want sound and valid arguments. And that's the point: we need to know what the premises are

what these answers require for exporting effect direction or effect size and tendencies are a far more powerful tool more widely applicable: Tendencies can hold and be of use across a wide range of circumstances where ESEV Answer 1 fails; they also underwrite condition EDEV 1b when it holds yet can be of use even where it fails; and they do not depend, as EDEV 2 and ESEV 2 do, on getting the weights of various subpopulations right in order to be a reliable tool for predicting direction of changes in the outcome.

Let us turn then to this alternative route, which involves exporting not probabilistic facts but causal facts. Doing so requires that we be careful in how we formulate causal claims. In particular it is important for this purpose to distinguish three different kinds of causal claim.

35.3 Three kinds of causal claim

The distinctions that matter for our discussion are those among

1. *It-works-somewhere claims:* T causes O somewhere under some conditions (e.g. in study population X administered by method M).
2. *Tendency claims:* T has a (relatively) stable tendency to promote O.
3. *It-will-work-for-us claims:* T would cause O in 'our' population θ administered as it would be administered.

35.3.1 T causes O somewhere

This is just the kind of claim that method-of-difference studies can provide evidence for; and it is important information to have. In saying this I follow, for instance, Curtis Meinert[11] when he says: 'There is no point in worrying whether a treatment works the same or differently in men and women until it has been shown to work in someone.'[12]

It-works-somewhere claims are the kind of claim that medical and social sciences work hard to establish with a reasonably high degree of certainty. But what makes these claims evidence for effectiveness claims: T will cause O for us? I have reviewed the standard answer: external validity. My alternative is tendency claims: T has a (relatively) stable tendency to promote O.

for a valid argument; only then can we get on with the serious job of seeing to what degree they can be warranted.

[11] Meinert is a prominent expert on clinical trial methodology and outspoken opponent of the US NIH diversity act demanding studies of subgroups because they generally cannot be based on proper RCT design. I agree with him about the importance of knowing it works somewhere. But my point in this chapter is that that knowledge is a tiny part of the body of evidence necessary to make reasonable predictions about what will work for us.

[12] Quoted from Epstein (2007).

35.3.2 T has a stable tendency to promote O

What are tendencies?

I have written a lot about the metaphysics, epistemology and methodology of tendencies already.[13] Here I hope to convey a sense of what they are and what they can do with a couple of canonical examples. For instance,

- Masses have a stable tendency to attract other masses.
- Aspirins have a relatively stable tendency to relieve headaches.

The driving concept in the logic of tendencies is that of a *stable contribution.* A feature, like having a mass, has a stable tendency when there is a fixed contribution that it can be relied on to make whenever[14] it is present (or properly triggered), where contributions do not always (indeed in many areas seldom) result in the naturally associated behaviours. The contribution from one cause can be – and often is – offset by contributions from features as well as unsystematic interferences. The mass of the earth is always pulling the pin towards it even if the pin lifts into the air because the magnet contributes a pull upwards. What actually happens on a given occasion will be some kind of resultant of all the contributions combining together plus any unsystematic interferences that may occur.

Reasoning in terms of contributions is common throughout the natural and social sciences and in daily life. Consider the California class-size reduction failure.[15] Here is a stripped down version of the widely accepted account of what went wrong.

There were well conducted RCTs in Tennessee showing that small class sizes improved reading scores there (that is, providing evidence for an it-works-somewhere claim). But when California cut its class sizes almost in half, little improvement in scores resulted. That is not because there was a kind of holisitc effect in Tennessee where the result depended on the special interaction among all the local factors there. Rather, so the story goes, the positive contribution of small class size was *offset* by the negative contributions of reduced teacher quality and inadequate classroom and backup support. These latter resulted because the programme was rolled out statewide over the course of a year. This created a demand for twice as many teachers and twice as many classrooms that couldn't be met without a dramatic reduction in quality. The positive contribution of small class size was not impugned by these results but possibly even borne out: The presumption seems to be that scores would have been even worse had the poorer quality teaching and accommodation been introduced without reducing class sizes as well.

[13] Cf. Cartwright (1989) and (2007a).

[14] Though note that some tenancies can be purely probabilistic and also the range of application can be limited.

[15] Bohrnstedt *et al.* (2002).

The reasoning is just like that with a magnet and gravity acting together on a pin.

Tendency claims are thus a natural conduit by which it-works-somewhere claims come to count as evidence for it-will-work-for-us claims. It should be noted however that a stable tendency to contribute a given result is not in any way universally indicated by the fact that a feature like class size participates in causing that result somewhere. Nevertheless, if a result is to be exported from a study to help predict what happens in a new situation, it can seldom be done by any other route.

The big problem for tendency logic

The central problem for reasoning involving tendencies is that we do not have good systematic accounts of what it takes to establish such claims. We have nice histories of establishing particular claims, especially in physics, but little explicit methodology. This contrasts, for instance, with it-work-somewhere claims. We have a variety of well-known well-studied methods for establishing these, methods for which we have strong principled accounts of how they are supposed to work to provide warrant for their conclusions and of where we must be cautious about their application. Recently, for instance, there has been a great deal of attention and debate devoted to Mill's-method-of-difference studies and to the advantages and disadvantages of various methods for ensuring that the requisite conditions are met that allow them to deliver valid conclusions. But if I am right that tendencies are the chief conduit by which it-works-somewhere claims come to support it-will-work-for-us, this attention focuses on only a very small part of the problem. For an it-works-somewhere claim is at best a single rock in the kind of foundation needed to support a tendency claim.

So I want to plead for more systematic work to lay out the kinds of studies and types of evidence that best support tendency claims. As best I can tell ultimately we need a theory to establish tendency claims, though admittedly often we will have to settle for our best stab at the important relevant features of such a theory. That's because contributions come in bundles and are characterized relative to each other. We only have good evidence that gravity is still working when the pin soars into the air because we can 'subtract away' the contribution of the magnet and thus calculate that gravity is still exerting its pull. To do that we need to have an idea both about what other factors make what other contributions and what the appropriate rule of composition for them is.[16]

[16] Note though the tension here: Most advocates of RCTs like them because, they claim, no substantive theory is required to do what they purport to do – i.e. establish an 'it-works-somewhere' claim.

Of course we most often have to proceed to make it-will-work-for-us predictions without a well-developed theory. In that case we make our bets. My point is that we must be clear what we are betting on and what evidence is available to back up the bet, even what kind of further evidence we should be setting out to learn. Are we betting on, and using the logic of, stable tendencies, and if so, to what extent does our evidence back us up in this? Or are we betting on facts about identical causal laws and correct distributions of other causal factors between study and target populations, and if so, to what extent does our evidence support that?

Tendencies versus external validity

My overall message is that sometimes there are tendencies to be learned about. Where there is a stable tendency, this provides a strong predictive tool for a very great range of different kinds of target populations. It naturally does not tell us what the observed result will be unless we know there are no unsystematic interferences at work, we have good knowledge of the contributions that will be made by the other causal factors present and we can estimate how these contributions combine, which is very seldom the case outside the controlled environment of a physics laboratory. But when we know a tendency claim we can make a prediction about the direction of change. Whatever the result would have been, if the cause is added the new result will differ by just the amount predictable from the contribution. But beware. The comparison we can make is with what the result would have been *post implementation*, just subtracting the effect of T itself. So, even restricting ourselves just to claims about direction of change, we still have not arrived at an 'it will work for us' claim, as I have characterized that.

Let us return to a comparison of tendencies versus external validity – predicting that 'the same' effect, either effect size or effect direction, will hold in the target as in the study population.

- Neither can be taken for granted.
- Both require a great deal of evidence to warrant them, though of different kinds.
- With respect to **effect direction**:
  - **Stable tendencies**: Post-implementation effect direction can be predicted from knowledge that T has a stable tendency to promote O (that it makes, say, a known contribution) without requiring knowledge of the distribution of other causal factors in the target.
  - **External validity**:
    - Recall by contrast that under **EDEV 2** the distribution of causally homogeneous subpopulations must be 'right' in order for the effect direction to be the same in the target as in the study population; and

of course for cases in which some set of right conditions holds, it takes considerable background knowledge of what the other causal factors are and what the target situation is like to be warranted in assuming they do.

- T has a stable tendency to promote O implies **EDEV 1.b**.
- What about **EDEV 1.a**? I have not gone into the issue of the range across which a cause must make the same contribution in order to be labelled as a tendency. Obviously there is no firm answer. What matters is that there should be good reasons to back up whatever range is presupposed in a given application. Many well-known tendencies, however, can survive a change in the other causes that affect the same outcome. Philosophers keen on modularity as a mark of genuine causation often insist that this is a widespread feature and it is often supposed in science as well. For instance most of us are familiar from elementary economics with exercises to calculate what happens if the demand laws change while the contribution to exchange from the supply side stays fixed, and vice versa. When that's the case tendency reasoning can provide predictions of effect direction that EDEV 1 cannot, though of course the assumptions that a tendency is stable across changes in other causal laws needs good arguments to back it up.

- With respect to **effect size**:
  - **Stable tendencies**. Effect size can be calculated when the contributions of all major tendencies present in a situation are known, or reasonably approximated, along with the appropriate rule of combination. This is typically what we demand from an engineering design but can surely never be supposed for social and economic policies for effects on crime, education or public health. Various narrow medical cases are generally thought of as lying in between these extremes.
  - **External validity**. It is seldom the case that the target and study populations have the same causal laws and same distribution of causal factors, and even more rare that we should be warranted in supposing so. So if the external validity of effect size is our primary method for learning something about target populations from Mill's method-of-difference studies, these studies will be of very little use to us.
- Use of the logic of tendencies is epistemically demanding. But so is external validity, only in different ways. Tendency knowledge, where available, can do more than traditional external validity reasoning and is far more widely applicable. Moreover tendency logic is well established to work well in a variety of domains. So it is wasteful and capricious to refuse to use this logic when evidence is available for it. Of course often some evidence will be available but not enough to clinch our conclusions. That

is the human condition and it applies in spades to external validity reasoning as well. When clinching evidence is missing, we had best proceed with caution and, if we can, hedge our bets.

35.3.3 It will work for us

Counterfactuals: Case-specific versus general-purpose causal models

Julian Reiss and I[17] each argue that it-will-work-for-us claims are best supported by case-specific causal models. It is not unusual among causal theorists nowadays to urge that these kinds of claims are best evaluated via causal models. After all, these are singular counterfactual claims: T would cause O if it were implemented in our population as it would be implemented there. The central difference between our claims and many others is the emphasis on 'case-specific' – i.e. on models built specifically for the counterfactual at hand, as it will be implemented.

For contrast consider the models of Judea Pearl, who has developed what must be the most detailed and thorough semantics for causal counterfactuals now available.[18] In Pearl's semantics counterfactuals are, as I advocate, evaluated on the basis of a causal model. I think I can explain the kinds of difficulties that face the use of general-purpose as opposed to case-specific models by reference to Pearl's models, without laying out details of his approach.

Causal models for Pearl are of a very specific form. The form connects neatly with our general probabilistic methods for discovering 'it-works-somewhere' claims; and this is both their strength and their weakness. For the somewhere is never here. Even if – contrary to what can ever realistically happen – a study encompasses the entire target population, the population of the study is not literally the same as the one about which future predictions are made. One may suppose that the same causal model will describe the 'same' population in the future as in the past but that is a strong assumption of external validity and it should have evidence, reason and argument to back it up.[19]

Reiss and I both stress that a causal model for evaluating 'it-will-work-for-us' claims needs to be built to the case at hand – for the given cause as, where

17 Reiss (2007); Cartwright (2007b).

18 Pearl (2000).

19 It should be noted that this is not just a reappearance of Hume's problem of induction. For the problem itself presupposes that there are general principles of some kinds at work in nature and even that we can find out about them, understand how they work and predict what kinds of conditions are required for a system to continue to operate as before. This is how we can often be confident that our interventions will not be successful because they will shift the arrangements of causes at work or undermine the operating principles. A better label for the problems for invariance I raise here is 'Mill's problem of induction' since it is the kind of worry that he described in arguing that economics cannot be an inductive science. (Mill 1836; for further discussion see Cartwright, forthcoming.)

and when it will implemented. A causal model for the system as it has been functioning or for similar systems is neither necessary nor sufficient. It is not sufficient because implementations of a cause often bring about importantly relevant changes, not only in the arrangement of other causes but also in the basic governing causal principles. It is not necessary because, as with external validity, *the same* as before or as elsewhere is the wrong idea. How the system has behaved so far or how 'similar' systems behave can be a clue to what will happen when the cause is implemented, but only a clue. We often have reason to suppose that it is the central clue; often we have reason to think it is not because we know how easily the system of laws or the arrangement of causes at work in our case might be disturbed. 'The same' causal model is just as much a hypothesis about a future case as is any new causal model proffered in its stead.

In the ideal a case-specific causal model to evaluate a specific it-will-work-for-us counterfactual will contain two essential ingredients:

- a list of 'all' the causes (or all the ones that can have a significant effect on the outcome) that will be present once the targeted cause is implemented
- a tool for calculating what happens with respect to the targeted effect when these all act together.

With this information we can predict the effect.

The trouble with causal models of this form is that we are seldom in a position to produce them with anything like a high degree of reliability. It is thus a good thing that for many kinds of predictions they are not necessary. Sometimes there are 'shortcut' models, or what following Gerd Gigerenzer[20] we might call 'cheap heuristics', that predict approximately enough the same result, sometimes even provably so, without mirroring the causal narrative that will unfold in nature as an ideal case-specific causal model does. Alternatively, sometimes there are good partial models that predict aspects of the effect, for example, estimates of effect size difference. Moreover when we are lucky an already constructed model laying out the causal laws that have governed the system till now or that govern similar systems can be taken over wholesale to serve for the specific case. But to repeat, the case-specific model that we get by this strategy is as much in need of justification as any other.

Tendencies and causal models for it-will-work-for-us claims

Tendency claims play an important role in constructing causal models for evaluating singular causal counterfactuals. Where causes act with stable tendencies we can be in a powerful situation with respect to either full or partial models because in this case the causes contribute by a systematic rule that we can learn about and encode in our theories. Otherwise prediction is

[20] Gigenerzer *et al.* (1999).

more piecemeal and local and though we often do it well, there is little good philosophical work on how to do so. So where tendencies can be relied on,[21] these will be a huge help in constructing a causal model for the evaluatsion of a specific causal counterfactual. Even if not all the causes present have a stable tendency so there are unsystematic interferences, if the targeted one has a known contribution then it may at least be possible to calculate an effect direction or even an effect difference. And certainly tendency claims are the central way by which it-works-somewhere claims can come to count as evidence that it will work for us.

Where we know of no stable tendencies then we are more at sea. I take it that we are often good at local detailed causal reasoning but that we need a great deal more concerted research on what strategies are reasonable to pursue in these areas. What matters, I believe, is to recognize the epistemic and ontological situation we are in when we want to judge if a treatment will work for us and do the best we can, hedging our bets and recognizing when we are making heroic assumptions in constructing our causal model and when not.

Given the limitations of tendency logic it is important to recognize that Pearl's semantics, and others like it, presuppose tendencies.[22] Pearl's causal models consist of a set of causal claims in functional form, one for each effect under study, with a dependent variable as effect and the independent variables as causes, plus a probability measure over the exogenous variables (i.e. those variables representing quantities not caused by other quantities represented in the set under study). To evaluate the counterfactual 'T would produce O for us'[23] Pearl substitutes for the law in the model for $t (t = f(x, y, z, \ldots))$, $t = T$, leaving all other laws in the model the same. This represents setting the value of t 'surgically', as should be done in a method-of-difference study. The assumption that this is always possible for any cause in the model is called *modularity*. The value of O that results is ultimately calculated from the law in the model for O, a law of the form $O = g(r, s, t, \ldots)$.

What we should note is that the general assumption that a system of laws is modular presupposes that the causes in that model have stable tendencies, stable at least across all the uses to which the model and its accompanying semantics is put. The contribution of a cause to an effect is given by

[21] But be careful. Many tendencies are conditional: They hold relative to an underlying structure that gives rise to them. So in using them we are betting on the stability of the underlying structure – in my language, a 'nomological machine'; and, as always, it is best to have as much evidence as possible to decide which way and how much to bet. (For a longer discussion see Cartwright, forthcoming, and 1989.)

[22] Although James Woodward (2004) does not offer a detailed semantics for counterfactuals, he is another causal theorist who makes very strong modularity assumptions, hence very strong tendency assumptions.

[23] Here I suppose that T and O are specific values that some random variable, t, o, can take.

the term in which that cause appears in the law for the effect; the rule of combination, by the functional form. Consider for instance the law, acc = $GM/r^2 + \varepsilon q_1 q_2/r^2$, for the acceleration of a particle of charge $q_2$in the vicinity of the Earth (of mass M) and of another particle of charge of q_1. The mass of the Earth makes a stable contribution of the size of its mass (M) multiplied by the acceleration of gravity G and the inverse of the square of the distance of its centre of mass from the particle. This adds vectorially with the contribution that the charge q_1 makes, which is its size multiplied by $\varepsilon q_2/r^2$. Ask now 'Would setting $q_1 = Q_1$ increase the particle's acceleration?' To answer, following Pearl, calculate acc = $GM/r^2 + \varepsilon Q_1 q_2/r^2$, substituting for the other values in this equation the values they take in the situation at hand. The assumption that the mass of the earth continues to contribute in exactly the same way as the value of the charge is changed is to suppose that the mass has a stable tendency. Similarly, to assume that the functional form for the electrostatic term stays the same, and indeed the overall functional form for acceleration does too, is to assume that charge has a stable tendency.[24] So to assume modularity for changes under every variable in the model is to make very strong tendency assumptions.[25]

I obviously have no quarrel with tendencies, having defended them for well over two decades. But we need to keep clearly in mind the lesson of Section 35.3.2. Causes often act holistically; tendencies cannot be taken to be the rule. Mill himself felt that the logic of tendencies applied in physics and in political economy but not in chemistry or more generally in the study of society[26] and Julian Reiss argues that they are not all that common even in political economy.[27] Nor are the conventional methods by which we test causal claims sufficient to establish tendency claims, especially not the wide-ranging claims about tendencies presupposed in taking a causal model to be modular. And I should stress that this is true not only for the method-of-difference

[24] This can be a misleading example because these tendencies are, or are often supposed to be basic, hence universal. As mentioned in footnote 21, most tendencies, however, depend instead on some stable underlying structure to give rise to and maintain them. So they are stable across changes that affect only arrangements in the superstructure, not necessarily across those that affect the substructure.

[25] Again, there is a serious caution to be urged. I said that Pearl's equations were of a familiar kind that we have rules for how to estimate and sufficient conditions (as with instrumental variable models or others I describe in Cartwright 2007) for determining if they can be interpreted causally. But neither these standard methods nor the sufficient conditions I know about warrant the modularity assumptions necessary to use the equations as instructed to draw counterfactual conclusions. This remark is essentially a repeat of my two-fold point that the equations, given their prescribed use in warranting counterfactual predictions, presuppose tendencies and that tendencies need a good deal more evidence to be warranted than that provided by the standard methods that warrant it-works-somewhere conclusions.

[26] Mill (1836). This would have placed Mill in the later methodenstreit (the battle of methods) more on the side of Schmoller and the holists, as opposed to Menger and those who believed in the wide applicability throughout the social sciences of the analytic method.

[27] Reiss (2007).

methods discussed in this chapter but for a wide variety of other valid methods for causal inference as well, including various econometric methods and many that trace causal pathways.

Aside on representation

This brings me to a point about representation that is somewhat more complex than the issues I have discussed so far, but one that matters to the question of how causal models help in the evaluation of it-will-work-for-us claims. Pearl, faced with challenges like mine to strong modularity assumptions, maintains that when the model is not modular that just means it is misspecified; that is, we haven't written down the right model. Whether he is right or not depends on how one conceives of his causal models. One way is to start with an independent notion of 'causal law', one that meshes at least reasonably well with our accepted methods for testing/establishing causal laws. Then one can consider how this model can be used (if at all!) to evaluate singular causal counterfactuals. If we read Pearl this way then it looks as if he offers a semantics that should allow us to evaluate any counterfactual with any variable from the model[28] in the antecedent and any variable from the model in the consequent.

This I believe is what Pearl is generally taken to be doing; and as a strategy it has exactly the problems I have described here. First, we do not have sufficient reason to take tendencies to be the rule, or even the fallback position. The causal laws governing situations are often holistic so that they are not much of a guide about what happens when the whole causal complex is no longer the same. Second is the point I have mentioned but not developed in any detail here,[29] that causal laws generally depend on some underlying structure that gives rise to and maintains them and many of the ways we implement antecedents in counterfactuals can undermine this structure in ways that destroy the very causal laws we hope to use to evaluate the counterfactual.

A usual fix for these problems is to try to extend the variables in the model. In this case the new variables would have to include descriptions of the possible underlying structures that could arise from any method of implementing a change on any variable in the model plus all new variables implicated in causal relations that the various new substructures would give rise to. Of course this is no fix for the problem of holism. As a fix for the problem that causal laws as we usually think of them and test for them depend on vulnerable substructures to support them, it seems impossible. Moreover, it is a cheat. We cannot define a proper variable whose values are the unending open-ended array of possible substructures that could exist once we start to

[28] Actually the semantics is stronger than that for it allows a mix of variables in the antecedent.

[29] For more on this point see my various discussions of nomological machines (to be found in the two references from footnote 21 plus further references in those).

intervene;[30] and if we could, it certainly would not be a random variable of the kind required in Pearl's models: Neither nature nor we supply a probability measure over any such array of possibilities.

The second way to interpret the causal laws in a model is to backread the 'causal laws' for a situation from the proffered semantics and the set of counterfactuals true for a given set of features in that situation. That is, the causal laws are whatever they have to be to allow the semantics to give correct results for the counterfactuals. This interpretation fits more closely with the claim that if the models aren't modular then they are misspecified. Probably it is easy to show that a model of Pearl form can be created that gives the correct results for any targeted counterfactual. But there is no guarantee that such a model can be created for an arbitrary collection of true counterfactuals over features under consideration, let alone a full set of them.

My own version of a causal model falls between these two. It is a model purpose built for evaluating a particular counterfactual as it would be implemented. Write down the causes of the targeted effect that will be in place given the implementation and consider what together they produce. The strength of this proposal is that it is sure to produce correct answers if we can carry it off. This is just the flip side of its chief weakness: We do not have set procedures for doing this and often are at sea.

One may wish for more. Indeed a referee for this paper expresses just this: 'We are told that we need to model the causal situation with tendencies but there is little detail on what such models would look like.' I am happy that sometimes such models will look like Pearl models, and that we could then use Pearl's semantics to generate counterfactuals with them. What I do not accept is that we can give much advice about how to build the model. I have spent a lot of time studying very successful models in physics – like models for lasers or for the gyroscopes that reveal precession due to space-time coupling in the Stanford Gravity-probe experiment, and also studying promising models in economics that are less predictively successful but are not disasters. The most I can say is that the modelling enterprise and importantly the enterprise of figuring out how good these models are *ex ante* – before they are used for prediction – seems to have no fixed rules and little good substance-neutral advice. But that I think is not only a fundamental fact about evidence; it is the human condition, better to be acknowledged and managed than denied or ignored.

30 John Worrall, in the referee's comments, suggests that many people think that the array of structures that could exist is not open-ended. I suppose they take a view of the world reflected in Wittgenstein's *Tractatus:* crudely, there are a fixed number of features in the world and the possible facts are exactly all the combinations of all the possible values of all the possible features. If I had to indulge in metaphysics, this is not one I would go for. But even if it were true, this does little in aid of establishing that there are random variables to represent this vast array since that requires reason to believe that there is a proper probability measures over it. And where does that come from?

35.4 Conclusion

My focus here has been on Mill's method-of-difference studies and what they can teach us about whether proposed interventions will have targeted effects when implemented as they would in fact be implemented (i.e. 'it-will-work-for-us', claims). These methods, I have argued, can establish claims of the form 'It works somewhere.' But it's a long road from 'It works somewhere' to 'It will work for us'.

The central problem I raise is that we do not have very good methodological guides for how to traverse this road. I argue that 'external validity' is generally a dead end: it seldom obtains and, because it depends so delicately on things being the same in just the right ways, it is even rarer that we can have reasonable warrant that it obtains. Instead tendency claims are the chief conduits by which 'it-works-somewhere' claims come to be evidence that a proposed intervention will work for us. This narrows the problem but does not solve it. For we do not have good explicit methodologies for how to establish tendency claims. Nor do we have explicit methodologies for how to use them to build case-specific models for evaluating whether the proposed intervention will work. And if I am right about how predicatively successful models are usually built even in physics, we haven't much reason to think any such methodology will be forthcoming.

What then is the role of the highly vaunted Mill's method-of-difference studies, including the current favourite, the RCT, in providing evidence that T will work for us to promote O? The ideal RCT can show that T works somewhere; a real RCT is one fallible indicator in what is hopefully a far fuller evidence base that T works somewhere. That T works somewhere can be a part, albeit a small part, of an evidence base to support T's capacity to contribute to O. That T has a capacity to promote O can serve as part, again probably only a small part, of the evidence that supports the case-specific causal model that is the eventual base for our predictions about whether T will work for us. So it is indeed a long road and most often an insecure one. But it is better to understand and acknowledge that than to presuppose heroic assumptions without admission, without examination, without evidence and without all the hedging that responsible betting calls for.

References

Atkins, D., Best, D., Briss, P.A. *et al.* [GRADE Working Group] (2004). Grading quality of evidence and strength of recommendations, *BMJ* 328 (7454): 1490 (19 June), doi:10.1136/bmj.328.7454.1490.

Bohrnstedt, G.W. and Stecher, B.M. (eds.) (2002). 'What we have learned about class size reduction in California', California Department of Education.

Cabinet Office Performance and Innovation Unit (2000). *Adding It Up: Improving Analysis & Modelling in Central Government*, London: HMSO.

Cartwright, N. (2009). How to do things with causes, Presidential Address, *Proceedings and Addresses of the APA*, (Vol. 83: 2).

Cartwright, N. (2007a). Causal laws, policy predictions and the need for genuine powers in N. Cartwright's *Causal Powers, What Are They? Why Do We Need Them? What Can and Cannot be Done with Them?*, Contingency and Dissent in Science Series, London: Centre for Philosophy of Natural and Social Science, LSE. Also in Handfield, T. (ed.), (2009). *Dispositions and Causes*, Oxford: Clarendon Press.

Cartwright, N. (2007b). *Hunting Causes and Using Them: Studies in Philosophy and Economics*, New York: Cambridge University Press.

Cartwright, N. (1989). *Nature's Capacities and their Measurement*, New York: Oxford University Press.

Cartwright, N. (1983). *How the Laws of Physics Lie*, New York: Oxford University Press.

Epstein, S. (2007). *Inclusion: The Politics of Difference in Medical Research*, Chicago: University of Chicago Press.

Gigerenzer, G., Todd, P.M., ABC Research Group (1999). *Simple Heuristics that Make us Smart*, New York: Oxford University Press.

Ludwig, J., Kling, J., Duncan, G., Katz, L., Kessler, R. and Sanbonmatsu, L. (2008). What can we learn about neighborhood effects from the moving to opportunity experiment?, *American Journal of Sociology*, 114, 144–88.

Mill, J. S. (1836 [1967]). 'On the definition of political economy and on the method of philosophical investigation in that science', reprinted in *Collected Works of John Stuart Mill*, Vol. 4, Toronto: University of Toronto Press.

Pearl, J. (2000). *Causality: Models, Reasoning and Inference*, Cambridge: Cambridge University Press.

Reiss, J. (2007). *Error in Economics: The Methodology of Evidence-Based Economics*, London: Routledge.

Sackett, D.L., Straus, S.E., Richardson, W.S., Rosenberg, W.M. and Haynes, R.B. (2000). *Evidence-Based Medicine: How to Practice and Teach EBM* (second edition), Edinburgh: Churchill Livingstone.

Salmon, W. (1971). *Statistical Explanation and Statistical Relevance*. Pittsburgh: University of Pittsburgh Press.

SIGN (Scottish Intercollegiate Guidelines Network) (2008). *SIGN 50: A Guideline Developer's Handbook (Revised edition, January 2008)*. Edinburgh: SIGN Executive.

US Department of Education Institute of Education Sciences National Center for Education Evaluation and Regional Assistance (2003). *Identifying and Implementing Educational Practices Supported by Rigorous Evidence: A User Friendly Guide. http://www.ed.gov/rschstat/research/pubs/rigorousevid/rigorousevid.pdf.*

Woodward, J. (2004). *Making Things Happen*, Oxford: Oxford University Press.

Worrall, J. (2007). Why there's no cause to randomize, *BJPS*, 58, 451–88.

PART VII
Causality and mechanisms

·36·

The idea of mechanism

Stathis Psillos

Abstract

In this chapter, I disentangle two ideas of mechanism and point to the key problems they face. Section 36.2 offers an outline of the mechanical conception of mechanism, as this was introduced in the seventeenth century and developed later on. Section 36.3 presents Poincaré's critique of mechanical mechanism in relation with the principle of conservation of energy. The gist of this critique is that mechanical mechanisms are too easy to get to be informative, provided that energy is conserved. Section 36.4 motivates the quasi-mechanical conception of mechanism and traces it to Kant's *Critique of Judgement* and to C.D. Broad's critique of pure mechanism. Section 36.5 reconstructs Hegel's critique of the idea of quasi-mechanism, as this was developed in his *Science of Logic*. Hegel's problem, in essence, was that the unity that mechanisms possess is external to them and that the very idea that *all* explanation is mechanical is devoid of content. Section 36.6 brings together Poincaré's problem and Hegel's problem and concludes that though mechanisms are not the building blocks of nature, the search for mechanisms is epistemologically and methodologically welcome.

36.1 Introduction

When we think about mechanisms, there are two general issues we need to consider. The first is broadly epistemic and has to do with the understanding of nature that identifying and knowing mechanisms yields. The second is broadly metaphysical and has to do with the status of mechanisms as building blocks of nature (and in particular, as fundamental constituents of causation). These two issues can be brought together under a certain assumption, which has had long historical pedigree, namely that nature is fundamentally mechanical.

What exactly does it mean to say that nature is *mechanical*? What is the content of this thesis? This assumption has had no concrete ahistorical conceptual content. Rather, its content has varied according to the dominant conception of nature that has characterized each epoch. Nor has it been the case that the very idea of mechanism has had a fixed and definite content. Even if in the seventeenth century and beyond, the idea of mechanism had something to do with matter in motion subject to mechanical laws, *current* conceptions of mechanism have only a very loose connection with this. A mechanism, nowadays, is virtually *any* relatively stable arrangement of entities such that, by engaging in certain interactions, a function is performed or an effect is brought about. To call a structure a mechanism is simply to *describe* it in a certain way – focusing on the steps or processes through which an effect is brought about.

This broad understanding of mechanism is typical of the new mechanical philosophy, as it is sometimes called, that has started to become a vocal, if not the dominant, approach to causation and explanation.[1] Take a very typical characterization of mechanism by Bechtel and Abrahamsen (2005, p. 423):

> (M) A mechanism is a structure performing a function in virtue of its component parts, component operations, and their organization. The orchestrated functioning of the mechanism is responsible for one or more phenomena.

On this conception,[2] a mechanism is *any* structure which is identified as such (that is as possessing a certain causal unity) via the function it performs. Moreover, a mechanism is a *complex* entity whose behaviour (that is, the function it performs) is determined by the properties, relations and interactions of its parts. This priority of the parts over the whole – and in particular, the view that the behaviour of the whole is determined by the behaviour of its parts – is the distinctive feature of this broad account of mechanism.

It will be helpful and accurate to distinguish between two concepts of mechanism – or, if you like, between two ideas of mechanism. We may call the first *mechanical* mechanism and the second *quasi-mechanical* mechanism. The first conception of mechanism is narrow: mechanisms are configurations of matter in motion subject to mechanical laws (*the laws of mechanics*). It is this conception that has been associated with the rise and dominance of the mechanical conception of nature in the seventeenth century. The key features of this conception are nicely captured by Margaret Wilson (1999, p. xiii, note 1):

> The mechanism characteristic of the new science of the seventeenth century may be briefly characterised as follows: Mechanists held that all macroscopic bodily phenomena result from the motions and impacts of submicroscopic particles, or corpuscles, each of which can be fully characterised in terms of a strictly limited range of (primary) properties: size, shape, motion and, perhaps, solidity and impenetrability.

As already noted, the second conception of mechanism is broader. A quasi-mechanical mechanism is *any* arrangement of parts into wholes in such a

[1] For defences of mechanical approaches to causation and explanation see Machamer, Darden and Craver (2000), Glennan (2002; 2008) and Craver (2007). Craver and Darden (2005) offer a nice summary/survey of recent conceptions of mechanism. For a critique of the mechanistic perspective, see Psillos (2004).

[2] For similar accounts of mechanism, see Machamer *et al.* (2000).

way that the behaviour of the whole depends on the properties of the parts and their mutual interactions. Rom Harré (1972, p. 116) has called this kind of mechanism *generative* mechanism. The focus is not on the mechanical properties of the parts, nor on the mechanical principles that govern the behaviour of the parts and determine the behaviour of the whole. Instead, the focus is on the causal relations there are between the parts and the whole. Generative mechanisms are taken to be the bearers of causal connections.[3] It is in virtue of them that the causes are supposed to produce the effects. There is a concomitant conception of mechanical explanation as a kind of decompositional explanation: an explanation of a whole in terms of its parts, their properties and their interactions. This second conception is, arguably, associated with Kant's idea of mechanism in his third critique.

In this chapter, I will disentangle these two ideas of mechanism and point to the key problems they face. Section 36.2 will offer an outline of the mechanical conception of mechanism, as this was introduced in the seventeenth century and developed later on. Section 36.3 will present Poincaré's critique of mechanical mechanism in relation with the principle of conservation of energy. The gist of this critique is that mechanical mechanisms are too easy to get to be informative, provided that energy is conserved. Section 36.4 will motivate the quasi-mechanical conception of mechanism and will trace it to Kant's *Critique of Judgement* and to C.D. Broad's critique of pure mechanism. Section 36.5 will reconstruct Hegel's critique of the idea of quasi-mechanism, as this was developed mainly in his *Science of Logic*. Hegel's problem, in essence, was that the unity that mechanisms possess is external to them and that the very idea that *all* explanation is mechanical is devoid of content. Section 36.6 will bring together Poincaré's problem and Hegel's problem and conclude that though mechanisms are not the building blocks of nature, the search for mechanisms is epistemologically and methodologically welcome.

36.2 Mechanical mechanism

In the seventeenth century, the mechanical conception of nature was taken to be a weapon against the Aristotelian view that each and every explanation was not complete unless some efficient *and* some final cause were cited. The emergent mechanical philosophy placed in centre-stage the new science of mechanics and left Aristotelian physics behind. Accordingly, the call for a mechanical explanation of phenomena has had definite content: all natural phenomena are *produced* by the mechanical interactions of the parts of matter according to mechanical laws.

The broad contours of the mechanical conception of nature were not under much dispute, at least among those who identified themselves as mechanical

[3] As (Harré 1972, p. 118) has put it: 'not all mechanisms are mechanical'.

philosophers. The key ideas were that all natural phenomena are explicable mechanically in terms of matter in motion; that efficient causation should be understood, ultimately, in terms of *pushings* and *pullings*; and that final causation should be excised from nature.[4] Though definite, this conception was far from monolithic. As Marie Boas (1952) has explained in detail, there had been different and opposing conceptions as to the structure of matter (atomistic vs corpuscularian); the reality of the void (affirmation of the existence of empty space vs the plenum); the primary qualities of matter (solely extension vs richer conceptions that include solidity, impenetrability and other properties). And yet, the unifying idea was that all explanation is mechanical explanation and proceeds in terms of matter and motion. As Robert Boyle put it, matter and motion are 'the two grand and most catholick principles of bodies' (quoted by Boas, p. 468).

Part of the appeal of the mechanical conception of nature was that it stood against a rival framework for the explanation of natural phenomena and fared better than it. For Boyle, for instance, at stake were not the details of what he called the *mechanical hypothesis*, but its being superior to its Aristotelian rival. This was judged by Boyle to be the case on the basis of the fact that the mechanical hypothesis possessed virtues such as consistency, simplicity, comprehensiveness and applicability to the phenomena that outrun its rival.

With Newton, the content of the mechanical conception of nature was altered and broadened.[5] The category of *force* was firmly introduced alongside the traditional mechanical categories of *matter* and motion. Actually, though this category was not strictly speaking new, it was for the first time set in a mechanical framework in which it was measured by the *change* in the quantity of motion it could generate. But Newton insisted that his concept of force was mathematical (cf. *Principia*, Book I, Definition VIII). Mechanical interactions were enriched to include attractive and repulsive forces between particles. Mechanical explanation was taken to consist in the subsumption of phenomena under Newton's laws.

Capitalizing on Gregor Schiemann's enlightening (2008), it can be argued that even within what I have called the mechanical conception of mechanism, there have been two distinct senses of mechanism, one wide and another narrow. The wide sense takes it that matter in motion is the ultimate cause of all natural phenomena. As such, mechanism covers everything, but its content

[4] To be sure, most mechanical philosophers did find a role for final causation via God's design of the world, but crucially, this design was precisely that of a *mechanism*. More specifically, mechanical philosophers denied the presence in nature of immanent final causes such as Aristotelian forms. Indeed, an important characteristic of the mechanical conception of nature was its denial of *forms* as part of the acceptable ontology.

[5] Not necessarily to the eyes of his contemporaries. To some (e.g. Leibniz) Newton had just abandoned the principles of mechanical philosophy, especially in light of the admission of action at a distance.

is quite unspecific, since there is no commitment to specific laws or principles that govern the workings of the mechanism. The narrow sense of mechanism, on the other hand, has it that mechanisms are governed by the *laws of mechanics*, as enunciated paradigmatically by Newton and Lagrange. Mechanics becomes privileged because it offers universal structural principles. But then, the form of the mechanical conception of nature depends on the details of the principles of mechanics and the content of the concept of mechanical mechanism is specified by the historical development of mechanics.

Schiemann draws an important distinction between monistic and dualistic conceptions of mechanics and, consequently, of mechanisms. On the monistic conception, there is only one fundamental mechanical category; on the dualistic conception, there are two fundamental categories. The monistic conception is further divided into two sub-categories: one takes matter to be the fundamental mechanical concept (called materialist, by Schiemann) while the other takes force to be the single fundamental mechanical category (called dynamic, by Schiemann). Huygens and Descartes had materialist conceptions of mechanical mechanism, while Leibniz and Kant had dynamic conceptions. The dualist conception of mechanical mechanism admits two distinct fundamental mechanical concepts – matter *and* force. Newton was a dualist in this sense and so was Helmholtz, according to Schiemann. Helmholtz's case is particularly instructive since he enunciated the principle of conservation of energy. It is precisely this principle that, as we shall see in the next section, holds the key to the very possibility of a mechanical explanation of all phenomena.[6]

With the emergence of systematic theories of heat, electricity and magnetism, one of the central theoretical questions was how these were related to the theories of mechanics. In particular, did thermal, electrical and magnetic phenomena admit of mechanical explanations?

This question was addressed in two different ways. One, developed mostly in Britain, was by means of building of mechanical *models*. These models were meant to show (a) the realizability of the system under study (e.g. the electromagnetic field) by a mechanical system; and (b) the possible inner structure and mechanisms by means of which the physical system under study operates. The other way was developed mostly in continental Europe and was the construction of abstract mechanical *theories* under which the phenomena under study were subsumed and explained. These theories were mechanical because they started with principles that embodied laws of mechanics and offered explanation by deductive subsumption. This tradition scorned the construction of mechanical models (especially of the wheels-and-pulleys

[6] As Schiemann (2008, p. 90) notes, what made the principle of conservation of energy special, at least for Helmholtz, was that energy can 'be used directly for measuring things (particularly mechanical work and heat) and their conserving properties can be examined experimentally in physical processes'.

form that many British scientists of the time were fond of). But even within this model-building tradition, especially in its mature post-Maxwellian period, mechanical models were taken to be, by and large, heuristic and illustrative devices – the focus being on the development of systematic theories (mostly based on abstract theoretical principles such as those of Lagrangian dynamics) under which the phenomena under study were subsumed and explained. Joseph Larmor (1894, p. 417) drew this division of labour clearly when he noticed

> (t)he division of the problem of the determination of the constitution of a partly concealed dynamical system, such as the aether, into two independent parts. The first part is the determination of some form of energy-function which will explain the recognised dynamical properties of the system, and which may be further tested by its application to the discovery of new properties. The second part is the building up in actuality or in imagination of some mechanical system which will serve as a model or illustration of a medium possessing such an energy function.

36.3 Poincaré's problem

How exactly was the idea of a mechanical explanation to be rendered? The problem here was not so much related to the nature of explanation as to what principles count as *mechanical*. In 1900, Henri Poincaré addressed the International Congress of Physics in Paris with the paper 'Relations entre la Physique Expérimentale et de la Physique Mathématique' (cf. 1900; this paper was reproduced as chapters 9 and 10 of his 1902). He did acknowledge that most theorists had a constant predilection for explanations borrowed from mechanics. Historically, these attempts had taken two particular forms: either they traced all phenomena back to the motion of molecules acting-at-a-distance in accordance to central force-laws; or, they suppressed central forces and traced all phenomena back to the contiguous actions of molecules that depart from the rectilinear path only by collisions. 'In a word' Poincaré said, 'they all [physicists] wish to bend nature into a certain form, and unless they can do this they cannot be satisfied'. (*ibid.*) And he immediately queried: 'Is nature flexible enough for this?'

The answer is positive, but in a surprising way. Poincaré's ground-breaking contribution to this issue was the proof of a theorem that a necessary and sufficient condition for a complete mechanical explanation of a set of phenomena is that there are suitable experimental quantities that can be identified as the kinetic and the potential energy such that they satisfy the principle of conservation of energy.[7] Given that such energy functions can be specified,

[7] The details of the proof (as well as further discussion of Poincaré's conception of mechanical explanation) are given in my (1995).

Poincaré proved that there will be *some* configuration of matter in motion (that is, a configuration of particles with certain positions and momenta) that can underpin (or model) a set of phenomena. As he put it:

In order to demonstrate the possibility of a mechanical explanation of electricity, we do not have to preoccupy ourselves with finding this explanation itself; it is sufficient to know the expressions of the two functions T and U which are the two parts of energy, to form with these two functions the equations of Lagrange and, afterwards, to compare these equations with the experimental laws (1890/1901, p. viii).

Poincaré presented these results in a series of lectures on light and electromagnetism – delivered at the Sorbonne in 1888 and published as *Électricité et Optique* in 1890 – which primarily aimed to deliver Maxwell's promise, i.e. to show that electromagnetic phenomena could be subsumed under, and represented in, a suitable mechanical framework. As Poincaré put it, he aimed to show that 'Maxwell does not give a mechanical explanation of electricity and magnetism; he confines himself to showing that such an explanation is possible' (1890/1901, p. iv). In effect, Poincaré noted that once the first part of Larmor's foregoing division of labour is dealt with, the second part (the construction of configurations of matter in motion) takes care of itself. Maxwell's achievement, according to Poincaré, was precisely this and he 'was then certain of a mechanical explanation of electricity' (1902, p. 224).

The irony was that Poincaré's demonstration had the following important corollary: if there is one mechanical explanation of a set of phenomena, i.e. if there is a possible configuration of matter in motion that can underpin a set of phenomena, there is an *infinity* of them. And not just that. Another theorem proved by the French mathematician Gabriel Königs suggested that for any material system such that the motions of a set of masses (or material molecules) is described by a system of linear differential equations of the generalized coordinates of these masses, these differential equations (which are normally attributed to the existence of forces between the masses) would be satisfied even if one replaced all forces by a suitably chosen system of *rigid connections* between these masses. Indeed, Heinrich Hertz (1894) had made use of this result to develop a system of mechanics that did away with forces altogether.

Poincaré thought that these formal results concerning the multiplicity of mechanical configurations that could underpin a set of phenomena described by a set of differential equations were natural. They were only the mathematical counterpart of the well-known historical fact that in attempting to form potential mechanical explanations of natural phenomena, scientists had chosen several theoretical hypotheses, e.g. forces acting-at-a-distance, retarded potentials, continuous or molecular media, hypothetical fluids, etc. Poincaré was sensitive to the view that even though some of these attempts had been discredited in favour of others, more than one potential mechanical model of, say, electromagnetic phenomena were still available (cf. 1900, pp. 1166–1167).[8]

So, the search for a *complete* mechanical explanation of electromagnetic phenomena was heavily underdetermined by possible configurations of matter in motion. Different underlying mechanisms could all be taken to give rise to the laws of electromagnetic phenomena. By the same token, though the possibility of a mechanical explanation of electromagnetic phenomena is secured, the empirical facts alone could not dictate any choice between different mechanical configurations that satisfy the same differential equations of motion. The choice among competing underlying mechanisms (possible configurations of matter in motion) was heavily underdetermined by the empirical facts. How then can one choose between these possible mechanical configurations? How can one find the correct complete mechanical explanation of electromagnetic phenomena? For Poincaré this was a misguided question. As he said 'The day will perhaps come when physicists will no longer concern themselves with questions which are inaccessible to positive methods and will leave them to the metaphysicians' (1902, p. 225). His advice to his fellow scientists was to content themselves with the possibility of a mechanical explanation of all conservative phenomena and to abandon hope of finding the true mechanical configuration that underlies a particular set of phenomena. He (1900, p. 1173) stressed:

> We ought therefore to set limits to our ambition. Let us not seek to formulate a mechanical explanation; let us be content to show that we can always find one if we wish. In this we have succeeded.

According to Poincaré, the search for mechanical explanation (i.e. for a configuration of matter in motion) of a set of phenomena is of little value not just because this search is massively underdetermined by the phenomena under study but mainly because this search sets the wrong target. What matters, for Poincaré, is not the search of mechanism *per se*, but rather the search for *unity* of the phenomena under laws of conservation. Understanding is promoted by the unification of the phenomena and not by finding mechanical mechanisms

[8] The turning point in Poincaré's thinking about mechanics is in his review of Hertz's (1894) for *Revue Générale des Sciences*. Concerning the 'classical system', which rests on Newton's laws, Poincaré agreed with Hertz that it ought to be abandoned as a foundation for mechanics (cf. 1897, p. 239). Part of the problem was that there were no adequate definitions of force and mass. But another part was that Newton's system was incomplete precisely because it passed over in silence the principle of conservation of energy (cf. 1897, p. 237). Like Hertz, Poincaré was more sympathetic to the 'energetic system', which was based on the principle of conservation of energy and Hamilton's principle that regulates the temporal evolution of a system (cf. 1897, pp. 239–240). According to Poincaré (1897, pp. 240–241) the basic advantage of the energetic system was that in a number of well-defined cases, the principle of conservation of energy and the subsequent Lagrangian equations of motion could give a full description of the laws of motion of a system.

that bring them about. As he said 'The end we seek (...) is not the mechanism. The true and only aim is unity'. (*ibid.*).

One may question the status of the law of conservation of energy as a mechanical principle. But that's beside the point. For the point is precisely that there is no fixed characterization of what counts as mechanical. It may well be that Poincaré's notion of mechanical explanation is too wide from the point of view of physical theory, since it hardly excludes any phenomena from being subject to mechanical explanation. Still, and this is quite important, it does block certain versions of vitalism that stipulate new kinds of forces. As is well known, in the twentieth century, the search for mechanisms and mechanical explanations was taken to be a weapon against vitalism. One key problem with vitalist explanations (at least of the sort that C.D. Broad has dubbed *substantial vitalism*) is that they are in conflict with the principle of conservation of energy and in *this* sense, they cannot be cast, even in principle, as mechanical explanations.

The significance of Poincaré's problem for the mechanical conception of mechanism can hardly be overestimated. But we should be careful to note exactly what this problem is. It is not that mechanical mechanisms are unavailable or non-existent. It is not that nature is *not* mechanical. Hence, it is not that mechanical explanation – that is, explanation in terms of mechanical mechanisms – is impossible. On the contrary, Poincaré has secured its very possibility, thereby securing, as it were, the victory of traditional mechanical philosophy over Aristotelianism. Rather, the problem for the mechanical conception of mechanism that Poincaré has identified is that mechanical mechanisms are *too* easy to get, provided nature is conservative. Under certain plausible assumptions that involve the principle of conservation of energy, the call for mechanical explanation is so readily satisfiable that it ceases to be genuinely informative.

36.4 Quasi-mechanical mechanisms

In his (1969, p. 216), A.C. Ewing drew a distinction between two conceptions of mechanical necessity in Kant's *Third Critique*. The first is related to what I have called the mechanical conception of mechanism: a determination of the properties of a whole by reference to matter in motion, and in particular by the mechanical properties of its parts and the mechanical laws they obey. The second, which Ewing calls 'quasi-mechanical', is still a determination of the properties of the whole by reference to the properties of its parts, but with no particular reference to mechanical properties and laws. This quasi-mechanical conception of mechanism is broader than the mechanical conception since there is no demand that the laws that govern the behaviour of the parts, or the properties of these parts, are mechanical – at least in the strict sense associated with the mechanical conception.

Peter McLaughlin (1990) has developed a similar account Kant's conception of mechanical explanation, according to which the mechanism of nature is a form of causation, whose differentia is that it takes it that the whole is determined by its parts. Thus understood, a mechanical explanation is a kind of de-compositional explanation: an explanation of a whole in terms of its parts, their properties and their interactions. McLaughlin bases his account on the following point made by Kant in his *Critique of Judgement* (1790/2008, p. 408):

> Now where we consider a material whole, and regard it as in point of form a product resulting from the parts and their powers and capacities of self-integration (including as parts any foreign material introduced by the co-operative action of the original parts), what we represent to ourselves in this way is a mechanical generation of the whole.

Accordingly, what renders a structure a mechanism is the fact that it possesses a reductive unity: its behaviour is determined by the properties its part have 'on their own, that is independently of the whole' (McLaughlin 1990, p. 153).

This is not the place to discuss in any detail whether this was indeed Kant's own conception.[9] The key point is that if this conception is viable at all (and, as the current mechanistic turn demonstrates, it is), then the concept of mechanism is not tied to mechanics; nor to the operation of specifically mechanical laws; nor to the ultimate determination of the behaviour of mechanism by reference to mechanical properties and interactions. Rather, the mechanism is *any* complex entity which exhibits reductive stability and unity in the sense that its behaviour is determined by the behaviour of its parts.

Kant, to be sure, contrasted mechanical explanation to teleological explanation. In its famous antinomy of the teleological power of judgement, he contrasted organisms to mechanisms. *Qua* material things, organisms (like all material things) should be generated and governed by merely mechanical laws. And yet, some material things (*qua* organisms, and hence natural purposes, as Kant put it) 'cannot be judged as possible according to merely mechanical laws (judging them requires an entirely different law of causality, namely that of final causes)'. The defining characteristics of an organism – that is of a non-mechanism – are two: (a) the whole precedes its parts and, ultimately, determines them; and (b) the parts are in reciprocal relations of cause and effect. Famously, Kant claimed that the very idea of non-mechanism (organism) is regulative and not constitutive – we have the right to proceed *as if* there were organisms (non-mechanisms) but this is not something that can

[9] There are competing views on this. Hannah Ginsborg takes it that Kant's conception of mechanism is closely tied to his account of forces and mechanical laws. For her, according to Kant, 'we explain something mechanically when we explain its production as a result of the unaided powers of matter as such' (2004, p. 42). For an attempted synthesis of Ginsborg's and McLaughlin's views, see Breitenbach (2006).

be known or proved, though Kant did think that this regulative principle is a *safe* presupposition, not liable to refutation by the progress of science.

This contrast of mechanism and non-mechanism suggests that the key feature of mechanism – what really sets it apart from organism – is the priority of the parts over the whole in the constitution of the mechanism and the determination of its behaviour.[10] It is also worth noting that it is precisely this contrast that C.D. Broad (1925) has had in mind in his own critique of mechanism.

Broad mounted an attack on what he called 'the ideal of Pure Mechanism'. This is an extreme and purified version of what I have called the mechanical conception of mechanism. Broad's Pure Mechanism is a worldview, which he (1925, p. 45) characterizes thus:

The essence of Pure Mechanism is

(a) a single kind of stuff, all of whose parts are exactly alike except for differences of position and motion;

(b) a single fundamental kind of change, viz., change of position (...);

(c) a single elementary causal law, according to which particles influence each other by pairs; and

(d) a single and simple principle of composition, according to which the behaviour of any aggregate of particles, or the influence of any one aggregate on any other, follows in a uniform way from the mutual influences of the constituent particles taken by pairs.

The gist of Pure Mechanism is that it is an ontically reductive thesis and in particular a reductive thesis with a very austere reductive basis of a single kind of fundamental particle, a single kind of change and a single causal law governing the interaction of the fundamental particles. Broad contrasted this view with two others. The first is what he called emergent vitalism. This is the view that living organisms and their behaviour cannot be fully and exhaustively determined by the properties and behaviour of their component parts, as these would be captured by the ideal of Pure Mechanism. Emergent vitalism is also opposed to a view we have already noted in Section 36.3, viz.,

[10] In her (2004), Ginsborg takes it that *qua* natural purposes, organisms are non-machine-like (and hence mechanically inexplicable) in the sense that 'they are not assemblages of independent parts, but that they are instead composed of parts which depend for their existence on one another, so that the organism as a whole both produces and is produced by its own parts, and is thus in Kant's words 'cause and effect of itself'' (2004, p. 46). This way to read Kant's account of organism distinguishes it from mechanism in two senses. (a) Organism cannot be explained in terms of the powers of matter as such; and (b) organism is such that its parts depend on the whole and cannot 'exist independently of the whole to which they belong' (2004, p. 47). Hence, what renders mechanism distinctive is precisely the fact that its unity and behaviour is determined by its parts, as they are independently of their presence in the whole. For a useful attempt to synthesise Kant's antinomy in the light of modern evolutionary biology, see Walsh (2006).

substantial vitalism: that living organisms are set apart from mechanism by an extra element (a kind of life-conferring force) that they share while pure mechanisms do not. In denying substantial vitalism, emergent vitalism puts emphasis on the structural arrangement of the whole vis-à-vis its parts and on the interaction among the parts when they are put together in a whole. A certain whole W may consist of constituents A, B, C placed in a certain relation R(A, B, C). There is emergence – emergent properties – when A, B, C cannot determine, even in principle, the properties of R(A, B, C).

Broad (1925, p. 61) put this point in terms of the lack of an in principle deducibility of the properties of R(A, B, C) 'from the most complete knowledge of the properties of A, B, and C in isolation or in other wholes which are not of the form R(A, B, C)'. This way to put the matter might be unfortunate, since what really matters is the metaphysical determination (or its lack thereof) of the whole by its parts and not deducibility *per se* – which is dependent on the epistemic situation we might happen to be in. But what matters for our purposes is Broad's thought that the denial of Pure Mechanism need not lead to the admission of spooky forces and mysterious powers, associated with substantial vitalism.

Still, our main concern here is not the opposition of Pure Mechanism to emergent vitalism, but rather its opposition to what Broad rightly took it to be a milder form of mechanism. This form, which Broad associated with what he called *Biological Mechanism*, is committed to the view that the behaviour of a whole (and of a living body in particular) is determined by its constituents, their properties and the laws they obey, but relies on a broader conception of what counts as a constituent and what laws are admissible. As Broad (1925, p. 46) put it:

> Probably all that he [a biologist who calls himself a 'Mechanist'] wishes to assert is that a living body is composed only of constituents which do or might occur in non-living bodies, and that its characteristic behaviour is wholly deducible from its structure and components and from the chemical, physical and dynamical laws which these materials would obey if they were isolated or were in non-living combinations. Whether the apparently different kinds of chemical substance are really just so many different configurations of a single kind of particles, and whether the chemical and physical laws are just the compounded results of the action of a number of similar particles obeying a single elementary law and a single principle of composition, he is not compelled as a biologist to decide.

This is, clearly, what we have called a quasi-mechanical conception of mechanism, and as Broad rightly notes, this kind of conception is enough to set the mechanist biologist apart from the emergent vitalist. The controversy need not be put, nor is it useful to be put, in terms of the ideal of Pure Mechanism.[11]

[11] In his very useful (2005), Garland Allen notes that 'operative, or explanatory mechanism refers to a step-by-step description or explanation of how components in a system interact to yield

Enough has been said, I hope, to persuade the reader that there is a distinct quasi-mechanical idea of mechanism, which – to recapitulate – proclaims a form of determination of a whole by its parts, their properties and interactions, as these would occur independently of their presence in the whole. With this in mind, let us now see what the key problem of this quasi-mechanical conception of mechanism is.

36.5 Hegel's problem

Long before Poincaré's critique of mechanical mechanism, Georg Hegel had, in his *Science of Logic*, attacked the idea that all explanation must be mechanical. According to James Kreines (2004), Hegel argued that making mechanism an absolute category – applicable to everything – obscures the distinction between explanation and description and hence undermines itself.

Hegel's writings on mechanism are rather cryptic (and perhaps, obscure). Essentially, he took the characteristic of mechanism to be that it possesses only an external unity. Its constituents (the objects that constitute it) retain their independence and self-determination, although they are parts of the mechanism. As he put it in his *The Encyclopaedia Logic* (1832/1991, p. 278) 'the relation of mechanical objects to one another is, to start with, only an external one, a relation in which the objects that are related to one another retain the semblance of independence'. And in his *Science of Logic* (2002, 711) he stressed:

> This is what constitutes the character of *mechanism*, namely, that whatever relation obtains between the things combined, this relation is one *extraneous* to them that does not concern their nature at all, and even if it is accompanied by a semblance of unity it remains nothing more *than composition, mixture, aggregation* and the like.

The determinant of the unity of a mechanism, or as Hegel put it 'the *form* that constitutes [its] difference and combines [it] into a unity' is 'an external, indifferent one; whether it be a *mixture*, or again an *order*, a certain *arrangement* of parts and sides, all these are combinations that are indifferent to what is so related' (2002, p. 713). And elsewhere, he stressed that being external, the unity of the mechanism 'is essentially one in which no *self-determination* is manifested' (2002, p. 734).

On Kreines's reading of Hegel's critique of mechanism, Hegel raised a perfectly sensible and quite forceful objection to the view that *all* explanation

a particular outcome (...)' (cf. 2005, p. 261). He contrasts this with what he calls 'philosophical mechanism' which he takes it to assert that living things are material entities. He then offers an instructive historical account of approaches to biological mechanism in the early twentieth century (and their opposition to vitalism), emphasising that 'the form that Mechanistic thinking took in the early twentieth century (...) differed from earlier (eighteenth and nineteenth-century) mechanistic traditions. It was physico-chemical not merely mechanical (...)' (2005, p. 280).

is mechanical explanation; that the only mode of explanation is mechanical; that to explain X is to offer a mechanical explanation of it.

Hegel's argument against the idea of mechanism – *qua* an all-encompassing explanatory concept – goes like this. Mechanistic explanation proceeds in terms of breaking an object down to its parts and of showing its dependence on them and their properties and relations. Explanation, then, amounts to a certain de-composition of the explanandum, viz., of a composite object whose behaviour is the result of the properties of, and interactions among, its parts. But there are indefinitely many ways to decompose something to parts and to relate it and its behaviour to them. For the call for explanation to have any bite at all, there must be some principled distinction between those decompositions that are merely descriptions of the explanandum and those decompositions that are genuinely explanatory. In particular, some decomposition – that which offers the mechanical *explanation* – must be privileged over the others, which might well reflect only pragmatic criteria or subjective interests. But how is this distinction to be drawn within the view that all explanation is mechanical? If all explanation is indeed mechanical, and if mechanical explanation amounts to decomposition, no line can be drawn between explanation and description – no particular way to decompose the explanandum is privileged over the others by being mechanical; mechanical as opposed to what? All decompositions will be equally mechanical and equally arbitrary. Hence, there will be no difference between explanation and description.

Hegel was pushing this line of argument in order to promote his own organic view of nature and, in particular, to reinstate a teleological kind of explanation – one that explains the unity of a composite object in terms of its internal purposeful activity.[12] But the point he makes is very general. In essence, Hegel's problem is that something external to the mechanism (considered as an aggregate of parts) is necessary for understanding how mechanistic explanation is possible. His general point is that the unity of a mechanism is not just a matter of arranging a set of elements into a whole; nor is it just a matter of listing their properties and mutual relations. Nor is it determined by the parts of the mechanism, as they are independently of their occurrence within the mechanism. There are indefinitely many ways to arrange parts into wholes, or to decompose wholes into parts. Most of them will be arbitrary since they will *not* be explanatorily relevant. The unity of the mechanism comes from something external to it, viz., from its function – from what it is meant to be a mechanism *for*. The function that a mechanism performs is something external to the description of the mechanism. It is the function that fixes a criterion of explanatory relevance. Some descriptions of

[12] For an informative and intelligible account of Hegel's organic world view, see Beiser (2005, chapter 4).

the mechanism are explanatorily relevant while others are not because the former and not the latter explain how the mechanism performs a certain function.

Let me illustrate this point with a couple of examples. Consider a toilet flush – a very simple mechanism indeed. What confers unity to it *qua* mechanism is the function it performs. As a complex entity, it can be decomposed into elements in indefinitely many different ways. Actually, in all probability, there is a description of it in terms of the interactions of the molecules of water and their collisions with the walls of the tank, etc. What fixes the explanatorily relevant description is surely the function it performs. Or consider telephone conversation through which some information is passed over from one end to the other – a very simple social mechanism. What confers unity to it *qua* mechanism is its function, viz., to transfer information between two ends. In all probability, there is a description of this mechanism in terms of the interactions of sound waves, collisions of particles, triggering of nerve-endings, etc. But this description is explanatorily irrelevant when it comes to explaining how this simple social mechanism performs its function. Notice that a point brought out by these examples, and certainly a point that Hegel had in mind, is that the truth of a description (supposing that it is to be had) does not necessarily render this description explanatorily relevant.

We could sum up Hegel's problem like this: first the function, then the mechanism.[13] The functional unity of the mechanism determines, ultimately, which of the many properties that the constituents of the mechanism have are relevant to the explanation of the performance and function of the whole. Hegel (1832/1991, p. 275) did think that mechanism is a form of objectivity, claimed that it is applicable to areas other than 'the special physical department from which it derives its name' but denied that it is an 'absolute category', that is constitutive of 'rational cognition in general'.

36.6 Concluding thoughts

Qua thinkers, Hegel and Poincaré were as different as chalk and cheese. Yet, they both point – with different philosophical arguments – towards a decline of the mechanistic worldview. It's not that there are no mechanisms. Actually, mechanisms, in the broad sense of stable arrangements of matter in motion, are ubiquitous. But it does not follow from this that nature has a definite mechanical structure (or, if that's too strong, that we can know which definite mechanical structure is the one actually characterizing nature). This

[13] This is indeed something that many modern mechanists have come to accept – but it is certainly not universally acknowledged among the new mechanists.

is, in essence, Poincaré's problem. How the mechanisms are individuated is a matter external to them – what counts as a mechanism, where it starts and where it stops, what kind of parts are salient and what kind of properties are relevant depend on the function they are meant to perform. The unity of the mechanisms is not intrinsic but extrinsic to them. This is, in essence, Hegel's problem. But even after a function has been determined, there are indefinitely many ways to configure mechanical mechanisms that perform it; that is, to offer a mechanical model (a configuration of matter in motion) that performs it. This is a corollary of Poincaré's problem.[14] Nature, even if it is mechanical, does not fix the boundaries of mechanisms. When it comes to the search for mechanisms, *anything* can count as a quasi-mechanism provided it performs a function that it is meant to explain. This is a corollary of Hegel's problem.

So, are mechanisms the ultimate building blocks of nature? The answer is both positive and negative. It is positive, given that the world is governed by conservation laws. But it is negative, given that mechanisms are functionally individuated: there are many ways to skin the cat!

Does the search for mechanism improve understanding? The answer is unequivocally positive. The description of a mechanism is a theoretical description and, as such, it tells a story as to how the phenomenon under study is brought about – if the story is true, our understanding of nature is enhanced. Insofar as mechanisms are taken to be functionally individuated stable explanatory structures (whose exact content and scope may well vary with our best conception of the world) which enhance our understanding of how some effects are brought about or are the realizers of certain functions, they can play a useful role in the toolkit of explanation.

Acknowledgements

Many thanks to two anonymous readers for extremely useful and sensitive comments and to Phyllis McKay Illari, Federica Russo and Jon Williamson.

References

Allen, G.F. (2005). Mechanism, vitalism and organicism in late nineteenth and twentieth-century biology: The importance of historical context, *Studies in the History and Philosophy of Biological and Biomedical Sciences* **36**: 261–283.

[14] Hegel was confident that 'not even the phenomena and processes of the physical domain in the narrower sense of the word (such as the phenomena of light, heat, magnetism, and electricity, for instance) can be explained in a merely mechanical way (i.e. through pressure, collision, displacement of parts and the like' (1832/1991, p. 195). Poincaré proved him wrong on this.

Bechtel, W. & Abrahamsen, A. (2005). Explanation: A mechanist alternative, *Studies in the History and Philosophy of Biological and Biomedical Sciences* **36**: 421–441.

Beiser, F. (2005). *Hegel*, New York: Routledge.

Boas, M. (1952). The establishment of the mechanical philosophy, *Osiris* **10**: 412–541.

Breitenbach, A. (2006). Mechanical explanation of nature and its limits in Kant's *Critique of Judgement, Studies in the History and Philosophy of Biological and Biomedical Sciences* **37**: 694–711.

Broad, C.D. (1925). *Mind and its Place in Nature*, London: Routledge and Kegan Paul.

Craver, C. (2007). *Explaining the Brain: Mechanisms and the Mosaic Unity of Neuroscience*, New York: Oxford University Press.

Craver, C. & Darden, L. (2005). Introduction, *Studies in the History and Philosophy of Biological and Biomedical Sciences* **36**: 233–244.

Ginsborg, H. (2004). Two kinds of mechanical inexplicability in Kant and Aristotle, *Journal of the History of Philosophy* **42**: 33–65.

Glennan, S. (2002). Rethinking mechanical explanation, *Philosophy of Science* **69**: S342–S353.

Glennan, S. (2008). Mechanisms, in S. Psillos and M. Curd (eds) *The Routldege Companion to Philosophy of Science*, London: Routledge, pp. 376–384.

Harré, Rom (1972). *The Philosophies of Science: An Introductory Survey*, Oxford: Oxford University Press.

Hegel, G.W.F. (1832/1991). *The Encyclopaedia Logic, Part I of the Encyclopaedia of Philosophical Sciences with the Zusätze*, (T. F. Geraets, W. A. Suchting & H. S. Harris (trans.) Indianapolis & Cambridge: Hackett Publishing Company.

Hegel, G. W. F. (2002). *The Science of Logic*, London: Routledge.

Hertz, H. (1894). *The Principles of Mechanics Presented in a New Form*, (first English Trans. 1899, reprinted by Dover, 1955), New York: Dover Publications.

Ewing, A.C. (1969). *Kant's Treatment of Causality*, Hamden Conn.: Archon Books.

Kant, I. (1790/2008). *Critique of Judgement*, N. Walker & J.C. Meredith (trans.), Oxford: Oxford University Press.

Kreines, J. (2004). Hegel's critique of pure mechanism and the philosophical appeal of the logic project, *European Journal of Philosophy* **12**: 38–74.

Larmor, J. (1894). A dynamical theory of the electric and luminiferous medium (Part I), *Philosophical Transactions of the Royal Society* **185**: 719–822; reprinted in J. Larmor *Mathematical and Physical Papers*, Vol. 1, Cambridge: Cambridge University Press (1929).

Machamer, P., Darden, L. & Craver, C. (2000). Thinking about mechanisms, *Philosophy of Science* **67**: 1–25.

McLaughlin, P. (1990). *Kant's Critique of Teleology in Biological Explanation*, Lewiston NY: Edwin Mellon Press.

Poincaré, H. (1890/1901). *Électricité et Optique: La Lumière et les Théories Électromagnétiques*, 2nd edition, Paris: Gathier-Villairs.

Poincaré, H. (1897). Les Idées de Hertz sur la Mécanique, *Revue Générale des Sciences* **8**: 734–743; reproduced in *Ouevres de Henri Poincaré*, VII, 231–250, (1952) Paris: Gauthier-Villars.

Poincaré, H. (1900). Relations entre la Physique Expérimentale et de la Physique Mathématique, *Revue Générale des Sciences* **11**: 1163–1165.

Poincaré, H. (1902). *La Science et L'Hypothése* (1968 reprint), Paris: Flammarion.

Psillos, S. (1995). Poincaré's conception of mechanical explanation, in J.-L. Greffe, G. Heinzmann & K. Lorenz (eds), *Henri Poincaré: Science and Philosophy*, Berlin: Academie Verlag & Paris: Albert Blanchard.

Schiemann, G. (2008). *Hermann von Helmholtz's Mechanism: The Loss of Certainty*, Berlin: Springer.

Walsh, D.M. (2006). Organisms as natural purposes: The contemporary evolutionary perspective, *Studies in the History and Philosophy of Biological and Biomedical Sciences* **37**: 771–791.

Wilson, M. (1999). *Ideas and Mechanism*, Princeton: Princeton University Press.

·37·

Singular and general causal relations: A mechanist perspective

Stuart Glennan

Abstract

What is the connection between general causal relations, like the relation between heating butter and butter melting, and singular causal relations, like the relation between my heating butter on my stove last night and that butter melting? The generalist view holds that singular causal relations obtain because they are instances of a general causal regularity or law. The singularist view holds the converse position. Singular causal relations can obtain even if they are not instances of causal regularities or laws, and what makes causal generalizations true, when they are true, is that they correctly describe a pattern of singular instances of causally related events.

In this chapter I make a case for the singularist view of causal relations from the perspective of a mechanistic account of causation. I explore the role of causal generalizations in the mechanistic approach, as well as in the related process and manipulability approaches to causation. I argue that, notwithstanding the centrality of such generalizations to describing mechanisms and explaining causal relationships, the most plausible metaphysical view is that singular rather than general causal relations are fundamental.

37.1 Introduction

Causal claims can be divided into two kinds – singular (token) and general (type). For instance, we claim generally that heating butter causes it to melt and, regarding a single case, that the heat that I applied to the butter in my kitchen yesterday evening caused it to melt. One of the central puzzles about the nature of causation concerns the relationship between these sorts of claims and the causal relations that underlie them. There seem to be two main options. What I shall call the generalist view holds that singular causal relations obtain because they are instances of a general causal regularity or law. What makes it true that the heat I applied to the butter last night caused it to melt is that there is a general causal law that whenever butter is heated past a certain point it will melt. What I shall call the singularist view holds the converse position. Singular causal relations can obtain even if they are not instances of causal regularities or laws, and what makes causal generalizations true, when they are true, is that they correctly describe a pattern of singular instances of causally related events. It is a basic fact that last night my heating the butter caused it to melt, and the general claim that heating butter causes it to melt is true only because it happens that in most or all of the individual cases, heating butter causes it to melt. To put the matter succinctly, the generalist holds that general causal relations make singular causal claims true, while the singularist holds that singular causal relations that make general causal claims true.[1]

My aim in this chapter is to make a case for the singularist view from the perspective of a mechanical theory of causation (Glennan 1996, 1997, 2010a, 2010c), and to explain what, from this perspective, causal generalizations mean, and what role they play within the mechanical theory. Prior to making this argument, it is important to clarify the relationship between the singularist/generalist distinction and another distinction widely discussed in the contemporary literature on causality. It is now commonly held that there are two concepts of cause – or at least that our causal assertions make two different sorts of claims (Hall, 2004, Hitchcock 2007, Godfrey-Smith 2010, Glennan 2010c). On the one hand, causes are said to produce or bring about effects. On the other hand, causes are said to depend upon, be relevant to, or make a difference to their effects. The case for thinking of these concepts as distinct is that there appear to be instances in which something can be a cause in one of these senses and fail to be a cause in the other. Two phenomena that illustrate this point are overdetermination and causation by omission. A paradigmatic example of overdetermination is a prisoner being executed by firing squad. In such a case a particular soldier's shot produces a wound that causes death, but the soldier's shot does not make a difference to the prisoner's death because the other soldiers' shots were each sufficient to cause the death. Had the first soldier's shot not hit the prisoner, the prisoner still would have died. In cases of causation by omission, on the other hand, we appear to have causes that make a difference but that are not productive. Suppose I rear-end a car because I fail to brake. In such a case my failure to brake clearly made a difference to the occurrence of the accident. Had I braked, I would not have hit the car. But my omission cannot be said to have produced the collision, because my omission is not, properly speaking, an event or occurrence. What produced

[1] Two other papers which explicitly discuss the relationship between singular and general causal claims are Hitchcock's (1995) and Russo and Williamson (2011). What I call singularist and generalist approaches, Russo and Williamson call bottom-up and top-down strategies and Hitchcock calls Humean (or neo-Humean) and generalization approaches. A third option holds that the truth of singular and general causal claims is independent of one other. This position has been argued for principally by Eells (1991). I shall not discuss this position directly, but I hope to argue indirectly against it by providing a singularist account that addresses the concerns that motivate it.

the collision was the forward momentum of my car, which was produced by my earlier pressing of the accelerator.

In what follows, I shall refer to these two concepts of cause as productivity and relevance. The significance of the productivity/relevance distinction for our discussion of singular and general causes is that the singularist view seems to fit more naturally with the productivity approach, while the generalist view seems to more naturally make sense of relevance or difference-making. The connection between these two distinctions is evident in an early paper by Sober (1984) that makes a case for there being two concepts of cause. Sober's two concepts of cause are not causal productivity and causal relevance but are rather token causation and property (or type) causation. Nonetheless, Sober makes the case that token and type causation are distinct by appealing to the difference between productivity and relevance. Sober does this in the context of probabilistic theories of causality. Most probabilistic theories of causality are type-level difference-making accounts. Probabilistic theories of causality typically assert that causes must raise the probabilities of their effects. The standard problem for difference-making accounts is that there appear to be singular causal processes in which events that lower the probability of an outcome nonetheless are productively connected to the effect – sometimes called the problem of 'doing it the hard way'. Many of these examples involve golf balls and squirrels. Suppose I have hit a putt that is heading cleanly toward the hole and there is a high probability of the ball going in. As the ball rolls towards the hole, a squirrel runs up and kicks it off its path, but fortuitously the ball ricochets off an acorn that has just dropped to the green and bounces into the hole. Sober argues that the correct way to analyze this case is to argue that 'the kind of kick' the squirrel made is a type-level negative cause of holing putts, in the sense that such kicks are negatively probabilistically relevant to holing putts, but that the particular squirrel kick is a token cause of holing this particular putt, because the particular process by which the putt made it to the hole 'traces back' to the squirrel's kick.

Notwithstanding Sober's argument, causal relevance is not essentially a property or type-level notion. Difference-making can be understood in the single case by appeal to counterfactuals. The case of omission discussed above is an example of just such a case. When I say that my failure to brake was causally relevant to my rear-ending the car, I am talking about an omission that made a difference in a particular case. But while intuitive appeal to counterfactuals allows us to make sense of singular causal relevance claims, it remains to be seen whether we can understand these singular counterfactuals in a way that does not implicitly make reference to general causal claims. The central claim of this paper is that the mechanical theory will provide a way to meet this challenge.

In the next section of the chapter we will review some traditional arguments for and against singularism. The remainder of the chapter is concerned with describing the relationship between three sorts of theories of causation: process theories, mechanical theories and manipulability theories. Section 37.3 contrasts process and mechanical theories, arguing that both theories support a singularist account of causal relations, but that mechanical theories are superior to process theories in their treatment of the problem of causal relevance. In Section 37.4, I examine the relationship between mechanical theories and manipulability theories. The upshot of this examination is that the two sorts of theories should not be thought of as competitors but as describing different aspects of the nature of causal relations. In the final section of the chapter I make the case that the mechanical theory as it has been developed in the previous sections really supports a singularist metaphysics.

37.2 Preliminary arguments for and against singularism

The modern singularist view begins as a reaction to Hume's regularity theory of causation. Hume's belief that all knowledge of matters of fact derives ultimately from impressions, combined with his view that we have no impression of a necessary connection in a single case, leads to his view that singular causal claims are true because they are instances of regular patterns of association. So, for instance, billiard ball *a*'s striking billiard ball *b* causes billiard ball *b* to move, because (1) billiard ball *a* did strike billiard ball *b*, (2) billiard ball *b* did start to move, and (3) *ceteris paribus*, whenever one billiard ball strikes another, the second begins to move. The motivation for his regularity theory is essentially epistemological and pragmatic. Singular causal sequences are instances of regular causal sequences, because, unless a singular causal sequence is an instance of a regular causal sequence, it would be impossible to recognize it as causal. Moreover, if cause–effect sequences are instances of regularities, then it is possible to predict and control effects by observing or manipulating their regular causes.

Two of the most widely discussed singularist critiques of Hume come from Ducasse ([1926] 1993) and Anscombe ([1971] 1993). Both reject the epistemological strictures that suggest it is not possible to observe causal relations in the single case. Ducasse argues that the problem arises from Hume's presumption that the connection between causally related events was some third entity, analogous to the relata. He writes:

> Hume's view that no connection between a cause and its effect is objectively observable would be correct only under the assumption that a 'connection' is an entity of the same sort as the terms themselves between which it holds, that is, for Hume and his followers, a sense impression.... [But] the fact is the causal connection is not a sensation at all, but a relation.... We observe it whenever we perceive that a certain change is the only one to have taken a place immediately before, in the immediate environment of another. (Ducasse [1926] 1993)

In describing causality as a relation, Ducasse is suggesting that it is like contiguity. When we observe two people sitting beside each other, we do not observe the first person, the second person and some third thing – 'besideness', but rather we just observe that the first person is beside the second person. Similarly, Ducasse argues, when one event causes a second, we observe the first event, the second event, and the fact that they are causally related.

Anscombe's argument is Wittgensteinian in character. People learn to use a variety of specific causal concepts – her examples are '*scrape, push, wet, carry, eat, burn, knock over, keep off, squash, make* (e.g. noises, paper boats), *hurt*' (Anscombe [1971] 1993, 93). Only when one has mastered specific causal concepts is one able to master the highly general concept of cause. For Anscombe, as for Wittgenstein, acquisition of a concept involves mastery of certain techniques. These techniques are parts of language games in which a variety of words – nouns, verbs and other kinds – are connected to behaviours and social practices. There is no 'observation' apart from mastery of these techniques, and once such mastery is achieved, observation of causal connections is no more or less problematic than observations of the objects that enter into them.

These arguments about how causal knowledge is acquired are helpful to the singularist's case but not decisive. The singularist's position is a metaphysical rather than an epistemological one, and the possibility of acquiring singular causal knowledge is neither necessary nor sufficient for establishing this metaphysical position. Davidson's position, for instance, is that Ducasse was correct that it was possible to know that a singular causal relation obtained without knowledge of a general causal law, but that the fact that a singular causal relation obtained entailed that there exists some law, even if we do not know what it is. ([1967] 1993, pp. 84–5).

Both Anscombe and Ducasse believe however that there are metaphysical or conceptual grounds for singularism. Anscombe's singularism stems from an observation she finds

> so obvious as to seem trite.... Causality consists in the derivedness of an effect from its causes. This is the core, the common feature, of causality in its various kinds. Effects derive from, arise out of, come of, their causes. For example, everyone will grant that physical parenthood is a causal relation. Here the derivation is material, by fission. Now analysis in terms of necessity or universality does not tell us of this derivedness of the effect; rather it forgets about that (Anscombe [1971] 1993, p. 92)

Anscombe here articulates the productive conception of cause. The fetus was produced by the interaction of one egg and one sperm, and the arrival of the baby in the world was produced by the act of labour. It may be the case that this is how all babies are conceived and most babies come into the world, but these general facts need not be true for the singular causal claims to hold. Greek myths tell us that the goddess Athena had a most unusual birth – springing fully armored from Zeus' head after Hephaestus cracked it open with an axe. The story is doubtless false, but there is nothing inconceivable in such a singular birth; the fact that in general whacking heads with axes is not a way to produce children does not entail that the story of Athena's birth is wrong. Ducasse echoes this point about the relation between singular causal claims and causal generalizations:

> … [T]he cause of a particular event [is defined] in terms of but a single occurrence of it, and thus in no way involves the supposition that it, or one like it, ever has occurred before or ever will again. The supposition of recurrence is thus wholly irrelevant to the meaning of cause; that supposition is relevant only to the meaning of law. And recurrence becomes related at all to causation only when a law is considered which happens to be a generalization of facts themselves individually causal to begin with. (Ducasse [1926] 1993, p. 129)

While Ducasse and Anscombe have made a strong case for the singularist intuition, in my view neither of them has provided an adequate explanation of just what this singular causal relationship is. Ducasse does offer a reductive definition of cause, but, for reasons I shall not explore, it seems wholly inadequate. Anscombe does not attempt to define what she means by determination. Presumably this is because she feels that this relation is both unanalysable and directly observable. While I do not doubt that in an ordinary sense we are quite capable of observing causal relationships, I do not think Anscombe's conclusion will do. While our conception of causality may originate in our typically successful observations of ordinary things pushing and scraping, identifying causal relationships can be far more complex. In the first place, there are circumstances – like magic shows – where our observations of ordinary causal relationships can be quite off the mark. Secondly, we often make causal inferences without observing a causal relation – as when we infer from a patient's symptoms that they have interacted with an infectious agent. An advocate of a singularist approach must then say something more about the nature of the causal relation.

While the arguments of Anscombe and Ducasse provide *prima facie* grounds for doubting Hume's view and for adopting a singularist perspective, the singularist perspective faces some important difficulties. Hitchcock's (1995) discusses one difficulty, which concerns the semantic relationship between singular and general causal claims. What Hitchcock calls the generalization strategy supposes that singular causal claims are basic, and that general causal claims should be analysed as generalizations over these singular causal claims. For instance, the general causal claim that smoking causes cancer is true because it generalizes over true singular causal claims – that Emily's smoking caused her to get cancer, that Edward's smoking caused him to get cancer, and so on. The problem with this strategy is there is not a simple relationship between generalizations and their instances. In probabilistic causal generalizations it is too strong to suppose that every instance of the singular causal relation obtain. Notwithstanding the general causal connection, not all who smoke get lung cancer. Perhaps one might treat a generalization like this as an existential one, but Hitchcock argues that there are true causal generalizations that have *no* instances. His example (*ibid.*, p. 265) is that eating one kilogram of uranium 235 causes death. This generalization, he claims, 'is true in virtue of certain features of human physiology and the physics of nuclear chain reactions; however, no one has ever died in this unusual way and it is unlikely that anyone ever will.' (*ibid.*). Hitchcock's example cannot be analysed either as:

(x)(x's ingesting 1 kg of U235 causes x to die)

or

(∃x)(x's ingesting 1 kg of U235 causes x to die).

Because no one has ingested 1 kg of U235, the first claim is vacuous and the second claim is false.

I would argue here that while Hitchcock's objection undercuts a natural sort of analysis of the relationship between singular and general causal claims, it doesn't undermine singularism as such. The problem here though doesn't have to do with singularism but with the fact that general causal claims have counterfactual import. Hitchcock's example is plausibly analysed as:

(x)(if x were to ingest 1 kg of U235, x's ingesting 1kg of U235 would cause x to die).

The singularist can maintain that the truth of this general counterfactual claim would depend upon the truth of singular counterfactual claims – for instance, that if Emily were to ingest 1 kg of U235 her ingesting it would cause her to die.

Russo and Williamson (2011) raise an epistemological objection to singularism, arguing that the singularist (or as they call it, the bottom-up) causal metaphysics is hard to square with actual practices of inference in the sciences. They divide the sorts of evidence available into evidence of (singular) mechanisms and evidence of (general) difference-making, and claim that in the health sciences at least to establish a cause one must have evidence of both types. Their argument is based upon the analysis of causal inference in the case of autopsy:

To determine that *Alf's heart attack was a cause of his death*, the medical practitioner needs to have evidence both that there is a viable biological mechanism linking heart-attack and death *and* that the heart attack made a difference to his death.... At the generic level, in order to establish that *pneumonia is a cause of death in hospital patients*, those conducting an academic autopsy need to be aware of evidence *both* of a mechanism linking pneumonia and death, *and* that pneumonia makes a significant difference to death in the population in question.... The proponent of the mechanistic analysis cannot explain why, in cases where there is excellent mechanistic evidence, evidence of difference-making is also required.

Contrary to Russo and Williamson, I think the mechanistic analysis can easily explain why evidence of difference-making is required. In their first case, the 'viable biological mechanism linking heart attack and death' is a generic description of a mechanism. Not all heart attacks cause death. Whether one does depends upon the details of the heart attack, the state of the victim's other vital systems and the place and circumstances of the heart attack. The fact that heart attacks on some occasions are linked via a physiological mechanism to a person's death makes a heart attack a *prima facie* candidate for the cause of a particular death. But to establish the heart attack as a cause of death in Alf's case, one would have to show that it made a difference in this case. One would have to show that had Alf not had the heart attack, he would not have died. If, for instance, Alf was suffering from sepsis and the sepsis brought about a failure of a number of organs including the heart, and that, given all these conditions, the heart attack did not make a difference to the death.[2] Similar arguments could be offered in response to other cases to show that a diversity of epistemic methods does not undermine the metaphysical position of the singularist.

37.3 The mechanical approach and the process approach[3]

Probably the most prominent attempt to provide a positive account of the nature of singular causal relations involves what I call *process theories*.[4] Process theories assert that a cause is related to an effect via a series of processes and interactions. Processes are world lines of objects that propagate causal influence through space-time, while interactions involve intersections of these world lines in which properties of the processes are changed. Here is a simple example: Gretchen's throwing a baseball causes the window to break because the motion of her throw (an interaction of the ball and arm processes) leads to the flight of the ball (a process) that leads to the impact with the window

[2] Identifications of cause of death would seem to involve a host of pragmatic factors, including social and legal conventions. How does one choose between proximal and distal causes, between environmental or internal causes, between various overdetermining causes? I am inclined to think that there is no objective answer to the question of what is 'the cause of death'.

[3] The material in this section summarizes an analysis of the relationship between mechanical and process approaches to causation and explanation that I have developed elsewhere (Glennan 2002, 2010a, 2010c). My discussion of process theories is of necessity very brief. For a more nuanced discussion of types of process theories, as well as rejoinders to concerns about problems of causal relevance see Dowe (2000, 2008, 2010).

[4] Process theories are sometimes called transference accounts (cf. Craver 2007) or causal-mechanical accounts, or physical accounts.

(an interaction between the ball process and the window process) that produces the break. Note here that the process is question is a particular process involving particular objects at a particular place in space and time.

Process theories emerge historically as a response to difficulties with probabilistic theories of causality. Salmon (1980) suggests such a theory is required as a response to probability-lowering cause cases like the squirrel/golf ball scenario, and Sober's (1984) suggestions about token causation are also in keeping with this view. The view has been most thoroughly developed in the work of Wesley Salmon (1984, 1994) and, more recently, in Dowe's conserved quantity theory (2000). On Dowe's conserved quantity theory, causal processes are world lines of objects that possess conserved quantities, while causal interactions are intersections of causal processes in which conserved quantities are exchanged (Dowe, 2000, p. 90). Conserved quantities are things like mass-energy, linear momentum and charge (*ibid.*, p. 91).

While process theories offer the promise of yielding a theory of singular causation, they appear to fall victim to a series of objections involving causal relevance. A first class of objections involve cases where there are processes and interactions linking irrelevant events and properties to an effect. In Hitchcock's (1995) well-known example, a chalked cue strikes a ball, changing both the colour of the ball and its linear momentum. The marking of the ball with the chalk is an interaction, but it is causally irrelevant to the outcome of the shot. A second class of objections involves causation by omission or prevention. In cases of omission, it is the non-occurrence of some potential preventing cause that causes (or at least allows) some effect to occur. For instance, my failure to turn off the alarm when I walked in the door caused the police to come to my house. In cases of prevention, the occurrence of some event prevents another event from occurring. For instance, my catching the vase as it topples off the shelf prevents it from breaking on the floor. The problem with omission and prevention is that either the putative cause (in omission) or the putative effect (in prevention) is a non-occurrence. These non-occurrences are problematic for process theories because there can be no set of processes that link non-occurring omissions to effects or preventive events to non-occurring effects. A third important objection concerns what might be called the reductionist character of process theories. Process theories are typically theories of physical causation, in the sense that they seek to identify properties of causal connection in terms of concepts drawn from current physics. But the great majority of causal claims made both in ordinary and scientific discourse involve events and processes not described in the language of physics. We seem to have good evidence for the truth of causal claims in biology, psychology, economics, history, etc. that do not make any reference to the exchange of conserved quantities or any such concept drawn from physical theory. In fact, there is a long history arguing for the autonomy of these higher-level causal claims from physical theory (e.g. Fodor 1974, Kitcher 1984). These arguments suggest that the 'gory details' at the physical level are irrelevant to the truth of causal-explanatory claims in the higher-level sciences.

In the remainder of this section I will argue that the mechanical approach to causation is a singularist approach that avoids problems with causal relevance. While the mechanical approach to causation is quite different from the process approach, readers would be excused for thinking that processes and mechanisms come down to very much the same thing. In my first paper on the subject (Glennan 1996), I argued that, roughly speaking, two events were causally connected just in case there was an intervening mechanism. This sounds very much like the process theory. Matters are not helped by the fact that some process theorists have characterized their approach as mechanistic. Salmon (1984), for instance, calls his approach to explanation 'causal-mechanical'.

The difference between the process theory and the mechanical theory lies in their rather different conceptions of what a mechanism is. For the process theorist, a mechanism just is a process of the sort described by their theory. To the mechanical theorist, however, a mechanism is a *system*. To get a sense of what this distinction amounts to, consider two widely discussed attempts to characterize a mechanism.

> Mechanisms are entities and activities organized such that they are productive of regular changes from start or set-up to finish or termination conditions. (Machamer *et al.* 2000, p. 3).

and

> A mechanism for a behavior is a complex system that produces that behavior by the interaction of a number of parts, where the interactions between parts can be characterized by direct, invariant, change-relating generalizations (Glennan 2002, p. S344).

Machamer, Darden and Craver argue that mechanisms are *organized* and that they are productive of *regular changes*. Glennan argues that mechanisms are *systems* of interacting parts, where these interactions are characterized by *generalizations*. In both of these cases we see that mechanisms are systems that have a certain degree of stability. If we consider, for instance, the circulatory mechanism in vertebrates, this system contains a number of parts – heart, arteries, capillaries, veins, and blood – that are stable in their organization and operate in a regular way over the lifetime of an organism. A second important and shared feature of these conceptions of mechanisms is their hierarchical character. The parts that comprise a mechanism may themselves be complex systems whose capacities and dispositions are explained by the regular operations of the parts' own parts. Within the circulatory system, a part of that system – say the heart – will have parts (valves, chambers, and so forth) and these parts will have parts, and the parts at each level will have characteristic activities and interactions that are productive of the behaviour of the mechanism of which they are a part. On this systems conception of mechanism, causal processes are understood as instances of the operation of mechanical systems. The circulation of blood through a particular animal's body is a continual process that results from the operation of the circulatory system.[5]

Both the regular and hierarchical nature of this approach to mechanisms contrasts with the view of process theories. The process theory focuses on a single process at a single point in space and time, as in our example of Gretchen throwing the ball and breaking the window. The collection of entities – Gretchen, the ball and the window – do not in any ordinary sense form a system. They do not act in a regular way to produce a repeatable behaviour. The Gretchen-breaking-the-window process is also not hierarchical, because the properties that are required to establish that Gretchen's throw caused the window to break are basic physical properties – exchanges of conserved quantities like momentum. It is the regular and hierarchical nature of mechanisms that provide the resources to address relevance problems. Consider first the example of the chalked cue stick. If we treat the situation in terms of the account of mechanisms given in Glennan (2002) we consider the chalk, the cue stick, the player, the balls, the table and the pocket as parts of a system describe the direct invariant change-relating generalizations that describe interactions between these parts. Some of the generalizations involved in this case would be in the form of equations describing the change in momentum of a part as a function of the momentum of a previous part at the moment of their impact. Other generalizations would describe the effect of the table on the ball as it rolled along the table and was slowed by friction. The chalking would not be part of the description of the system because changes in the chalking would not produce changes in the motion of the ball.[6]

The hierarchical character of the mechanical approach is important in avoiding the objection leveled against process accounts that the account of causation is overly reductive. Unlike the process theories, which seek to identify a physical criterion like exchange of conserved quantities that characterizes all physical interactions, on the hierarchical mechanical approach,

[5] It is not plausible to suppose that all causal processes involve regular operations of a mechanical system of this sort. The squirrel who kicks the ball against the acorn into the hole is a clear example of such an irregular causal process. Processes of this sort are examples of what I call ephemeral mechanisms (Glennan 2010b). A complete mechanistic account of causation needs to explain what these ephemeral connections are, how they are related to mechanisms on the systems conception, and how one can provide an account of productivity and relevance for connections mediated by such mechanisms.

[6] We should note here that there is a simplification involved. I am no pool player, but I suppose that if chalking is anything more than a ritual affectation, it does have an effect on the trajectories of balls – presumably by making the cue surface less slippery and allowing the player to impart spin to cue balls. So, on a careful analysis, chalking may be relevant. What couldn't be relevant is any inadvertent colouring of the ball.

causal interactions can occur at multiple levels of organization. In a circulatory system for instance, one characterizes an interaction between the blood (as a fluid) and the heart as an interaction between parts that can be characterized in terms of change-relating generalizations describing the relationship between variables such as heart rate, blood pressure and rate of blood flow. What makes these relationships causal is that they can be described by these invariant generalizations of physiology. While there is a further mechanistic explanation of why these generalizations are true, the behaviour of the mechanism will be largely invariant with respect to changes in the structure of blood and tissue at the cellular and sub-cellular level.

Craver has argued that my version of the mechanistic approach does not in fact provide a suitable solution to the problem of causal relevance. He makes his case by providing a description of a particular mechanism that characterizes that mechanism in terms of a set of causally irrelevant properties. The mechanism in question is the mechanism of long-term potentiation (LTP)—a mechanism for strengthening the connection between pre- and post-synaptic neurons by rapidly stimulating pre-synaptic neurons (Craver 2007, 92) . Craver offers a 'bizarre description' of this mechanism:

> A glutamate molecule with molecular weight w crosses the synaptic cleft at velocity v, collides with a passing protein, alters the position of amino acids in the NMDA receptor, and lowers the concentration of Na+ in the intracellular fluid.

He goes on:

> This description includes a set of parts and mechanistically explicable interactions. Each stage is linked via a mechanism to its predecessor. Yet no one would claim this is a good explanation of LTP. This is because the putative explanation is composed of irrelevant features of the synapse. It is not the molecular weight of the glutamate molecule or its velocity that matter, but rather its conformation and charge configuration... (*ibid*., p. 92)

Craver's strategy is just an application of Hitchcock's chalked ball argument to a neuroscientific example. He claims that my approach cannot rule out this bizarre description, but I think this is incorrect. What it means to say that a feature like the velocity of a particular glutamate molecule is irrelevant is to say (counterfactually) that if the velocity of the molecule had been different, the mechanism would still have produced the same behaviour, and to say (actually) that the mechanism, which involves a large number of these molecules that will move across the cleft at a variety of velocities, will produce the behaviour it does in spite of these variations – both among molecules in the synapse at a particular time and between molecules travelling across the synapse at different times. The mechanism that produces long-term potentiation will utilize interactions that can be characterized by invariant change-relating generalizations. Change-relating generalizations describe functional relations between two or more variables where an ideal intervention on one variable will bring about a change in another variable. According to Craver's 'correct' description of the LTP mechanism for instance, the binding of glutamate to an NMDA receptor changes the conformation of the receptor in order to open a channel for Ca^{2+} (*ibid.*, p. 70). This is an interaction between glutamate molecules and the receptor in which one change – binding to the receptor – produces another change – opening a channel. The reason that irrelevant characteristics, like the velocity of the glutamate model, are not included in a description of the interactions is that the behaviour of the mechanisms (and the generalizations describing interactions between its parts) are invariant under interventions that change these characteristics.

What appears to have happened here is that in his criticism Craver has appealed implicitly to the Salmon-Dowe conception of an interaction, as opposed to the sort that I advocate, in which the interactions must be interactions that are part of a mechanism that produce a particular behaviour and must interact in accordance with invariant change-relating generalizations. While the version of the mechanistic account I favor addresses the causal relevance problem, potential objections to the mechanical theory remain. First of all, it may be, because of the centrality of my appeal to invariant change-relating generalizations in characterizing what constitutes a mechanism, that the mechanistic theory is in reality just a version of Woodward's counterfactual-based manipulability theory. Second, it may also appear that the prominent appeal to generalizations solves the relevance problem only by moving away from the singularist stance. I shall address the first of these concerns in Section 37.4 of this chapter and the second in Section 37.5.

37.4 The mechanical approach and the manipulability approach

There has been considerable discussion in the literature about the relationship between the mechanistic approach and counterfactual approaches to causation. In Glennan (1996) I argued that the mechanistic approach explained the truth conditions for counterfactuals in a way that was more epistemically and scientifically helpful than that of Lewis (1973); but in Glennan (2002) I suggested that Woodward's counterfactual account of invariant generalizations was essential to characterizing an interaction between parts of a mechanism. Machamer and Bogen have argued that a mechanistic approach to causation allows one to avoid counterfactuals altogether, a point that has been criticized by Psillos, Woodward and myself (Bogen 2004, Glennan 2010a, Psillos 2004, Woodward 2004). Woodward has suggested that one can give a definition of a mechanism, or at least a mechanical model, in terms of manipulability criteria (Woodward 2002). For myself, I have come to believe that the mechanical

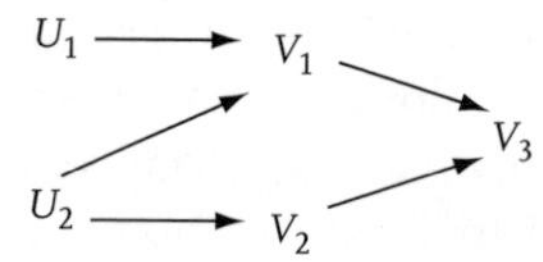

Fig. 37.1 A directed acyclic graph.

theory and the manipulability theory – at least as it is advocated by Woodward (2003), Pearl (2000), and Spirtes, Glymour and Scheines (2000) – are not really rival theories, but rather highlight different features of a unified conception of causation. The manipulability account emphasizes procedures for discovery, prediction and control. The mechanical account provides different sorts of resources for discovery and prediction, a metaphysical underpinning for the manipulability approach and an enriched conception of causal explanation.

To understand how the manipulability theory and the mechanical theory are related, it is necessary to examine how causal relationships are typically represented in the manipulability theory. I base my exposition on Pearl (2000) and Woodward (2003), which I take to represent different pieces of a single approach to causation[7] Pearl and Woodward assume that causal relationships can be represented as relations between variables, where variables can take Boolean values representing the occurrence or non-occurrence of an event, integer valued variables representing discrete states, or continuous quantities representing different values of variable magnitudes. Causal relationships between variables are represented by causal models (or structural equation models). Causal models consist of a set of exogenous variables, a set of endogenous variables, and for each endogenous variable, a function from some subset of the variables (endogenous or exogenous) – its parents – to that variable (Pearl 2000, p. 203). A causal model will determine a directed acyclic graph (DAG). For instance, Figure 37.1 shows a DAG involving two exogenous and three endogenous variables:

This graph would be determined by a causal model involving functions of the following variables:

$$\begin{aligned} v_1 &= f_1(u_1, u_2) \\ v_2 &= f_2(u_2) \\ v_3 &= f_3(v_1, v_2). \end{aligned}$$

Causal models provide a way of factoring dependence relationships among variables, so that the value of a variable depends only upon some subset of

[7] Woodward (2003, p. 38) acknowledges his debt to Pearl. As Woodward sees it, his theory complements Pearl's. While Pearl is more focused on questions of inference, Woodward is more concerned with providing explicit definitions of notions like being a total or contributing cause, and more generally with relating Pearl's approach to causation to the philosophical literature.

other variables – its Markovian parents. Parents are clearly represented in the DAG notation. We can see, for instance, that V_3's parents are V_1 and V_2, and that conditional on its parents, the value of V_3 will be independent of U_1 and U_2.

I claim that a causal model is a representation of a mechanism in the sense described in the mechanical theory. I will argue for this by way of an example. In Glennan (1996), one of the examples I used of a mechanism was a toilet. A toilet is a mechanism for a certain behaviour, which for purposes of illustration we can describe as follows: When the handle is pulled, water is released from the storage tank into the bowl, and the storage tank is refilled. Here is how the mechanism works. Pulling the handle (H) pulls a chain (C) which opens the flapper valve at the bottom of the tank (B_1). The open flapper valve allows the water to empty out of the tank (T_1), which has two effects. First, it causes the bottom valve to close (B_2) and second it causes a float to drop (F_1). The dropping of the float opens the float valve (V_1) allowing water to enter the tank. The opening of the float valve together with the closing of the flapper valve causes the tank to fill (T_24) which causes the float to rise (F_2) and the float valve to close (V_2). Figure 37.2 shows the DAG.

This graph is a representation of a system of parts whose interactions can be characterized by a set of direct invariant change-relating generalizations. These generalizations will be in the form of equations characterizing the relation between each endogenous variable and its parent or parents (the f_i's). This representation does lose some information one might want in the model of a mechanism. First, the model characterizes the mechanism with a set of binary-valued variables. In fact the system contains some continuously varying magnitudes – the level of the water, the amount that the valve is opened, and so on. The causal modelling approach can however handle quantitative variables. Perhaps more importantly, the variables here do not represent parts (like the float valve) but rather changes in the state of the parts (like the opening or closing of the float valve). This is connected to the fact that the DAG representation does not illustrate the cyclical nature of the toilet mechanism, in which a part like a valve begins in a closed position, is opened, and is closed again.

If a causal model is to be a representation of a mechanism, it requires another feature, which Woodward calls *modularity*. If a model is modular, it must be possible in principle to intervene in order to change the value of a

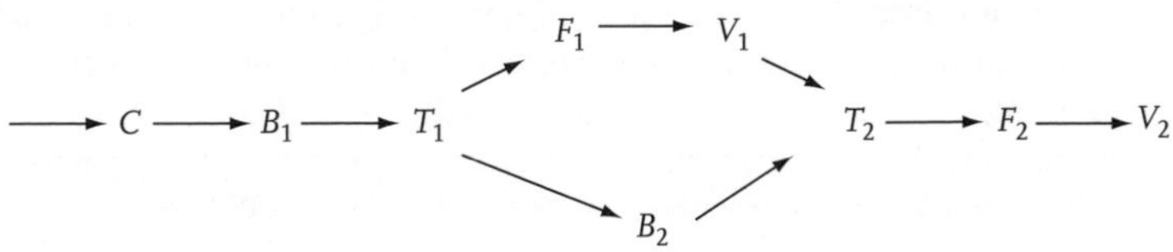

Fig. 37.2 A DAG for a toilet mechanism.

dependent variable without altering any of the other functional relationships in the model. Modularity so defined is a property of models, but it corresponds to an important property of the mechanisms modeled. In the case of the toilet, for instance, the modularity condition implies that one should be able to intervene on the state of one part and not thereby alter any of the functional relationships downstream of that part. For instance, one ought to be able to intervene on the chain, pulling it up, and not thereby interfere with the functional relationship between the water level and the float. Woodward offers the following motivation for the modularity requirement:

> It is natural to suppose that if a system of equations correctly and fully represents the causal structure of some system, then those equations should be modular. One way of motivating this claim appeals to the idea that each equation in the system should represent the operation of a distinct causal mechanism. (Correlatively, each complete set of arrows directed into each variable in a directed graph should also correspond to a distinct mechanism.) If we make the additional plausible assumption that a necessary condition for two mechanisms to be distinct is that it be possible in principle to interfere with the operation of one without interfering with the operation of the other and vice versa, we have a justification for requiring that systems of equations that fully and correctly represent causal structure should be modular. (Woodward 2003, p. 48)

While I have suggested that the causal model as a whole is, given the modularity assumption, a model of a mechanism, here Woodward (cf. Pearl 2000, sections 1.3, 7.2.4) suggests that a single structural equation, representing the causal relationship between a node and its parent nodes, represents a 'distinct causal mechanism'. So, continuing with our example, an equation describing how pulling the chain relates to opening the flapper valve represents a distinct causal mechanism. Thus, it is clear in this instance that what Woodward and Pearl mean by a causal mechanism is what I have called an interaction between parts.

In another article, where Woodward responds directly to the systems account of mechanism he understands mechanisms more in the manner of Glennan (1996) and Machamer, Darden and Craver (2000). He offers the following description of a necessary condition for a causal model to be a model of a mechanism:

> **(MECH)** a necessary condition for a representation to be an acceptable model of a mechanism is that the representation (i) describe an organized or structured set of parts or components, where (ii) the behavior of each component is described by a generalization that is invariant under interventions, and where (iii) the generalizations governing each component are also independently changeable, and where (iv) the representation allows us to see how, in virtue of (i), (ii) and (iii), the overall output of the mechanism will vary under manipulation of the input to each component and changes in the components themselves. (Woodward 2002, p. S375).

In this characterization, the whole causal model (or directed graph) is the representation of a mechanism. My supposition is that when Woodward says that the behaviour of each part is characterized by an invariant generalization, he really means that there are generalizations (perhaps multiple) describing both activities (behaviour of the part) and interactions (relations between the behaviour of one part and the behaviour of directly connected parts). Given this definition, his view of what constitutes a mechanical model essentially coincides with Glennan (2005). His view is similar to that of Machamer, Darden and Craver, except that he insists, contrary to Machamer (2004) and Bogen (2004) that the characterization of activities and interactions between parts of mechanisms requires counterfactuals.

It may be significant that Woodward says both that causal models as a whole and equations representing the relation between parts of a mechanical system are both representations of mechanisms. This suggests that Woodward is in agreement with the account I sketched earlier regarding the hierarchical character of mechanisms, and specifically with a thesis I have defended elsewhere (Glennan 1996, 1997) that the generalizations describing the interactions between parts of mechanisms are in most cases mechanically explicable. Consider again the toilet mechanism. The chain connecting the lever to the flapper valve is a part of the toilet mechanism, but it is also itself a mechanism. The chain has parts (links) whose properties and interactions explain the behaviour of the chain as a whole. Thus, an invariant change-relating generalization describing how pulling on the top of the chain will change the position of the bottom of the chain, will be a mechanically explicable generalization. Similarly, we can treat each link of the chain as a part of the chain and we can describe how each link is connected via a mechanically explicable change-relating generalization.

A final important similarity between the mechanical and manipulationist approach concerns the understanding of the semantics and epistemology of counterfactuals. In Glennan (1996) I argue that one of the virtues of the mechanical approach over Lewis' counterfactual approach is that it provides an unproblematic way to understand and evaluate counterfactuals by reference to mechanisms. Given a model of a mechanism that exhibits the functional dependence of variables that represent the mechanism's parts and their properties, one evaluates a counterfactual claim by using the model to calculate what would happen if one were to intervene and fix the value of a variable to the antecedent of the counterfactual. For instance, in the case of the toilet one knows that if the chain were broken then the tank would not empty, because, if one were to intervene and break the link between C (the chain being pulled) and B_1 (the flapper valve opening), then all the events downstream of C would not occur.

Judea Pearl (2000, chapter 7) has developed a complete analysis of what he calls 'structure-based counterfactuals' that formalizes this approach in terms of structural equation models and his 'do operator'. Like Glennan (1996), Pearl sees this analysis of providing an analysis of the truth conditions of counterfactuals that does not rely on the metaphysically extravagant and cognitively/epistemologically problematic possible-worlds semantics of Lewis:

In contrast with Lewis' theory, counterfactuals are not based on an abstract notion of similarity among hypothetical worlds; instead they rest directly on the mechanisms (or 'laws' to be fancy) that produce those worlds and on the invariant properties of those mechanisms. Lewis' elusive 'miracles' are replaced by principled minisurgeries, $do(X = x)$ which represent the minimal change (to a model) necessary for establishing the antecedent $X = x$ (for all [values of exogenous variables] u) (Pearl 2000, p. 239)[8]

Pearl appears to understand the term 'mechanism' here as Woodward does in his discussion of modularity – as productive relationships between parts of systems that can be manipulated by interventions. It is interesting here how Pearl equates mechanisms and laws. It suggests a deflationary view of laws in which laws are simply descriptions of dependence relations between parts of particular systems rather than exceptionless universals. Pearl's view appears, like Woodward's, to be consistent with the position argued for in Glennan (1996, 1997) that laws are mechanically explicable.

These observations about the relationship between my analysis of causality and the manipulationist counterfactual approach of Woodward and Pearl should suffice to show how the two approaches are interconnected. The mechanical approach relies on the counterfactual approach because there is no way to define interactions between parts of mechanisms except by appeal to counterfactual-supporting generalizations. The counterfactual approach relies on the mechanical approach because the truth-conditions for counterfactuals depend upon the structure of mechanisms.

Psillos (2004) has also argued for an account of causation that seeks to 'harmonize' mechanisms and counterfactuals, and most of what I have said here is consistent with Psillos' explication of the relation between mechanisms and counterfactuals. Psillos has, however, argued that 'there is a sense in which the counterfactual approach is more basic than the mechanistic one in

[8] The formal analysis is not without difficulties. In particular, it may strike readers as problematic that the truth of a counterfactual claim is relativized to a model of a mechanism rather than to the mechanism itself. Of course if we are to appeal to mechanisms to make judgments about the truth of counterfactuals, we must inevitably rely on our models of these mechanisms, but we would like the truth itself not to depend on our representation. A second issue has to do with background conditions. For Pearl, a counterfactual 'If it were the case that $X = x$, then it would be the case that $Y = y$' will be true only if the model calculates $Y = y$ for all values of background variables U. This may be too strict. For instance, consider the counterfactual 'If the flapper valve weren't to close properly, water would keep running into the bowl'. Intuitively, this seems to be true, but its truth depends upon a certain background condition remaining constant, namely that the water supply to the toilet is kept on. If the water supply were included as an exogenous variable in the model, then the counterfactual would not be true. These complications do not seem to me to undermine the structural approach, but rather to be an inevitable consequence of the vagueness of counterfactuals.

that a proper account of mechanisms depends on counterfactuals while counterfactuals need not be supported (or depend on) mechanisms' (Psillos 2004, p. 288). To complete our analysis of the relationship between the mechanical and manipulationist approaches, we need to assess Psillos' claim.

The feature of my account upon which Psillos bases his claim is my claim that *most but not all* causal generalizations (or laws) are mechanically explicable. According to my account, the relationship between causes, causal generalizations and mechanisms is this: Two events are causally connected when there is an intervening mechanism. An intervening mechanism consists of a number of interacting parts, and these interactions are truly interactions (as opposed to accidental correlations) because they are described by invariant change-relating generalizations, which support counterfactuals. But if these generalizations are mechanically explicable, then what ultimately makes it true that the parts interact is that these interactions are produced by the operation of further, lower-level mechanisms. These mechanisms will in turn be systems of parts interacting in accordance with invariant change-relating generalizations, and these generalizations may too be mechanically explicable. Ultimately, however, one will reach a level where the parts of a mechanism interact, but where there is no further mechanism that explains this interaction. These are the fundamental interactions. What makes it the case that these relationships are truly interactions? The answer would seem to be that there is some basic, mechanically inexplicable, counterfactual dependence between events, perhaps one that holds in virtue of a fundamental law. As Psillos sees it, 'the presence of a mechanism is *part* of a metaphysically sufficient condition for the truth of certain counterfactuals; the fully sufficient condition includes some facts about the fundamental laws that, ultimately, govern the behavior of the mechanism' (*ibid.*, p. 310).

Psillos is correct that the mechanical approach cannot eliminate counterfactuals, and because of this that it cannot provide a complete and reductive analysis of causal relations. This fact does not, however, entail Psillos' asymmetry claim. In the first place, it appears that counterfactuals really do need mechanisms. We have seen in the above analysis of the manipulability theory that causal models are models of mechanisms and that in Pearl's analysis of the semantics of counterfactuals, their truth conditions depend upon the structure of mechanisms. At least on Pearl's analysis, and arguably on Woodward's, what makes a certain counterfactual claim true is that there is a mechanism that would respond in a certain way to a manipulation. And like the mechanical theory, the manipulability approach faces a charge of a prima facie circularity and uses the same strategy to respond to that circularity. Woodward, as he himself notes (2003, pp. 103–107), defines causal relations in terms of the outcomes of possible interventions, and interventions are themselves kinds of causing. Woodward's response to this potential objection is to argue that the circularity is not vicious. To determine whether two variables X and Y are causally related, we must know something about other causal relationships (e.g. between an intervention I and X), but not about the causal relationship between X and Y we are seeking to establish. True enough, but how do we know that an intervention I causes a change in X? Presumably we know this because there is a mechanism connecting I and X. This will involve further variables (representing further parts and interactions), and how can we know that these variables are connected? By further interventions of course! Just as the mechanist must ultimately run out of nested mechanisms, Woodward must ultimately get to an intervention that cannot be further analysed in terms of interventions on further mechanisms. One is left with the brute claim that if one were to intervene on X a change in Y would result. But this is not an analysis of the counterfactual dependence of X and Y; it is just a restatement of it. Thus I would contend that the truth of causal claims according to the manipulability theory will depend upon an unanalysed notion of counterfactual dependence.

Psillos' argument for the asymmetry of mechanisms and counterfactuals is really an epistemological one (Psillos 2004, pp. 315–317). It is possible to construct a perfectly randomized experiment that establishes a causal connection between a treatment and a control. One can establish this connection without having any idea of what the mechanism is, and indeed, one could establish this even if there is no mechanism, but just a brute pattern of causal dependence. But one should not let this fact mislead one into thinking that the manipulationist approach has provided a metaphysical grounding for causal relations any more than the mechanistic account has. I do not think Woodward would object to this characterization of the situation, because he is emphatic that his analysis of causation is not reductive. It does not seek to ground the truth of causal claims in some ultimately non-causal state of affairs, but rather to explain the relationship between certain causal claims and others. This seems like a wise idea, especially given Woodward's focus on causal explanation, but, like the mechanical theory, Woodward's theory leaves crucial metaphysical questions unanswered. These have to do with the ultimate truth grounds for claims of counterfactual dependence at the level of fundamental physics, where the notion of causal interaction cannot be explicated by appeal to further mechanisms. How we understand these truth grounds will turn out to have a crucial impact on our understanding of our original question – whether causal claims are ultimately singular.

37.5 The mechanical approach and the grounding of singular causal claims

We are now finally in a position to make the argument that the mechanical approach supports a singularist view of causation. The basic reason why

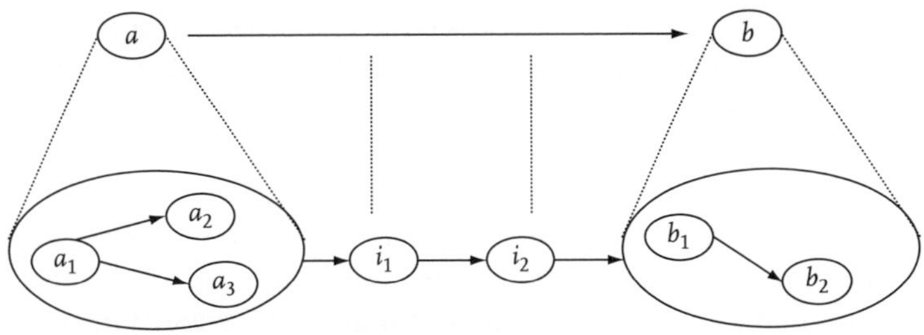

Fig. 37.3 A hierarchical mechanism.

the mechanical approach is a singularist one is that it suggests that causal interactions are mediated by mechanisms, and mechanisms are particular systems of interacting parts, where these interactions occur at a particular place and time. On this view, causal generalizations are generalizations about the behavior of mechanisms, and they are true because mechanisms do or would behave in the way described on actual or hypothetical occasions. The problem that remains is that our definition of mechanism frequently makes reference to causal generalizations, and the suspicion will arise that the truth of singular causal claims depends ultimately on the truth of these generalizations, especially the non-mechanically explicable generalizations upon which the causal productivity of mechanisms would ultimately seem to depend.

Because the central issue concerns the implications of the hierarchical character of mechanisms for the status of singular causal relations, it will be helpful to have an abstract representation of a hierarchical mechanism, as in Figure 37.3.

The top of the diagram is a representation of two parts of a mechanism interacting. The dotted lines represent constitutive relations and the solid arrows represent causal interactions. The constitutive lines indicate that the part a is itself a mechanism with three parts and part b is a mechanism with two parts, where these parts interact as indicated by the arrows. The interaction between a and b may involve the operation of a further mechanism, as indicated by the parts i_1 and i_2. Although not pictured in this diagram, we should imagine that the parts in the lower level of the diagram themselves have parts, and that the arrows representing interactions between these parts are themselves complex mechanisms with subparts. This hierarchy of mechanisms can go a long way down but will ultimately bottom out with fundamental parts and interactions. Imagine the parts to be atoms or corpuscles, much in the way Descartes imagined, and that these parts have some set of basic properties (e.g. mass, charge), interacting with each other in a manner determined by these properties.[9]

[9] A problem that I can only allude to here has to do with the implications of quantum mechanics for how we understand fundamental interactions. The picture I offer of a fundamental inter-

To defend the singularist interpretation, we must explicate the role of causal generalizations in this picture. The definition of a mechanism in Glennan (2002) claims that interactions between parts 'can be characterized by direct, invariant, change-relating generalizations'. What exactly are these generalizations and what is their relation to the interaction? One reading would be to treat these generalizations as a form of causal law, and to argue that the causal interactions are *governed* by these causal laws. If this were the case, the singular causal claim that a change in a property of *a* produces a change in a property of *b* would be made true by the causal law. This reading would undermine the singularist account, but it is not consistent with the hierarchical character of mechanisms and the mechanical explicability of these generalizations. The hierarchical picture suggests a second reading in which change-relating generalizations are statements that 'characterize' the interaction, but the interaction itself involves the operation of the underlying mechanism and is not governed by the generalization. For example, there might be a change-relating generalization indicating that when I ingest caffeine, my motor activity increases. This generalization is true and reliable, but it simply characterizes the outcome of a complex metabolic mechanism.

Ontologically, the crucial point to observe is that mechanisms are not universals but particulars. They are structured collections of parts which occupy a certain region of space and which interact over a certain definite period of time. We characterize these mechanisms by generalizations because very often a mechanism's behaviour is repeatable. My body is a mechanism that on repeated occasions interacts with coffee, and coffee repeatedly and reliably has an effect on my behaviour. Not only does the very same mechanism exhibit repeatable behaviour, but particular mechanisms may be instances of types with consistent behaviour. My metabolic mechanisms are broadly similar to those of other human beings and, as a consequence, there will be generalizations (say about the ingestion of caffeine) that hold true of my body and the bodies of many others. But these generalizations are true in virtue of the fact that these mechanisms can and do operate in particular ways on particular occasions, rather than conversely. This understanding of these generalizations also accounts for why they always involve approximations and are only true *ceteris paribus*.

action is essentially a classical Cartesian/Newtonian one. Indeterminism in quantum mechanics raises some problems for this picture. More significant though are problems raised by the measurement problem and by violations of locality. The mechanistic picture seems to require bottom-level interactions that are local and have definite properties independent of measurement. I don't have anything constructive to say about this problem. I can only offer as consolation the fact that except under special conditions quantum mechanical peculiarities wash out as one gets past sub-atomic scales. Wherever this point is, we can treat it as the fundamental level with respect to the hierarchy of mechanisms.

This explanation of the role of causal mechanisms is available so long as the generalizations are mechanically explicable, but here we come to what may seem the key metaphysical issue. If mechanisms are truly going to explain how one event produces another, all of the interactions between parts, at all levels in the hierarchy of mechanisms, will need to be genuinely causally productive. If it were to turn out that these interactions at the fundamental level were not truly interactions, then none of the putative causal relations mediated by mechanisms would be genuine (cf. Psillos 2004, Craver 2007).

We are now concerned with an interaction between two (or more) parts at the bottom of a mechanistic hierarchy. These parts interact (by hypothesis) in accordance with a change-relating generalization or law. But how are we to understand the relationship between the generalization and this interaction? There seem to be three main metaphysical possibilities:

(1) **Humean lawlessness** – The interaction nothing more than an instance of a pattern that is described by a generalization.

(2) **Nomological determination** – The interaction is governed by the generalization (law).

(3) **Singular determination** – The interaction is a singular case of causal determination and any generalizations describing interactions are true in virtue of there being a general pattern of such singular instances.

The first view is the position that fits most naturally with the Mill–Ramsey–Lewis (MRL) view of laws.[10] On this view, laws are statements that provide the best balance of simplicity and strength in characterizing events within the world. If there are (as there appear to be in our world) a relatively small number of kinds of fundamental-level parts and some relatively simple generalizations describing how these parts behave in relation to each other, then these generalizations would be obvious candidates for MRL laws. This small set of laws, together with a much larger volume of information about how these parts are organized into hierarchies of mechanisms, will provide a simple and powerful description of the pattern of events in the world. I borrow the term 'Humean lawlessness' from Stephen Mumford (2004), who argues persuasively, that the MRL laws aren't truly laws, because they supervene on particulars of the actual world. They describe a pattern, but they do not create or explain the pattern. Such a view is anti-realist with respect both to laws and to causes.

The second view – nomological determination – holds that fundamental interactions are governed by laws. This view requires some form of nomological realism about fundamental laws, such as the Armstrong–Dretske–Tooley (ADT) view. A law on this view is some third metaphysically real entity, apart

[10] There has been a good deal of recent literature on alternate interpretations of laws of nature, including the MRL and ADT views. Two helpful introductions are Psillos (2003) and Carroll (2008).

from the particular events, which makes it true that one event produces the second. Causal relations are real but subordinate to nomological relations.

The third view, singular determination, holds that there are genuine interactions between parts at the bottom of the mechanistic hierarchy, but that these parts are not governed by laws. In calling these interactions genuine, I am suggesting that the relationship is a modal one. We can express the modality of the relationship counterfactually: When a change in *a* produces a change in *b*, it follows (with the usual caveats about overdetermination, etc.) that if *a* had not changed, *b* would not have changed. But the counterfactual locution should be understood not as a claim about non-actual worlds, but a claim about the determining power of *a* in this world. The singular determination view is the view that is consistent with Anscombe's and Ducasse's arguments for singular determination, in the sense that it supports the basic idea that whether a particular event (or change in an object) causes another is at bottom a fact about the relationship between these two events and is ultimately independent of any facts about other events.

Because the mechanistic approach to causation requires that there be genuine causal connections between parts all of the way down the mechanistic hierarchy, it appears that the question of whether or not the mechanistic approach is a genuinely singularist one would essentially depend upon which of these metaphysical options is correct. If Humean lawlessness is operative at the fundamental level, then mechanisms are simply parts of the pattern of events in the world and they cannot imbue relations between those events with any genuine sort of causal necessity. Alternatively, if fundamental interactions are *governed* by fundamental laws, then the truth of all claims about productive relations between entities at any level in a mechanistic hierarchy will depend upon these laws. Finally, if causal relationships between events at the fundamental level involve singular determination, then so too will events at all levels of the mechanistic hierarchy.

Each of these metaphysical possibilities is genuine in the sense that each of them is consistent with the pattern of objects and events that both science and common experience reveal in the world. If this is the case, which of them should we accept? One approach would be to reject this question as meaningless on the grounds that the options are not empirically distinguishable. While I am not entirely unsympathetic to this sort of anti-metaphysical response, I do think there are arguments that may lend credence to one or more of these metaphysical positions. I cannot hope to survey the many arguments that have been offered in support or criticism of these positions. I can only offer here some explanation of why I think the metaphysic of singular determination fits naturally with the mechanistic approach to causation that I have argued for.

In the first place, it is difficult to reconcile our intuitions about manipulation, which are central to the mechanistic view, with the Humean view. On the Humean view there are no such things as genuine modal relationships (or necessary connections as Hume would say) between events. Moreover, singular counterfactual claims are not really claims about what would have happened in a single case. Causal and counterfactual expressions are elliptical ways of talking about complicated patterns in the experience of the actual world. Because of this, manipulations or interventions are not modally effective ways to change the world; they are simply part of the pattern of the actual world. Manipulation, like all other forms of causing, is shown to be a fiction of the human mind. What patterns or regularities there are in the world just are. There are no explanatory principles to account for them. I concur with Mumford's summary judgment of this sort of metaphysical view – 'irrefutable, but neither compelling, appealing nor intuitive' (Mumford 2004, p. 33). What I think makes it unappealing and non-intuitive is that it is inconsistent with the belief that we manipulate things and cause things in the single case.

The main argument in favour of the nomological determination view is simply that it is implausible to suppose that the order and regularity that we find in nature would exist without laws.[11] Suppose, for the sake of illustration, that patterns of gravitational attractions between bodies are correctly described by the generalizations we call Newton's laws of motion and the law of universal gravitation, and furthermore that these generalizations are not mechanically explicable. Statements of these laws collectively entail certain claims about regular patterns of behavior in the objects – for instance, that satellites will travel around planets in elliptical orbits. It seems quite reasonable, the defender of nomological determination would contend, that these objects behave as they do because they are governed by these general laws. If these laws do not govern their instances, there would be no explanation for the existence of this regularity in nature.

In response to this argument, the singularist must contend that in each particular interaction, the change in one entity produces a change in the other entity, and the fact of this productive relationship does not depend in any way on a general relationship between properties or instances. In the gravitational case, each body acts in each instant on the other body, producing accelerations which, over time, produce elliptical orbits. We do live in a world in which fundamental interactions fit within general patterns, but from this it does not follow that it is in virtue of falling under those patterns that the productive relationship holds. It is possible that we could live in a higgledy-piggledy world in which causes determined effects but in which these determination relations were not regular.[12]

[11] This is what Mumford (2004, chapter 5) refers to as the nomological argument.

[12] Not all will agree. Heathcote (1991, pp. 63–73), from whom I borrow the term 'higgledy-piggledy world', conclude that this situation , while conceivable, is not possible, and that there is an a posteriori necessary connection between laws and causes.

Woodward's view of the relationship between invariant generalizations and counterfactuals appears to support the singularist view. Woodward suggests that generalizations used in explanations are explanatory only if those generalizations support a particular sort of interventionist counterfactual – which he calls a 'same object' counterfactual. If a generalization characterizes an interaction between parts of a mechanism, the relevant counterfactuals that must be true would have to do with what would happen if one were to intervene to change a property of one of those very same parts. It is explanatorily irrelevant whether or not those generalizations hold counterfactually of other objects.[13]

One potential objection to the singularist interpretation concerns the relevance of properties to causation. A proponent of nomological determination would point out that causal relations appear to hold in virtue of properties of the related events. For instance, the ringing of the alarm clock caused me to wake in virtue of its loudness. Thus, if one takes laws to be relations between properties, it may seem natural to infer that causes depend upon laws. The singularist, however, can argue that on a particular occasion a causal relation between events may hold in virtue of certain properties of those events but not hold to a nomological theory of causation. The singularist does not deny the importance of properties in characterizing causal relations, but insists that it is not essential to causal relations that the relationship between cause and effect be the same on different occasions.

This argument is moreover bolstered by the mechanistic view on the nature of properties and their relations to objects. I have argued (Glennan 2010c) that much of the literature on causation and laws suffers from a property bias – a tendency to think of causal relations as relations between properties without recognizing how properties themselves depend upon particulars. Consider for example a property of butter – whether it is in a solid or liquid state. Butter is not a basic substance, but a combination of a number of different types of fats – saturated and unsaturated. Whether a fat is saturated or not is in turn dependent upon the molecular structure of the fat. When we say that heating the butter will cause the butter to melt, this is, on the face of it, a change-relating generalization involving properties – changing the temperature of the butter will change the solid/liquid state of the butter. But the properties themselves are not basic facts about the substance butter, but depend instead upon the particular structure of the butter – the molecules that make it up and their arrangement, as well as the arrangement and bonding of submolecular structures within these fat molecules. One consequence of this is that there is no such property as *the* melting point of butter. Different samples of butter will have different kinds and proportions of fat molecules which will have different consequences for their interaction with heat.

[13] I am grateful to an anonymous referee for this observation. See Woodward (2003, section 6.9.)

This point about properties applies only to higher-level properties. If an object has fundamental properties like mass and charge whose presence or causal role cannot be explained by reference to the organization and interaction of parts of that individual, then we cannot show how those properties depend upon particulars. For this reason, the observation about property bias is not decisive at the fundamental level. In favour of the singularist interpretation of fundamental interactions we can only say that it has the advantage of providing a consistent picture of the role of properties and laws in characterizing causal relations.

Where does this all leave us with regard to our original question about the relationship between singular and general causal relations? If the argument has succeeded it has shown that a mechanistic approach to causation is consistent with a singularist causal metaphysics. This is so even though causal generalizations are part and parcel of the apparatus we use to describe and manipulate mechanisms and to formulate causal explanations. Moreover, the singularist picture is the simplest one for the defender of a mechanistic approach to causation, because it fits most naturally with the view that the causal mechanisms which are the truth-makers for causal claims operate at particular locations in space and time. One of the virtues of the mechanistic approach to causation is that it at once fits nicely with a singularist metaphysics and explains the centrality of causal generalization to our epistemic and explanatory practices.

References

Anscombe, G.E.M. [1971] 1993. Causality and determination. In *Causation.*, eds. Ernest Sosa, Michael Tooley, pp. 88–104. Oxford: Oxford University Press.

Bogen, Jim. (2004). Analysing causality: The opposite of counterfactual is factual. *International Studies in the Philosophy of Science* 18 (1): 3–26.

Carroll, John W. (2008). Laws of nature. In *Stanford Encyclopedia of Philosophy.*, ed. Edward N. Zalta. Fall 2008 Edition.

Craver, Carl F. (2007). *Explaining the Brain: What a Science of the Mind–Brain Could Be.* New York: Oxford University Press.

Davidson, Donald. [1967] 1993. Causal relations. In *Causation.*, eds. Ernest Sosa, Michael Tooley, 75–87. Oxford: Oxford University Press.

Dowe, Phil. (2010). Causal process theories. In *Oxford Handbook of Causation.*, eds. Helen Beebee, Christopher Hitchcock and Peter Menzies. New York: Oxford University Press.

Dowe, Phil. (2008). Causal processes. In *The Stanford Encyclopedia of Philosophy (fall 2008 edition).*, ed. Edward Zalta. URL = http://plato.stanford.edu/archives/fall2008/entries/causation-process/:.

Dowe, Phil. (2000). *Physical Causation.* Cambridge Studies in Probability, Induction, and Decision Theory. Cambridge: Cambridge University Press.

Ducasse, Curt J. [1926] 1993. On the nature and observability of the causal relationship. In *Causation.*, eds. Ernest Sosa, Michael Tooley. Oxford: Oxford University Press.

Eells, Ellery. (1991). *Probabilistic Causality*. Cambridge, MA: Cambridge University Press.

Fodor, Jerry. (1974). Special sciences (or: The disunity of science as a working hypothesis). *Synthese* 28 (2): 97–115.

Glennan, Stuart S. (2010a). Mechanisms. In *Oxford Handbook of Causation.*, eds. Helen Beebee, Christopher Hitchcock and Peter Menzies, pp. 315–325. New York: Oxford University Press.

Glennan, Stuart S. (2010b). Ephemeral mechanisms and historical explanation. *Erkenntnis* 72 (2): 251–255.

Glennan, Stuart S. (2010c). Mechanisms, causes and the layered model of the world. *Philosophy and Phenomenological Research* 81 (2): 362–381.

Glennan, Stuart S. Modeling mechanisms (2005). *Studies in the History and Philosophy of the Biological and Biomedical Sciences* 36 (2): 375–388.

Glennan, Stuart S. (2002). Rethinking mechanistic explanation. *Philosophy of Science* 69, (3 Supplement): S342–53.

Glennan, Stuart S. (1997). Capacities, universality, and singularity. *Philosophy of Science* 64 (4) (D): 605–26.

Glennan, Stuart S. (1996). Mechanisms and the nature of causation. *Erkenntnis* 44 (1) (Ja): 49–71.

Godfrey-Smith, Peter. (2010). Causal pluralism. In *Oxford Handbook of Causation.*, eds. Helen Beebee, Christopher Hitchcock and Peter Menzies. New York: Oxford University Press.

Hall, Ned. (2004). Two concepts of causation. In *Causation and Counterfactuals.*, eds. John Collins, Ned Hall and L. A. Paul, pp. 225–276. Cambridge, MA: Bradford Book/MIT Press.

Heathcote, Adrian, and D. M. Armstrong. (1991). Causes and laws. *Noûs* 25 (1): 63–73.

Hitchcock, Christopher. (2007). How to be a causal pluralist. In *Thinking about Causes: From Greek Philosophy to Modern Physics.*, eds. Gereon Woters, Peter Machamer, pp. 200–221. Pittsburgh: University of Pittsburgh Press.

Hitchcock, Christopher Read. (1995). Discussion: Salmon on explanatory relevance. *Philosophy of Science* 62 (2) (June): 304–20.

Hitchcock, Christopher Read. (1995). The mishap at Reichenbach fall: Singular vs general causation. *Philosophical Studies: An International Journal for Philosophy in the Analytic Tradition* 78 (3) (June): 257–91.

Kitcher, Philip. (1984). 1953 and all that: A tale of two sciences. *Philosophical Review* 93 (JL): 335–74.

Lewis, David. (1973). Causation. *Journal of Philosophy* 70: 556–567.

Machamer, Peter. (2004). Activities and causation: The metaphysics and epistemology of mechanisms. *International Studies in the Philosophy of Science* 18 (1): 27–39.

Machamer, Peter, Lindley Darden, and Carl F. Craver. (2000). Thinking about mechanisms. *Philosophy of Science* 67 (1): 1–25.

Mumford, Stephen. (2004). *Laws in Nature*. New York: Routledge.

Pearl, Judea. (2000). *Causality: Models, Reasoning, and Inference*. Cambridge: Cambridge University Press.

Psillos, Stathis. (2004). A glimpse of the secret connexion: Harmonizing mechanisms with counterfactuals. *Perspectives on Science* 12 (3): 288–319.

Psillos, Stathis. (2003). *Causation and Explanation.* Montreal: McGill-Queen's University Press.

Russo, Federica, and Jon Williamson (2011). Generic Versus Single-Case Causality: The Case of Autopsy. *European Journal for Philosophy of Science. DOI: 10.1007/s13194–010–0012–4.*

Salmon, Wesley C. (1997). Causality and explanation: A reply to two critiques. *Philosophy of Science* 64 (3) (S): 461–77.

Salmon, Wesley C. (1994). Causality without counterfactuals. *Philosophy of Science* 61 (2) (Je): 297–312.

Salmon, Wesley C. (1984). *Scientific Explanation and the Causal Structure of the World.* Princeton: Princeton University Press.

Salmon, Wesley C. (1980). Probabilistic causality. *Pacific Philosophical Quarterly* 61, 50–74.

Sober, Elliott. (1984). Two concepts of cause. *PSA: Proceedings of the Biennial Meeting of the Philosophy of Science Association* 1984, (Volume Two: Symposia and Invited Papers): pp. 405–24.

Spirtes, Peter, Clark N. Glymour, and Richard Scheines. (2000). *Causation, Prediction, and Search.* Lecture Notes in Statistics. 2nd edn. Vol. 81. Boston: MIT Press.

Woodward, James. (2004). Counterfactuals and causal explanation. *International Studies in the Philosophy of Science* 18, (1): 41–72.

Woodward, James. (2003). *Making Things Happen: A Theory of Causal Explanation.* Oxford: Oxford University Press.

Woodward, James. (2002). What is a mechanism? A counterfactual account. *Philosophy of Science* 69 (3 Supplement): S366–77.

·38·

Mechanisms are real and local

Phyllis McKay Illari and Jon Williamson

Abstract

Mechanisms have become much-discussed, yet there is still no consensus on how to characterize them. In this chapter, we start with something everyone is agreed on – that mechanisms explain – and investigate what constraints this imposes on our metaphysics of mechanisms. We examine two widely shared premises about how to understand mechanistic explanation: (1) that mechanistic explanation offers a welcome alternative to traditional laws-based explanation and (2) that there are two senses of mechanistic explanation that we call 'epistemic explanation' and 'physical explanation'. We argue that mechanistic explanation requires that mechanisms are both real and local. We then go on to argue that real, local mechanisms require a broadly *active* metaphysics for mechanisms, such as a capacities metaphysics.

38.1 Introduction

Mechanisms have become much-discussed in the current philosophy literature, to begin to match the long-enduring interest in mechanisms in the sciences. Yet there is still no consensus as to the best way to characterize mechanisms. A brief glance at only three major papers attempting to characterize mechanisms illustrates:

Machamer, Darden and Craver: 'Mechanisms are entities and activities organized such that they are productive of regular changes from start or set-up to finish or termination conditions.' (Machamer, Darden and Craver, 2000, p. 3)

Glennan: 'A mechanism for a behavior is a complex system that produces that behavior by the interaction of a number of parts, where the interactions between parts can be characterized by direct, invariant, change-relating generalizations.' (Glennan 2002, p. S344)

Bechtel and Abrahamsen: 'A mechanism is a structure performing a function in virtue of its component parts, component operations, and their organization. The orchestrated functioning of the mechanism is responsible for one or more phenomena.' (Bechtel and Abrahamsen 2005, p. 423)

In this chapter, we will start with one thing everyone – including both philosophers and scientists – agrees on about mechanisms: mechanisms explain. We will investigate what constraints this imposes on a metaphysics of mechanisms. In Section 38.2 we examine two important premises about mechanistic explanation shared by many in the mechanisms debate: that mechanisms offer a form of explanation distinct from laws-based explanation, and that there two senses of explanation, that we call epistemic explanation and physical explanation. In Section 38.3 we argue that both kinds of explanation require real mechanisms, and in Section 38.4 we argue that scientific explanation using mechanisms requires that mechanisms must be local. In Section 38.5 we argue that real, local mechanisms require what we call a broadly *active* metaphysics, such as a capacities metaphysics, rather than a more passive best-system laws-based metaphysics. In Section 38.6 we deal with two possible objections to our view.

Dialectically, we take our argument to be important because the general trend in characterizing mechanisms in the literature is to use a more passive metaphysics, which our argument shows to be at odds with the basic reasons we examine in Section 38.2 that are often given for developing an account of mechanisms at all. Further, if you think, as many do, that mechanistic explanation is *causal* explanation, then everything argued here transfers to this important species of causal explanation.

38.2 Mechanistic explanation

There are various views about the nature of mechanistic explanation which are sometimes implicit, sometimes explicit in the mechanisms literature. In this section, we will pick out two views which, while not universal, are widely shared by those writing about mechanisms. They are first that mechanistic explanation is a distinct alternative to laws-based explanation, and secondly that there is something which we will call 'physical' explanation, where the mechanism in the world produces the phenomenon of interest. We agree with both of these deep premises of thinking about mechanisms and mechanistic explanation. We will argue that they impose constraints on a plausible metaphysics of mechanisms.

The first view is that mechanistic explanation is a new form of explanation that is distinct from traditional laws-based approaches to explanation – and far more promising as an account of explanation across much of the sciences. As Torres writes: 'The mechanistic model of explanation represents an appealing alternative to classical covering-law (CL) models' (Torres forthcoming, Section 38.1). There are three independent reasons for this, which influence different philosophers in different ways. All three reasons are important.

The first reason is simple: mechanistic explanation as distinct from law-based explanation fits the *practice* of the special sciences in a way that law-based explanation fails quite drastically to do. And the special sciences now amount to a great deal of science.[1] They proceed by identifying phenomena requiring explanation, and decomposing them. They look for the parts of the mechanism underlying the phenomenon, and start trying to figure out what they do. They work out how the parts go together where, so that what the parts can do in conjunction changes, and ultimately the way the organized behaviour of the whole mechanism produces the phenomenon becomes clear. What such scientists do not do is look for laws and try to build up any kind of explanation resembling a law-based one. Bechtel and Abrahamsen are influenced by this: 'The received view of scientific explanation in philosophy (the deductive-nomological or D-N model) holds that to explain a phenomenon is to subsume it under a law .... However, most actual explanations in the life sciences do not appeal to laws in the manner specified in the D-N model.' (Bechtel and Abrahamsen 2005, p. 421–2)

The second reason for thinking an alternative to traditional laws-based explanation is vital is also simple, and has been extensively argued elsewhere. There are no laws – no exceptionless non-accidental generalisations – in the special sciences and no reason to suppose there ever will be.[2] Those 'laws' that do exist in the special sciences are nothing more than generalizations backed up by an understanding of the mechanisms underlying those laws.[3] This is the direction of explanation between laws and mechanisms that many scientists adhere to. It seems to us that a reductive faith that there will ultimately be discoverable laws everywhere has been seriously empirically undermined by the huge success of science without laws – such as the extraordinary amount now known about the detailed mechanisms of protein synthesis – and should be abandoned. Leuridan mentions this reason for preferring mechanistic explanation: 'If there are no strict laws, there are no D-N explanations. Hence the mechanicist alternative, which states that explanation involves *mechanistic models* (i.e. descriptions of mechanisms) instead of strict laws, might be very welcome.' (Leuridan forthcoming Section 38.1, emphasis in original)

The third reason is that even if fundamental laws were available in the special sciences – exceptionless non-accidental generalizations that we could treat as brute nomological facts about the universe – it is hard to see how they could explain in the way that mechanisms explain. Brute nomological

[1] The special sciences span from chemistry to psychology, economics and social science. Some areas traditionally thought of as physics might still best be thought of as special sciences, as James Ladyman has argued in private communication. He suggests optics, cosmology and solid-state physics. We don't mean to imply that mechanistic explanation is *absent* from physics, just that it *is* ubiquitous in the special sciences.

[2] Cartwright has been very influential here. She even argues that many of the classic universal laws of fundamental physics describe only idealized, closed systems, so are in fact *not* universal laws when applied to the real world. See Cartwright (1983).

[3] Glennan (2002) discusses this extensively. Note that idealised models are a different issue, quite distinct from physical mechanisms – consider that they are *models*. See also Mitchell (2007).

facts themselves call for explanation, which is precisely why scientists do try to explain laws – and they often use mechanisms to try to do this. Glennan (2002) is concerned about this issue, concluding that at least the fundamental laws of physics are not mechanically explicable, or explicable at all. It seems to us that this is acceptable if it is assumed that there is a final, fundamental, lowest, level of physics. But so far we have not discovered such a final level, and we don't have empirical reason to believe there must be one. Mechanisms offer a more accurate characterization of genuine special science explanation. This reason clearly influences Bechtel and Abrahamsen: 'For present purposes we leave laws in place as statements of particularly robust and general phenomena. However, we suggest that explanation is to be found in the mechanisms that account for these laws, not in the laws themselves.' (Bechtel and Abrahamsen 2005, p. 422, footnote 1)

So the first view that mechanistic explanation is a promising and much-needed alternative to laws-based explanation is widely shared and important.[4]

The second view is sometimes only implicit in the literature, but it is clearly present. It is the view that there are two different kinds of mechanistic explanation in line with Salmon's more general distinction between epistemic and 'ontic' explanation (Salmon 1998a, 1998b). The first, epistemic, sense is of explanation as a human practice, aimed at increasing understanding of the world. It often involves the passing of information between human beings. Although it is aimed at understanding the world, it is highly sensitive to the cognitive abilities and background knowledge of human beings.[5] The second, ontological, sense of explanation, particularly important in the scientific practice of explanation using mechanisms, we will call physical explanation. This second sense of explanation is the sense in which mechanisms explain the phenomena they explain by being responsible for them. This happens whether human beings understand what is going on or not.[6] Knowledge of the mechanisms involved in physical explanation might in some cases be beyond our cognitive capacities. There is certainly no a priori reason to be sure we will always be able to know such mechanisms.

[4] It is not universal. We will show later that careful study of Psillos' views suggests that he believes mechanistic explanation collapses to laws-based explanation. And see Leuridan (forthcoming) for an interesting examination of mechanisms and laws-based explanation using Sandra Mitchell's understanding of 'pragmatic' laws, which are intended to be available in the special sciences.

[5] It is in this sense of explanation that the explanation a biologist gives his small son is very different from the one he will give in a research seminar. It is also the sense in which an explanation which is strictly speaking *false* might often be the best way of explaining something to a particular audience. One might say, for example, 'Mummy is getting fat because she has a baby in her tummy'. Strictly speaking Mummy is not getting fat, nor is the baby in her stomach. Thanks to Julia Tanney for suggesting this useful example.

[6] We will discuss the tricky cases of social and psychological mechanisms in Sections 38.3 and 38.4.

The two senses of explanation are naturally intertwined, of course. If the epistemic sense of explanation is to succeed in increasing understanding of the *world*, rather than merely making up interesting stories about it, the stories had better be describing the mechanisms in the world. In the other direction, as soon as we start trying to *describe* the physical mechanism that produces the phenomenon, we begin to abstract, to prioritise salient details and so on, so that our *description* takes on some epistemic features. But while these two senses of explanation are so intertwined that they can be difficult to separate, they are not the same.[7] This can be clearly seen by realising that they are subject to different constraints. For example, an epistemic mechanistic explanation should be perspicuous to its audience, while the correctness of a description of a physical mechanism producing a phenomenon does not depend on whether its audience understands that description. A second important difference is that the epistemic sense of explanation allows for the imposing of normative desiderata that would be inappropriate to apply to physical mechanisms. For example, one might reasonably attempt to make epistemic mechanistic explanations modular – very crudely, organised so that you can wiggle one bit of the mechanism without affecting the working of other bits.[8] But you couldn't possibly demand that every physical mechanism be modular. They may be or they may not. These differences exist because in the epistemic sense of explanation it is the *description* of the mechanism that explains, while in the physical sense, the *mechanism itself* does the explaining.

This distinction is implicit in much of the literature on mechanisms, and explicit in some papers. In many papers, an explicit distinction is made between mechanisms as they are in the world, and the models, schema or descriptions of mechanisms that we construct to explain the phenomena.[9] In the construction of models or schema we are involved in an explanatory project that is epistemic. This notion is much explored by Bechtel and Abrahamsen, among others, and of course has a very important place in science. But Machamer, Darden and Craver, Bechtel and Abrahamsen, and Glennan all recognize the deeper relation between mechanisms and phenomena. The mechanism as it is in the world is responsible for the phenomena we observe.

[7] Achinstein explores the relation between explaining acts and worldly explainers. See Achinstein (1983).

[8] Woodward does this. His section 5 is all about modularity, and he writes: 'The basic idea that I want to defend is that the components of a mechanism should be independent in the sense that it should be possible in principle to intervene to change or interfere with the behavior of one component without necessarily interfering with the behavior of others.' (Woodward 2002, p. S374).

[9] See Glennan (2002), Machamer, Darden and Craver (2000), and Bechtel and Abrahamsen (2005). Craver goes on to make the distinction thoroughly explicit. He talks of explanations as texts, models or representations used to convey information. He goes on: 'Other times, the term explanation refers to an objective portion of the causal structure of the world, to the set of factors that bring about or sustain a phenomenon.' (Craver 2007, p. 27).

Machamer, Darden and Craver phrase it as the mechanism being 'productive of regular changes'; Glennan says that the mechanism 'produces that behavior'; and Bechtel and Abrahamsen say that mechanisms are 'responsible for one or more phenomena'. They all mean much the same thing.

Bechtel and Abrahamsen don't want to call this kind of thing explanation. We do, following both scientific and common practice. Sometimes a request for explanation is a request to identify the responsible portion of the world. For example, when asking why the Sun looks larger at the horizon, the requested explainer is something about the world, even though it is described in terms of something epistemic – a story about the world. So this physical sense of explanation is a genuine sense of explanation, and it is this physical sense of explanation that we are most concerned with in this paper. It comes in at least two varieties – decompositional and etiological. In the first case, mechanisms explain phenomena by being the lower-level entities and activities that are organized to produce the higher-level observed phenomena. In the etiological case, mechanisms explain events by being the detailed causal history leading up to these events. Although a decompositional mechanistic explanation for a one-off event could be sought, we will focus on the case where mechanisms are the decomposition of a regularly occurring phenomenon. There are interesting relations between these slightly different kinds of mechanistic explanation which we do not have space to explore here.[10]

In conclusion, scientific practice allows for two distinct understandings of mechanistic explanation: epistemic, where the description of the mechanism explains, and physical, where the mechanism itself explains. The literature on mechanisms has followed this, and the view that physical explanation exists, where the mechanism produces the phenomenon of interest, is widely shared. Similarly, the view that mechanistic explanation is a crucial alternative to laws-based explanation is a driving force of the mechanisms literature, widely shared in it.

38.3 Mechanistic explanation requires mechanisms to be real

For now, we put aside the view that mechanistic explanation is distinct from laws-based explanation, although it will become crucial to the overall argument again in Section 38.6. In this section and the next, we focus on teasing out two implications for the metaphysics of mechanisms arising from the use of mechanisms in explanation: mechanisms must be real in this section and local, in Section 38.4.

For mechanisms to yield physical explanations, they must be real. That is, at least some of the mechanisms posited by scientists must exist as worldly

[10] See Glennan (2002) and Craver (2007) for interesting discussion.

entities responsible for the phenomena we take them to explain.[11] This fits scientific practice, matching how scientists treat mechanisms in mechanistic explanation. The mechanism of protein synthesis is a system in the world that produces proteins; natural selection is a complex of worldly processes that produces the adaptation of a population to its environment.[12] It also fits the interesting uses to which mechanisms might be put, in causal inference, for example. Leuridan and Weber (2011, this volume) suggest that identification of an underlying mechanism can be important in addressing the problem of external validity – working out whether a causal relation established in one population will apply in another. If knowing the underlying mechanism is to help, it is because you have identified a worldly entity that you can with reasonable confidence expect to be present/absent or more or less similar in altered circumstances. Cartwright (2006 and forthcoming) suggests that knowing social mechanisms is important to making social policy decisions, which seems to make the same assumption. Physical explanation is the most important kind of mechanistic explanation, because epistemic explanation is parasitic upon it. If epistemic explanations are to explain, rather than merely being stories, there must be real mechanisms to describe. So both kinds of mechanistic explanation require real mechanisms.

So the mechanisms literature is implicitly or explicitly committed to mechanisms being real. The claim is seldom made so baldly: perhaps it has seemed too obvious to comment on. A more detailed understanding of how mechanisms are real is slightly tricky. There are two problems, which we will take in turn: non-'physical' mechanisms, and the functional individuation of mechanisms.

The problem of non-'physical' mechanisms arises because not all mechanisms are completely independent of what people believe of them and how people describe them – the existence of social and psychological mechanisms means that that independence comes in degrees. Whatever social and psychological mechanisms are, they are partly constituted by people, and some may include the beliefs of people as components. For example, enough people believing that the economy is going into recession, and behaving accordingly, might be a necessary step in the economy entering recession. In terms of traditional debates about the existence of the external world, the dependence of such mechanisms on minds, their representational or constructed nature, might be thought to be an unacceptable mind-dependence of such mechanisms, undermining their reality.

[11] The claim that mechanisms are real does *not* commit us to some variety of physical reductionist thesis. We believe there is serious reason to doubt whether science really proceeds by classical reduction. See for example Craver (2007) for an argument that the field of neuroscience has not advanced in ways remotely resembling the reductive paradigm.

[12] Skipper and Millstein argue that natural selection is not a mechanism – at least not one that fits the characterizations of mechanisms offered by MDC and by Glennan. We disagree. See our examination of natural selection in Illari and Williamson (2010).

However, this does not mean that social and psychological mechanisms must be merely explanatory schemas or models, thoroughly and essentially dependent on human minds, and so forever restricted to merely *epistemic* explanation. So long as people, and their minds, exist, are part of the ontology of the world, such mechanisms exist and are susceptible to scientific investigation. Such mechanisms are still worldly entities. Further, although models and schemas might under some circumstances themselves be mechanisms, they cannot themselves be part of the mechanisms that they describe. This view also contrasts with the view that such explanatory posits are merely instruments – instruments for making accurate predictions about the behaviour of observables.[13] An instrumentalist view renders science less explanatory. There is no commitment to anything that can explain why the predictions work.

The second problem for the reality of mechanisms is that mechanisms are at least partially functionally individuated. This is a light sense of 'function', where the function of a thing is the role it plays in a system. That role does not require a selective history. Mechanisms are conceived of as the *mechanism for* producing a particular behaviour, where the behaviour is identified as a phenomenon of interest by scientists, when characteristically producing that behaviour is the role of that mechanism in that system.

This partial functional individuation of mechanisms is clear in scientific practice. Take for example the discussions of the mechanism of protein synthesis in three biochemistry textbooks. The notion of function is ubiquitous: all three texts took the function of protein synthesis to be the decoding of the information in DNA to produce proteins, and many lower-level mechanisms involved were functionally described.[14] For example, regulatory mechanisms are understood in terms of ensuring that the right proteins come out. All three texts talk about some kind of repair mechanisms, and they all mean mechanisms whose function is to correct various kinds of malfunction.

This kind of functional individuation of mechanisms is also recognized in the philosophical literature. Both Glennan, and Bechtel and Abrahamsen, are explicit that mechanisms have functions. For Bechtel and Abrahamsen a mechanism is a structure 'performing a function' (2005, p. 423), while for Glennan a mechanism is a mechanism 'for a behaviour' (2002, p. S344). Machamer, Darden and Craver's characterization does not explicitly make a function an essential aspect of a mechanism, but they don't explicitly rule it out either. Independent work by Darden and by Craver makes the link between a

[13] Instrumentalism has a long intellectual history, being traceable to work of Mach and Duhem, among others. See van Fraassen (1980) for a more recent discussion.

[14] See Adams *et al.* (1992), Voet and Voet (2004), and Whitford (2005). For example: 'Uncovering the cellular mechanisms resulting in sequential transfer of information from DNA (our genes) to RNA and then to protein represents one of major achievements of biochemistry in the 20th century.' (Whitford 2005, p. 247) Or consider: 'How do genes function, that is, how do they direct the synthesis of RNA and proteins, and how are they replicated?' (Voet and Voet 2004, p. 92).

mechanism and its explanandum phenomenon explicit (see Darden 2006 and Craver 2007).

The possible problem for the reality of mechanisms is probably clear. A function is required for something to be a mechanism. But the role something plays in a system seems to be a matter of description – the description of the system it is in. The same object individuated structurally, such as the heart, can have different functions according to the description of the system it is in. It might have the function of pumping blood when considered as part of the circulatory system, or the function of making a thump-thump noise when considered as part of a system for comforting a newborn baby. Then functions seem to be not wholly in the world, but set by the description of the system. This is not a spurious unscientific example. In scientific practice, the very same thing individuated structurally can sometimes have one function, sometimes another – and sometimes has no function at all.

Consider an example of this from a discussion of branch migration, the moving of the crossover point at which two molecules of DNA exchange excised strands of DNA. Voet and Voet write: 'such a process moves forward and backward at random and, moreover, is blocked by as little as a single mismatched base pair. In *E.coli*, and most other bacteria, branch migration is an ATP-dependent unidirectional process that is mediated by two proteins ... ' (Voet and Voet 2004, p. 1189.) In the context of protein synthesis as having the function of decoding DNA to produce proteins, branch migration doesn't have a function. It just happens, being a nuisance that the cell has to have various mechanisms to fix. That is why it is described as a process here.

The same responsiveness to description occurs for other kinds of mechanisms. Take natural selection. When it's the mechanism for adaptation, then it has a function – the bringing about of fit between organism and environment – and so is properly described as a mechanism. But when something else is being described and natural selection is just something happening in the background, then it doesn't have a function, and is described as a process. There are many other examples showing that both biochemists and evolutionary biologists consistently use 'mechanism' to describe only something that has a function, using the word 'process' when there is no function. So mechanisms are mechanisms *for*, while processes are just processes in themselves. Mechanisms, of course, are still what is used in explanation because once in the domain of explanation, you are thinking in terms of a behaviour or phenomenon to be explained. Once you identify that phenomenon, then you are looking for something that has a function – the mechanism that produces that phenomenon, which is thought of as the mechanism *for* that phenomenon.

This looks like a more serious problem for the reality of mechanisms than it really is. It is true that what function a mechanism performs can vary according to our *description* of the system it is in. But this is because many entities *really do belong to more than one system*. The heart really does circulate blood, and it really does make a thump-thump noise which is comforting to newborns because it is familiar. When we consider different systems we will naturally identify a different role-function for the heart, but this is because of the variation between systems, not because of our *description* of that system. The role an entity plays in any particular system is not a matter of description – it is a worldly fact. Functionally individuated mechanisms are real.

In conclusion, both scientific practice and much of the philosophical literature on mechanisms is committed to real mechanisms, either in physical explanation, or as items to be described in epistemic explanation. That is, mechanisms exist as worldly entities that are responsible for the phenomena they produce. The mechanism itself is different from any model, schema or other description or representation of *that* mechanism, and the mechanism itself is real. Mechanisms are physical explainers, while representations of mechanisms are epistemic explainers. This is the first important constraint on a metaphysics of mechanisms.

38.4 Mechanistic explanation requires mechanisms to be local

The second interesting constraint on a metaphysics of mechanisms comes from a feature of the *kind* of explanation you get when you get a physical mechanistic explanation. The explanation is local to the phenomenon produced. Mechanistic explanations use parts organized to produce the phenomenon, and may also look to containing systems to set the phenomenon, but both parts and systems are local to the phenomenon. This is true of scientific practice, and is also reflected in the philosophical literature on the issue.

Mechanisms, recall, are individuated by their explanandum phenomena. Mechanistic explanation begins by identifying a phenomenon, usually a regularly occurring one, to be explained. What then occurs is the process of examining the area local to the explanandum phenomenon to see how the phenomenon is produced. This is how characteristic mechanistic explanation proceeds.[15] Take protein synthesis as an example. Scientists have a phenomenon that they wish to explain – the production of proteins. What they want to know about is the underlying mechanisms that produce proteins, and the

[15] This understanding of decompositional mechanistic explanation is uncontroversial in the mechanisms literature. See for example Bechtel and Abrahamsen: 'The quest to understand the mechanism responsible for a given phenomenon requires decomposing the responsible system.' (2008, p. 560) They also discuss the importance of looking upwards to higher-level mechanisms. For this chapter, we set aside etiological mechanistic explanation. We suspect that it will exhibit the same kind of locality as decompositional mechanistic explanation, but lack space to establish that here.

conditions in the cell at the time. So scientists examine the details of the cells where proteins are produced. What is going on when proteins are produced? What kinds of molecules and so on are involved, and what do they do? They discover that there is a stage where DNA is copied, a stage where (various kinds of) RNA are made from DNA, and a stage where protein is made from mRNA. In fact, they discover vastly more information than this about all of these stages. But what they are doing certainly involves decomposing into local parts, and investigation of the local cell conditions. They look for the lower level stuff that produces the higher-level stuff, and the conditions under which this happens. In this sense, a great deal of what is described at great length in biochemistry textbooks just *is* protein synthesis.

Precisely because of the functional individuation of mechanisms, the extent of the locality you expect in a mechanism is set by the phenomenon the mechanism produces. Protein synthesis is very constrained, happening within a single cell. This is where the proteins appear, and where we look for the mechanisms that produce them. Compare natural selection, which happens within entire populations – much larger entities than single cells. Nevertheless, if you are interested in how natural selection has produced a particular distribution of trait types in a population, you study the history of the happenings in *that* population. You will also have to study the environment of that population, but that is also local – you don't look anywhere else. Investigating social mechanisms might lead you to look at social relations in an entire country, or even internationally. Investigating emission mechanisms for gravitational waves would lead you to look at something much bigger – the behaviour of binary pulsars, or even the movement of entire galaxies. But in each case the phenomenon sets the extent of appropriate locality. It isn't that mechanisms are always the size of 200 molecules, or three galaxies, it is that they are constrained in locality by where the phenomenon is that they are thought to produce. Other regions of space are not investigated, since they are not considered relevant.

Locality is implicit in much philosophical discussion of mechanisms, and explicit in some cases. Although he doesn't use the word 'locality', Craver, for example, gives the matter some extended discussion. Craver thinks that mechanisms are generally hierarchically nested, and mechanisms at different levels fall in (certain sorts of) part-whole relations. He writes 'The primary difference is that LM levels [levels in the multilevel mechanism for spatial memory] are relationships between a whole and its parts, while levels of processing are relationships between distinct items.' (Craver 2007, p. 178) Clearly, parts of wholes are local to those wholes, so Craver thinks that lower-level mechanisms are local to higher-level ones.

Craver also examines the issue of what, in the local area of the explanandum phenomenon, might be *left out* of the mechanism for the phenomenon. Not everything local is relevant. Writing again of his spatial memory case study, he claims: 'Components at lower levels are organized to make up the behaviours at higher levels, and lower- and higher-level items stand in relationships of mutual manipulability...' (Craver 2007, p. 170) By mutual manipulability, he means, very crudely, that if you wiggle one, the other wiggles, and vice versa. Not all parts of the cell, for example, are components of the mechanism of protein synthesis in this sense. And this is generally true of mechanistic explanation. If you are investigating social relationships the behaviour of the atoms or molecules of people are irrelevant, and a star collapsing will emit gravitational waves quite independently of the behaviour of any nearby life. While this shows that not everything that is local will be part of the mechanism, it is true that the parts of the mechanism will be local to where the phenomenon they produce exists. Mechanisms come in different sizes because the phenomena they produce come in different sizes. The extent of mechanisms that produce various phenomena is part of what is learned by scientists as the mechanisms for the phenomena are better understood. One of the things that is surprising about various subatomic phenomena in physics, for example, is how non-local their causes and effects might be.[16] Nevertheless scientists in different fields come to have a good idea of how dispersed the mechanisms they are looking for are likely to be. Once that is established, they do not look further afield. It is, if you like, an empirical discovery of any mechanism what extent of local space-time is relevant to the explanation. In all cases, that locality is limited.

Our discussion of the locality of mechanisms sails confusingly close to debates elsewhere in philosophy. It is worth pausing to make clear that the locality claim is a claim about mechanistic explanation in scientific practice and how it is understood in the mechanisms literature. We are not here concerned with metaphysical debates about causality, such as the claim that causal relations are intrinsic.[17] The locality of mechanisms might imply that a precise copy of a mechanism is still a mechanism, which is an interesting parallel of some ways of understanding the intrinsicality claim. But this debate is not what concerns us here. We are also not concerned with alternative *scientific* claims about locality. A familiar locality claim from physics is that there can be no covariation of physical properties that are spacetime separated, where the covariation would demand that the propagation of any causal influence would have to be faster than the speed of light. This might be recast in less scientific terms as a doctrine of no action at a distance, which confusingly enough seems to be what some philosophers mean by locality when they talk about

[16] In this way, quantum non-locality isn't a counterexample to our locality claim, since quantum phenomena are still local to the system. What's surprising is how spread out in space that system can be.

[17] Ned Hall characterises this informally: 'Intrinsicness: The causal structure of a process is determined by its intrinsic, non-causal character (together with the laws).' (Hall 2004, p. 225).

causality.[18] If these are basic laws of physics, then no doubt mechanisms will not violate them. But such laws require support from physics more generally – examining how mechanisms are used in explanation cannot establish that nothing moves faster than the speed of light! Since we are looking for a claim supported by the use of mechanisms in explanation, it should be clear that this is not the locality claim we have in mind.

To reiterate, the claim is that in decompositional mechanistic explanation, the mechanism that produces the phenomenon of interest is looked for and discovered in the area of the phenomenon produced. This makes perfect sense of the idea that the mechanisms discovered are the mechanisms underlying the phenomenon. Other regions of space and time are not considered relevant.

Before ending this section, we will raise three possible problems for the locality of mechanisms: the functional individuation of mechanisms, the existence of non-'physical' mechanisms, and omissions. We will argue that none puts paid to locality, although examining them does yield a better grasp of the locality claim.

The first problem is that mechanisms having functions potentially damages the locality of a mechanism. If functions are relative to a description of a system, and that description, along with the wider system, is not part of the world where the mechanism is, then mechanisms are not wholly local. However, just as for the reality of mechanisms, this problem can be resolved by recognising that functions are more accurately thought of as objective, worldly relations between a mechanism and the (perhaps several) different systems it is in. Consider the heart again, as structurally individuated. Its role as blood-circulator is set by the higher-level system of blood circulation it is a part of, while its role of baby-comforter is set by that different system. But this is not a matter of description, just a relation between the heart and two different (and local) systems it really is a part of. This is not a serious problem for locality any more than it is for reality.

Again, the possibility of non-'physical' mechanisms requires careful thought. We have pointed out already that locality is always relative to the phenomenon explained, so that some non-'physical' mechanisms, such as social mechanisms, might be very large indeed. But it is also true that how thoroughgoing the locality is, how deeply it extends, varies from mechanism to mechanism, affecting effective strategies of decomposition.

Craver recognizes this: 'Localization is one of the most fundamental spatial constraints on interlevel integration (Bechtel and Richardson 1993). Not all mechanisms have easily localized *components*, but when they do, the location

[18] Ned Hall, for example, uses the following sense of locality: 'Locality: Causes are connected to their effects via spatiotemporally continuous sequences of causal intermediates.' (Hall 2004, p. 225).

of different processes can be crucial to understanding a mechanism that incorporates them.' (Craver 2007, p. 261–2) This is a particularly interesting issue for psychological mechanisms, where there is great debate on localization strategies. There is disagreement, for example, on whether specific psychological functions are localized to particular parts of the brain, or processed in a way far more dispersed through the brain.[19] Nevertheless, this becomes an issue once the description of the phenomenon is underway, which is once the locality of the mechanism is already set. In these cases, both sides of the debate still agree that the mechanisms producing psychological phenomena are in the brain. That is local enough.

There is a far trickier problem case.[20] Some decompositional psychological mechanisms cite content-bearing mental states as part of the mechanism. For example, the behaviour of a rat in a maze might routinely require explanation in terms of false beliefs about where food is – where of course the *content* of the false belief is crucial to the explanation. If mental content is external, depending on the history of the interaction of the organism with its environment, then this particular decompositional mechanism is not local. The phenomenon may arise in the brain, but some of its constituents are elsewhere.

This is of course a controversial case. It seems to us there are three options. The first is to say that psychological mechanisms are just quite different from other mechanisms in this way. This is a position one might be driven to, but is ad hoc as a first move. The second option is a denial of externalism about content. All that is necessary for mechanisms is the narrow content of any content-bearing mental state, which supervenes only on the local brain state. Our third option is to take externalism about content more seriously – perhaps as seriously as the extended mind thesis does. Then, we re-construe the appropriate extent of mental phenomena, so that external items are appropriate parts for decomposition of mental events.[21] These are interesting issues worthy of wider consideration, but we lack space to develop them here. We have said enough to indicate the defence of our locality thesis, because on either the second or the third story, the case doesn't violate locality.

The third problem is difficult. The problem is that if omissions are part of any mechanism, which they routinely seem to be, then that mechanism will not be local. The mechanism might depend on features thoroughly nonlocal to the phenomenon it produces – perhaps on features that have no location at all. The problem of locating omissions is well known from the vast literature on omissions and *causation*.[22] To illustrate, suppose I, your

[19] See the debate about core realizers. Advocates include Block (2005) and Wilson (2004). These are very interesting issues we intend to examine at length elsewhere.

[20] Thanks to Anthony Everett for raising this point.

[21] See the forthcoming Menary collection for extensive discussion of the extended mind thesis.

[22] See for example Schaffer (2004) and Dowe (2004) for an interesting symposium.

neighbour, promise to water your plants when you are on holiday. I don't, and they die.[23] It seems my omitting to water the plants causes their death. But where is my omission located? Is it in your house where I fail to turn up? Or is it wherever I am, busy not watering plants? There seems to be no satisfactory answer. Yet it is not impossible that omissions are routinely part of some mechanisms. In the context of causation, Schaffer discusses the example of a gunshot through the heart causing death, but it isn't difficult to see the circulatory system and the internal workings of a gun that he describes as mechanisms. Schaffer writes: 'But heart damage only causes death by negative causation: heart damage (c) causes an absence of oxygenated blood flow to the brain ($\sim d$), which causes the cells to starve (e).' Later he considers the gun: 'But trigger pullings only cause bullet firings by negative causation: pulling the trigger (c) causes the removal of the sear from the path of the spring ($\sim d$), which causes the spring to uncoil, thereby compressing the gunpowder and causing an explosion, which causes the bullet to fire (e).' (Schaffer 2004, p. 199) Examples in biochemistry are numerous. Cells routinely alter which enzymes they produce in response to which metabolites are available. A cell stops producing lactase, for example, in response to the *absence* of lactose diffusing into the cell's cytoplasm.

Our view is that omissions are not a problem for the locality of mechanisms in the same way that they are for causation. Causal relations are often subject to absences. Just as my failure to water your plants caused their death, so the Queen's failure to water them caused their death, and so on. There is no location for these omissions. But the kinds of omissions that are routinely part of mechanisms are locatable, and they are local. This is clear in all three examples above. The deoxygenated blood is part of the circulatory system within the body; the removal of the sear from the path of the spring is part of the gun mechanism, and the alteration of gene expression within the cell is in response to differing levels of metabolites within that cell. All of these are in the familiar area where we would attempt a decomposition of the relevant phenomena.

Causation and mechanisms are different here because difference-making is important to causation and not to mechanisms. Often, the conviction that difference-making is crucial to causation is decisive on thinking that absences cause. This is because in the causation case an absence often stands in a difference-making relation to the effect. To illustrate, it is missing the bus that *made the difference* to my lateness, not the empty bus-stop when I got there.[24] That is why it is the *absence* that causes my lateness, rather than the positive

[23] This useful example is originally Helen Beebee's. See for example her (2004).

[24] We take this important point to be a central view of David Lewis' later work. See Lewis (2004). For further discussion of the idea of difference-making and its relation to causation, see Hall (2004). Difference-making itself might be tracked in different ways, such as using simple counterfactual notions, invariance relations, or correlations.

story about where the missing bus actually is. This idea of difference-making is plausibly very important to causation, but it isn't central to a mechanism. So such external omissions are not in the same way relevant to mechanisms, being outside the relevant mechanisms. This means that only local omissions are relevant to mechanisms.

In conclusion, and in spite of some challenging problems, it is a genuine feature of mechanisms that they are local. This is a kind of locality that scientific practice is committed to in how it explains using real mechanisms, and a locality that is, albeit sometimes implicitly, reflected widely in the mechanisms literature. This is the second important constraint on a metaphysics of mechanisms. Mechanisms are both real and really there, and the right metaphysics of mechanisms must respect that.

38.5 Reality and locality require an active metaphysics

The problem for a good metaphysics of mechanisms is to characterize the *interactions* in mechanisms. Recall that all characterizations of mechanisms have two components, with something about the parts of mechanisms, and something about how the parts interact. The metaphysics of the parts of mechanisms has been uncontroversial. They have had different names – 'entities', 'parts' and 'component parts' in the characterizations above – but they have not been much discussed. What is controversial is how to characterize what the parts of mechanisms *do*, the activities or interactions of the parts. It is not surprising that this is controversial, since the interactions are more interesting metaphysically. Unlike the parts, or entities, which are actual, understanding mechanisms also involves understanding what those entities will do in non-actual situations. Scientists seem to know of many mechanisms what they would do when the initial conditions are changed in various ways.

It is in characterizing the interactions of parts of mechanisms that the reality and locality of mechanisms becomes relevant, because not all metaphysical approaches to the interactions allow mechanisms to be real and local. Understanding the metaphysics of mechanisms on this level is now a philosophical problem with no immediate bearing on scientific method, of course. It does, however, bear on our understanding of science. Since both scientific practice and many philosophical treatments of mechanisms are committed to their locality and reality, our argument should be of wide interest.

There are broadly two approaches to characterizing the interactions in mechanisms, and we will argue that all the approaches to metaphysics that allow mechanisms to be real and local lie within only one of these approaches. We call these broad approaches 'passive' and 'active'. Passive approaches characterize interactions using laws or some counterfactual notion or other – either relatively simple counterfactuals, or their more sophisticated cousin the invariance relation. They then use either a best-system laws grounding for such counterfactual claims, or a modally realist grounding.[25] We call this approach 'passive' because broadly the grounding for counterfactual claims is just patterns of the objects in this or in other worlds.

Passive approaches contrast with active approaches. These latter approaches give an account of interactions in terms of the capacities, powers or activities of entities. So active approaches include Machamer, Darden and Craver's activities approach, where activities are varied things that entities can engage in, like bonding, breaking, pushing and coiling. Also included are Nancy Cartwright's capacities approach, which claims that capacities are properties of objects, and Carl Gillett's powers approach, which follows Shoemaker in individuating properties by their causal powers, so that having a property implies action in certain conditions.[26] We will argue that only active approaches give a local characterization of a mechanism.

We take the points we make about the metaphysical systems to be well-understood aspects of these systems. They also have well-known problems, which we shall not repeat here. Nevertheless the significance of these points for thinking about the metaphysics of mechanisms needs to be spelled out. Dialectically, this is interesting to the mechanisms literature because – outside of traditional metaphysics – passive approaches are generally regarded as metaphysically less problematic than active approaches. This yields a dialectical reason for philosophers working on philosophy of science, but not inclined to do traditional metaphysics, to plump for passive characterizations of interactions in mechanisms. In the core mechanisms literature, only Machamer, Darden and Craver are trying to use an active approach to characterize interactions in mechanisms, but the novelty and apparent extremity of their approach is off-putting to many. Recent work by Gillett using a powers approach is interesting too, but has yet to be picked up extensively in the mechanisms literature. We wish to oppose this dialectical trend towards passive approaches by pointing out that there are well-worked-out metaphysical systems available that do a better job for mechanisms than passive approaches.

Using some counterfactual account or other of the interactions in a mechanism has been popular. This is the approach of Woodward, Psillos and later Glennan. We will raise concerns about the status of this claim later on, but suppose for now that it is a claim about the nature of mechanisms.

[25] These familiar ideas derive from Ramsey, through significant development by David Lewis. Ramsey discusses the possibility of us systematizing our knowledge in a deductive system, suggesting that the laws are the axioms of such a system. See Ramsey (1990, p. 143). We focus on the best-system laws grounding for counterfactuals as it is Stathis Psillos' explicit view (see Psillos 2004), and include modal realism as the major alternative.

[26] See Cartwright (1989), Machamer *et al.* (2000) and Gillett (2006). There is considerable variation within the active tradition, with Machamer, Darden and Craver's essentially dynamic activities the most 'active'. We will come to some of these distinctions later.

Whether such a view yields real, local mechanisms or not depends on what you take to ground counterfactuals. Take the passive views first. Suppose the truth-conditions for counterfactuals – whether simple or sophisticated – are grounded in a best-system account of laws of nature, where the best-system is judged by the simplicity and strength of laws.[27] If this is so, then mechanisms are not local. Mechanisms depend on two kinds of non-local features. The first is features of many other places and times in this world, those necessary to determine the laws of nature. As we have said in the previous section, this is in tension with the actual practice of mechanistic explanation in the sciences, which examines only local regions of spacetime in constructing mechanistic explanations. The second kind of feature is the simplicity and strength of laws that establish what is the best system of laws. No such features are local, being dependent first on the entire universe, and second on the abstract concepts simplicity and strength, which cannot be located clearly at all. Such mechanisms might also fail to be real – depending on the status of laws. In general, an anti-realist account of laws clearly yields non-real mechanisms; while any passive realist account of laws still has truthmakers widespread in the universe for law, so mechanisms based on such laws will be non-local.[28]

Take the alternative modal realist account of truthmakers for counterfactuals, where the truth of any counterfactual claim depends on what happens to counterparts in nearby possible worlds.[29] Mechanisms involving such counterfactuals might well be real. But the situation regarding locality is worsened. This view makes the truth of counterfactual claims depend not only on what happens elsewhere in this world, but also on what happens in nearby possible worlds. This is the most radically non-local account of the interactions in mechanisms it is possible to have. In general, any passive metaphysical grounding for such counterfactual claims as part of mechanisms will yield non-local, and in some cases also non-real, mechanisms.[30]

[27] This is Psillos' view. He writes: 'The one [story] I favor ties the truth-conditions of counterfactual assertions to *laws of nature*.' He adds later: 'Laws are those regularities that are members of a coherent system of regularities, in particular, a system which can be represented as an ideal deductive axiomatic system striking a good balance between *simplicity* and *strength*.' (Both from Psillos 2004, p. 299).

[28] David Lewis seems committed to this, since he accepts the parallel implication for a laws-based account of causation: 'Like any regularity theory, the best-system analysis says that laws hold in virtue of patterns spread over all of space and time. If laws underlie causation, that means that we are wrong if we think, for instance, that the causal roles of my brain states here and now are an entirely local matter. That's an unpleasant surprise, but I'm prepared to bite the bullet.' (Lewis 1994, p. 479) Note that if the best-system view of laws is as a best system in a world consisting of four-dimensional spacetime, then mechanisms today will also depend on the future.

[29] No one in the mechanisms literature explicitly holds this view, to our knowledge. However, given the history of metaphysical theorizing about counterfactuals, it is too important an alternative to neglect.

[30] We do not discuss Woodward's test-conditions for the truth of counterfactuals because Woodward claims that his view is not a metaphysical one. See Woodward (2003). Nevertheless, if

The only prospect for a real and local metaphysics is some variety of active metaphysics, a metaphysics such as Machamer, Darden and Craver's activities, or Cartwright's capacities. We begin here with Cartwright's capacities view, since it is a more familiar real and local metaphysics. We will raise the issue of how far our arguments transfer to other active approaches later.

On the capacities approach, the interactions between parts of a mechanism are described in terms of the capacities of the entities in the mechanism. Cartwright holds that most general causal claims such as 'aspirins relieve headaches' or 'electromagnetic forces cause motions perpendicular to the line of action', are really ascriptions of capacities – the capacity to relieve headaches ascribed to aspirin, and the capacity to cause motions perpendicular to the line of action ascribed to electromagnetic forces. Capacities are properties, and their instances are real. [31] Cartwright allows for special science 'laws' of the kind we allow for – the rule of thumb, ceteris paribus generalizations that are produced in the special sciences; laws with exceptions that can be explained using a mechanism. But these laws arise out of the reasonably regular reactions of entities with similar capacities: the truthmakers for laws are capacities. She writes: 'It is not the laws which are fundamental, but rather the capacities. . . . Whatever associations occur in nature arise as a consequence of the actions of these more fundamental capacities.' (Cartwright 1989, p. 181)

Using capacities in an account of mechanisms allows what a mechanism is to be local. For Cartwright, the capacities of the entities in a mechanism are properties of the entities, not dependent on anything anywhere else in this world or any other. This nicely fits scientific practice since scientists in many domains spend a lot of time figuring out the capacities of the entities in mechanisms underlying the phenomena that are interesting to them. When scientists point to a mechanism and identify it as the mechanism responsible for certain phenomena, on this view they are pointing to something real and really *there*.

Cartwright ensures that her metaphysics is real and local. An ontology using entities and their powers is structurally similar to that of Cartwright. This approach might also yield real and local mechanisms. See the work of Carl Gillett (2006) and recent as yet unpublished work by Stephen Mumford, both using an active powers metaphysics in their approach to

you did take Woodward's position to be a metaphysical one, his invariance relations would be non-local, albeit more local than either a modal realist or best-system laws view. Invariance relations would still depend on what happened elsewhere and at other times in this world.

[31] Cartwright argues this extensively. See her (1989). She also argues against the possibility of describing away capacities in terms of regularities. She writes: 'One does not just say the acid and the base interact because they behave differently together from the way they behave separately; rather, we understand already a good deal about how the separate capacities work and why they should interfere with each other in just the way they do.' (Cartwright 1989, p. 165).

mechanisms.[32] Machamer, Darden and Craver are also trying to use an active approach as a good metaphysics for mechanisms, but the view remains as yet under-developed. We think it an interesting view worthy of the necessary development.[33]

These views are promising approaches to mechanisms, but note that not all active metaphysical approaches will do. Dispositional approaches are structurally similar to Cartwright's capacities approach and the powers approach, having a basic ontology of entities plus their dispositional properties, rather than entities and their capacities or powers. But the further detail of most dispositional approaches would create problems. First, many accounts of what dispositions are, are non-local. This would apply to either conditional or law-based accounts of dispositions. Second, the local approach to characterizing dispositions is the one which claims that dispositions are intrinsic properties of objects. This makes it an acceptable approach to those wanting a real and local metaphysics and willing to accept metaphysical claims about intrinsic properties, but we are not inclined to accept them. All science needs is clusters of capacities that stick around together for long enough and produce a phenomenon regularly enough for us to get interested in it and look for the mechanism. As Cartwright says of capacities: 'They do indeed endure; on the other hand, their characteristics may evolve naturally through time, and they may be changed in systematic, even predictable, ways as a consequence of other factors in nature with which they interact.' (1989, p. 157)

In conclusion, those wishing for a local and real metaphysics of mechanisms should not use counterfactual notions grounded in laws or other possible worlds in their characterization of mechanisms. There are alternative available metaphysics, along the capacities or active powers lines, which are real and local. These aspects of the capacities or active powers metaphysics are well-known in the core metaphysics literature, of course, but their existence seems to have been largely ignored in the mechanisms literature.

38.6 Objections: Laws, capacities and fundamental explainers

In this section we introduce two major objections to our line of argument. The first objection claims that capacities or powers cannot explain anything at

[32] One might hold that powers are a better prospect for the metaphysics of mechanisms than dispositions, since dispositions can be seen as structural properties of static objects, whereas powers are more dynamic. On this point, see also Machamer (2004). See Illari and Williamson (forthcoming) for further discussion of the dynamic nature of activities.

[33] We argue elsewhere that Machamer, Darden and Craver's activities-entities dualism compares well to Cartwright's entities-capacities ontology, on various criteria for ontology. See Illari and Williamson forthcoming.

all. We argue that they can, and do a job more suited to the special sciences use of mechanisms than laws or counterfactuals. The second objection is an argument attempting to show that on laws or counterfactuals-based stories, mechanisms can still be local. We argue that whether this succeeds depends on precisely what you are claiming in characterizing mechanisms. On one of the cases we identify, there is no serious disagreement, on the other account we argue that mechanisms remain non-local.

38.6.1 Capacities cannot explain

Recall that the major reason for introducing mechanisms is to explain. The most uncontroversial claim about mechanisms is that they explain, whether in a physical or in an epistemic way. It might be objected to our arguments that positing an active metaphysics such as capacities or powers as explainers, particularly as fundamental explainers, is illegitimate, because capacities and powers do not explain anything. To say A produced B because it has a capacity or power to produce B, or engaged in the activity which brings about B, explains nothing – it might be thought a mere assertion of 'dormitive virtue'.

The first thing to notice in this debate between capacities or powers, and laws or counterfactuals, is that a parallel complaint can be made about laws or counterfactuals. To say that A produced B because there is a law that A produces B might also be thought to explain nothing. The only 'explanation' offered by a law is the recognition that things just tend to happen that way, that things like A tend to produce things like B. Intuitions on whether to prefer something like capacities or something like laws as fundamental explainers do seem to vary.

There is something odd in considering a law or a capacity, alone, as a complete explanation. The explanations we do get tell us so much more. Consider the explanation of how the cell produces proteins. Neither the claim that it has a capacity to do so, nor the claim that cells like it also produce proteins, tell us much. *Mechanisms* explain in terms of lower-level entities and their capacities, powers, activities or some such item of an active metaphysics. Mechanisms as a whole are neither just capacities, nor just laws. Mechanistic explanation generally starts with a regularity: the identification of a phenomenon requiring explanation – usually a regularly occurring phenomenon. In the case of protein synthesis, distinguishing between kinds of proteins produced, and so further dividing the explanandum phenomenon is important. Mechanistic explanation then proceeds by identifying the parts that make up the phenomenon – the production of each protein, and what those parts do, and can do under similar circumstances. To see this as the identification of the entities present, and the capacities or powers that those entities have, seems natural. Lawlike regularities can be useful in describing mechanisms, but as we have explained, this is consistent with an active metaphysics account of mechanisms. On this view lawlike regularities are not fundamental. For those influenced by the widespread concern of those in the mechanisms literature that there are no special science laws, which we identified as a key reason for turning to consideration of mechanisms, this mechanistic view of explanation is far superior.

38.6.2 Mechanisms using laws or counterfactuals can still be local

There may be a deeper objection to our argument that using passive grounds for counterfactual notions to characterize mechanisms renders them non-local. This is as follows: the status of mechanisms as mechanisms depends only on the natural properties that the mechanisms have. Although these natural properties depend on the laws of nature, or other possible worlds, nevertheless the natural properties are local. Thus, the mechanisms themselves are local. The broad idea seems to be to push the underlying non-local metaphysics into the background, and insist that in general, say, protein synthesis depends only on the natural properties of the molecules in the cell, while natural selection depends only on the natural properties in the population and its local environment. This, the claim would go, yields the required locality in spite of the fact that these natural properties only produce anything or interact with anything in virtue of things widely spread in time and space. Thus mechanisms are constitutively local since what makes a mechanism a mechanism is these natural properties, the laws being merely some kind of inert background conditions.

This is an interesting possibility, which raises the issue of what a characterization of mechanisms is intended to do. Over the course of this chapter we have identified a number of differences in approaches to mechanisms. A genuine possibility here is that some in the debate are not concerned to make any claims about the nature of mechanisms – that is, no claims that really impinge on the metaphysics of mechanisms.[34] This might be a way to defend Woodward, Psillos and Glennan. Perhaps they are merely trying to give an account of mechanisms that will let you *pick them out*, so that you can discriminate mechanisms from non-mechanisms – rather than illuminate what they *actually are*. Perhaps the best characterisation uses counterfactuals or invariance relations because they best let you pick out the mechanisms.

It is possible to read both Woodward and Glennan as merely characterizing how you pick out mechanisms. Woodward's paper is titled, 'What is a mechanism? A counterfactual account,' but his abstract summarizing his argument is less clear. He writes: 'This paper presents a counterfactual account of what a mechanism is. Mechanisms consist of parts, *the behavior of which conforms to generalizations that are invariant under interventions*, and which are modular in the sense that it is possible in principle to change the behavior of one part independently of the others. Each of these features can be captured by

[34] This might be partially due to their primary concern being with epistemic explanation.

the truth of certain counterfactuals.' (Woodward 2002, S366, emphasis added) Perhaps he is only claiming something about what the parts of mechanisms typically do, not what they are. Glennan is also open to this interpretation. He writes: '"Interaction" is a causal notion that *must be understood in terms of* the truth of certain counterfactuals. The stipulation that these interactions can be characterized by invariant, change-relating generalizations is meant to capture the relevant counterfactual truth claims.' (Glennan 2002, S344, emphasis added) If you read 'must be understood in terms of' fairly lightly, this may not be a claim about a deeper metaphysics.

If this is indeed the aim Woodward, Psillos and Glennan have in mind, then their accounts would fit that aim. Mechanisms do typically exhibit a stability which can very naturally be characterized using various counterfactual notions. If this is the claim then it is perfectly legitimate to say that a counterfactual characterization using a laws or modal realist grounding for counterfactuals is compatible with a local deeper metaphysics. A characterization of this sort says nothing about metaphysics. When serious questions about the nature of mechanisms arise, the non-local metaphysics used merely to pick out mechanisms can fade quietly into the background.

That is one possibility. However, it seems that Woodward, Psillos and Glennan all have at least some intention of arguing for a claim about the nature of mechanisms. In some form or other, they all argue that some counterfactual notion is essential or ineliminable in characterizing mechanisms. This might well be taken as a stronger claim about the nature of mechanisms, beyond any claim about a handy way to pick them out. The quotes from Woodward and Glennan above are certainly open to this stronger reading. Psillos must be read in this stronger way.[35] He argues extensively that counterfactuals are indispensable to a characterization of mechanisms, repeating this kind of claim at several points in his paper. He summarizes the thesis he has argued for towards the end of the paper: 'mechanisms need counterfactuals; but counterfactuals do not need mechanisms. In other words, mechanistic causation requires counterfactual dependence but not conversely. It is in this sense, that the counterfactual approach is more basic than the mechanistic.' (Psillos 2004, p. 315) To claim that the counterfactual approach is more basic than the mechanistic does look like a metaphysical claim in this chapter.

If the claims of any of the three are claims about the basic – ineliminably basic – metaphysics for mechanisms, then it is a claim about the nature of mechanisms. It is not clear how such a claim can fade into the background to allow mechanisms to be considered constitutively local. It is claimed that the nature of mechanisms is to have the natural properties that they do lead to the interactions that they do only in virtue of the laws of nature, or facts about

[35] Note that Psillos could defend himself by retreating into a general anti-realism. But then we think his general anti-realist views will support his views on mechanisms, not vice versa.

other possible worlds. But then what mechanisms *are* is non-local. Depending on the detail of the further claims, mechanisms might also turn out to be non-real.

Another thought is that Psillos, Woodward and Glennan are in various ways concerned with understanding causation, as well as mechanisms. Perhaps the claim that a non-local metaphysics doesn't make *causal* claims non-local is plausible in a way that the parallel claim for mechanisms is not. In considering causal claims, we immediately focus on salient causes, and standardly assume vast amounts of stable background conditions. Perhaps in this context the claim that natural properties are local, treating a passive metaphysics as background, is not unreasonable. But in the context of physical mechanistic explanation – not *epistemic* mechanistic explanation – we are not in the same situation. For mechanisms, the entire structural background is crucial to what a mechanism is. On this metaphysical view, such mechanisms are non-local.

Perhaps the objection discussed in this section could be read as intending merely to deflect the counterintuitiveness of mechanistic explanation turning out to be non-local.[36] If this is so, it seems to fail. Non-local mechanistic explanations of this sort are committed to the core claim that the behaviour of other people like me in this world, or my counterparts in other possible worlds, is in some way relevant to explaining why I bump into lampposts. This is precisely the counterintuitive claim that we deny. A mechanistic explanation of my clumsiness depends only on facts about me. There is a metaphysics available without such a counterintuitive consequence – a metaphysics of capacities, powers or activities, which should be preferred.

In conclusion, capacities can explain and mechanisms are still non-local on any claim that counterfactual notions are part of their nature, their metaphysics. The active metaphysics of capacities or powers clearly comes out better, thoroughly satisfying both locality and reality.

38.7 Conclusion

We have argued extensively that both scientific practice and much of the mechanisms literature is committed to mechanisms being both real and local. We further argued that if mechanisms are to be real and local, so that they can be used in physical explanations of phenomena, in a form distinct from laws-based explanation, they require an active metaphysics such as Cartwright's capacities approach, a powers approach, or an activities approach.

We have framed all our arguments here about mechanistic explanation. We believe that exploring mechanistic explanation will be illuminating to meta-

[36] We thank an anonymous reviewer for raising this possibility.

physical debates about causation, but not necessarily in a simple way, meaning that it is best to understand mechanistic explanation thoroughly, and then go on explicitly to consider its relation to causation and causal explanation. For those interested in scientific methodology more than metaphysical debates, it is probably sufficient to note that scientists involved in mechanistic explanation of the sort we describe see themselves as straightforwardly involved in causal explanation. See for example the work of an evolutionary biologist: 'The main purpose of evolutionary biology is to provide a rational explanation for the extraordinarily complex and intricate organization of living things. To explain means *to identify a mechanism that causes evolution and to demonstrate the consequences of its operation*.' (Bell 1997, p. 1, emphasis added.) Or consider the view of a biochemist: 'Uncovering the cellular mechanisms *resulting in sequential transfer* of information from DNA (our genes) to RNA and then to protein represents one of major achievements of biochemistry in the 20th century.' (Whitford 2005, p. 247, emphasis added) Both clearly see themselves as investigating causes. From this point of view, all our arguments apply straightforwardly to this important variety of causal explanation.

Acknowledgements

We would like to thank the Leverhulme Trust for funding this work on mechanisms. We are also grateful to three anonymous referees for extensive comments leading to improvement of the work. We would also like to thank numerous colleagues at Kent, Bristol, and internationally for discussion of ideas used here. Remaining errors are, of course, our own.

References

Peter Achinstein (1983). *The Nature of Explanation*, OUP, Oxford.

Roger L. Adams, John T. Knowler and David P. Leader (1992). *The Biochemistry of the Nucleic Acids* (11th edn), Chapman and Hall, London.

William Bechtel and Adele Abrahamsen (2005). Explanation: A mechanist alternative, in *Studies in the History and Philosophy of the Biological and Biomedical Sciences*, 36, 421–41.

William Bechtel and Adele Abrahamsen (2008). From reduction back to higher levels, In B.C. Love, K. McRae and V.M. Sloutsky (eds.), *Proceedings of the 30th Annual Conference of the Cognitive Science Society* (pp. 559–64), Austin, TX, Cognitive Science Society.

Helen Beebee (2004). Causing and nothingness, in L.A. Paul, E.J. Hall and J. Collins (eds.) *Causation and Counterfactuals*, MIT Press, Cambridge, MA, pp. 291–308.

Graham Bell (1997). *Selection: The Mechanism of Evolution*, Chapman and Hall, New York.

Ned Block (2005). Two neural correlates of consciousness, in *Trends in Cognitive Sciences*, Vol. 9, 2, 46–52.
Nancy D. Cartwright (1983). *How the Laws of Physics Lie*, Clarendon Press, Oxford.
Nancy D. Cartwright (1989). *Nature's Capacities and their Measurement*, Clarendon Press, Oxford.
Nancy D. Cartwright (2006). Evidence-based policy: Where is our theory of evidence?, in *Selected Proceedings of the German Analytic Philosophy Conference*, Berlin, September 2006.
Nancy D. Cartwright (2009). 'Causality, invariance and policy', in Harold Kincaid and Don Ross (eds.) *The Oxford Handbook of Philosophy of Economics*. Oxford, OUP, pp. 410–23.
Carl Craver (2007). *Explaining the Brain*, Clarendon Press, Oxford.
Lindley Darden (2006). *Reasoning in Biological Discoveries*, CUP, Cambridge.
Phil Dowe (2004). Causes are physically connected to their effects: Why preventers and omissions are not causes, in C. Hitchcock (ed.) *Contemporary debates in Philosophy of Science*, Blackwell, Oxford.
Bas van Fraassen (1980). *The Scientific Image*, Clarendon Press, Oxford.
Carl Gillett (2006). The metaphysics of mechanisms and the challenge of the new reductionism, in M. Schouten & H.L. de Jong (eds.) *The Matter of the Mind*, Blackwell, Oxford.
Stuart Glennan (2002). Rethinking mechanistic explanation, in *Philosophy of Science*, 69 (September 2002), S342–S353.
Ned Hall (2004). Two concepts of causation, in L.A. Paul, E.J. Hall and J. Collins (eds.) *Causation and Counterfactuals*, MIT Press, Cambridge, MA, pp. 225–76.
Phyllis McKay Illari and Jon Williamson (forthcoming) 'In defense of activities'.
Phyllis McKay Illari and Jon Williamson (2010). 'Function and organization: Comparing the mechanisms of protein synthesis and natural selection' in *Studies in the History and Philosophy of the Biological and Biomedical Sciences*, 41, 279–291.
Bert Leuridan (2010). Can mechanisms really replace laws of nature?, in *Philosophy of Science* 77, pp. 317–340.
Bert Leuridan and Erik Weber (2011). The IARC, and mechanistic evidence, in Phyllis McKay Illari, Federica Russo and Jon Williamson (eds.) *Causality in the Sciences*; Oxford, Oxford University Press.
David Lewis (1994). Humean supervenience debugged, *Mind*, 412, 471–90.
David Lewis (2004). Void and object, in L.A. Paul, E.J. Hall and J. Collins (eds.) *Causation and Counterfactuals*, MIT Press, Cambridge, MA, pp. 227–90.
Peter Machamer, Lindley Darden and Carl Craver (2000). Thinking about mechanisms, *Philosophy of Science* 67 (March 2000) 1–25.
Peter Machamer (2004). Activities and causation: The metaphysics and epistemology of mechanisms, *International Studies in the Philosophy of Science*, 18:1 (March 2004) 27–39.
Richard Menary (2010). *The Extended Mind*, Ashgate MIT Press, Cambridge, Massachusetts.
Sandra D. Mitchell (1997). Pragmatic laws, *Philosophy of Science*, 64(4), supplement, S468–S479.
Stathis Psillos (2004). A glimpse of the secret connexion: Harmonizing mechanisms with counterfactuals, *Perspectives on Science*, 12(3), 288–319.

Frank P. Ramsey (1990). Law and causality, in D.H. Mellor (ed.), *FP Ramsey: Philosophical Papers*, CUP, Cambridge.

Mark Ridley (2004). *Evolution* (3rd edn). Blackwell, Oxford.

Wesley C. Salmon (1998a). Comets, pollen, and dreams: Some reflections on scientific explanation, in his *Causality and Explanation*, OUP, Oxford, pp. 50–67.

Wesley C. Salmon (1998b). Scientific explanation: Three basic conceptions, in his *Causality and Explanation*, OUP, Oxford, pp. 320–32.

Jonathan Schaffer (2004). Causes need not be physically connected to their effects: The case for negative causation, in C. Hitchcock (ed.), *Contemporary debates in Philosophy of Science*, Blackwell, Oxford.

Robert A. Skipper (Jr.) and Roberta L. Millstein (2005). Thinking about evolutionary mechanisms: Natural selection, *Studies in History and Philosophy of Biological and Biomedical Sciences*, 36, 327–347.

Phillip J. Torres (2009). A modified conception of mechanisms, in *Erkenntnis* DOI 10.1007/s10670-008-9125-y. 71, 233–51.

Donald Voet and Judith G. Voet (2004). *Biochemistry*, John Wiley, New York.

David Whitford (2005). *Proteins: Structure and Function*, John Wiley, Sussex.

Robert A. Wilson (2004). *Boundaries of the Mind*, CUP, Cambridge.

James Woodward (2002). What is a mechanism? A counterfactual account, *Philosophy of Science* 69, S366–S377.

James Woodward (2003). *Making Things Happen: A Theory of Causal Explanation*, OUP, Oxford.

·39·

Mechanistic information and causal continuity

Jim Bogen and Peter Machamer

Abstract

Contrary to Griffiths and others, we claim that properly conceived, the notion of *information* is important to the philosophical and scientific understanding of causal continuities that connect the steps of some important biological processes. Far from being 'little more than a metaphor that masquerades as a theoretical concept' (as Sarkar claims), we believe the relevant notion of information can be well enough understood to qualify as a useful and perfectly acceptable scientific concept. As our title suggests, this paper's treatment of information develops from Machamer, Darden, and Craver's mechanistic account of causally productive causal processes. We suppose that what we are calling mechanistic information can be understood in terms of goals served by mechanisms, and the influence on connections among the initial and final stages of their operation. We use the examples of Crick's early conception of gene expression and a sensory-motor reflex in the leech to illustrate our account and to contrast ours to some familiar ideas of information including Shannon and Weaver's, Millikan's teleosemantic notion, and Crick's own conception of information transmission as pattern replication.

39.1 Introduction

Since the middle of the twentieth century neuroscientists, evolutionary theorists, molecular geneticists and other biologists, have talked as though *information* and *information flow* are important explanatory notions. However, some influential recent literature in philosophy of science disagrees. Paul Griffiths says that although

> ... there is a genetic code by which the sequence of DNA bases in the coding regions of a gene corresponds to the sequence of amino acids in the primary structure of one or more proteins, ... the rest of 'information talk' in biology is no more than a picturesque way to talk about correlation and causation. (Griffiths 2001, p. 395)

In conceding that 'there is a genetic code' all Griffiths commits himself to is a simple (though degenerate) mapping relationship between sequences of

DNA codon bases and amino acids on protein precursors. In rejecting the rest of 'information talk' he denies that explanations and descriptions of biological phenomena couched in terms of information flow can tell us anything that cannot be said simply by talking about causes and correlations. He endorses Sahotra Sarkar's claim that no matter how much working biologists talk about information

> ...there is no clear, technical notion of 'information' in molecular biology. It is little more than a metaphor that masquerades as a theoretical concept and...leads to a misleading picture of possible explanations in molecular biology. (Sarkar 1996, p. 187)

We agree with the biologists.[1] Ideas about information have been and continue to be important to the development and articulation of exemplary explanations of some fundamental biological phenomena.[2]

Standard biology textbooks invoke information to answer questions about biological processes they seek to explain. We maintain that information talk in biological explanations cannot always or often be satisfactorily replaced by descriptions of correlations and non-informational causal connections. Crick's *On Protein Synthesis* is a classic illustration (Crick 1958). Although we don't agree with much of what Crick says about information flow, we follow him in thinking that the continuities of some fundamental biological processes do depend upon information storage and transmission. This chapter's main examples of this are DNA expression and a sensory-motor reflex which functions to move leeches away from things that press against them.

The poet Frank O'Hara described the development of his poems as taking the form: I do this, I do that (http://findarticles.com/p/articles/mi_m1248/is_2_88/ai_59450177/). The continuity of a causal process might be analogously described by saying that this causes (or contributes to the production of) this, and that causes (or contributes along with such and such other factors

[1] The debate over whether *information* is a helpful concept in biology often focuses on the role of genes and genetic expression in evolutionarily significant processes involving the emergence and transmission of the adaptive effects of heritable traits (see Smith 2000, 2000a Sterelny 2000, Godfrey-Smith 2000, Sarkar 2000). A further focus is on the question whether genotypes and phenotypes are coupled closely enough to support controversial versions of genetic and evolutionary determinism. We agree with Griffiths' skepticism about the idea that 'genes' code or transmit information about phenotypes, but we disagree with him about the role of information in the synthesis of protein precursors and in some other biological processes.

[2] Sarkar grants that the concept of information can be play a useful heuristic role 'in the construction of some scientific entity' but denies that it 'explicitly occurs in that entity'. (Sarkar 2000 p. 209). If this means that it was useful for biologists like Crick to think about information in constructing their theories, we agree. We don't know how Sarkar distinguishes ideas which occur explicitly and do real scientific work from those that don't. But we maintain contrary to Sarkar that over and above its heuristic value *information* does do important explanatory work. Furthermore we maintain that information transmission plays a causally significant role in the biological processes it is invoked to explain.

to the production of) that. Frank O'Hara style causal descriptions can tell us all there is to know about how a number of biological processes develop from step to step. As we discuss in Section 39.3 and 39.4, the Krebs cycle is a case in point. But they fail to capture important facts about connections between the steps that take the mechanism of protein synthesis from the transcription of a segment of DNA to the production of a polypeptide string of amino acids. The same holds for connections between the steps of the leech reflex whose operation is initiated by something pressing on the organism's body and completed (if all goes well) by muscle contractions that move the leech away from the source of the pressure. In cases like these, causal factors at work in the initial steps of the process exert a strong influence on the development of the process and the result that completes it. The kind of influence they exert distinguishes these processes from processes whose continuities can be explained without appeal to information. In the remainder of this chapter we set out the notion of information that we take to be appropriate to the explanation of the continuities of DNA expression, the leech escape reflex and other biological mechanisms whose initial causal factors make similar contributions to their continuities.

39.2 Processes and mechanisms

The processes we consider in this chapter – informational and non-informational alike – are operations of mechanisms in the sense of Machamer *et al.* (MDC) (2000). An MDC mechanism is an arrangement of entities which engage in an ordered sequence of activities. The activities that entities engage in move the mechanism from an initial or start-up condition through one or more steps[3] to a result that marks the end of its successful operation. The activities move the mechanism forward by initiating, sustaining, modifying and damping the activities of other entities and in some cases, by incorporating or producing new component entities and eliminating or modifying preexisting ones. Mechanistic explanations answer questions about the entities, the activities and the parts they play in producing results to be explained.[4] As MDC observe, ideal mechanistic explanations

> ... exhibit productive continuity without gaps from the set-up to termination conditions. Productive continuities are what make the connection between stages intelligible. (Machamer, Darden and Craver 2000, p. 3)

[3] Steps are individuated in terms of causal sub-processes. We lack space to say more.

[4] For details see Machamer, Darden, Craver (2000) *passim*, and Darden (2006 pp. 1–12, 13–98, 271–312).

All mechanisms

> ...have productive continuity from one stage to the next...[such that] entities and activities of one stage give rise to the next stage...but few mechanisms have information flow through multiple stages of the [operation of the] mechanism. (Darden 2006, p. 283)

Whether information ought to be invoked to explain how a mechanism operates depends in part on the interpretive stance taken by its investigators. For example, molecular biologists found it useful to interpret certain features of DNA expression in terms of information when biochemical interpretations proved unfruitful (Darden 2006, p. 280 ff.). But the correctness of their explanatory claims depends on facts about the makeup and the operation of the mechanism that obtain independently of interpretive or explanatory strategies. One crucial difference between the operations of mechanisms that do, and those that do not involve information turns on what we call the *reach* of causal influences exerted by initial factors. To illustrate what this means we will sketch some differences between DNA expression (an informational process) and the Krebs cycle (a non-informational process).

A second crucial difference is that the continuity of an informational mechanism is a function of its teleological structure. Informational biological mechanisms operate for the sake of achieving or promoting goals of the organism (or one or more of its component subsystems) to which they belong.[5] Information, as we think of it, consists of the causal influences that achieve or promote relevant goals. When the system is in good working order and the mechanism functions as it should, the information an entity or activity transmits (i.e. the causal influence it exerts on other components of the mechanism) contributes to the production of a result that achieves or promotes the goal for the sake of which the mechanism operates. We discuss this below in Section 39.6. But mechanisms are subject to different kinds of malfunction. In some malfunctions – e.g. where interfering factors keep the mechanism from operating or from moving all the way to its final step – information plays no significant role. In others, e.g. the expression of mutant DNA responsible for cystic fibrosis, information contributes to malfunction by moving the mechanism toward a result that prevents the achievement of its normal goal. We discuss this in Section 39.7.

[5] Information may figure in the operation of artificial as well as natural mechanisms. The goals that the mechanisms (help) satisfy may be natural or imposed on them by humans or other organisms that use them. We focus on natural mechanisms and ignore what they do to satisfy the goals of human and other users. For example, what we have to say should apply to fermentation as a metabolic process in yeast, but not as a step in intentionally producing wine.

39.3 Reach

To illustrate the notion of reach, consider how DNA codons direct the selection and arrangement of amino acids to form protein precursor polypeptides. The production of amino acid strings begins when the bases on a DNA segment bind weakly to their complements[6] to produce a string of nucleotides which is then detached to form a strand of pre-mRNA. Later the pre-mRNA strand is cut and spliced to produce a strand of mRNA. Still later, amino acids attached to ribosomes decorated with the complements of mRNA bases are carried and attached to mRNA strands. The bases to which ribosome bases bond are the complements of codon bases on the DNA segment that is expressed. As a result the amino acids carried by the ribosomes are arranged on the polypeptide to stand in the same spatial relations as the bases on the DNA codons. Because the bases bind weakly only to their complements, DNA codon bases exert a direct influence on the production of pre-mRNA and a strong indirect influence that extends to the products of subsequent steps of polypeptide construction.

To make the notion of reach more vivid, compare the extensive influence of DNA codon bases on polypeptide construction to the weak influence the oxaloacetate molecule that interacts with other chemicals to begin each round of the Krebs (citrate) cycle.[7] Each round of the cycle consists of eight chemical reactions. Each reaction uses chemicals supplied by a number of different mechanisms to produce a molecule that serves as a substrate for the next step in the cycle. Many of the mechanisms that supply chemicals to interact with the substrates operate more or less independently of one another. And they are part of the Krebs cycle only to the extent that some of its byproducts contribute to the production of energy carriers and other vital molecules needed to sustain their operation. Thus, none of the substrates of Krebs cycle reactions exert an influence of any considerable reach on subsequent steps.

In slightly more detail, a number of different chemicals from different sources enter into the first two steps of the cycle. Oxaloacetate (the substrate for step one) does little by itself to limit the number of results that can be produced in step two. Moreover, the influence of oxaloacetate diminishes from step to step as the new substrates are produced. For example what goes on in step 5 in Figure 39.1 depends upon the chemical behaviour of succinyl-CoA, the substrate supplied by step 4, together with additional chemicals (including

[6] For example, cytosine bonds weakly to guanine, and adenine bonds weakly to thymine or uracil (Alberts *et al.* 2002, p. 302).

[7] The production of citrate from oxaloaetate is customarily singled out as the first step of the cycle because of its place in the metabolic process that takes the organism from food intake and breakdown to the production of physiologically useful energy carriers. (For details see Alberts *et al.* 2002, pp. 95–108, 126–7.) But treating any other step as the beginning of the cycle would make no difference to what we have to say here.

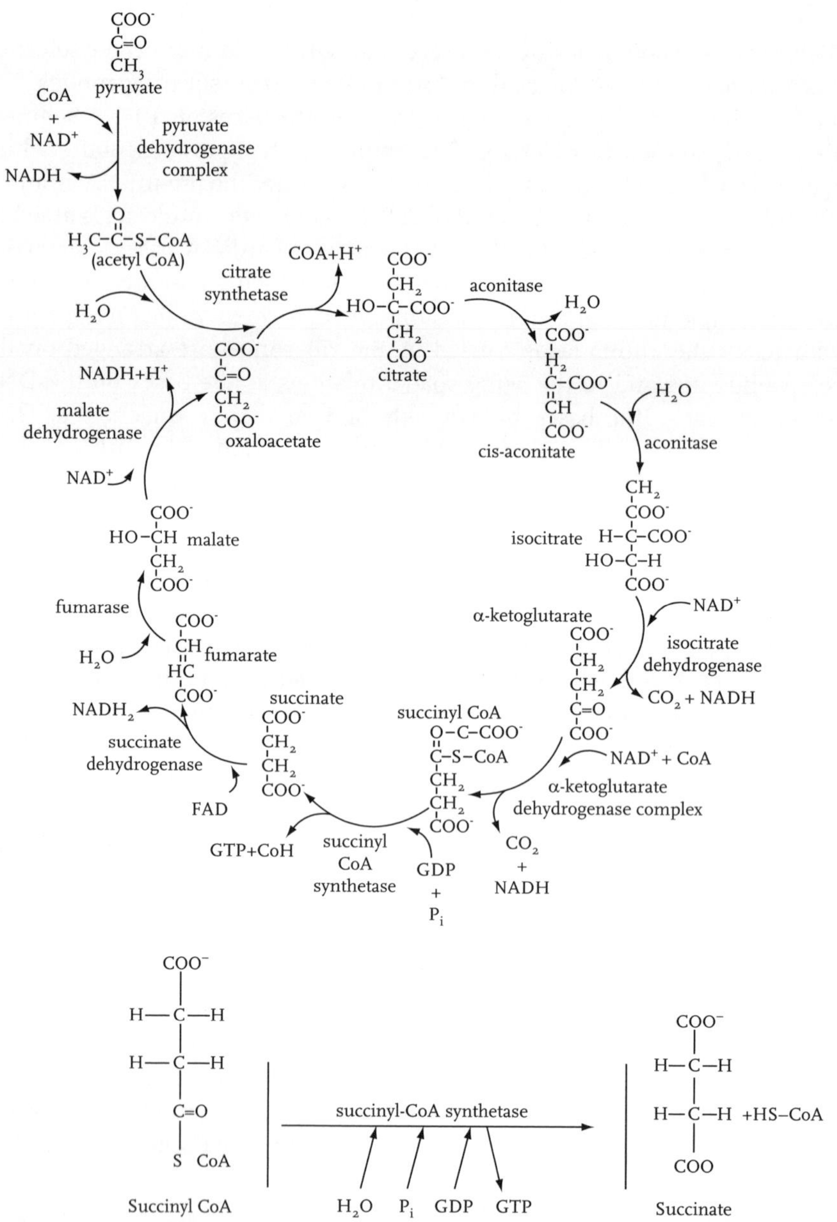

Fig. 39.1 Top: The Krebs cycle, courtesy of Georgia State Chemistry Department, http://chemistry.gsu.edu/Glactone/PDB/Proteins/Krebs/Krebs.html. Bottom: Step five of the Krebs cycle.

GDP, water, inorganic phosphate molecules and synthetase enzyme, for example) that are made available by interactions that do not belong to the cycle itself.

These chemicals work together to produce succinate, the substrate for step 6. The chemical makeup of succinyl-CoA places some constraints on what molecules can be produced at subsequent steps. But these constraints are so weak that a great many different interactions producing a great many different molecules would occur if different enzymes and reactants were present instead of the ones that are normally available to help move the cycle forward. According to Peter Wipf

> For the 9 small molecules[8] involved in the citric acid cycle, any good chemist could draw you upwards from 200 different reactions giving different products from the specific enzyme-mediated processes. For somebody trained in the art of synthesis, the number would go up to maybe 5,000. (Personal correspondence)

The burden of deciding which of these molecules can be produced falls mainly to the influence of the reactants and enzymes that various mechanisms provide after the completion of step 1.

We characterize *reach* in terms of strength and independence of influence as follows:

39.3.1 Strength of influence

The strength of an entity's or activity's influence depends on how many alternative results it rules out or renders significantly improbable in subsequent steps in the operation of the mechanism.[9] The more downstream outcomes it leaves open, the weaker is its influence.

39.3.2 Independence of influence

This is a robustness condition. The less a factor's influence depends on background conditions over which it has no control, and the greater the range of different background conditions under which it can produce or contribute to the production of the same downstream results, the more independent is its influence on the operation of the mechanism.[10] Because bases bind to their complements independently of the background conditions under which

[8] The nine small molecules are the substrates that begin each step together with acetyl CoA.

[9] By 'alternative results', we mean outcomes which are physically, chemically, anatomically and physiologically possible rather than outcomes which are just logically possible or conceivable from the vantage point of the armchair from which philosophers think about 'nearby possible worlds'. A detailed account of what this amounts to must await a further paper.

[10] We distinguish independence from strength because a factor, X, that influences the production of results at a number of different steps may do so only in connection with different factors, $y_1, \ldots, y_n$, at every step. If the y factors decide which results are produced and X has no control over which of the y are available at any given step, X's influence can be strong but not independent.

binding occurs, the influence DNA codons exert on the construction of RNA strands and polypeptide chains is considerably more independent than the influence that the oxaloacetate molecule exerts on the production of molecules in the Krebs cycle.[11]

39.3.3 Reach

The reach of an entity's or an activity's influence depends on how many steps are strongly influenced by it and how independently it influences them. It should be clear that the reach of the influence exerted by DNA base sequences extends to the ordering of the amino acids. (We discuss the reach strength and independence of initial causal factors in the leech's pressure escape mechanism below in Section 39.5 and note 20.)

While the reach of initial factors is important, it is not sufficient to distinguish informational from non-informational continuities. Consider the example of a very smooth block that slides down a very smooth plane in response to a gentle tap, strikes a domino at the bottom of the incline and knocks it over.[12] Absent interfering causes, the tap that starts it on its way exerts a strong and independent influence on its direction and velocity at every step of its slide and on the fall and final position of the domino, but this is by no means an information system. A Frank O' Hara style description of the relevant causal sequence suffices to explain what happens from the tap to the domino's falling over and coming to rest. Similarly, there is no information in the process that takes a drinker from alcohol ingestion to intoxication even though the alcohol exerts an influence of great reach. Furthermore, the notion of reach applies only trivially to causal influences in mechanisms that move from start up to end states in a single step. The teleological structure of DNA expression and the leech reflex is what distinguishes them from the falling dominos.

39.4 Teleological structure and reach differentiate the Krebs mechanism from informational mechanisms

To recapitulate, we've seen that the mechanisms responsible for the leech reflex and for DNA expression operate to satisfy needs for the organisms they belong to. These mechanisms are activated by factors that indicate what results they must produce in order to serve their purposes. Thus in response to pressure, sensory neurons engage in activities whose reach extends far enough through the operation of the reflex mechanism to direct it toward the

[11] We are indebted to Jim Woodward for conversation on this topic and for making available a draft of a paper in progress on causality in biology that develops a related idea.

[12] We stipulate that the block, the incline, and the domino have not (as perhaps they could have) been incorporated into some sort of Rube Goldberg mechanism whose proper functioning depends upon the way the domino falls.

production of teleologically appropriate muscular activity. A crucial difference between the pressure escape reflex and the Krebs cycle is that in comparison to the factors that set it in motion, the factors that begin the Krebs cycle have very little influence on what happens downstream. But reach doesn't explain why the operation of the block-incline-domino mechanism is non-informational. The tap that starts the block on its way toward the domino has a great deal to do with the fall of the domino. With regard to mechanistic information, the crucial difference between the escape reflex and the block-incline-domino mechanism is teleological. The block-incline-domino mechanism does not belong to a system with goals for the block to promote by knocking over the domino. DNA expression differs from the operation of the block-incline-domino mechanism in the same way – by virtue of operating to help satisfy needs of the organism.[13]

Someone will want to know how we can accommodate what we've just been saying to the fact that the Krebs cycle benefits organisms by producing energy carrying GTP and NADH, and other beneficial molecules including the precursor of energy carrying ATP (Alberts *et al.* 2002, pp. 92, 102, 106). It is arguable that this is why the Krebs cycle survived natural selection. We are indebted to Lindley Darden for commenting that philosophers who identify the goals of biological mechanisms with functions for which they were selected might conclude from this that the Krebs cycle operates for the sake of supplying the organism with certain vital molecules. But even so, biologists and biochemists seldom invoke information to explain how the cycle operates as they do in connection with DNA expression. According to us that is as it should be because the molecules that figure in its first step have next to no influence on the direction the cycle takes in moving toward the result that marks its final step. As we said, reactants that originate outside of the mechanism do much of the work in selecting products to be produced, e.g. at steps 5 through 8 of the cycle. As a result, the strength and independence of the influence of oxaloacetate molecules diminishes to relative insignificance as the cycle moves on.

Furthermore, although the Krebs cycle plays a teleological role it does not have the kind of teleological structure that is required for a mechanistic informational process. When all goes well the operations of the leech reflex and the DNA expression mechanisms end with the very result whose teleological appropriateness was indicated by entities and activities belonging to their first steps. By contrast, the oxaloacetate molecule whose production completes the Krebs cycle benefits the organism only as a substrate for the first step of its next round. In contributing to the continuation of the cycle, the

[13] See previous note. We are indebted to Ken Schaffner for pointing out how often biologists describe biological functions in terms of needs. For just a few examples see Alberts *et al.* (2002, p. 380), and Lewin (1994, p. 418).

benefit oxaloacetate provides is no different from that of the products of every other step. The vital molecules the Krebs cycle produces for the organism are released as byproducts before the cycle reaches its final step.

Jason Byron objected in discussion that the Krebs cycle resembles DNA expression more closely than we acknowledge. The molecules that interact with oxaloacetate to begin the cycle are obtained from carbohydrates whose ingestion is typically increased by hunger which indicates a need for energy carried by GTP and NADH. Thus if hunger tracks energy closely enough, and sugars are metabolized efficiently enough, hunger could regulate the rate at which the Krebs cycle mechanism produces needed energy carriers.[14] But this doesn't imply that the Krebs cycle has the teleological structure exemplified by DNA expression and the leech reflex. Over and above the fact that GTP, FADH, etc., are byproducts rather than results completed by the cycle's final step, we've seen that the molecules that set the cycle in motion don't even do much to direct the cycle toward the production of its beneficial byproducts. Thus Frank O'Hara style causal accounts can tell us all we need to know about the continuity of the Krebs cycle. But trying to explain the continuity of polypeptide construction or the pressure escape reflex without appeal to the goals their mechanisms serve, and connections between their goals and the control that first step factors impose on their operation would be like trying to explain the continuity of a chess game without reference to the object of the game or the influence of tactical considerations.

39.5 Mechanistic information

We use the term 'mechanistic information' in connection with the notion of information that we think is appropriate for explaining continuities in the operation of goal directed mechanisms that Frank O'Hara descriptions cannot account for. We turn now to our characterization of mechanistic information, beginning with the first of several comparisons to other notions of information.

As Lindley Darden explains, Crick believed that information flows from DNA to direct protein precursor construction and that information flow consists of pattern replication. Amino acid strings are encoded after the manner of Morse code messages by sequences of DNA codon bases (Darden 2006, p. 282–3). The pattern into which these bases are arranged is instantiated on RNA strands by codons whose bases are the complements of the bases in their

[14] For the sake of the argument, ignore people like us who eat when they have no physiologically good reason to do so. Byron's example was an athlete who eats after exercise in response to hunger caused in part by energy depletion.

DNA codon counterparts. According to Crick the same pattern is replicated from step to step until its final instantiation – a string of amino acids that will be folded to complete the protein encoded in the DNA segment.[15] We agree with Crick in thinking of DNA information in terms of '... the specification of the amino acid sequence of the protein' that has been selected for synthesis (Crick 1958, p.144; Cp.p.153). But we disagree with his ideas about information flow. As we said, DNA segments are expressed in order to supply the organism (or one or more of its subsystems) with proteins it needs. So-called housekeeping genes are expressed more or less continuously to provide proteins the organism needs at all times, e.g. to support cell metabolism. Other DNA segments are expressed occasionally to satisfy temporary needs for proteins to support special functions (e.g. muscle contractions and relaxations in the execution of a dance step). The construction of a protein precursor actually does begin with the replication of a pattern; a DNA segment is selected to serve as the template for the production of a strand of pre-mRNA whose nucleotide bases stand to one another in the same spatial relations as their DNA complements. Anti-codons are sequences of nucleotides. Like codons, each anti-codon is a sequence of three nucleotides. Anti-codon nucleotides complement and stand in the same special relations to one another as those of the corresponding DNA codons. Thus pre-mRNA anti-codons instantiate the same pattern as codons. But pre-mRNA is cut and spliced to produce mRNA, and in the process molecules standing in between anti-codons are removed. As a result anti-codons that were spatially separated on the pre-mRNA strand are adjacent to one another on the mRNA strand. Suppose CUCAGCGU-UACCAU are the bases on a string of anti-codon nucleotides that comprises a part of the mRNA strand. As it stands this string could function as any one of several different sequences of anti-codons, each of which codes for a different arrangement of amino acids.[16] Because the string is ambiguous with regard to alternative nucleotide sequences, it is neither plausible nor illuminating to think of it as the replication of any specific DNA codon pattern. Further failures of pattern replication result from a variety of editing, erasing and other processes. If all goes well, a pattern once filled by DNA bases will be replicated

[15] For simplicity we ignore the influence of DNA bases which do not encode polypeptide amino acid sequences but contribute instead to the production of non-coding RNA molecules that help move the mechanism forward through a number of interactions including the selection of sites for cutting, splicing, and editing. DNA base sequences play roughly the same role in producing non-coding RNA molecules as they do in polypeptide construction. See e.g. Mendes, Soares and Valcárcel (2006).

[16] The string could be read as CUC, AGC, GUU, and ACC followed by two members, AU of an anti-codon whose third base is located to the right of our string. Alternatively, the leftmost C could be read as the last base of an anti-codon whose other bases are to the right of the string. If so, the complete anti-codons would be UCA, GCG, UUA, CCA, followed by U. Another alternative would be CA followed by CAG, CGU, UAC, and CAU (Alberts *et al.* 2002, p. 336).

by an amino acid sequence completed in its final step. But that pattern is by no means unambiguously instantiated at every step of the process.

The leech's pressure escape reflex provides even more vivid examples of information transmission without pattern replication. Its pressure sensing neurons synapse on interneurons. The interneurons synapse on the motor neurons that drive muscle fibres. Each interneuron receives inputs from sensory neurons that originate at different locations on the leech's body. Because they are subjected to different pressure stimuli they fire at different rates or intervals. To produce information to transmit to motor neurons, interneurons must process and resolve the multiple inputs they receive from sensory neurons (Churchland and Sejnowski 1992, p. 34). As a result temporal organizations of sensory and interneuron spike trains are too different to be usefully construed as instances of the same spatial or temporal pattern.[17]

Teleology is essential to our alternative to Crick's pattern replication story. As Ruth Millikan emphasized, in order to understand what makes biological information *information* one must consider its role in directing DNA expression toward the production of precursors of needed proteins (Millikan 1993, p. 186). As things stand before a DNA segment is selected for transcription the mechanism can be set in motion toward the production of many different protein precursors. The selection and transcription of a specific DNA segment in response to a specific need drastically reduces the number of results that could otherwise have been produced and promotes the production of results that move the mechanism toward the assembly of the teleologically appropriate protein precursor. If all goes well the activities that entities engage in at subsequent steps eliminate more and more teleologically inappropriate results until the appropriate amino acid string has been assembled. Had a segment encoding a different protein precursor been selected, the mechanism would have eliminated different possibilities and promoted different results.

We will use the term 'function indicator' to refer to features like temporal organizations of leech reflex spike trains and arrangements of codon bases of entities and activities in mechanisms like DNA expression and the leech reflex. Like Crick's replicated patterns, they help direct the operation of the mechanism toward teleologically appropriate results. But they do not supply direction after the manner of a series of pattern replications, or a navigational map, an architectural plan, a diagram, a recipe, or any other sort of representation of the relevant result. Nor do they convey semantic content. Instead, they help move the mechanism toward its goal by determining or limiting

[17] We ignore highly contrived, mathematical definitions of higher level variables produced just for the purpose of describing trivially similar patterns that are of no particular functional significance.

the causal influences that the entities or activities they belong to can bring to bear on other components of the mechanism. For example firing rates or temporal distributions of action potentials determine what causal influence a pre-synaptic spike train can bring to bear on a post-synaptic neuron or muscle fibre to move the leech reflex toward its goal of getting the leech away from a putatively harmful environmental influence. The teleological significance of the relevant temporal features and the way in which they guide the reflex depend upon their causal history. When the reflex mechanism operates properly under conditions conducive to its serving its purpose, the temporal organization of sensory spike trains varies with the locations of pressure sensors relative to the part of the leech's body that is pressed. As a result, the function indicating temporal features of sensory spike trains vary with location and intensity of the pressure source and therefore with the direction in which the reflex must move the leech to serve its purpose. The causal influence that sensory spike trains exert on interneuron spiking depends upon their function indicators; interneurons don't respond in the same way to sensory inputs that differ with regard to their temporal organization. When all goes well sensory neurons are thus constrained by their function indicators in such a way as to promote interneuron electrical activity that moves the mechanism closer to its goal. The same holds for function indicators on interneuron and motor neuron spike trains. Thus upstream function indicators influence the production of downstream function indicators. This, together with the constraints function indicators impose on the activities that entities engage in accounts for the goal directed reach of first step factors in the leech reflex. This is how sensory spike trains provide the organism with information about the location of the harmful stimulus that it needs in order to move away from that stimulus. In this sense it is information about what direction to move to avoid a pressure source.

In our other example DNA segments are expressed to supply proteins needed by the organism or one or more of its subsystems on a regular or temporary basis. The expression mechanism is set in motion by transcription factors and other molecules made available and set to work in response to specific needs. The part they play in selecting and beginning the transcription of DNA segments that encode teleologically appropriate protein precursors is roughly analogous to the role pressure stimuli play in setting up teleologically appropriate spike trains in leech pressure sensors. Nucleotide bases and their spatial arrangements are the function indicators that belong to DNA codons. Their contribution to transcription is analogous to the contribution of teleologically significant temporal features of leech pressure sensor spike trains. We call such things function indicators because when all goes well they and the constraints they place on the operation of their bearers are indicative of the functions that entities and activities must carry out in order for the mechanism to serve its purpose.

Mechanistic information is the causal influence that entities and activities at one step in the operation of a mechanism exert to select teleologically appropriate results for production in one or more subsequent steps. In short, mechanistic information is selective causality. The mechanistic information transmitted by a segment of DNA or RNA consists in its contribution to such processes as assembling, cutting and splicing strings of nucleotides. In this case mechanistic information is transmitted through bonding and bond breaking. The DNA and RNA segments are constrained by their function indicators to transmit information that moves the mechanism toward the production of precursors of needed proteins. It is information for protein production.[18] Function indicators constrain the influence of spike trains at every step of the leech's pressure avoidance reflex to promote the selection of results that move the mechanism toward the teleologically appropriate muscular responses.[19] This is information for an avoidance response.

More generally, the mechanistic information an entity or activity transmits is the causal influence it exerts on other entities or activities to select teleologically appropriate results for production and to prevent or discourage teleologically inappropriate results at one or more subsequent steps. The mechanistic information an entity or activity receives is the teleologically significant causal influence the relevant entity or activity exerts on it. To store mechanistic information is to have the ability to exert a teleologically significant influence on the selection of downstream results.

[18] As an anonymous referee pointed out, the sketch of protein precursor construction we use to illustrate our account ignores recent investigations of complicated, widespread, and diverse contributions of small non-coding RNA molecules to protein synthesis. In addition to setting the mechanism of DNA expression in motion in response to the organism's needs, non-coding RNAs and other molecules participate in promotion, inhibition, splicing, reassembling, editing, and other functions. Under their influence, the same DNA segments can be expressed to construct more than one polypeptide, and contribute to the synthesis of more than one protein (see Mendes, Soares and Valcárcel 2006). It is plausible, as Richard Burian suggests, that the most the genome contains is instructions about how the mechanism of protein construction is

> ... to respond when the information it contains is unpacked in specific contexts and settings... [T]he contingencies that go into when and how that information is unpacked, and how it is processed before or during its use cannot be specified by DNA alone. (Burian, personal communication.)

We suppose that our simplified, textbook style picture is faithful enough relative to specific settings and contexts in which DNA is typically expressed to illustrate our ideas about mechanistic information. We lack the space to consider whether or how mechanistic information figures in the contributions of non-coding RNA and other molecules to polypeptide construction in any given context.

[19] The influence of sensory neuron spike trains extends throughout the operation of the reflex mechanism even though it loses some of its independence because each interneuron resolves and fires in response to inputs from more than one sensory neuron.

39.6 Some differences between mechanistic, Shannon–Weaver, semantic, and teleo-semantic conceptions of information

Some weak points of analogy hold between the selection of results produced in mechanistic informational processes and the communication of a message by Shannon–Weaver signal transmission. Like Shannon and Weaver we think of information in connection with the reduction of uncertainty. But Shannon–Weaver information is a measure of uncertainty as to which of a number of alternative possible messages or signals has been chosen for transmission, or which message or signal is to be received. If noise interferes with the signal to increase uncertainty, Shannon–Weaver information increases in the sense that more possibilities remain open (Shannon and Weaver 1998, p.19). Thus Shannon–Weaver information *decreases* as more and more of the transmitted signal reaches the receiver intact. By contrast mechanistic information is a causal influence that decreases uncertainty with regard to which of a number of alternative results a mechanism is to produce.

An important disanalogy is that mechanistic information can and often does do its work without benefit of any biological counterpart to a Shannon–Weaver signal (e.g. a sequence of electrical pulses) that conveys a message by traveling through a channel from a transmitter to a receiver. For example, we saw that no single sequence of electrical of electrical spikes moves (or is duplicated) intact from pressure sensors through interneurons and motor neurons to muscles in the leech reflex. Thus we reject Crick's characterization of a 'flow of information' specifying an amino acid as a flux analogous to a flow of energy or a flow of matter (Crick 1958, pp.133–4). We maintain that his talk of information flow should be replaced by descriptions of causal and teleological relations between function indicators and mechanistic information.

Shannon–Weaver information has no semantic meaning.

> ... [T]wo messages, one of which is heavily loaded with meaning and the other of which is pure nonsense can be exactly equivalent... as regards [Shannon–Weaver] information. (Shannon and Weaver 1998, p. 8)

Mechanistic information lacks semantic content for a different reason: neither mechanistic information nor function indicators are, or function as symbols. An instance of mechanistic information is meaningful in the sense that it selects results to move a mechanism toward its goal. But in doing so it functions as the causal influence it is, not as a symbolic representation, e.g. a description, recipe, plan, map, or set of instructions.[20] One can of course use semantically meaningful expressions to describe the results that information selects. However, that is no reason to think that mechanistic information is,

[20] Pace Stegman (2005).

or consists of symbols belonging to a language. Nor is it any reason to think that mechanistic information has syntax, signification, a pragmatics, or an inferential role. One can produce a semantic representation of a home run pitch and the features that explain how the batter hit it into the stands. But that is no reason to think the pitch has semantic content (e.g. that it expresses instructions) for the bat to receive and respond to. Similar considerations hold, of course, for function indicators.[21]

Ruth Millikan's early bio-semantic account of information proposed that informational content depends upon the evolutionary history of the mechanism in whose operation it figures (Millikan 1993, pp. 83–102). Our main objection to this is that it rules out attributing information or identifying its significance to the continuity of mechanisms whose evolutionary history is unknown or irrelevant. Moreover, biologists typically don't rest their cases for claims about the function of a mechanism or the role of information on theories about the adaptive value or evolutionary history of it or similar mechanisms. Crick knew that DNA expression and genetic coding were important to variation, adaptation and natural selection. But his account of the role of genetic information in protein synthesis neither implies nor assumes any specific account of how the mechanism of DNA expression evolved. The same holds for investigations of sensory-motor reflexes since Sherrington. Investigators typically don't need to find out how a mechanism evolved in order to develop or test accounts of its function and its use of biological information. Indeed, they sometimes rely on what they know on independent grounds about a mechanism's function and the purposes it serves to draw conclusions about its adaptive value or evolutionary history.

[21] The fact that semantic notions don't fit the processes by which mechanistic information contributes to the selection of results to be produced argues against construing biological information on the model of Millikan's inferential account of how signs represent. Ignoring details, Millikan thinks that a sign, *s*, carries information about something, *r*, only if there is a 'natural connection' between *s* and *r* such that a system that is properly attuned to the connection can do something analogous to inferring *r* from *s*. For example, she thinks a mitten she found on the path outside her house carries information about her daughter's whereabouts for anyone who knows enough to infer that her daughter's having come home and dropped it on the way in is the best explanation of how the glove got there (Millikan 2004, p. 37). Millikan characterizes the inferential process as a matter of 'tracking' the connection between the glove, and the daughter or her whereabouts *(ibid.)*. Although there certainly are cases in which signs are said to carry information about something by virtue of what we can infer from them, it sheds no light on biological information to think of DNA codons or leech spike trains as signs from which the relevant mechanisms draw conclusions about what results they should produce. Such inferences would be impossible unless the spike trains or codons carried information with semantic content or the expression or reflex mechanisms produced semantically meaningful descriptions of the codons or spike trains to draw conclusions from. If any tracking goes on in DNA expression or leech bending, the tracking must be understood non-inferentially in terms of function indicators and the teleologically significant influences they enable their bearers to exert.

A further difficulty for bio-semantics arises from the idea that a mechanism can use information only for purposes it was naturally selected to serve. Adapting an example of Rick Grush's, imagine two leeches with similar nervous systems and other body parts living in similar environments with similar predators, food sources, etc. and similar needs for self maintenance. Suppose their nerves and muscles behave in the same way to produce the same responses to the same pressure stimuli. According to bio-semantics, they could not transmit the same information if their neuro-muscular systems did not evolve in the same way. Suppose one of them is an artificial leech designed and assembled by engineers who gave no thought to how or whether its nerves and muscles might function to benefit it. According to bio-semantics its sensory neurons could not carry information even though they behave in exactly the same way as their counterparts in the natural leech and benefit both leeches by helping them escape environmental perils (Cf. Grush 2001, p. 166ff).

Mechanistic information is subject to no such difficulties. Even though it is to be understood teleologically, the goals it serves need not be fixed by any evolutionary history. Many biological mechanisms function to promote goals that are learned or acquired during the individual organism's career rather than having been evolutionarily conferred on the species it belongs to. They often serve to promote the satisfaction of temporary desires.[22] Factors whose presence is susceptible to evolutionary explanation often serve mechanisms without determining the purposes they function to serve.[23] And as we said, investigators may have to learn how a biological mechanism functions and what goals it serves before they can begin to draw conclusions about its evolutionary history.[24]

[22] A number of mechanisms belonging to Venus and Serena Williams function from time to time to promote the goal of winning a tennis tournament. Evolutionary psychology and sociobiology to the contrary, there is no good reason to think that evolutionary history accounts for their pursuit of that goal.

[23] For example, voltage gated channels which control the flow of electrical currents carried by K^+ ions through neuron membranes are similar enough to voltage gated K^+ channels in certain bacteria membranes to suggest that the former evolved from the latter. But the mechanisms these channels belong to have remarkably different functions (Jiang, et al 2003).

[24] An anonymous reviewer asks whether one can reject evolutionary accounts as we do without treating goals and proper functioning as relative to investigators' interests in such a way that there is no objective fact of the matter as to which of a number of possible alternatives is the goal that a mechanism operates to promote. Biologists often approach such questions by asking what contribution the mechanism of interest makes to functions (involved in processes as various as nutrition, respiration, hair and fingernail growth, body temperature regulation, and disease resistance) which can be identified without appeal to evolutionary history. For example, we submit that there are facts of the matter about the proper functioning of neuronal systems such that there are objective but non-evolutionarily based answers to such questions as whether temperature regulation is the main function of the brain as Aristotle thought (Aristotle, 1991, pp. 1015–18),

39.7 Mechanistic information and malfunction

We have been describing the role of information in prototypically successful operations of mechanisms like DNA expression and leech bending. But some mechanisms malfunction to produce results that make no contribution to, or run contrary to the achievement of goals they would promote if they were functioning as they should. The expression of the mutant gene responsible for cystic fibrosis is an example of the role that mechanistic information can play in the kind of malfunction we have in mind.

Cystic fibrosis is caused by a mutation in a DNA segment that encodes the precursor of a protein involved in transporting chlorine ions through cell membranes. Without that protein, chlorine ions are trapped inside lung cells where they promote an unhealthy accumulation of mucous. In sweat ducts chlorine ions are trapped outside cells. There they attract sodium ions to produce an abnormally high concentration of salt in perspiration. Both cystic fibrosis mutants and their counterparts in healthy subjects are expressed to supply a protein the organism needs. Mutant cystic fibrosis proteins could further the attainment of the organism's goals if they could get to the plasma membranes of the cells that need them. But structural abnormalities in the mutant proteins trigger a quality control mechanism that discards them before they reach their destinations (Alberts *et al.* 2002, p. 631, 728). Thus the result that completes the operation of the expression mechanism in a cystic fibrosis patient prevents rather than promotes the attainment of the goal the properly functioning mechanism serves. But in many respects information plays the same role in the malfunctioning mechanism that it would play if the mechanism were functioning properly. In both cases DNA segments are selected for transcription in response to the organism's need for Cl^- transport proteins. In both cases they transmit information to move the mechanism toward the satisfaction of the same need. Thus the teleological structures of the healthy and the malfunctioning mechanism are surprisingly similar. The only difference is that because the cystic fibrosis gene encodes a defective protein, the result of its expression fails to promote the goal for the sake of which the mechanism operates.

39.8 Conclusion

We have tried to describe the ways in which and the reasons for considering some mechanisms as carrying mechanical information. Basically, the claim is that some mechanisms carry information about upstream stages that is

or whether the lung's main contribution to an animal's life is respiration. How such facts are established and how disputes about them are adjudicated is a topic for another paper.

used to produce what is needed by the organism (or system) downstream. The causal specificity of these selective influences produce what is needed to fulfil a goal or purpose. We have tried to detail these functions in terms of reach (independence and strength of influence), and most importantly the teleology of the information carrying mechanism. It is because of these features that mechanistic information supplies a unique from of continuity for the working of certain mechanisms.

Acknowledgements

We began work on the ideas in this chapter a long time ago in conversation with Lindley Darden to whom we are greatly indebted for this and for helpful criticisms over the past several years. We are also indebted to Megan Delehanty, Fidel Mejia, Jim Woodward, Jack Vickers, Ken Schaffner and to audiences who discussed earlier versions and related talks at the University of Pittsburgh, the University of Maryland, the University of Calgary, the College of William and Mary, to Emi Iwatani, Jason Byron, and other students in a seminar we taught on mechanistic explanation, Anjan Chakravartty and an anonymous referee for helpful suggestions, and to Deborah Bogen.

References

B. Alberts, A. Johnson, J. Lewis, M. Raff, K. Roberts and P. Walter, (2002). *The Cell, 4th edition,* New York, Garland.

Aristotle (1991). *Parts of Animals* Bk. II, chpt 7, in J. Barnes, ed., *Complete Works of Aristotle,* vol., Princeton, Princeton University Press.

F. Crick, (1958). On protein synthesis, *Symposium of the Society of Experimental Biology,* 12: 138–163.

P.S. Churchland and T.J. Sejnowski, (1992). *The Computational Brain,* Cambridge, MIT Press.

L. Darden, (2006). Flow of information in biological mechanisms, *Biological Theory,* 1 (3): 280–287.

L. Darden, (2006a). *Reasoning in Biological Discoveries, Essays on Mechanisms, Interfield Relations, and Anomaly Resolution,* Cambridge: Cambridge University Press.

F.I. Dretske, (1983). *Knowledge and the Flow of Information,* Cambridge, MIT Press.

P. Godfrey-Smith, (2000). Information arbitrariness and selection: Comments on Maynard-Smith, *Philosophy of Science,* 67 (2): 202–207.

P.E. Griffiths, (2001). Genetic information: A metaphor in search of a theory, *Philosophy of Science,* 68 (3): 394–412.

R. Grush, (2001). The semantic challenge to computational neuroscience, in P.K. Machamer, R. Grush, and P. McLaughlin, eds. *Theory and Method in the Neurosciences,* Pittsburgh, University of Pittsburgh Press, pp. 155–172.

Y. Jiang, A. Lee, J. Chen, V. Rutta, M. Cadene, B.T. Chait, and R. MacKinnon, (2003). X-ray structure of a voltage-dependant K^+ channel, *Nature*, 423, 33–41.

C. Koch, (1999). *Biophysics of Computation, Information Processing in Single Neurons*, New York, Oxford University Press.

B. Lewin, (1994). *Genes V*, Oxford, Oxford University Press.

P. Machamer, (1977). Teleology and selective processes, in R. Colodny, ed., *Logic, Laws, and Life: Some Philosophical Complications*, Pittsburgh Series in Philosophy of Science, University of Pittsburgh Press, pp. 129–142.

P. Machamer, L. Darden and C. Craver, (2000). Thinking about mechanisms, *Philosophy of Science*, 67 (1): 1–25.

L.M. Mendes Soares and J. Valcárel, (2006). The expanding transcriptome: The genome as 'the Book of Sand', *The European Molecular Biology Organization Journal*, 25: 923–931.

R.G. Millikan, (1993). Biosemantics, in R.G. Millikan, *White Queen Psychology and Other Essays for Alice*, Cambridge, MIT, pp. 83–102.

R.G. Millikan, (2004). *Varieties of Meaning: The 2002 Jean Nicod Lectures*, Cambridge, MIT Press.

S.D. Mitchell, (2009). *Komplexitäten. Warum Wir Erst Anfangen, die Welt zu Verstehen*, Frankfurt, Suhrkamp Verlag, a German language version of the expanded.

S.D. Mitchell, (2009). *Unsimple Truths: Science, Complexity, and Policy*, Chicago, University of Chicago Press.

A. Rosenberg, (2006). *Darwinian Reductionism*, Chicago, University of Chicago Press.

W.C. Salmon, (1984). *Scientific Explanation and the Causal Structure of* the *World*, Princeton, Princeton University Press.

S. Sarkar, (2000). Information in genetics and developmental biology: Comments on Maynard Smith, *Philosophy of Science*, 67 (2): 208–213.

C.E. Shannon and W. Weaver, (1963). *The Mathematical Theory of Communication*, Urbana and Chicago, University of Illinois Press.

J.M. Smith, (2000). The concept of information in biology, *Philosophy of Science*, 67 (2): 177–195.

J.M. Smith, (2000a). Reply to commentaries, *Philosophy of Science*, 67 (2): 214–218.

U. Stegman, (2005). Genetic information as instructional content, *Philosophy of Science*, 72: 425–443.

K. Sterelny, (2000). The 'Genetic Program' Program: A commentary on Maynard Smith on information in biology, *Philosophy of Science*, 67 (2): 195–201.

E. Werner, (2007). All systems go, *Nature*, 446: 493–4.

·40·

The causal-process-model theory of mechanisms

Phil Dowe

Abstract

Wesley Salmon and I have argued that causation and causal explanation need to appeal to causal processes understood in terms of conserved quantities. This has the consequence of ruling out absence 'causation' as being genuine causation. Carl Craver has argued persuasively that absences are crucial in causal explanations in neuroscience, and so he gives an account of mechanisms in terms of causal relevance where the latter is understood along the lines of causal modelling. This allows for absences to be causes and hence to feature in causal explanations, but it is not compatible with the claim that causal explanation needs to appeal to causal processes understood in terms of conserved quantities. I therefore offer an account of mechanisms, in particular the role of causal relevance in mechanisms, which can respect the theory that causation involves causal processes understood in terms of conserved quantities, but which also allows absences to figure in causal explanation.

40.1 Introduction

In some areas of science, the right correlations are routinely taken to indicate causality. In other areas, it's common enough that a scientist might be unwilling to infer causality from correlations without also knowing *how* one factor is responsible for another. Some philosophers take this to show that correlations are insufficient for causation. They might alos take this to indicate that scientific explanation, in those areas of science at least, involves appeal to mechanisms. I agree, for some areas of science, that this indicates both that correlations are insufficient for causation and that mechanisms are the basis for scientific explanation.

In what follows I survey two attempts to give an account of mechanistic explanation, namely those due to Salmon (1984) and Craver (2007). The question to be addressed is: what theory of mechanisms can account for the idea that scientists sometimes seek mechanisms that underlie correlations. I will assume that the answer cannot be 'more correlations'.

40.2 Salmon's mechanistic theory of explanation

Wesley Salmon draws the following lesson, notably in *Scientific Explanation and the Causal Structure of the World* (1984). There are two tiers to scientific explanation. The first tier involves set of statistical relevance relations. But this is not enough to guarantee causation. We need a second tier, which involves exhibiting the causal connections. Together, this provides an account of explanation which is 'ontic' rather than epistemic. Hempel's account of explanation as arguments, for example, makes explanation an epistemic matter, but Salmon's account in terms of objective facts of statistical relevance and causal connection makes for an ontic account.

According to Salmon's Statistical Relevance Criterion, explanations involve first an assemblage of facts, facts which are statistically relevant to the thing to be explained. An event C is statistically relevant to another event E if the probability of E is affected by whether or not C occurs: C is statistically relevant to E iff $\mathrm{P}(E|C) \neq \mathrm{P}(E)$ and C is statistically irrelevant to E iff $\mathrm{P}(E|C) = \mathrm{P}(E)$. For Salmon these probabilities are to be understood as relative frequencies.

An explanation of an event E involves four steps. First, we begin with the prior probability of E – the likelihood of that event relative to an appropriate reference class (R). This will be of the form $\mathrm{P}(E|R)$. Second, we need to find relevant partitions of this reference class. A partition is relevant if the probability of E is different in the relevant cells (parts of the reference class). For example, if the probability that a female will die of lung cancer is the same as the probability that a male will, the male/female partition is not relevant to death by lung cancer. However, whether an individual smoked or not will be relevant if the probability of the individual getting cancer if they smoke is different to the probability that they get cancer if they don't smoke. Third, we need to know the posterior probabilities, that is, the probabilities in each cell after we make the partition. We need to know the probability of getting cancer if one smokes, and if one doesn't smoke; ie $\mathrm{P}(E|S)$ and $\mathrm{P}(E|\sim S)$. Finally, we need to locate the individual in one of these cells. Then, if we are satisfied that there are no further relevant partitions, the explanation involves citing all the factors in the definition of that cell.

Salmon gives a fictional example of an American teenager Albert who is convicted of the offence (O) of stealing a car. Albert is a male from an urban environment. Take the class of American teenagers (T) as the reference class. Dividing it into male (M) and female (F) happens to be a relevant partition (it's more likely that teenage males will commit offences in America than female teenagers) and so does dividing into urban (U) and rural (R) background (city bred teenagers are more likely to commit offences than those who are rural bred). This gives four cells in the reference class: male-urban, male-rural, female-urban, female-rural. The probabilities for each will be different. We select the relevant cell: male-urban, and the fact that the probability in that cell is higher than the prior probability – provided there are no further relevant partitions – explains the teenager's being convicted. One of the advantages of this approach as Salmon sees it is that it allows for low probability explanations. Explanatory factors raise the probability of the thing they explain, but don't necessarily make it probable.

Exhibiting the statistically relevant facts is the first step in explanation, but this needs to be added to, by exhibiting the causal links between the fact to be explained and the statistically relevant facts. When we have statistically relevant facts linked by a causal process to the thing to be explained, then we have a satisfactory explanation. Salmon therefore turns his attention to the question of causality. His contribution here is twofold: he offers a persistent line of criticism against the probabilistic theories of causation, and he offers his own positive account.

The essence of the probabilistic theory of causality is the idea that a cause raises the probability of its effect (Suppes 1970). Salmon's argument against this theory concerns counterexamples where a particular causal chain contains elements that do not stand in the probability raising relation. In an example due originally to Deborah Rosen a golfer slices her shot, but by sheer fortune hits a tree branch, and the ball bounces back onto the green and into the hole. The slice lowered the chance of a hole in one, but in fact caused it.

Salmon's positive account treats causation primarily as a property of individual processes. Salmon proposes to overcome traditional difficulties with determining the nature of the causal relation by treating causation as primarily a characteristic of continuous processes rather than as a relation between events. This treatment involves two elements, the production and the propagation of causal influence. The latter is achieved by causal processes. Salmon's views on how to characterize causal processes underwent various changes which are not relevant to our purposes. In his 1997 *Philosophy of Science* paper Salmon presents the following revised theory of causality:

> A causal process is the world line of an object that transmits a non-zero amount of a conserved quantity at each moment of its history (each spacetime point of its trajectory) (Salmon 1997, p. 468)

The concept of transmission is to be understood by the following definition:

> A process transmits a conserved quantity between A and B ($A \neq B$) if and only if it possesses [a fixed amount of] this quantity at A and at B and at every stage of the process between A and B without any interaction in the open interval (A, B) that involve an exchange of that particular conserved quantity (Salmon 1997, p. 463)

Thus, Salmon appeals to a special kind of regularity which involves the possession of a fixed amount of a conserved quantity at every spacetime point of the process (for my attempt to do the same, in the context of the causal theory of explanation see Dowe 1992).

So a causal process is a worldline of an object that possesses a conserved quantity, a causal interaction is an intersection of worldlines involving exchange of conserved quantities. For one event to stand in a causal relation to another they must be connected by a set of causal processes and interactions. This, however, gives us a necessary condition, but not a sufficient condition. We also need to say in what respects an object at one time is causally relevant to an object at another time. (For a discussion of my attempt to satisfy this requirement, and objections to that see Dowe, 2007.) However, that it is only a necessary condition won't matter in this context because Salmon conjoins the causal processes-interactions requirement with the statistical relevance requirement, which should rule out irrelevances.

What will matter is the problem of chance lowering causes, as identified by Salmon himself, and used by him to rule out the chance-raising theory of causation. While it is true that a process theory of causation will not be open to an objection from chance-lowering causes, Salmon's theory of explanation will be, because it conjoins the process theory with a statistical relevance requirement. This means we cannot appeal to chance-lowering causes to explain their effects.

In any case, Salmon's account doesn't go far enough to give us an account of explanations or mechanisms. First, the account seems geared to explaining particular events, whereas mechanisms provide general explanations. Explaining why Albert was convicted no doubt needs to establish that he was an actual cause of the offense, and this may well be done in terms of causal processes. But a mechanism of the sort appealed to for example in medical sciences would provide a general explanation.

Take the following example from lipid metabolism research:

> The association between abdominal fat accumulation and risk of chronic diseases, including type II diabetes and coronary heart disease, has long been recognized. Insulin resistance may be a key factor in this link. Many studies have pointed to an association between insulin resistance and intra-abdominal fat accumulation (visceral obesity). However there is no clear proof of a causal link between visceral fat accumulation and insulin resistance. In assessing the probability of a causal link, it is useful to consider potential mechanisms. One such potential causal link is the release of non-esterified fatty acids from visceral fat into the portal vein, so that they have direct effects on hepatic metabolism. Visceral fat has been shown in many studies to exhibit a high rate of lipolysis compared with subcutaneous fat depots. However, if the idea that visceral fat releases fatty acids into the portal vein at a high rate is examined critically, a number of difficulties appear. Not least of these is the fact that continued high rates of lipolysis should lead to the disappearance of the visceral fat depot, unless these high rates of fat mobilization are matched by high rates of fat deposition. There is far less evidence for high rates of fat deposition in visceral adipose tissue, and some contrary evidence. Evidence for high rates of visceral lipolysis in vivo from studies involving catheterization of the portal vein is not strong. If this potential link is discounted, then other reasons for the relationship between visceral fat and

insulin resistance must be considered. One is that there is no direct causal link, but both co-correlate with some other variable. A possibility is that this other variable is subcutaneous abdominal fat, which usually outweighs intra-abdominal fat several-fold. Subcutaneous fat probably plays the major role in determining systemic plasma non-esterified fatty acid concentrations, which are relevant in determining insulin resistance. (Frayn 2000, p. S71)

Note that correlations by themselves are open to at least two causal interpretations, a direct causal connection and a common causal connection: two factors are the effects of a common cause. There are of course plenty of techniques to distinguish the two interpretations in terms of correlations. But these are neither conclusive, nor representative of how scientists in certain areas approach the question. Scientists rather seek mechanisms, as Salmon urges.

But in this example, we want a mechanism to explain insulin resistance in general in the first instance, not an explanation of what occurred in one particular instance. Mechanisms explain how things work, and this entails saying how various *alternatives* operate. Mechanisms typically have more than one possible value of an input. If we are, for example, to appeal to the effects of non-esterified fatty acids in the portal vein on hepatic metabolism, we cannot simply trace the connections between single values of the relevant variables.

And finally, to explain how something works we sometimes need to account for the fact that something of interest didn't happen (Woodward 2003, pp. 227ff). For example, the idea that high rates of lipolysis lead to the *disappearance* of the visceral fat depot, not to mention the notion of insulin resistance itself, seems to involve absences rather than positive occurrences (see below for more examples). But there's no causation by or of absences according to the causal process theory, so there must be more to a mechanism than just an actual set of actual causal processes and interactions.

40.3 Craver's account of mechanisms

Carl Craver's *Explaining the Brain: Mechanisms and the Mosaic Unity of Science* (2007) provides an account which does each of these things, with a specific orientation to neuroscience. According to Craver a mechanism is a set of entities with associated productive activities organised so as to constitute some phenomenon. 'Activities' is a 'filler term' for a set of causal components, meaning that to count as a mechanism the entities that constitute the phenomenon must be linked by "causal relevance". To count as a mechanism the entities must also be linked with the phenomenon they constitute by 'constitutive relevance', which in turn is understood in terms of mutual manipulability: change the phenomenon and you change the parts, change a part and you change the phenomenon.

Drawing on Salmon's distinction between 'ontic' and 'epistemic' explanation, a distinction to which he subscribes, Craver notes an ambiguity in the word 'explanation':

> Sometimes explanations are texts – descriptions, models, or representations of any sort that are used to convey information from one person to another. Explanatory texts are the kinds of things that are spoken and written, and drawn. They are the kinds of things that can be more or less complete and more or less accurate. They are representations. Other times explanations are objective portions of the causal structure of the world, the set of factors that bring about or sustain a phenomenon (call them objective explanations). What explains the accident? The ice on the road, the whiskey, the argument, the tears, and the severed brake cables. What explains the release of neurotransmitters? The action potential, Ca^{2+} influx, vesicular binding, and fusion. There are mechanisms (the objective explanations) and there are their descriptions (explanatory texts). Objective explanations are not texts; they are full-bodied things. They are facts, not representations. They are the kinds of things that are discovered and described. There is no question of objective explanations being 'right' or 'wrong,' or 'good' or 'bad'. They just are. (Craver 2007, p. 27)

Causes and mechanisms, then, are "things in the world". A second implication of explanations being ontic – the first is that they aren't arguments – is that they aren't representations.

According to Craver mechanisms are typically multi-level, requiring at the level of description what he calls the 'mosaic'. This includes multi-level causation and in particular the fact that higher levels can be causal. As he sees it this ties in with integrative neuroscience, where mechanisms, and hence explanations, involve multiple levels and cross multiple fields so that entities dealt with in different fields jointly constitute explanations. This notable aspect of Craver's account will not be dealt with in this chapter.

To give an account of causal relevance Craver draws heavily on Woodward (2003): a 'variable *X* is a cause of variable *Y* in conditions *W*, if and only if it is possible in conditions *W* to change the value of *Y* with an ideal intervention that changes the value of *X*' (Craver 2007, p. 94), where an *ideal* intervention *I* on *X* with respect to *Y* is a change in the value of *X* that changes *Y*, if at all, *only via* the change in *X*. More specifically, this requirement implies that:

(I1) *I* does not change *Y* directly;

(I2) *I* does not change the value of some causal intermediate *S* between *X* and *Y* except by changing the value of *X*;

(I3) *I* is not correlated with some other variable *M* that is a cause of *Y*; and

(I4) *I* acts as a 'switch' that controls the value of *X* irrespective of *X*'s other causes, *U*. (Craver 2007, p. 96)

Craver provides a number of detailed examples. One (Craver 2007, pp, 65–72) is a certain type of 'Long-Term Potentiation' (LTP), where changes to a synapse are produced which are thought to underlie certain kinds of learning. Two neurons are involved, the pre-synaptic and post-synaptic. Firing of the pre-synaptic neuron releases the neurotransmitter glutamate, which crosses the synapse and binds to the receptors on the post-synaptic cell. There are two types of receptors relevant here, NMDA receptors (because they are responsive to N-methyl-D-aspartate) and AMPA receptors (because they are responsive to α-amino-3-hydroxyl- 5-methyl-4-isoxasolepropionic acid). The function of the NMDA receptor is to create a Ca^{2+} selective channel. However, if the post-synaptic cell is polarized, the Ca^{2+} channel is blocked by large Mg^{2+} ions. The function of the AMPA receptor is to depolarize the post-synaptic cell and thereby repel the Mg^{2+} ions from the channel, enabling Ca^{2+} to flow through the MNDA receptor. The increase in Ca^{2+} concentrations in the post-synaptic neuron in turn activates a number of intracellular biochemical pathways including those responsible for the production of proteins used to alter the structure of the synapse. The firing of the pre-synaptic neuron can be induced experimentally, and this intervention is correlated with changes in synaptic efficiency (see Craver 2007, p. 67). This is already an impressive amount of detail, and I've only summarized the bare bones of Craver's example.

At first pass, the idea of singular productive activities looks like it might provide the right 'ontic something' that underlies and explains correlations. But there are worries, ironically, about both whether the account really is ontic in the relevant sense, and also whether it can be said to provide that something beyond correlations which explain correlations. Both concerns focus on the Woodwardian explication of causal relevance. It's true that there's more to Craver's mechanisms than causal relevance – they also exhibit constitutive relevance (although arguably the same worries re-emerge) and organization. If the concerns have merit then it would seem the latter must be what really does the work. I don't want to sound too pessimistic about the prospects of its doing so, as there is much to commend the account in Craver's book including the detailed workings of examples. My point is that the account of causal relevance *per se* raises these concerns. And it is apparent that Craver intends the notion of causal relevance to indeed carry the burden (i.e. to be that ontic something which explains correlations). He says:

> In saying that activities are *productive*, I mean that they are not mere correlations... and most fundamentally, that they can potentially be exploited for the purposes of manipulation and control. (Craver 2007, p. 6)

> More explicitly: what it means to say that one stage of a mechanism is productive of another (as I suggest in Machamer *et al.* 2000; Craver and Darden 2001; Darden 2002), and to say that one item (activity, entity, or property) is relevant to another, is to say, at least in part, that one has the ability to manipulate one item by intervening to change another. (Craver 2007, pp. 93–4)

Talk of 'one having the ability' should not be taken too literally. As for Woodward, this is a way of appealing to certain counterfactuals, as outlined above. However, as there is no account of the truth conditions of the counterfactuals, but rather an appeal to experimentation to discern their truth, Craver is open to the charge that the account of causal relevance and hence mechanisms cannot provide that something beyond correlations. It is simply a particular subset of correlations (not the 'mere' correlations) that the account rests on. I say 'might be open to the concern' because, given the absence of an account of causation *per se*, and indeed of the truth conditions for the counterfactuals, the theory is sufficiently incomplete to allow that it may be elaborated in such a way that obviates the concern.

The second concern, again tied up with the appeal to causal modelling, is that the account seems to make mechanisms epistemic rather than ontic, contrary to Craver's stated aim. The causal modelling approach to causation, and the manipulability approach derived from (or inherent in) it, involves models which are *abstractions* (see Menzies 2004, pp. 154–7). There are many ways to abstract, and thus, as is well known, causal modeling makes causation (here causal relevance) model relative. This in itself makes the notion of causal relevance – a key component in a mechanism on Craver's account – look rather more like a representation than an objective feature of the world.

But whether that is so or not, it's clear that we never will get beyond correlations on this account. To illustrate these concerns, let's return to Craver's example. We could simply model some of the phenomena in LTP by the following model:

$S = 1$ if the pre-synaptic neuron is stimulated, $S = 0$ if not.

$R = 1$ if there is an increase in post-synaptic response time, $R = 0$ if not.

$\Pr(R = 1 | S = 1) > \Pr(R = 1 | S = 0)$.

S is causally relevant to R by the definition: an intervention on S raises the probability of $R = 1$, and only does so via S. But all we've done is encode the correlations that we should explain in an adequate explanation. So let's give a slightly more detailed model:

$S = 1$ if the pre-synaptic neuron is stimulated, $S = 0$ if not.

$C = 1$ if the concentration of Ca^{2+} in the post-synaptic neuron are above a certain threshold, $C = 0$ if not.

$M = 1$ if Mg^{2+} ions block the Ca^{2+} channel, $M = 0$ if not.

$\Pr(C = 1 | S = 1) > \Pr(C = 1 | S = 0)$.

$\Pr(M = 0 | S = 1) > \Pr(M = 0 | S = 0)$.

$\Pr(C = 1 | M = 1) = 0$.

$\Pr(C = 1 | M = 0) \neq 0$.

Causal modelling would represent this on a two-pathway graph with an arrow from S to C, an arrow from S to M and an arrow from M to C. An intervention on S would change the values of M and C, and an intervention on M would change the value of C. This nicely captures the way in which experiments inform our understanding of causation and mechanisms. But, again, all we have done is encode more correlations. This is inadequate if our task is figuring out what it means to move beyond correlations to a mechanism.

I suggest that we rather appeal to causal processes to begin to capture the force of Craver's impressive examples. What carries the argument is the appeal to processes such as the movement of neurotransmitters, and the flow of Ca^{2+} ions. These processes qualify as causal processes on the account set out above because glutamate and Ca^{2+} ions have conserved quantities like mass and charge. Charge in particular is most pertinent to explanations in neuroscience. Craver objects that the conserved quantity theory drives us to fundamentalism: 'CQ also presents a view of causation tailor-made for physicalist/fundamentalist metaphysics. If causal interactions are exchanges of conserved quantities, and if conserved quantities are found only at the fundamental level, then all causation is located at the fundamental level.' (Craver 2007, p. 77). However, this misreads the account set out above. The CQ theory appeals to the trajectories of objects possessing conserved quantities. A steel ball possesses mass and charge as much as an electron does, and so too for a Ca^{2+} ion.

However, we still have the problem of absence causation. In the example just discussed, both $S = 1$ and $M = 0$ cause $C = 1$. $M = 0$ is the absence of Mg^{2+} ions in the Ca^{2+} channel, and Craver makes a convincing case for admitting absences such as this one into explanatory mechanisms in neuroscience (e.g. 2007, pp. 80–81). I think this has to be accepted. But absences are not causes and the CQ theory rules them out as causes. But the problem with absence causation is not just that the CQ theory rules them out. On a counterfactual theory of causation (for example Lewis 2004), absence causation violates relativity (Dowe 2009). (Ironically, Craver's initial characterization of a mechanism as involving singular productive mechanisms looks like it would rule out absences: how can something's not occurring be 'productive'.)

Nevertheless, Craver also runs into trouble trying to say exactly how absences figure in explanations. Or more specifically, why they don't when they don't. The problem concerns what Peter Menzies calls 'profligate causation' (Menzies 2004, p. 145). If absences are causes then there are far more causes than we expect intuitively. Craver notes

> A ... problem raised against the acceptance of negative causes is that there are too many of them, and most negative causes are of no use for understanding explanation in neuroscience. As Dowe (2004) and Beebee (2004) argue, many instances of negative causation run counter to our common sense, scientific, and theoretical uses of the

concept of 'cause', and there is no available account of negative causation that allows in all and only the intuitively satisfactory instances. (Craver 2007, p. 83)

It would seem, then, that a satisfactory account of mechanistic explanation should tell us when we can and cannot appeal to absence causation. Craver's response is this:

The extravagant cases of negative causation can be handled in a number of ways. Some negative causes are too improbable or abnormal to be included in explanatory texts or even counted as causes. Others are ruled out by, for example, legal, moral, and epistemic factors that determine the salience of a fact in a particular discussion (see Beebee 2004).... Consider a neuroscientific example: is the gasoline in my car's tank a cause of the instance of LTP in the Petri dish? It is likely true that if I had doused the dish with the gasoline, then the cells would not induce LTP, but it seems odd to think of the absence of gasoline as a cause of LTP. Although I do not have a general formula for ruling out non-explanatory causes of this sort, it is clear enough that gasoline is neither normally part of cells nor part of their extracellular environment. Gasoline is not part of the set-up or background conditions under which the cell normally operates. It is not a cellular constituent. Gasoline levels do not vary as the mechanism works. The distinction between intuitive and counterintuitive cases is a psychological distinction that is drawn on a number of different grounds in different epistemic contexts. (Craver 2007, p. 85)

This appears to make what counts as a mechanisms dependent on an individual's or a group's psychology. But perhaps a better interpretation is to take the above response to apply not to explanations but to explanatory texts, and objective explanations just do involve numerous absences, all of which are actually causally relevant. Given that Craver refers here to 'normal operation', certain approaches (Menzies 2004, Hitchcock 2007), which attempt to deal with the problem of profligate absence causation by appeal to a notion of normal operation might be of use to Craver here, if one could establish an objective notion of 'normal operation' for neuroscience. However, this is not the direction I want to take.

40.4 Causal process mechanisms

Drawing together some of the lessons from our discussion of Salmon's and Craver's accounts, we shall require the following desiderata of an account of a mechanism:

1. Mechanisms should explain correlations, setting out for example how one correlate causes the other.
2. Mechanisms should encode alternatives.
3. Mechanisms should include absences.
4. Mechanistic explanation should be ontic not epistemic.

A mechanism is not necessarily named as a phenomenon, but often is when used to explain. A mechanism can but need not involve sub-mechanisms. Mechanisms are glued together by causation.

We start with a model of a type of situation. We have a choice of the variables, *U, V, W,...* and a choice of a partition of each variable into incompatible and jointly exhaustive values: u_1, u_2, u_3, v_1, v_2, etc. There are laws which apply to this kind of situation. Based on the laws, for some set of values of initial variables (yet to be defined) each pair of values of a variable pair, either will or won't be connected in the right way by causal processes and interactions. (This account would also work with other physical connection theories, such as that of Fair 1979, but for my reasons for rejecting that account see Dowe 1995. For an attempt to combine causal modeling and causal processes in an account of causation, see Handfield *et al.* 2008.)

To generate a *Causal Process Model* we write down all values that are thus connected, with an arrow indicating the connections:

$u_i \rightarrow v_j$

$v_j \rightarrow w_k$

etc.

When two variable values are connected by such an arrow, the antecedent is a cause and the consequent is an effect (the process theory of causation). The Relevance Condition on a mechanism requires that some value of every variable in the model is connected one way or another to some value of every other variable. Where a causal connection depends on some variable value which is not a cause of the variable value named as the antecedent (compare Hitchcock 2001), write down the required value before the antecedent, say:

$w_k, u_i \rightarrow v_j.$

The motivation for this technique is to make transparent what conditions would interrupt the causal connection. We remove indirect connections by the following rule:

Non redundancy: Remove any connection in the set that can be generated by Transitivity from other connections in the set.

Transitivity: If $u_i \rightarrow v_j$ and $v_j \rightarrow w_k$ then $u_i \rightarrow w_k$.

This minimal set of connections together with the variable value set constitutes the causal process model. (Whether causation is transitive is a controversial matter. For the case that it is see Lewis 2004; for the case that it isn't see Hitchcock 2001. I side with the former, but won't argue for that here.)

Take as an example the standard late preemption case. Billy and Suzi throw rocks at a glass bottle, Suzi's throw is slightly stronger and will smash the glass, but were Suzi not to throw, Billy's throw would smash the glass. Let $B(1,0)$ stand for Billy's throwing or not, $S(1,0)$ for Suzi's throwing or not, and $G(1,0)$ the glass bottle being smashed or not. We model the following causal connections:

$$s_1 \rightarrow g_1$$
$$s_{0,} b_1 \rightarrow g_1.$$

This is not a case that we would normally call a mechanism, but the analysis will apply to any simple backup mechanism where a signal S and its back-up B both typically fire, and when they both do S causes effect G, but when S doesn't fire, B causes H. The mechanism represented by the Causal Process Model enables us to explain various outcomes, depending on the input values of S and B. We then understand how the mechanism works in general, because we know the possible causal processes, and how their operation depends, or not, on other variables. So mechanisms explain in both the particular and general sense. They explain particular events, and they explain how in general a system works. In the particular case, the explanation can appeal to actual causal connections that connect the actual values, and possible connections that connect non-actual values. In the general case, explanations appeal to possible connections between possible values.

Any thus defined mechanism will be an approximation in an important sense. If a mechanism is a system of connections that would hold given certain values of certain variables, then there will always be another more detailed mechanism which the first mechanism approximates. That the system of connections would obtain given certain values of certain variables holds only on the assumption that other interfering factors – factors not modelled – are not present. A more detailed mechanism would include some of these factors. In practice no model would be so completely detailed as to avoid such an assumption. As Mill pointed out, 'a special enumeration of... the negative conditions... of any phenomenon... would generally be very prolix' (Mill 1843, pp. 370–1). Nevertheless mechanisms explain correlations between the variables in the mechanism. We are interested in certain stable correlations which arise because other possible interfering factors are held fixed; perhaps because they are rare or easily controlled in an experimental situation.

On this account explaining by appeal to mechanisms is an ontic matter. Actual causal connections are things that are 'in the world' no matter how they are represented, and possible causal connections are guaranteed by the laws. Is this claim threatened by the fact that it is model (mechanism) relative which interfering factors are included? No. First, whether a particular actual causal connection obtains or not is not model relative. It's true that what causal connections one appeals to in order to explain something depends on the choice of model; i.e. on which mechanism one appeals to. That doesn't mean the explanation is not ontic. Second, it is model relative which interfering factors a non-actual causal connection depends on. But again, this does not mean the account is not ontic, just that different mechanisms specify differently – more or less completely – the conditions under which the connection would hold.

The discussion of interfering factors leads us into the question of how absences enter into mechanisms and explanations. Take our model of LTP

$S = 1$ if the pre-synaptic neuron is stimulated, $S = 0$ if not.

$C = 1$ if the concentration of Ca^{2+} in the post-synaptic neuron is above a certain threshold, $C = 0$ if not.

$O = 1$ if the Ca^{2+} channel is open, $O = 0$ if not.

$M = 1$ if Mg^{2+} ions block the Ca^{2+} channel, $M = 0$ if not.

$P = 1$ if the post-synaptic cell is polarized, $P = 0$ if not.

$N = 1$ if there is sufficient Ca^{2+}outside the post-synaptic cell, $N = 0$ if not. A causal process model is then

$S(1,0)$, $C(1,0)$, $O(1,0)$, $M(1,0)$, $P(1,0)$, $N(1,0)$ and

$s_1 \rightarrow o_1$

$s_1 \rightarrow p_0$

$o_1, p_1 \rightarrow m_1$

$m_{0,}, o_1, n_1 \rightarrow c_1.$

The opening of the Ca^{2+} channel, and the absence of Mg^{2+} ions each allow the Ca^{2+} flow, but do not cause it. Nevertheless, they are part of the mechanism, and enter into the explanation. (Compare my account of prevention and omission, originally in Dowe 1999.) So, in a particular case the absence of Mg^{2+} ions might help explain the Ca^{2+} concentration (the explanation appeals to an omission). In another case the presence of Mg^{2+} ions explains the absence of adequate Ca^{2+} concentration (the explanation appeals to a prevention).

What about the problem of profligate omissions? Craver's position is that some absences enter into mechanisms, others do not. The absence of gasoline in the Petri dish is not part of the mechanism of LTP. The problem then is that the definition of the *mechanism* becomes interest relative. On my account there are two mechanisms, one that includes the absence of gasoline, and another, an approximation to the first, which omits that interfering factor. Gasoline floods don't normally occur in the brain, and are controlled for in experimental situations, so the second mechanism is what we appeal to in neuroscience. That mechanism explains the correlations that obtain because of the general absence of that interfering factor. But that certain causal connections in that mechanism would occur under certain conditions is true only because we assume the absence of that interfering factor.

References

Beebee, H. (2004). Causing and nothingness, in J. Collins, N. Hall and L. Paul (eds.), *Causation and Counterfactuals,* Cambridge, Mass.: MIT Press, pp. 291–308.

Craver, C. (2007). *Explaining the Brain: Mechanisms and the Mosaic Unity of Neuroscience,* Oxford: Oxford University Press.

Dowe, P. (1992). An empiricist defence of the causal account of explanation, *International Studies in the Philosophy of Science* 6: 123–128.

Dowe, P. (1995). What's right and what's wrong with transference theories, *Erkenntnis* 42: 363–374.

Dowe, P. (1999). Good connections: Causation and causal processes, in H. Sankey (ed.), *Causation and Laws of Nature,* Dordrecht: Kluwer, pp. 247–63.

Dowe, P. (2004). Causes are physically connected to their effects: Why preventers and omissions are not causes in C. Hitchcock (ed.) *Contemporary Debates in Philosophy of Science,* London Blackwell, pp. 189–96.

Dowe, P. (2007). Causal processes, in E. Zalta (ed.), *Stanford Encyclopedia of Philosophy.* Stanford University. http://plato.stanford.edu/entries/causation-process/causation-process.html; ISSN 1095-5054, revised edition.

Dowe, P. (2009). Absences, possible causation, and the problem of non-locality *Monist* 92: 24–41.

Fair, D. (1979). Causation and the flow of energy, *Erkenntnis* 14: 219–50.

Frayn, K. (2000). Visceral fat and insulin resistance – causative or correlative? *British Journal of Nutrition* 83, Suppl. 1: S71–S77.

Handfield, T., Twardy, C., Korb, K., and Oppy, G. (2008). The metaphysics of causal models: Where's the biff?, *Erkenntnis* 68: 149–168.

Hitchcock, C. (2001). The intransitivity of causation revealed in equations and graphs, *Journal of Philosophy* 98: 273–299.

Hitchcock, C. (2007). Prevention, preemption, and the principle of sufficient reason, *Philosophical Review* 116: 495–532.

Lewis, D. (2004). Causation as influence, in J. Collins, N. Hall and L. Paul, (eds.), *Causation and Counterfactuals,* Cambridge, Mass.: MIT Press, pp. 75–106.

Menzies, P. (2004). Difference-making in context, in J. Collins, N. Hall and L. Paul, eds., *Causation and Counterfactuals,* Cambridge, Mass.: MIT Press, pp. 139–80.

Mill, J.S. (1843). *A System of Logic,* London: Longman, Book III Chapter V.

Pearl, J. (2000). *Causality: Models, Reasoning, and Inference,* Cambridge: Cambridge University Press.

Salmon, W. (1984). *Scientific Explanation and the Causal Structure of the World,* Princeton: Princeton University Press.

Salmon, W. (1994). Causality without counterfactuals, *Philosophy of Science* 61: 297–312.

Salmon, W. (1997). Causality and explanation: A reply to two critiques, *Philosophy of Science,* 64: 461–77.

Salmon, W. (1998). *Causality and Explanation*, New York: Oxford University Press.
Spirtes, P., Glymour, C., and Scheines, R. (2000). *Causation, Prediction, and Search. Springer Lecture Notes in Statistics*. 2nd revised edition, Cambridge, MA: MIT Press.
Suppes, P. (1970). *A Probabilistic Theory of Causality*, Amsterdam : North Holland.
Woodward, J. (2003). *Making Things Happen*, New York: Oxford University Press.

·41·

Mechanisms in dynamically complex systems

Meinard Kuhlmann

Abstract

In recent debates mechanisms are often discussed in the context of 'complex systems' which are understood as having a complicated compositional structure. I want to draw attention to another, radically different kind of complex system, in fact one that many scientists regard as the only genuine kind of complex system. Instead of being compositionally complex these systems rather exhibit highly non-trivial dynamical patterns on the basis of structurally simple arrangements of large numbers of nonlinearly interacting constituents. The characteristic dynamical patterns in what I call 'dynamically complex systems' arise from the interaction of the system's parts largely irrespective of many properties of these parts. Dynamically complex systems can exhibit surprising statistical characteristics, the robustness of which calls for an explanation in terms of underlying generating mechanisms. However, I want to argue, dynamically complex systems are not sufficiently covered by the available conceptions of mechanisms. I will explore how the notion of a mechanism has to be modified to accommodate this case. Moreover, I will show under which conditions the widespread, if not inflationary talk about mechanisms in (dynamically) complex systems stretches the notion of mechanisms beyond its reasonable limits and is no longer legitimate.

41.1 Introduction

In recent debates mechanisms are often discussed in the context of 'complex systems', with certain biological examples, for instance concerning biochemical processes, as paradigmatic cases. Complex systems of this kind often have a complicated compositional structure. I want to draw the attention to the fact that there is still another, radically different kind of complex system, in fact one that many scientists – in particular in the physical sciences – regard as the only genuine kind of complex system. Instead of being compositionally complex these systems rather exhibit highly non-trivial *dynamical* patterns, on the basis of structurally simple arrangements of large numbers of nonlinearly interacting constituents. To be sure, I want to call this kind '*dynamically* complex systems'. The characteristic dynamical patterns in dynamically complex systems arise from the interaction of the system's parts largely irrespective of many properties of these parts. One example, which has been studied extensively in statistical physics, is the ferromagnet with a surprising dynamical behaviour despite the fact that it consists of nothing more than a simple array of numerous identical dipoles. Analyses of dynamically complex systems are by no means limited to physics. For instance, it is common practice to model socio-economic contexts by using dynamical multi-agent systems, which deal with 'microscopic' agents with a very simple individual behaviour in a very simple arrangement.

Whereas for a compositionally complex system it is usually feasible to predict its behaviour once the compositional structure and the behaviour of its parts is known, this is completely different in the case of dynamically complex systems. Here the knowledge of the compositional structure, e.g. agents with only two possible actions arranged on a square lattice, together with the knowledge of the behaviour of its parts in isolation as well as in simple composites, often allows for hardly any straightforward predictions of the dynamical behaviour of a given complex system. Nevertheless, an ensemble of similar complex systems can exhibit surprising statistical characteristics, the robustness of which calls for an explanation in terms of underlying generating mechanisms. Thus, not only for compositionally complex systems, but also for dynamically complex systems, the identification of generating mechanisms is essential in order to explain their often surprising behaviour.

However, I want to argue, dynamically complex systems are not sufficiently covered by the available conceptions of mechanisms. Whereas for mechanisms in compositionally complex systems the decomposition into modules is an essential and non-trivial task, it is usually largely a non-issue for dynamically complex systems. Instead, the recognition and detailed (statistical) analysis of dynamical patterns that are to be explained becomes one main task, besides the identification of generating mechanisms. The most important novelty in dynamically complex systems is the fact that the material nature of the mechanisms' parts in dynamically complex systems is irrelevant in a far more drastic way than in many classical biological mechanisms, for instance. Structurally similar dynamical patterns can occur in materially completely diverse phenomena such as traffic jams, avalanches, earthquakes, tsunamis and financial market crashes. In each of these cases one has a system with a large number of elements, which displays a surprising macroscopic behaviour that results purely from the *local* interaction of the system's components. Due to the resulting predominance of structural over material considerations in complex systems research, which is underlined by the formation of numerous interdisciplinary projects and even whole scientific fields, mechanisms in dynamically complex systems must be construed in a more abstract structural fashion.[1] Despite these and other differences, it is still appropriate to talk about 'mechanisms' both for compositionally as well as for dynamically complex systems, since, among other things, in both cases the interaction of parts and the robustness regarding the resulting behaviour of the composite system are essential, albeit these features need to be filled in in a different way.

Many widely used notions in complex systems research, such as complexity, emergence and mechanisms, are notoriously and to some extent inevitably vague. Among other things this vagueness manifests itself in various lists, e.g. of emergent phenomena or nonlinear mechanisms, whose items are neither on the same level nor situated in any clear conceptual hierarchy. It is one of my aims in this paper to go some way towards a clarification, either by characterizing some notions or by highlighting certain hidden ambiguities. My main goal is to explore whether and how the notion of a mechanism has to be modified in the case of mechanisms in (dynamically) complex systems. Moreover, I will show under which conditions the widespread, if not inflationary talk about mechanisms in (dynamically) complex systems stretches the notion of mechanisms beyond its reasonable limits and is no longer legitimate. Thus, I want to explore the boundary of the notion of mechanisms by giving reasons for distinguishing warranted from unwarranted claims about successful mechanistic explanations. To this end I will carve out a minimal notion of mechanisms that allows theorists, first, to incorporate many complex systems analyses into the mechanistic programme and, second, to say in which cases claims for a mechanistic explanation are at least premature. I will present two detailed examples, one of which I see on the (yet) unwarranted and the other one on the warranted side of the boundary of a minimal notion of mechanisms.

[1] In short, I use the term 'structural' in the philosophy of physics fashion (structural versus material) and not the philosophy of biology one (structural versus functional). For philosophers of biology, 'functional descriptions' abstract from everything other than input–output relations or the function that a part has in a given whole, whereas 'structural descriptions' go beyond functional ones by referring to the inner material structure that underlies or may underlie a given functional relation. In contrast to that, the philosophy of physics' usage of 'structural' refers to abstracting from any particular material entities. This usually means that one focuses on abstract mathematical structures, such as the harmonic oscillator or certain symmetries or dynamical patterns like bifurcations in chaotic systems. In physical contexts the focus on abstract mathematical structures often reveals decisive insights into essential features of a physical system that are invisible with a detailed material description.

41.2 Dynamical complexity

41.2.1 Compositional versus dynamical complexity

The most important distinction in my analysis is that between compositional and dynamical complexity. What I call *compositional* complexity is also discussed under the labels *structural, combinatorial* or *detail* complexity.[2] Alternatively, one could also talk about *set-up* complexity because the complexity is due to a complicated organization of the set-up. Note that my usage of the term 'compositional complexity' should not be understood in the sense of complex rules of composition, which play a role in the discussion of emergence.[3] The *compositionally* complex systems I have in mind are typically linear systems which obey the principle of superposition, i.e. the behaviour of the compound system is the summation of the behaviours of its component parts in the sense that the system behaviour can be predicted by the traditional reductionist procedure of identifying components and characterising their individual input–output behaviours. In compositionally complex systems the complexity resides in the large number of relevant variables that characterise the component parts together with the detailed organization which is one out of very many possible combinations of the component parts. I call such a system compositionally complex because the individual behaviours of its parts and the detailed way how these parts are organized in the compound system are decisive for the overall system behaviour. Change the behaviour or the input of one of its parts or change the relation of two parts and you will in general change the behaviour of the whole system. Thus in compositionally complex systems many micro details have a measurable (linear) effect on the studied behaviour of the whole system.

In contrast, for dynamically complex systems very few parameters are usually sufficient to describe the behaviour of the whole system one is interested in. In most cases the vast majority of micro details is irrelevant, in the sense that a change of most microscopic variables as well as a change of most interrelations of the component parts will have no effect at all on the overall system behaviour.[4] Dynamical complexity is characterized by the fact that

[2] For instance, see Sterman (2000 p. 21) and Érdi (2007 p. 1). Moreover, I should point out that there is also a divergent technical notion of 'compositional complexity' as 'a measure of bias in the sequence composition' (Zvelebil/Baum 2007 p. 151), which is used in particular in the analysis of DNA sequences.

[3] See Hüttemann (2004, chapter 3) and Hüttemann and Terzidis (2000).

[4] The occurrence of the so-called butterfly effect, a well-known characteristic feature of dynamically complex systems is not in conflict with how I characterized dynamical complexity and explains the addition 'In most cases'. In some cases, tiny variations of the initial conditions are sufficient to generate a drastic effect for the whole system. But even this effect does not occur for compositionally complex systems where, due to their linearity, similar variations of the initial conditions always lead to similar effects for the whole system.

even in compositionally simple systems with simple (but nonlinear) rules that determine the dynamics the resulting time series can be unexpectedly complex. For instance, a nonlinear double pendulum, i.e. a pendulum with another pendulum attached to it, exhibits complex chaotic behaviour, due to the non-linearity of the rules that determine its dynamics, while the compositional complexity is as low as one can think. Thus, dynamical complexity arises from the nonlinear interactions of the subunits over time.

The example of the double pendulum allows me to forestall a possible misunderstanding. Dynamically complex systems also do have components, no less than compositionally complex systems. However, most facts about the nature of these components as well as their initial arrangement have no bearing on the complex system behaviour one wants to explain. In the case of the double pendulum, for instance, it makes no difference for the complexity of its behaviour how long the two pendulums are, out of which material they are made and in which initial state they are arranged. To put it another way, *dynamically* complex systems don't need to look differently from *compositionally* complex systems. They may still have recognizable components which behave and interact in a regular fashion and thereby give rise to a particular behaviour of the whole system. However, knowledge about the detailed nature of these components and the way how they are organized in the whole does not render the complexity of the system behaviour understandable.

Strictly speaking, a system exhibits either compositional or dynamical complexity only *with respect to* a certain behaviour to be explained. That is, one and the same system can be compositionally complex with respect to the behaviour of, e.g. one quantity and dynamically complex with respect to the behaviour of another quantity, or not complex at all with respect to the behaviour of still another quantity. For instance, the community of financial market traders may be compositionally complex with respect to the money they spend on travelling, dynamically complex with respect to the stock market prices they generate and not complex at all with respect to their collective weight. The dependence of complexity on the particular quantity or phenomenon one is studying is reflected by the fact that the issue of compositional versus dynamical complexity as well as the issue of mechanisms in such systems play a major role in the broader context of explanations. And the quality of an explanation in turn depends on that aspect of a phenomenon one seeks to understand. Thus complexity, mechanisms and explanations are pragmatic matters, which depend decisively on one's explanatory interests.

Talk about 'dynamical complexity' in complex systems' research has a certain further ambiguity since the term refers to two intimately connected but still different perspectives, strictly speaking. The first meaning is dynamically emerging complexity, i.e. an unpredictable complexity in the system behaviour that arises while the system evolves in time, although the rules for the interactions between its components are very simple (albeit nonlinear). The second meaning is complexity displayed in the statistical characteristics of the dynamics, i.e. it refers to a complex measurable phenomenology of the dynamics. Roughly the ambiguity concerning 'dynamical complexity' is the difference between process and result. Examples for such a complex coming about are the *endogenous formation of* abrupt changes and extreme events through the nonlinear interaction of the system's subunits without an abrupt or extreme external influence. An example for a 'complex result' is statistical self-similarity, e.g. of the fluctuations of some quantity (see below). If some given dynamics is statistically self-similar, it is often referred to as 'fractal dynamics'. But of course, while the system evolves, fractality cannot be recognized. The fractality involved here shows up only in the statistical analysis of the data set of the whole time series, which results from the dynamics of the system. To a certain degree the ambiguity of 'dynamical complexity' is already inherent in the term 'dynamics', which is often referred to almost as an object, whereas the paraphrase of 'dynamics' as the 'evolution of a system in time' exhibits the procedural character.

Dynamical complexity in the first sense emerges only in the temporal evolution, i.e. in the dynamics of a compound system without any need for complex initial conditions. In other words, even if the set-up of the system is very simple, its dynamics can exhibit an unpredictable complexity. In addition, the composition of the system may, in concrete cases, also be complicated or, if one wishes to say so, 'complex', but this compositional complexity is not responsible for the dynamically emerging complexity I am addressing. In order to isolate and understand how complexity arises it is therefore advisable to make the assumptions about the initial set-up of the system as simple as possible. Although some kind of dynamical complexity in the second sense, i.e. a complex statistical phenomenology of the dynamics, could in principle result from the compositional complexity of the initial set-up or complex influences from the system's environment, there are very powerful and subtle methods for discriminating complex statistical characteristics that most likely emerged only in the temporal evolution of compositionally simple systems. One of the main reasons behind this assessment is the experience with other systems that are well understood and where a complex dynamical phenomenology emerged purely endogenously by the (nonlinear) interaction of the system's otherwise simple subunits.

41.2.2 Dynamical complexity: From data analysis to mechanisms

Often, significant complex dynamical patterns are very difficult to identify because they are hidden beneath other regular or random processes (see the 'DFA method' below). Since the non-trivial identification of certain characteristic features in the dynamics of a system is taken as a strong indicator for corresponding underlying 'mechanisms', it becomes hard to disentangle description and analysis on the one side and explanation on the other side. Interestingly, Goldberger (2006), for instance, makes no clear distinction between phenomena (which are to be explained) and mechanisms (on which the explaining is usually taken to rest) when he presents his list of 'nonlinear/complexity-related mechanisms and phenomena in physiology'.[5] Since, on the one hand, dynamically emerging complexity necessarily results in complexity displayed in the statistical characteristics of the dynamics and, on the other hand, the occurrence of such statistical characteristics is in turn the most important indication of underlying dynamically emerging complexity (of the corresponding kind), these two aspects are often identified without further reflection about their difference.

As one can see in the later examples, the ambiguity of the expression 'dynamical complexity' (i.e. process versus result) is, to a certain degree, transferred to the way the term 'mechanism' is used in complex systems' research. For instance, there is, side by side, talk about 'feedback mechanisms' and 'fractal mechanisms', although these two issues, feedback and fractality, are not on the same conceptual level. Whereas feedback is a process that can occur in the evolution of a system, fractality is (in this context) a characteristic property of the statistics of a time series. Thus, strictly speaking, fractality itself is not a process in time at all – although it may be a strong indicator for a certain underlying process or mechanism that generates fractality. I can see two options now. Either one 'simply' points out that the expression 'mechanism' is inappropriate in such cases. Or one explores under which conditions one can make sense of this common use. I will go for the second option.

As mentioned above, investigations of dynamically complex systems are – due to the crucial role of structural considerations – often pursued in interdisciplinary research groups. Econophysics, for instance, is a relatively young special science between physics and economics that tries to analyse and explain economical phenomena by using models, techniques and analytical tools from physics.[6] Although the possibility of econophysics first appears puzzling, it is 'simply' grounded on the insight that important properties of, e.g. financial markets can be understood if one adopts a complex systems framework. The same reasoning stands behind many other, mostly computer-aided analyses of, e.g. traffic flow, opinion dynamics, social networks, avalanches, earthquakes, turbulences, tsunamis, etc., and in a more general fashion in chaos theory, game theory or the theory of self-organization. In these diverse contexts one observes similar dynamical patterns, which is

[5] Among other things, Goldberger's list (p. 469) contains abrupt changes (e.g. bifurcations, intermittency and other bursting behaviours, bistability/multistability, phase transitions), complex periodic cycles and quasiperiodicities, nonlinear oscillations (e.g. limit cycles, phase-resetting, entrainment phenomena, pacemaker annihilation) and scale-invariance (diffusion limited aggregation, fractal and multifractal scaling, long-range correlations, self-organized criticality).

[6] See, for instance, Mantegna and Stanley (2000) and Johnson *et al.* (2003).

seen as an indication that they are generated by structurally similar mechanisms. For this reason, it is often possible to use the same methods, models and analytical strategies, many of which were first devised in physics. Although in the case of econophysics the use of analytical tools from physics is particularly dominant, the general reasoning, the concepts, and the strategies are very similar in various other investigations of complex systems. Outstanding examples for the success of econophysics are the analysis and description of financial market crashes by using the advanced physical theory of phase transitions, where the common characteristic is a sudden occurrence of a comprehensive change of the state of affairs.

Scale-invariance/self-similarity, power-law behaviour and the closely connected occurrence of 'universal behaviour' and criticality are important indications that one is dealing with a dynamically complex system. The existence of long-range correlations in fluctuating quantities is particularly interesting because it indicates that there may be an underlying long-memory process, i.e. that the fluctuation at a given time depends on what has happened at earlier times. By contrast, a Gaussian random walk process exhibits no long-range correlations because each change of the respective quantity is an independent event, so to say, which is not affected by previous changes. Long-range correlations are an implication of statistical self-similarity, which in turn is tantamount to power-law behaviour. The equivalence of statistical self-similarity and power-law behaviour is primarily a mathematical issue. A power law looks the same everywhere, i.e. if you take a small piece of a power-law tail and inflate it, it is identical with a larger piece of the initial curve. This is different, in particular, for exponential functions like the Gaussian, which drops sharply towards zero for small values already and then lies almost on the x-axis. The next point to be explained is the connection between statistical self-similarity and long-range correlations. For random processes like coin tossing there is no correlation between, in this case, two coin tosses. There is a fifty-fifty chance for either side in each toss. Even if you had heads ten times in a row, there is still a fifty percent chance for heads in the eleventh toss. And if you look at the probability distribution for many samples of 10 consecutive coin tosses you will get a completely different result than for samples of 100 consecutive coin tosses. The probability distributions for sequences of independent random variables are not self-similar, i.e. they have different statistical properties on different scales. This is very different, say, for (healthy) heartbeat intervals or stock prices. If there are days with drastic stock price movements there is a much higher chance for still more days with large changes in the immediate future than in quiet times (even though there is no correlation between the direction of these changes, i.e. up or down, which is the reason why it is not easy to exploit this knowledge). Thus changes of stock prices have a memory.

Another important and closely connected point is that for independent random processes (e.g. coin tosses) there is a negligible probability that very many subunits (e.g. individual tosses) all do the same, which would lead to extreme events like 100 consecutive times heads. If such an extreme event happens, there is either an external cause, e.g. a magnetic heads-detecting device, or the subunits most likely interacted with a coordinating effect. Thus statistical facts about a complex system can supply strong reasons for specific inferences about the existence or non-existence of underlying interactions between the system's parts with a collective effect.

Scientists in complex systems research make every effort to discover power-law and therefore scale-invariant relations because it has far-reaching implications for the behaviour of the system under investigation. Most importantly, under certain conditions, most microscopic details become irrelevant for the dynamics of the system on the macroscopic level. As one can learn from studies of so-called critical phenomena in statistical physics, the occurrence of scale-invariance and hence self-similarity is the deeper reason why diverse systems can exhibit very similar or even identical behaviour, a fact that physicists call 'universal behaviour'.[7] 'Universality' in this sense can be explained via the method of renormalization involving iterative coarse graining, which in turn would not be possible without self-similarity. Thus there is a direct road from power-law behaviour, scale-invariance and self-similarity to understanding why certain universal structural mechanisms can account for phenomena in physics as well as in economics.[8] More specifically, power-law behaviour allows applying 'scaling methods', which were first devised in physics, in very different contexts such as economics.

In the context of dynamically complex systems, and also in my two examples in the next section, statistical self-similarity is of particular importance. *Spatial* self-similarity is relatively well-known from branching trees, snowflakes and coastlines which display spatial structures of the same type on small and large scales (see Figure 41.1). The temporal kind of self-similarity shown in the right diagram is more abstract. It refers to the statistical properties of a temporal process, namely the probability distribution of the deviations of some quantity from one time step to another. For this reason one talks about 'statistical self-similarity'.

One practically important aspect of scale-invariance stems from the fact that it is a symmetry principle. As in the well-known spatio-temporal cases, e.g. translational, rotational or Galilean invariance, the corresponding symmetry principles often allow for simple and elegant solutions of otherwise

[7] See Binney *et al.* (1992) for a detailed account of the physical and Batterman (2002) of the philosophical perspective.

[8] Newman (2005) cautions that the mechanism discovered for critical phenomena is only one among various different mechanisms generating power-law behaviour.

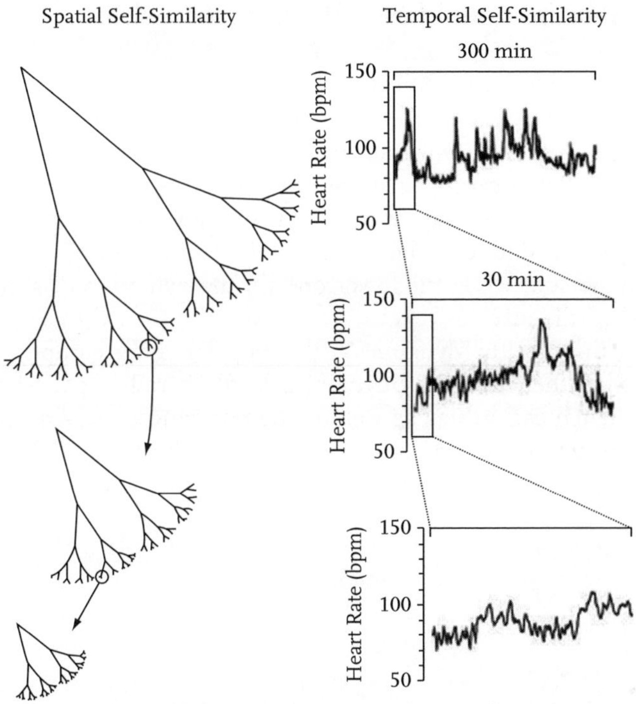

Fig. 41.1 Schematic comparison of spatial (left diagram) and dynamical or temporal (right diagram) self-similarity. In both cases, zooming repeatedly into parts of the initial spatial structure/dynamics reveals structures/dynamics with similar or even identical characteristics as the initial one. Reprinted from Goldberger *et al.* 2002. (Copyright (2002) National Academy of Sciences, U.S.A.)

intractable problems. For instance, symmetry considerations often make it possible to derive important aspects of a system's dynamics without solving the underlying equations of motion. In other words, in certain important respects the dynamical behaviour of a system can be understood by abstract reasoning concerning its symmetries without any detailed knowledge about the behaviour of its fundamental constituents. The significance of these facts for econophysics, for instance, is straightforward. The application of physics to economical issues is, to a large extent, possible because financial market behaviour exhibits invariances that allow neglecting certain micro details, thus making way for analytical methods and explanatory models developed in physics, in order to understand the behaviour of systems in those special circumstances where, just as in financial markets, many micro details lose their relevance in a sharply specified sense.

In the following section I will exemplify the significance of scale-invariance in two different concrete contexts. More importantly, I will discuss the connection with mechanisms in dynamically complex systems.

41.3 On heartbeat and financial market crashes

41.3.1 Congestive heart failure

Traditionally, cardiologists have described the normal heart beat activity as a regular sinus rhythm. However, in contrast to our subjective impression and to the traditional cardiologists' assumption, interbeat intervals normally fluctuate, even for individuals at rest, in a complex way, which appears to be erratic. The upper time series (a) in Figure 41.2 shows the heart rate dynamics of a healthy person, while the lower one (b) the dynamics of a person with congestive heart failure.

Although the 'healthy dynamics' exhibits a far more complex pattern of variability than the 'unhealthy dynamics' with its rather periodic temporal structure, their mean values as well as their variances are almost identical. Thus the unexpectedly irregular behaviour of heart beat activity defies conventional methods of analysis, which only work for stationary or 'well-behaved'

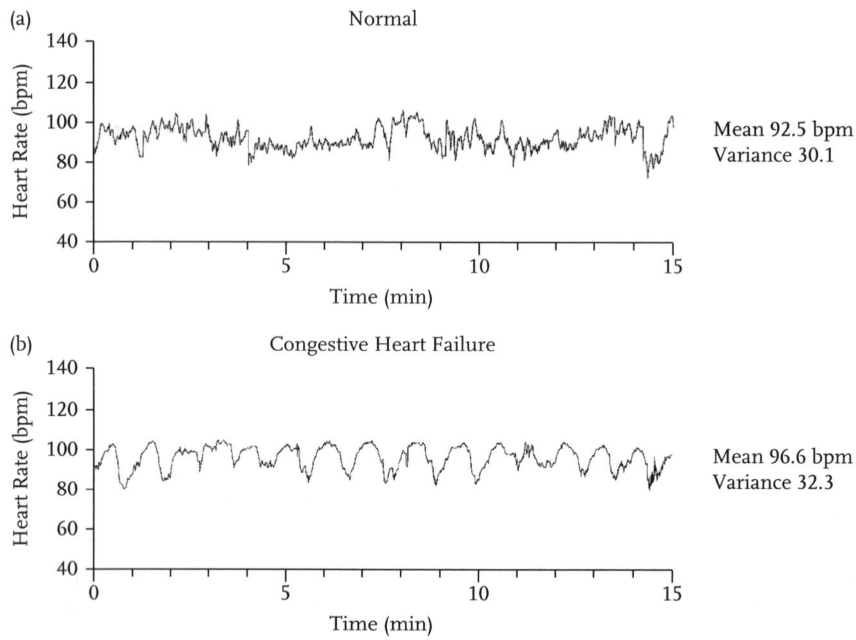

Fig. 41.2 Comparison of (a) healthy and (b) unhealthy heart rate dynamics. Contrary to one's expectation, healthy heart rate activity is far more complex, and seemingly erratic, than the relatively regular heart rate activity of unhealthy individuals. Only subtle data analysis is the first important step towards explaining this surprising difference. Reprinted from Ivanov *et al.* (2002) (with kind permission from Springer Science+Business media).

time series. In order to analyse such data sets with fluctuations on multiple time scales one needs sophisticated techniques of 'fractal analysis', one of which is depicted in Figure 41.3.

The detrended fluctuation analysis, or short 'DFA method', is very useful in revealing to what extent there are so-called long-range correlations in a given non-stationary time series, where non-stationarity means that the statistical

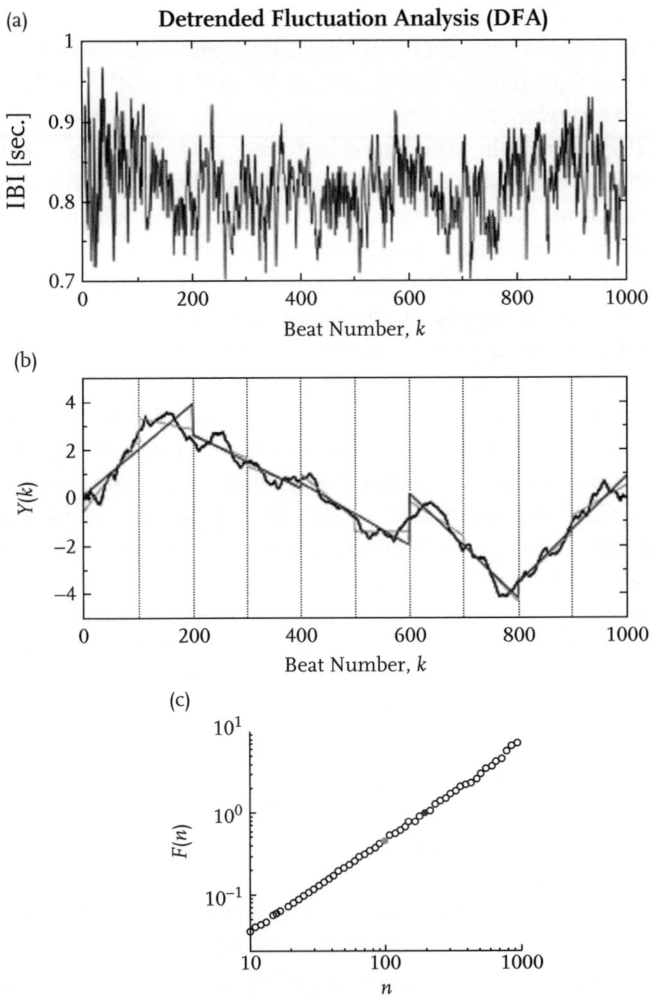

Fig. 41.3 The method of detrended fluctuation analysis (DFA) allows uncovering hidden dynamical patterns, which are strong indicators for specific underlying mechanisms. The above diagrams depict the stepwise isolation of a power law in a seemingly erratic time series of heart beat (see main text for further explication). Reprinted from Goldberger *et al.* (2002). (Copyright (2002) National Academy of Sciences, U.S.A.)

properties of the time series, such as the mean value, vary with time.[9] In such cases one needs a sophisticated method since linear or higher order polynomial trends in the data often lead to the spurious detection of long-range correlations. In the first step of the DFA method, applied to heart beat dynamics, a time series of interbeat intervals such as the one in diagram A of Figure 41.3 with N beats in total) is integrated, where the average interbeat interval RR_{ave} is subtracted from each interbeat interval $RR(i)$, so that one gets $y(k) = \Sigma_{i=1}^{k}[RR(i) - RR_{ave}]$ as the integrated time series (see diagram B in Figure 41.3).[10] Note that if the interbeat intervals were all equal, then $RR(i) - RR_{ave}$ and thereby $y(k)$ would vanish and the DFA method would be redundant. Moreover, if the interbeat interval time series were stationary, i.e. if the interbeat intervals would randomly fluctuate around a constant mean value, then $y(k)$ would also vanish, again making the DFA method redundant. Thus, a *non-vanishing* integrated time series $y(k)$ reveals fluctuations that are not evenly distributed around some mean value, i.e. it makes the non-stationarity of the time series visible and quantifiable. In particular, it allows detecting whether the heart beat intervals temporarily tend into one direction, e.g. becoming either shorter or longer. That this happens is well-known to everyone by first-hand experience.

The DFA method allows investigators to extract and put aside these 'trends' (therefore the name '*detrended* fluctuation analysis') in order to get an undisturbed view into hidden statistical patterns which may indicate certain underlying processes. This removal of trends is done in the following way. The integrated time series gets divided into equal boxes and in each box a 'least squares line' is fitted to the data, which is taken to represent the trend in that box (see diagram B in Figure 41.3). In the next step, the integrated time series gets *detrended* by subtracting the local trend in each box of the chosen size n. Eventually, one calculates the root-mean-square fluctuation of the integrated and detrended time series and repeats the same procedure over all time scales, i.e. all box sizes, in order to determine the relationship between the average fluctuation $F(n)$ and the respective box size n. If, on a double logarithmic plot one finds a linear relationship, this indicates the presence of power law scaling and thereby of a fractal structure of the time series.[11] In this case the fluctuations can be characterized by a scaling exponent, which is the slope of the line in diagram C of Figure 41.3. At this stage it may be helpful to consult Section 41.2.2 again, where I already explained the more general implications of these issues.

[9] The DFA method was first presented by Peng *et al.* (1994) for the analysis of DNA nucleotides.

[10] 'RR' stands for Scipione **R**iva-**R**occi, who invented the traditional procedure for measuring blood pressure.

[11] In a double logarithmic plot, i.e. if the logarithm of both the x- and the y-coordinate of a function $y(x)$ is taken, a power law $y(x) = x^{a}$ becomes a straight line.

Thus using the DFA method one finds that the fluctuations of the healthy (but not the unhealthy) heart rate dynamics are statistically self-similar, i.e. the statistical properties of heart rate fluctuations are identical on different time scales. Thus healthy heart rate *regulation* generates statistically self-similar fluctuations, which is tantamount to long-range correlations in the time series.[12] Goldberger, one of the pioneers of 'fractal physiology', offers the following conclusion:

A defining feature of healthy function is adaptability, the capacity to respond to often unpredictable stimuli. [...] Fractal physiology, exemplified by long-range correlations in heartbeat and breathing dynamics, may be adaptive for at least two reasons [...]: (1) long-range correlations serve as an organizing *mechanism* [my emphasis, M. K.] for highly complex processes that generate fluctuations across a wide range of time scales and (2) the absence of a characteristic scale may inhibit the emergence of very periodic behaviors that greatly narrow system responsiveness. This hypothesis is supported by findings from life-threatening conditions, such as chronic heart failure where the breakdown of fractal correlations is often accompanied by the emergence of a dominant mode [...] The paradoxical appearance of highly ordered dynamics with pathologic states ("disorders") exemplifies the concept of complexity loss (decomplexification) in aging and disease [...]. Physiologic stability appears to relate in part to complex patterns of variability that incorporate long-range correlations [...] The opposite of a fractal (scale-free) system [...] is one dominated by a characteristic frequency [...]. (Goldberger 2006)

Goldberger directly (although tentatively) interprets the long-range correlations in the time series of healthy heart beat, as found by means of the DFA method, as an 'organizing *mechanism*' that helps to generate fluctuations across many time scales, which secure the responsiveness of a healthy heart to unpredictable influences.

What is it that justifies Goldberger's (and others') hypothesis that a mechanism has been identified? The first, very important point is the robustness of the statistical characteristics that the DFA method allows one to identify in healthy heartbeat dynamics. For different healthy individuals in the same context (e.g. sleep or wake phase) one always finds the same characteristics. Second, there has been some transfer of knowledge from structurally similar situations.[13] In particular there is a large body of experience with complexly fluctuating quantities in condensed matter physics. From these cases one

[12] See Binney *et al.* (1992) for the connection between the statistical self-similarity of fluctuations and the existence of long-range correlations.

[13] The following quote by condensed matter physicists indicates the fundamental significance of this point: What makes continuous phase changes especially interesting is the scale-freedom of the fluctuations at [...]. Not only is the creation of long-range structure by short-range intermolecular forces intriguing, but any example of scale-freedom is worthy of close examination since this phenomenon occurs in several physical systems that are inadequately understood. (Binney *et al.* 1992, p. 30)

knows that certain statistical characteristics of fluctuations imply long-range correlations, which arise purely endogenously by the nonlinear interaction of the system's subunits in the absence of any coordinating external force. If this inference is justified, it suggests an (endogenous) mechanism that leads to correlations between (spatially or temporally) distant subunits of the system. As a third point, a mechanistic interpretation of the results of data analysis such as Goldberger's always rests on specific contextual knowledge. For instance, one needs to know, I want to argue, that the responsiveness of heartbeat to unpredictable influences is an important ability of healthy individuals, so that one can surmise that there is some corresponding mechanism. Despite of these points in favour of Goldberger's mechanistic interpretation the question is not yet conclusively answered whether he is in fact justified in claiming the discovery of a mechanism. In my evaluation such an interpretation is not sufficiently grounded as long as *not even a sketch* has been supplied as to how the component parts may, by their compositional and interactive organization, generate the phenomenon of interest. I will come back to this point in the final discussion. Before doing so I want to present another example which is closely related to the first one in some structural aspects, whereas materially we will be concerned with a completely different subject matter. Moreover, the situation will be different regarding the legitimacy of talking about mechanisms.

41.3.2 Financial markets

My second candidate for a mechanism in a dynamically complex system occurs in so-called microscopic models of financial markets, within the context of econophysics (see above). Since mechanisms are always relative to some behaviour that is to be explained, it is necessary to describe what it is that econophysicists want to explain and why they rate complex systems theories, as developed in condensed matter physics, as the appropriate framework for this end. The endogenous formation of financial market crashes, i.e. without any particular external causes, is one particularly well-known example for a characteristic dynamical pattern that calls for an explanation. Put more generally, financial markets experience far more large changes and extreme events, like crashes and bubbles, than one would traditionally expect for random processes, such as Brownian motion. Econophysics often talk about 'fat tails' in the probability distributions for price changes in assets like stocks or commodities, since the corresponding functions stay way above the x-axis much longer than in Gaussian probability distributions for random variables like body size or IQ. Another closely connected example for a characteristic dynamical feature in financial markets that calls for an explanation is the so-called *volatility clustering*, i.e. the tendency of quiet and turbulent market periods to cluster together in packages. These characteristics of financial markets indicate that the interaction between market participants is of crucial significance. That is, the best explanation for the high probability of extreme events in financial markets involves the assumption that financial markets are complex systems with nonlinearly interacting constituents, just as many other composite systems that show a similar tendency for the endogenous formation of extreme events in the absence of any dramatic external causes.[14]

One main research activity in econophysics is the construction of so-called microscopic models of financial markets[15] that reproduce the observed statistical features of market movements (e.g. fat tailed return distributions, clustered volatility, crashes) by employing or inventing highly simplified models with large numbers of agents (market participants).[16] Thereby one tries to understand the main statistical characteristics of observed probability distributions in terms of underlying random processes, e.g. random walk. The relevant parts of physics that are used to build microscopic models of financial markets are usually models and methods from condensed matter physics and statistical physics. Microscopic models of financial markets are highly idealized as compared to what they are meant to model. Often all agents have identical properties or there are very few subgroups. Another option is to have a set of agents with random variations. The interaction between agents is usually modelled as extremely simple, like 'do what your nearest neighbour does'.

A paradigm case of a microscopic model for financial markets arose from the collaboration of the economist T. Lux and the physicist M. Marchesi.[17] Their stochastic multi-agent model rests on the empirical fact that the universal characteristics of price change statistics (fat tails, clustered volatility) are structurally similar to scaling laws in physics. In physics, scaling laws arise from the interaction of a large number of interacting units, e.g. particles, where most microscopic details are irrelevant. This structural similarity of observed phenomena in physics and finance suggests an equally similar explanation. In the Lux–Marchesi model there are two types of traders, 'fundamentalists' and 'noise traders' (or 'chartists').[18] Whereas *fundamentalists* are rational traders who base their action on the comparison of the fundamental value p_f of the traded asset (e.g. stocks, bonds or currencies) and the actual market price p, the behaviour of *noise traders* only depends on the current price trend and the opinion of other traders. A crucial feature of the setting used by

[14] See Mantegna and Stanley (2000, p. 5), Sornette (2003, p. 15) and Schweitzer (2003, section 1.1).

[15] In the last decade economists and physicists investigated various microscopic (or 'agent-based') models of financial markets, for instance the Kim–Markowitz, the Levy–Levy–Solomon, the Cont–Bouchaud, the Solomon–Weisbuch, the Lux–Marchesi, the Donangelo–Sneppen and the Solomon–Levy–Huang model. See Samanidou *et al.* (2007) for a review of these models.

[16] See Voit (2001) and Johnson *et al.* (2003) as well as Casti (1997) for the wider background.

[17] See Lux and Marchesi (1999).

[18] The coinage of and the distinction between 'fundamentalists' and 'noise traders' is not due to Lux and Marchesi, but is established in economics.

Lux and Marchesi refers to the dynamics for the fundamental value p_f, more precisely its relative (logarithmic) changes between two time steps, which are assumed to be Gaussian random variables. This assumption is decisive for the Lux–Marchesi approach because it means that changes of p_f cannot be the reason for the typical statistical features of financial assets like fat tails and clustered volatility, which the model is meant to reproduce in its dynamics. Figure 41.4 shows the result of one 'computer simulation run'.

The intuitively most compelling impression of the result can be gained by comparing the time developments of the market price p (upper curve) and the fundamental value p_f (lower curve), first, with each other, and, second, with respect to their statistical properties. The most interesting statistical property is the frequency of price changes from one time step to another. The crucial point of the result is that the time developments of the market price and the fundamental value are very similar whereas, at first sight surprisingly, their statistical properties differ remarkably. The two lower diagrams show the relative price changes which are extracted from the time developments shown

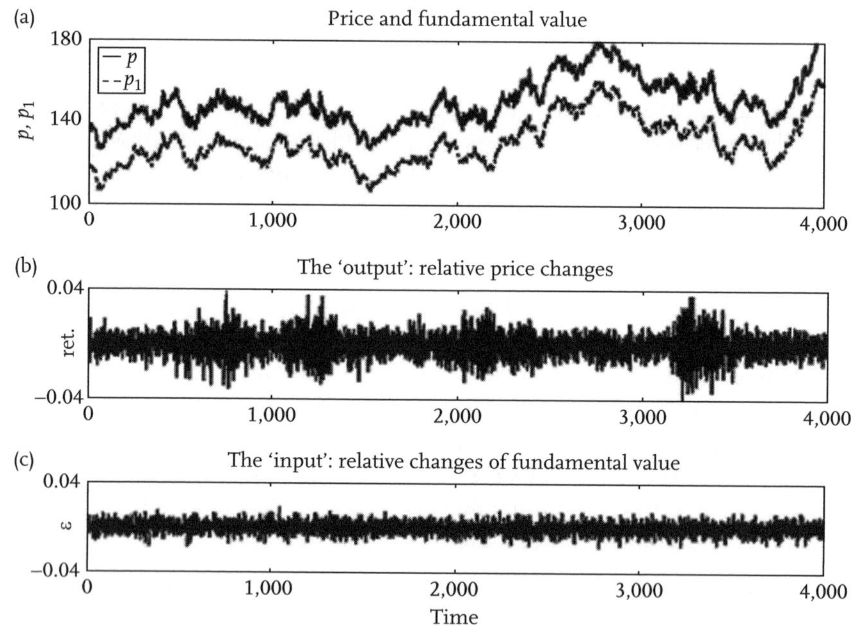

Fig. 41.4 Result of (a) typical simulation run for a stochastic multi-agent model of a financial market. Most importantly, in comparison to the assumed input series in diagram (c), the resulting time series in diagram (b) shows far more large changes as well as packages or clusters of very volatile asset prices. Both of these characteristics must have been produced purely endogenously by the interaction of the market participants. (Reprinted from Lux and Marchesi 1999, with permission by Nature.)

in the first diagram. Only after this extraction does the difference between the distribution of changes of the market price and the fundamental value become visible. Although the market price tracks the fundamental value in average it deviates significantly on a short time scale, allowing for the typical 'extreme events' and the clusters of high volatility which are observed in real markets. Lux and Marchesi conclude that the market is efficient in the sense that the market price follows the fundamental value. This does not apply to the short term, however, where the relative changes of the market price deviate from the normal distribution, which was assumed for the relative changes of the fundamental value.

In their analysis Lux and Marchesi also use the DFA method which I introduced above for the investigation of heart rate dynamics. The analysis of the scaling properties (in particular the extraction of critical exponents) shows that the exponents for the exogenous input series (i.e. the random changes of the fundamental price p_f) do not allow for fluctuations of the order of empirically observed price changes. Lux and Marchesi show that the emergence of a power-law distribution of price changes is produced by changes from quiet to volatile periods, which are due to transitions of agents from one group to another, more precisely from fundamentalists to noise traders. This behaviour, which is sometimes called 'switching', will play an important role in my own analysis of how Lux and Marchesi contribute to the scientific explanation of financial market behaviour. Another result of Lux and Marchesi is that a system loses its stability when the number of noise traders exceeds a certain critical value, and they observed so-called 'on-off intermittency', i.e. the fact that instabilities are recurrent but only temporary. Eventually, it should be stressed that the qualitative results of Lux and Marchesi are very robust since temporary instability (high volatility) occurs for a wide range of parameter values.

Again, let me ask the question, whether it is justified to say that certain mechanisms have been found? Similarly as in the heart beat example, the first important point in favour of talking about mechanisms is the robustness of the statistical characteristics which have been identified in financial market dynamics. For Lux and Marchesi's microscopic models of a financial market, fat-tailed probability distributions and volatility clustering are stable characteristics of their computer simulations that do not depend on any particular parameter values or any particular initial configuration. And the other two points in my evaluation of the heart beat example, namely about the transfer of knowledge from structurally similar situations as well as about additional contextual knowledge, are also in place. But in the financial market case, compared to the heart beat example, much more has been said about component parts (the agents) and their compositional and interactive organization (different groups of traders, certain rules for buying and selling behaviour). And one also gets a clearer picture about how the compositional and interactive organization of the component parts can generate the phenomenon of interest. Nevertheless, this is still much less than in classical mechanisms. In comparison to the well-known multifacetedness and irrationality of real financial markets the described model of a financial market seems ludicrously simple. This observation seems a crucial point to me to which I will come back in my final evaluation.

Summing up, for both the heart beat and the financial market example, one strong point in favour of talk of mechanisms is the fact that certain subtle statistical characteristics of the respective dynamics can be shown to arise in a robust way. However, although the reference to the level of component parts and their compositional and interactive organization also remains more or less vague in both cases, when compared to the complexity of the phenomena that are investigated, the sketch of the 'microscopic' processes are far more convincing in the financial market case. In the next section I will argue that this difference is in fact crucial regarding claims about mechanistic explanations.

41.4 Complex systems and mechanical philosophy

How could the potential mechanisms in dynamically complex systems be incorporated into the program of mechanical philosophy? Thorough answers are hard to be found. A promising and very recent answer is presented in Bechtel and Abrahamsen (2011). Although Bechtel and Abrahamsen come closest to how I see the matter there remain some diverging points. One of their central claims is that despite of terminological differences[19] there is a consensus about the crucial steps of what they call a 'basic mechanistic explanation', namely '(1) the identification of the working parts of the mechanism, (2) the determination of the operations they perform, and (3) an account of how the parts and operations are organized so that, under specific contextual conditions, the mechanism realizes the phenomenon of interest.'[20] Bechtel and Abrahamsen concede that, as it stands, the basic notion of mechanistic explanation is too limited to account for the insights into the complex dynamics of biological mechanisms which have been achieved by complexity theories. However, they argue, there is no need to supplant the mechanistic philosophy of science by the new paradigm of complex systems modelling since it is possible and preferable to correct and thus modify the mechanistic approach appropriately by incorporating the relevant ideas of complex systems theories. The resulting notion of what they call 'dynamic

[19] For instance, Machamer *et al.* (2000) stress the dichotomy of 'entities' and 'activities', whereas Glennan (1996, 2002) emphasizes the interaction between the parts of a mechanism.

[20] See Bechtel and Abrahamsen (2011); all following quotes come from this paper, unless otherwise stated.

mechanistic explanation' is meant to recognize the 'previously neglected temporal dynamics and the implications for our understanding of how operations are orchestrated in real time' or in other words, the 'temporal dynamics that orchestrate the functioning of biological mechanisms'.

I agree with Bechtel and Abrahamsen that the idea of mechanistic explanations should not be given up in favour of complex systems theories since many tools and concepts of complex systems theories can and should be integrated into the more comprehensive conception of mechanistic explanations. However, I think the phenomenon of dynamical complexity that I am focussing on cannot be fitted into the existing theories of mechanisms by *adding* insights into the complexity of the dynamics of mechanisms, by recognizing how operations are 'orchestrated in real time', to use a phrase by Bechtel and Abrahamsen. Instead, I claim that in the case of what I call dynamically complex mechanisms, understanding the robust dynamical patterns of the system is in fact the core task of the researcher whereas the identification of parts, operations, and their organization loses much weight, although it is not completely lost. Detailed analyses of the parts and the operations of these parts and the organization, including the detailed organization of their interactions, are of minor interest, since they have, to a surprisingly high degree, no effect on the dynamical characteristics of interest. They constitute the set-up but nothing much is understood if only the parts, their interaction behaviour and the initial arrangement of the whole system are specified. Rather, one has to identify the dynamical patterns of the compound system of interacting parts. When complex systems researchers try to understand by which mechanisms these dynamical patterns are generated they usually make certain assumptions about the parts of the complex system, their interaction behaviour, and the basic arrangement of the whole system so that it displays the phenomenon one wants to be explained. But finding an appropriate set of assumptions as such by no means exhausts the identification and understanding of mechanisms in dynamically complex systems. Moreover, often the parts in dynamically complex systems can change their nature completely while the complex system evolves in time, as one can quickly see in the econophysics example. Traders can switch from one group to another, which is in fact a crucial characteristic of the model. But if the parts are taken to be identified by their behaviour, then one is forced to say that that there isn't even a stable set of entities in a dynamically complex system.

So what should we conclude concerning the question how complex systems research and mechanical philosophy are related? Bechtel and Abrahamsen offer the following conclusion:

> Dynamic mechanistic explanation stands in contrast not only to purely mechanistic explanation but also to theoretical inquiries that emphasize complex dynamics in living systems conceived abstractly – at best neglecting but in some cases explicitly rejecting

the mechanistic project. Artificial life research, for example, is conducted on a plane removed from research on actual biological mechanisms. While accounts oriented purely to complexity or dynamics can make unique and valuable contributions, they provide no understanding of how the dynamic relations are actually realized in living systems if they do not get anchored to component parts and operations of actual mechanisms. That is, they are empty. We contend that complexity and dynamical systems theory find their best use as tools in a more integrated endeavor. (Bechtel and Abrahamsen, 2011)

Again, I agree with Bechtel and Abrahamsen that an understanding of complex dynamics does and should not supplant mechanistic explanations. However, as I have shown in the two examples above, mechanisms in dynamically complex systems are not appropriately covered by the standard notions of mechanisms. To some extent this judgement is in agreement with Bechtel and Abrahamsen, but a closer look reveals important differences. In order to see that it is helpful to consult two other writings by Bechtel, together with Richardson:

The interactions *between* subsystems become increasingly important as the units engage in more complex modes of interaction, such as [...][different] kinds of feedback [...]. [...] Thus, emergence is a consequence of complex interaction. Different models are needed to characterize the interactions between the components in a complexly organized system than are needed to characterize the behavior of the independent components. With emergent phenomena, it is the *interactive organization* [my emphasis, M.K.], rather than the component behaviour, that is the critical explanatory feature. (Bechtel and Richardson 1992, p. 285)

In what Bechtel and Richardson (1993) call 'integrated systems' the behaviour of the whole system is mostly determined by the (nonlinear) interaction of its components. And in still more extreme cases 'the activities of the parts seem to be different in kind from, and so far simpler than those performed by the whole. The parts can be so simple, in fact, that they do not seem to contribute anything of interest to understanding the behavior of the whole; in some cases it is possible to destroy or disable much of the system without significantly affecting performance.'[21] While Bechtel and Richardson think of network models of cognition something similar applies to microscopic models of financial markets, for instance. Here the statistical patterns are not altered by adding or removing however many specific traders, as long as we are still dealing with a large number of nonlinearly interacting heterogeneous subunits.

For a better understanding of mechanisms in dynamically complex systems I think the eventual shift away from classical mechanistic thinking is not radical enough as long as the basic idea remains that we have a certain number of working parts, say, A, B, C, D, E, F and G, each of which can

[21] Bechtel and Richardson (1993, p. 202f).

perform certain operations and which interact with each other in a nonlinear way thus giving rise, for instance, to self-sustained oscillations like in the circadian rhythm. I do not intend to reject this analysis in the cases Bechtel in particular is studying but I want to point out that not all explanations in terms of nonlinear mechanisms are appropriately represented by this description. In those dynamically complex systems I am focusing on in this chapter it is inappropriate to emphasize the identification of particular working parts and certain operations they perform. The kind of nonlinear mechanisms I scrutinize work largely irrespective of the detailed individual natures of the subunits that are involved and their initial compositional as well as their detailed interactive organization in the whole system.

Since it is apparent from complex systems that are microscopically well-understood that most micro details can be irrelevant (relative to one's explanatory target), it is only consistent that detailed investigations about the organization of mechanisms become less important in the case of dynamically complex systems. Instead, the focus is shifted towards studying the dynamics. For instance, it is of great interest, under which conditions the dynamics is robust and in which cases instabilities occur. However, if the attention is *exclusively* directed towards analysing the (statistical) characteristics of the dynamics, then the point is reached where, in my judgement, the researchers' use of the term 'mechanism' is no longer justified since it transgresses the limits of even the most minimal notion of mechanisms. For example, one can sometimes find talk about 'fractal mechanisms', although fractality is a feature that only refers to the statistics of time series. Without any further knowledge, all one is warranted to say in this case is that the statistical characteristics one has found indicate certain underlying mechanisms. But no mechanism has been identified unless at least some indication has been given about how an interaction of subunits may be involved to generate the phenomenon of interest. The inevitably vague qualification 'at least some indication' is of crucial importance. Requiring more than that would, in my view, make too many complex systems studies non-explanatory. And more importantly, the valuable explanatory perspective of complex systems theories would be diminished if its structural focus were given up, where specific material details are deliberately faded out.

The difficulties in fitting the potential mechanisms in dynamically complex systems into the existing notions of mechanisms prompts the question whether one should talk about mechanisms at all in this case. Is there any need or at least are there good reasons for construing explanations for the behaviour of dynamically complex systems in terms of mechanisms? A first strong indication that one should indeed be talking about mechanisms is that the term is ubiquitous in actual analyses of complex systems. For instance, there is an extensive discussion about the 'mechanisms' that generate power law behaviour, which is closely connected with scale-invariance (see above).[22] Although in some cases the term 'mechanism' seems inappropriate, widespread terminology among scientists should be taken seriously. Nevertheless, this point alone does not yield a conclusive justification. A closer look at the econophysics example from above already provides a firmer basis. In the actual scientific practice of econophysics there are clearly two different areas of research. On the one hand we have statistical analyses which provide the systematic identification of explananda. But on the other hand, many investigations are concerned with the formulation of explanatory microscopic models, i.e. models that reproduce or 'generate' the observed phenomena, in particular their statistical characteristics. And it seems that the underlying conception of explanation is mechanistic, partly because there are no established laws on the micro level that would allow invoking the covering law model, for instance. Moreover, to a certain degree, interdisciplinary approaches such as econophysics rest on the transfer of mechanistic models from one scientific field to another, e.g. from condensed matter physics to economics. Eventually, there are two further reasons why a construal of explanations in complex systems theories in terms of mechanisms is attractive. First, it supplies important means in order to answer questions concerning the reducibility of complex systems behaviour. Naturally, this point is only attractive if one is interested in micro reductions. Second, mechanistic explanations are arguably the best way towards finding effective interventions and many investigations in complex systems research have this goal.

41.5 Towards a more structural notion of mechanisms

Structural explanations that rest, for instance, on basic symmetries independently of any particular ontology have a long and successful history in physics. Elementary particle physics lives on considerations where symmetry principles are the cornerstones. With the advent of the statistical mechanics of complex systems and modern computing, structural explanations spread into various fields far beyond fundamental physics, at first within physics, eventually into almost each science. Today, the same analytical techniques, concepts, models and explanatory strategies are applied across radically different sciences such as physics, biology, economics and social science. Apparently, the success of this transfer does not rest on a common ontology – unless one wants to reify structures, which I do not advocate. In a sense these sciences have the same underlying ontology since, for instance, market traders, human hearts and ocean waves ultimately all consist of elementary particles. But this common fundamental ontology is not the reason why the same explanatory

[22] See, for instance, Newman (2005) and Sornette (2006, chapter 14).

strategies can successfully be applied. In the context of complex systems theories the reason is the observable fact that there are structural similarities in the dynamics of compound systems with completely different kinds of subunits. These structural similarities can be classified in terms of certain dynamical patterns that can in turn be represented and discriminated in a mathematically precise and subtle way.

A concrete example for structural similarities in the dynamics of extremely diverse complex systems is probably more helpful than a thousand words. Ferromagnets have the surprising ability to form a macroscopic magnetization if the temperature falls below a certain threshold. Detailed analyses revealed, roughly sketched, that the underlying mechanism involves the endogenous, i.e. not externally coordinated, parallel alignment of neighbouring dipoles (spins) across the whole piece of matter, whereas the dipoles are irregularly oriented for higher temperatures. Physicists talk of a phase transition, which results in long-range correlations of otherwise uncorrelated dipoles (and of course in self-similarity, power laws, and all that). Reasoning in structural analogies helped enormously to understand that something very similar happens in financial markets. Here as well it is the mutual interaction between traders (analogous to dipoles) and their ability to change the neighbour's trading behaviour (analogous to the orientation of the dipoles) that is crucial for understanding the endogenous formation of large changes and even comprehensively collective behaviour (e.g. financial market crashes). Once this structural analogy is understood it allows for far-reaching explanatorily valuable conclusions without the need for detailed analyses of the micro details. What still needs to be done, however, is a convincing proof that the analogy actually holds. That is, sufficient grounding in the actual situation of financial markets must be supplied, for otherwise one just has an interesting speculation. But insisting on a complete specification of the microscopic situation would spoil the explanatory efficacy of this approach, since one of its crucial characteristics is the insight into the irrelevance of most micro details.

Today, complex systems with large numbers of nonlinearly interacting subunits have a similar significance as analytically tractable systems in the past. The behaviour of complex systems is much harder to understand and to predict than the behaviour of simpler classical systems. Nevertheless, for good reasons complex systems theorists firmly believe that – bearing in mind the much higher complexity of the subject matter – they can do more than just describe similarities of dynamical patterns. For instance, one can show under which conditions the statistical characteristics of dynamical patterns are robust and how these patterns arise on the basis of nonlinear interactions of subunits – subunits that need not be described more than in a rough structural way. Moreover, in some cases it can precisely be said at which point a system may lose its stability. This is less than in the classical cases since the further development cannot be accurately predicted, but still something explanatorily helpful can be said, e.g. for purposes of intervention.

Summing up one can say that complex systems theories can contribute substantially to the explanation of when and why certain structural dynamical patterns[23] are generated in a robust way by the nonlinear interaction of the system's parts, even if these parts and their compositional and interactive organization in the whole system are only roughly sketched. Therefore, I think it is justified to say that complex systems theories supply *mechanistic explanations*, provided a sufficient grounding in concrete interacting parts is supplied. In many cases, more detailed and concrete grounding may be desirable, but still a large number of cases will remain, where more details will be very hard to supply without deteriorating the explanatory efficacy. However, if one is willing to follow this step, the notion of mechanisms must be modified or adapted in a rather drastic way. Even the more sophisticated gloss that the understanding of mechanisms comprises the identification of parts, of the input–output behaviours of these parts and how the compositional and interactive organization can bring about the phenomenon of interest has an inappropriate focus in the case of dynamically complex systems. Here, the emphasis lies not on the identification of *material* parts, their detailed behaviours and the initial set-up of the whole system but on identifying the structural conditions for the robust generation of characteristic dynamical patterns. To this end, a very simple description of the lower-level organization can be sufficient for a mechanistic explanation, and sometimes even the best one can do.

Acknowledgements

I wish to express my gratitude to William Bechtel, Carl Craver, Stuart Glennan, Manfred Stöckler and two anonymous referees for very helpful comments on an earlier draft of this chapter.

References

Batterman, R. W. (2002): *The Devil in the Details*. Oxford: Oxford University Press.

Bechtel, W., and A. Abrahamsen (2005): Explanation: A mechanist alternative. *Studies in History and Philosophy of Biological and Biomedical Sciences*, 36: 421–441.

Bechtel, W., and A. Abrahamsen (2011): Complex biological mechanisms: Cyclic, oscillatory, and autonomous. In Collier, J., and C.A. Hooker (eds.): *Handbook of the Philosophy of Science, Vol. 10: Philosophy of Complex Systems*. New York: Elsevier.

[23] By 'structural dynamical patterns' I mean patterns independently of any particular ontology.

Bechtel, W. and R.C. Richardson (1992): Emergent phenomena and complex systems. In Beckermann, A., Flohr, H., and J. Kim (eds.): *Emergence or Reduction? Essays on the Prospects of Nonreductive Physicalism*. Berlin, New York: Walter de Gruyter.

Bechtel, W., and R. C. Richardson (1993): *Discovering Complexity: Decomposition and Localization as Strategies in Scientific Research*. Princeton: Princeton University Press.

Binney, J. J., Dowrick, N. J., Fisher, A. J. and M. E. J. Newman (1992): *The Theory of Critical Phenomena: An Introduction to the Renormalization Group*. Oxford: Clarendon Press.

Casti, J. L. (1997): *Would-Be Worlds – How Simulation is Changing the Frontiers of Science*. New York: John Wiley & Sons.

Érdi, P. (2007): *Complexity Explained*. Berlin: Springer.

Glennan, S. S. (1996): Mechanisms and the nature of causation, *Erkenntnis* 44: 49–71.

Glennan, S. S. (2002): Rethinking mechanistic explanation, *Philosophy of Science* 69: S342–S353.

Goldberger, A.L. (2006): Giles F. Filley Lecture. Complex systems. *Proceedings of the American Thoralic Society*. 3: 467–72.

Goldberger, A.L., Amaral, L.A.N., Hausdorff, J.M., Ivanov, P.Ch., Peng C.K., and H.E. Stanley (2002): Fractal dynamics in physiology: Alterations with disease and aging. *Proceedings of the National Academy of Science*, 99: 2466–2472.

Hüttemann, A. (2004): *What's Wrong With Microphysicalism?* London: Routledge.

Hüttemann, A., and O. Terzidis (2000): Emergence in physics, *International Studies in the Philosophy of Science* 14: 267–281.

Ivanov, P.Ch., Goldberger, A.L., and H.E. Stanley (2002): Fractal and multifractal approaches in physiology. In Bunde, A., Kropp, J., und H.J. Schellnhuber (ed..): *The Science of Disaster: Market Crashes, Heart Attacks, and Climate Disruptions*, Berlin: Springer.

Ivanov, P.Ch., Hu, K., Hilton, M.F., Shea, S.A., and H.E. Stanley (2007): Endogenous circadian rhythm in human motor activity uncoupled from circadian influences on cardiac dynamics. *Proceedings of the National Academy of Science* 104[52]: 20702–20707.

Johnson, N. F., Jefferies, P., and P. M. Hui (2003): *Financial Market Complexity: What Physics Can Tell Us about Market Behaviour*, Oxford: Oxford University Press.

Lux, T., and M. Marchesi (1999): Scaling and criticality in a stochastic multi-agent model of a financial market, *Nature* 397: 498–500.

Lux, T., and M. Marchesi (2000): Volatility clustering in financial markets: A microsimulation of interacting agents, *International Journal of Theoretical & Applied Finance* 3: 675–702.

Machamer, P., Darden, L., and C. Craver (2000): Thinking about mechanisms, *Philosophy of Science* 67: 1–25.

Mantegna, R. N., and H. E. Stanley (2000): *An Introduction to Econophysics: Correlations and Complexity in Finance*, Cambridge: Cambridge University Press.

Newman, M. E. J. (2005): Power laws, Pareto distributions and Zipf's law. *Contemporary Physics* 46: 323–351

Peng, C.K., Buldyrev, S.V., Havlin, S., Simons, M., Stanley, H.E., and A.L. Goldberger (1994): Mosaic organization of DNA nucleotides. *Physical Review E* 49: 1685–1689.

Peng, C.K., Havlin, S., Stanley, H.E., and A.L. Goldberger (1995): Quantification of scaling exponents and crossover phenomena in nonstationary heartbeat time series *Proceedings of the NATO Dynamical Disease Conference,* edited by L. Glass, Chaos 5: 82–87.

Samanidou, E., Zschischang, E., Stauffer, D., and T. Lux (2007): Agent-based models of financial markets, *Reports on Progress in Physics* 70: 409–450.

Sornette, D. (2003): *Why Stock Markets Crash: Critical Events in Complex Financial Systems,* Princeton: Princeton University Press.

Sornette, D. (2006): *Critical Phenomena in Natural Sciences – Chaos Fractals, Selforganization and Disorder: Concepts and Tools,* Berlin: Springer.

Sterman, J. (2000): *Business Dynamics: Systems Thinking and Modeling for a Complex World.* New York: Irwin McGraw-Hill.

Voit, J. (2001): *The Statistical Mechanics of Financial Markets,* Berlin: Springer.

Woodward, J. (2003): *Making Things Happen – A Theory of Causal Explanation.* Oxford: Oxford University Press.

Zvelebil, M.J., and J.O. Baum (2007): *Understanding Bioinformatics.* Garland Science.

·42·

Third time's a charm: Causation, science and Wittgensteinian pluralism

Julian Reiss

Abstract

Pluralism about causation seems to be an attractive option as the term seems to defy analysis in terms of necessary and sufficient conditions. This chapter examines a specific form of conceptual pluralism about causation, one that has been termed 'Wittgensteinian'. I will present three such accounts in detail. All three accounts share the rejection of attempting to define 'cause' in terms of necessary and sufficient conditions, and they regard instances of causal relationships to share family resemblance at best. After criticizing and rejecting two earlier accounts, I will develop an alternative that, to the best of my knowledge, does not suffer from the deficiencies of its fellows and is more firmly grounded in some of Wittgenstein's ideas about meaning.

42.1 Introduction

Pluralism about causation is an attractive option. All theories of causation face counterexamples and all attempts to fix them lead to new counterexamples. Though, as always in philosophy, guarantees are hard to come by, there is ample *prima facie* evidence that there is no single essential property or set of essential properties that is shared among all causal relations. In response, a growing number of philosophers have considered pluralistt stances towards causation (Anscombe 1971; Campaner and Galavotti 2007; Cartwright 1999; 2007; De Vreese 2006; Godfrey-Smith 2009; Hall 2004; Hitchcock 2003; Longworth 2006aa, b; Psillos forthcoming; Reiss 2009; Russo and Williamson 2007; Weber 2007).

Pluralism about causation is, however, more of an assortment of ideas than a definite theory.[1] Most fundamentally, one can distinguish pluralism about causation at three different levels:

[1] In a recent survey paper, for instance, Chris Hitchcock distinguishes no less than nine forms of pluralism (Hitchcock 2007).

- *evidential pluralism*: the thesis that there are more than one reliable ways to find out about causal relationships;
- *conceptual pluralism*: the thesis that 'cause' and its cognates has more than one meaning; and
- *metaphysical pluralism*: the thesis that there is no one kind of thing in the world that makes a relationship causal.

This chapter is concerned with a specific form of conceptual pluralism about causation, one Chris Hitchcock terms 'Wittgensteinian' (Hitchcock 2007, pp. 216–7). I will present three such accounts in detail. All three accounts share the rejection of attempting to define 'cause' in terms of necessary and sufficient conditions, and they regard instances of causal relationships to share family resemblance at best. After criticizing and rejecting two already existing accounts, I will develop an alternative that, to the best of my knowledge, does not suffer from the deficiencies of its fellows and is more firmly grounded in some of Wittgenstein's ideas about meaning.

42.2 Wittgensteinian pluralism, takes one and two

Wittgenstein famously claimed that we cannot give a definition of the concept 'game'. He asks us whether all games – board games, card games, ball games, Olympic games – had something in common and observes that although some kinds of games have some characteristics in common there is no one characteristic or set thereof common to all instances games. Hence, we cannot define 'game' in terms of necessary and sufficient conditions (Wittgenstein 1953, §66). Instead, he argues, 'we see a complicated network of similarities overlapping and criss-crossing: sometimes overall similarities, sometimes similarities of detail'. Further, 'I can think of no better expression to characterize these similarities than "family resemblance"; for the various resemblances between members of a family: build, features, colour of eyes, gait, temperament, etc. etc. overlap and criss-cross in the same way. – And I shall say, "games" form a family' (§66–7).

Although the focus of her paper is an attack on two Humean dogmas – that causes necessitate their effects and that causal relations are not observable – Elizabeth Anscombe presents an account of causation that understands 'cause' as analogous to 'game' (Anscombe 1971 [1992]). She explains (*ibid.* p. 93; emphasis original),

> The word 'cause' itself is highly general. How does someone show that he has the concept *cause*? We may wish to say: only by having such a word in his vocabulary. If so, then the manifest possession of the concept presupposes the mastery of much else in language. I mean: the word 'cause' can be *added* to a language in which are

already represented any causal concepts. A small selection: *scrape, push, wet, carry, eat, burn, knock over, keep off, squash, make* (e.g. noises, paper boats), *hurt*. But if we care to imagine languages in which no special causal concepts are represented, then no description of the use of a word in such languages will be able to present it as meaning cause.

If such causatives or 'thick causal verbs' (Cartwright 2004)[2] are understood as constituting the meaning of 'cause', the account faces various problems. To see these, let us define:

Wittgensteinian Pluralism X causes Y if and only if X stands in relation $r \in R$ to Y, where each element of R can be described using a causative in Anscombe's sense.

An immediate problem with this formulation is that causal relations are typically transitive but it is hard to describe the resulting relation using a causative. Consider the following example. *A child upsets a glass of milk. The milk flows on the table, creating a white puddle. Observing the puddle alarms a parent who rushes to fetch a cloth and wipe it off.* It is perfectly meaningful to say that the child (or the child's action) caused the cloth to be milky. But the child didn't *wet* or *stain* or *soak* the cloth. A possible solution would be the following amendment:

*Wittgensteinian Pluralism** X causes Y if and only if X stands in relation $r \in R$ to Y, or such that there is a chain of relations $Xr_1C_1r_2C_2 \ldots C_{n-1}\ r_nY$ with $r_1, r_2, \ldots r_n \in R$, where each element of R can be described using causative in Anscombe's sense.

In this formulation there may remain problems regarding transitivity because it builds transitivity into the concept of cause and not all causal relations are transitive (see for instance McDermott 1995). I will not pursue difficulties relating to the transitivity of causation any further here because they are not specific to the Wittgensteinian account at stake here.

There are, however, two objections that require closer attention. The first is that this proposal limits causation to cases where there is an active agent, mechanism or process that produces the effect, and not all cases in which 'cause' is used meaningfully involve such an agent, mechanism or process. The second objection is that the account fails to provide a criterion to distinguish genuine causatives from non-causal transitive verbs.

The first objection concerns cases of causation by absences. Absences can figure in causal claims both on the side of the cause as well as on the side

[2] Hitchock (2007) regards Cartwright's theory as a form of Wittgensteinian pluralism. This theory is one of physical causation rather than meaning and therefore not necessarily subject to the criticisms raised here.

of the effect. Cases of the former type are omissions. For instance, Billy's failure to water the plants caused their wilting. Cases of the latter type are preventions. For instance, Suzy's catch caused the ball not to hit the window; it prevented the shattering of the window. In neither case can the abstract 'cause' be substituted by a more concrete causative. Whatever Billy did when he failed to water the plants, he did not *desiccate*, *dehydrate* or *dehumidify* them. Billy did not act, he failed to act. Likewise, Suzy (or Suzy's catch), while stopping the *ball*, did nothing to the *window*.

Proponents of process or mechanistic theories of causation bite the bullet and deny that omissions and preventions are genuine cases of causation. Phil Dowe, for one, uses a (counterfactual) concept of pseudo-causation to describe such cases (Dowe 2000). Peter Machamer thinks that these are not cases of causation, but that can be causally explained (Machamer 2004, 35f.):

> Non-existent activities cannot cause anything. But they can, when other mechanisms are in place, be used to explain why a given mechanism did not work as it normally would, and why some other mechanism became active. Failures and absences can be used to explain why another mechanism, if it had been in operation, would have disrupted the mechanism that actually was operating. Maybe we should draw a distinction and say they are causally relevant rather than causally efficacious. They are not, to use an old phrase, true causes.

But such responses cut no ice when the meaning of causal claims is at stake. Neither ordinary language nor the language of science makes a difference to whether the causal relation involves 'presences', i.e. entities that can act and be acted upon or absences of such entities. Below I will discuss in detail an example from the health sciences that involves causation by absences at the generic level. In some cases it may not even be clear whether or not a relatum is present or absent, and causal language can be used to describe the case perfectly meaningfully (Schaffer 2004).

The second objection was that the Anscombe account lacks a criterion to distinguish causatives from non-causal verbs. How do we demarcate verbs that belong in the category used to describe the relation *R* from those which don't? Certainly not all *verbs* belong in this category. Even though many causal processes are involved in someone walking, we don't describe a causal relation by saying 'Billy is walking'. Nor are all *transitive verbs* causal: 'Billy measures five foot nine' does not describe a causal relation. There are many relations that are non-causal and that can be described using transitive verbs: '*A* entails *B*', '5 and 7 sum up to 12', 'H_2O consists of two hydrogen and one oxygen molecules'; 'The fall in the barometer reading predicts the storm'.

It seems to be the case that once we discover that a certain transitive verb applies to some situation, it is an *additional* discovery that this verb belongs to the set of causal verbs. Moreover, there are numerous verbs that can have causal and non-causal meanings: *determine, induce, fix, lead to, depend on*. And perhaps this phenomenon is more wide-spread than seems at first sight. Many verbs have numerous meanings, only some of which are causal in the way required for Anscombe's account to work. '*To scrape*' means '(1a) *to remove from a surface by usually repeated strokes of an edged instrument*' (causal) or '(1b) *to make (a surface) smooth or clean with strokes of an edged instrument or an abrasive*' (causal) but also '(2a) *to grate harshly over or against*' (non-causal); '*to carry*' means '(1) *to move while supporting*' (causal) but also '(14b) *to provide sustenance for ⟨land carrying 10 head of cattle⟩*' (non-causal); '*to eat*' means '(3a) *to consume gradually*' (causal) but also '(1) *to take in through the mouth as food*' (non-causal).[3] Thus, for every verb we have to discover that it can be used causally and for some we have to discover in addition that it is used causally on a given occasion.

A potential way out is to say that certain cases of causal verbs are paradigm cases, and whether or not a new verb is causal is determined by its family resemblance with paradigm cases. This, however, is an unpromising route. Take, for the sake of the argument, Anscombe's verbs: *scrape, push, wet, carry, eat, burn, knock over, keep off, squash, make* (*e.g.* noises, paper boats) and *hurt* as paradigms, and *yield* as a yet-to-be-determined case. How *could* we say that 'yield' bears a family resemblance to, say, 'scrape'? Any two things are similar and dissimilar in many, perhaps indefinitely many ways. There simply is no sense in which two things are similar to each other *simpliciter*. Rather, things are similar *with respect to* some feature or another. 'Yield', then, is supposed to be similar to 'scrape' with respect to its causal content, but how do we determine that without having an independent grasp on the concept of cause?

An alternative to Anscombe's theory, also Wittgensteinian in spirit, is to regard causation as a cluster concept. For the concepts of ordinary language, we apply one or the other of the standard tests for causality. To take an example, consider the claim 'Jim used a blanket to *smother* the fire'. First of all, presumably on this occasion we mean by this something like 'Jim used a blanket to suppress the fire by excluding oxygen' (*cf.* definition (2c) from Merriam–Webster). Did Jim's action cause the fire to end? Yes: Had Jim not thrown the blanket over the fire, it would have persisted; Jim's action increased the probability of the fire's death; covering a fire with a blanket is an effective strategy to end it; there is a regularity between covering fires with blankets and their end; there is a mechanism by which the blanket kills the fire; and so forth. Unless the case answers positively to some or all of these tests (I will discuss the details of how many tests have to be satisfied in the next section), we do not have a case of causation. Hence, satisfying the tests is basic for causation, not the application of a verb that's presumed to be causal.[4]

[3] If that is not convincing, '*ingest*' and '*absorb*' can very clearly be used causally and non-causally. All definitions are taken from the Merriam-Webster online dictinary www.merriam-webster.com. Accessed on 27.10.2009.

[4] Stathis Psillos makes a very similar point about the Machamer–Darden–Craver (MDC) notion of 'activity', focusing on the counterfactual test (Psillos 2004, p. 314; emphasis original): 'Activities,

Are we committing a fallacy here, mistaking test for identity or truth conditions? I don't think so. If '*X* causes *Y*' is true if and only if '*X R Y*' is true, where *R* is a relation (or an activity or capacity) described by a thick causal verb, then we need some principled way of telling which verbs do describe relationships that are causal. And this cannot be done, or so I've been trying to argue, unless we have an independent concept of cause. The tests I've mentioned are meant to help us in determining which transitive verbs are causal, not to define causation.

Francis Longworth has developed this proposal in detail. He regards causation as a cluster concept, by which he means the following (Longworth 2006a, p. 112f.):

Cluster concept. There are a number of features that are relevant to, or 'count towards' an individual's being an instance of the concept. X is a cluster concept if and only if the following conditions are jointly satisfied:

(1) The presence of the entire set of features (the 'cluster set') is sufficient for the concept to be applied.

(2) No feature is necessary.

(3) At least one feature from the cluster set must be instantiated.

Longworth suggests that (perhaps, among others) the following features are members of the cluster set (Longworth forthcoming; this is a paraphrase):

- *Counterfactual dependence* ('*E* counterfactually depends on *C*');
- *Lawlike regularity* ('There is a law such that "whenever *C*, then *E*"');
- *Manipulability* ('Changing *C* is an effective strategy to change *E*');
- *Probability raising* ('$P(E|C\&K) > (C|K)$, where *K* is a set of background factors');
- *Mechanism* ('There is a local physical process from *C* to *E*');
- *Responsibility* ('*C* is [morally] responsible for *E*').

Counterexamples to univocal theories of causation show that none of these features is necessary for causation. For example, cases of redundant causation

such as bonding, repelling, breaking, dissolving etc., are supposed to embody causal connections. But, one may argue that causal connections are distinguished, at least in part, from non-causal ones by means of counterfactuals. If "*x* broke *y*" is meant to capture the claim that "*x caused y* to break," then "*x* broke *y*" must issue in a counterfactual of the form "if *x* hadn't struck *y*, then y would have broken." So talk about activities is, in a sense, disguised talk about counterfactuals'. Notice that Psillos doesn't say '*x* broke *y*' *means* '*x* caused *y* to break', leaving open the possibility of extra content.

Though the authors seem to disagree, I believe that the MDC notion of 'activity' is very close to Cartwright's notion of thick causal verbs in that thick causal verbs describe activities. Hitchcock makes a similar observation (2007, p. 300), pointing out that a difference lies in the fact that MDC use activities as building blocks for their more fundamental notion of a mechanism.

demonstrate the non-necessity of *counterfactual dependence*, in indeterministic cases that of *lawlike regularity* and so forth. However, some subsets of the cluster set are sufficient, e.g. counterfactual dependence and responsibility; production and responsibility; and dependence holding fixed some *G* and responsibility.

Longworth argues that his cluster theory is superior to other accounts in that it explains the truth of five theses regarding the concept of causation (2006a, p. 100; the discussion of how the cluster theory meets these desiderata occurs on pp. 119ff.):

1. *Counterexamples*: There are many extant univocal theories of causation and all of them have counterexamples.
2. *Disagreement*: There are some cases about which individuals disagree in their intuitive causal judgements.
3. *Vagueness*: There are borderline cases of causation.
4. *Error*: Individuals' intuitions are sometimes clearly mistaken.
5. *Degrees of Typicality*: Some cases of causation appear to be 'better' or more typical examples of the concept than others.

Univocal theories must fail because they inflate a single feature of causation into a necessary and sufficient condition; hence, there are counterexamples. Disagreements and vagueness obtain because it is not always clear what precise subset of criteria is sufficient for the application of the concept. Individuals' intuitions are sometimes mistaken because they take the fact that the envisaged scenario has one feature from the cluster set as sufficient to apply the concept while closely analogous cases (which have that and only that feature) are judged differently. Degrees of typicality, naturally, stem from the fact that scenarios have smaller and larger numbers of features from the cluster set.

42.2.1 Understood as account of our ordinary concept of causation

Longworth's account is successful. I know of no case of causation that has *none* of the mentioned features. Whether or not a case that has some but not other features is judged as causation depends on the subsets of the cluster set we take to be sufficient. Longworth does not give a final answer to that question but this flexibility is an advantage of the account. Language is in flux and the subsets of features that are taken to be sufficient for causation and how important the satisfaction of each criterion is each may change over time.

According to this theory, then, 'cause' is ambiguous, vague, gives rise to disagreements in individuals' judgements as well as occasional error, and it comes in degrees. But what seems advantageous from the point of view of our ordinary concept of causation may turn out to be unfavourable for science and policy. For science and policy we require concepts that have a definite meaning and clear conditions of application. Disagreements, so they arise, should be resolvable with reference to an external standard, not individuals' intuitions.

Perhaps it is not a problem for our ordinary concept of causation that some people believe that the father's inattention was a cause of the child's drowning while others think that it was only a quasi-cause because the there was no physical process of the appropriate kind; or that, for some, the fact that a murderer's parents met at a ball in Vienna is a cause of her criminal deed while for others this thought appears ridiculous. For science and policy having clear answers to such questions matters greatly. In determining whether the father should be held liable for his child's accident, we don't only have to know whether certain normative considerations apply but also whether he was *causally* responsible for the accident. And it won't do to answer the question whether he was causally responsible with 'according to some intuitions yes, according to others, no'. Nor will it do to answer 'in some sense, yes; in another, no'.

The account that I develop in the three sections that follow might answer the question 'does X cause Y?' with 'in some sense, yes; in another, no', depending on the case. But unlike other forms of conceptual pluralism, this one has a methodology built into it how disagreements can be resolved. One could say that it makes cause *unspecific* rather than *ambiguous*. 'Cause' here is an unspecific term that is specified by what I will call an 'inferential analysis': an analysis of what set of propositions the claim in which 'cause' occurs is inferentially connected with. So let us now look at what causation has to do with inference.

42.3 Causation and inference

To develop my own Wittgensteinian account of causation I need to digress for a moment. My account builds on the idea that causation and inference are intimately related. This is most easily seen in Hume's theory of causation because within that theory causation and inference are the two sides of the same medal.

In Hume's theory, for any two independent, spatially contiguous and temporally ordered events A and B, if one knows that A causes B, one is entitled to infer B upon observing A. And if one is entitled to infer B upon observing A, one knows that A causes B. The problem is only that one cannot know that A causes B because one cannot see it. Concomitantly, one is never entitled to infer B upon observing A because the future might not resemble the past. The problems of causation and induction thus collapse into one.

But they do so only because Hume held a regularity view of causation, and that view is well known to be false. Without the regularity view, the relation between causation and inference is less tight. Few of us hold that an effect *must* follow its cause – an effect might fail to follow its cause for instance because an intervening factor prevents it from doing so or because the cause is indeterministic. Therefore, an observer of the cause is not entitled to infer the effect (but rather something weaker such as 'the probability of the effect is high' or '*ceteris paribus*, the effect will obtain'). Likewise, few of us hold that if an agent is indeed in the position to infer a later event from an earlier that the earlier event *must* be the cause of the later – for instance because the relation may be due to a common cause such that earlier and later event are epiphenomena. Knowing that *A* is regularly followed by *B* then does not entitle a language user to infer that *A* causes *B* (but rather something weaker such as the disjunctive proposition "'*A* causes *B*" or "*A* and *B* share a common cause" or "there is some non-causal reason for the association between *A* and *B*"'). More tenuously than in Hume, causation and inference are nevertheless related.

An inferentialist theory of the meaning of causal claims explains simply and elegantly why this should be so. Inferentialist theories of meaning hold, roughly, that the meaning of an expression is given by its inferential connections to other expressions. According to some interpreters, Wittgenstein held such a theory in the period between the *Tractatus* and developing the theory of meaning as use in the *Philosophical Investigations*. For instance, in his *Remarks on the Foundation of Mathematics* he says (quoted from Peregrin 2006, p. 2):

> The rules of logical inference cannot be either wrong or right. They determine the meaning of the signs . . . We can conceive the rules of inference – I want to say – as giving the signs their meaning, because they are rules for the use of these signs.

Building on this idea I propose the following for causal claims. The meaning of a causal claim is constituted by the system of propositions with which it is inferentially connected; that is, the system comprised of those propositions that entitle a language user to infer the causal claim as well as those she is entitled to infer from it.

Let us call such a system an 'inferential system for causal claim CC' or short 'inferential system-CC'. An inferential system-CC can roughly be divided into inferential base, inferential target and the causal claim CC itself. The inferential base (for CC) comprises all those propositions *from which* a language user is entitled to infer CC. The inferential target (of CC) comprises all those propositions that a language user is entitled to infer *from* CC.

Scientists seldom establish causal claims for their own sake but rather because they take them to be conducive to the more ultimate goals of science such as scientific explanation, policy and prediction (to give some examples). If a causal claim together with the relevant background knowledge entitles a user to infer a scientific explanation, a policy claim or a prediction, then these latter propositions constitute what I call the inferential target of the causal claim. In concrete terms, consider a claim such as 'aflatoxin is hepatocarcinogenic' ('exposure to aflatoxin causes liver cancer'). An epidemiologist might be interested in explaining the population-level correlation between aflatoxin exposure and liver cancer and thus whether it is due to the carcinogenicity of the substance; a policy maker in inferring 'controlling aflatoxin is an effective strategy to reduce mortality'; finally, a person exposed to aflatoxin in knowing whether consumption of aflatoxin will lead to (an increased chance of) liver cancer *in him* and thus in prediction. Below, I will illustrate the kinds of propositions that must be part of the inferential base in order for a language user to be entitled to these inferences in the context of this case.

Here I will say no more about inferential systems-CC in general save two brief remarks. First, the inferences that form the connections between the propositions contained in it are material rather than formal inferences. Formal models of inference (such as *modus ponens*), as the name suggests, are valid in virtue of their form and independently of the propositions that they take as arguments. Material inferences, by contrast, are valid due to the content of the propositions. To illustrate, consider John Norton's example of contrasting the two inferences 'Some samples of the element bismuth melt at 271°C, therefore all sample of the element bismuth melt at 271°C' and 'Some samples of wax melt at 91°C, therefore all samples of wax melt at 91°C' (Norton 2003, p. 649). It is subject and domain specific (or as Norton calls it, 'material') background knowledge that entitles a language user to the former but not the latter inference. In this case, that background knowledge includes the empirical generalization that chemical elements tend to share physical properties and the fact that bismuth is an element whereas wax is a generic name for a variety of substances. Importantly, proponents of theories of material inference hold that it is not the case that there must be implicit premisses that turn the material argument into a formally valid one once made explicit. Rather, the inferences are licensed by the material facts concerning the subject matter of the propositions involved (Norton 2003; Brigandt forthcoming).

Second, I use the rather clumsy formulation 'inferences a language user is entitled to' in an attempt to strike a balance between a descriptive and prescriptive perspective on meaning. It is clearly the case that ordinary folk as much as sophisticated scientists sometimes make mistakes when inferring a causal claim from evidence or some other claim in the inferential target from a causal claim. It would therefore be incorrect to take those inferences language users actually make as the basis for meaning. On the other hand, there aren't many hard-and-fast rules that philosophers can use to prescribe scientists and ordinary folk what inferences they should and shouldn't make. The best guide to what's doable and what isn't is scientific practice and therefore I won't make highly general claims about what a language user is entitled to. Instead, in the next section I will show how tightly inferential base and target are connected on the basis of a brief analysis of two brief case studies.

42.4 An inferentialist analysis of two causal claims

In this section I consider the kinds of material inferences a user is entitled to make when she knows, first, that 'aflatoxin causes liver cancer' and second, that 'lack of sunlight causes multiple sclerosis'. In particular I will ask under what conditions knowing the causal claim entitles the user to infer (a) a more specific causal claim; (b) a claim about explanation; (c) a claim about policy; (d) a claim about prediction; and (e) a mechanistic claim.

42.4.1 Is aflatoxin carcinogenic *in humans*?

The carcinogenicity of aflatoxin is more like Norton's wax example than his bismuth example in that there is a great deal of variability of the toxicity of substances among different species and populations in general. Aflatoxin turns out to be carcinogenic in human populations but the inference could only be made on the basis of population-specific evidence.[5] Thus, in general, when the causal claim concerns the toxicity of a substance, language users are entitled to inferences about a given population only when the inferential base contains evidence claims about just that population.

42.4.2 Does the carcinogenicity of aflatoxin explain the (human) population-level correlation between the substance and incidence of liver cancer?

It turns out that the inferential base for the human population specific causal claim contains mostly evidence regarding the mechanism of its operation. That is, it contains a claim such as 'There exists a pathway through which aflatoxin produces cancerous growths in liver cells'. For at least two reasons this claim does not entitle to infer the explanatory claim. First, the existence of one or several mechanisms through which aflatoxin causes and therefore increases the chance of liver cancer is compatible with the existence of further mechanisms through which aflatoxin prevents the disease. In this particular case, it is implausible that there should exist a pathway such that exposure to aflatoxin is actually beneficial (e.g. Steel 2008, p. 116). But this is an additional claim the inferential base must contain, which in no way follows from the claim about the carcinogenicity of aflatoxin.

Second, the population-level association is likely to be confounded. In the given case it is infection with the hepatitis-B virus (HBV) that may be responsible for the association. Populations subject to high exposure to aflatoxin are

[5] Steel (2008) argues that the example is a case of successful extrapolation from a claim about animal models (in particular Fischer rats) to humans. I am doubtful whether he is right (Reiss forthcoming). But even if we go along with Steel, the reasoning he presents depends in large part on evidence regarding the *human* metabolism. The important point is that causal claims about toxicity are almost always population specific.

also populations where HBV prevalence is high, and HBV is a known cause of liver cancer. Moreover, HBV is known to *interact* with aflatoxin but in ways that are not fully appreciated (Wild and Ruggero 2009). That is, the carcinogenicity of aflatoxin itself depends on whether or not the compound is co-present with other causes of liver cancer, and it may be the case that even though aflatoxin causes liver cancer in some humans, in populations also affected by HBV aflatoxin is causally irrelevant for cancer (or is even a preventative) so that the association is entirely due to the carcinogenicity of HBV. It is thus no surprise that in one and the same article we can read the following statements: 'Aflatoxins, which are the metabolites of some *Aspergillus* species, are among the most potent hepatocarcinogens known'; 'Several ecological studies have shown a correlation between liver cancer incidence and aflatoxin consumption at the population level, but findings are not entirely consistent'; and 'Case-control studies with dietary questionnaires or biomarkers of recent exposure to aflatoxin have also provided inconsistent results' (Henry *et al.* 1999, p. 2453).

Thus, it may or may not be that the association between exposure to aflatoxin and liver cancer incidence can be explained by the causal claim. Hence the inference cannot be made on the basis of the causal claim alone. In addition, knowledge about other pathways through which the compound affects liver cancer as well as about confounders and modes of interaction is required.

42.4.3 Is control of aflatoxin an effective strategy to reduce mortality of the affected populations?

The usual approach to controlling aflatoxin exposure is to set standards for a maximum level of contamination of finished food products. According to the best available estimates lowering the standard does indeed achieve a small reduction of liver cancer incidence (*ibid.*). However, for two reasons setting stricter contamination standards is not considered a good strategy to reduce mortality. First, higher food standards will lead countries to limit the import of affected products, which may mean that the least contaminated foods and feeds are exported, leaving the more highly contaminated products in the most affected countries. Second, it may lead to food shortages in those countries (*ibid.*). Thus, controlling aflatoxin is not an effective strategy to reduce mortality in the affected populations because the intervention, while decreasing mortality along one path – through aflatoxin consumption and liver cancer – *in*creases mortality along another, viz. food deprivation.

Such an intervention would certainly be 'ham-fisted', to use Elliott Sober's term (Sober 2009). A ham-fisted intervention is one that affects the target variable through pathways that do not go through the cause variable of interest. But there is no guarantee that there exist interventions that are not ham-fisted. Nor is there a guarantee that an intervention that affects, if at all, the effect (mortality) only through the cause (exposure to aflatoxin) leaves the causal relation intact. Especially in the social sciences interventions might be structure altering and therefore unable to be exploited for policy purposes. Again, therefore, a claim about policy can only be inferred when a number of additional pieces of knowledge is contained in the inferential base.

42.4.4 Does exposure to aflatoxin predict liver cancer in the individual case?

Just as there is much variability between species, there is often much variability within a single species. Therefore, whether the causal claim is relevant for an individual depends on whether or not the individual belongs to the precise population for which the causal claim has been established. In the aflatoxin case, the toxicity of the substance depends on details of the metabolism that are widely shared among humans, hence establishing carcinogenicity for some humans is likely to be relevant for all humans (and this, once more, is an additional proposition that has to be part of the inferential base if a prediction is to be made). However, even if that is the case, three possible circumstances may drive a wedge in between the truth of the causal claim and successfully using the claim for prediction. First, even if aflatoxin is toxic in most humans, some may have a rare genetic make-up that makes them immune to aflatoxin (that this is not an idle possibility is demonstrated by the fact that some species such as mice are immune). Second, even if a given individual is susceptible to aflatoxin, intervening factors may prevent the causal relation from realising. People might swallow antidotes or die before aflatoxin has made its way through the metabolism. Third, even if the individual is susceptible and nothing intervenes, the cause may fail to produce its effect because the mechanism operates indeterministically. None of these possibilities can be excluded without additional evidence.

Let us now examine a case in which a causal claim has been established by means of epidemiological – that is, probabilistic – data. It has long been known that there is a characteristic pattern in the global distribution of multiple sclerosis (MS): high latitude is associated with a high risk for MS (Kurtzke 1977). But it is difficult to disentangle genetic factors and various environmental factors such as nutrition and culture. Strong evidence that sunlight exposure is the relevant factor came from a quasi natural experiment in Australia. Australia presents a very favourable case for causal analysis because it displays enormous latitudinal spread and climatic variation at the same time as genetic and cultural homogeneity (van der Mai *et al.* 2001, p. 169; references suppressed):

> In Australia, a more than sixfold increase in age-standardized MS prevalence has been demonstrated from tropical Queensland to Tasmania. Within Europe and the United States, there is also an at least two- to threefold gradient of increasing MS prevalence

with increasing latitude. These geographical differences were initially interpreted to represent environmental influences which varied by latitude, such as climatic factors, dietary characteristics and infectious agents. More recent analyses indicate that geographical MS variation, at least in North America, may result from a complex interplay of genes and environment. The marked Australian latitudinal gradient found in the national prevalence survey of 1981 is unlikely to be explained by genetic factors only, because the gradient is evident even among UK and Irish immigrants to Australia, a population subgroup that is predominantly Caucasian. These findings together with the large latitudinal spread across the continent, stretching from 10° to 44° South in latitude, and a uniform health care system provide a good opportunity to examine the relationship between latitude-related factors and MS.

[...]

The aim of this study was to conduct an ecological analysis of the extent to which UVR [ultraviolet radiation] levels might explain the regional variation of MS in Australia. We contrasted the relationship between UVR and MS prevalence with that of UVR and melanoma incidence, because the latter association has previously been demonstrated to be causal.

42.4.5 Is there a mechanism from (lack of) sunlight to multiple sclerosis?

Let us suppose then that it is true that lack of sunlight causes MS.[6] The first thing to note is what has been established is a probabilistic causal claim. That is, in a certain population (caucasians, say), lack of sunlight increases the probability of MS, holding fixed other causes of MS. Many of the limitations described above hold here too. For instance, the claim is population relative and without population-specific evidence no inferences can be made about a hitherto unexamined population. Above I also argued that a mechanistic causal claim does not license an inference regarding the corresponding population-level probabilistic claim. Here let me ask the reverse question: does a population-level probabilistic causal claim entail anything about mechanisms? My answer is once more no but the reasoning requires some elaboration.

When some time passes between the occurrence of a cause and the onset of an effect, it is plausible to assume that there exist some intermediaries that transport the causal message from cause to effect. In the type of biomedical cases I have been talking about, there lie long stretches of time between cause and effect, often many years. There is some evidence, for instance, that

[6] If it is indeed the case, as I believe it is, that this causal hypothesis is widely accepted in the biomedical community, the vitamin-D/MS link provides an interesting case study against the so-called Russo–Williamson thesis according to which both mechanistic as well as probabilistic evidence is required to establish a causal claim (Russo and Williamson 2007). Whereas parts of the vitamin-D metabolism are understood fairly well, the etiology of MS is still completely unknown (e.g. Ramagopalan and Giovannoni 2009).

sunlight exposure during age 6–15 is an important risk factor associated with MS (van der Mei 2003). The onset of the disease typically occurs much later, between the ages 20 and 40 (van Amerongen *et al.* 2004).

Sunlight is required for the skin to metabolise vitamin-D3. UV-B radiation photolyses provitamin D3 to previtamin D3, which, in turn, is converted by a thermal process to vitamin-D3. Vitamin-D3 is biologically inactive but when converted into $1, 25-(OH)_2D$, the hormonally active form of vitamin-D, involved in an abundance of biological functions including calcium homeostasis, cell differentiation and maturation and, most relevantly, immune responses. How precisely $1, 25-(OH)_2D$ affects MS is unknown but studies with mice have shown that the hormone successfully prevents the onset of experimental autoimmune encephalomyelitis (EAE), which is recognized as a useful animal model for MS (van Etten *et al.* 2003). Moreover, there is some evidence that vitamin D interacts with the major genetic locus which determines susceptibility to MS (Ramagopalan *et al.* 2009).

None of this shows, however, that there is a mechanism from sunlight exposure to onset of MS. It is the *lack* of sunlight that causes vitamin-D *deficiency*. As vitamin D is an important *preventer* of MS, it is the *absence* of vitamin D that causes MS. Now, one might call this a (sketch for a) mechanism. But it is important to see the differences between the causal relations involved in this example and those involved in other cases such as the aflatoxin case that was described above. Exposure to aflatoxin causes cancer through a series of intermediate stages, all of which contain markers that have a clear (and, in fact, unique) association with the toxin. At least in principle, therefore, the causal effect of aflatoxin on liver cells could be learned by both forward as well as backward chaining. Forward chaining uses the early stages of a mechanism to make inferences about the types of entities and activities that are likely to be found downstream and backward chaining reasons conversely from the entities and activities in later stages about entities and activities appearing earlier (Darden 2002, p. 362). Forward chaining thus would start with the consumption of aflatoxin, examine the various stages of its metabolism and eventually establish an effect of an aflatoxin metabolite on liver cells. Backward chaining proceeds by examining these cells, asking what could possibly have caused the characteristic mutation and then backtracking further. As the mechanism is fully present in each individual in which aflatoxin has caused liver cancer, it could (again, in principle) be discovered on the basis of a single individual.

The role of sunlight is not analogous to a chemical compound making its way through the human metabolism. Sunlight is a factor that enables the skin to synthesise vitamin D, which, after several transformations, plays an active role in regulating immune responses among other things. There would be no use in attempting forward or backward chaining in an individual suffering from MS. Even if that individual were deficient in vitamin D, there would be no sense in which 'lack of sunshine' could be regarded as 'the' cause of the deficiency, analogously to the sense in which exposure to aflatoxin is 'the' cause of the presence of its various metabolites in the blood stream. We might say that lack of sunlight was among the causes of the vitamin-D deficiency because of the truth of the counterfactual 'had the individual been more exposed to sunlight, her vitamin-D levels would have been higher'. But alternative antecedents (e.g. 'had the individual eaten more oily fish' or 'had the individual taken dietary supplements') also make the counterfactual true and with it the associated causal claims. Such counterfactual claims we judge in turn on the basis of population-level epidemiological – i.e. probabilistic – data.

Aflatoxin is an entity that damages liver cells by way of various activities the compound and its metabolites engage in. Nothing analogous is true in the sunlight/MS case. Using the well-known Machamer–Darden–Craver definition of a mechanism according to which 'Mechanisms are entities and activities organized such that they are productive of regular changes from start or set-up to finish or termination conditions' (Machamer *et al.* 2000, p. 3), it is straightforward to conclude that there is a mechanism in the former but not in the latter case.[7]

Another way of describing the difference is the following. If it is true that at the population level aflatoxin causes liver cancer, then there must be some individuals whose liver cancer was brought about by aflatoxin. But it is not the case that if at the population level lack of sunlight causes MS, there must be some individuals whose MS was brought about by lack of sunlight. When a mechanism is present, a causal generalization entails something about singular causal relations. When no mechanism is present, there is no such entailment either.

42.5 Re-enter Wittgenstein

Even the more patient among the readers might have wondered by now what these musings about inference have to do with Wittgenstein, pluralism and Wittgensteinian pluralism. Let us look at Wittgenstein first.

Wittgenstein is famous for having remarked that 'the sense of a proposition is the method of its verification' in a conversation with the Vienna Circle (McGuinness 1985, p. 352). But apparently he himself expressed out-

[7] This is not to deny that there is something *similar* to a mechanism at the type level. It is certainly true that the variable 'exposure to sunlight' is causally relevant to the variable 'vitamin-D level', which in turn is relevant to the variable '1,25-(OH)2D', which, finally, is relevant to the risk of MS. One way to put my point is to say that that if we want to call that a mechanism we can infer at best a mechanism of this type but not a mechanism of the type that mediates the influence of aflatoxin on liver cancer.

rage when the 'verification principle' was attributed to him (Anscombe 1995, p. 405) and at least according to some interpretations (e.g. Medina 2001; Peregrin 2006) held an *inferentialist* theory of meaning in the period between the *Tractatus* and developing the theory of meaning as use in the *Philosophical Investigations*. For instance, in his *Remarks on the Foundation of Mathematics* we can read (quoted from Peregrin 2006, p. 2):

> The rules of logical inference cannot be either wrong or right. They determine the meaning of the signs... We can conceive the rules of inference – I want to say – as giving the signs their meaning, because they are rules for the use of these signs.

According to this theory, then, the meaning of an expression is given by the role it plays in our inferential practises. On this view, then, there is a perfectly natural and simple explanation why causation and inference are so intimately related: the meaning of a causal claim is given by its inferential role.

How do we know with what other expressions a given expression is inferentially connected? This is where in Wittgenstein's theory of verification comes in. José Medina explains its role as follows (Medina 2001, p. 308; emphasis is Medina's):

> That the verificationism of the Satzsystem view is at the service of an inferentialist semantics becomes explicit when Wittgenstein remarks that the import of asking of a proposition 'What is its verification?' is that 'an answer gives the meaning by showing the relation of the proposition to other propositions. That is, it shows *what it follows from and what follows from it*. It gives the grammar of the proposition.' [Wittgenstein 1979: 19–20] So, for Wittgenstein, verificationism seems to be a heuristic tool that enables us to analyze the content of propositions in terms of their *inferential use*.

Thus, whereas the meaning of an expression is given by its inferential connections with other expressions in a system of propositions, its method of verification determines what these inferential connections are. This latter point is precisely what I've argued in the preceding section: the method of verifying a causal claim – of evidentially supporting it – determines with what other claims it is inferentially related.

Moreover, it is easy to see how this theory of meaning leads to a form of pluralism about causation. If its inferential connections to other propositions constitute the meaning of a causal claim and the kinds of propositions from which a causal claim can be inferred and those that can be inferred from a causal claim differ from claim to claim, the case for pluralism has been made. Very roughly, we can define identity conditions for causal claims as follows. Suppose the term 'cause' is used on two different occasions and it is not known whether it has the same meaning on both occasions. Two such claims would have the form 'X α-causes Y' and 'Z β-causes W'. We can then say that 'α-causes' has the same meaning as 'β-causes' (on these occasions) to the extent that 'X α-causes Y' is inferentially connected to the same kinds of propositions regarding the relation between X and Y as 'Z β-causes W' is inferentially connected to propositions regarding the relation between Z and W. If, to give a fictional example, both 'X α-causes Y' and 'Z β-causes W' have been established by RCTs and both license claims about effective strategies (such as 'promoting X is an effective means to raise the chance of Y' and likewise for Z and Y), then 'α-causes' means the same as 'β-causes' (on these occasions).

There is no guarantee that the kinds of propositions found in inferential base and target are the same for different instances of 'cause'.[8] Different methods of supporting a causal claim license different kinds of inference: this is just what the previous section aimed to establish. Therefore, the meaning of 'cause' in 'Aflatoxin causes liver cancer' and 'Lack of sunlight causes MS' differs – as these claims differ both with respect to the kinds of propositions in their inferential base as well as those in their inferential target.

42.6 Conclusions

The advantages of the account proposed here over its two Wittgensteinian competitors are easy to see. Unlike Anscombe's account inferentialism has no difficulty with cases of causation by absence, as was shown in the discussion of the causal claim about lack of sunlight and MS. The issue whether or not a given transitive verb is a genuine causative simply doesn't arise.[9] Unlike Longworth's account, inferentialism doesn't make causal claims ambiguous or vague or both. There is a definite set of propositions with which any causal claim is inferentially related. True, we might not always have a very clear idea of what these sets are. But this is a question of epistemology, not of semantics.

Finally, inferentialism has an answer to Jon Williamson's challenge: 'If one can't say much about the number and kinds of notions of cause then one can't say much about causality at all' (Williamson 2006, p. 72). It is certainly the case that the type of pluralism entailed by an inferentialist theory of meaning is of the indeterminate variety in that number and kinds of notion of cause are not

[8] Though if the Russo–Williamson thesis were true, researchers in the health sciences did indeed always require both difference-making evidence and evidence about mechanistic connections in order to establish causal claims, and in addition the kinds of propositions one is entitled to infer from causal claims were also the same, then conceptual monism about causation in the health sciences, which they favour, would be supported. I do not think that that thesis is true, and I think that my second case can serve as a counterexample (footnote 5) but it is interesting to note that the thesis (plus one further assumption) entails conceptual monism under an inferential conception of meaning.

[9] An issue that does arise is the parallel one of *justifying* the inferences among base, causal claim and target. But this is one we ought to leave to science. As I claimed above, the best guide to what works and what doesn't is scientific practise, and there is no reason why this area should be exempt from the general principle.

fixed once and for all times. But, as the inferentialist analyses of section four have shown, there is a great deal one can say about causality.

Acknowledgements

Though developed and motivated independently, the view on causation I present here resembles some work on scientific representation by my former colleagues in Madrid, Mauricio Suárez (Complutense University) and Jesús Zamora Bonilla (UNED). I received financial support from two projects of the Spanish ministry of education, FFI2008-01580 and CONSOLIDER INGENIO CSD2009-0056.

References

Anscombe, Elizabeth (1971) [1992]. *Causality and Determination*, Cambridge: Cambridge University Press.

Anscombe, Elizabeth (1995). Ludwig Wittgenstein (Cambridge Philosophers II), *Philosophy* **70**: 395–407.

Brigandt, Ingo (2010). Scientific reasoning is material inference: Combining confirmation, discover, and explanation, *International Studies in the Philosophy of Science* 24(1): 31–43.

Campaner, Raffaella and Maria Carla Galavotti (2007). Plurality in causality, in Peter Machamer and Gereon Wolters (eds), *Thinking About Causes: From Greek Philosophy to Modern Physics*, Pittsburgh (PA): University of Pittsburgh Press: pp. 178–99.

Cartwright, Nancy (1999). *The Dappled World*, Cambridge: Cambridge University Press.

Cartwright, Nancy (2003). Causation: One word, many things, *Philosophy of Science* **71**: 805–19.

Cartwright, Nancy (2007). *Hunting Causes and Using Them*, Cambridge: Cambridge University Press.

Darden, Lindley (2002). Strategies for discovering mechanisms: Schema instantiation, modular subassembly, forward and backward chaining, Supplement to *Philosophy of Science* 69(3), vol. 2: S354–S365.

De Vreese, Leen (2006). Causal pluralism and scientific knowledge: An underexposed problem, *Philosophica* **77** (1): 125–150.

Dowe, Phil (2000). *Physical Causation*, New York: Cambridge University Press.

Godfrey-Smith, Peter (2009). Causal pluralism, in Helen Beebee, Peter Menzies and Christopher Hitchcock (eds), *Oxford Handbook of Causation*, Oxford: Oxford University Press: pp. 326–37.

Hall, Ned (2004). Two concepts of causation, in John Collins, Ned Hall and Laurie Paul (eds), *Causation and Counterfactuals*, Cambridge (MA): MIT Press: pp. 225–76.

Henry, Sara Hale, F. Xavier Bosch, Terry C. Troxell and P. Michael Bolger (1999). Reducing liver cancer – Global control of aflatoxin, *Science* **286**(5449): 2453–4.

Hitchcock, Christopher (2003). Of Humean bondage, *British Journal for the Philosophy of Science* **54**(1): 1–25.

Hitchcock, Christopher (2007). How to be a causal pluralist, in Peter Machamer and Gereon Wolters (eds), *Thinking About Causes: From Greek Philosophy to Modern Physics*, Pittsburgh (PA): University of Pittsburgh Press: pp. 200–21.

Kurtzke, John (1977). Geography in multiple sclerosis, *Journal of Neurology* **215**(1): 1–26.

Longworth, Francis (2006a). *Causation, Counterfactual Dependence and Pluralism*. PhD Thesis. University of Pittsburgh.

Longworth, Francis, (2006b). Causation, Pluralism and Responsibility. *Philosophica* **77**(1): 45–68.

Machamer, Peter (2004). Activities and causation: The metaphysics and epistemology of mechanisms, *International Studies in the Philosophy of Science* **18**(1): 27–39.

Machamer, Peter, Lindley Darden and Carl Craver (2000). Thinking about mechanisms, *Philosophy of Science* **67**(1): 1–25.

Mackie, John (1974). *The Cement of the Universe: A Study of Causation*, Oxford: Oxford University Press.

McDermott, Michael (1995). Redundant causation, *British Journal for the Philosophy of Science* **46**: 523–44.

McGuinness, Brian (1985). Wittgenstein and the Vienna Circle, *Synthese* **64**: 351–8.

Medina, José (2001). Verificationism and inferentialism in Wittgenstein's philosophy, *Philosophical Investigations* **24**(4): 304–313.

Norton, John (2003). A material theory of induction, *Philosophy of Science* **70**(4): 647–670.

Peregrin, Jaroslav (2006). Meaning as inferential role, *Erkenntnis* **64**: 1–35.

Psillos, Stathis (2004). Glimpse of the secret connection: Harmonizing mechanisms with counterfactuals, *Perspectives on Science* **12**(3): 288–319.

Psillos, Stathis, forthcoming, Causal pluralism, unpublished manuscript, University of Athens.

Ramagopalan, Sreeram and Gavin Giovannoni (2009). Can we predict MS?, *The Lancet Neurology* **12**(8): 1077–79.

Ramagopalan, Sreeram, Narelle Maugeri, Lahiru Handunnetthi, Matthew Lincoln, Sarah-Michelle Orton, David Dyment, Gabriele DeLuca, Blanca Herrera, Michael Chao, Dessa Sadovnick, George Ebers and Julian Knight (2009). Expression of the multiple sclerosis-associated MHC class II allele HLA-DRB1*1501 is regulated by vitamin D, *PLoS Genetics* **5**(2): e1000369. doi:10.1371/journal.pgen.1000369.

Reiss, Julian (2009). Causation in the social science: Evidence, inference, and purpose, *Philosophy of the Social Sciences* **39**(1): 20–40.

Reiss, Julian, forthcoming, Review of Daniel Steel's Across the Boundaries: Extrapolation in Biology and Social Science, *Economics and Philosophy*.

Russo, Federica and Jon Williamson (2007). Interpreting causality in the health sciences, *International Studies in the Philosophy of Science* **21**(2): 157–70.

Schaffer, Jonathan (2004). Causes need not be physically connected to their effects: The case for negative causation, in Christopher Hitchcock (ed.), *Contemporary Debates in Philosophy of Science*, Oxford: Blackwell: pp. 197–216.

Sober, Elliott (2009). Reichenbach's cubical universe and the problem of the external world, *Synthese* (online first).

Steel, Daniel (2008). *Across the Boundaries: Extrapolation in Biology and Social Science*, Oxford: Oxford University Press.

van Amerongen, B.M., C.D. Dijkstra, P. Lips, C.H. Polman (2004). Multiple sclerosis and vitamin D: an update, *European Journal of Clinical Nutrition* **58**(8): 1095–109.

van Etten, E., D.D. Branisteanu, L. Overbergh, R. Bouillon, A. Verstuyf and C. Mathieu (2003). Combination of a 1,25-dihydroxyvitamin D(3) analog and a bisphosphonate prevents experimental autoimmune encephalomyelitis and preserves bone, *Bone* **32**: 397–404.

van der Mei, I., A.-L. Ponsonby, L. Blizzard, and T. Dwyer (2001). Regional variation in multiple sclerosis prevalence in Australia and its association with ambient ultraviolet radiation, *Neuroepidemiology* **20**: 168–74.

van der Mei, I., A.-L. Ponsonby, T. Dwyer, L. Blizzard, R. Simmons, B.V. Taylor, H. Butzkueven, and T. Kilpatrick (2003). Past exposure to sun, skin phenotype, and risk of multiple sclerosis: case-control study, *British Medical Journal* **327**: 1–6.

Weber, Erik (2007). Conceptual tools for causal analysis in the social sciences, in Federica Russo and Jon Williamson (eds), *Causality and Probability in the Sciences*, London: College Publications, pp. 192–207.

Wild, Christopher and Ruggero Montesano (2009). A model of interaction: Aflatoxins and hepatitis viruses in liver cancer aetiology and prevention, *Cancer Letters* **286**: 22–8.

Williamson, Jon (2006). Causal pluralism versus epistemic causality, *Philosophica* **77**(1): 69–96.

Wittgenstein, Ludwig (1953). *Philosophical Investigations*, Trans. Elizabeth Anscombe. Oxford: Blackwell.

Wittgenstein, Ludwig (1979). *Wittgenstein's Lectures, 1932–35*, edited by Alice Ambrose, Oxford: Blackwell.

Index